T0184435

# Lecture Notes in Computer Science     11302

*Commenced Publication in 1973*
Founding and Former Series Editors:
Gerhard Goos, Juris Hartmanis, and Jan van Leeuwen

## Editorial Board

David Hutchison
  *Lancaster University, Lancaster, UK*
Takeo Kanade
  *Carnegie Mellon University, Pittsburgh, PA, USA*
Josef Kittler
  *University of Surrey, Guildford, UK*
Jon M. Kleinberg
  *Cornell University, Ithaca, NY, USA*
Friedemann Mattern
  *ETH Zurich, Zurich, Switzerland*
John C. Mitchell
  *Stanford University, Stanford, CA, USA*
Moni Naor
  *Weizmann Institute of Science, Rehovot, Israel*
C. Pandu Rangan
  *Indian Institute of Technology Madras, Chennai, India*
Bernhard Steffen
  *TU Dortmund University, Dortmund, Germany*
Demetri Terzopoulos
  *University of California, Los Angeles, CA, USA*
Doug Tygar
  *University of California, Berkeley, CA, USA*
Gerhard Weikum
  *Max Planck Institute for Informatics, Saarbrücken, Germany*

More information about this series at http://www.springer.com/series/7407

Long Cheng · Andrew Chi Sing Leung
Seiichi Ozawa (Eds.)

# Neural
# Information Processing

25th International Conference, ICONIP 2018
Siem Reap, Cambodia, December 13–16, 2018
Proceedings, Part II

 Springer

*Editors*
Long Cheng (iD)
The Chinese Academy of Sciences
Beijing, China

Andrew Chi Sing Leung
City University of Hong Kong
Kowloon, Hong Kong SAR, China

Seiichi Ozawa
Kobe University
Kobe, Japan

ISSN 0302-9743         ISSN 1611-3349   (electronic)
Lecture Notes in Computer Science
ISBN 978-3-030-04178-6         ISBN 978-3-030-04179-3   (eBook)
https://doi.org/10.1007/978-3-030-04179-3

Library of Congress Control Number: 2018960916

LNCS Sublibrary: SL1 – Theoretical Computer Science and General Issues

© Springer Nature Switzerland AG 2018
This work is subject to copyright. All rights are reserved by the Publisher, whether the whole or part of the material is concerned, specifically the rights of translation, reprinting, reuse of illustrations, recitation, broadcasting, reproduction on microfilms or in any other physical way, and transmission or information storage and retrieval, electronic adaptation, computer software, or by similar or dissimilar methodology now known or hereafter developed.
The use of general descriptive names, registered names, trademarks, service marks, etc. in this publication does not imply, even in the absence of a specific statement, that such names are exempt from the relevant protective laws and regulations and therefore free for general use.
The publisher, the authors, and the editors are safe to assume that the advice and information in this book are believed to be true and accurate at the date of publication. Neither the publisher nor the authors or the editors give a warranty, express or implied, with respect to the material contained herein or for any errors or omissions that may have been made. The publisher remains neutral with regard to jurisdictional claims in published maps and institutional affiliations.

This Springer imprint is published by the registered company Springer Nature Switzerland AG
The registered company address is: Gewerbestrasse 11, 6330 Cham, Switzerland

# Preface

The 25th International Conference on Neural Information Processing (ICONIP 2018), the annual conference of the Asia Pacific Neural Network Society (APNNS), was held in Siem Reap, Cambodia, during December 13–16, 2018. The ICONIP conference series started in 1994 in Seoul, which has now become a well-established and high-quality conference on neural networks around the world. Siem Reap is a gateway to Angkor Wat, which is one of the most important archaeological sites in Southeast Asia, the largest religious monument in the world. All participants of ICONIP 2018 had a technically rewarding experience as well as a memorable stay in this great city.

In recent years, the neural network has been significantly advanced with the great developments in neuroscience, computer science, cognitive science, and engineering. Many novel neural information processing techniques have been proposed as the solutions to complex, networked, and information-rich intelligent systems. To disseminate new findings, ICONIP 2018 provided a high-level international forum for scientists, engineers, and educators to present the state of the art of research and applications in all fields regarding neural networks.

With the growing popularity of neural networks in recent years, we have witnessed an increase in the number of submissions and in the quality of submissions. ICONIP 2018 received 575 submissions from 51 countries and regions across six continents. Based on a rigorous peer-review process, where each submission was reviewed by at least three experts, a total of 401 high-quality papers were selected for publication in the prestigious Springer series of *Lecture Notes in Computer Science*. The selected papers cover a wide range of subjects that address the emerging topics of theoretical research, empirical studies, and applications of neural information processing techniques across different domains.

In addition to the contributed papers, the ICONIP 2018 technical program also featured three plenary talks and two invited talks delivered by world-renowned scholars: Prof. Masashi Sugiyama (University of Tokyo and RIKEN Center for Advanced Intelligence Project), Prof. Marios M. Polycarpou (University of Cyprus), Prof. Qing-Long Han (Swinburne University of Technology), Prof. Cesare Alippi (Polytechnic of Milan), and Nikola K. Kasabov (Auckland University of Technology).

We would like to extend our sincere gratitude to all members of the ICONIP 2018 Advisory Committee for their support, the APNNS Governing Board for their guidance, the International Neural Network Society and Japanese Neural Network Society for their technical co-sponsorship, and all members of the Organizing Committee for all their great effort and time in organizing such an event. We would also like to take this opportunity to thank all the Technical Program Committee members and reviewers for their professional reviews that guaranteed the high quality of the conference proceedings. Furthermore, we would like to thank the publisher, Springer, for their sponsorship and cooperation in publishing the conference proceedings in seven volumes of *Lecture Notes in Computer Science*. Finally, we would like to thank all the speakers, authors, reviewers, volunteers, and participants for their contribution and support in making ICONIP 2018 a successful event.

October 2018

Jun Wang
Long Cheng
Andrew Chi Sing Leung
Seiichi Ozawa

# ICONIP 2018 Organization

## General Chair

Jun Wang      City University of Hong Kong,
Hong Kong SAR, China

## Advisory Chairs

Akira Hirose      University of Tokyo, Tokyo, Japan
Soo-Young Lee      Korea Advanced Institute of Science and Technology,
South Korea
Derong Liu      Institute of Automation, Chinese Academy of Sciences,
China
Nikhil R. Pal      Indian Statistics Institute, India

## Program Chairs

Long Cheng      Institute of Automation, Chinese Academy of Sciences,
China
Andrew C. S. Leung      City University of Hong Kong, Hong Kong SAR,
China
Seiichi Ozawa      Kobe University, Japan

## Special Sessions Chairs

Shukai Duan      Southwest University, China
Kazushi Ikeda      Nara Institute of Science and Technology, Japan
Qinglai Wei      Institute of Automation, Chinese Academy of Sciences,
China
Hiroshi Yamakawa      Dwango Co. Ltd., Japan
Zhihui Zhan      South China University of Technology, China

## Tutorial Chairs

Hiroaki Gomi      NTT Communication Science Laboratories, Japan
Takashi Morie      Kyushu Institute of Technology, Japan
Kay Chen Tan      City University of Hong Kong, Hong Kong SAR,
China
Dongbin Zhao      Institute of Automation, Chinese Academy of Sciences,
China

## Publicity Chairs

Zeng-Guang Hou              Institute of Automation, Chinese Academy of Sciences,
                                            China
Tingwen Huang              Texas A&M University at Qatar, Qatar
Chia-Feng Juang            National Chung-Hsing University, Taiwan
Tomohiro Shibata           Kyushu Institute of Technology, Japan

## Publication Chairs

Xinyi Le                   Shanghai Jiao Tong University, China
Sitian Qin                 Harbin Institute of Technology Weihai, China
Zheng Yan                  University Technology Sydney, Australia
Shaofu Yang                Southeast University, China

## Registration Chairs

Shenshen Gu                Shanghai University, China
Qingshan Liu               Southeast University, China
Ka Chun Wong               City University of Hong Kong,
                                            Hong Kong SAR, China

## Conference Secretariat

Ying Qu                    Dalian University of Technology, China

## Program Committee

Hussein Abbass             University of New South Wales at Canberra, Australia
Choon Ki Ahn              Korea University, South Korea
Igor Aizenberg             Texas A&M University at Texarkana, USA
Shotaro Akaho              National Institute of Advanced Industrial Science
                                            and Technology, Japan
Abdulrazak Alhababi       UNIMAS, Malaysia
Cecilio Angulo             Universitat Politècnica de Catalunya, Spain
Sabri Arik                 Istanbul University, Turkey
Mubasher Baig              National University of Computer and Emerging
                                            Sciences Lahore, India
Sang-Woo Ban               Dongguk University, South Korea
Tao Ban                    National Institute of Information and Communications
                                            Technology, Japan
Boris Bačić                Auckland University of Technology, New Zealand
Xu Bin                     Northwestern Polytechnical University, China
David Bong                 Universiti Malaysia Sarawak, Malaysia
Salim Bouzerdoum           University of Wollongong, Australia
Ivo Bukovsky               Czech Technical University, Czech Republic

| | |
|---|---|
| Ke-Cai Cao | Nanjing University of Posts and Telecommunications, China |
| Elisa Capecci | Auckland University of Technology, New Zealand |
| Rapeeporn Chamchong | Mahasarakham University, Thailand |
| Jonathan Chan | King Mongkut's University of Technology Thonburi, Thailand |
| Rosa Chan | City University of Hong Kong, Hong Kong SAR, China |
| Guoqing Chao | East China Normal University, China |
| He Chen | Nankai University, China |
| Mou Chen | Nanjing University of Aeronautics and Astronautics, China |
| Qiong Chen | South China University of Technology, China |
| Wei-Neng Chen | Sun Yat-Sen University, China |
| Xiaofeng Chen | Chongqing Jiaotong University, China |
| Ziran Chen | Bohai University, China |
| Jian Cheng | Chinese Academy of Sciences, China |
| Long Cheng | Chinese Academy of Sciences, China |
| Wu Chengwei | Bohai University, China |
| Zheru Chi | The Hong Kong Polytechnic University, SAR China |
| Sung-Bae Cho | Yonsei University, South Korea |
| Heeyoul Choi | Handong Global University, South Korea |
| Hyunsoek Choi | Kyungpook National University, South Korea |
| Supannada Chotipant | King Mongkut's Institute of Technology Ladkrabang, Thailand |
| Fengyu Cong | Dalian University of Technology, China |
| Jose Alfredo Ferreira Costa | Federal University of Rio Grande do Norte, Brazil |
| Ruxandra Liana Costea | Polytechnic University of Bucharest, Romania |
| Jean-Francois Couchot | University of Franche-Comté, France |
| Raphaël Couturier | University of Bourgogne Franche-Comté, France |
| Jisheng Dai | Jiangsu University, China |
| Justin Dauwels | Massachusetts Institute of Technology, USA |
| Dehua Zhang | Chinese Academy of Sciences, China |
| Mingcong Deng | Tokyo University of Agriculture and Technology, Japan |
| Zhaohong Deng | Jiangnan University, China |
| Jing Dong | Chinese Academy of Sciences, China |
| Qiulei Dong | Chinese Academy of Sciences, China |
| Kenji Doya | Okinawa Institute of Science and Technology, Japan |
| El-Sayed El-Alfy | King Fahd University of Petroleum and Minerals, Saudi Arabia |
| Mark Elshaw | Nottingham Trent International College, UK |
| Peter Erdi | Kalamazoo College, USA |
| Josafath Israel Espinosa Ramos | Auckland University of Technology, New Zealand |
| Issam Falih | Paris 13 University, France |

| | |
|---|---|
| Bo Fan | Zhejiang University, China |
| Yunsheng Fan | Dalian Maritime University, China |
| Hao Fang | Beijing Institute of Technology, China |
| Jinchao Feng | Beijing University of Technology, China |
| Francesco Ferracuti | Università Politecnica delle Marche, Italy |
| Chun Che Fung | Murdoch University, Australia |
| Wai-Keung Fung | Robert Gordon University, UK |
| Tetsuo Furukawa | Kyushu Institute of Technology, Japan |
| Hao Gao | Nanjing University of Posts and Telecommunications, China |
| Yabin Gao | Harbin Institute of Technology, China |
| Yongsheng Gao | Griffith University, Australia |
| Tom Gedeon | Australian National University, Australia |
| Ong Sing Goh | Universiti Teknikal Malaysia Melaka, Malaysia |
| Iqbal Gondal | Federation University Australia, Australia |
| Yue-Jiao Gong | Sun Yat-sen University, China |
| Shenshen Gu | Shanghai University, China |
| Chengan Guo | Dalian University of Technology, China |
| Ping Guo | Beijing Normal University, China |
| Shanqing Guo | Shandong University, China |
| Xiang-Gui Guo | University of Science and Technology Beijing, China |
| Zhishan Guo | University of Central Florida, USA |
| Christophe Guyeux | University of Franche-Comte, France |
| Masafumi Hagiwara | Keio University, Japan |
| Saman Halgamuge | The University of Melbourne, Australia |
| Tomoki Hamagami | Yokohama National University, Japan |
| Cheol Han | Korea University at Sejong, South Korea |
| Min Han | Dalian University of Technology, China |
| Takako Hashimoto | Chiba University of Commerce, Japan |
| Toshiharu Hatanaka | Osaka University, Japan |
| Wei He | University of Science and Technology Beijing, China |
| Xing He | Southwest University, China |
| Xiuyu He | University of Science and Technology Beijing, China |
| Akira Hirose | The University of Tokyo, Japan |
| Daniel Ho | City University of Hong Kong, Hong Kong SAR, China |
| Katsuhiro Honda | Osaka Prefecture University, Japan |
| Hongyi Li | Bohai University, China |
| Kazuhiro Hotta | Meijo University, Japan |
| Jin Hu | Chongqing Jiaotong University, China |
| Jinglu Hu | Waseda University, Japan |
| Xiaofang Hu | Southwest University, China |
| Xiaolin Hu | Tsinghua University, China |
| He Huang | Soochow University, China |
| Kaizhu Huang | Xi'an Jiaotong-Liverpool University, China |
| Long-Ting Huang | Wuhan University of Technology, China |

| | |
|---|---|
| Panfeng Huang | Northwestern Polytechnical University, China |
| Tingwen Huang | Texas A&M University, USA |
| Hitoshi Iima | Kyoto Institute of Technology, Japan |
| Kazushi Ikeda | Nara Institute of Science and Technology, Japan |
| Hayashi Isao | Kansai University, Japan |
| Teijiro Isokawa | University of Hyogo, Japan |
| Piyasak Jeatrakul | Mae Fah Luang University, Thailand |
| Jin-Tsong Jeng | National Formosa University, Taiwan |
| Sungmoon Jeong | Kyungpook National University Hospital, South Korea |
| Danchi Jiang | University of Tasmania, Australia |
| Min Jiang | Xiamen University, China |
| Yizhang Jiang | Jiangnan University, China |
| Xuguo Jiao | Zhejiang University, China |
| Keisuke Kameyama | University of Tsukuba, Japan |
| Shunshoku Kanae | Junshin Gakuen University, Japan |
| Hamid Reza Karimi | Politecnico di Milano, Italy |
| Nikola Kasabov | Auckland University of Technology, New Zealand |
| Abbas Khosravi | Deakin University, Australia |
| Rhee Man Kil | Sungkyunkwan University, South Korea |
| Daeeun Kim | Yonsei University, South Korea |
| Sangwook Kim | Kobe University, Japan |
| Lai Kin | Tunku Abdul Rahman University, Malaysia |
| Irwin King | The Chinese University of Hong Kong, Hong Kong SAR, China |
| Yasuharu Koike | Tokyo Institute of Technology, Japan |
| Ven Jyn Kok | National University of Malaysia, Malaysia |
| Ghosh Kuntal | Indian Statistical Institute, India |
| Shuichi Kurogi | Kyushu Institute of Technology, Japan |
| Susumu Kuroyanagi | Nagoya Institute of Technology, Japan |
| James Kwok | The Hong Kong University of Science and Technology, SAR China |
| Edmund Lai | Auckland University of Technology, New Zealand |
| Kittichai Lavangnananda | King Mongkut's University of Technology Thonburi, Thailand |
| Xinyi Le | Shanghai Jiao Tong University, China |
| Minho Lee | Kyungpook National University, South Korea |
| Nung Kion Lee | University Malaysia Sarawak, Malaysia |
| Andrew C. S. Leung | City University of Hong Kong, Hong Kong SAR, China |
| Baoquan Li | Tianjin Polytechnic University, China |
| Chengdong Li | Shandong Jianzhu University, China |
| Chuandong Li | Southwest University, China |
| Dazi Li | Beijing University of Chemical Technology, China |
| Li Li | Tsinghua University, China |
| Shengquan Li | Yangzhou University, China |

| | |
|---|---|
| Ya Li | Institute of Automation, Chinese Academy of Sciences, China |
| Yanan Li | University of Sussex, UK |
| Yongming Li | Liaoning University of Technology, China |
| Yuankai Li | University of Science and Technology of China, China |
| Jie Lian | Dalian University of Technology, China |
| Hualou Liang | Drexel University, USA |
| Jinling Liang | Southeast University, China |
| Xiao Liang | Nankai University, China |
| Alan Wee-Chung Liew | Griffith University, Australia |
| Honghai Liu | University of Portsmouth, UK |
| Huaping Liu | Tsinghua University, China |
| Huawen Liu | University of Texas at San Antonio, USA |
| Jing Liu | Chinese Academy of Sciences, China |
| Ju Liu | Shandong University, China |
| Qingshan Liu | Huazhong University of Science and Technology, China |
| Weifeng Liu | China University of Petroleum, China |
| Weiqiang Liu | Nanjing University of Aeronautics and Astronautics, China |
| Dome Lohpetch | King Mongkut's University of Technology North Bangoko, Thailand |
| Hongtao Lu | Shanghai Jiao Tong University, China |
| Wenlian Lu | Fudan University, China |
| Yao Lu | Beijing Institute of Technology, China |
| Jinwen Ma | Peking University, China |
| Qianli Ma | South China University of Technology, China |
| Sanparith Marukatat | Thailand's National Electronics and Computer Technology Center, Thailand |
| Tomasz Maszczyk | Nanyang Technological University, Singapore |
| Basarab Matei | LIPN Paris Nord University, France |
| Takashi Matsubara | Kobe University, Japan |
| Nobuyuki Matsui | University of Hyogo, Japan |
| P. Meesad | King Mongkut's University of Technology North Bangkok, Thailand |
| Gaofeng Meng | Chinese Academy of Sciences, China |
| Daisuke Miyamoto | University of Tokyo, Japan |
| Kazuteru Miyazaki | National Institution for Academic Degrees and Quality Enhancement of Higher Education, Japan |
| Seiji Miyoshi | Kansai University, Japan |
| J. Manuel Moreno | Universitat Politècnica de Catalunya, Spain |
| Naoki Mori | Osaka Prefecture University, Japan |
| Yoshitaka Morimura | Kyoto University, Japan |
| Chaoxu Mu | Tianjin University, China |
| Kazuyuki Murase | University of Fukui, Japan |
| Jun Nishii | Yamaguchi University, Japan |

| | |
|---|---|
| Haruhiko Nishimura | University of Hyogo, Japan |
| Grozavu Nistor | Paris 13 University, France |
| Yamaguchi Nobuhiko | Saga University, Japan |
| Stavros Ntalampiras | University of Milan, Italy |
| Takashi Omori | Tamagawa University, Japan |
| Toshiaki Omori | Kobe University, Japan |
| Seiichi Ozawa | Kobe University, Japan |
| Yingnan Pan | Northeastern University, China |
| Yunpeng Pan | JD Research Labs, China |
| Lie Meng Pang | Universiti Malaysia Sarawak, Malaysia |
| Shaoning Pang | Unitec Institute of Technology, New Zealand |
| Hyeyoung Park | Kyungpook National University, South Korea |
| Hyung-Min Park | Sogang University, South Korea |
| Seong-Bae Park | Kyungpook National University, South Korea |
| Kitsuchart Pasupa | King Mongkut's Institute of Technology Ladkrabang, Thailand |
| Yong Peng | Hangzhou Dianzi University, China |
| Somnuk Phon-Amnuaisuk | Universiti Teknologi Brunei, Brunei |
| Lukas Pichl | International Christian University, Japan |
| Geong Sen Poh | National University of Singapore, Singapore |
| Mahardhika Pratama | Nanyang Technological University, Singapore |
| Emanuele Principi | Università Politecnica elle Marche, Italy |
| Dianwei Qian | North China Electric Power University, China |
| Jiahu Qin | University of Science and Technology of China, China |
| Sitian Qin | Harbin Institute of Technology at Weihai, China |
| Mallipeddi Rammohan | Nanyang Technological University, Singapore |
| Yazhou Ren | University of Science and Technology of China, China |
| Ko Sakai | University of Tsukuba, Japan |
| Shunji Satoh | The University of Electro-Communications, Japan |
| Gerald Schaefer | Loughborough University, UK |
| Sachin Sen | Unitec Institute of Technology, New Zealand |
| Hamid Sharifzadeh | Unitec Institute of Technology, New Zealand |
| Nabin Sharma | University of Technology Sydney, Australia |
| Yin Sheng | Huazhong University of Science and Technology, China |
| Jin Shi | Nanjing University, China |
| Yuhui Shi | Southern University of Science and Technology, China |
| Hayaru Shouno | The University of Electro-Communications, Japan |
| Ferdous Sohel | Murdoch University, Australia |
| Jungsuk Song | Korea Institute of Science and Technology Information, South Korea |
| Andreas Stafylopatis | National Technical University of Athens, Greece |
| Jérémie Sublime | ISEP, France |
| Ponnuthurai Suganthan | Nanyang Technological University, Singapore |
| Fuchun Sun | Tsinghua University, China |
| Ning Sun | Nankai University, China |

| Norikazu Takahashi | Okayama University, Japan |
| Ken Takiyama | Tokyo University of Agriculture and Technology, Japan |
| Tomoya Tamei | Kobe University, Japan |
| Hakaru Tamukoh | Kyushu Institute of Technology, Japan |
| Choo Jun Tan | Wawasan Open University, Malaysia |
| Shing Chiang Tan | Multimedia University, Malaysia |
| Ying Tan | Peking University, China |
| Gouhei Tanaka | The University of Tokyo, Japan |
| Ke Tang | Southern University of Science and Technology, China |
| Xiao-Yu Tang | Zhejiang University, China |
| Yang Tang | East China University of Science and Technology, China |
| Qing Tao | Chinese Academy of Sciences, China |
| Katsumi Tateno | Kyushu Institute of Technology, Japan |
| Keiji Tatsumi | Osaka University, Japan |
| Kai Meng Tay | Universiti Malaysia Sarawak, Malaysia |
| Chee Siong Teh | Universiti Malaysia Sarawak, Malaysia |
| Andrew Teoh | Yonsei University, South Korea |
| Arit Thammano | King Mongkut's Institute of Technology Ladkrabang, Thailand |
| Christos Tjortjis | International Hellenic University, Greece |
| Shibata Tomohiro | Kyushu Institute of Technology, Japan |
| Seiki Ubukata | Osaka Prefecture University, Japan |
| Eiji Uchino | Yamaguchi University, Japan |
| Wataru Uemura | Ryukoku University, Japan |
| Michel Verleysen | Universite catholique de Louvain, Belgium |
| Brijesh Verma | Central Queensland University, Australia |
| Hiroaki Wagatsuma | Kyushu Institute of Technology, Japan |
| Nobuhiko Wagatsuma | Tokyo Denki University, Japan |
| Feng Wan | University of Macau, SAR China |
| Bin Wang | University of Jinan, China |
| Dianhui Wang | La Trobe University, Australia |
| Jing Wang | Beijing University of Chemical Technology, China |
| Jun-Wei Wang | University of Science and Technology Beijing, China |
| Junmin Wang | Beijing Institute of Technology, China |
| Lei Wang | Beihang University, China |
| Lidan Wang | Southwest University, China |
| Lipo Wang | Nanyang Technological University, Singapore |
| Qiu-Feng Wang | Xi'an Jiaotong-Liverpool University, China |
| Sheng Wang | Henan University, China |
| Bunthit Watanapa | King Mongkut's University of Technology, Thailand |
| Saowaluk Watanapa | Thammasat University, Thailand |
| Qinglai Wei | Chinese Academy of Sciences, China |
| Wei Wei | Beijing Technology and Business University, China |
| Yantao Wei | Central China Normal University, China |

| | |
|---|---|
| Guanghui Wen | Southeast University, China |
| Zhengqi Wen | Chinese Academy of Sciences, China |
| Hau San Wong | City University of Hong Kong, Hong Kong SAR, China |
| Kevin Wong | Murdoch University, Australia |
| P. K. Wong | University of Macau, SAR China |
| Kuntpong Woraratpanya | King Mongkut's Institute of Technology Chaokuntaharn Ladkrabang, Thailand |
| Dongrui Wu | Huazhong University of Science and Technology, China |
| Si Wu | Beijing Normal University, China |
| Si Wu | South China University of Technology, China |
| Zhengguang Wu | Zhejiang University, China |
| Tao Xiang | Chongqing University, China |
| Chao Xu | Zhejiang University, China |
| Zenglin Xu | University of Science and Technology of China, China |
| Zhaowen Xu | Zhejiang University, China |
| Tetsuya Yagi | Osaka University, Japan |
| Toshiyuki Yamane | IBM, Japan |
| Koichiro Yamauchi | Chubu University, Japan |
| Xiaohui Yan | Nanjing University of Aeronautics and Astronautics, China |
| Zheng Yan | University of Technology Sydney, Australia |
| Jinfu Yang | Beijing University of Technology, China |
| Jun Yang | Southeast University, China |
| Minghao Yang | Chinese Academy of Sciences, China |
| Qinmin Yang | Zhejiang University, China |
| Shaofu Yang | Southeast University, China |
| Xiong Yang | Tianjin University, China |
| Yang Yang | Nanjing University of Posts and Telecommunications, China |
| Yin Yang | Hamad Bin Khalifa University, Qatar |
| Yiyu Yao | University of Regina, Canada |
| Jianqiang Yi | Chinese Academy of Sciences, China |
| Chengpu Yu | Beijing Institute of Technology, China |
| Wen Yu | CINVESTAV, Mexico |
| Wenwu Yu | Southeast University, China |
| Zhaoyuan Yu | Nanjing Normal University, China |
| Xiaodong Yue | Shanghai University, China |
| Dan Zhang | Zhejiang University, China |
| Jie Zhang | Newcastle University, UK |
| Liqing Zhang | Shanghai Jiao Tong University, China |
| Nian Zhang | University of the District of Columbia, USA |
| Tengfei Zhang | Nanjing University of Posts and Telecommunications, China |
| Tianzhu Zhang | Chinese Academy of Sciences, China |

| Ying Zhang | Shandong University, China |
| Zhao Zhang | Soochow University, China |
| Zhaoxiang Zhang | Chinese Academy of Sciences, China |
| Dongbin Zhao | Chinese Academy of Sciences, China |
| Qiangfu Zhao | University of Aizu, Japan |
| Zhijia Zhao | Guangzhou University, China |
| Jinghui Zhong | South China University of Technology, China |
| Qi Zhou | University of Portsmouth, UK |
| Xiaojun Zhou | Central South University, China |
| Yingjiang Zhou | Nanjing University of Posts and Telecommunications, China |
| Haijiang Zhu | Beijing University of Chemical Technology, China |
| Hu Zhu | Nanjing University of Posts and Telecommunications, China |
| Lei Zhu | Unitec Institute of Technology, New Zealand |
| Pengefei Zhu | Tianjin University, China |
| Yue Zhu | Nanjing University, China |
| Zongyu Zuo | Beihang University, China |

# Contents – Part II

**Stability Analysis**

## Optimization

## Supervised Learning

# Other Neural Network Models

# Improved Kernel Density Estimation Self-organizing Incremental Neural Network to Perform Big Data Analysis

Wonjik Kim[1(✉)] and Osamu Hasegawa[1,2]

[1] Department of Systems and Control Engineering, Tokyo Institute of Technology, Tokyo, Japan
{kim.w.ab,hasegawa.o.aa}@m.titech.ac.jp
[2] Inc.SOINN, Cureindo-Building 405, Turuma8-4-30, Tamachi, Tokyo, Japan
oh@soinn.com

**Abstract.** Plenty of data are generated continuously due to the progress in the field of network technology. Additionally, some data contain substantial noise, while other data vary their properties in according to various real time scenarios. Owing to these factors, analyzing big data is difficult. To address these problems, an adaptive kernel density estimation self-organizing neural network (AKDESOINN) has been proposed. This approach is based on the kernel density estimation self-organizing incremental neural network (KDESOINN), which is an extension of the self-organizing incremental neural network (SOINN). An SOINN can study the distribution using the input data online, while KDESOINN can estimate the probability density function based on this information. The AKDESOINN can adapt itself to the changing data properties by estimating the probability density function. Further, the experimental results depict that AKDESOINN succeeds in maintaining the performance of KDE-SOINN, while depicting an ability to adapt to the changing data.

**Keywords:** Neural network · Kernel density estimation · Data analysis
Self-organizing incremental neural network

## 1 Introduction

Due to the expansion of network communications, data are generated continuously. Such data are called big data, and there have been many attempts to analyze and apply them to various research fields [1, 2, 11, 16, 24].

Laney derived three concepts of the big data characteristics, which are as follows [18]:

(1) Data Volume: Massive amounts of data that continue to grow after being generated.
(2) Data Velocity: Increasing numbers of networks are generating data continuously, which means that the data generation velocity is very high.
(3) Data Variety: Data in a pool can be of different types, such as time-series, real environment, artificial environment, textual, and image data.

© Springer Nature Switzerland AG 2018
L. Cheng et al. (Eds.): ICONIP 2018, LNCS 11302, pp. 3–13, 2018.
https://doi.org/10.1007/978-3-030-04179-3_1

These three characteristics that are considered by [18] are known as the 3Vs, and are taken into consideration while dealing with big data.

In machine learning and data analysis research, it is necessary to estimate the probability density. However, it is difficult to estimate the probability density of big data due to the three reasons [22].

First, the density estimator for big data must be nonparametric because of the data volume. Further, we observe that parametric methods are effective for handling fixed data, because it is possible to tune the parameters of the method to obtain an optimal performance. However, the volume of big data is not observed to be constant. Therefore, the volume of big data cannot be analyzed in advance in order to obtain optimal parameters for the density estimator. However, we observe that the nonparametric density estimator is not troublesome, since analyzing and constructing a big data model beforehand is not necessary for a nonparametric density estimator.

Second, the density estimator for big data must use online learning methods due to the observed data velocity. In big data, massive amounts of data grow quickly until the total size of data becomes gigantic. Online learning methods can be sequentially updated using the growing data.

Third, the density estimator for big data must be robust. Data that are collected from real environments often contain noise, which could cause overfitting and decrease performance. Thus, robust methods are required to deal with data that contain noise.

Further, we observe that robustness is defined differently across various fields [9, 13]. In this study, we define robustness as 'a function that provides almost the same results as learning data without noise when learning with noisy data.' [22]. Further, we observe that there are two types of noise. The first type is the noise that is generated by the environment, but that is not related to the objective distribution. Thus, this type of noise needs to be eliminated. The second type is observed to be related to variance and fluctuation. Therefore, this type of noise must be preserved.

The kernel density estimation self-organizing incremental neural network (KDE-SOINN) method [22] satisfies all the three conditions for dealing with big data and is further observed to be robust to noise. However, it cannot adapt to a changing environment. Due to the variety of big data, the structure of data is likely to vary at any instance. Therefore, an ability to adapt to the observed variation of data is required. In this study, we propose a revised KDESOINN method to solve this problem. Further, our proposed method has been termed adaptive KDESOINN (AKDESOINN) in this paper.

## 2 Related Works

### 2.1 Kernel Density Estimation

Kernel Density Estimation (KDE) is a typical nonparametric density estimation approach [23]. The methodology of KDE process is presented in Algorithm 1

**Algorithm 1.** Kernel Density Estimation

(1) **Require**: training samples $\{x_i | x_i \in \mathbb{R}^d, i = 1, 2, \ldots, N\}$, $K$ : kernel function, $H$ : bandwidth matrix

(2) $\hat{p}(x) = \frac{1}{N} \sum_{i=1}^{N} K_H(x - x_i)$

For the kernel function $K$, the Gaussian kernel is often used in an identical manner as that in (1)

$$K_H(x - \mu) = \left(1 / \sqrt{(2\pi)^d |H|}\right) * \exp\left(-(x - \mu)^T H^{-1}(x - \mu)/2\right) \qquad (1)$$

H in algorithm 1 is a parameter, which influences the performance of the estimation function. Further, attempts have been made to optimize the estimation function [8, 10]. KDE has been investigated using several methods such as by method of setting the number of kernels [3], gradient descent method [19], and online clustering method [17].

## 2.2 Self-organizing Incremental Neural Network

In the field of artificial intelligence, artificial neural networks have been recently proposed. They are usually classified into two groups, namely, supervised and unsupervised learning [25].

SOINN is an unsupervised learning method that is driven by growing neural gas [4]. There are several kinds of SOINN, including two-layer [5], enhanced [6], and adjusted SOINN [7]. Since the adjusted SOINN has less parameters than that of the other SOINNs, it is generally used in applied research [12, 14, 15].

While SOINN learns from the training data, it constructs a data network through competitive learning. Various nodes are added or deleted from the network or they may update their location. Further, the edges are added or deleted in a similar manner as the nodes. Thus, the SOINN network is updated in order to approximate the distribution using the added input data.

The flowchart of the adjusted SOINN is depicted in Fig. 1, and its procedural flow is presented in Algorithm 2

**Algorithm 2.** Adjusted SOINN process

(1) **Require**: A: set of all neurons. $C \subset A \times A$: set of all edges. $N_i$: set of all neighbors of neuron $i$. $W_i$: weight of neuron $i$. $\lambda$: time period to delete redundant neurons. $age_{max}$: parameter to delete edges.

(2) **if** first time of input **then**

(3)  $A \leftarrow c_1, c_2$; randomly pick up two vectors from training data to initialize the neuron set.

(4)  $C \leftarrow \emptyset$

(5) **end if**

(6) **while** input data $\xi$ exist **do**

(7)  $s_1 \leftarrow argmin_{c \in A} \|\xi - W_c\|$: find out the winner.

(8)  $s_2 \leftarrow argmin_{c \in A \setminus s_1} \|\xi - W_c\|$: find out second winner.

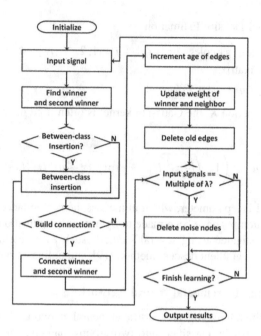

**Fig. 1.** Flowchart of SOINN

(9)  calculate similarity thresholds $T_{s_1}, T_{s_2}$. If $i$ got neighbors, $T_i$ is the distance to the farthest neighbor, else the distance to the nearest neuron.

(10)  **if** $\|\xi - W_{s_1}\| > T_{s_1}$ or $\|\xi - W_{s_2}\| > T_{s_2}$ **then**

(11)  $A \leftarrow A \cup \xi$: insert $\xi$ as a new neuron.

(12)  **else**

(13)  **if** $(s_1, s_2) \notin C$: there is no edge between the winner and second winner, **then**

(14)  $C \leftarrow C \cup (s_1, s_2)$: add new edge into the network

(15)  **end if**

(16)  $age_{(s_1,s_2)} \leftarrow 0$: reset the age of $(s_1, s_2)$

(17)  $age_{(s_1,i)} \leftarrow age_{(s_1,i)} + 1 (\forall i \in N_{s_i})$: increase age of edges connected with the winner by 1.

(18)  $\Delta W_{s_i} = \epsilon(t_{s_1})(\xi - W_{s_1}), \Delta W_i = \epsilon(100t_i)(\xi - W_i)(\forall i \in N_{s_i}), \epsilon(t) = \frac{1}{t}$

(19)  using $vartriangle W_{s_i}, \Delta W_i$ to adjust the winner and its neighbors

(20)  delete edges whose age is larger than $age_{max}$

(21)  among these neurons which the edge deleted in last step connected to, delete neurons having no neighbors.

(22)  **end if**

(23)  **if** input data number becomes $n \times \lambda (n \in N^+)$ **then**

(24)  Delete neurons having less than one neighbor

(25)  **end if**

(26)  **end while**

## 2.3 KDESOINN

KDESOINN is an extended version of the adjusted SOINN [22]. It determines the structure of the network using each kernel in the node of a local network that is located near the node. Additionally, it estimates the probability function using the sum of the kernels. In the adjusted SOINN, only the Euclidean distance is used for calculating the similarity thresholds. Conversely, KDESOINN calculates the threshold using Algorithm 3.

**Algorithm 3.** KDESOINN threshold calculation

(1) **Require**: A: set of all neurons. $\xi$: new sample data. $P_i$: set of nodes connected to node $i$. $\rho$: parameter for threshold. $\varpi_i \in \mathbb{R}^d$: positional vector of node $i$. $t_i$: number of wins of node $i$ in competitive learning. $I$: identity matrix. $\Theta_i$: threshold region of node $i$.

(2) calculate $\gamma_i = \begin{cases} \min_{p \in P_i} \|w_p - w_i\| & (P_i \neq \phi) \\ \min_{p \in A\{i\}} \|w_p - w_i\| & (otherwise) \end{cases}$

(3) $T_{P_i} \leftarrow \sum_{i \in P_i} t_i$

(4) $C_i \leftarrow \frac{1}{T_{P_i}} \sum_{p \in P_i} t_p (w_p - w_i)(w_p - w_i)^T$

(5) $M_i \leftarrow C_i + \rho \gamma_i I$

(6) threshold region $\Theta_i = (\xi - w_i)^T M_i^{-1}(\xi - w_i) \leq 1$

KDESOINN can divide clusters more effectively than the adjusted SOINN. The entire process of KDESOINN is presented in Algorithm 4

**Algorithm 4.** KDESOINN process

(1) **Require**: A: set of all neurons. $C \subset A \times A$: set of all edges. $N_i$: set of all neighbors of neuron $i$. $W_i$: weight of neuron $i$. $\lambda$: time period to delete redundant neurons. $age_{max}$: parameter to delete edges. $P_i$: set of nodes connected to node $i$. $\rho$: parameter for threshold. $t_i$: number of wins of node $i$ in competitive learning. $I$: identity matrix. E(G): set of edges in graph G.

(2) **if** first time of input **then**

(3) $A \leftarrow c_1, c_2$; randomly pick up two vectors from training data to initialize the neuron set.

(4) $C \leftarrow \emptyset$

(5) **end if**

(6) **while** input data $\xi$ exist **do**

(7) $s_1 \leftarrow argmin_{c \in A} \|\xi - W_c\|$: find out the winner.

(8) $s_2 \leftarrow argmin_{c \in A \setminus s_1} \|\xi - W_c\|$: find out second winner.

(9) calculate similarity thresholds $\Theta_{s_1}, \Theta_{s_2}$ by algorithm 3.

(10) **if** $(\xi - W_{s_1})^T M_{s_1}^{-1}(\xi - W_{s_2}) > 1$ or $(\xi - W_{s_2})^T M_{s_2}^{-1}(\xi - W_{s_2}) > 1$ **then**

(11) $A \leftarrow A \cup \xi$: insert $\xi$ as a new neuron.

(12) **else**

(13) **if** $(s_1, s_2) \notin C$: there is no edge between the winner and second winner, **then**

(14) $C \leftarrow C \cup (s_1, s_2)$: add new edge into the network

(15) **end if**

(16)  $age_{(s_1,s_2)} \leftarrow 0$: reset the age of $(s_1,s_2)$

(17)  $age_{(s_1,i)} \leftarrow age_{(s_1,i)} + 1(\forall i \in N_{s_i})$: increase age of edges connected with the winner by 1.

(18)  $\Delta W_{s_i} = \epsilon(t_{s_1})(\xi - W_{s_1}), \Delta W_i = \epsilon(100 t_i)(\xi - W_i)(\forall i \in N_{s_i}), \epsilon(t) = \frac{1}{t}$

(19)  using $vartriangle W_{s_i}, \Delta W_i$ to adjust the winner and its neighbors

(20)  delete edges whose age is larger than $age_{max}$

(21)  among these neurons which the edge deleted in last step connected to, delete neurons having no neighbors.

(22)  **end if**

(23)  **if** input data number becomes $n \times \lambda(n \in N^+)$ **then**

(24)  delete neurons having no neighbor

(25)  create a k-NN graph $G$ whose set of nodes is A.

(26)  $C \leftarrow C \cup \{(i,j)|(i,j) \in E(G), (j,i) \in E(G)\}$

(27)  **end if**

(28)  **end while**

(29)  create a k-NN graph $G$ whose set of nodes is A.

(30)  $C \leftarrow C \cup \{(i,j)|(i,j) \in E(G), (j,i) \in E(G)\}$

## 3  Proposed Method

To improve KDSOINNs ability of adapting to the changing data, algorithm 5 was used after line 10 of Algorithm 4

**Algorithm 5.** Adaptive step

(1)  **Require**: $s_1$: first winner. $s_2$: second winner. $\eta$: parameter for adapting. $\xi$: new sample data.

(2)  $D_{s_1} \leftarrow |s_1 - \xi|, D_{s_2} \leftarrow |s_2 - \xi|$

(3)  update $s_1 \leftarrow s_1 + \frac{\eta D_{s_2}}{D_{s_1}+D_{s_2}}(\xi - s_1)$

(4)  update $s_2 \leftarrow s_2 + \frac{\eta D_{s_1}}{D_{s_1}+D_{s_2}}(\xi - s_2)$

By applying algorithm 5, SOINN can adapt to the data as they change with time. $\eta$ is the adaptation parameter. Further, if $\eta$ is observed to be equal to 0, the performance of AKDESOINN is observed to be exactly the same as that of KDESOINN. If $\eta$ is observed to be bigger than 1, it is possible that it can fit over $\xi$. To avoid overfitting and low performance, it is recommended to set $\eta$ within the range of 0 to 1. The entire process of AKDESOINN is presented in algorithm 6.

**Algorithm 6.** AKDESOINN process

(1)  **Require**: A: set of all neurons. $C \subset A \times A$: set of all edges. $N_i$: set of all neighbors of neuron $i$. $W_i$: weight of neuron $i$. $\lambda$: time period to delete redundant neurons. $age_{max}$: parameter to delete edges. $P_i$: set of nodes connected to node $i$. $\rho$: parameter for threshold. $\eta$: parameter for adapting. $t_i$: number of wins of node $i$ in competitive learning. $I$: identity matrix. E(G): set of edges in graph G.

(2)  **if** first time of input **then**

(3)  $A \leftarrow c_1, c_2$; randomly pick up two vectors from training data to initialize the neuron set.

(4)  $C \leftarrow \emptyset$

(5)  **end if**

(6)  **while** input data $\xi$ exist **do**

(7)  $s_1 \leftarrow argmin_{c \in A} \xi - W_c$: find out the winner.

(8)  $s_2 \leftarrow argmin_{c \in A \setminus s_1} \xi - W_c$: find out second winner.

(9)  calculate similarity thresholds $\Theta_{s_1}, \Theta_{s_2}$ by Algorithm 3.

(10)  **if** $(\xi - W_{s_1})^T M_{s_1}^{-1} (\xi - W_{s_2}) > 1$ or $(\xi - W_{s_2})^T M_{s_2}^{-1} (\xi - W_{s_2}) > 1$ **then**

(11)  $D_{s_1} \leftarrow |s_1 - \xi|, D_{s_2} \leftarrow |s_2 - \xi|$

(12)  update the location of $s_1 \leftarrow s_1 + \frac{\eta D_{s_2}}{D_{s_1} + D_{s_2}} (\xi - s_1)$

(13)  update the location of $s_2 \leftarrow s_2 + \frac{\eta D_{s_1}}{D_{s_1} + D_{s_2}} (\xi - s_2)$

(14)  $A \leftarrow A \cup \xi$: insert $\xi$ as a new neuron.

(15)  **else**

(16)  **if** $(s_1, s_2) \notin C$: there is no edge between the winner and second winner, **then**

(17)  $C \leftarrow C \cup (s_1, s_2)$: add new edge into the network

(18)  **end if**

(19)  $age_{(s_1, s_2)} \leftarrow 0$: reset the age of $(s_1, s_2)$

(20)  $age_{(s_1, i)} \leftarrow age_{(s_1, i)} + 1 (\forall i \in N_{s_i})$: increase age of edges connected with the winner by 1.

(21)  $\Delta W_{s_i} = \epsilon(t_{s_1})(\xi - W_{s_1}), \Delta W_i = \epsilon(100 t_i)(\xi - W_i)(\forall i \in N_{s_i}), \epsilon(t) = \frac{1}{t}$

(22)  using $vartriangle W_{s_i}, \Delta W_i$ to adjust the winner and its neighbors

(23)  delete edges whose age is larger than $age_{max}$

(24)  among these neurons which the edge deleted in last step connected to, delete neurons having no neighbors.

(25)  **end if**

(26)  **if** input data number becomes $n \times \lambda (n \in N^+)$ **then**

(27)  delete neurons having no neighbor

(28)  create a k-NN graph $G$ whose set of nodes is A.

(29)  $C \leftarrow C \cup \{(i,j) | (i,j) \in E(G), (j,i) \in E(G)\}$

(30)  **end if**

(31)  **end while**

(32)  create a k-NN graph $G$ whose set of nodes is A.

(33)  $C \leftarrow C \cup \{(i,j) | (i,j) \in E(G), (j,i) \in E(G)\}$

## 4  Experimental Study

In order to compare AKDESOINN with other methods, experimental evaluations were performed to evaluate the robustness, calculation time, accuracy, and adaptation ability of SOINN, KDESOINN, and AKDESOINN. The experimental environment comprised MATLAB2017b that was used on a personal computer having an eight-core CPU at 3.40 GHz and 16.0 GB RAM.

## 4.1  Fixed Gaussian Distribution

Initially, we evaluated the performance of the proposed method using a fixed Gaussian distribution. Specific details regarding the experiment are described in Table 1.

**Table 1.**  Information of experiment 1

| Description | Details |
|---|---|
| Gaussian distribution 1 | $\mu = (1, 1), \sigma = 0.25$ |
| Gaussian distribution 2 | $\mu = (0, 0), \sigma = 0.25$ |
| Gaussian distribution 3 | $\mu = (-1, -1), \sigma = 0.25$ |
| Uniform distribution | Range of $(-2, -2)$ to $(2, 2)$ |
| Number of data | 1000 in each distribution Total number: 4000 |
| $\lambda$ | 200 |
| $age_{max}$ | 50 |
| $\rho$ | 0.1 |
| H | 1 |

The experiment was repeated 100 times. Further, the Jensen-Shannon divergence was used to compare the accuracy [20], and the results are presented in Table 2.

**Table 2.**  Result of the 100 trials of experiment 1

| Description | Mean | Variance | Total computation time [s] |
|---|---|---|---|
| SOINN | 4.40E−02 | 3.63E−05 | 4.42E01 |
| KDESOINN | 1.03E−02 | 8.68E−07 | 1.09E02 |
| AKDESOINN | 1.10E−02 | 1.71E−06 | 1.27E02 |

According to Table 2, KDESOINN is observed to depict the most effective performance in this experiment. Further, AKDESOINN depicts better performance than SOINN.

## 4.2  Changing Gaussian Distribution

To evaluate the adaptation performance to the changing data, a Gaussian distribution was used in the experiment. Specific details of the experiment are provided in Table 3.

The experiment was repeated 100 times, and the results were compared using the Jensen-Shannon divergence, in a similar manner as that in experiment 1. The results of the comparison are presented in Table 4.

According to Table 4, AKDESOINN was observed to be the most effective in experiment 2 in terms of mean value

Table 3. Information of experiment 2

| Description | Details |
|---|---|
| Gaussian distribution 1 | $\mu = (1, 1)$ |
| Gaussian distribution 2 | $\mu = (0, 0)$ |
| Gaussian distribution 3 | $\mu = (-1, -1)$ |
| Sigma | $\sigma : 0.25 \rightarrow 0.15$ with $0.01$ interval |
| Uniform distribution | Range of $(-2, -2)$ to $(2, 2)$ |
| Number of data | 100 in each distribution<br>Total number 4400 |
| $\Lambda$ | 200 |
| $age_{max}$ | 50 |
| $\rho$ | 0.1 |
| H | 1 |

Table 4. Result of the 100 trials of experiment 2

| Description | Mean | Variance | Total calculation time [s] |
|---|---|---|---|
| SOINN | 6.88E−02 | 7.72E−05 | 52.2E01 |
| KDESOINN | 2.04E−02 | 2.10E−06 | 1.33E02 |
| AKDESOINN | 1.88E−02 | 3.13E−06 | 1.73E02 |

## 5  Conclusion

In this study, AKDESOINN was proposed, not only as a robust fast online nonparametric density estimator, but also as an adaptive method to the changing data. KDESOINN is a method combining both KDE and SOINN and is known to outperform the existing nonparametric density estimators in terms of robustness, calculation cost, and accuracy. The revised KDESOINN algorithm was successful in adapting to the changing data without depicting any performance loss.

For future studies, we could analyze the application of AKDESOINN to meteorological data. Because of the extraordinary climatic conditions, analysis models need to be updated constantly [21]. Because AKDESOINN can adapt its model online with analyzing the data, AKDESOIN may be effective in addressing this problem of constant change of large amounts of climatic data. Apart from the meteorological data, AKDESOINN could be extensively applicable to other fields that require nonparametric density estimation with a suitable adaptation ability to the changing data.

# References

1. Amimi, S., Ilias Gerostathopoulos, I,. Prehofer, C.: Big data analytics architecture for real-time traffic control. In: 5th IEEE International Conference on Models and Technologies for Intelligent Transportation Systems, pp. 710–715 (2017)
2. Anker, S., Asselbergs, F.W., Brobert, G., Vardas, P., Grobbee, D.E., Cronin, M.: Big data in cardiovascular disease. Eur. Hear. J. **38**(24), 1863–1865 (2017)
3. Deng, Z., Chung, F.L., Wang, S.: FRSDE: fast reduced set density estimator using minimal enclosing ball approximation. Pattern Recognit. **41**(4), 1363–1372 (2008)
4. Fritzke, B.: A growing neural gas network learns topologies. In: Advances in Neural Information Processing Systems, vol. 7, pp. 625–632, MIT Press, USA (1995)
5. Furao, S., Hasegawa, O.: An incremental network for on-line unsupervised classification and topology learning. Neural Netw. **19**(1), 90–106 (2006)
6. Furao, S., Ogura, T., Hasegawa, O.: An enhanced self-organizing incremental neural network for online unsupervised learning. Neural Netw. **20**(8), 893–903 (2007)
7. Furao, S., Hasegawa, O.: A fast nearest neighbor classifier based on self-organizing incremental neural network. Neural Netw. **211**(10), 1537–1547 (2008)
8. Hall, P., Sheather, S.J., Jones, M.C., Marron, J.S.: On optimal data-based bandwidth selection in kernel density estimation. Biometrika **78**(2), 263–269 (1991)
9. Huber, P.J., Ronchetti, E.M.: Robust Statistics. International Encyclopedia of Statistical Science, pp. 1248–1251. Springer Press, Berlin (2011)
10. Jones, M.C., Marron, J.S., Sheather, S.J.: A brief survey of bandwidth selection for density estimation. J. Am. Stat. Assoc. **91**(433), 401–407 (1996)
11. John, W.S.: Big data: a revolution that will transform how we live, work, and think. Int. J. Advert. **33**(1), 181–183 (2014)
12. Kawewong, A., Pimup, R., Hasegawa, O.: Incremental learning framework for indoor scene recognition. In: Proceedings of the 27th AAAI Conference on Artificial Intelligence, pp. 496–502. Bellevue (2013)
13. Kim, J., Scott, C.D.: Robust kernel density estimation. J. Mach. Learn. Res. **13**, 2529–2565 (2012)
14. Kim, W., Hasegawa, O.: Prediction of tropical storms using self-organizing incremental neural networks and error evaluation. In: Liu, D., Xie, S., Li, Y., Zhao, D., El-Alfy, E.S. (eds.) Neural Information Processing. ICONIP 2017. LNCS, vol. 10636, pp. 846–855. Springer Press, Cham (2017). https://doi.org/10.1007/978-3-319-70090-8_86
15. Kim, W., Hasegawa, O.: Time series prediction of tropical storm trajectory using self-organizing incremental neural networks and error evaluation. J. Adv. Comput. Intell. Intell. Inform. **22**(4), 465–474 (2018)
16. Rob, K., Gavin, M.: What makes Big Data, Big Data? Exploring the ontological characteristics of 26 datasets. Big Data & Society **3**(1) (2016)
17. Kristan, M., Leonardis, A., Skočaj, D.: Multivariate online kernel density estimation with Gaussian kernels. Pattern Recognit. **44**, 2630–2642 (2011)
18. Laney, D.: 3D data management: controlling data volume, velocity, and variety. META group research note 6, 1 (2001)
19. Bottou, L.: Large-scale machine learning with stochastic gradient descent. In: Lechevallier, Y., Saporta, G. (eds,) Proceedings of COMPSTAT 2010, pp. 177–186. Springer Press, Heidelberg (2010). https://doi.org/10.1007/978-3-7908-2604-3_16
20. Lin, J.: Divergence measures based on the Shannon entropy. IEEE Trans. Inf. Theory **37**(1), 145–151 (1991)

21. Lobell, D.B., et al.: Prioritizing climate change adaptation needs for food security in 2030. Science **319**(5863), 607–610 (2008)
22. Nakamura, Y., Hasegawa, O.: Nonparametric density estimation based on self-organizing incremental neural network for large noisy data. IEEE Trans. Neural Netw. Learn. Syst. **28**(1), 8–17 (2017)
23. Parzen, E.: On estimation of a probability density function and mode. Ann. Math. Stat. **33**, 1065–1076 (1962)
24. Chris, P.P.: Big knowledge from big data in functional genomics. Emerg. Top. Life Sci. **1**(3), 245–248 (2017)
25. Zurada, J.M.: Introduction to Artificial Neural Systems. West, St. Paul (1992)

# HISBmodel: A Rumor Diffusion Model Based on Human Individual and Social Behaviors in Online Social Networks

Adil Imad Eddine Hosni[1,2], Kan Li[1(✉)], and Sadique Ahmed[1]

[1] School of Computer Science, Beijing Institute of Technology,
100081 Beijing, China
{hosni.adil.emp,likan}@bit.edu.cn, ahmad01.shah@gmail.com
[2] Ecole Militaire Polytechnique, Bordj El-Bahri, Algiers 16046, Algeria

**Abstract.** This paper attempts to address the rumor propagation problem in online social networks (OSNs) and proposes a novel rumor diffusion model, named the HISBmodel. Its originality lies in the consideration of various human factors such as the human social and individual behaviors and the individuals' opinions. Moreover, we present new metrics that allow accurate assessment of the propagation of rumors. Based on this model, we present a strategy to minimize the influence of the rumor. Instead of blocking nodes, we propose to launch a truth campaign to raise the awareness to prevent the influence of a rumor. This problem is formulated from the perspective of a network inference using the survival theory. The experimental results illustrate that the HISBmodel depicts the evolution of rumor propagation more realistic than classical models. Moreover, Our model highlights the impact of human factors accurately as proven in the studies of the literature. Finally, these experiments showed the outstanding performance of our strategy to minimize the influence of the rumor by selecting precisely the candidate nodes to diminish the influence of the rumor.

**Keywords:** Rumor propagation · Humans individual behaviors
Humans social behaviors · Rumor influence minimization

## 1 Introduction

Online social networks (OSNs) have changed the Internet ecosystems in which they have infiltrated every aspect of human lives. These OSNs, such as Twitter, Facebook, and Weibo[1], are becoming increasingly popular each day. They provide users with a variety of online social activities (e.g., content publishing, friendship creation, and commenting), permitting more extensive, fast, and real-time information access. Moreover, a psychological study showed that when individuals are approved in OSNs by the content they shared (e.g., posts, tweets, comments),

---

[1] http://weibo.com/.

© Springer Nature Switzerland AG 2018
L. Cheng et al. (Eds.): ICONIP 2018, LNCS 11302, pp. 14–27, 2018.
https://doi.org/10.1007/978-3-030-04179-3_2

it activates a section of the brain called the nucleus accumbens. This section is activated when an individual is doing pleasurable things (such as winning money, etc.) [17]. Eventually, this is a formal proof that individuals are influenced by OSNs to spread any information including rumors. The propagation of rumors in an OSN often results in a faster and wider spread than ever before. This propagation can quickly change public opinion [8], create social panic, and even end up with chaos [6]. Undoubtedly, rumors represent a threat to society, which must be stopped as soon as it is detected, so as to reduce its influence.

Various works address the problem of rumor influence minimization (RIM). The major lacks of these works consist in the weakness of the diffusion models to capture the human social activities in an OSN. Modeling the propagation of rumors is increasingly essential to overcome its adverse effect. An efficient model must reflect a real propagation of a rumor and consider the crucial aspects involved in the process to permit a better understanding of this phenomenon. Mastering the problem is half the solution; this allows us to set up strategies to limit the spread of undesirable information and test them to ensure that they can be applied in the real world. The large size and complex topology of OSNs, as well as various users' characteristics, make this problem even more challenging.

Realizing the significant role of a diffusion model in the problem of propagation of the rumor, the attention in this paper has been pointed to it. Therefore, we propose a novel rumor propagation model, named the HISBmodel, considering the human individual and social behaviors in OSNs. Accordingly, we introduce an individual behavior formulation describing the attractiveness of users towards a rumor analogous to a damped harmonic motion. We establish rules of rumor transmission between a pair of individuals. Additionally, we incorporate the opinion of the individuals in the propagation process. Then, we present new metrics that allow an accurate assessment of the impact of the rumor spreading in an OSN. Based on this model, we propose the truth campaign strategy for the RIM problem which aims to raise individuals' awareness and prevent the spread of the rumor. The survival theory is used to analyze the likelihood of the nodes that could be selected as candidates by the strategy. Finally, experiments were performed on real datasets to illustrate the realism of our model to depict the evolution of rumor propagation. Furthermore, experiments are conducted to show the performance of our strategy to minimize the influence of the rumor.

The paper is organized as follows. In Sect. 2, we introduce related work and the preliminary knowledge. Section 3 describes the proposed rumor propagation model. In Sect. 4, we present the proposed RIM strategy. In Sect. 5, we display the results of experiments. Finally, we conclude this paper in Sect. 6.

## 2   Related Work and Preliminary Knowledge

### 2.1   Rumor Propagation Problem in OSNs

By carefully scrutinizing the existing work about rumor propagation problems, we can categorize these works based on their designing assumptions in two classes: macroscopic and microscopic approaches.

**Macroscopic Approaches.** The first category is mainly based on the Epidemic models [5] where research engaging rumor propagation under these models started during the 1960s. Quite recently, research has tended to focus on the role of human individual and social behaviors, such as the forgetting and remembering mechanism [24–26], the hesitating mechanism [22], the education level [1], the trust mechanism [21], the herd mentality [20], and the role of social reinforcement [15,19] in the spreading of online behavior. Nevertheless, these models are macroscopic approaches, in which they consider all the users have identical characteristics with the same abilities to influence others. The variety of the human being and the complexity of the human behaviors beyond our imagination is neglected, which are the primary challenge to model such a phenomenon as the spreading of rumors. Additionally, the spread of rumors and epidemics is different, in that individuals have the ability to accept the rumor and decide to spread it. However, in epidemics, individuals do not play a role in the infection process.

**Microscopic Approaches.** The known models in this category are the independent cascades (IC) model, the linear threshold (LT) model [12], the energy model [10], and Galam's Model [8]. The IC and LT are most known models of this category where they are the milestone for different information diffusion problems in general and RIM in particular. Works investigated the blocking nodes or link [13,18] strategies to limit the spread of undesirable information. While, other researchers have proposed to initiate a truth (good) campaign that fights the false (bad) campaign [2,4,7,11]. Accordingly, researchers have proposed to improve the IC [4] and LT [2] models to a Multi-Campaign Model. Research in this direction focuses on designing strategies to limit the spread of rumors where little attention has been paid to the propagation model. Furthermore, these models still lack in some major details, which they failed to capture some aspects of the human social and individual behaviors.

### 2.2 Damped Harmonic Motion

For better clarity of the paper, a brief introduction to the damped harmonic motion is given in this subsection. In mechanics and physics, a damped harmonic motion is a type of periodic or oscillation motion, often described in terms of a ball linked to a horizontal spring on a table. The immobile state is the equilibrium state. When the ball is moved from this state, it stretches the spring, which results in a restoring force acting directly proportional to the displacement. The ball starts a back and forth motion with frequency $\omega$. The damped harmonic motion experiences friction in which dissipating forces would eventually damp the motion with parameter $\beta$ to the point that the oscillation would no longer occur. Thus, if a system moves under the combined influence of a linear restoring force and a resisting force, the motion is described by the equation [16]

$$x(t) = A_0 e^{-\beta t} \cos(\omega t + \delta), \tag{1}$$

where $x(t)$ is the position of the system at time $t$, $\delta$ is the phase of the motion determining the starting point at $t = 0$, and $A_0$ is the amplitude of the oscillation.

# 3 Proposed Rumor Propagation Model

In this section, the proposed model will be detailed where the light will be spotted in three major details: the individual behavior towards a rumor, individuals opinion formulation and the established rumor transmission rules.

## 3.1 Individual Behavior Toward a Rumor Formulation

While analyzing the individual behavior in OSN, we were inspired by a model of physics that fits the description of the behaviors. We use the analogy that the attraction of an individual to a rumor is similar to an oscillator systems when it is displaced from its equilibrium position. The individual's attraction to the rumor is initially large and then exhibits a gradual downtrend [10]. Likewise, the amplitude of the motion is high in the beginning, then decreases gradually, depending on the damping parameter. The damping parameter represents, in this case, the individuals' background knowledge (IBK) about the rumor, which can define the abilities of an individual to evaluate the trustworthiness of a rumor [1]. However, due to the hesitating mechanism (HM), an individual can eventually have a latent time before spreading the rumor which is relatively related to the degree of doubt of an individual toward a rumor [22] which is analogous to the phase of the system. Furthermore, during the propagation process, individuals can cease and restart transmitting the rumor due to the forgetting-remembering (FR) factor, which has been studied by various works [24–26]. This parameter represents the periodicity of an individual to switch between the forgetting and remembering phases. Finally, for non-negative values of the individual's attraction, we consider absolute value of the formula. An individual's attraction to the rumor is presented as follows

$$A(t) = A_{int}e^{-\beta t}|\cos(\omega t + \delta)|, \tag{2}$$

where $A(t)$ is the attraction of the individual to the rumor at the time $t$, $A_{int}$ is the initial attraction to the rumor, $\beta$ represents the IBK, the FR factor $\omega$ represents the period of forgetting and remembering and $\delta$ is the HM factor.

## 3.2 Individual Opinion Formulation

During the propagation process of a rumor, individuals identify the same rumor differently. Individuals can confirm, refute, question, or talk about matters of interest. The existing rumor propagation models did not consider this aspect, which has been proved by [3] its significant role in the diffusion of information. In this work, we adopt an additive model to estimate $B_v$, the opinion of an individual $v$ toward the rumor. Each opinion correspond to a range of values, where

$B_v \in [-\infty, 0]$ is denying, $B_v \in [0, 10]$ is neutral, $B_u v \in [10, 20]$ is questioning, and $B_v \in [20, \infty]$ is supporting. This formulation is based on the effect of herd behavior on the propagation of rumors [20]. It refers to the fact that individuals are influenced in OSN by other users. The studies have shown that most people exhibit a herd mentality phenomenon where in some cases it easily causes people to follow blindly other and adopt their opinions. However, when individuals receive the same information more than one time, it may not affect them as the first time due to the information redundancy. Therefore, we define the opinion of an individual $v$ as follows:

$$B_v(t) = \sum_{u \in \mathbb{N}^v} \sum_{j=1}^{n} \frac{B_u(t-1)}{j}, \quad \text{for} \quad t > 0 \tag{3}$$

where $\mathbb{N}^v$ is the set of neighbors the individual $v$. $n$ indicating the number of times $v$ received the rumor from a single neighbor.

### 3.3   Human Social Interactions Rules

In this section, attention is brought to human interactions, and the emphasis is on: when will the rumor be sent? when will it be accepted?' Thus, we define the rumor transmission rules between two pairs of nodes. Inspired by the work of [10,18], these rules are evaluated in two steps: sending probability and acceptance probability. We define rumor transmission probability between two pairs of nodes, $u$ and $v$ as follows

$$p_{u,v}(t) = p_u^{send}(t) p_{v,u}^{acc}. \tag{4}$$

Firstly, the sending probability estimates the chances of an individual to send the rumor to his neighbors. We assume that it depends on the individual's attraction to the rumor. Thus, we definedthe sending probability as $A(t)/A_{int}$, where the probability of a node $u$ to send the rumor at time $t$ is

$$p_u^{send}(t) = e^{-\beta t} |\cos(\omega t + \delta)|. \tag{5}$$

The acceptance probability evaluates the chances of an individual to accept the rumor from his neighbor. We consider high-degree nodes had greater chances to send the rumor and a greater ability (authority) to influence other nodes, but they cannot be easily influenced. Accordingly, we definedbalanced weighted probability after considering the impact of the influence of both the sender $u$ and the receiver $v$ and the effect of the social reinforcement, which is as follows

$$p_{v,u}^{acc} = \frac{1}{1 + d_v/d_u}, \tag{6}$$

where $p_{v,u}^{acc}$ is the acceptance probability of the node $v$ from $u$, and $d_u$ and $d_v$ are the connection degrees of nodes $v$ and $u$ respectively.

## 3.4   Rumor Propagation Process

Based on the above analysis and analogy, we can simulate the progress of rumor propagation by the HISBmodel. Consider a population of $N$ users of an OSN represented by a graph $G(V, E, C)$. Set of nodes $V$ represents the users, where $|V| = N$ and the set of edges $E$ refers to the relationship between users. $C$ is a set of individuals' characteristics, where $\forall c_i \in C$, $c_i = (\omega_i, \beta_i, \delta_i)$ represents the characteristics of the individual $v_i$. It starts with a set of spreaders with different opinions assigned to each node $v$ randomly $(B_v(0))$. Each time a node accepts a rumor, he has a single chance to transmit the rumor to their neighbors, according to Eq. 5. During this process, if an ignoramus accepts the rumor based on the probability of Eq. 6, she/he will be a spreader and obeys the behavior toward the rumor Eq. 2; simultaneously, its opinion is updated. Any node can accept more then one rumor, but he have a single chance to spread each accepted rumor. When a spreader's attraction to the rumor fades away, he becomes a stifler. Finally, the spreading process ends when rumor popularity fades $R(t) \simeq 0$.

$$R(t) = \sum_{v \in V} A_v(t) \cdot d_v, \tag{7}$$

where $R(t)$ is the accumulative attraction of all the individuals considering their authority, $d_v$ the connection degrees of $v$. The rumor popularity is a metric proposed by our model illustrating evolution of the rumor considering the impact of each user. Moreover, this model highlights the evolution of the opinion of individuals and gives their final proportion. Considering the individuals with a negative opinion do not contribute to the spread of the influence of the rumor. We evaluate the impact of the rumor by considering the number of individuals who believe the rumor which it will be exploited by our RIM strategy.

## 4   Rumor Influence Minimization Strategy Formulation

Based on the proposed propagation model, we present truth campaign strategies to minimize the influence of rumor. This strategy aims to raise the level of awareness of individuals to prevent the adoption of the rumor by starting a campaign of truth to fight the rumor. We adopt the addictive survival model inspired by [9] to analyze and select the most influential nodes to initiate the campaign of truth. We formulate the problem as follows: we assume that the rumor was propagated through the OSN and it is detected at time $t = t_{det}$. At this time, we consider the population $V$ is divided into two sets $V = V_{B+} + V_{B-}$ where $V_{B-}$ is set of individuals who have a negative opinion and $V_{B+}$ is the rest of the population. Then, our goal is to select $k$ most influential nodes from $V_{B+}$ to propagate a negative opinion about the rumor to prevent the individuals from adopting the rumor. We adopt the survival theory to analyze the likelihood of a node $v$ getting infected at each time step $t$ during the propagation process. The survival function is given as follows

$$S(t) = Pr(T > t) = 1 - F(t), \tag{8}$$

---

**Algorithm 1.** Truth campaign strategy

---
**Input**: $G(V, E, C), V_{B+}, V_{B-}, k$.
**for** $i \leftarrow 1 : k$ **do**

$\quad\left\lfloor \begin{array}{l} u = \arg\max\limits_{v \in V_{B+}} \left[ f_{V_{B-} \cup \{v\}}(t_{det}) - f_{V_{B-}}(t_{det}) \right]; \\ V_{B+} = V_{B+} \backslash u; \\ V_{B-} = V_{B-} \cup \{u\}; \end{array} \right.$

**Output**: $V_{B-}$

---

where $F(t)$ is the cumulative distribution function. $T$ is a continuous random variable representing the infection of individuals by a rumor; $t$ is a specified constant. Thus, the survival function represents the probability that individuals survive the infection after the observation deadline $t$. Then, the density function $f(t)$ is given as follows

$$ f(t) = \frac{dF(t)}{dt} = -S'(t). \tag{9} $$

The distribution of $T$ can be defined as well by the hazard rate function of occurrence of the event $h(t)$ which is a conditional probability that the event will occur during the interval $[t, t+dt]$ given that it has not occurred before time $t$, defined as

$$ h(t) = \lim_{dt \to 0} \frac{\Pr\{t \leqslant T < t+dt | T \geqslant t\}}{dt} = \frac{f(t)}{S(t)} = -\frac{S'(t)}{S(t)}. \tag{10} $$

Accordingly, we propose an additive survival model where the probability of node $u$ getting activated is the summation of the propagation probabilities represented by the Eq. 4. Therefore, the hazard rate of the nodes getting infected by a node $v$.

$$ h_v(t) = \sum_{u \in \mathbb{N}^v} p_{v,u}(t) = p_v^{send}(t) \sum_{u \in \mathbb{N}^v} p_{v,u}^{acc}. \tag{11} $$

Then survival function is defined from Eqs. 10 and 11

$$ S(t) = e^{-\int_0^t h(\tau)d\tau}. \tag{12} $$

Thus yields the cumulative distribution function

$$ F_v(t) = 1 - e^{-\int_0^t p_v^{send}(\tau) \sum_{u \in \mathbb{N}^v} p_{v,u}^{acc} d\tau} = 1 - \prod_{u \in \mathbb{N}^v} e^{-p_{v,u}^{acc} \int_0^t p_v^{send}(\tau)d\tau}. \tag{13} $$

Then the likelihood function of nodes getting infected by $v$ is given as follow

$$ f_v(t) = \frac{dF_v(t)}{dt} = \sum_{u \in \mathbb{N}^v} p_{v,u}^{acc} p_v^{send}(t) \prod_{w \in \mathbb{N}^v} e^{-p_{v,w}^{acc} \int_0^t p_v^{send}(\tau)d\tau}. \tag{14} $$

From Eq. 14, we could generalize the likelihood function of the nodes getting infected given as follows

$$f_V(t) = \prod_{v \in V : A_v(t) > 0} \sum_{u \in \mathbb{N}^v} p_{v,u}^{acc} p_v^{send}(t) \prod_{w \in \mathbb{N}^v} e^{-p_{v,w}^{acc} \int_0^t p_v^{send}(\tau)d\tau} . \qquad (15)$$

A greedy algorithm was designed based on Eq. 15 presented in Algorithm 1. The objective function is maximize the likelihood of the nodes getting infected by the truth campaign. The objective function is given as follows.

$$\max_V f_{V_{B^-}}(t_{det}) \quad . \qquad (16)$$

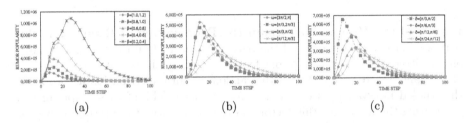

(a)               (b)               (c)

**Fig. 1.** The impact of the human factors on the rumor popularity.

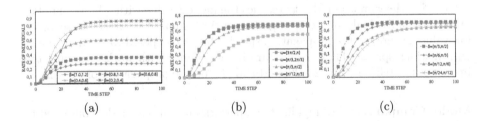

(a)               (b)               (c)

**Fig. 2.** The impact of the human factors on the rumor final size.

## 5   Experiments

In this section, experiments have been conducted to highlight the performance of the HISBmodel as well as the efficiency RIM strategy. Three datasets crawled from real OSNs [14] have been employed. These datasets are detailed in Table 1.

**Table 1.** Data sets description.

| No. | Network | Type | Nodes | Edges | $\langle k \rangle$ |
|---|---|---|---|---|---|
| 1 | Facebook | Undirected | 4,039 | 88,234 | 21.84 |
| 2 | Twitter | Directed | 4,546 | 280 ,846 | 39.53 |
| 3 | Slashdot | Directed | 77,360 | 905,468 | 54.55 |

## 5.1  Model Performance

We propose in this section two sets of experiments to highlight the performance of the proposed model. The following experiments are conducted in Twitter network [14] where for each scenarios we select ten (10) initial spreaders chosen randomly; we ran the simulations 500 times to avoid randomness.

**Impact of the Human Factors on the Propagation of Rumors.** In this part, we illustrate the impact of the IBK, FR, and HM on the rumor spreading on a Twitter network by varying one factor and fixings the two others. The values of $\omega$, $\beta$ and $\delta$ were assigned with the random uniform distribution according to the studied interval. Figures 1 and 2 display the results of these experiments, illustrating the impact of this factors on the popularity of the rumor as well as the ratio of infected individuals. An overall view, the results show that these factors have an important impact on: (1) the popularity of the rumor,(2) the speed of propagation and (3) the ratio of the infected individuals(see Figs. 1 and 2). Moreover, We see that IBK has a greater impact on the propagation of the rumor compared to the two factors. We can see in Figs. 1(a) and 2(a) for low values of $\beta$, the rumor popularity reaches the highest value with a larger number of infected individuals and long propagation time. Moreover, we can observe for low values $\omega$ and $\delta$ of the propagation speed is faster compared to the high values with a wider spread. These results are in accordance with works studying the impact of the IBK [1], FR [24–26] and HM [22] on the propagation of the rumor.

**Model Comparison.** To highlight the performance of our model compared to other models, we compare the trends of the rumor propagation between three models: the HISBmodel, the IC Model [18], and the Epidemic Model. We select the classical SIR models to be the baseline for the epidemic models where we set the initial conditions according to [25]. For the IC model, we select the model of [18] as the baseline which is an improved IC model considering the individual tendency. Finally, for the proposed model, we assigned a uniform probability for $\beta \in [0.2, 1.2]$, $\omega \in [\pi/12, \pi]$ and $\delta \in [\pi/24, \pi/2]$. The results are displayed in Fig. 3(a) showing the evolution of the normalized density of spreaders representing the trends of rumor propagation for the three models. Furthermore, The literature has proved that the evolution of the rumor propagation in OSN has a rising and falling pattern. This pattern does not present a stable or straightforward shape; it grows in a fast way and fades in a slowly fluctuating way

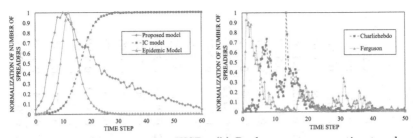

(a) Comparison between the HISB-model, IC model, and Epidemic model.   (b) Real rumors propagation trends .

**Fig. 3.** Comparison among HISBmodel, traditional models and real rumor propagation.

[10, 23]. To confirm this assumption, we display in Fig. 3(b) the trends of two real-world rumor propagation in twitter [27] about the "Charlie Hebdo attacks[2]" and "Ferguson Event[3]". The results show that: firstly, the epidemic model describes the spread of rumors going through two stages of growth and declining in the same way and a spiky tendency. Secondly, the IC Model simulates only the rising stage of the propagation this is due to the assumption that once an individual is activated, she/he remains in this states forever. Lastly, the proposed model represents the spread of the rumor with rapid growth and a slow decline, fluctuating in two stages. This pattern of evolution fits not only the trend of rumor propagation describes in the literature but also a real-world rumor propagation as shown in Fig. 3(b). Consequently, among these models, the proposed model depicts the pattern of rumor propagation the more realistically providing additional proof about the performance of the proposed model.

## 5.2   Performance of the Proposed Strategy

This section aims to evaluate the performance of the proposed RIM strategy under our model on the three networks listed in Table 1. The algorithms presented for comparison are as follows: (1) **Classic Greedy Algorithm** (CGA), which is used as the baseline based on a descendant order of nodes degree, (2) the **DRIMUX** of [18] which is a blocking node strategy, (3) **the truth campaign strategy** (TCS) and (4) **Natural propagation** (NP) to illustrate the difference between these algorithms. For more clarity, we note the ratio of the nodes selected as targets as RNSasT and the final ratio of individuals who believe the rumor as FRB. We run simulations for different scenarios by varying the detection time of the rumor $t_{det} = \{2, 4, 8, 12, 15\}$ as well as the RNSasT $k = \{5\%, 10\%, 15\%, 20\%\}$ where the initial conditions are set as the previous section (see Sect. 5.1). Then, each simulations are repeated 500 times in order to avoid randomness where the results presented in Table 2.

---

[2] https://en.wikipedia.org/wiki/CharlieHebdoshooting.

[3] https://en.wikipedia.org/wiki/Fergusonunrest.

**Table 2.** The final ratio of the believers of the rumor after the introduction of the rumor propagation strategies.

| $t_{det}$ | OSNs | Twitter | | | | Facebook | | | | Slashdot | | | |
|---|---|---|---|---|---|---|---|---|---|---|---|---|---|
| | RNSasT | 0.05% | 0.10% | 0.15% | 0.20% | 0.05% | 0.10% | 0.15% | 0.20% | 0.05% | 0.10% | 0.15% | 0.20% |
| 2 | NP | 0.5752 | 0.5752 | 0.5752 | 0.5752 | 0.6374 | 0.6374 | 0.6374 | 0.6374 | 0.5274 | 0.5274 | 0.5274 | 0.5274 |
| | CGA | 0.4798 | 0.4145 | 0.3127 | 0.2214 | 0.4675 | 0.2840 | 0.1448 | 0.0954 | 0.3604 | 0.3017 | 0.2624 | 0.2333 |
| | DRIMUX | 0.4451 | 0.3651 | 0.2537 | 0.1714 | 0.4335 | 0.2125 | 0.1015 | 0.0860 | 0.2008 | 0.1757 | 0.1478 | 0.1297 |
| | TCS | **0.1266** | **0.1223** | **0.0957** | **0.0952** | **0.1201** | **0.1002** | **0.0896** | **0.0832** | **0.1700** | **0.1487** | **0.1329** | **0.1265** |
| 4 | NP | 0.5752 | 0.5752 | 0.5752 | 0.5752 | 0.6374 | 0.6374 | 0.6374 | 0.6374 | 0.5274 | 0.5274 | 0.5274 | 0.5274 |
| | CGA | 0.4974 | 0.4326 | 0.3636 | 0.3208 | 0.4721 | 0.3413 | 0.2139 | 0.1533 | 0.4618 | 0.4215 | 0.3887 | 0.3587 |
| | DRIMUX | 0.4671 | 0.4139 | 0.2572 | 0.2116 | 0.3905 | 0.2751 | 0.1556 | 0.1331 | 0.3718 | 0.3397 | 0.3030 | 0.2818 |
| | TCS | **0.2454** | **0.2383** | **0.2169** | **0.1891** | **0.1728** | **0.1582** | **0.1327** | **0.1269** | **0.3096** | **0.2798** | **0.2712** | **0.2550** |
| 8 | NP | 0.5752 | 0.5752 | 0.5752 | 0.5752 | 0.6374 | 0.6374 | 0.6374 | 0.6374 | 0.5274 | 0.5274 | 0.5274 | 0.5274 |
| | CGA | 0.5142 | 0.4537 | 0.3999 | 0.3687 | 0.4896 | 0.3615 | 0.3036 | 0.2357 | 0.4934 | 0.4665 | 0.4411 | 0.4321 |
| | DRIMUX | 0.4821 | 0.4315 | 0.3677 | **0.2614** | 0.4466 | 0.3565 | 0.2360 | **0.1924** | 0.4227 | 0.4017 | 0.3850 | 0.3727 |
| | TCS | **0.3829** | **0.3660** | **0.3498** | 0.3261 | **0.2550** | **0.2184** | **0.2165** | 0.2077 | **0.3469** | **0.3312** | **0.3044** | **0.2940** |
| 12 | NP | 0.5752 | 0.5752 | 0.5752 | 0.5752 | 0.6374 | 0.6374 | 0.6374 | 0.6374 | 0.5274 | 0.5274 | 0.5274 | 0.5274 |
| | CGA | 0.5316 | 0.4806 | 0.4545 | 0.3935 | 0.4710 | 0.4163 | 0.3334 | 0.2961 | 0.5021 | 0.4903 | 0.4753 | 0.4669 |
| | DRIMUX | 0.4878 | 0.4348 | 0.4192 | **0.3409** | 0.4068 | 0.3591 | **0.2055** | **0.1508** | 0.4402 | 0.4290 | 0.4196 | 0.4159 |
| | TCS | **0.4197** | **0.4007** | **0.3796** | 0.3792 | **0.3355** | **0.3206** | 0.3076 | 0.2850 | **0.3590** | **0.3579** | **0.3462** | **0.3470** |
| 15 | NP | 0.5752 | 0.5752 | 0.5752 | 0.5752 | 0.6374 | 0.6374 | 0.6374 | 0.6374 | 0.5274 | 0.5274 | 0.5274 | 0.5274 |
| | CGA | 0.5455 | 0.5032 | 0.4697 | 0.4464 | 0.4711 | 0.4213 | 0.3746 | 0.3355 | 0.5077 | 0.4994 | 0.4877 | 0.4837 |
| | DRIMUX | 0.4945 | 0.4607 | **0.4299** | **0.4268** | 0.4237 | 0.3956 | **0.3520** | **0.2864** | 0.4486 | 0.4385 | 0.4314 | 0.4288 |
| | TCS | **0.4520** | **0.4457** | 0.4355 | 0.4280 | **0.3865** | **0.3716** | 0.3677 | 0.3582 | **0.3730** | **0.3695** | **0.3620** | **0.3585** |

Table 2 shows that the three strategies reduce significantly the impact of the rumor in the early-stage detection of the rumor as compared to the late-stage detection. Similarly, an important decrease of the impact of the rumor is seen for higher values RNSasT. More precisely, for RNSasT $k = 0.05\%$ and $0.10\%$, we see that the TCS performs the best among the strategies in the three networks in which it reduces significantly the impact of the rumor when it is discovered in the early-stage $t_{det} = 2, 4$ compared to middle and late-stage detection $t_{det} = 8, 12, 15$. This is because when the rumor is detected in the early-stages the selected nodes will contribute significantly to diminish the influence of the rumor even when RNSasT is low. When RNSasT $k = 0.15\%$ and $0.20\%$, we perceive that the TCS has better result among the strategies in the three OSN for $t_{det} = 2, 4, 8$ However, in the late-stage detection $t_{det} = 12, 15$ we can see that DRIMUX have a better performance than TCS in some cases (see Table 2). Presumably, this observation is explained by the fact that, when the rumor is detected at a relatively late-stage, it would have already infected a large portion of the nodes in the entire network and individuals will already lose interest about the rumor. Moreover, the high values of RNSasT are an ideal scenario which is not always the case in real life. Despite this, we can still state that the TCS has the best performance among these strategies. Moreover, since the Slashdot is the densest network followed by Twitter, Facebook illustrated by their average degree of connection per node (see Table 1). The rumor will spread faster in Slashdot compared to other networks where this feature will improve the result of TCS by amplifying the propagation of the truth campaign. Thus, the TCS accomplishes the best results in all the scenarios in Slashdot. The evidence from these results confirms the outstanding performance of the proposed strategy in selecting target nodes accurately based on the HISBmodel.

# 6    Conclusions

This work addresses the rumor propagation problem in online social networks (OSNs). We propose a novel rumor propagation model based on humans individual and social behaviors in OSNs called the HISBmodel. This model is based on a physics theory, where its originality lies in taking into account various human factors as the individuals' opinions, social influence, and behaviors. Based on this model, we propose a truth campaign strategy to minimize the influence of the in OSNs from the perspective of a network inference using the survival theory. The experiments part illustrates that our model depicts the evolution of rumors propagation more realistically than other models; it reproduces as well all the trends features of rumor propagation mentioned in literature and real-world rumor propagation. Finally, we illustrate the performance of our strategy to minimize the influence of the rumor, in which we showed that the HISBmodel allows an accurate selection of target nodes.

**Acknowledgment.** The Research was supported in part by National Basic Research Program of China (973 Program, No.2013CB329605).

# References

1. Afassinou, K.: Analysis of the impact of education rate on the rumor spreading mechanism. Phys. A Stat. Mech. Appl. **414**, 43–52 (2014)
2. Borodin, A., Filmus, Y., Oren, J.: Threshold models for competitive influence in social networks. In: Saberi, A. (ed.) WINE 2010. LNCS, vol. 6484, pp. 539–550. Springer, Heidelberg (2010). https://doi.org/10.1007/978-3-642-17572-5_48
3. Bredereck, R., Elkind, E.: Manipulating opinion diffusion in social networks. In: Proceedings of the 26th IJCAI (2017)
4. Budak, C., Abbadi, A.E.: Limiting the spread of misinformation in social networks. Distribution, pp. 665–674 (2011)
5. Daley, D., Kendall, D.: Epidemics and rumours. Nature **204**(4963), 1118 (1964)
6. DiFonzo, N., Bordia, P., Rosnow, R.L.: Reining in rumors. Organ. Dyn. **23**(1), 47–62 (1994)
7. Fan, L., Lu, Z., Wu, W., Bhavani, T., Ma, H., Bi, Y.: Least cost rumor blocking in social networks. In: IEEE 33rd International Conference on Distributor Computer Systems (2013)
8. Galam, S.: Modelling rumors: the no plane pentagon french hoax case. Phys. A Stat. Mech. Appl. **320**(7603), 571–580 (2003)
9. Gomez-Rodriguez, M., Leskovec, J.: Modeling information propagation with survival theory. In: Proceedings of the 30th ICML (ICML-13), pp. 666–674 (2013)
10. Han, S., Zhuang, F., He, Q., Shi, Z., Ao, X.: Energy model for rumor propagation on social networks. Phys. A Stat. Mech. Appl. **394**, 99–109 (2014)
11. He, X., Song, G., Chen, W., Jiang, Q.: Influence blocking maximization in social networks under the competitive linear threshold model. Education p. Technical report CoRR abs/1110.4723 (2011)
12. Kempe, D., Kleinberg, J., Tardos, É.: Maximizing the spread of influence through a social network. In: KDD, p. 137 (2003)
13. Kimura, M., Saito, K., Motoda, H.: Minimizing the spread of contamination by blocking links in a network. In: AAAI, pp. 1175–1180 (2008)
14. Leskovec, J., Krevl, A.: SNAP datasets: stanford large network dataset collection, June 2014
15. Ma, J., Li, D., Tian, Z.: Rumor spreading in online social networks by considering the bipolar social reinforcement. Phys. A: Stat. Mech. Appl. **447**, 108–115 (2016)
16. Marion, J.B., S.T.T.: Classical dynamics of particles and systems. Thomson (2003)
17. Meshi, D., Morawetz, C., Heekeren, H.R.: Nucleus accumbens response to gains in reputation for the self relative to gains for others predicts social media use. Front. Hum. Neurosci. **7**, 439 (2013)
18. Wang, B., Chen, G., Fu, L., Song, L., Wang, X.: Drimux: dynamic rumor influence minimization with user experience in social networks. IEEE Trans. Knowl. Data Eng. **29**(10), 2168–2181 (2017)
19. Wang, H., Deng, L., Xie, F., Xu, H., Han, J.: A new rumor propagation model on SNS structure. In: Proceedings of IEEE International Conference Granular Computing GrC 2012 (2012)
20. Wang, J., Wang, Y.Q., Li, M.: Rumor spreading considering the herd mentality mechanism. In: Control Conference, 2017 36th Chinese, pp. 1480–1485. IEEE (2017)
21. Wang, Y.Q., Yang, X.Y., Han, Y.L.: Rumor spreading model with trust mechanism in complex social networks. Commun. Theor. Phys. **59**(4), 510 (2013)

22. Xia, L.L., Jiang, G.P., Song, B., Song, Y.R.: Rumor spreading model considering hesitating mechanism in complex social networks. Phys. A Stat. Mech. Appl. **437**, 295–303 (2015)
23. Yang, J., Leskovec, J.: Patterns of temporal variation in online media. Time **468**, 177–186 (2011)
24. Zhao, L., Cui, H., Qiu, X., Wang, X., Wang, J.: SIR rumor spreading model in the new media age. Phys. A Stat. Mech. Appl. **392**(4), 995–1003 (2013)
25. Zhao, L., Wang, J., Chen, Y., Wang, Q., Cheng, J., Cui, H.: SIHR rumor spreading model in social networks. Phys. A Stat. Mech. Appl. **391**(7), 2444–2453 (2012)
26. Zhao, L., Wang, Q., Cheng, J., Chen, Y., Wang, J., Huang, W.: Rumor spreading model with consideration of forgetting mechanism: A case of online blogging LiveJournal. Phys. A Stat. Mech. Appl. **390**(13), 2619–2625 (2011)
27. Zubiaga, A., Hoi, G.W.S., Liakata, M., Procter, R.: Pheme dataset of rumours and non-rumours. (2016)

# Multi-scale Feature Decode and Fuse Model with CRF Layer for Boundary Detection

Zihao Dong[1], Ruixun Zhang[2], Xiuli Shao[1(✉)], Huichao Li[1], and Zihan Yang[1]

[1] College of Computer and Control Engineering, NanKai University, Tian Jin, China
{1120170132,2120160395}@mail.nankai.edu.cn, shaoxl@nankai.edu.cn,
pink_edward@126.com
[2] MIT Laboratory for Financial Engineering, Cambridge University, Cambridge,
MA, USA
zhangruixun@gmail.com

**Abstract.** The key challenge for edge detection is that salient edge is difficult to detect due to the complex background. To improve the resolution and accuracy of salient edge effectively, we propose a novel method of edge detection called MSDF (Multi Scale Decode and Fusion) based on deep structured multi-scale features in this paper. The decoding layer of MSDF can fuse the adjacent features of the DNN multi-scale and increase the correlation between the features. In the fusion of different scale's information, the traditional method of up-sample based on deconvolution is not used and Subpixel [16] algorithm is adopted to improve the resolution of the convolution layer's output image. We also build a new Conditional Random Fields (CRF) model with CRF-RNN layer to reduce the number of irrelevant features and eliminate the weak correlation information while retaining the important structural attributes. Extensive experiments on BSDS500 [1] dataset and the larger NYUD [17] dataset show that the effectiveness of the proposed model and of the overall hierarchical framework.

**Keywords:** Edge detection · HED [20] · Multi-scale feature
Subpixel · CRF

## 1 Introduction

Edge detection is to extract the salient edge from the input image with complex background accurately. It can be applied to image segmentation, visual saliency, object detection and image retrieval. Convolutional Neural Networks (CNNs) have become a recent trend to improve the state of varieties of methods and training models.

CNN is no longer dependent on the manual design of researchers for learning image features unlike traditional methods. Though training large quantities of image data, it tends to get general features such as intensity, depth, and texture.

© Springer Nature Switzerland AG 2018
L. Cheng et al. (Eds.): ICONIP 2018, LNCS 11302, pp. 28–40, 2018.
https://doi.org/10.1007/978-3-030-04179-3_3

These features are usually used to get a classifier by supervised learning of edge detection, and the classifier will predict the edge and non-edge pixels. But edge detection is still a very challenging problem and remains unsolved due to the facts that: (i) Edge features are very sparse, we hardly distinguish the relevant features and irrelevant features of the edges. (ii) Some methods based on leaning multi-scale features have a rough treatment of different scale features, which result in the lower correlation between scales, and a simple average of these edge maps of multi-scale will output low-quality edges.

This paper mainly has two parts. The first one is to propose a new edge detection method called MSDF (Multi Scale Decode and Fusion) based on learning multi-scale features. MSDF can extract the edge features of 5 scales in VGG-16 network [18] using Holistic Edge Detector (HED) [20], and combine the bottom-up decoding architecture with subpixel method [16], which is designed for learning clear edge features. Finally, the multi-scale edge extracted by HED is fused with the decoded edge to obtain the detection result of MSDF. The second one is to build a end-to-end training CRF model based on MSDF edge detector for weakening unrelated features and reducing the error rate.

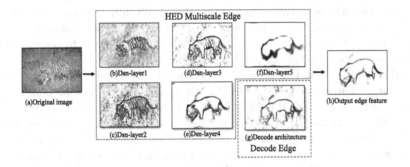

**Fig. 1.** Multi-scale extraction and fusion of MSDF algorithm.

In Fig. 1, we show a visualization of MSDF multi-scale edges. One can clearly see that the resolution of decoded edge features is higher than HED multiscale including low-level and high-level, but it contains some fine details that do not appear in other layers. MSDF model will output edge map that closed to Ground Truth through fusing feature maps from each stage.

The remainder of this paper is structured as follows. In Sect. 2, the related work is introduced in edge detection. Section 3 describes the overall framework of the proposed method. Section 4 introduces the CRF model structure based on CRF-RNN layer. Section 5 presents the experiment and application in BSDS500 and NYUD datasets. Section 6 concludes this paper.

## 2    Related Work

This paper focuses on edge detection in images with complex background. In edge detection algorithms proposed in the literature, we divide them into four groups, including the local edge feature detectors, edge classification methods using machine learning, and CNN-based training methods.

Typically, some traditional methods focus on extracting local cues of brightness, colors, gradients and textures, or other manually designed features. As for the local edge feature detectors, Arbelaez et al. [1] combine multiscale local brightness, color and texture cues into a globalization framework using spectral clustering to detect salient edge. Lim et al. [9] defines the method of sketch token to represent the structure estimation of local edges, then uses random forest to complete the classification of different image patch in the sketch token. Among machine learning-based methods, Dollar et al. [2] present structured edge detector SE to construct a structured decision tree for edge detection, where PCA is used to achieve data dimensionality reduction and the random forest is used to capture the structured information.

In recent years, CNN-based method began to be applied and achieve impressive performance in the edge detection. Shen et al. [15] use k-means clustering method to classify image patches and multi-class shape patches is extracted through a 6-layer CNN network, then structured random forest is used to further classify the edge to obtain more discriminative features. Xie and Tu [20] propose a deep learning model combining full convolutional network (FCN) [11] and deeply supervised nets [7] to detect edges. Based on the deep learning model of [20], some methods focus on improving the network structure to generate better features for performing the pixel-wise prediction, such as RCF [10] and CED [19], other methods add some useful components to achieve better accuracy, such as Deep Boundary [6] and COB [13], these components include multi-scale, extern training data with PASCAL Context dataset [14] and Normalized Cuts [1]. In addition to the above-mentioned researches, some works have begun to consider a new type of network that can go deeper without the problem of vanishing/exploding gradients and use feature parameters effectively. Taking Dense Net as an example, [5] has successfully applied it to the area of object segmentation and edge detection.

To explore the object level contours, [8,21,22] focus on how to add other mechanisms in VGG-16 network to weaken the influence of unrelated feature information to obtain edge map. Liao et al. [8] proposed an edge correlation graph (CCG) to predict the validity of candidate edges and convert the CCG segmentation information into feature mapping function to obtain an effective CRF model. Yang et al. [22] develop an object-centric edge detection method using efficient fully convolutional encoder-decoder network, this method uses a refinement method based on dense CRF to concern with the imperfect edge annotations from polygons. Xu [21] proposed the AG-CRFS model, which mainly adopts two methods of CRF and attention mechanism to produce more rich and complementary representations.

# 3   MSDF Model

## 3.1   Network Architecture

**Overview.** The overall network architecture of the proposed MSDF model is illustrated in the left part of Fig. 2. The MSDF model leverages a bottom-up decoded pathway to complete the fusion of multi-layer features. It consists of two sub-networks: (i) a HED subnet that extracts top-down multi-scale features, (ii) a decoder subnet to decode bottom-up feature information. Each branch produces different edge maps at different layers of this network, such as HED dsn-layer (1–5) and decoded convolutional layer. In decoded subnet, each module fuses a bottom-up feature map from its bottom layer with a top-down feature map from the convolutional layer in VGG-16 FCN. The two subnets are fused finally, we average them to generate the final edge map.

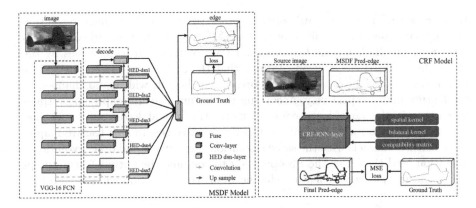

**Fig. 2.** Architecture of the proposed multi-scale deep learning MSDF model (left) and CRF model with CRF-RNN layer (right).

**Top-down Scale Feature Extraction.** The top-down architecture is based on that of VGG-16, we cut all the fully connect layers and pool5 layer. The multi-scale features can be learned, including low-level and object-level. In Fig. 2, some parts of this model are omitted: a conv-layer with kernel size $(1 \times 1)$ and channel depth 1 follows each conv-layers (convolutional layers) of VGG-16 network, then a deconvolutional layer is used to up-sample this scale feature map, finally edge map of input image will be obtained from HED dsn-layer.

**Bottom-up Decoder Subnet.** The modules of decoder are associated with each conv-layer in VGG-16 Net. Each conv-layer of this subnet up-samples the map by a factor $(2\times)$ and fuses with the left layers of VGG-16 FCN. For bottom to up pathway, the number convolutional layer is set to 1024 for the bottom layer, this value is reduced by 2 times from bottom layer to top layer. For example, the first Conv-layer on the top of the bottom layer will have 256 feature channels and the second, third bottom-up module will have 128 and 64 feature channels respectively. In this sub-net, up-sampling and fusion is important for better resolution and correlation of edges.

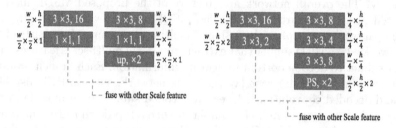

**Fig. 3.** The assumption of two different fuse methods of HED and decoder sub-net.

**Up-sample of Decoder Sub-net.** In order to improve the image resolution after up-sampling, the subpixel method is used to perform standard convolution processing on the low-scale features, and then rearrange feature values by the phase shift (PS) [16] operation. Assume that the scale of a convolution feature map is $h \times w \times c$, where $h$, $w$ and $c$ represent the height, width, and channels of the feature map, respectively. According the output feature map being $r$ times larger than it, the size of feature map using the traditional deconvolution method is $(r \times h) \times (r \times w) \times c$. The subpixel method firstly generates a feature map of $r^2 \times c$ feature channels with the same resolution through a standard convolutional operation, and then uses the PS operation to combine the output feature maps into a feature map with $c$ feature channels, but the resolution will increase by $r$ times.

**Fusion Method of Decoder Subnet.** HED uses a simple fusion strategy that directly completes the stitching of two feature matrices, but the premise of this strategy is that the number of channels of the input feature map must be the same. In the decoder subnet of MSDF model, the number of feature channels of different scales is not the same, splicing two feature matrices directly will lead to lost some low-dimensional features. Therefore, this paper reduces the dimensions of high-scale features by increasing the number of convolutional layers and has the same number of channels as low-scale features. In this way, the matrix splicing method of HED can be used to fuse the feature map between different scales. Please see Fig. 3 for the fusion method details. We assume that the features of $(3 \times 3, 16)$ convolutional layer and $(3 \times 3, 8)$ convolutional layer need to be fused, 3 is kernel size, (8 and 16) are number of channels, $(\frac{w}{2} \times \frac{h}{2})$ is the size of feature map. The left is the strategy of HED, it only concatenate two different feature map by additional $(1 \times 1, 1)$ convolutional layer and upsample layer. The right is our fusion method, we add some refinement modules such as convolutional layers and PS operation above to realize the fuse between two different feature maps. From this case, our method can retain more information of low-dimensional features.

## 3.2 Learning and Deployment

The image edge detection problem can be expressed as a general mathematical form. This paper uses $X_n$ to represent the image in data layer, which is generally transformed into a multidimensional matrix, $Y_n$ denotes the set of edge feature pixels of the Ground Truth $Y_n = \{y_j^{(n)}, j = 1, ..., |X_n|\}$ Where $|X_n|$ represents the number of pixels in the image n. $P(y_j = 1|X_n)$ is defined to represent the possibility of predicted pixel labeled by annotator. Here, 0 means the pixel of edge predicted by the model is not the Ground Truth label, and 1 means the pixel of edge predicted by the model is labeled by all annotators. Because BSDS500 has five different annotations for each image, which leads to controversial issues, a threshold $\mu$ is defined here, we set it to 0.7 depending on training data in loss function. When $P(y_j = 1|X_n)$ is greater than 0 and less than $\mu$, the pixel is controversial and should not be considered. The image training data will be represented as $S = \{(X_n, Y_n), n = 1, 2, ..., N\}$, where $N$ is the number of training images.

We compute the loss functions of the top-down architecture and the decoder subnet at every pixel with respect to the Ground Truth pixel label as.

When $P(y_j = 1|X_n) \geq \mu$,

$$l_{side}(W, X_n) = -\beta \sum_{j \in y^+} \log P(y_j = 1|X_n, W) - (1 - \beta) \sum_{j \in y^-} \log P(y_j = 0|X_n, W)$$

(1)

When $0 < P(y_j = 1|X_n) < \mu$, $l_{side}(W, X_n) = 0$.

In which

$$\beta = \frac{|Y_-|}{|Y_-| + |Y_+|}; 1 - \beta = \frac{|Y_+|}{|Y_-| + |Y_+|}$$

(2)

$Y_+$ is an edge label and $Y_-$ is a non-edge label.

The total loss function of the multi-scale edge detection model also needs to include the fusion loss. This paper assumes that the top-down architecture has the same loss function as the decoder subnet, the value of $W$ will be learned and updated during the training process. Therefore, our improved loss function can be formulated as:

$$l(W, X_n) = \sum_{i=1}^{N} (\sum_{j=1}^{M} l_{side}^j(W, X_n) + l_{fuse}(W, X_n))$$

(3)

where $N$ is the number of training images, and $M$ is the total number of HED dsn-layers and the layers of decoder subnet (here is set to 6). $l_{fuse}(W, X_n)$ represents the loss function of the fusion layer.

## 4 CRF Model Based on MSDF

### 4.1 CRF Model Architecture

This paper applies a new CRF model to connect with the network structure of MSDF model. It can be trained end-to-end utilizing the usual back-propagation algorithm and mainly used to post-process the edge maps predicted by MSDF model. The input of CRF model is composed of the original image and the final edge maps predicted by MSDF method. It estimates the CRF parameters on each edge and is adopted to fulfill the optimization of estimation. In the right part of Fig. 2, our CRF model is made up of CRF-RNN Layer with the weights of spatial kernel and bilateral kernel [23] which depend on the number of classes. Finally, the weights of this network are optimized to solve a regression task by Mean Squared Error (MSE) function, where the objective to reduce the error between predicted edge map and ground truth. Due to the CRF model can make full use of the features of adjacent scales and predict the validity of local edges, our method will eliminate most blurry and noisy boundaries significantly.

**CRF-RNN Layer.** [23] combines the strengths of CNN and CRF-based probabilistic graphical modeling to imply end to end training in semantic segmentation. In our model, CRF-RNN method is adopted to solve the optimization problem of edge detection. We use the function $f(U, Q_{in}, X_n, \theta)$ to denote the transformation done by one mean-field iteration: the image $X_n$, the unary potential function $U$, the marginal distribution $Q_{in}$ and the CRF parameters $\theta$. The unary potential function $U$ is the multi-scale fuse layer's output from VGG-16 network. Each iteration takes $Q_{in}$ value from the previous iteration to achieve multiple mean-field iterations. The CRF parameters $\theta$ includes class numbers, iteration numbers $(i)$, and other hyperparameters. So, the behavior of CRF-RNN layer is given by the following equations:

$$H_1(t) = \begin{cases} sigmoid(U), & i = 0. \\ H_2(t-1), & 0 < t < i. \end{cases} \qquad (4)$$

where $H_1(t)$ and $H_2(t-1)$ are hidden states. We use $sigmoid(U)$ instead of $softmax(U)$ to solve binary classification problem.

$$H_2(t) = f(U, H_1(t), \theta), \quad 0 < t < i. \qquad (5)$$

Output is get by Eq. (5) operation when $t$ achieve to mean-field iterations $T$ (set to 5):

$$Y(t) = H_2(t), \quad t = T \qquad (6)$$

In order to be compatible with predicted result of MSDF, we build the CRF network to solve a regression problem and employ the MSE function to reduce the training error. Using the predicted results of Eq. 6, we can obtain the loss value computed by MSE, which is shown as follows:

$$l_{CRF} = \sum_{i=1}^{N} ||\hat{Y}_i - Y(t)_i||^2, \quad t = T, N = batchsize \qquad (7)$$

with $Y_i$ the $ith$ the ground truth, $Y(t)_i$ the $ith$ predicted edge map of Eq. (6), $N$ the batch size.

In this way, two independent end-to-end models (MSDF and CRF model) are trained for edge detection.

**Training with External Data.** The number of training image in BSDS500 dataset is only 300, so the author in [20] used 16 rotations and flipping of 200 images used in train set and 100 images used in val set. But this method will result in generating some different scale images. We have used a new method called random-cropping to achieve dataset augmentation. Our modification to this boundary have been to random generate $X$ and $Y$, which will help to crop $256 \times 256$ image in this coordinate. We also flip the image at 16 different angles, leading to an augmented training set with 30000 training images.

## 5  Experiments

**Dataset.** We perform model training and performance evaluation on the BSDS500 [1] and NYUD [17] datasets. BSDS500 is a dataset widely used in image segmentation and image edge detection. It is composed of 200 training images, 100 verification images, and 200 test images. Each image is manually identified and there is a corresponding Ground Truth label. We rotated and scale 200 training pictures and 100 verification pictures to expand the size of the training set, and use 200 test pictures as performance evaluation. The NYUD dataset consists of 1449 images containing three parts: RGB, Depth, and acceleration data. In order to facilitate the model training, we divide the NYUD dataset into two parts of RGB and HHA. We found that the training of MSDF model on HHA part is unstable making it difficult to draw any conclusion, only RGB part is selected. There are 381 images in the RGB training set, 414 in the validation set, and 654 images in the test set.

**Implementation Details.** Our MSDF and CRF model build on the publicly available code of HED [20], using TensorFlow as backend with a single NVIDIA GTX1070 GPU. In this MSDF model, we initialize the conv-layers of encoder subnet with Gaussian random distribution with fixed mean (0.0) and variance (0.01) and the conv-layers of top-down subnet with constant weight (0.2) and variance (0). The hyper parameters, including the initial learning rate, weight decay and momentum, are set to $3e - 5$, $1e - 6$ and 0.7, respectively. We use AdamW optimizer (see details in next section) as the method of gradient optimization with the parameter $\varepsilon = 1e - 3$. The learning rate is set as $3e - 5$ at first 30000 iterations and further decreased to $1e - 6$ for another 10000 iterations with an epoch size of 100. In BSDS500 training, we randomly cropped an original $321 \times 481$ image of size $20 \times 30$ and rotated it $90°$ for data augmentation during training. In NYUD training, we randomly cropped an original $425 \times 560$ image of size $26 \times 35$, rotated it $90°$, $180°$, $270°$, and flipped it vertically and horizontally for data augmentation during training.

**ADAMW Optimization.** In order to recover the original formulation of weight decay regularization by decoupling the weight decay, we use a new adaptive gradient method called AdamW [12] as the method of gradient optimization instead of Adam. This method improves Adam's performance in three practices: decoupling weight decay, formal analysis of weight decay vs $L2$ regulation and cosine annealing and warm restarting.

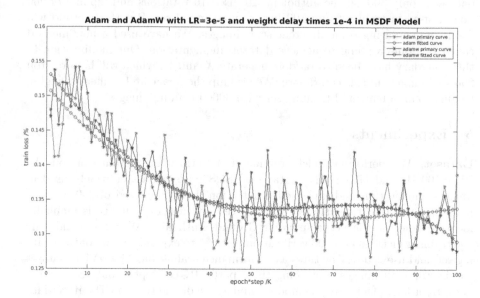

**Fig. 4.** Loss curves obtained by MSDF+CRF model trained by Adam and AdamW on BSDS500 dataset.

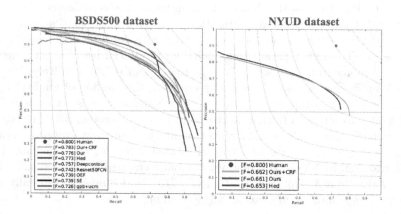

**Fig. 5.** The evaluation results on BSDS500 dataset and NYUD dataset. The left one is the P-R curve in BSDS500 and the right one is the P-R curve in NYUD-RGB. (Color figure online)

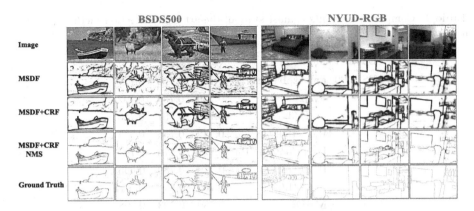

**Fig. 6.** Some examples of MSDF and MSDF+CRF. From left to right: BSDS500 and NYUD. From top to bottom: origin image, MSDF edge map, MSDF+CRF edge map, MSDF+CRF edge map with NMS, and Ground Truth.

According to the experiment of [12], the hyper parameters including the initial learning rate and weight decay, are set to $3e-5$ and $1e-4$, respectively. The learning rate is set as $3e-5$ at first 30000 iterations and further decreased to $1e-6$ for another 10000 iterations with an epoch size of 100.

Figure 4 shows the results for same settings of weight decay of Adam and AdamW. Importantly, the use of weight decay $1e-4$ in Adam did not yield as good results as in AdamW, the train loss of AdamW fitted curve is lower than the Adam Curve in 100 epoch and has the faster convergence.

**Evaluation Metrics.** The evaluation metrics used here are P/R (precision/recall) curves, ODS (optimal dataset scale), OIS (optimal image scale), AP (average precision), and F-measure parameters for predicting edge matching degree. ODS divides the fixed parameters of all images. OIS selects the optimal segmentation parameters for each image, and F-measure is calculated by the precision and recall:

$$F-measure = \frac{2 \times precision \times recall}{precision + recall} \tag{8}$$

**Comparisons Against Other Methods.** Our model is compared to other better methods in Table 1. The expansion method of dataset (details as the section of implementation details) is different from other algorithms such as HED, so the evaluation results based on our model structure and the reports of other methods in original papers exist obvious variances. The following observations can be made: (i) Our method (MSDF+CRF) outperforms all compared models. For BSDS500 data, ODS F-measure of MSDF is 1% higher than HED, which indicates the decode part of our method plays a main role for improving the accuracy of boundary detection. But AP of COB is the best performance, the reason may be that the decoding part of MSDF model adds some noise to the

**Table 1.** Comparative results against state-of-the-art edge detection method performance.

| DataSet | Methods | ODS | OIS | AP |
|---------|---------|-----|-----|-----|
| BSD500 | gpb+spb+ucm [1] | 0.726 | 0.760 | 0.727 |
| | SE [2] | 0.739 | 0.759 | 0.792 |
| | OEF [3] | 0.739 | 0.761 | 0.720 |
| | Resnet50FCN | 0.742 | 0.755 | 0.585 |
| | DeepContour [15] | 0.757 | 0.776 | 0.790 |
| | HED [20] | 0.770 | 0.789 | 0.645 |
| | COB [13] | 0.782 | 0.801 | **0.824** |
| | **MSDF** | 0.780 | 0.793 | 0.798 |
| | **MSDF+CRF** | **0.783** | **0.802** | 0.801 |
| NYUD | HED [20] | 0.653 | 0.665 | 0.559 |
| | **MSDF** | **0.661** | **0.674** | **0.565** |
| | **MSDF+CRF** | **0.662** | **0.675** | **0.567** |

detection process, which is the direction we need to further study. For NYUD data, when compared with HED, ODS F-measure of MSDF is higher 0.8% than it, the performance of MSDF+CRF is almost same as the single MSDF model. (ii) In Fig. 5, the precision-recall curves of our methods are also higher than HED's. (iii) The poor result of Resnet50 [4] suggests that the performance of DNN is not better as deep as possible, that's why we choose VGG-16 net as the based network. (iv) ODS F-measure of Our model in NYUD is lower than BSDS500's, because images in NYUD dataset are larger than images in BSDS500 dataset.

**Qualitative Results.** Example edge detection results of the proposed model are shown in Fig. 6. The results suggest that the model is more robust over different kinds of images. It can be seen that MSDF model can improve global resolution of the image edges, and the CRF model has a better effect on eliminating irrelevant features, our edge detection method is fairly close to the annotated edge of Ground Truth, especially the edges detected in the second column, Our method almost eliminates the irrelevant effects of complex backgrounds.

## 6  Conclusion

This paper proposes a new CNN structure MSDF based on learning multi-scale features. The architecture mainly includes the top-down scale feature extraction and the new proposed decoder subnet, the decoder subnet shows a new method to balance the fusion and up-sampling to carry out edge detection. Finally, the paper uses a new CRF model based on MSDF to link relevant information together. The implementation code of MSDF model will be available at https://github.com/zihaodong/Multi-Scale-Decode-Feature-.

# References

1. Arbelaez, P., Maire, M., Fowlkes, C., Malik, J.: Contour detection and hierarchical image segmentation. IEEE Trans. Pattern Anal. Mach. Intell. **33**(5), 898–916 (2011)
2. Dollár, P., Zitnick, C.L.: Fast edge detection using structured forests. IEEE Trans. Pattern Anal. Mach. Intell. **37**(8), 1558–1570 (2015)
3. Hallman, S., Fowlkes, C.C.: Oriented edge forests for boundary detection. In: Proceedings of the IEEE Conference on Computer Vision and Pattern Recognition, pp. 1732–1740 (2015)
4. He, K., Zhang, X., Ren, S., Sun, J.: Deep residual learning for image recognition. In: Proceedings of the IEEE Conference on Computer Vision and Pattern Recognition, pp. 770–778 (2016)
5. Hou, Q., Liu, J., Cheng, M.M., Borji, A., Torr, P.H.: Three birds one stone: a unified framework for salient object segmentation, edge detection and skeleton extraction. arXiv preprint arXiv:1803.09860 (2018)
6. Kokkinos, I.: Pushing the boundaries of boundary detection using deep learning. arXiv preprint arXiv:1511.07386 (2015)
7. Lee, C.Y., Xie, S., Gallagher, P., Zhang, Z., Tu, Z.: Deeply-supervised nets. In: Artificial Intelligence and Statistics, pp. 562–570 (2015)
8. Liao, Y., Fu, S., Lu, X., Zhang, C., Tang, Z.: Deep-learning-based object-level contour detection with CCG and CRF optimization. In: 2017 IEEE International Conference on Multimedia and Expo (ICME), pp. 859–864. IEEE (2017)
9. Lim, J.J., Zitnick, C.L., Dollár, P.: Sketch tokens: a learned mid-level representation for contour and object detection. In: Proceedings of the IEEE Conference on Computer Vision and Pattern Recognition, pp. 3158–3165 (2013)
10. Liu, Y., Cheng, M.M., Hu, X., Wang, K., Bai, X.: Richer convolutional features for edge detection. In: 2017 IEEE Conference on Computer Vision and Pattern Recognition (CVPR), pp. 5872–5881. IEEE (2017)
11. Long, J., Shelhamer, E., Darrell, T.: Fully convolutional networks for semantic segmentation. In: Proceedings of the IEEE Conference on Computer Vision and Pattern Recognition, pp. 3431–3440 (2015)
12. Loshchilov, I., Hutter, F.: Fixing weight decay regularization in adam. arXiv preprint arXiv:1711.05101 (2017)
13. Maninis, K.K., Pont-Tuset, J., Arbeláez, P., Van Gool, L.: Convolutional oriented boundaries: from image segmentation to high-level tasks. IEEE Trans. Pattern Anal. Mach. Intell. **40**(4), 819–833 (2018)
14. Mottaghi, R., et al.: The role of context for object detection and semantic segmentation in the wild. In: Proceedings of the IEEE Conference on Computer Vision and Pattern Recognition, pp. 891–898 (2014)
15. Shen, W., Wang, X., Wang, Y., Bai, X., Zhang, Z.: Deepcontour: a deep convolutional feature learned by positive-sharing loss for contour detection. In: Proceedings of the IEEE Conference on Computer Vision and Pattern Recognition, pp. 3982–3991 (2015)
16. Shi, W., et al.: Real-time single image and video super-resolution using an efficient sub-pixel convolutional neural network. In: Proceedings of the IEEE Conference on Computer Vision and Pattern Recognition, pp. 1874–1883 (2016)
17. Silberman, N., Hoiem, D., Kohli, P., Fergus, R.: Indoor segmentation and support inference from RGBD images. In: Fitzgibbon, A., Lazebnik, S., Perona, P., Sato, Y., Schmid, C. (eds.) ECCV 2012. LNCS, vol. 7576, pp. 746–760. Springer, Heidelberg (2012). https://doi.org/10.1007/978-3-642-33715-4_54

18. Simonyan, K., Zisserman, A.: Very deep convolutional networks for large-scale image recognition. arXiv preprint arXiv:1409.1556 (2014)
19. Wang, Y., Zhao, X., Huang, K.: Deep crisp boundaries. In: Proceedings of the IEEE Conference on Computer Vision and Pattern Recognition, pp. 3892–3900 (2017)
20. Xie, S., Tu, Z.: Holistically-nested edge detection. Int. J. Comput. Vision **125**(1–3), 3–18 (2017)
21. Xu, D., Ouyang, W., Alameda-Pineda, X., Ricci, E., Wang, X., Sebe, N.: Learning deep structured multi-scale features using attention-gated CRFs for contour prediction. In: Advances in Neural Information Processing Systems, pp. 3961–3970 (2017)
22. Yang, J., Price, B., Cohen, S., Lee, H., Yang, M.H.: Object contour detection with a fully convolutional encoder-decoder network. In: Proceedings of the IEEE Conference on Computer Vision and Pattern Recognition, pp. 193–202 (2016)
23. Zheng, S., et al.: Conditional random fields as recurrent neural networks. In: Proceedings of the IEEE International Conference on Computer Vision, pp. 1529–1537 (2015)

# Sentimental Analysis for AIML-Based E-Health Conversational Agents

David Ireland, Hamed Hassanzadeh, and Son N. Tran[(⊠)]

The Australian E-Health Research Centre, CSIRO, Brisbane, QLD 4026, Australia
{d.ireland,hamed.hassanzadeh,son.tran}@csiro.au

**Abstract.** Conversational agents or chat-bots are emerging in various applications including finance, education and e-health. Recent research has highlighted the importance of the consistency between the response of the chat-bot and the sentiment of the input utterance. This is quite challenging as detecting the sentiment of an utterance often depends on the context and timing of the conversation. Moreover, whereas humans have complex repair strategies, encoding these for human-computer interaction is problematic. This paper presents five sentiment prediction models for conversational agents that are trained on a large corpus of smart-phone application reviews and their sentiment ranks obtained from the Google playstore. These models are tested on collected, real-life conversations between a human and a machine. It is found that positive utterances are classified with a high accuracy but classifying negative utterances is still challenging.

**Keywords:** Sentiment analysis · E-health chatbots

## 1 Introduction

Artificial intelligent conversation agents or chat-bots are natural language processing systems designed to mimic human communication. Despite having existed for several decades, it is only recently that the technology has spawned a growing list of applications and commercial opportunities. The first chat-bot was developed in 1966 by Weizenbaum and named *Eliza* which famously simulated a Rogerian psychotherapist [19]. Eliza was able to generate responses with surprising intelligibility by simply parsing the input text and substituting key words into pre-stored phrases

One particularly exciting application with a chat-bot is in the electronic-health (e-health) domain. With smart-phones becoming ubiquitous, autonomous remote monitoring applications have been developed that log and analyse multi-domain data related to our health and well-being. Many people living with neurological conditions like autism, Parkinson's disease or dementia have difficulties communicating or conversing. Studies have shown people with Parkinson's disease (PwPD) have problems with conversation initiation, turn-taking, topic management, word-retrieval, and memory [2]. People on the autism spectrum

© Springer Nature Switzerland AG 2018
L. Cheng et al. (Eds.): ICONIP 2018, LNCS 11302, pp. 41–51, 2018.
https://doi.org/10.1007/978-3-030-04179-3_4

often have problems understanding the nuances of social interaction [13]. There is growing literature suggesting chat-bots may provide some form of therapy for these conditions as they allow the user to practise in the privacy of their own homes without feeling frustrated, judged or embarrassed whilst providing encouragement and individualised feedback [13]. Moreover, chat-bot technology can monitor progress and the impact of difficulties with communication, helping the individual as well as health professionals and researchers. Chat-bots have been reported in the literature for health related applications. Examples include health behavior change for obesity and diabetes [18]; disease self-management [1]; and health education for adolescents on topics related to sex, drugs and alcohol [3].

There are a number of approaches that apply online resources to improve the quality of the responses of the conversational agents. However, none of them, to the best of our knowledge, has employed available app reviews to enhance a chat-bot engine. For example, Huang et al. [6] developed a knowledge base of pairs of question-answers using discussion forums. Similar to our approach, they investigated the application of Support Vector Machines (SVM) to extract related replies to the topics of the threads and then rank them based on their qualities. The topic-replies pairs then formed the main component of their chat-bot. This is contrary to our system that uses the acquired knowledge from sentiment analysis as a supportive feature to produce high quality utterances. Kirange and Deshmukh [12] presented an emotion classifier using SVM. They classified human's emotions (e.g., anger, disgust, sadness, etc.) when reading news headlines. Although not directly used for chat-bots, they argued that the automatic recognition of such emotions can help conversational agents to produce more realistic and human-like responses. This paper studies models that could potentially be used to predict a sentiment giving an utterance by the user. The first is a novel deterministic dictionary using a directed graph model while the others are the well-known Naive Bayes classifier, support vector machine (SVM), convolutional neural networks (CNN), and Long-short term memory (LSTM). These models are trained using a large, 100 MB data-set of smartphone application reviews from the Android playstore. This data-set was obtained by the authors via web-crawler program and is available online [9].

**Fig. 1.** Overview of the chatbot system showing the three stages. The speech utterance is captured by the smartphone; this is converted to text where it is fed into Harlie's brain. A response is generated which is subsequently converted to a digital voice that is spoken back to the user.

## 2   Harlie an E-Health Chat-Bot

A chat-bot named HARLIE (Human and Robot Language Interaction Experiment) was introduced recently that was designed for people who have troubles with communication [7, 10]. Harlie was programmed to have conversations with participants using full voice interaction on a range of topics including family, sports, hobbies and daily living. Whilst conversing, Harlie analysed the dialog and acoustic signals of the users voice looking for speech irregularities, response times, vowel articulation and other speech metrics. The Harlie study has hitherto logged more than 350 conversations with a large range of people having none or some form of neurological condition including Parkinson's disease and autism [10]. Interaction via Harlie is done completely by voice. Harlie operated via case-based reasoning and textual pattern matching algorithms in particular the use of a standardised computer language referred to as artificial intelligence mark-up language (AIML). Figure 1 shows the main components of the Harlie chat-bot showing a speech-to-text module which is processed to derive a suitable response which is then converted to an acoustic signal and spoken back to the user.

How the text input is processed to construct a meaningful response has been the main research question of chat-bot researchers for some time. The most prominent technique for the last decade has been via case-based reasoning and textual pattern matching algorithms in particular the use of a standardised computer language referred to as artificial intelligence mark-up language (AIML). Two examples of AIML code are given in Figs. 2a and b. Harlie uses AIML to converse with a user in a way that is both non-deterministic and meaningful. An AIML based chat-bot is comprised of a collections of AIML files for various topics and situations.

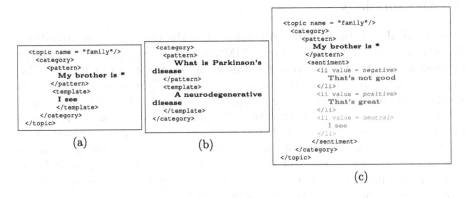

**Fig. 2.** Example AIML code (a) from the Harlie chat-bot, and (b) showing the input stimuli *pattern* and the chat-bot response, and (c) from the proposed extensions to feature responses back on sentiment of the human utterance. The * in the XML pattern tag denotes a wildcard for all of the utterance.

AIML was developed in early 2000 [17] and is based on the common extensible markup language (XML) which utilises tags to identify commands and specific input stimuli and responses. AIML is based on basic units of dialogue formed from user input patterns and respective chat-bot response. The basic units in AIML are called categories; within categories are a *pattern* and *template*. The pattern tag defines a possible user input while the template tag gives the response. Creation of AIML content requires no advanced computer programming skill and this has likely contributed to the significant amount of content already released by bot-masters. As of 2017, a large collection of AIML sets have been released under the GNU public license making them freely available. These sets make up approximately 97,431 categories in topics including, politics, religion, science, sex, sports, history, food and geography. A simple example of AIML code given in Fig. 2b, when the input stimuli is *"What is Parkinson's disease"*, the chat-bot responds with a *"A neurodegenerative disease"*.

This simplistic example consists of a direct response only; more advanced AIML techniques are often needed to formulate a response that is human-like and non-deterministic. An exhaustive list of AIML features is beyond the scope of this paper, however, the most useful features are: (i) The ability to learn new responses and alter existing responses whilst interacting with a human during run-time; (ii) Wild card searches allowing for responses to be generated when an incomplete match of the pattern occurs; (iii) A template that includes multiple responses that are randomly chosen during run-time; (iv) Scope for internal variables that allow the chat-bot to store information that may be later accessed; (v) A dedicated topic variable that maintains the current topic allowing category to be activated conditionally on the current topic; (vi) Internal data-processing, conditional statements and tests that are not visible to the user; (vii) Recursion allowing different patterns to target a single template tag.

## 3    Methodology

Despite AIML's features, issues still arise when the user utterance contains a fragment of text not considered in the AIML code-base. For instance, an excerpt of problematic AIML code from the Harlie chat-bot is given in Fig. 2a. This module is activated when, whilst speaking about family, the user mentions his brother and uses an adjective(s) not explicitly found. A wildcard is substituted for the unknown combination of words. The programmed response was neutral however this often did not match the sentiment of what the user said. This often resulted in termination of the conversation. We propose an extension to the AIML standard as given in Fig. 2c where the returned response is based upon the sentiment of what the user said. How the sentiment heuristic could be implemented is discussed in the rest of this article.

### 3.1    Graph-Based Sentiment Model

The first model considered in this work uses a dictionary approach comprised of a graph data structure approach referred to as the *graphmaster* that is typically used in chat-bots to store AIML code according to the corresponding stimulus utterance (pattern). The graphmaster holds all possible human utterances (AIML pattern tag) in the form of a rooted, directed graph $G(n, w)$ where $n$ is the node and $w$ is a word. The set $S^n = \{w : \exists m | G(n, w) = m\}$ is the set of words forming the branches from node $n$. If $r$ is the root node, $S_r$ is a collection of all the first words in the set of patterns.

The graphmaster stores AIML patterns along a path from a root node $r$, to a terminal node $t$, where the AIML template is stored. Let $w_1, w_2, \cdots, w_k$ be the sequence of $k$ words the procedure for inserting a pattern. An example of a excerpt of such a graph is given in Fig. 3. Here nodes of the graph are represented by circles where a shaded circle indicates the node contains a payload which is typically the AIML code to be executed to generate the response. In the proposed model however, this also holds a sentiment level denoted $S$. The graphmaster achieves considerable compression by sharing common prefixes. For example, "do you have the time" and "do you have time for chatting" share the common nodes "do you have". This is quite advantageous when traversing large data-sets of text.

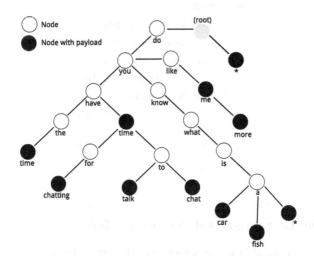

**Fig. 3.** Example of rooted, directed graph that contains *payloads* which is commonly AIML code but also could include sentimental values for the particular nodes of the graph.

Nodes store single words or special characters such as wildcards; terminal nodes store a payload with is typically AIML code that generates a response. Given an utterance denoted as a sequence of $k$ words as $\boldsymbol{w} = \{w_1, w_2, \cdots, w_k\}$, to insert the pattern and payload, the graphmaster checks if $m = G(r, w_1)$ where $r$ is the root node; if it does, the process continues the traversing the graph for the remaining words $w_2, \cdots, w_k$. When the process, at index $i$ encounters $\exists n | G(n, w_i)$ to be undefined, a new node for each of the remaining words

$w_i, \cdots, w_k$ is created. This particular method, satisfies all conditions mentioned in the introduction and has the following advantages: (i) Response times are measurable in milliseconds; (ii) Online learning is supported i.e. the model doesn't need to be taken offline and retrained. (iii) It re-uses algorithms and data structures already existing in most common chat-bots e.g. Siri.

**Finding a Match.** Given an utterance by the user, a best-match is obtained when a successful transverse of the graphmaster has occurred with the maximum number of words that comprise the utterance. Using for example, "this is actually not bad", Table 1 shows the order of phrases that are tested with the graphmaster. Matching in the graphmaster for a particular utterance is typically done using a backtracking, depth first search. Once a successful transverse has resulted in a payload being found (sentiment score) the process is aborted and the sentiment returned to the chat-bot system.

**Table 1.** Order of phrases to search a match in the graph. * denotes a wildcard which consists of zero or more words.

| Order | Phrase |
|-------|--------|
| 1 | * this is actually not bad * |
| 2 | * is actually not bad * |
| 3 | * this is actually not * |
| 4 | * actually not bad * |
| 5 | * this is actually * |
| 6 | * not bad * |
| 7 | * this is * |
| 8 | * bad * |
| 9 | * this * |

## 3.2 Machine Learning-Based Sentiment Model

Two machine learning algorithms are employed in order to evaluate the proposed graph-based method: support vector Machines and Naive Bayes.

**Naive Bayes Sentiment Model.** The Naive Bayes Classifier (NBC) is also used as a potential sentiment model. The literature shows it is surprising effective despite it's apparent simplicity. It also has the advantage of not considering the dependence of word sequences therefore the model might be potentially more robust to missing words caused by the speech to text process. The NBC was generally a 1-gram model for all except a set of *special words* where a 2-gram model was used. This allowed the correct modeling of negation for example, "not-good", "not-bad" etc. The probability the utterance belongs to sentiment class $S$ is computed as: $P(S|\boldsymbol{w}) = P(S) \prod_i^M P(w_i|S)$, where $S$ is the class, $w_i$ is the i-th word in the utterance. If a particular word is not found in any class it is ignored. If the utterance has no words contained in any of the classes, a sentiment with the highest prior, $P(S)$ is returned otherwise the class with the highest probability is returned.

**Support Vector Machine.** A support vector machine (SVM) is another machine learning algorithm that is employed to create a sentiment model. SVM is a *kernel*-based approach (i.e., it uses a function to map a space of non-linearly separable data points onto a new linearly separable space) and has been extensively used in the literature [14,15]. The devised SVM model for sentiment analysis uses syntactical and statistical features of utterances in order to learn and classify their sentiments. The syntactical features include words and their n-grams (2 to 4-grams) and the statistical features derived from calculated term frequency–inverse document frequency (tf-idf).

**Convolutional Neural Network.** A Convolutional Neural Network (CNN) is a feedforward neural network, in which, the information flows from the input to an intermediate computation for defining one or more functions and finally to the output [4]. The input to the CNNs is a grid-like data, for example, time-series data that is considered as a 1D vector, or images that are 2D grid of pixels, or as in document classifications, vector representation of words that results in a matrix for each document. In CNNs, at least one layer uses convolution (a mathematical operation) to perform matrix multiplication between the input and multiple weight matrices (or filters). Convolutional layers are typically connected to pooling (or subsampling) layers that are then followed by fully connected (or output) layer [16]. The CNN network in this work follows the architecture proposed for text classification in [11].

**Long-Short Term Memory.** Long-short term memory (LSTM) [5] is a recurrent neural network equipped with memory gates for controlling and memorizing information given in the past. In this application for sentiment analysis, we use each sentence as an input for the LSTM and connect the last hidden state with an output layer for prediction. The words in each sentence are converted into 300-dimensional vectors using pre-trained word2vecs from GoogleNews[1].

### 3.3   Data Corpus

**Training.** The training corpus used in this work was compiled by the authors using a web-crawler that scanned the Google Playstore for reviews of smartphone applications across a broad range of categories. More then one million smartphone applications were scanned resulting in 3,163,878 unique reviews. The review title and corresponding score (ranging from 1–5) was extracted. This data-set is freely available at [9]. The raw data consisting of review title and score was further processed to eliminate conflicting utterances with different scores. In this instance, reviews that had the same utterance but different scores were averaged to find the total score. This is was converted into a *positive* and

---

[1] https://github.com/mmihaltz/word2vec-GoogleNews-vectors.

**Table 2.** Table of comparisons between the five proposed methods of sentiment classifaction evaluated on the Android playstore collection and the Harlie dialogs. Performance is determined by the recall, precision and F1-scores for both positive and negative utterances.

| Testing dataset | Model | Performance | | | | | |
|---|---|---|---|---|---|---|---|
| | | Positive | | | Negative | | |
| | | Recall | Precision | F1-Score | Recall | Precision | F1-Score |
| Android Playstore | Graph | 0.8158 | **0.9838** | 0.8943 | 0.567 | 0.1089 | 0.1877 |
| | Naive-Bayes | 0.90452 | 0.8862 | 0.8952 | 0.5894 | 0.6247 | 0.6028 |
| | SVM | 0.8896 | 0.9512 | **0.9113** | **0.7073** | 0.4872 | 0.5735 |
| | CNN | **0.9385** | 0.8907 | **0.9140** | 0.5651 | **0.7086** | **0.6288** |
| | LSTM | 0.9103 | 0.9072 | 0.9028 | 0.5547 | 0.6892 | 0.6111 |
| Harlie Dialogs | Graph | 0.9641 | 0.9605 | **0.9694** | 0.1233 | 0.1048 | 0.1185 |
| | Naive-Bayes | 0.9624 | 0.9223 | 0.9451 | 0.1197 | 0.2215 | 0.1592 |
| | SVM | 0.9656 | 0.9348 | 0.9525 | 0.1455 | **0.2377** | 0.1703 |
| | CNN | 0.8654 | **0.9701** | 0.9148 | **0.4228** | 0.1268 | **0.1950** |
| | LSTM | 0.9632 | 0.9508 | 0.9152 | 0.2441 | 0.1136 | 0.1885 |

*negative* sentiment according to the following schema for sentiment $S$ where $r$ denotes the average review score for the utterance $w$:

$$S(w) = \begin{cases} \texttt{positive,} & \text{if } r > 2.5. \\ \texttt{negative,} & \text{if } r \leq 2.5. \end{cases} \tag{1}$$

The rationale behind the binarisation of sentiments is that, in reality, a chatbot generally deals with negative and neutral utterances in a similar way (i.e., by giving sympathetic responses). After processing the data consists of 3,163,878 unique utterances. Table 3 shows the statistics of the final corpus.

**Table 3.** Size of the retrieved application reviews.

| | Positive | Negative |
|---|---|---|
| Training Set | 1,765,011 (80%) | 449,705 (20%) |
| Testing Set | 755,701 (80%) | 193,371 (20%) |

**Real-Life.** In order to gauge the effectiveness of the five models on real-life conversation data, the Harlie dialogs of conversation between the chat-bot and the generation population are used. These are available online at [8]. As this text is unclassified, a human manually labeled each utterance positive and negative. An utterance was labeled negative if it referred to any of the following: insults and profanity; expresses dislike or dissatisfaction; melancholy, depression, sickness, exhaustion or death; illegal activities; and communication breakdown or misunderstandings. An utterance was labeled positive if it did not fit the negative criteria. This corpus consists of 3225 (95%) positive and 149 (5%) negative utterances.

## 4   Results

Table 2 shows the results of the graph-based and machine learning-based approaches over both Android App reviews and Harlie Dialogs corpora. It can be observed that the graph-based approach achieved highest precision utterances on Android reviews and ranked second on Harlie Dialogs corpora with 0.9838 and 0.9605, respectively. The CNN sentiment model showed highest recall on identifying positive utterances in Android reviews with 0.9385, while all five approaches gained ~96% recall for such utterances in Harlie dialogs. In terms of F1-Score, the SVM and CNN models achieved better performance for classifying positive utterances in Android reviews with 0.9113 and 0.9140 respectively, and the Graph-based approach was the best on predicting positive utterances in Harlie Dialogs.

All five approaches exhibited lower performance on identifying negative utterances in both corpora. CNN showed the highest precision with 0.7086 on Android reviews and SVM achieved slightly better precision on Harlie's negative utterances with 0.2377. The SVM model performed better than other approaches in identifying negative utterances in Android reviews with relatively higher recall of 0.7073. Overall, the CNN model worked slightly better for negative utterances in both Android reviews and Harlie's negative utterances with 0.6288 and 0.1950 F1-Scores respectively.

## 5   Discussion

The experimental results show that the proposed graph-based approach and the two Machine Learning-based models are highly confident in predicting positive utterances in both written and conversational corpora (i.e., the Android App reviews and Harlie's transcribed conversations, respectively). However, both corpora exhibit an unbalanced distribution of positive and negative samples; as shown in Table 2, 80% of the utterances in both corporate positive. This characteristic considerably affects the performance of Machine Learning algorithms as they need abundant samples of all target classes to develop discriminative patterns. In addition, the graph-based approach showed the lowest performance in classifying negative sentiment. This is mostly likely caused by insufficient training data for negative utterances. As shown in Table 2, the graph-based approach achieved the best F1-Score over the positive utterances in the Harlie dialogs corpus. Since this corpus was derived from real chat-bot conversations, it can provide a better evaluation of the sentiment models and their estimated performance as supportive engines inside chat-bots to produce more reliable and accurate responses. All five approaches achieved similar results on identifying positive utterances. However, the performance of these approaches varied for predicting negative utterances: Naive-Bayes, SVM, CNN and LSTM models showed almost similar performance, while the Graph-based approach achieved lower performance, especially on Android reviews.

# 6  Conclusions

The role of chat-bots in the future is difficult to predict. It is however, strongly believed, there will be a profound impact as previous technology jumps (graphical user interfaces, web browsers and touchscreens etc.). In order for humans to develop a strong personal relationship with an AI agent, quantifying the sentiment of the user is paramount. This paper has discussed the potential for future applications particularly in the emerging e-health domain. A large data corpus of pre-classified utterances was used to build five competing models to detect the sentiment of an utterance. How well these models adapted to real-life human-computer machine interactions was presented. All five methods were highly accurate in recognizing positive sentiment but had significant reduced accuracy in detecting a negative utterance.

# References

1. Bickmore, T., Gruber, A., Picard, R.: Establishing the computer-patient working alliance in automated health behavior change interventions. Patient Educ. Couns. **59**(1), 21–30 (2005)
2. Saldert, C., Ferm, U., Bloch, S.: Semantic trouble sources and their repair in conversations affected by parkinson's disease. Int. J. Lang. Commun. Disord. **49**(6), 710–721 (2014)
3. Crutzen, R., Peters, G., Portugal, S.D., Fisser, E.M., Grolleman, J.: An artificially intelligent chat agent that answers adolescents' questions related to sex, drugs, and alcohol: an exploratory study. J. Adolesc. Health **48**, 514–519 (2011)
4. Goodfellow, I., Bengio, Y., Courville, A., Bengio, Y.: Deep Learning, vol. 1. MIT Press, Cambridge (2016)
5. Hochreiter, S., Schmidhuber, J.: Long short-term memory. Neural Comput. **9**(8), 1735–1780 (1997)
6. Huang, J., Zhou, M., Yang, D.: Extracting chatbot knowledge from online discussion forums. IJCAI **7**, 423–428 (2007)
7. Ireland, D., Liddle, J., McBride, S., Ding, H., Knuepffer, C.: Chat-bots for people with parkinson's disease: science fiction or reality? In: Driving Reform: Digital Health is Everyones Business. vol. 214. Studies in Health Technology and Informatics (2015)
8. Ireland, D.: Harlie dialog data (2015). https://github.com/djireland/Dialog-Harlie-2016
9. Ireland, D.: Classified utterances from the google playstore (2016). https://github.com/djireland/labelled_utterances
10. Ireland, D., et al.: Hello harlie: enabling speech monitoring through chat-bot conversations. In: Driving Reform: Digital Health is Everyones Business. vol. 227. Studies in Health Technology and Informatics (2016)

11. Kim, Y.: Convolutional neural networks for sentence classification. In: Proceedings of the 2014 Conference on Empirical Methods in Natural Language Processing (EMNLP), pp. 1746–1751 (2014)

12. Kirange, D., Deshmukh, R.: Emotion classification of news headlines using SVM. Asian J. Comput. Sci. Inf. Technol. **2**(5), 104–106 (2013)

13. Milne, M., Luerssen, M., Lewis, T., Leibbrand, R., Powers, D.: Designing and evaluting interactive agents as social skills tutors for children with autism spectrum disorder. In: Conversation Agents and Natural Language Interaction, chap. 2, pp. 23–50. Information Science Reference, Hershey PA 17033 (2011)

14. Moraes, R., Valiati, J.F., Neto, W.P.G.: Document-level sentiment classification: an empirical comparison between svm and ann. Expert. Syst. Appl. **40**(2), 621–633 (2013)

15. Mullen, T., Collier, N.: Sentiment analysis using support vector machines with diverse information sources. In: Conference on Empirical Methods in Natural Language Processing, vol. 4, pp. 412–418 (2004)

16. Suk, H.I.: An introduction to neural networks and deep learning. In: Deep Learning for Medical Image Analysis, pp. 3–24. Elsevier (2017)

17. Wallace, R.: The elements of aiml style (2003)

18. Watson, A., Bickmore, T., Cange, A., Kulshreshtha, A., Kvedar, J.: An internet-based virtual coach to promote physical activity adherence in overweight adults: randomized controlled trial. J. Med. Internet Res. **14**(1), 1–12 (2012)

19. Weizenbaum, J.: Eliza a computer program for the study of natural language communication between man and machine. Commun. ACM **9**(1), 36–45 (1966)

# Hashtag Recommendation with Attention-Based Neural Image Hashtagging Network

Gaosheng Wu, Yuhua Li$^{(\boxtimes)}$, Wenjin Yan, Ruixuan Li, Xiwu Gu, and Qi Yang

School of Computer Science and Technology,
Huazhong University of Science and Tecnnology, Wuhan 430074, China
{wugaosheng,idcliyuhua,ywj,rxli,guxiwu,ayang7}@hust.edu.cn

**Abstract.** With the increasing number of microblog users, the hashtag recommendation task has become an important component in social media. Most hashtag recommendation related methods get relative low precisions, because hashtags are not necessarily related to the content of tweets, which makes hashtag recommendation more challenging. In this work, we propose a new sequence-to-sequence method named attention based neural image hashtagging network (A-NIH) to model sequence relationship between social images and hashtags. To the best of our knowledge, this is the first work that applies attention mechanism to the image-only hashtag recommendation tasks. Our experimental results on the real-world social image dataset shows that our model performs better than the state-of-the-art methods.

**Keywords:** Image hashtag recommendation · Deep learning
Attention mechanism

## 1 Introduction

In the real-world social network, users spontaneously generate several hashtags [4] (see in Fig. 1) for users' contents (such as tweets, user's personal social images), so assigning hashtags to the social content is a kind of social behavior. Hashtags can be used as category labels, topics, and it also can be used for information retrieval and help people to understand social content or convey the social meaning.

There have been a lot of works devoted to the hashtag recommendation task, but most of them focus on hashtag recommendation for tweets. Very few works pay attention to hashtag recommendation with only the given social images. But in some social networks, such as Instagram, Flicker, etc, people tend to share more pictures, so there is a large number of images existing in these social networks. The social images contain much semantic information that we can make use of for recommendation tasks.

© Springer Nature Switzerland AG 2018
L. Cheng et al. (Eds.): ICONIP 2018, LNCS 11302, pp. 52–63, 2018.
https://doi.org/10.1007/978-3-030-04179-3_5

**Fig. 1.** An example of the social image hashtags.

Automatically generating a list of hashtags of the social image is a very challenging task, because hashtags are related to user preference, and it is not necessarily related to the content of the images. The relation between the hashtags and images can be called weak relations. Most approaches mainly focus on text twitter hashtag recommendation and get very low precision. Inspired by image captioning task, we propose a novel attention based neural image hashtagging network (A-NIH) to model hashtag's inner relationship. The main contribution of this work can be summarized as follows:

(a) Considering the sequence relationship among the hashtags, we use a novel sequence-to-sequence neural network framework to model the dependencies of the hashtags.

(b) We propose an attention-based image to hashtags model to perform image hashtag recommendation task. To the best of our knowledge, this is the first work that applies attention mechanism to the image-only hashtag recommendation tasks.

(c) Experimental results on the real-world dataset shows that our method performs better than the state-of-the-art methods.

The remaining of this paper is organized as follows. Section 2 describes related works. The framework and adopted methods are discussed in Sect. 3. In Sect. 4 we present the dataset used to evaluate the effectiveness of our model and experimental results obtained. Section 5 concludes the paper.

## 2   Related Works

We briefly review hashtag recommendation in the context of hashtag & tag recommendation and attention based hashtag recommendation.

## 2.1  Hashtag & Tag Recommendation

Most recent works have not illustrated the relationship between hashtag recommendation and tag recommendation, we empirically define the relationships here:

It is obvious that the hashtag [4] is a kind of metadata tag used in social networks such as Twitter and other microblogging services. In social network, users create and use hashtags by placing the number sign or pound sign # in front of a word or a short phrases. This is a kind of social behavior. Compared with hashtags, tags have more general concepts, and tags [1] can be a word or a short phrase, too. Tags have been widely used in many more applications, such as, QA websites (Stack Overflow), video websites (Youtube), etc.

In essential, they are in same concepts, but used in relatively different occasions. Tags can occur anywhere for information retrieval or information categorizing, while hashtags are specially used for microblogging, some musical websites and so on.

Except for applying deep learning in hashtag recommendation, there are lots of methods for hashtag recommendation. For the textual and visual information of a microblog, Gong et al. [5] proposed a generative method to incorporate textual and visual information. Wang et al. [19] proposed a Tensor factorization method to obtain to top ranked tags which are suitable for image visual contents and users' interest. Surendra et al. [14] proposed to recommend hashtags for hyperlinked tweets, and formulated the hashtag recommendation problem as a learning to rank problem. Kuntal et al. [3] proposed a word embedding based method to recommend twitter hashtags, by deriving the embeddings of hashtags using the context words appearing with the hashtag.

However, these methods are not easy to be applied in image-only hashtag recommendation tasks. So far, the public available dataset for social image hashtag recommendation is the HARRISON [12] dataset, and Jocic et al. [8] employed the dataset to do the image tagging task. The model used only 50 predefined hashtags, evaluated the performance with precision, recall and F1-measure, which is not truly a model for original benchmark improvement.

## 2.2  Attention Based Hashtag Recommendation

Attention mechanism has been successfully applied to the computer vision, machine translation, etc. Attention mechanism [11,20] is originated from the human's attention. Humans often focus on the typical part of an image or a text, then draw the conclusions. Similarly, the neural network generate better results through paying more attention to the specific parts of the input.

In the field of hashtag recommendation, many attention based methods have been proposed. Zhang et al. [21] proposed a co-attention Network, but the attention network needs to make use of the image and the corresponding tweet as the input, which makes it not easy to use other datasets that don't cover information about the tweets. Huang et al. [7] proposed a Hierarchical Attention network to model users' history information, attention mechanism can help the model select appropriate histories. Li et al. [10] proposed a topic attention based LSTM model for text tweets hashtag recommendation, the attention mechanism incorporates the topic modeling into LSTM for hashtag recommendation. Yuyun et al. [6] proposed an attention-based convolutional neural network for hashtag recommendation. These attention mechanism is applied only in text datasets.

Motivated by above successful applications, in this work, we adopt the attention mechanism to model the social images for a better representation.

## 3   The Proposed Method

In this work, we formulate the hashtag recommendation task as a multi-class classification problem. The input of our network is a raw social image, and the output is a sequence of hashtags. Inspired by the neural image captioning model, we call the model as the neural image hashtagging (A-NIH) model.

The overall framework is illustrated in Fig. 2. A-NIH mainly consists of two parts, namely, encoder with sequential attention and decoder for hashtag recommendation. In the rest of this section, we will present the two parts in detail.

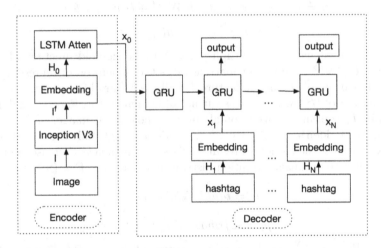

**Fig. 2.** The overall structure of neural image hashtagging model (A-NIH). The Inception V3 extracts the features of the social images, and then we map all the hashtags and image features to the same dimension. In order to model the sequential dependency of an image, LSTM attention model is introduced to capture the information. The unrolled GRU units generate the outputs, sharing the same parameters.

### 3.1   Encoder with Sequential Attention

The encoder is a very deep convolutional neural network (CNN). Since training a very deep convolutional network on a large dataset is a very time consuming task, we use public pretrained model as the nuclear part of the encoder. Then we directly use the encoder to extract features without training the CNN.

Our choice for the CNN is the Inception V3 [17] network. Experimental results show that Inception V3 network can get a higher precision than VGG [15] network. The features are embedded with same size equal to the hashtag embedding (described in next subsection). The sequential attention is applied in the encoded feature. The process can be formulated as Eqs. (1)–(2):

$$I^f = CNN(I) \tag{1}$$

$$x_0 = LSTM(I^f)^{att} \tag{2}$$

where I represents the input image, CNN is Inception V3 network. $I^f = (v_1, v_2, .., v_N)$ is the feature vector. LSTM processes it sequentially. The transition equations of LSTM are defined as Eqs. (3)–(7):

$$i_t = \sigma(W_i I^f + U_i h_{t-1} + b_i) \tag{3}$$

$$f_t = \sigma(W_f I^f + U_f h_{t-1} + b_f) \tag{4}$$

$$o_t = \sigma(W_o I^f + U_o h_{t-1} + b_o) \tag{5}$$

$$c_t = f_t \odot c_{t-1} + i_t \odot tanh(W_c I^f + U_c h_{t-1} + b_c) \tag{6}$$

$$h_t = o_t \odot tanh(c_t) \tag{7}$$

where $\odot$ is the element-wise multiplication, $\sigma$ is the sigmoid function, all $W \in \mathbb{R}^{d \times N}$ and $U \in \mathbb{R}^{d \times d}$ are weight matrixes, d is the dimension of representation, n is the dimension of the image feature $I^f$, all $b \in \mathbb{R}^d$ are bias vectors.

The output of LSTM layer is a sequence of hidden vectors $[h_1, h_2, ...h_d]$. Each annotation $h_t$ contains information about the whole input image.

Taking all hidden states $[h_1, h_2, ...h_d]$, we adopt a single fully connected neural network to project the hidden states into a fixed feature space, then use a softmax layer to generate an image attention distribution, shown in Eqs. (8)–(9):

$$\overline{h_i} = tanh(W_{\overline{v}} h_i + b_{\overline{v}}) \tag{8}$$

$$p_i = softmax(W_v \overline{h_i} + b_v) \tag{9}$$

where $W_{\overline{v}}, W_v \in \mathbb{R}^{d \times N}$ are the weight matrixes, all $b_{\overline{v}}, b_v \in \mathbb{R}^{d \times d}$ represent the bias vectors, tanh is the activation function, $\overline{h_i}$ is the candidate hidden state at time step i, $p_i$ is the attention probability. Based on the attention probability $p_i$ of the feature vector, the new representation of the image $x_0$ is conducted as the weighted image vector:

$$x_{0,i} = p_i h_i \tag{10}$$

The $x_0 = \{x_{0,1}, x_{0,2}...x_{0,d}\}$ will be used in decoder described in the next subsection.

## 3.2   Decoder for Hashtag Recommendation

The decoder is a GRU network. We first map the hashtag into a fixed feature space, combined with the image feature as the inputs. The GRU will model the hashtag-hashtag relationships and the image-hashtag relationships. The process can be formulated as Eqs. (11)–(12):

$$x_t = W_e H_t, t \in \{1, ..., N\} \tag{11}$$

$$\hat{h}_{t+1} = GRU(x_t), t \in \{0, ..., N\} \tag{12}$$

where $H = (H_0, H_1, ..., H_N)$ is the hashtag sequence. We represent each hashtag as a one-hot vector $H_t$ of dimension equal to the size of hashtag dictionary. $W_e \in \mathbb{R}^{d \times N}$ is the weight of the hashtag at time step t, d is the size of the fixed feature space, N is the maximum number of the hashtags of the social image, $x_t$ is the hashtag embedding feature at time step t.

For each position $x_t$, given the previous output $\hat{h}_{t-1}$, GRU use the update gate $z_t$ and reset gate $r_t$ to generate next output $\hat{h}_t$. The transition equations of GRU is defined as Eqs. (13)–(16):

$$r_t = \sigma(W_r x_t + U_r \hat{h}_{t-1}) \tag{13}$$

$$z_t = \sigma(W_z x_t + U_z \hat{h}_{t-1}) \tag{14}$$

$$\overline{h}_t = tanh(W_{\overline{h}} x_t + U_{\overline{h}} r_t \odot h_{t-1}) \tag{15}$$

$$\hat{h}_t = (1 - z_t)\hat{h}_{t-1} + z_t \overline{h}_t \tag{16}$$

where $W_r, W_z, W_{\overline{h}} \in \mathbb{R}^{d \times L}$ and $U_r, U_z, U_{\overline{h}} \in \mathbb{R}^{d \times d}$ are weight matrixes (same with the previous subsection), L is the output size of each GRU, $\sigma$ is the sigmoid function, $tanh$ is the nonlinear function, $\overline{h}_t$ is the candidate hidden state at time step t.

Note that we denote a special start word with $S_0$. Both the image and hashtags are mapped to the same feature space. The image attention $x_0$ combined with the special start word $S_0$ is only input once at t=0 to inform the GRU about the image contents.

Taking the output of the GRU $\hat{h} = \{\hat{h}_1, \hat{h}_2, ..., \hat{h}_N\}$, we adopt a single fully connected network at each time step. The softmax function will generate the distribution probability of each hashtag:

$$P_t = softmax(W_i \hat{h}_t) \tag{17}$$

where $P_t$ is the probability of each hashtag at time step t, $W_i \in \mathbb{R}^{d \times L}$ is the weight matrix, L is the dimension of the hashtag dictionary.

## 3.3  Training

Our loss function is the sum of the negative log likelihood of the correct hashtag word at each time step as follows:

$$J = - \sum_{t=0}^{N} \log p(H_t | I, H_0, ..., H_{t-1}; \theta) \qquad (18)$$

where $H_t$ is the hashtag at time step t, I is the image (described in the previous subsection), $\theta$ represents the parameters of our proposed network. The above loss is minimized w.r.t. all the parameters of encoder and decoder.

The network parameters are updated using stochastic gradient descent with the adam [9] update rule. L2 regularization and Dropout [16] regularization are used to improve the generalization ability, and to prevent the model from being overfitted.

## 3.4  Hashtag Recommendation

We perform hashtag recommendation as follow: for the given the image, we first extract the feature vector of the image, then the attention mechanism is used to generate the attention based feature, GRU networks generate the sequence of the recommended hashtags. Since the output words are often repetitive and we don't need to care about the order of the hashtags, so we simply use greedy search (in Fig. 3) to filter out the repeated words, and get the best top K hashtags.

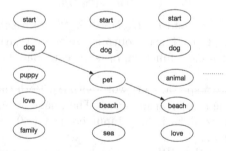

**Fig. 3.** An example of the customized greedy search. The greedy search finds the best one path with the highest probability. If one path contains repetitive hashtag, we prune it.

## 4   Experiments

We conduct the experiments on the real-world image dataset (HARRISON) in social network collected by Minseok et al. [12]. HARRISION has a total of 57,383 images and approximately 260,000 hashtags. Each image has an average of 4.5 associated hashtags (minimum 1 and maximum 10 associated hashtags). After removing infrequently used hashtags, there are 997 most frequently used hashtags reserved for the dataset. We split 90% dataset for training and 10% for testing. The size of training dataset is 51644, the size of the testing set is 5739. To evaluate the performance, we use precision@K, recall@K, accuracy@K defined in [12]. For the comparison of the benchmark [12], we set K to 1 for precision and 5 for recall and accuracy.

### 4.1 Experiments Settings

We take the image with size 229 x 229 and channel RGB as visual input. Our encoder employs the Inception V3 [17] network pretrained on ImageNet 2012 classification [13] challenge dataset. We regard avg_pool's output as the visual features of images. One LSTM attention layer is stacked on top of the image feature layer, with dimensions of 128, the decoder GRU network generate the hashtags. Finally, all of the implementations are trained on one NVIDIA Titan Xp using Keras [2] deep learning framework and all the source code of our models will be released upon the publication of this work.

### 4.2 Baseline Model

In this section, state-of-the art baseline models are described in detail.

**VGG Based Model.** This model was proposed by Minseok et al. [12]. The model employs two base networks, VGG-Scene [18] pretrained on Place Database [22] and VGG-Object [15] as the visual feature extractor, then concats the features, feeds them into a fully connected network. It regards the hashtag recommendation task as a multi-label classification problem and gets the recommended hashtags with top K posterial probability.

**Neural Image Captioning.** The neural image generater was proposed by Oriol Vinyals et al. [20]. The model uses VGG pretrained on Imagenet model to extract image features. The image features are fed into the LSTM to predict each hashtag.

**Table 1.** Evaluation results of different methods on the evaluation dataset

| Model | Precision@1 | Recall@5 | Accuracy@5 |
|---|---|---|---|
| VGG based model | 0.3002 | 0.2174 | 0.5250 |
| Neural Image captioning (NIC) | 0.3232 | 0.1290 | 0.3554 |
| Neural Image hashtaging (A-NIH) | **0.3296** | **0.2408** | **0.5975** |

## 4.3   Comparison Experiments

Table 1 shows the results of our experiments. We compare our model with the neural image captioning model and the benchmark VGG based model. A-NIH gets the best performance in terms of the precision@1, accuracy@5, recall@5. Here is the detail, our approach provides improvements of 2.94% in precision@1, 2.34% in recall@5, and 7.25% in accuracy@5 over the VGG based model, which is the state-of-the-art method for this task. The baseline model used two base networks for feature extraction, our model only employs single CNN for feature extraction. Compared with the NIC, A-NIH achieves an improvement of 0.64% in precision@1, 11.18% in recall@5, 24.21% in accuracy@5.

## 4.4   Model Component Impact Analysis

Images contain the sequence information. We employ RNN to capture the sequence information of hashtags and the corresponding images. But how neural network impacts the final results, we replace parts of the A-NIH with other baseline models and get VGG + LSTM, VGG + GRU, Inception V3 + LSTM, Inception V3 + GRU, Inception V3 + GRU + attention models for component impact analysis. The following part will illustrate the details. Since training the model is a time-consuming task, the following experiments are all trained 100 epoches with same experimental settings, then we choose the model with the lowest validation loss for parameter analysis.

**Table 2.** Model component impact analysis for A-NIH

| Model | Precision@1 | Recall@5 | Accuracy@5 |
|---|---|---|---|
| VGG+LSTM | 0.2933 | 0.1655 | 0.4168 |
| VGG+GRU | 0.2962 | 0.1729 | 0.4307 |
| Inception V3+LSTM | 0.3084 | 0.1748 | 0.4503 |
| Inception V3+GRU | 0.3171 | 0.1849 | 0.4719 |
| Inception V3+GRU+atten | **0.3177** | **0.2128** | **0.5200** |

Table 2 shows the model component impact results for A-NIH, VGG represents the VGG-16 network. We can see, the feature representation of inception V3 is better than the VGG-16 feature representation, and GRU recommender is better than the LSTM recommender, the attention mechanism can promote better performance.

Figure 4 shows some examples of good and bad hashtag recommendation of the A-NIH and VGG based baseline model, GT represents ground truth, the first two examples are the good recommendation examples, the third example is the bad one. Considering the sequence information of hashtags and images, A-NIH can recommend more correct hashtags than the VGG based baseline

| A-NIH | VGG based Model | GT |
|-------|-----------------|-----|
| yellow | orange | morning |
| green | instadaily | coffee |
| home | photography | yellow |
| coffee | flower | flower |
| love | school | mood |

| A-NIH | VGG based Model | GT |
|-------|-----------------|-----|
| wedding | igdaily | wedding |
| makeup | instagood | bride |
| beauty | fashion | funtimes |
| hair | tired | goodtimes |
| model | picoftheday | silly |

| A-NIH | VGG based Model | GT |
|-------|-----------------|-----|
| instafood | night | word |
| yummy | bored | live |
| food | swag | gold |
| foodie | follower | day |
| dinner | school | new |

**Fig. 4.** Examples of good and bad hashtag recommendation of the A-NIH and VGG based baseline model

model. Because of the weak relations between the hashtags and social images, the performance is still not good. As for the last example, both the two models don't recommend any correct hashtags, this is because that the ground truth hashtags are unrelated to the image. So this task still needs to be improved in the future.

For the sequential attention, we give empirical explanation. Distribution of the image feature vector is different from the hashtag feature vector. If we directly use the image feature to do the hashtag sequence task, owing to the inconsistency of the image feature and hashtag feature, the image feature will bring noise for the sequence task. Attention mechanism can change the distribution of the image feature so that it can be fit for the sequence task.

## 5   Conclusions

In this paper, we propose a novel attention based model (A-NIH) for improving the performance of the image hashtag recommendation. Inception v3 plays as the encoder for image feature extraction, the GRU plays the decoder for hashtag recommendation. Our experimental results show that the attention mechanism can filter out the image noise to some extent and the proposed method outperform the recent state-of-the-art baseline model on the real-world social image dataset. Our future work will try to construct the semantic network of hash-tag and visualize it for semantic and statistic analysis, what's more, we seek to visualize the attention mechanism feature to further analyse the relationship between social images and hashtags.

**Acknowledgments.** The authors wish to thank the anonymous reviewers for their helpful comments, and we gratefully acknowledge the support of NVIDIA Corporation with the donation of the Titan X Pascal GPU used for this research, what's more, we feel thankful that authors can provide baseline models for us so that we can conduct the experiments smoothly. This work is supported by the National Key Research and Development Program of China under grants 2016QY01W0202 and 2016YFB0800402, National Natural Science Foundation of China under grants 61572221, U1401258, 61433006, 61772219 and 61502185, Major Projects of the National Social Science Foundation under grant 16ZDA092, Science and Technology Support Program of Hubei Province under grant 2015AAA013, Science and Technology Program of Guangdong Province under grant 2014B010111007 and Guangxi High level innovation Team in Higher Education Institutions–Innovation Team of ASEAN Digital Cloud Big Data Security and Mining Technology.

# References

1. Berendt, B., Hanser, C.: Tags are not metadata, but "just more content" - to some people. In: Proceedings of the First International Conference on Weblogs and Social Media, ICWSM 2007, Boulder, Colorado, USA, 26–28 March 2007 (2007)
2. Chollet, F., et al.: Keras (2015). https://github.com/fchollet/keras
3. Dey, K., Shrivastava, R., Kaushik, S., Subramaniam, L.V.: Emtagger: a word embedding based novel method for hashtag recommendation on twitter. In: 2017 IEEE International Conference on Data Mining Workshops, ICDM Workshops 2017, New Orleans, LA, USA, 18–21 November 2017, pp. 1025–1032 (2017)
4. Efron, M.: Hashtag retrieval in a microblogging environment. In: Proceeding of the 33rd International ACM SIGIR Conference on Research and Development in Information Retrieval, SIGIR 2010, Geneva, Switzerland, 19–23 July 2010, pp. 787–788 (2010)
5. Gong, Y., Zhang, Q., Huang, X.: Hashtag recommendation for multimodal microblog posts. Neurocomputing **272**, 170–177 (2018)
6. Gong, Y., Zhang, Q.: Hashtag recommendation using attention-based convolutional neural network. In: Proceedings of the Twenty-Fifth International Joint Conference on Artificial Intelligence, IJCAI 2016, New York, NY, USA, 9–15 July 2016, pp. 2782–2788 (2016)
7. Huang, H., Zhang, Q., Gong, Y., Huang, X.: Hashtag recommendation using end-to-end memory networks with hierarchical attention. In: COLING 2016, 26th International Conference on Computational Linguistics, Proceedings of the Conference: Technical Papers, 11–16 December 2016, Osaka, Japan, pp. 943–952 (2016)
8. Jocic, M., Obradovic, D., Malbasa, V., Konjo, Z.: Image tagging with an ensemble of deep convolutional neural networks. In: 2017 International Conference on Information Society and Technology, ICIST Workshops 2017, New Orleans, LA, USA, 18–21 November 2017, pp. 13–17 (2017)
9. Kingma, D.P., Ba, J.: Adam: A method for stochastic optimization. CoRR abs/1412.6980 (2014)

10. Li, Y., Liu, T., Jiang, J., Zhang, L.: Hashtag recommendation with topical attention-based LSTM. In: COLING 2016, 26th International Conference on Computational Linguistics, Proceedings of the Conference: Technical Papers, 11–16 December 2016, Osaka, Japan, pp. 3019–3029 (2016)

11. Mnih, V., Heess, N., Graves, A., Kavukcuoglu, K.: Recurrent models of visual attention. In: Advances in Neural Information Processing Systems 27: Annual Conference on Neural Information Processing Systems 2014, Montreal, Quebec, Canada, 8–13 December 2014, pp. 2204–2212 (2014)

12. Park, M., Li, H., Kim, J.: HARRISON: A benchmark on hashtag recommendation for real-world images in social networks. CoRR abs/1605.05054 (2016)

13. Russakovsky, O., et al.: Imagenet large scale visual recognition challenge. Int. J. Comput. Vis. **115**(3), 211–252 (2015)

14. Sedhai, S., Sun, A.: Hashtag recommendation for hyperlinked tweets. In: The 37th International ACM SIGIR Conference on Research and Development in Information Retrieval, SIGIR 2014, Gold Coast, QLD, Australia, 06–11 July 2014, pp. 831–834 (2014)

15. Simonyan, K., Zisserman, A.: Very deep convolutional networks for large-scale image recognition. CoRR abs/1409.1556 (2014)

16. Srivastava, N., Hinton, G.E., Krizhevsky, A., Sutskever, I., Salakhutdinov, R.: Dropout: a simple way to prevent neural networks from overfitting. J. Mach. Learn. Res. **15**(1), 1929–1958 (2014)

17. Szegedy, C., Vanhoucke, V., Ioffe, S., Shlens, J., Wojna, Z.: Rethinking the inception architecture for computer vision. In: 2016 IEEE Conference on Computer Vision and Pattern Recognition, CVPR 2016, Las Vegas, NV, USA, 27–30 June 2016, pp. 2818–2826 (2016)

18. Wang, L., Guo, S., Huang, W., Qiao, Y.: Places205-vggnet models for scene recognition. CoRR abs/1508.01667 (2015)

19. Wang, Y.: Image tag recommendation algorithm using tensor factorization. J. Multimed. **9**(3), 416–422 (2014)

20. Xu, K., et al.: Show, attend and tell: Neural image caption generation with visual attention. In: Proceedings of the 32nd International Conference on Machine Learning, ICML 2015, Lille, France, 6–11 July 2015, pp. 2048–2057 (2015)

21. Zhang, Q., Wang, J., Huang, H., Huang, X., Gong, Y.: Hashtag recommendation for multimodal microblog using co-attention network. In: Proceedings of the Twenty-Sixth International Joint Conference on Artificial Intelligence, IJCAI 2017, Melbourne, Australia, 19–25 August 2017, pp. 3420–3426 (2017)

22. Zhou, B., Lapedriza, À., Xiao, J., Torralba, A., Oliva, A.: Learning deep features for scene recognition using places database. In: Advances in Neural Information Processing Systems 27: Annual Conference on Neural Information Processing Systems 2014, Montreal, Quebec, Canada, 8–13 December 2014, pp. 487–495 (2014)

# CocoNet: A Deep Neural Network for Mapping Pixel Coordinates to Color Values

Paul Andrei Bricman[1] and Radu Tudor Ionescu[2(✉)]

[1] George Coşbuc National College, 29-31 Olari, Bucharest, Romania
paubric@gmail.com
[2] University of Bucharest, 14 Academiei, Bucharest, Romania
raducu.ionescu@gmail.com

**Abstract.** We propose a deep neural network for mapping the 2D pixel coordinates in an image to the corresponding RGB color values. The neural network is termed *CocoNet*, i.e. **co**ordinates-to-**co**lor **net**work. During the training process, the neural network learns to encode the input image within its layers, i.e. it learns a continuous function that approximates the discrete RGB values sampled over the discrete 2D pixel locations. At test time, given a 2D pixel coordinate, the neural network will output the RGB values of the corresponding pixel. By considering every 2D pixel location, the network can actually reconstruct the entire learned image. We note that we have to train an individual neural network for each input image, i.e. one network encodes a single image. Our neural image encoding approach has various low-level image processing applications ranging from image denoising to image resampling and image completion. Our code is available at https://github.com/paubric/python-fuse-coconet.

**Keywords:** Neural networks · Image encoding · Image denoising
Image resampling · Image restoration · Image completion
Image inpainting

## 1 Introduction

After the success of the AlexNet [16] convolutional neural network (CNN) in the ImageNet Large Scale Visual Recognition Challenge (ILSVRC) [27], deep neural networks [10,30,33] have been widely adopted in the computer vision community to solve various tasks ranging from object detection [7,19,25,26,31] and recognition [10,26,30] to face recognition [23] and even image difficulty estimation [13]. In the usual paradigm, a deep neural network (DNN) [8] is trained on thousands of labeled images in order to learn a function that classifies the input images into multiple classes, according the labels associated to the images. In this paper, we propose a different paradigm in which the neural network is trained on a single image in order to memorize (encode) the image by learning a function that

© Springer Nature Switzerland AG 2018
L. Cheng et al. (Eds.): ICONIP 2018, LNCS 11302, pp. 64–76, 2018.
https://doi.org/10.1007/978-3-030-04179-3_6

maps the 2D pixel coordinates to the corresponding values in the Red-Green-Blue (RGB) color space. At test time, given a 2D pixel coordinate, the neural network will output the approximate RGB values of the corresponding pixel. By considering every 2D pixel location, the network can actually reconstruct the entire learned image at the original scale. By taking pixel locations at different intervals, the network can also reconstruct the image at various scales. We coin the term *CocoNet* for our neural network, i.e. coordinates-to-color network. To the best of our knowledge, we are the first to propose a neural approach for encoding images individually, by learning a mapping from the 2D pixel coordinate space to the RGB color space. Our novel paradigm enables the application of neural networks for various low-level image processing tasks, without requiring a training set of images. Applications include, but are not limited to, image encoding, image compression, image denoising, image resampling, image restoration and image completion. In this paper, we focus on only three of these tasks, namely image denoising, image upsampling (super-resolution) and image completion. First of all, we present empirical evidence that CocoNet is able to surpass standard image denoising methods on the CIFAR-10 data set [15]. Second of all, we show that our approach can obtain better upsampling results compared to bicubic interpolation on a set of images usually employed in image super-resolution, known as Set5 [3]. We also present qualitative results for the task of image completion on Set5.

The rest of this paper is organized as follows. We present related works in Sect. 2. We describe our approach in Sect. 3. We present the results of our image denoising, image upsampling and image completion results in Sect. 4. Finally, we draw our conclusion and discuss future work in Sect. 5.

## 2 Related Work

To the best of our knowledge, there are no previous scientific works that propose to learn a mapping of the pixel coordinates to the corresponding pixel color values using neural networks. However, there are numerous neural models that learn a mapping from image pixels to a set of classes [10,16,30,33] or from pixels to pixels [1,2,4–6,12,14,18,20,21,24,28,29,35,39–42]. The neural models that map pixels to pixels are usually applied on tasks such as image compression [1,2,4,6, 21,24,35], image denoising and restoration [20,39,42], image super-resolution [5, 14,18,20,28,29,39,40], image completion [12,41] and image generation [11,36]. Since our approach, CocoNet, can be applied to similar tasks, we consider that the neural models that map pixels to pixels are more closely related to ours. It is important to mention that, different from all these state-of-the-art models, CocoNet does not require a training set of images, as it operates on a single image at once. This represents both an advantage and a disadvantage of our approach. CocoNet can be applied to a given task even if a training set is not available, but models trained a set of images can learn important patterns that generalize well from one image to another. Since CocoNet does not use a set of training images (it just learns a continuous representation of the test image it is applied on), we believe that it is unfair to compare it with supervised models that benefit from the access to a set of training images. We next present related (but supervised) models for each individual task.

**Image Denoising and Restoration.** Image restoration is concerned with the reconstruction of an original (uncorrupted) image from a corrupted or incomplete one. Typical corruptions include noise, blur and downsampling. Hence, image denoising and image super-resolution can be regarded as sub-tasks of image restoration. There are models that address multiple aspects of image restoration [20,39,42], but there are some models that deal with specific sub-tasks, e.g. image super-resolution [5,14,18,20,28,29,34,39,40]. Zhao et al. [42] propose a deep cascade of neural networks to solve the inpainting, deblurring, denoising problems at the same time. Their model contains two networks, an inpainting generative adversarial network (GAN) and a deblurring-denoising network. Wu et al. [39] propose a novel 3D convolutional fusion method to address both image denoising and single image super-resolution. To address the same tasks, Mao et al. [20] describe a fully convolutional encoding-decoding framework composed of multiple layers of convolution and deconvolution operators.

**Image Super-resolution.** Super-resolution techniques reconstruct a higher-resolution image from a low-resolution image. Tang et al. [34] employ local learning and kernel ridge regression to enhance the super-resolution performance of nearest neighbor algorithms. Dong et al. [5] propose a CNN model that learns an end-to-end mapping between the low and high-resolution images. Kim et al. [14] propose a deeply-recursive convolutional network with skip connections and recursive-supervision for image super-resolution. Ledig et al. [18] present a GAN model for image super-resolution capable of inferring photo-realistic natural images for $4 \times 4$ upscaling factors. Sajjadi et al. [28] employ fully-convolutional neural networks (FCN) in an adversarial training setting. Yamanaka et al. [40] propose a highly efficient and faster CNN model for single image super-resolution by using skip connection layers and network in network modules.

**Image Compression.** Image compression is applied to digital images in order to reduce the cost for storing and transmitting the images. Lossy image compression methods usually obtain better compression rates, but they introduce compression artifacts, which are not desirable in some cases, e.g. medical images. Baig et al. [1] study the design of deep architectures for lossy image compression. Interestingly, they show that learning to inpaint (from neighboring image pixels) before performing compression reduces the amount of information that must be stored. Ballé et al. [2] describe an image compression method based on three sequential transformations (nonlinear analysis, uniform quantization and nonlinear synthesis), all implemented using convolutional linear filters and nonlinear activation functions. Cavigelli et al. [4] present a deep CNN with hierarchical skip connections and a multi-scale loss function for artifact suppression during image compression, while Minnen et al. [21] combine deep neural networks with quality-sensitive bit rate adaptation using a tiled network. Both Cavigelli et al. [4] and Minnen et al. [21] report image compression results that are better than the JPEG standard. Dumas et al. [6] address image compression using sparse representations, by proposing a stochastic winner-takes-all auto-encoder in which image patches compete with one another when their sparse representation is computed. Prakash et al. [24] design a technique that makes JPEG content-aware by training a deep CNN model to generate a map that highlights semantically-salient regions so that they can be encoded at higher quality as compared to background regions. Toderici et al. [35] present several recurrent neural network (RNN) architectures that provide variable compression rates during deployment without requiring retraining. Their approach is able to outperform JPEG across most bit rates.

**Image Completion.** Image completion or image inpainting refers to the task of filling in missing or corrupted parts of images. Iizuka et al. [12] employ a FCN to complete images of various resolutions by filling in missing regions of arbitrary shape. To train their model, the authors use global and local context discriminators that distinguish real images from completed ones. Yang et al. [41] propose a multi-scale neural patch synthesis approach based on joint optimization of image content and texture constraints. The approach of Yang et al. [41] has two important advantages, namely it preserves contextual structures and produces high-frequency details.

**Image Generation.** Perhaps more difficult than image completion is the task of generating entire images that look natural to the human eye. Pixel Recurrent Neural Networks [36] sequentially predict the conditional distribution over the possible pixel values in an image along the two spatial dimensions, given the scanned context around each pixel. The Pixel RNN model consists of up to twelve fast two-dimensional Long Short-Term Memory (LSTM) layers. Huang et al. [11] propose a novel generative model named Stacked Generative Adversarial Networks (SGAN), which is trained to invert the hierarchical representations of a bottom-up discriminative network. They first train each model in the stack independently, and then train the whole model end-to-end.

## 3  Method

### 3.1  Intuition

The human brain is able to solve an entire variety of tasks ranging from object detection to language understanding, by learning to detect, recognize and understand complex patterns in the input data provided by senses such as sight, hearing or touch. Since the brain has an impressive capacity in solving such complex tasks, researchers have tried to mimic the brain capacity through neural networks [22], which are a mathematical model of the brain. Most neural network

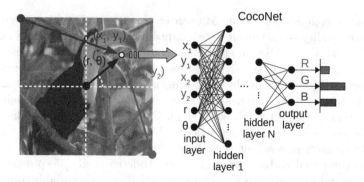

**Fig. 1.** The generic architecture of CocoNet. The input is given by the pixel coordinates represented in three different coordinate systems (two Cartesian systems and one polar system). The depth of the network can vary, depending on the application. Upon training, the output layer reproduces (with some degree of approximation) the RGB color values for the pixel coordinates provided as input. Best viewed in color. (Color figure online)

models [10,16,17] are designed to solve classification tasks, but, as the human brain, neural networks also have the capacity to memorize data. For classification problems, memorizing the training data (overfitting) is not desired, as it affects the generalization capacity of the neural model. On the other hand, memory is an important function of the brain and it is useful in solving various tasks, for example in remembering (mapping) important locations in the world, e.g. your home or the nearby grocery store. Although neural networks with memorization capacity have been designed before [9,32] to solve tasks such as question answering and language modeling, these models are based on an external memory tape. Since our main goal is to memorize the provided input data, we exploit the intrinsic memorization capacity of the neural networks, in a different manner than all previous works. As the human brain is able to remember various details corresponding to important locations in the world, we design a neural network that is able to remember the color values corresponding to pixel coordinates in an image.

## 3.2 CocoNet

We propose a deep neural network approach that learns a mapping function $f$ from the pixel coordinates in an image to the corresponding RGB color values. The input layer of our neural network is actually composed of 6 neurons which receive 6D pixel coordinate-related information in the form of a pair of polar coordinates $(r, \theta)$ with the origin in the center of the image and two pairs of Cartesian coordinates $(x_1, y_1)$ and $(x_2, y_2)$. One pair of Cartesian coordinates has the origin in a corner (bottom right) of the image that is diagonally opposite from the corner (top left) taken as origin for the other pair of Cartesian coordinates. It is clear that, from a single pair of coordinates, we can derive every other pair

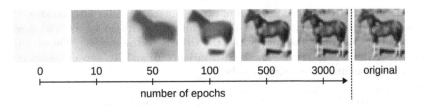

0        10        50        100        500        3000    |    original

number of epochs

**Fig. 2.** An image from the CIFAR-10 data set [15] reconstructed (memorized) by CocoNet, at various stages during the training process. After 3000 epochs, the memorized image is nearly identical to the original (input) image. Best viewed in color. (Color figure online)

of coordinates through simple transformations. Although the original input is just 2D, i.e. $(x_1, y_1)$, we observed that CocoNet learns a better mapping if we provide the pixel coordinates using multiple coordinate systems. For this reason, we choose to describe each pixel in three different coordinate systems, as shown in Fig. 1. The input and the output values (targets) are normalized to the $[0, 1]$ interval, for a more efficient optimization during training. The mapping function $f$ learned by our neural network is defined as follows:

$$f : \mathbb{R}^6 \longrightarrow [0, 1] \times [0, 1] \times [0, 1]. \tag{1}$$

The generic architecture of our deep neural network consists of fully-connected layers throughout the entire model. Depending on the application, the hidden layers can have a different architecture in terms of depth and width. For image compression, we need to use a thinner architecture with a reduced number of hidden layers. We can employ deeper and wider architectures for image denoising, resampling and completion. The output layer is identical in all applications, consisting of 3 neurons that output the RGB color values corresponding to the input pixel coordinates, as illustrated in Fig. 1. Hidden layers are based on the hyperbolic tangent (tanh) activation function, while the final layer is based on the sigmoid transfer function, since the output needs to be in the $[0, 1]$ interval.

Each training sample is formed of a 6D input feature vector (composed of pixel coordinates expressed in three different coordinate systems) and a 3D target vector (composed of RGB color values). In this formulation, our neural network can learn to map pixel coordinates to color values for a single image. Each pixel in the image can be taken as training sample, and, for most applications, we consider all pixels as training samples. However, for image completion, we intentionally leave out a region (patch) of the input image that needs to be later reconstructed by the network (at test time). We train the neural network using learning rates between $10^{-4}$ and $10^{-5}$, which provide a more stable convergence during training. We use the mean squared error (MSE) loss function and we optimize the parameters using the Adam stochastic gradient descent algorithm. During training, CocoNet inherently learns a continuous function $f$ from a discrete input image (represented through pixels). Since the learned mapping is continuous, CocoNet can naturally reproduce the image at any scale, at test

**Table 1.** The PSNR and SSIM metrics of various image denoising filters (mean, Gaussian, median, bilateral) versus a CocoNet architecture with 15 layers. The reported PSNR and SSIM values are averages on the CIFAR-10 test set [15]. The results are obtained for two Gaussian noise levels with variance 10 and 20, respectively. The best result for each metric is highlighted in bold.

| Method | Noise variance = 10 | | Noise variance = 20 | |
|---|---|---|---|---|
| | PSNR | SSIM | PSNR | SSIM |
| Noisy image | 28.26 | 0.8928 | 22.36 | 0.7437 |
| 3 × 3 mean filter | 25.47 | 0.8598 | 24.56 | 0.8216 |
| 5 × 5 mean filter | 21.80 | 0.6768 | 21.60 | 0.6629 |
| 3 × 3 Gaussian filter | 27.33 | 0.9057 | 25.83 | 0.8583 |
| 5 × 5 Gaussian filter | 24.96 | 0.8436 | 24.35 | 0.8160 |
| 3 × 3 median filter | 26.51 | 0.8732 | 24.72 | 0.8120 |
| 5 × 5 median filter | 22.97 | 0.7269 | 22.43 | 0.6943 |
| 3 × 3 bilateral filter | 31.24 | 0.9475 | 26.46 | 0.8629 |
| 5 × 5 bilateral filter | 30.60 | 0.9413 | **27.49** | 0.8902 |
| CocoNet (15 layers) | **31.25** | **0.9503** | 27.22 | **0.8925** |

time. Therefore, our neural network model can be used for image resampling, without requiring any additional developments. Another important aspect is that the neural network implicitly learns a mapping function $f$ that is smooth, hence it will automatically reduce noisy pixel values. Therefore, CocoNet can also be used for image denoising, without any additional changes.

For a better understanding of how CocoNet learns to memorize the input image within its own layers and weights, we provide a memorized image at various intermediate stages of the training process, in Fig. 2. After initializing the network with random weights (close to zero), we can observe that it produces an image that is almost entirely white. After 10 epochs, the network learns to approximate the average color of the input image. After 100 epochs, we can already distinguish the main object in the image (a horse). After completing the training process, the image reproduced by the network is very close to the original image. There are very small differences that are indistinguishable to the naked eye. We can observe that during the training process, CocoNet starts by memorizing a smooth version of the input image, which is gradually refined with increasingly finer details. This comes to support our statement that the neural network tends to learn a continuous and smooth mapping function $f$.

# 4   Experiments

## 4.1   Data Sets

We use the CIFAR-10 data set [15] in the image denoising experiments. The CIFAR-10 data set consists of 60 thousand $32 \times 32$ color images which are divided into 10 mutually exclusive classes, with 6000 images per class. There are 50 thousand training images and 10 thousand test images.

We perform image resampling and completion experiments on Set5 [3], a commonly used data set for image super-resolution [14,18,40]. The Set5 data set consists of 5 images with various sizes between 228 and 512 pixels.

## 4.2 Evaluation

The image denoising and resampling results are evaluated against the original images, in terms of the peak signal-to-noise ratio (PSNR) and the structural similarity index (SSIM). The PSNR is the ratio between the maximum possible power of a signal and the power of corrupting noise that affects the fidelity of its representation. Many researchers [37,38] argue that PSNR is not highly indicative of the perceived similarity. The SSIM metric [38] aims to address this shortcoming by taking texture into account.

## 4.3 Image Denoising Results

For image denoising, we consider several filters (mean filter, Gaussian, median and bilateral) as baseline denoising methods. For each filter, we consider filter sizes of $3 \times 3$ and $5 \times 5$. The baselines methods are compared with a CocoNet architecture with 15 layers, each of 200 neurons. To generate input images, we apply Gaussian noise over the CIFAR-10 test images, using two noise levels with variance 10 and 20, respectively. Table 1 presents the results of the considered baseline filters versus CocoNet. The $3 \times 3$ filters perform better for images with noise variance 10, while the $5 \times 5$ filters perform better for images with noise variance 20. In terms of the SSIM metric, CocoNet surpasses all filters for both noise variances. Although the bilateral filter obtains a slightly better PSNR value for noisy images with variance 20, it seems that CocoNet is better correlated with the similarity perceived by humans, according to the SSIM metric. Figure 3 shows that the output images of CocoNet are indeed more similar to the original CIFAR-10 images than all filters.

## 4.4 Image Resampling Results

For image resampling, we consider a scale factor of $4\times$, a standard choice for image super-resolution approaches [3,14,18,34,40]. Although we use the same scale factor, it is not fair to compare our approach with image super-resolution approaches [14,18,28,34,40] that are trained on both low-resolution (downsampled) and high-resolution (upsampled) patches sampled from the input image or from additional training images, since we train our method solely on the downsampled image. Nevertheless, we compare our CocoNet architecture, that consists of 10 layers each of 200 neurons, with a single-image super-resolution approach [34] and bicubic interpolation. The corresponding results are presented in Table 2. In terms of PSNR, we obtain better results for two images (*butterfly*

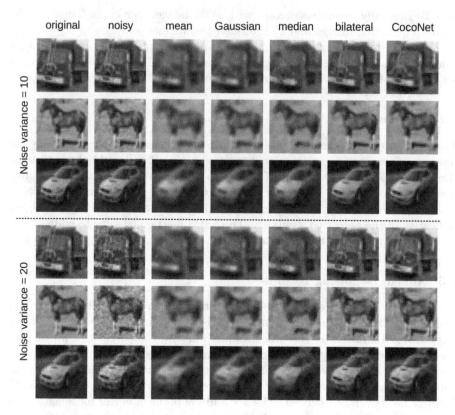

**Fig. 3.** Three original, noisy and denoised images selected from the CIFAR-10 data set [15]. The output images resulted after applying various filters (mean, Gaussian, median, bilateral) are compared with the output images produced by a CocoNet architecture with 15 layers. Best viewed in color. (Color figure online)

**Table 2.** The PSNR and SSIM metrics for image upsampling on Set5 [3] using two baseline methods (bicubic interpolation, Tang et al. [34]) versus a CocoNet architecture with 10 layers. The upsampling scale factor is 4×. The best result for each image in the data set and each metric is highlighted in bold.

| Image | Bicubic interpolation | | Tang et al. [34] | | CocoNet (10 layers) | |
|---|---|---|---|---|---|---|
| | PSNR | SSIM | PSNR | SSIM | PSNR | SSIM |
| Baby | 29.39 | **0.8162** | **29.70** | - | 26.88 | 0.7146 |
| Bird | 26.84 | **0.8265** | **27.84** | - | 26.71 | 0.7816 |
| Butterfly | 19.93 | 0.6779 | 20.61 | - | **20.92** | **0.7229** |
| Head | 28.48 | 0.6786 | **29.83** | - | 29.26 | **0.7016** |
| Woman | 24.17 | 0.7979 | 24.46 | - | **25.33** | **0.8150** |

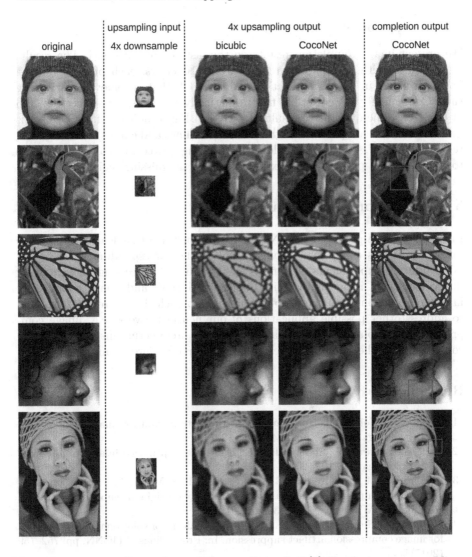

**Fig. 4.** Image upsampling and completion results on Set5 [3]. For the image upsampling experiments, the original images are first downsampled by a scale factor of $4\times$. For the image completion experiments, a random patch (surrounded by a red bounding box) from the original image is removed during training. For both applications, the CocoNet architecture has 15 layers. Best viewed in color. (Color figure online)

and *woman*). In terms of SSIM, we obtain better results than bicubic interpolation for three images. The upsampling results of bicubic interpolation and CocoNet are depicted in Fig. 4. In the *bird*, the *butterfly* and the *woman* images, we notice that the lines produced by CocoNet are smoother than those resulted after bicubic interpolation.

## 4.5   Image Completion Results

We present image completion results on the last column of Fig. 4. The results are obtained with a CocoNet architecture of 10 layers, each of 200 neurons. For each image in Set5, we select a random patch that is removed at training time. Consequently, CocoNet has to automatically complete the respective patch at test time. The qualitative results depicted in Fig. 4 indicate that CocoNet is generally able to complete the high-level shapes (lines and curves) in the missing regions. However, without any additional training sources, CocoNet is not able to generate (or reproduce) the texture details of the missing regions.

# 5   Conclusion and Further Work

In this paper, we proposed CocoNet, a neural approach that learns to map the pixel coordinates (in a single input image) to the corresponding RGB values. We conducted experiments on various low-level image processing tasks, namely image denoising, image upsampling (super-resolution) and image completion, in order to demonstrate the usability of our approach. In future work, we aim to address other possible applications of our neural network, e.g. image compression. We also aim to try out a convolutional architecture instead of a fully-connected one (the preliminary results in this direction are promising).

# References

1. Baig, M.H., Koltun, V., Torresani, L.: Learning to inpaint for image compression. In: Proceedings of NIPS, pp. 1246–1255 (2017)
2. Ballé, J., Laparra, V., Simoncelli, E.P.: End-to-end optimized image compression. In: Proceedings of ICLR (2017)
3. Bevilacqua, M., Roumy, A., Guillemot, C., Alberi-Morel, M.: Low-complexity single-image super-resolution based on nonnegative neighbor embedding. In: Proceedings of BMVC, pp. 135.1–135.10 (2012)
4. Cavigelli, L., Hager, P., Benini, L.: CAS-CNN: A deep convolutional neural network for image compression artifact suppression. In: Proceedings of IJCNN, pp. 752–759 (2017)
5. Dong, C., Loy, C.C., He, K., Tang, X.: Learning a deep convolutional network for image super-resolution. In: Fleet, D., Pajdla, T., Schiele, B., Tuytelaars, T. (eds.) ECCV 2014. LNCS, vol. 8692, pp. 184–199. Springer, Cham (2014). https://doi.org/10.1007/978-3-319-10593-2_13
6. Dumas, T., Roumy, A., Guillemot, C.: Image compression with stochastic winner-take-all auto-encoder. In: Proceedings of ICASSP, pp. 1512–1516 (2017)
7. Girshick, R.: Fast R-CNN. In: Proceedings of ICCV, pp. 1440–1448 (2015)
8. Goodfellow, I., Courville, A., Bengio, Y.: Deep Learning. MIT Press (2016). http://www.deeplearningbook.org
9. Graves, A., Wayne, G., Danihelka, I.: Neural Turing Machines. arXiv preprint arXiv:1410.5401 (2014)
10. He, K., Zhang, X., Ren, S., Sun, J.: Deep residual learning for image recognition. In: Proceedings of CVPR, pp. 770–778, June 2016

11. Huang, X., Li, Y., Poursaeed, O., Hopcroft, J., Belongie, S.: Stacked generative adversarial networks. In: Proceedings of CVPR, pp. 5077–5086 (2017)
12. Iizuka, S., Simo-Serra, E., Ishikawa, H.: Globally and locally consistent image completion. ACM Trans. Graph. **36**(4), 107 (2017)
13. Ionescu, R.T., Alexe, B., Leordeanu, M., Popescu, M., Papadopoulos, D., Ferrari, V.: How hard can it be? Estimating the difficulty of visual search in an image. In: Proceedings of CVPR, pp. 2157–2166, June 2016
14. Kim, J., Kwon Lee, J., Mu Lee, K.: Deeply-recursive convolutional network for image super-resolution. In: Proceedings of CVPR, pp. 1637–1645 (2016)
15. Krizhevsky, A.: Learning Multiple Layers of Features from Tiny Images. Technical report, University of Toronto (2009)
16. Krizhevsky, A., Sutskever, I., Hinton, G.E.: ImageNet classification with deep convolutional neural networks. In: Bartlett, P., Pereira, F., Burges, C., Bottou, L., Weinberger, K. (eds.) Proceedings of NIPS, pp. 1106–1114 (2012)
17. LeCun, Y., Bottou, L., Bengio, Y., Haffner, P.: Gradient-based learning applied to document recognition. Proc. IEEE **86**(11), 2278–2324 (1998)
18. Ledig, C., et al.: Photo-realistic single image super-resolution using a generative adversarial network. In: Proceedings of CVPR, pp. 4681–4690 (2016)
19. Liu, W., et al.: SSD: single shot MultiBox detector. In: Proceedings of ECCV, pp. 21–37 (2016)
20. Mao, X., Shen, C., Yang, Y.B.: Image restoration using very deep convolutional encoder-decoder networks with symmetric skip connections. In: Proceedings of NIPS, pp. 2802–2810 (2016)
21. Minnen, D., et al.: Spatially adaptive image compression using a tiled deep network. In: Proceedings of ICIP, pp. 2796–2800 (2017)
22. Montavon, G., Orr, G.B., Müller, K.-R. (eds.): Neural Networks: Tricks of the Trade. LNCS, vol. 7700, 2nd edn. Springer, Heidelberg (2012). https://doi.org/10.1007/978-3-642-35289-8
23. Parkhi, O.M., Vedaldi, A., Zisserman, A., et al.: Deep face recognition. In: Proceedings of BMVC, pp. 6–17 (2015)
24. Prakash, A., Moran, N., Garber, S., DiLillo, A., Storer, J.: Semantic perceptual image compression using deep convolution networks. In: Proceedings of DCC, pp. 250–259 (2017)
25. Redmon, J., Divvala, S., Girshick, R., Farhadi, A.: You only look once: unified, real-time object detection. In: Proceedings of CVPR, pp. 779–788 (2016)
26. Ren, S., He, K., Girshick, R., Sun, J.: Faster R-CNN: towards real-time object detection with region proposal networks. In: Proceedings of NIPS, pp. 91–99 (2015)
27. Russakovsky, O., et al.: ImageNet large scale visual recognition challenge. Int. J. Comput. Vis. **115**(3), 211–252 (2015)
28. Sajjadi, M.S., Schölkopf, B., Hirsch, M.: Enhancenet: single image super-resolution through automated texture synthesis. In: Proceedings of ICCV, pp. 4501–4510 (2017)
29. Shi, W., et al.: Real-time single image and video super-resolution using an efficient sub-pixel convolutional neural network. In: Proceedings of CVPR, pp. 1874–1883 (2016)
30. Simonyan, K., Zisserman, A.: Very deep convolutional networks for large-scale image recognition. In: Proceedings of ICLR (2014)
31. Soviany, P., Ionescu, R.T.: Optimizing the trade-off between single-stage and two-stage deep object detectors using image difficulty prediction. In: Proceedings of SYNASC, pp. 1–6 (2018)

32. Sukhbaatar, S., Weston, J., Fergus, R., et al.: End-to-end memory networks. In: Proceedings of CVPR, pp. 2440–2448 (2015)
33. Szegedy, C., et al.: Going deeper with convolutions. In: Proceedings of CVPR, pp. 1–9, June 2015
34. Tang, Y., Yan, P., Yuan, Y., Li, X.: Single-image super-resolution via local learning. Int. J. Mach. Learn. Cybern. **2**(1), 15–23 (2011)
35. Toderici, G., et al.: Full resolution image compression with recurrent neural networks. In: Proceedings of CVPR, pp. 5306–5314 (2017)
36. Van Oord, A., Kalchbrenner, N., Kavukcuoglu, K.: Pixel recurrent neural networks. In: Proceedings of ICML, pp. 1747–1756 (2016)
37. Wang, Z., Bovik, A.C.: Mean squared error: Love it or leave it? A new look at signal fidelity measures. Signal Process. Mag. **26**(1), 98–117 (2009)
38. Wang, Z., Bovik, A.C., Sheikh, H.R., Simoncelli, E.P.: Image quality assessment: from error visibility to structural similarity. IEEE Trans. Image Process. **13**(4), 600–612 (2004)
39. Wu, J., Timofte, R., Van Gool, L.: Generic 3D convolutional fusion for image restoration. In: Chen, C.-S., Lu, J., Ma, K.-K. (eds.) ACCV 2016. LNCS, vol. 10116, pp. 159–176. Springer, Cham (2017). https://doi.org/10.1007/978-3-319-54407-6_11
40. Yamanaka, J., Kuwashima, S., Kurita, T.: Fast and accurate image super resolution by deep CNN with skip connection and network in network. In: Proceedings of ICONIP, pp. 217–225 (2017)
41. Yang, C., Lu, X., Lin, Z., Shechtman, E., Wang, O., Li, H.: High-resolution image inpainting using multi-scale neural patch synthesis. In: Proceedings of CVPR, pp. 6721–6729 (2017)
42. Zhao, G., Liu, J., Jiang, J., Wang, W.: A deep cascade of neural networks for image inpainting, deblurring and denoising. Multimed. Tools Appl. **77**(22), 29589–29604 (2018)

# BCMLP: Binary-Connected Multilayer Perceptrons

Ningqi Luo[✉], Binheng Song, Yinxu Pan, and Bin Shen

Graduate School at Shenzhen, Tsinghua University, Beijing, China
lnq16@mails.tsinghua.edu.cn

**Abstract.** Sparse connection has been used both to reduce network complexity and sensitivity with input perturbations in multilayer perceptrons as well as artificial neural networks. We propose a novel binary-connected multilayer perceptrons where arbitrary node is connected with the only two nodes of previous layer. The sensitivity of this model is discussed both in theoretical methods and simulation experiments. Comparisons with related works show that our scheme achieves the least amount of parameters, the lowest deviation to input perturbations, and the highest accuracy in the noisy classification task.

**Keywords:** Sparse connection · Multilayer perceptrons
Sensitivity analysis

## 1 Introduction

Multilayer perceptrons (MLPs) and artificial neural networks (ANNs) have been successfully applied across an extraordinary range of problem domains. However, most of these models are fully connected which consume considerable storage, memory bandwidth, and computational resources. Many new network structures and learning methods are proposed to tackle these problems.

Sparse connection [1–4] is one of the schemes. It could be used to reduce hardware requirements as well as improve generalization capabilities. The often cited empirical requirement [5] is that the minimum number of training patterns $N_{min}$ required for good generalization performance as $N_{min} = U/\varepsilon$ where $U$ is the total number of independent network weights and biases, and $\varepsilon$ is the fraction of classification errors permitted on the test data. Sparse connection reduces the amount of parameters $U$ as well as promotes model's generalization capabilities when $N_{min}$ is limited. Thus we present a novel sparse connected MLPs called binary-connected multilayer perceptrons (BCMLP) where every node in each layer is only connected with two nodes from the previous layer (Sect. 3). Meanwhile BCMLP has the same property with full-connected MLPs that every node in the output layer is connected with all input nodes (Proposition 2).

Sensitivity evaluation [6–8,10] reflects the degree of its output deviation caused by its input perturbations. And it is an important measure for evaluating

© Springer Nature Switzerland AG 2018
L. Cheng et al. (Eds.): ICONIP 2018, LNCS 11302, pp. 77–88, 2018.
https://doi.org/10.1007/978-3-030-04179-3_7

the MLPs' performance, such as error-tolerance and generalization ability etc. Sensitivity evaluation of MLPs [9,10] is under the assumption that all the inputs and weights are independently and identically distributed (i.i.d.) with uniform distribution which weakens its applications to real problems. This assumption of input could be verified in BCMLP (Proposition 1). And we evaluate the sensitivity of BCMLP with mathematical methods (Sect. 4) as well as simulation experiments (Sect. 5). The agreement between the theoretical results and experimental results is quite good.

Our contributions are: (1) We proposed a novel binary-connected multilayer perceptrons in Sect. 3. (2) We proposed an algorithm for evaluating BCMLP's sensitivity in Sect. 4. (3) Comparisons in Sect. 5 show that BCMLP is more robust to input perturbations than related works.

## 2    Related Work

### 2.1    Sparse Connection

LocallyConnected1D and 2D layers [11] take $k$ nearest neighborhoods in the previous layer as input. Convolutional neural networks (CNNs) [12] is locally-connected with sharing weights. Tield CNNs [13] without sharing weights is able to learn complex invariances (such as scale and rotational invariance) beyond translational invariance. Gregor [14] learns sparse feature in a locally-connected network. Bruna et al. [1] proposed deep locally connected networks which restrict attention to sparse "filters" with receptive fields given by $k$ neighborhoods to get locally connected networks, thus reducing the number of parameters in a filter layer.

Network pruning has been used to reduce network complexity and overfitting. An early approach to pruning was biased weight decay [2]. Optimal Brain Damage [15] and Optimal Brain Surgeon [16] prune networks to reduce the number of connections based on the Hessian of the loss function and suggest that such pruning is more accurate than magnitude-based pruning such as weight decay. Han et al. [3] learns connections for efficient neural networks by pruning redundant connections and retraining the network to fine tune the weights of the remaining connections. Dropout [4] and DropConnect [17] randomly drop units (along with their connections) from the neural network during training to prevent units from co-adapting too much. During training, dropout samples are from an exponential number of different sparse connected networks.

### 2.2    Sensitivity Evaluation

The sensitivity evaluation [6–8,10] reflects the degree of its output deviation caused by its input perturbations. For the arbitrary $i^{th}$ neuron in the $l^{th}$ layer, with given weights $W_i^l$, bias $b_i^l$, input deviation $\Delta X^{l-1}$ and activation function $f(\cdot)$, its sensitivity is defined as the mathematical expection of output deviation with respect to all possible $X^{l-1}$. The sensitivity $s_i^l$ is expressed as

$s_i^l = E_{X^{l-1}}(|f((X^{l-1} + \Delta X^{l-1})^T W_i^l + b_i^l) - f((X^{l-1})^T W_i^l + b_i^l)|)$. $E_{X^{l-1}}(\cdot)$ is the mathematical expectation by taking $X^{l-1}$ as statistical variable. Obviously, $s_i^l$ can be regarded as a function of $\Delta X^{l-1}$, independent of specific $X^{l-1}$. The sensitivity of entire MLPs is the sensitivity of its output layer $\mathcal{S} = (s_1^L, s_2^L, ..., s_N^L)^T$ where $L$ is amount of layers and $N$ is amount of neurons per layer.

The sensitivity of MLPs' output to its inputs' perturbations is an important measure for evaluating the MLPs' performance, such as error-tolerance and generalization ability [18–21] etc. It could be found that a variety of studies on the computation of sensitivity have been carried out. Methods [22–25] analyzed the sensitivity of Madalines to weight perturbation. In the model [26, 27], the sensitivity was defined as the variance of the output deviation caused by input perturbation with antisymmetric squashing activation functions. Zeng et al. [6, 28] established a hypercube model instead of the hypersphere model with a heuristic algorithm to accurately compute the sensitivity of Adalines and Madaline. Zeng et al. [7] investigated the sensitivity behavior of an ensemble of MLPs, and then proposed an approach to quantify the sensitivity of an individual MLPs by establishing a hyperrectangle model [8]. Later on, Yang [9] and Yang [10] analyzed the sensitivity of MLPs under the assumption that all the inputs and weights are independently and identically distributed (i.i.d.) with uniform distribution which weakens its applications to real problems. To tackle this problem, the assumption of inputs in our model could be verified in proof 1.

## 3  BCMLP

This section introduces the definition and properties of our binary-connected multilayer perceptrons (BCMLP). Table 1 illustrates symbols used in our model. As shown in Fig. 1, BCMLP is $L$ layers perceptrons with $N$ nodes per layer which is sparse connected. And every node in the $l^{th}$ layer is connected with two nodes from the $(l-1)^{th}$ layer where $1 \leq l \leq L$.

**Table 1.** Symbols and their meanings.

| Symbols | Meanings |
|---------|----------|
| $N$ | Number of neurons in each layer |
| $L$ | Number of layers in BCMLP |
| $(i, l)$ | The $i^{th}$ neuron in the $l^{th}$ layer |
| $x_i^l$ | The output of node $(i, l)$. And $x_i^0$ is the input of BCMLP |
| $w_{ij}^l$ | Weight of linkage from node $(j, l-1)$ to $(i, l)$ |
| $b_i^l$ | Bias of node $(i, l)$ |
| $Z_i^l$ | Collection of all nodes which are connected with $(i, l)$ |

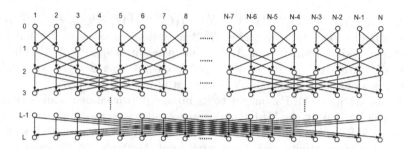

**Fig. 1.** The architecture of binary-connected MLPs.

**Definition 1.** *The linkage rule of BCMLP's structure. There are only two nodes in the $(l-1)^{th}$ layer which are connected with $(i, l)$ where $1 \leq i \leq N$, $1 \leq l \leq L$ and $L = log_2N$. One is $(i, l-1)$ and the other is $(i + (-1)^{\lfloor \frac{i-1}{2^{l-1}} \rfloor} \cdot 2^{l-1}, l-1)$*

**Proposition 1.** *If nodes B $(b, l-1)$ and C $(c, l-1)$ are both connected with node A $(a, l)$, $\{(x, y)|(x, y) \in Z_b^{l-1}, (x, y) \in Z_c^{l-1}, y \leq l-1\} = \emptyset$. Which means the collection of nodes connected with A before the $l^{th}$ layer is divided into two parts by B and C.*

*Proof.* We assume that nodes in the $l^{th}$ layer are divided into $2^{L-l}$ parts, and A $(a, l)$ satisfies the rule $P \cdot 2^l < a \leq (P+1) \cdot 2^l$, $0 \leq P < 2^{L-l}$ and $P \in N$. A is connected with B $(b, l-1)$ and C $(c, l-1)$ in the $(l-1)^{th}$ layer. Might as well assume $b < c$. And known that $P \cdot 2^l < b < c \leq (P+1) \cdot 2^l$. According to definition 1, $|b - c| = |i + (-1)^{\lfloor \frac{i-1}{2^{l-1}} \rfloor} \cdot 2^{l-1} - i| = 2^{l-1}$, so we can infer that $2P \cdot 2^{l-1} < b \leq (2P+1) \cdot 2^{l-1} < c \leq (P+1) \cdot 2^l$. The indexes of any nodes in the previous layers of B is limited in $(2P \cdot 2^{l-1}, (2P+1) \cdot 2^{l-1}]$, and the range of C is $((2P+1) \cdot 2^{l-1}, (P+1) \cdot 2^l]$. Which means any $b^* \in \{x|(x, y) \in Z_b^{l-1}, y \leq l-1\}$ and any $c^* \in \{x|(x, y) \in Z_c^{l-1}, y \leq l-1\}$ are satisfy $2P \cdot 2^{l-1} < b^* \leq (2P+1) \cdot 2^{l-1} < c^* \leq (P+1) \cdot 2^l$. So the Proposition 1 is true.

**Proposition 2.** *Every node in the output is connected with all input nodes. Which means $|\{x|(x, 0) \in Z_i^L\}| = N$ where $1 \leq i \leq N$ and $L = log_2N$. Each input node $(i, 0)$ affects all output nodes $(i, L)$.*

*Proof.* We first define a counting function $S(i, l)$ where $1 \leq i \leq N$ and $1 \leq l \leq L$. It returns the amount of nodes in the input layer which have a linkage with node $(i, l)$, where $S(i, l) = |\{(x, y)|(x, y) \in Z_i^l, y = 0\}|$. And $S(i, 1) = 2$ according to the Definition 1. According to Proposition 1, if node B $(b, l-1)$ and C $(c, l-1)$ are both connected with node A $(a, l)$, any previous nodes of B and any previous nodes of C are independent. So $S(a, l) = S(b, l-1) + S(c, l-1)$. B and C also satisfy the constraint rule in Proposition 1, so the $S(i, l)$ could be solved by the recurrence formula and $S(a, l) = 2^{l-1} + 2^{l-1} = 2^l$. Therefore $S(a, L) = 2^L = N$. There are $N$ distinct nodes in the input layer. We can infer that output node $(a, L)$ connects with all input nodes. All nodes of output layer satisfy the same

**Algorithm 1. BCMLP Process**

1: **function** FORWARD COMPUTATION
2:     $x^0 \leftarrow input$
3:     $w, b \leftarrow initialization$
4:     **for** $l = 1 \rightarrow L$ **do**
5:         **for** $i = 1 \rightarrow N$ **do**
6:             $j = i + (-1)^{\lfloor \frac{i-1}{2^{l-1}} \rfloor} \cdot 2^{l-1}$
7:             $x_i^l \leftarrow active(w_{ii}^l \cdot x_i^{l-1} + w_{ij}^l \cdot x_j^{l-1} + b_i^l \cdot 1)$
8:         **end for**
9:     **end for**
10:     $output \leftarrow x^L$
11:     **return** $output$
12: **end function**
13:
14: **function** BACKWARD PROPAGATION
15:     $\eta \leftarrow initialization$
16:     $Loss = sum[loss(i, L)] \; for \; i = 1 \rightarrow N$
17:     **for** $l = 1 \rightarrow L$ **do**
18:         **for** $i = 1 \rightarrow N$ **do**
19:             $p_i^l \leftarrow w_{ii}^l \cdot x_i^{l-1} + w_{ij}^l \cdot x_j^{l-1} + b_i^l \cdot 1$
20:             $\nabla p_i^l \leftarrow \frac{\partial Loss}{\partial x_i^l} \cdot \frac{\partial x_i^l}{\partial p_i^l}$
21:             **for** $param$ in $[w_{ij}^l, w_{ii}^l, b_i^l]$ **do**
22:                 $\nabla param \leftarrow \nabla p_i^l \cdot \frac{\partial p_i^l}{\partial param}$
23:                 $param^* \leftarrow param - \eta \cdot \nabla param$
24:             **end for**
25:         **end for**
26:     **end for**
27: **end function**

constraint rule in Proposition 1. In conclusion, all nodes $(i, L)$ of output connect with all input nodes. The Proposition 2 is true.

According to the Proposition 2, BCMLP has the same property with full-connected MLPs that every node in the ouput layer is connected with all input nodes. We could set the forward computation and backward propagation of BCMLP as Algorithm 1.

## 4   Sensitivity Analysis

In this section, we evaluate the sensitivity of BCMLP to input perturbations. BCMLP has $L$ layers and there are $N$ nodes in each layer where $L \geq 1$ and $N = 2^L$. For an arbitrary node $(i, l)$ in BCMLP where $1 \leq i \leq N$ and $1 \leq l \leq L$, its input vector $X_{ij}^{l-1}$ is $(x_i^{l-1}, x_j^{l-1})^T$ and weight vector $W_{ij}^l$ is $(w_{ij}^l, w_{ij}^l)^T$. Known that nodes $x_i^{l-1}$ and $x_j^{j-1}$ are independent according to Proposition 1 which means $\{(k, 0) | (k, 0) \in Z_i^{l-1} \cap Z_j^{l-1}\} = \emptyset$. So we could infer that the mathematical distribution of $x_i^{l-1}$ and $x_j^{l-1}$ are uncorrelated. We use sigmoid function $f(x) = 1/(1 + e^{-x})$ as the activation function of BCMLP. For the case of a perceptron, the measurement regards its bias as zero and weights as constant variable to reduce loss of generality. We define $g(\cdot)$ to calculate the output $x_i^l$ from input $(x_i^{l-1}, x_j^{l-1})^T$ as expressed in Eq. 1

$$x_i^l = g(x_i^{l-1}, x_j^{l-1}) = f(x_i^{l-1} \cdot w_{ii}^l + x_j^{l-1} \cdot w_{ij}^l). \tag{1}$$

For $X_{ij}^{l-1} = (x_i^{l-1}, x_j^{l-1})^T$ and their expected perturbation $\overline{\Delta X_{ij}^{l-1}}$ where $(\overline{\Delta x_i^{l-1}}, \overline{\Delta x_j^{l-1}})^T = (E(\Delta x_i^{l-1}), E(\Delta x_j^{l-1}))^T$. The corresponding deviation $\Delta x_i^l$ in $(i, j)$ is expressed as

$$\Delta x_i^l = g(x_i^{l-1} + \overline{\Delta x_i^{l-1}}, x_j^{l-1} + \overline{\Delta x_j^{l-1}}) - g(x_i^{l-1}, x_j^{l-1}) \tag{2}$$

Taylor series expansion for $g(x_i^{l-1} + \overline{\Delta x_i^{l-1}}, x_j^{l-1} + \overline{\Delta x_j^{l-1}})$ is

$$\sum_{k=0}^n \frac{1}{k!} (\overline{\Delta x_i^{l-1}} \frac{\partial}{\partial x_i} + \overline{\Delta x_j^{l-1}} \frac{\partial}{\partial x_j})^k g(x_i^{l-1}, x_j^{l-1}) + R_n \tag{3}$$

where $R_n = \frac{1}{(n+1)!} (\overline{\Delta x_i^{l-1}} \frac{\partial}{\partial x_i} + \overline{\Delta x_j^{l-1}} \frac{\partial}{\partial x_j})^{n+1} g(X_{ij}^{l-1} + \theta \overline{\Delta X_{ij}^{l-1}})$ and $0 < \theta < 1$. In the Eq. 3, the sub items $(\overline{\Delta x_i^{l-1}} \frac{\partial}{\partial x_i} + \overline{\Delta x_j^{l-1}} \frac{\partial}{\partial x_j})^k g(x_i^{l-1}, x_j^{l-1})$ from $k = 0$ to $n$ is equal to

$$\sum_{r=0}^k C_k^r \overline{\Delta x_i^{l-1}}^r \overline{\Delta x_j^{l-1}}^{k-r} \frac{\partial^k}{\partial x_i^r \partial x_J^{k-r}} g(x_i^{l-1}, x_j^{l-1}) \tag{4}$$

where

$$\frac{\partial^k}{\partial x_i^r \partial x_J^{k-r}} g(x_i^{l-1}, x_j^{l-1}) = w_{ii}^l{}^r w_{ij}^l{}^{k-r} f^k(x_i^{l-1} \cdot w_{ii}^l + x_j^{l-1} \cdot w_{ij}^l) \tag{5}$$

Here we ignore the Lagrange remainder $R_n$. And when $k = 0$, the sub item is $g(x_i^{l-1}, x_j^{l-1})$. So Eq. 2 could be transformed as

$$\Delta x_i^l = \sum_{k=1}^n \sum_{r=0}^k \frac{1}{k!} C_k^r w_{ii}^l{}^r w_{ij}^l{}^{k-r} \overline{\Delta x_i^{l-1}}^r \overline{\Delta x_j^{l-1}}^{k-r} f^k(x_i^{l-1} \cdot w_{ii}^l + x_j^{l-1} \cdot w_{ij}^l) \tag{6}$$

To evaluate expected deviation to input perturbations of an arbitrary node, it may not be enough to calculate at some specific value of $x_i^l$. We treat $x_i^l$ as a statistical variable ranging over all patterns. We give the interval of $x_i^l \in [a_i^l, b_i^l]$ and $w_i^l \in R$. Because of $0 \le f(x) \le 1$, the output interval of all layers is included in $[0, 1]$, and input layer can be normalized in $[0, 1]$ by the transformation: $(x_i^0 - a)/(b - a)$. Thus, we give the normalization: for $0 \le l \le L$ and $1 \le i \le N$, $X_{ij}^l = (x_i^l, x_j^l)^T \in [a_i^l, b_i^l]^2 \subseteq [0, 1]^2$. Now Eq. 6 can be expressed as

$$\int_{a_i^{l-1}}^{b_i^{l-1}} \int_{a_j^{l-1}}^{b_j^{l-1}} \varphi(X_{ij}^{l-1}) \sum_{k=1}^{n} \sum_{r=0}^{k} \frac{1}{k!} C_k^r w_{ii}^l{}^r w_{ij}^l{}^{k-r} \overline{\Delta x_i^{l-1}}^r \overline{\Delta x_j^{l-1}}^{k-r} f^k(X_{ij}^{l-1}{}^T W_{ij}^l) dX_{ij}^{l-1} \quad (7)$$

Where $\varphi(\cdot)$ is the density function. The mathematical distribution of $x_i^{l-1}$ and $x_j^{l-1}$ are uncorrelated according to Proposition 1. Under the assumption that $\varphi(x_i^l) = 1$, an approximate solution is given

$$\int_0^1 \int_0^1 \sum_{k=1}^{n} \sum_{r=0}^{k} \frac{1}{k!} C_k^r w_{ii}^l{}^r w_{ij}^l{}^{k-r} \overline{\Delta x_i^{l-1}}^r \overline{\Delta x_j^{l-1}}^{k-r} f^k(X_{ij}^{l-1}{}^T W_{ij}^l) dX_{ij}^{l-1} \quad (8)$$

The Eq. 8 is transformed as

$$\Delta x_i^l = \sum_{k=1}^{n} \sum_{r=0}^{k} \frac{1}{k!} C_k^r w_{ii}^l{}^r w_{ij}^l{}^{k-r} \overline{\Delta x_i^{l-1}}^r \overline{\Delta x_j^{l-1}}^{k-r} F^{k-2}(w_{ii}^l, w_{ij}^l) \quad (9)$$

$$F^{k-2}(w_{ii}^l, w_{ij}^l) = f^{k-2}(w_{ii}^l + w_{ij}^l) - f^{k-2}(w_{ii}^l) - f^{k-2}(w_{ij}^l) + f^{k-2}(0) \quad (10)$$

It is easy to calculate $f^{k-2}(x)$ functions with $f'(x) = f(x)[1 - f(x)]$. Especially $f^{-1}(x) = ln(1 + e^x)$ when $k = 1$, and $f(x) = 1/(1 + e^{-x})$ when $k = 2$.

**Definition 2.** *The sensitivity of single neuron* $(i, L)$ *to input perturbations. For input* $X_{ij}^{L-1} = (x_i^{L-1}, x_j^{L-1})^T$, *and their expected perturbations* $\overline{\Delta X_{ij}^{L-1}} = (\overline{\Delta x_i^{L-1}}, \overline{\Delta x_j^{L-1}})^T$. *The sensitivity of* $(i, L)$ *in the last layer is defined as the expectation of* $|\Delta x_i^L|$, *which is expressed as*

$$s_i = E(|\Delta x_i^L|) = E(|g(X_{ij}^{L-1} + \overline{\Delta X_{ij}^{L-1}}) - g(X_{ij}^{L-1})|) \quad (11)$$

$$\int_{a_i^{L-1}}^{b_i^{L-1}} \int_{a_j^{L-1}}^{b_j^{L-1}} \varphi(X_{ij}^{L-1})| \sum_{k=1}^{n} \sum_{r=0}^{k} \frac{1}{k!} C_k^r w_{ii}^L{}^r w_{ij}^L{}^{k-r} \overline{\Delta x_i^{L-1}}^r \overline{\Delta x_j^{L-1}}^{k-r} f^k(X_{ij}^{L-1}{}^T W_{ij}^L)| dX_{ij}^{L-1} \quad (12)$$

**Definition 3.** *BCMLP's sensitivity to input perturbations. With input* $X^0 = (x_1^0, x_2^0, ..., x_N^0)^T$, *and their expected perturbations* $\overline{\Delta X^0} = (\overline{\Delta x_1^0}, \overline{\Delta x_2^0}, ..., \overline{\Delta x_N^0})^T$. *The sensitivity of the entire BCMLP is defined as* $S = (s_1, s_2, ..., s_N)^T$. *And* $S$ *could be regarded as a function (in Algorithm 2) of* $\overline{\Delta X^0}$ *and* $W$, *independent of specific* $X^0$.

$\overline{\Delta X^0}$ is the expected input perturbations of the entire BCMLP. Firstly, compute $\Delta x_i^l$ by Eq. 7 from the $1^{st}$ layer to the $(L-1)^{th}$ layer. In the last layer, $s_i$ is calculated by calling formulation 12 and finally return $S$ as the sensitivity of the entire BCMLP. The simulation experiment of sensitivity in Sect. 5.2 shows that the agreement between the experimental results and theoretical results is quite good.

---

**Algorithm 2.** BCMLP Sensitivity

---

1: **function** CALCULATE SENSITIVITY($\overline{\Delta X^0}$,$W$)
2:     **for** $l = 1 \rightarrow L - 1$ **do**
3:         **for** $i = 1 \rightarrow N$ **do**
4:             $\overline{\Delta x_i^l} \leftarrow$ equation 7
5:         **end for**
6:     **end for**
7:     **for** $i = 1 \rightarrow N$ **do**
8:         $s_i \leftarrow$ equation 12
9:     **end for**
10:     **return** $S = (s_1, s_2, ..., s_N)^T$
11: **end function**

---

## 5    Experiments

In this section, we compare our scheme with related works in (1) amount of parameters, (2) sensitivity to input perturbations, and (3) noisy logic gates task. Our scheme is implemented in TensorFlow framework [29] and more experiment details is available at Github[1].

### 5.1    Amount of Parameters

While these large ANNs and MLPs are very powerful, their size consumes considerable storage, memory bandwidth, and computational resources. And full-connected layers occupy a large portion of the total parameters in these structures, for example the Alex Net [30], 5/6 of the total 60 million parameters in Alex Net are in full-connected layers.

Table 2 shows the amounts of parameters in different models where $N$ is from 2 to 128 and $L = log_2 N$. The amount of weights is $2N \cdot log_2 N$ and the amount of bias is $N \cdot log_2 N$ in BCMLP. The space complexity of BCMLP is $O(N \cdot log_2 N)$ which is much less than $O(N^2 \cdot log_2 N)$ of FC-MLPs. Known that

**Table 2.** Amount of parameters in fully-connected layer (FC-Layer), LocallyConnected1D ($k = 3$) [11], dropout ($p = 0.5$) [4], fully-connected MLPs (FC-MLPs) and BCMLP where $N$ is from 2 to 128 and $L = log_2 N$.

| N | 2 | 4 | 8 | 16 | 32 | 64 | 128 |
|---|---|---|---|----|----|----|-----|
| FC-Layer | 6 | 20 | 72 | 272 | 1056 | 4160 | 16512 |
| FC-MLPs | 6 | 40 | 216 | 1088 | 5280 | 24960 | 115584 |
| dropout | 6 | 40 | 216 | 1088 | 5280 | 24960 | 115584 |
| LocallyConnected1D | 6 | 28 | 90 | 248 | 630 | 1524 | 3570 |
| BCMLP | 6 | 24 | 72 | 192 | 480 | 1152 | 2688 |

---

[1] Available at https://github.com/luoningqi/bcmlp.

dropout does not discard any parameters actually. Dropout has the same amount of parameters with FC-MLPs. LocallyConnected1D's space complexity relies on $k$. When $N$ rose to 128, the amounts of parameter needed in BCMLP, dropout and LocallyConnected1D are 2688, 115584 and 3570 respectively.

## 5.2   Sensitivity Test

In the Sect. 4, the Algorithm 2 for computing the sensitivity of the entire BCMLP have been derived. In order to verify the above theoretical results as well as compare sensitivity with fully-connected MLPs (FC-MLPs), LocallyConnected1D [11] and dropout [4], a simulation experiment was taken. With input $X = [x_1, x_2, ..., x_N]$ which is uniform random in $[0, 1)$, weights $W \in [-10, 10)$ and input deviation $\Delta X$, we calculate sensitivity of FC-MLPs, LocallyConnected1D, dropout, simulation results and theoretical results (Algorithm 2) of BCMLP. For each pair of super parameters $(\Delta X, N)$, the average result is acquired from 10000 repeated trials.

As shown in Fig. 2, both red and purple results are higher than the blue or green result. The cyan results are higher than our results when $N \geqslant 16$ and

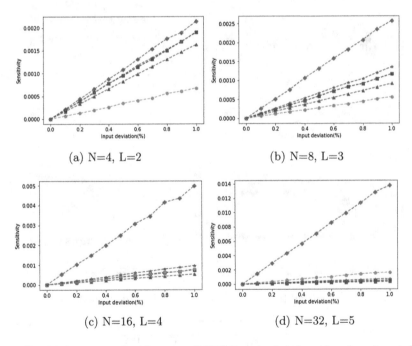

(a) N=4, L=2

(b) N=8, L=3

(c) N=16, L=4

(d) N=32, L=5

**Fig. 2.** Sensitivity comparison between BCMLP and related works where input deviation (%) is from 0.0 to 1.0 and $N$ is in $[4, 8, 16, 32]$. $-\blacklozenge-$ is sensitivity of FC-MLPs; $-\bigstar-$ is results of LocallyConnected1D ($k = 3$); $- \bullet -$ is results of dropout ($p = 0.5$); $-\blacksquare-$ is simulation results of BCMLP; $-\blacktriangle-$ is theoretical results of BCMLP. (Color figure online)

$L \geqslant 4$. The blue line is close to the green line in all 4 subfigs. With the input deviation increasing, the blue and green lines keeps a linear deviation caused of ignoring $R_n$ in Eq. 3. There are 2 verifications from this experiment: (1) BCMLP is less sensitive to input perturbations than related works. (2) The agreement between the experimental results and theoretical results is quite good.

### 5.3   Noisy Logic Gates

In this section, we compare FC-MLPs and BCMLP in a noisy classification task. Traing and testing data consist of vectors $(x_1, x_2, x_3, x_4)^T$ with the corresponding label $y = G(x_1, x_2, x_3, x_4) = (\neg x_1 \wedge x_2) \vee (x_3 \wedge \neg x_4)$ where $x$ and $y$ are both 0 or 1. $N$ and $L$ are 4 and 2 respectively, and batch size is 1. The initial weights are set as 1 or $-1$, and biases are all 0. An $N \rightarrow 1$ fully-connected layer follows with both FC-MLPs and BCMLP to output the results. The gradient descent optimizer is used to minimize the L1 loss. MLPs are trained 10000 steps for each different input deviation and learning rate. And the average accuracy is calculated from 100 testing instance in each step.

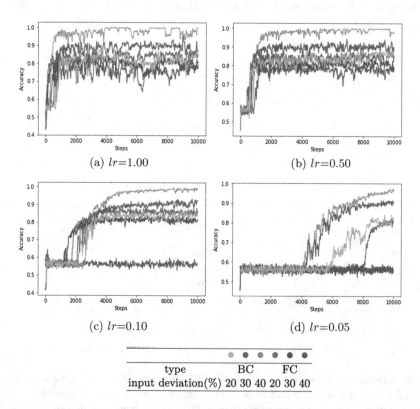

(a) $lr=1.00$       (b) $lr=0.50$

(c) $lr=0.10$       (d) $lr=0.05$

| type | BC | | | FC | | |
|------|----|----|----|----|----|----|
| input deviation(%) | 20 | 30 | 40 | 20 | 30 | 40 |

**Fig. 3.** Accuracy comparision between fully-connected and binary-connected MLPs on noisy classification task where input deviation (%) is from 20 to 40 and learning rate $lr$ is in $[1.00, 0.50, 0.10, 0.05]$.

In each pair of super parameters $(\Delta x, lr)$, BCMLP acquires higher accuracy than FC-MLPs. With $\Delta x$ increasing from 20% to 40%, the accuracy of BCMLP is deceasing from 95% to 85%, while the accuracy of FC-MLPs is deceasing from 85% to 75% as shown in Figs. 3a and b. The FC-MLPs are more sensitive to input perturbation. With $lr$ decreasing from 0.10 to 0.05, shown in Figs. 3a and d, the green and blue results are decreasing to 0.55. FC-MLPs urgently need a rectified learning rate. Comparison shows that our model is more robust in noisy logic gates task.

## 6    Conclusion

We have constructed binary-connected multilayer perceptrons (BCMLP) where every node is connected with two specific nodes of the previous layer. The algorithm for evaluating BCMLP's sensitivity is also proposed. Simulation experiments verify that the agreement between the theoretical sensitivity and experimental results is quite good. We compare our scheme with related works including fully-connected MLPs, LocallyConnected1D and dropout. Comparisons show that our scheme achieves the least amount of parameters, the lowest deviation to input perturbations, and the highest accuracy in the noisy classification task. We will further expand this model in parallel computing frameworks.

## References

1. Bruna, J., Zaremba, W., Szlam, A., LeCun, Y.: Spectral networks and deep locally connected networks on graphs. arXiv preprint 11312.6203 (2013)
2. Hanson, S.J., Pratt, L.Y.: Comparing biases for minimal network construction with back-propagation. In: NIPS, Denver, pp. 177–185 (1989)
3. Han, S., Pool, J., Tran, J., Dally, W.J.: Learning both weights and connections for efficient neural networks. In: NIPS, Montreal, pp. 1135–1143 (2015)
4. Srivastava, N., Hinton, G., Krizhevsky, A., Sutskever, I., Salakhutdinov, R.: Dropout: a simple way to prevent neural networks from overfitting. J. Mach. Learn. Res. **15**, 1929–1958 (2014)
5. Pierce, S.G., Ben-Haim, Y., Worden, K., Manson, G.: Evaluation of neural network robust reliability using information-gap theory. IEEE Trans. Neural Netw. **17**(6), 1349–1361 (2006)
6. Wang, Y., Zeng, X., Yeung, D.S., Peng, Z.: Computation of madalines' sensitivity to input and weight perturbations. Neural Comput. **18**(11), 2854–2877 (2006)
7. Zeng, X., Yeung, D.S.: Sensitivity analysis of multilayer perceptron to input and weight perturbations. IEEE Trans. Neural Netw. **12**(6), 1358–1366 (2001)
8. Zeng, X., Yeung, D.S.: A quantified sensitivity measure for multilayer perceptron to input perturbation. Neural Comput. **15**(1), 183–212 (2003)
9. Yang, S., Ho, C., Siu, S.: Sensitivity analysis of the split-complex valued multilayer perceptron due to the errors of the i.i.d. inputs and weights. IEEE Trans. Neural Netw. **18**(5), 1280–1293 (2007)
10. Yang, J., Zeng, X., Zhong, S.: Computation of multilayer perceptron sensitivity to input perturbation. Neurocomputing **99**(1), 390–398 (2013)

11. Keras documentation. http://keras-cn.readthedocs.io/en/latest/layers/locally_connected_layer/

12. LeCun, Y., Bottou, L., Bengio, Y., Haffner, P.: Gradient based learning applied to document recognition. Proc. IEEE **86**, 2278–2324 (1998)

13. Le, Q.V., et al.: Tiled convolutional neural networks. In: NIPS, Vancouver, pp. 1279–1287 (2010)

14. Gregor, K., LeCun, Y.: Emergence of complex-like cells in a temporal product network with local receptive fields. CoRR, abs/1006.0448 (2010)

15. LeCun, Y., Denker, J.S., Solla, S.A.: Optimal brain damage. In: NIPS, Denver, pp. 598–605 (1990)

16. Hassibi, B., Stork, D.G., et al.: Second order derivatives for network pruning: optimal brain surgeon. In: NIPS, Denver, p. 164 (1993)

17. Wan, L., Zeiler, M., Zhang, S., LeCun, Y., Fergus, R.: Regularization of neural networks using dropconnect. In: ICML, Atlanta, pp. 1058–1066 (2013)

18. Bernier, J.L., Ortega, J., Rojas, I., Prieto, A.: Improving the tolerance of multilayer perceptrons by minimizing the statistical sensitivity to weight deviations. Neurocomputing **31**, 87–103 (2000)

19. Yeung, D.S., Ng, W.W.Y., Wang, D., Tsang, E.C.C., Wang, X.: Localized generalization error model and its application to architecture selection for radial basis function neural network. IEEE Trans. Neural Netw. **18**, 1294–1305 (2007)

20. Ng, W.W.Y., Dorado, A., Yeung, D.S., Pedrycz, W., Izquierdo, E.: Image classification with the use of radial basis function neural networks and the minimization of localized generalization error. Pattern Recogn. **40**, 19–32 (2007)

21. Bernier, J.L., Ortega, J., Ros, E., Rojas, I., Prieto, A.: A quantitive study of fault tolerance, noise immunity and generalization ability of MLPs. Neural Comput. **12**, 2941–2964 (2000)

22. Piche, S.W.: The selection of weight accuracies for madalines. IEEE Trans. Neural Netw. **6**, 432–445 (1995)

23. Stevenson, M., Winter, R., Widrow, B.: Sensitivity of feed forward neural networks to weight errors. IEEE Trans. Neural Netw. **1**, 71–80 (1990)

24. Alippi, C., Piuri, V., Sami, M.: Sensitivity to errors in artificial neural networks: a behavioral approach. In: IEEE International Symposium on Circuits and Systems, Seatlle, pp. 459–462 (1995)

25. Cheng, A.Y., Yeung, D.S.: Sensitivity analysis of neocognitron. IEEE Trans. Syst. Man Sensitivity Anal. Neocognitron Part C **29**(2), 238–249 (1999)

26. Yeung, D.S., Sun, X.: Using function approximation to analyze the sensitivity of MLP with antisymmetric squashing activation function. IEEE Trans. Neural Netw. **13**(1), 34–44 (2002)

27. Ng, W.W.Y., Yeung, D., Ran, S.Q., Tsang, E.C.C.: Statistical output sensitivity to input and weight perturbations of radial basis function neural networks. In: IEEE International Conference on Systems, pp. 503–508 (2002)

28. Zeng, X., Wang, Y., Zhang, K.: Computation of adalines sensitivity to weight perturbation. IEEE Trans. Neural Netw. **17**(2), 515–519 (2006)

29. Abadi, M., et al.: Tensorflow: large-scale machine learning on heterogeneous distributed systems. arXiv preprint 1603.04467 (2016)

30. Krizhevsky, A., Sutskever, I., Hinton, G.: Imagenet classification with deep convolutional neural networks. In: NIPS, Nevada, pp. 1097–1105 (2012)

# Network of Recurrent Neural Networks: Design for Emergence

Chaoming Wang[1] and Yi Zeng[1,2,3(✉)]

[1] Institute of Automation, Chinese Academy of Sciences, Beijing 100190, China
oujago@gmail.com, yi.zeng@ia.ac.cn
[2] University of Chinese Academy of Sciences, Beijing 100190, China
[3] Center for Excellence in Brain Science and Intelligence Technology,
Chinese Academy of Sciences, Shanghai 200031, China

**Abstract.** *Emergence* plays an important role in Recurrent Neural Networks (RNNs). In order to *design for emergence*, we qualitatively and quantitatively design the recurrent neural network structures from the perspective of systems theory. From the qualitative viewpoint, we introduce two methodologies (*aggregation* and *specialization*) from systems theory to design the novel neural structure, and we name it as "Network Of Recurrent neural networks" (NOR). In NOR, RNNs are viewed as the high-level neurons and are used to build the high-level layers. Experiments on three predictive tasks show that under the same number of parameters, the implemented NOR models get superior performances than conventional RNN structures (e.g., vanilla RNN, LSTM and GRU). More importantly, from the quantitative perspective, we introduce an information-theoretical framework to quantify the information dynamics in recurrent neural structures. And the evaluation results show that several NOR models achieve similar or better emergent information processing capabilities compared with LSTM.

**Keywords:** Recurrent Neural Networks · Systems theory · Emergence

## 1 Introduction

In recent years, Recurrent Neural Networks (RNNs) [5,7,9] have been widely used in various applications. Traditionally, RNNs are directly used to build the final models. In this paper, we propose a novel idea called "Network Of Recurrent neural networks" (NOR) which utilizes existing RNNs to make the structure design of the higher-level layers. From a standpoint of systems theory [25], a recurrent neural network is a group composed of a number of interacting parts, and it actually is viewed as a *complex system*, or a *complexity* [19]. Every system is relative: it is not only the system of its parts, but also the part of a larger system. So, in NOR structures, RNN is viewed as the high-level neuron, then several high-level neurons are used to build the high-level layers rather than directly used to construct the final models.

© Springer Nature Switzerland AG 2018
L. Cheng et al. (Eds.): ICONIP 2018, LNCS 11302, pp. 89–102, 2018.
https://doi.org/10.1007/978-3-030-04179-3_8

Conventionally, there are three levels of structure in deep neural networks: *neurons*, *layers* and *whole nets* (or called *models*). From a perspective of systems theory, at each level of such increasing complexity, novel features that do not exist at lower levels emerge at higher levels [15]. For example, at the *neurons* level, single neuron is relatively simple and its generalization capability is comparatively very poor. But when a certain number of such neurons are accumulated into a certain elaborate structure by certain ingenious combinations, the *layers* at the higher level begin to get the unprecedented ability of classification and feature learning. More importantly, such new gained capability or property is deducible from but not reducible to constituent neurons at the lower level. That is, it is not a property of the simple superposition of all constituent neurons, and the whole is greater than the sum of its parts. In systems theory, such kind of phenomenon is known as *emergence* [10]. Emergent phenomena appear in various systems ranging from living beings to social networks [10]. A natural question is, what structure designs can make us achieve emergent computation in recurrent neural structures? Are there any mechanisms for emergence in systems theory which can lead us to locate the parametric regions where emergent computation are likely to occur?

Traditionally, it was assumed that emergence and complexity were solely the effects of nonlinearity of elements which was regarded, by Prigogine, as a precondition for a dynamic system [20]. While in 1993, Arthur identified another three mechanisms by which complexity tends to grow as systems evolve [1]:

- Mechanism 1: increase in *coevolutionary diversity*. The agent in the system seems to be a new agent class, type or species.
- Mechanism 2: increase in *structural sophistication*. The individual system steadily accumulates increasing numbers of new subsystems or parts.
- Mechanism 3: increase by "*capturing software*". The system captures simpler elements and learns to "program" them as "software" to be used as its own ends.

In this paper, with the guidance of the first two mechanisms, we introduce two methodologies to *design for emergence* in recurrent neural structures, which are named as *aggregation* and *specialization*. Aggregation and specialization are two natural operations for increasing complexity in complex systems [8]. The former is related to Arthur's second mechanism, in which traditional RNNs are aggregated and accumulated into a high-level layer in accordance with a specific structure, and the latter is related to Arthur's first mechanism, in which the RNN agent in the high-level layer is specialized as the agent that performs a specific function.

Based on these two methodologies, we design several NOR architectures and conduct experiments on three prediction tasks. Experimental results show that our implementations outperform conventional RNN structures (such as vanilla RNN, LSTM, and GRU) under the same number of parameters. More importantly, in order to quantify the emergent computational capabilities and gain more insights on how information is processed inside the implemented models, we adopt an information-theoretical framework [17] to measure information dynamics.

## 2 Background

### 2.1 Systems Theory and Emergence

*Systems Theory* was originally proposed for the theoretic investigation of biological phenomena [25]. In biology systems, there are several different levels which begin with the smallest units and reach to the most extensive category: molecule, cell, tissue, organ, $\cdots$, biosphere. Traditionally, such kinds of system at any level can be decomposed into a number of elementary components so that each component will be independently analyzed, and then added in a linear superposition manner to describe the overall system. However, Bertalanffy argued that such operation (i.e., simply breaking it down and then reforming it) cannot make us fully understand a system, because "the whole is greater than the sum of its parts" [25]. Instead, we need to apply a global and systematic perspective to underline its functionality [18]. In systems theory, such phenomenon is known as *emergence* [10] and it is often described in a two-level or multi-level architecture.

What design can achieve emergence in artificial systems and what mechanisms are behind emergence in natural systems have been widely discussed. There have already been many works trying to figure out the mechanisms behind emergence. One of the widely accepted is the repeated application and combination of two complementary forces or operations, which are named as folding and stretching in physics, merging and splitting in computer science, or cooperation and specialization in sociology [8]. The former mean a number of (sub-)agents are aggregated into a single agent. The latter mean the agents are separated from each other and each agent is constrained to a certain role or class. In this paper, we hypothesize that such two widely accepted operations can also be applied to organize and evolve the neural network structures.

### 2.2 Systems Theory and Recurrent Neural Networks

*Recurrent Neural Networks (RNNs)* are a class of deep neural networks that are able to process arbitrary sequences of inputs. Formally, given a sequence of vectors $\{x_t\}_{t=1\cdots T}$, vanilla RNN [7] is defined as $h_t = f(Wx_t + Uh_{t-1})$, where $W$ and $U$ are parameter matrices, and $f$ denotes the activation function. Afterwards, several improved RNNs are proposed to solve the training problems in vanilla RNN, such as Long Short-Term Memory [9] and Gated Recurrent Unit [5].

Actually, RNNs can behave chaotically. Some efforts have been made to analyze RNNs theoretically or experimentally from the perspective of systems theory. Sontag provided an exposition of research regarding system-theoretic aspects of RNNs with sigmoid activation function [23]. Bertschinger & Natschlager analyzed the *computation at the edge of chaos* in RNNs and calculated the critical boundary in parameter space where the transition from ordered to chaotic dynamics takes place [3]. Razvan et al. employed a dynamical systems perspective to understand the exploding/vanishing gradients problem in RNNs [21]. Boedecker et al. introduced two information-theoretical measurements to investigate information processing in Echo State Networks [4]. In this paper, we obtain methodologies from systems theory [1] to *design for emergence* and adopt an information-theoretical framework [17] to quantify emergence.

# 3   Network of Recurrent Neural Networks

We firstly give an overview of NOR structure. Then, based on two design methodologies, we propose four NOR models. Finally, three measurements [17] are introduced to quantify information dynamics of the proposed models.

## 3.1   Overall Architecture

As the high-level illustration shown in Fig. 1, NOR architecture is a three-dimensional spatial-temporal structure. It includes four components: I, M, S and O. Component I (input) and O (output) control the head and tail of NOR layer. Component S (subnetworks) is in charge of the spatial extension. Component M (memories) is responsible for the temporal extension. We formalize each component as follows:

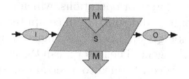

**Fig. 1.** The high level architecture of the Network of Recurrent Neural Networks (NOR).

*Component I*: It is responsible for tasks of data preprocessing and distribution. At each time-step $t$, component I may *copy* input data into $n$ duplicates and feed each of them into one single subnetwork in component S: $\{x_t^i\}_{i=1\cdots n} = C(x_t)$, where $C$ means copy function, and $x_t^i$ will be fed into the $i$-th subnetwork.

*Component M*: It manages all memories in the whole layer. Component M may *directly deliver* individual memory of each RNN neuron to the next time-step: $m_t^j = I(o_{t-1}^j)$, where $I$ means identity function, $m_t^j$ means the memory of $j$-th RNN neuron at time-step $t$ and $o_{t-1}^j$ means the output state of the $j$-th RNN neuron at time-step $t-1$.

*Component S*: Its responsibility is to manage the logic of each subnetwork and handle the interaction between them. Suppose component S is with in-degree $n$ and out-degree $m$ (i.e., component S receives $n$ inputs and produces $m$ outputs), the $k$-th output at time-step $t$ is generated by necessary inputs and memories $IN^k$ and $MEM^k$:

$$s_t^k = f(IN^k, MEM^k) \quad k \in \{1, \cdots, m\} \tag{1}$$

where $f$ means nonlinear interaction.

***Component O***: It integrates all outputs produced in component S into a whole. A simple method (named as *Concat-MLP*) is concatenating all outputs then using a MLP to measure the weight at each position: $o_t = g(W_{mlp} * [s_t^1; s_t^2; \cdots ; s_t^m])$, where $W_{mlp}$ is the weight of MLP and $g$ is the activation function. Pure pooling methods (such as *Mean-Pooling* and *Max-Pooling*) can be used to select optimal value in each position. *Attention method* [2] can also be employed to select the proper output.

## 3.2 Methodology I: Aggregation

According to Arthur's second mechanism, internal complexity can be increased by aggregation of sub-agents, which means a number of RNN agents are conglomerated into a single big system. In a concrete NOR layer designed by aggregation, component I first *copy* $x_t$ into $n$ duplications, component M *directly deliver* the memory of each RNN neuron from the last time-step to the current time-step, *Concat-MLP* method is used as component O. However, the number, the type and the interaction of the aggregated RNNs in component S determine the internal structure or inner complexity of the newly formed layer system. Thus, three kinds of topologies are presented here.

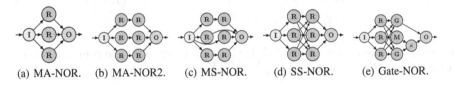

(a) MA-NOR.    (b) MA-NOR2.    (c) MS-NOR.    (d) SS-NOR.    (e) Gate-NOR.

**Fig. 2.** The sectional views of NOR layers. "I" is component I, "O" is component O, "R" denotes RNN agent, "G" denotes gate RNN agent, "M" means MLP agent, and "×" denotes element-wise multiplication.

**Multi-Agent.** The natural description of complex system is the Multi-Agent system. As shown in Fig. 2(a), a multi-agent NOR layer called MA-NOR is composed of three parallel RNNs. Figure 3(a) shows this layer unrolled into a full network. At time-step $t$, the $i$-th subnetwork of component S in MA-NOR is calculated as:

$$s_t^i = r(W^i x_t^i + U^i m_t^i) \tag{2}$$

where $r$ denotes the ReLU activation, and $W^i$ and $U^i$ are parameters of the $i$-th RNN neuron. While, each subnetwork may be more complex. For example, Fig. 2(b) shows another MA-NOR layer made up of three two-tier RNNs.

**Multi-Scale.** One way to increase the diversity in an aggregation structure is to use the Multi-Scale topology which can learn sequence dependencies in different timescales. Figure 2(c) shows a multi-scale NOR layer (MS-NOR) made up of four subnetworks, in which two of them are one-tier RNNs and the others are two-tier RNNs. Figure 3(b) shows it unfolded in two time-steps. In component S, two kinds of timescale dependencies can be learned, which are formalized as follows:

$$s_t^i = r(W^i x_t^i + U^i m_t^i) \qquad\qquad i \in \{2,3\} \tag{3a}$$

$$s_t^{j,1} = r(W^{j,1} x_t^j + U^{j,1} m_t^{j,1}) \qquad\qquad j \in \{1,4\} \tag{3b}$$

$$s_t^j = r(W^{j,2} s_t^{j,1} + U^{j,2} m_t^{j,2}) \qquad\qquad j \in \{1,4\} \tag{3c}$$

where $s_t^i$ denotes the output of the one-tier subnetwork, $s_t^j$ denotes the output of the two-tier subnetwork, and $s_t^{j,1}$ denotes the output of the RNN neuron at the first tier.

**Self-Similarity.** Repeated aggregation and high accumulation enables the emergence of the fractal and self-similar structure. As shown in Fig. 2(d), three paths are also used in SS-NOR (three-dimensional diagram can be observed in Fig. 3(c)). But after each path firstly learns its own intermediate representation, the second layers gather all intermediate representations of three paths to learn high-level abstract features. Such operation helps the model easy to share information. We formalize the cooperation process as follows:

$$o_t^{i,1} = r(W^{i,1} x_t^i + U^{i,1} m_t^{i,1}) \qquad\qquad i \in \{1,2,3\} \tag{4a}$$

$$s_t^j = r(W^{j,2} o_t^1 + U^{j,2} m_t^{j,2}) \qquad\qquad j \in \{1,2,3\} \tag{4b}$$

where $o_t^{i,1}$ is the output of the $i$-th RNN at first tier, $o_t^i$ is the concatenation of all first-tier RNN outputs, i.e., $[o_t^{1,1}; o_t^{2,1}; o_t^{3,1}]$.

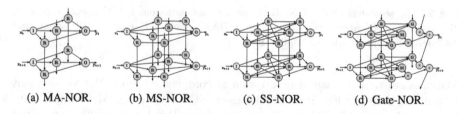

(a) MA-NOR.        (b) MS-NOR.        (c) SS-NOR.        (d) Gate-NOR.

**Fig. 3.** The unfolding of NOR layers in two time-steps.

### 3.3   Methodology II: Specialization

Specialization is related to Arthur's first mechanism. Based on the aggregation, specialization makes objects become objects of a certain class and make agents become agents of a certain type. The more such an agent becomes a particular class or type, the more it needs to delegate special tasks that it can not handle alone to other agents [8]. As a result, the effect of specialization is the existence of delegation and division of labor in the newly formed groups. Thus, the interactions in Eq. (1) can be rewritten as $s_t^k = g(f_1, \cdots, f_l, \cdots, f_L)$, where $f_l$ is the $l$-th specialized agent function, $g$ means the cooperation of all specialized agents, and $L$ is the number of specialized agents.

**Gate Specialization.** One way to specialize is the gate specialization. For example, a general RNN agent is specialized into two types of RNN agents, one is for gating and the other is for generalization. A concrete implementation which is named as Gate-NOR is shown in Figs. 2(e) and 3(d). We formalize it as follows:

$$s_t^{i,1} = r(W^{i,1}x_t^i + U^{i,1}m_t^{i,1}) \qquad i \in \{1,2,3\} \tag{5a}$$

$$g_c = \sigma(W^c s_t^1 + U^c m_t^c) \tag{5b}$$

$$g_s = \sigma(W^s s_t^1 + U^s m_t^s) \tag{5c}$$

$$h_t = g_c \odot tanh(W^1 s_t^1) + g_s \odot tanh(h_{t-1}) \tag{5d}$$

where $g_c$ and $g_s$ denote gate RNN agents, and $s_t^1$ is the concatenation of all first-tier RNN outputs, i.e., $[s_t^{1,1}; s_t^{2,1}; s_t^{3,1}]$. Gate-NOR is evolved from the SS-NOR, in which the medial second-tier RNN agent is separated into MLP unit $tanh(Wx_t)$ and state unit $tanh(h_{t-1})$, while other two second-tier RNN agents are specialized as two gate RNNs, in which the first gate $g_c$ is to control how much current information to let through and the second $g_s$ is to control how much information of previous state to let through.

**Relation with LSTM and GRU.** We consider LSTM [9] and GRU [5] as two special cases of "network of recurrent neural networks". For example, LSTM can be viewed as a special NOR which is made up of four RNNs, in which three of four RNNs are specialized as gate agents (i.e., input gate, forget gate and output gate) and the fourth RNN is specialized as an agent for generalization task. However, different from our implementations, there is only a public (not individual) memory in LSTM shared with each RNN agent.

### 3.4   Information-Theoretical Measurements

Information theory has already been proved to be a useful tool to design and analyze complex systems [17]. Lizier et al. provided the first complete framework which contains three measurements to quantify information dynamics in a complex system [17].

The first measurement is named as *Active Information Storage (AIS)* and is used to measure how much does a neuron's history (i.e., the storage of information by the neuron) influence its future state. It is defined as the mutual information between the $i$-th neuron's semi-infinite past state $x_{i,n-1}^{(k)} = \{x_{i,n-1}, \cdots, x_{i,n-k}\}$ and its future state $x_{i,n}$:

$$a(i,n) = \lim_{k \to \infty} \sum_{u_{i,n}} p(u_{i,n}) log_2 \frac{u_{i,n}}{p(x_{i,n-1}^{(k)})p(x_{i,n})} \qquad (6)$$

where $u_{i,n}$ denotes the state tuple $(x_{i,n-1}^{(k)}, x_{i,n})$.

The second measurement is *Transfer Entropy (TE)* which is used to measure how much does the current state of a source neuron influence the next state of a destination neuron. It is defined as the mutual information between the current state of the source neuron $y_{j,n-1}$ and the next state of the destination neuron $x_{i,n}$ conditioned on the history of the destination process:

$$t(i,j,n) = \lim_{k \to \infty} \sum_{w_{i,n}} log_2 \frac{p(x_{i,n}|x_{i,n-1}^{(k)}, y_{j,n-1})}{p(x_{i,n}|x_{i,n-1}^{(k)})} \qquad (7)$$

where $w_{i,n}$ is the state transition tuple $(x_{i,n}, x_{i,n-1}^{(k)}, y_{j,n-1})$.

The third measurement is *Separable Information (SI)* which measures synergy between information storage and transfer sources. It is defined as the independent sum of information between the information storage $a(i,n)$ and all information transfer contributors $t(i,j,n)$ at current state $n$ of the $i$-th neuron:

$$s(i,n) = a(i,n) + \sum_{j=1, j \neq i}^{h} t(i,j,n) \qquad (8)$$

where $h$ is the hidden number of a recurrent node. According to [17], the formulation of separable information is consistent with the description of emergence as "the whole is greater than the sum of its parts".

## 4  Task-Based Evaluations

In this part, we evaluate the performance of the presented neural structures on three prediction tasks: handwriting classification, named entity recognition and question classification. We compare all models under the same parameter numbers to validate the capacity of better utilizing the parametric space. In order to verify the reliability of the design, we conduct three comparative tests under total parameters of different orders of magnitude for each task (Table 1). Every experiment is repeated 20 times with different random initializations, then we report the mean results. It is worthy noting that our aim here is to compare the model performance under the same hyper-parameter settings, not to achieve best performance for one single model.

It has been reported that careful initialized vanilla RNN composed of ReLU activation (named as IRNN) can show excellent performance [13]. In our experiments, all basic RNN neurons are implemented with vanilla RNN applied with

**Table 1.** Number of hidden neurons for IRNN, GRU, LSTM, MA-NOR, MS-NOR, SS-NOR and Gate-NOR for each network size specified in terms of the number of parameters (weights).

| Task | # of Params. | IRNN | GRU | LSTM | MA-NOR | MS-NOR | SS-NOR | Gate-NOR |
|------|------|------|------|------|------|------|------|------|
| Handwriting classification | 150 k | 217 | 124 | 107 | 97 | 71 | 61 | 68 |
| | 350 k | 335 | 192 | 165 | 149 | 109 | 93 | 105 |
| | 650 k | 459 | 263 | 227 | 205 | 150 | 128 | 143 |
| Named entity recognition | 200 k | 197 | 86 | 67 | 74 | 54 | 53 | 58 |
| | 400 k | 319 | 148 | 119 | 122 | 88 | 83 | 91 |
| | 800 k | 497 | 244 | 199 | 193 | 139 | 126 | 140 |
| Question Classification | 100 k | 198 | 86 | 68 | 74 | 54 | 53 | 58 |
| | 200 k | 319 | 148 | 119 | 122 | 88 | 83 | 91 |
| | 400 k | 497 | 244 | 199 | 193 | 139 | 126 | 140 |

ReLU function. We keep the number of the hidden units the same over all RNN neurons in a NOR layer. Obviously, our baseline is a single giant IRNN model. Meanwhile, two improved popular RNNs (i.e., GRU [5] and LSTM [9]) are also chosen as our baselines.

In all experiments, models are regularized by dropout [26] with a rate of 0.5. Trainings are done through Adam optimizer [12]. In order to avoid overfitting and prevent unnecessary computation, early stopping is applied. For NLP tasks, we obtain the pre-trained Glove vectors [22] for the word embeddings. During training we fix all word embeddings and learn only the other parameters in all models.

### 4.1 Handwriting Classification

The handwriting classification task is with the MNIST benchmark [14], which consists of hand written digits 0–9. This dataset contains 60,000 training images and 10,000 testing images in total. We randomly select 5,000 training images as the validation set. MNIST images are $28 \times 28$ pixels large, so this yields a sequence of 28 bits vectors of dimension 28.

All test models use the same architecture: input layer $\rightarrow$ RNN/NOR layer $\rightarrow$ RNN/NOR layer $\rightarrow$ max-pooling layer $\rightarrow$ softmax layer. The initial learning rates of all models are set to 0.001. Three different network sizes are tested for each architecture, such that the number of parameters are roughly 150 k, 350 k and 650 k (Table 1). We set the minibatch size as 32, and use cross entropy criterion as the loss function.

The accuracy results of the experiments are shown in Table 2. It shows that NOR models get competitive performances compared with three widely used RNN structures under the same number of parameters. The accuracies of all models are improved as the network sizes grow. Among all NOR models, Gate-NOR gets the best.

**Table 2.** Evaluation results on three different prediction tasks.

| Model | Handwriting classification | | | Named entity recognition | | | Question classification | | |
|---|---|---|---|---|---|---|---|---|---|
| | 150k Params | 350k Params | 650k Params | 200k Params | 400k Params | 800k Params | 100k Params | 200k Params | 400k Params |
| IRNN | 98.97 | 99.01 | 99.03 | 85.09 | 85.32 | 85.57 | 92.73 | 93.22 | 93.62 |
| GRU | 98.91 | 99.03 | 99.06 | 84.93 | 85.37 | 85.78 | 92.33 | 93.34 | 93.74 |
| LSTM | 98.79 | 99.01 | 99.04 | 84.77 | 85.73 | 86.11 | 92.54 | 93.46 | 94.10 |
| MA-NOR | 99.06 | 99.10 | 99.13 | 85.80 | 86.10 | 86.21 | 92.73 | 93.56 | 93.73 |
| MS-NOR | 99.05 | 99.10 | 99.15 | 85.62 | 86.14 | 86.21 | 92.85 | 93.79 | 94.04 |
| SS-NOR | 99.07 | **99.12** | 99.16 | **86.06** | **86.20** | 86.24 | **93.01** | 93.62 | **94.12** |
| Gate-NOR | 99.08 | **99.12** | 99.18 | 85.89 | 86.12 | **86.27** | 92.84 | **93.81** | **94.12** |

## 4.2 Named Entity Recognition

Named Entity Recognition (NER) is a task that tries to identity the proper names of persons, organizations, locations, or other entities in the given text. We apply the CoNLL-2003 dataset [24] which consists of 14,987 sentences in the training set, and 3,466 sentences in the validation set and 3,684 sentences in the test set.

We adopt a widely used NER model (i.e., Bi-LSTM-CRF [11]) as the universal architecture of all test networks: embedding layer → Bi-RNN/Bi-NOR layer → CRF layer. Three hidden layer sizes are chosen such that the total number of parameters for the whole network is roughly 200 k, 400 k and 800 k (Table 1). The initial learning rate is set to 0.005 and the size of each minibatch is 20.

The accuracy results are summarized in Table 2. Again, our NORs perform much better than three baselines. Among them, SS-NOR and Gate-NOR get the best results.

## 4.3 Question Classification

Question classification is an important step in a question answering system which classifies a question into a specific type. For this task, we use the TREC [16] benchmark which divides all questions into 6 categories: location, human, entity, abbreviation, description and numeric. TERC provides 5,452 labeled questions in the training set and 500 questions in the test set. We randomly select 10% of the training data as the validation set.

All networks use the same architecture: embedding layer → RNN/NOR layer → max-pooling layer → softmax layer. Three hidden sizes are chosen such that the total number of parameters for any model is roughly 100 k, 200 k, 400 k (Table 1). All networks use 0.0005 learning rate and are trained to minimize the cross entropy error.

Table 2 shows the $F_1$ results. NOR models get comparable or even better results than baseline models. Among all NOR models, SS-NOR and Gate-NOR get the best.

## 5    Information-Theoretical Evaluations

In order to gain additional insights on information processing capabilities, we perform evaluation on three aforementioned information-theoretical measurements. We obtain the training set of MNIST dataset to stabilize the network, and the test set is used to calculate the measures. Different hidden numbers are used to keep the total parameter the same over all models. We replace ReLU as *tanh* activation function in all models in order to map all values into a range between $-1$ and 1. At the statistical phase, we *round* the hidden states to get discrete values $-1$, 0, and 1. In order to arrive at a single value, we take averages over all the neurons in a recurrent nodes. However, recurrent nodes for gate tasks are omitted in this evaluation. We choose three history length, i.e., $k = 1$, $k = 2$, and $k = 3$. All AIS, TE and SI results are averaged by 10 times, and plotted in Fig. 4, Fig. 5 and Fig. 6 respectively.

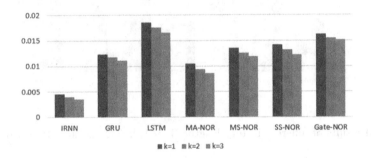

**Fig. 4.** AIS results of different history lengths ($k = 1$, $k = 2$ and $k = 3$) over different models.

As shown in Fig. 4, LSTM gets the best results, which demonstrates its effective capacity of information storage when evolving over time. More than 20 years of tests in various scenarios has proved LSTM is an effective network model to learn short-term memories. We come to the same conclusion from the perspective of information storage measurement. Meanwhile, as NOR structures evolve and become more complex, the information storage capabilities are growing. Figure 5 exhibits the information transfer evaluation results. Surprisingly, LSTM performs very poor, while the simple MA-NOR gets the high results. Figure 6 presents the interaction results between stored and transmitted information, which measures the emergent computational capabilities. We can observe that SS-NOR and Gate-NOR perform comparable with (and slightly better than) LSTM.

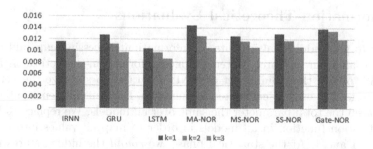

**Fig. 5.** TE results of different history lengths (k = 1, k = 2 and k = 3) over different models.

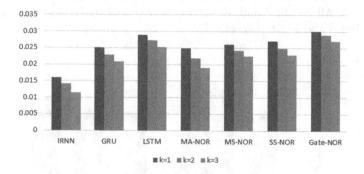

**Fig. 6.** SI results of different history lengths (k = 1, k = 2 and k = 3) over different models.

# 6    Conclusion

We introduced a novel neural network model named as "Network Of Recurrent neural networks" (NOR) which considers existing RNNs as high-level neurons and then utilizes RNN neurons to design higher-level layers. *Emergence* is an important goal to deign artificial systems [6]. In order to design for emergence, two efforts were made. Firstly, inspired from two mechanisms behind emergence [1], we proposed two kinds of methodologies (named as *aggregation* and *specialization*) to design different NOR topologies. Task-based evaluations demonstrated the good performances of implemented NOR models on prediction tasks. Secondly, we adopted an information-theoretical framework [17] to quantify emergent information processing capabilities. The information-theoretical evaluation results gave us more insights on how information is processed and led us to further narrow down the most appropriate models.

# References

1. Arthur, W.B.: On the evolution of complexity. In: Cowan, G.A., Pines, D., Meltzer, D.E. (eds.) Complexity: Metaphors, Models, and Reality. Advanced Book Classics, pp. 65–81. Westview Press, Cambridge (1999). Chapter 5
2. Bahdanau, D., Cho, K., Bengio, Y.: Neural machine translation by jointly learning to align and translate. arXiv preprint arXiv:1409.0473 (2014)
3. Bertschinger, N., Natschläger, T.: Real-time computation at the edge of chaos in recurrent neural networks. Neural Comput. **16**(7), 1413–1436 (2004)
4. Boedecker, J., Obst, O., Lizier, J.T., Mayer, N.M., Asada, M.: Information processing in echo state networks at the edge of chaos. Theor. Biosci. **131**(3), 205–213 (2012)
5. Cho, K., et al.: Learning phrase representations using RNN encoder-decoder for statistical machine translation. arXiv preprint arXiv:1406.1078 (2014)
6. Dessalles, J.L., Müller, J.P., Phan, D.: Emergence in multi-agent systems: conceptual and methodological issues. In: Phan, D., Amblard, F. (eds.) Agent-based Modelling and Simulation in the Social and Human Sciences, pp. 327–355. The Bardwell Press, Oxford (2007)
7. Elman, J.L.: Finding structure in time. Cogn. Sci. **14**(2), 179–211 (1990)
8. Fromm, J.: The Emergence of Complexity. Kassel University Press, Kassel (2004)
9. Hochreiter, S., Schmidhuber, J.: Long short-term memory. Neural Comput. **9**(8), 1735–1780 (1997)
10. Holland, J.H.: Emergence: From Chaos to Order. OUP, Oxford (2000)
11. Huang, Z., Xu, W., Yu, K.: Bidirectional LSTM-CRF models for sequence tagging. arXiv preprint arXiv:1508.01991 (2015)
12. Kingma, D.P., Ba, J.: Adam: a method for stochastic optimization. arXiv preprint arXiv:1412.6980 (2014)
13. Le, Q.V., Jaitly, N., Hinton, G.E.: A simple way to initialize recurrent networks of rectified linear units. arXiv preprint arXiv:1504.00941 (2015)
14. LeCun, Y., Bottou, L., Bengio, Y., Haffner, P.: Gradient-based learning applied to document recognition. Proc. IEEE **86**(11), 2278–2324 (1998)
15. Lehn, J.M.: Towards complex matter: supramolecular chemistry and self-organization. Eur. Rev. **17**(2), 263–280 (2009)
16. Li, X., Roth, D.: Learning question classifiers. In: Proceedings of the 19th International Conference on Computational Linguistics-Volume 1, pp. 1–7. Association for Computational Linguistics (2002)
17. Lizier, J.T., Prokopenko, M., Zomaya, A.Y.: A framework for the local information dynamics of distributed computation in complex systems. In: Prokopenko, M. (ed.) Guided Self-Organization: Inception. ECC, vol. 9, pp. 115–158. Springer, Heidelberg (2014). https://doi.org/10.1007/978-3-642-53734-9_5
18. Mele, C., Pels, J., Polese, F.: A brief review of systems theories and their managerial applications. Serv. Sci. **2**, 126–135 (2010)
19. Mitchell, M.: Complexity: A guided Tour. Oxford University Press, New York (2009)
20. Nicolis, G., Prigogine, I.: Self-organization in Nonequilibrium Systems: From Dissipative Structures to Order Through Fluctuations. Wiley (1977)
21. Pascanu, R., Mikolov, T., Bengio, Y.: On the difficulty of training recurrent neural networks. In: International Conference on Machine Learning, pp. 1310–1318 (2013)
22. Pennington, J., Socher, R., Manning, C.: Glove: global vectors for word representation. In: Proceedings of the 2014 Conference on Empirical Methods in Natural Language Processing (EMNLP), pp. 1532–1543 (2014)

23. Kárný, M., Warwick, K., Kůrková, V.: Recurrent neural networks: some systems-theoretic aspects. In: Kárnỳ, M., Warwick, K., Kůrková, V. (eds.) Dealing with Complexity, pp. 1–12. Springer, London (1998). https://doi.org/10.1007/978-1-4471-1523-6_1

24. Tjong Kim Sang, E.F., De Meulder, F.: Introduction to the CoNLL-2003 shared task: language-independent named entity recognition. In: Proceedings of the Seventh Conference on Natural Language Learning at HLT-NAACL, vol. 4, pp. 142–147. Association for Computational Linguistics (2003)

25. Von Bertalanffy, L.: General System Theory. G. Braziller, New York (1968)

26. Zaremba, W., Sutskever, I., Vinyals, O.: Recurrent neural network regularization. arXiv preprint arXiv:1409.2329 (2014)

# Computationally Efficient Radial Basis Function

Adedamola Wuraola$^{(\boxtimes)}$ and Nitish Patel

Department of Electrical and Computer Engineering, The University of Auckland,
Auckland, New Zealand
awur978@aucklanduni.ac.nz, nd.patel@auckland.ac.nz

**Abstract.** We introduced a Square-law based RBF kernel called SQuare RBF (SQ-RBF) which is computationally efficient and effective due to the elimination of the exponential term. In contrast to the Gaussian RBF, SQ-RBF requires smaller computational operation count and direct implementation without a call to higher order library. The derivative of the SQ-RBF is linear which will improve gradient computation and makes its applicability in multilayer perceptron neural network attractive. In experiments, SQ-RBF lead not only to faster learning but also requires significant low neurons than Gaussian RBF on networks. On an average, we recorded a speed-up in training time of about 8% for SQ-RBF based networks without affecting the overall generalizability of the network. SQ-RBF uses about 10% fewer neurons than Gaussian RBF hence making it very attractive.

**Keywords:** Activation function · Artificial Neural Networks · RBF
Function approximation

## 1 Introduction

In the field of Artificial Neural Network and machine learning, there is a constant need for improvement in cost-energy-performance. There is an exponent growth in data and the more data available for an ANN model during training, the better the inference (prediction) stage. This has led to several architectures being developed to be used during training as well as inference. Graphical Processing Units (GPUs) were able to provide speedups that were not possible when using Central Processing Units (CPUs). Recently, Google introduced an application specific unit dedicated to machine learning. This is referred to as Tensor Processing Units (TPUs) which has been described to be about 15X and 30X faster than GPUs and CPUs respectively [1]. This shows that computational speed is as important as performance accuracy when it comes to any Neural Network architecture.

Radial Basis Functions are one of the many architectures of Artificial Neural Network, commonly referred to as Radial Basis Function Neural Network (RBFNN). It was introduced by Broomhead and Lowe [2] in the late 80's and

© Springer Nature Switzerland AG 2018
L. Cheng et al. (Eds.): ICONIP 2018, LNCS 11302, pp. 103–112, 2018.
https://doi.org/10.1007/978-3-030-04179-3_9

has found significant applications and success in areas such as function approximation, interpolation, dynamic system design, classification. They perform excellently till date in any form of approximation problem, and their excellent approximation capabilities have been studied extensively ins literature [3,4]. The most important part of the RBFNN architecture is the hidden layer neurons commonly referred to as RBF kernels. As described in [5], the RBF kernels transform the input patterns (data) into a new feature space through the help of a strictly positive radially symmetric activation function. Researchers have been working to improve RBFNN training, more focus is on the training of RBFNN, in this work we shift focus to the center selection through the RBF kernel. The importance of the right RBF kernel was extensively illustrated in [5]. One of the most widely used RBF kernels is the Gaussian kernel.

The presence of the exponential term in the Gaussian function made its direct implementation on hardware not possible except through approximation methods. Therefore, the elimination of the exponential term will not only alleviate the computational intensity, but also lead to faster simulations. The introduction of GPU and other platforms for training Neural Networks has led to significant speedups in simulation time but as described in [6,7], the exponential function is not implemented with hardware on GPUs but with software library. With this, the speedup possible using GPUs is defeated. Over the years, there have been different approximation of the Gaussian RBF, and an exhaustive list can be found in [8]. The authors in [9] proposed a replacement for the Gaussian kernels which is adaptive by combining 3 activation functions namely sigmoid, multi quadratic and Gaussian activation functions. However, this new activation function still makes use of exponent term heavily.

The paper aims to present a computationally efficient and improved version of the Gaussian RBF activation function. This new function although having a similar convex shape as the Gaussian function is characterized by just multiplication and subtraction mathematical operators and it is capable of achieving better performance. The new function requires only multiplications, subtractions, and logical operations to obtain a bell-shaped Gaussian-type function. Moreover, the derivative of the function is purely combinatorial (comprising just addition and subtractions) which makes gradient computation easier when used in MLP architectures. The function proposed in this paper can replace the existing Gaussian AF through its well-formed bell-shaped non-linearity. Additionally, extensive experiments have been performed to show the performance and advantages of the proposed function.

## 2    RBF Networks and RBF Kernels

RBF networks are similar to MLP in that they are feedforward neuron networks architecture, but they are characterized with a single hidden layer whereas MLP networks can have more than one hidden layers. Each hidden layer unit is characterized by a radial activation function. The output layer, on the other hand, implements the weighted sum of the hidden node outputs. A survey carried out in [10] describe RBF networks as a current trend in the successful modeling of various industrial processes. The training of RBFNN is in two stages namely identification of the radial basis center for the RBF neurons in the hidden layers and the learning of the weights present in the hidden to the output layer. Several methods can be used for the center definition namely random sampling of input data, unsupervised clustering (commonly K-means) and Self Organising Map (SOM). On the other hand, the weights can be tuned using supervised linear perceptron training, backpropagation, Moore Penrose Pseudo Inverse, Least Mean Squares and some additional training methods as proposed in literature [10,11]. Like their MLP counterparts, the performance accuracy, convergence speed and generalizability of an RBFNN depends mostly on architecture, initialization heuristics, choice of activation function, regularization techniques and learning algorithms.

Among all the different types of activation functions for RBF networks, Gaussian function tends to be the most popularly used. Unfortunately, this function is characterized by an exponent term. Exponent function is generally computationally expensive due to a call to high order function term. As described in [5], Radial basis kernels fall under the family of functions whose value depend on the distance between the variable $x$ and a defined center point $c$. In other words, an RBF kernel $f(x, c) = ||x - c||$. There are various types of RBF kernel $f(x, c)$ as seen in literature, Table 1 shows some kernels with their associated mathematical expressions.

**Table 1.** Common and approximate RBF kernels

| RBF Kernel | Expression |
| --- | --- |
| Gaussian | $f(x) = \exp^{-(\beta x)^2}$ |
| Multiquadric | $f(x) = \sqrt{1 + (\beta x)^2}$ |
| Inverse multiquadric | $f(x) = \dfrac{1}{\sqrt{1+(\beta x)^2}}$ |
| Thin-plate spline | $f(x) = x^2 \log(\beta x^2)$ |
| $C^4$ Matern | $f(x) = exp^{-\beta x} \cdot (3 + 3\beta x + \beta x)^2$ |
| Approximate Gaussian | $f(x) = \dfrac{1}{1+(\beta x)^2}$ |
| Approximate Gaussian | $f(x) = \dfrac{1}{1+(\beta x)^4}$ |

The expressions in Table 1 all show a need for multiple computationally expensive operations. Although the two approximate Gaussian functions do not have exponent terms and square roots and are "simple" as described in literature [12]. However, the divisor term is not a power of 2 hence shift operation cannot be used, thereby requiring division which is resource intensive when compared to logical operation. The next section will show a new function that is computationally efficient and effective than all the RBF kernels listed in Table 1. The mappings of each of the functions listed in Table 1 are significantly different. A suitable normalizing technique is required such that the narrower functions are not adversely affected. There are many ways of normalizing the functions, but in this paper, we have opted to compare only the Gaussian RBF.

## 3   Square Non-linear Radial Basis Function (SQ-RBF)

We define a new convex kernel that makes use of a square-law and eliminates the exponential term present in Gaussian expression. We referred to this expression as SQ-RBF and is defined in (1).

$$f(x) = \begin{cases} 1 - \frac{x^2}{2} & : x \leq 1.0 \\ 2 - \frac{(2-x)^2}{2} & : 1.0 \leq |x| < 2.0 \\ 0 & : |x| \geq 2.0. \end{cases} \tag{1}$$

As reported in [13], a neural network model with an exponent in its activation function is about 5% slower than one without exponent term in the activation function. Although, SQ-RBF is similar in shape to Gaussian RBF as illustrated in Fig. 1 but in contrast to Gaussian kernel, SQ-RBF has smoother asymptotes which can be beneficial during training by using lesser hidden neurons. Furthermore, the approximate Gaussian kernels are shown in Table 1 is not computationally simple as SQ-RBF because of the presence of non-power of 2 divisors. SQ-RBF divisor is a power of two which is a shift operation and is not computationally expensive when compared with the former.

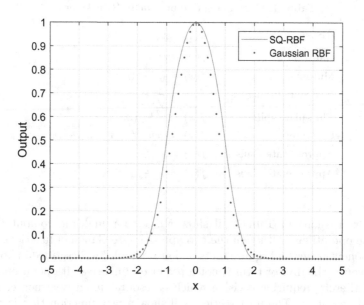

**Fig. 1.** The SQ-RBF and Gaussian RBF kernels

# 4 Experiments and Analysis

In this section, two experiments are presented to validate and demonstrate the effectiveness of the proposed function. Our proposed method is compared with the Gaussian function as the hidden layer activation function on series of benchmark tests and one new problem set. The benchmark tests as used in [10] are the SinE function approximation dataset, nonlinear dynamic system identification and finally the Mackey-Glass time series prediction dataset. Performance of the proposed function has been measured by the computational training time, the number of RBF kernel required, performance accuracy on both the training set and test set which is commonly used in literature to verify and demonstrate the effectiveness and performance of the proposed RBF kernel function. The performance accuracy was based on two criteria namely: number of neurons required to get to a specified MSE and the generalizability independent of the number of neurons. Therefore, the criteria for stopping experiment 1 is based on the MSE goal while for experiment 2 the stopping criteria are the number of RBF kernels. All the experiments were performed a total of 100 times, and the average was presented. We ran all the simulations in Matlab version 2017b Environment. Our system is a 64-bit windows 7 Enterprise with Intel Core i7 processor running at 3.40 GHz with 16.0 GB of installed memory (RAM).

## 4.1 SinE Function Approximation

A rapidly changing function named SinE is used in our first experiment to evaluate and compare the performance of the SQ-RBF and Gaussian RBF. SinE function is given as follows in (2)

$$y = 0.8 \exp(-0.2x) \sin(10x). \tag{2}$$

We defined a set $(x, y)$ of 500 data points for the training set and 150 data points for the testing samples. The value of the input $x$ was uniformly distributed in the interval $[0, 3]$ for the training and test samples.

**Experiment 1.** In this experiment, we defined a specified MSE on which the training stops, in order to know the number of RBF kernels (neurons) needed. The MSE goal is defined as the $0.01 * (\text{mean}(\text{variance}(\text{training target})))$. The kernel spread is given as 0.15. Table 2 shows the results. The results in bold perform better, and all the results are an average of 100 trials of network initialization.

**Experiment 2.** In this experiment, the stopping criteria are the number of RBF kernels. This number is defined as the size of the training samples. The same method was done as Experiment 1 to set up the remaining parameters. Table 3 shows the approximation results obtained with Gaussian and SQ-RBF kernels in terms of the error and computational time.

As shown in Tables 2 and 3, the time used in training an SQ-RBF based Network is smaller than Gaussian Function. The performance accuracy of the

**Table 2.** Experiment 1: Performance Comparison on SinE function

| RBF kernel | Training time (seconds) | Training MSE | Testing MSE | Number of neurons |
|---|---|---|---|---|
| Gaussian | 0.6189 | 0.8948 | **0.0060** | 90 |
| SQ-RBF | **0.5555** | **0.08934** | 0.0067 | **84** |

**Table 3.** Experiment 2: Performance comparison on SinE function

| RBF kernel | Training time (seconds) | Training MSE |
|---|---|---|
| Gaussian | 5.3898 | **2.93 × 10e−19** |
| SQ-RBF | **5.1141** | 6.75 × 10e−15 |

SQ-RBF is sometimes better and comparable to the Gaussian function. Finally, in terms of RBF kernels needed, SQ-RBF based Network requires less than Gaussian-based Network hence saving computational time as well as network size.

## 4.2   Nonlinear Dynamic System Identification

We evaluated the SQ-RBF on the identification of the non-linear dynamic system which is one of the commonly used benchmarks for approximation capability of RBF networks. The dataset was obtained from a device functioning as a hair dryer, for full description; readers are referred to [14]. There are ten inputs and one output in this model. We simulated the output using time $t = 1$ to 1000. We used 80% of the data as training and the remaining as test data.

**Experiment 1.** Using the same approach as session Sect. 4.1, we defined our target MSE as 0.01 ∗ (mean(variance(training target)) and the RBF kernel spread (width) as 1.8. Table 4 shows the average of 100 runs using the SQ-RBF and Gaussian RBFs. In terms of wall-clock time, SQ-RBF is about 14% faster computationally than the Gaussian kernel. Furthermore, SQ-RBF requires 26 fewer RBF kernels when compared with Gaussian without any detriment to the performance accuracy. This is a significant saving on both computational time and resources without any negative effect on the accuracy.

**Experiment 2.** Here the following parameters were changed due to the nature of this dataset. The kernel width remains the same; we constrain the numbers of RBF kernels to be 450 because the number of the target is large. The interest is on how long (in term of time) it takes to finish a training run to acquire the lowest error possible. Table 5 shows the results obtained.

**Table 4.** Experiment 1: Performance comparison on nonlinear dynamic system identification

| RBF kernel | Training time (seconds) | Training MSE | Testing MSE | Number of neurons |
|---|---|---|---|---|
| Gaussian | 5.5499 | 0.0065 | 0.0197 | 368 |
| SQ-RBF | **4.7737** | **0.0063** | **0.0162** | **342** |

**Table 5.** Experiment 2: Performance comparison on nonlinear dynamic system identification

| RBF kernel | Training time (seconds) | Training MSE |
|---|---|---|
| Gaussian | 11.0813 | $9.09 \times 10e-4$ |
| SQ-RBF | **10.82** | $\mathbf{7.49 \times 10e-4}$ |

### 4.3  Mackey-Glass Time Series Prediction

Mackey-Glass time series is one of the popular benchmark problems found in the literature for evaluating the performance of RBF networks. The Mackey-Glass series is non-periodic for $\tau$ greater than 17 and periodic otherwise. The initial values (for time $t$ less than $\tau$) were generated from a uniformly distributed pseudo-random numbers. Equation (3) is used to generate the remaining values. $\tau = 30$, 1500 data samples were extracted and used for training while 500 samples were used for testing.

$$\frac{ds}{dt} = \frac{0.9 * s(t) + (0.2 * s(t-\tau))}{1 + s(t-\tau)^{10}} \tag{3}$$

**Experiment 1.** Using the same approach as session Sect. 4.1, we defined our target MSE as $0.1 * (\text{mean}(\text{variance}(\text{training target}))$ and the RBF kernel spread (width) as 1.8. Table 6 shows the average of 100 runs using the SQ-RBF and Gaussian RBFs. In terms of wall clock time, SQ-RBF is about 8% faster computationally than the Gaussian kernel. Furthermore, SQ-RBF requires 19 fewer RBF kernels when compared with Gaussian without any detriment to the performance accuracy.

**Table 6.** Experiment 1: Performance comparison on Mackey-Glass time series prediction

| RBF kernel | Training time (seconds) | Training MSE | Testing MSE | Number of neurons |
|---|---|---|---|---|
| Gaussian | 20.2748 | 0.0092 | **0.0140** | 450 |
| SQ-RBF | **18.6580** | **0.0076** | 0.0173 | **431** |

**Experiment 2.** Here the following parameters were changed due to the nature of this dataset. The kernel width remains the same; we constrain the numbers of RBF kernels to be 450 because the number of the target is large. The interest is on how long (in term of time) it takes to finish a training run to acquire the lowest error possible. Table 7 shows the results obtained.

**Table 7.** Experiment 2: Performance comparison on Mackey-Glass time series prediction

| RBF kernel | Training time (seconds) | Training MSE |
|---|---|---|
| Gaussian | 20.1224 | 0.0091 |
| SQ-RBF | **19.5357** | **0.0060** |

### 4.4   Inverse Cosine Function Approximation

This dataset aims to use RBFNN to infer trigonometry angle. The repeating cosine wave is characterized with a magnitude of 1 and a phase offset of $pi$. We generated 2000 data points, using 1600 as training and the remaining 400 as the test set.

**Experiment 1.** Using the same approach as session 4.1, we defined our target MSE as 0.01 * (mean(variance(training target)) and the RBF kernel spread (width) as 0.15. Table 8 shows the average of 100 runs using the SQ-RBF and Gaussian RBFs. In terms of wall clock time, SQ-RBF is faster computationally than the Gaussian kernel.

**Table 8.** Experiment 1: Performance comparison on triangular function approximation

| RBF kernel | Training time (seconds) | Training MSE | Testing MSE | Number of neurons |
|---|---|---|---|---|
| Gaussian | 0.1603 | 0.0218 | 0.0213 | 3 |
| SQ-RBF | **0.1588** | **0.0189** | **0.0189** | 3 |

**Experiment 2.** Here the following parameters were changed due to the nature of this dataset. The kernel width remains the same, the number of epoch was left to be the number of samples in the dataset. The interest is on how long (in term of time) it takes to finish a training run to acquire the lowest error possible. Table 9 shows the results obtained. The numbers in bold shows where SQ-RBF network is better than the Gaussian RBF network. The SQ-RBF network is faster than the Gaussian RBF network without any decrease in accuracy as shown in Table 9.

**Table 9.** Experiment 2: Performance comparison on triangular function approximation

| RBF kernel | Training time (seconds) | Training MSE |
|---|---|---|
| Gaussian | 295.28 | $9.7208 \times -8$ |
| SQ-RBF | **260.94** | $\mathbf{4.4768 \times -9}$ |

# 5   Conclusion

In this paper, we proposed a computationally efficient and effective RBF kernel. This novel RBF kernel improves the training time without any detriment to the performance accuracy. We also recorded a consistent reduction in the number of RBF kernel required when using our function as to the Gaussian function. Two experiments were performed on 4 benchmarks and showed the generalizability of our function as well as convergence speed. On an average, we recorded a speed-up in training time of about 8% and decrease in the numbers of neurons to 10%. In the future, we aim to investigate the properties of this function that makes it use fewer neurons without causing adverse effect on the performance accuracy.

# References

1. Jouppi, N.P., et al.: In-Datacenter performance analysis of a tensor processing unit. In: Proceedings of the 44th Annual International Symposium on Computer Architecture, pp. 1–12. ACM, New York (2017)
2. Broomhead, D.S., Lowe, D.: Multivariable functional interpolation and adaptive networks. Complex Syst. **2**(3), 321–355 (1988)
3. Yojna, A., Singhal, A., Bansal, A.: A study of applications of RBF network. Int. J. Comput. Appl. **94**(2), 17–20 (2014)
4. Matthias, R., Eskofier, B.M.: An approximation of the Gaussian RBF kernel for efficient classification with SVMs. Pattern Recognit. Lett. **84**, 107–113 (2016)
5. Xu, B., Shen, F., Zhao, J., Zhang, T.: A self-adaptive growing method for training compact RBF networks. In: Liu, D., Xie, S., Li, Y., Zhao, D., El-Alfy, E.S. (eds.) ICONIP 2017. LNCS, vol. 10634, pp. 74–81. Springer, Cham (2017). https://doi.org/10.1007/978-3-319-70087-8_8
6. Reese, J., Zaranek, S.: GPU programming in MATLAB. MathWorks News and Notes, pp. 22–25 (2012)
7. Wuraola, A., Patel, N.: SQNL: a new computationally efficient activation function. IEEE International Joint Conference on Neural Network. IEEE (2018)
8. Duch, W., Jankowski, N.: Transfer functions: hidden possibilities for better neural networks. ESANN, pp. 81–94. De-facto, Brugge (2001)
9. Hoffmann, G.A.: Adaptive transfer functions in radial basis function (RBF) networks. In: Bubak, M., van Albada, G.D., Sloot, P.M.A., Dongarra, J. (eds.) ICCS 2004. LNCS, vol. 3037, pp. 682–686. Springer, Heidelberg (2004). https://doi.org/10.1007/978-3-540-24687-9_102
10. Meng, X., Rozycki, P., Qiao, J., Wilamowski, B.: Nonlinear system modeling using RBF networks for industrial application. IEEE Trans. Ind. Inform. **14**(3), 931–940 (2018)

11. Yanbing, L., Zhao, J., Xiao, Y.: C-RBFNN: a user retweet behavior prediction method for hotspot topics based on improved RBF neural network. Neurocomputing **275**, 733–746 (2018)
12. Włodzisław, D., Jankowski, N.: Survey of neural transfer functions. Neural Comput. Surv. **2**(1), 163–212 (1999)
13. Clevert, D.A., Unterthiner, T., Hochreiter, S.: Fast and accurate deep network learning by exponential linear units (ELUs). In: International Conference on Learning Representation (2016)
14. Lennart, L.: System Identification. Prentice-Hall, Boston (1998)

# A Model for Age and Gender Profiling of Social Media Accounts Based on Post Contents

Jan Kristoffer Cheng[(✉)], Avril Fernandez, Rissa Grace Marie Quindoza, Shayane Tan, and Charibeth Cheng

De La Salle University, Manila, Philippines
{jan_kristoffer_cheng,avril_fernandez,rissa_quindoza, shayane_tan}@dlsu.edu.ph, chari.cheng@delasalle.ph

**Abstract.** The growth of social networking platforms such as Facebook and Twitter has bridged communication channels between people to share their thoughts and sentiments. However, along with the rapid growth and rise of the Internet, the idea of anonymity has also been introduced wherein user identities are easily falsified and hidden. Hence, presenting difficulty for businesses to give accurate advertisements to specific account demographics. As such, this study searched for the best model to identify gender and age group of Filipino social media accounts through analyzing post contents. Two model structures for the classifier namely, the stacked/combined structure and the parallel structure were experimented on. Different types of features including those based on socio-linguistics, grammar, characters and words were considered. The results show that different model structures, features, feature reduction and classification algorithms apply to age classification and gender classification. For Facebook and Twitter, the best model for classifying age was Support Vector Classifier (SVC) with least absolute shrinkage and selection operator (Lasso) on a parallel model structure for Facebook, while a combined model structure is best for Twitter. For gender classification, the best model for Facebook used Ridge Classifier (RC), while the best model for Twitter used SVC, both utilizing Lasso on a parallel model structure. The features that were dominant in age classification for both Facebook and Twitter were word-based, socio-linguistic features and post time, while socio-linguistic features, specifically netspeak, were important in gender classification for both platforms. Based on the differences of the features affecting the performance of the models, Facebook and Twitter data must be analyzed separately as the posts found in these two platforms differ significantly.

**Keywords:** Profiling · Natural Language Processing
Machine learning · Information extraction · Language resources

© Springer Nature Switzerland AG 2018
L. Cheng et al. (Eds.): ICONIP 2018, LNCS 11302, pp. 113–123, 2018.
https://doi.org/10.1007/978-3-030-04179-3_10

# 1    Introduction

Businesses invest heavily on social media, with 92% of marketers claiming the importance of social media for their business [20], and social media ad spending expected to hit $35 billion this 2017 [1]. Businesses are highly concerned with the demographics of their target markets because it helps them understand their ideal customer and formulate their marketing strategies based on their target markets [22]. Online advertising also utilizes demographics in order to group audiences according to shared traits and focus an advertising campaign on these traits. However, businesses and advertisers who utilize the social media may find difficulty in accurately targeting account demographics due to the abundance of anonymous accounts that lack demographic information.

Gender and age are two of the basic information that compose the identity of a person, and are important for defining demographics. However, both can be easily hidden in social media. Exploring these information can be used in business intelligence where customer demographics are used for market studies in targeted advertising and product development.

Based on a 2016 report regarding digital statistics in the Philippines, Filipinos spend the most time on social media, with almost 4 h spent daily [1]. As one of the fastest growing countries in social media use and one of the top users of social media platforms [3], the demand to reduce anonymity in the country is high.

Language plays a vital part in profiling social media posts. Prior research have explored profiling social media posts written in English [4,11,12] and Spanish [9,16]. In the Philippines, the language in social media includes English, Filipino, Taglish, and the other Philippine languages and dialects. Taglish is the code-switching between the two most common language in the country, which are English and Tagalog. Posts written in Taglish normally follow the sentence structure of either English or Tagalog, then intersperses the sentence with words from the other language.

This paper focuses on profiling the age and gender of Filipino social media accounts. The research not only has significance in business but also provides insights on profiling using multilingual text, specifically in English, Filipino, and Taglish. The accounts to be profiled will be from the top active social networks used in the country, including Facebook with 26% and Twitter with 13% [1].

# 2    Related Works

Anonymity is a characteristic of cyberspace wherein one can hide one's identity [5]. The extent of existing anonymity on the Internet has already been proven in previous studies, such as the research conducted by Huffaker and Calvert [8] wherein they showed the high level of anonymity in teenagers. To determine if an anonymous account is owned by a teenager, an analysis on teenagers' use of language in an online setting was done by taking into account the amount of personal information disclosed by a teenager's blog, how they use emoticons in quantity and emotion displayed, how sexual identity is presented, and how words and its tone in a sentence's context are used to express feelings and ideas.

There have been numerous studies regarding age and gender classification, and the performance of their solutions differed primarily based on the selected feature sets and learning technique. Most of those feature sets are features under syntactical and grammatical features and socio-linguistic features such as function words both utilized by Cheng et al. [4] and Newman et al. [12], f-measures for formality measurement and POS sequence patterns both introduced by Mukherjee and Liu [11], stylistic features used by Rangel and Rosso [16], and other features such as POS n-grams.

There are also existing researches that explore different model structures for simultaneously profiling both age and gender. Rao et al. [17] explored the performance of a stack model using the predictions from both socio-linguistic and n-gram features along with their prediction weights and discovered that while the socio-linguistic model performed better than the n-gram model, the stacked model performed the best at an accuracy of 72.33%.

Various techniques and statistical classifiers were utilized by previous studies. Newman et al. [12] categorized the words of text written by men and women and analyzed these word categories using a Multivariate Analysis of Variance. The results show significant gender differences in language use and discovered the gender difference size was approximately twice as large for function words as for either nouns or verbs. Other studies such as Cheng et al. [4], Mukherjee and Liu [11], and Rao et al. [17] explored the use of Support Vector Machines (SVM) as one of their learning algorithms aside from Naive Bayes and Adaboost decision tree algorithm where the results showed SVM performing better in the gender classification process with an accuracy of at least 85% among other learning algorithms.

In terms of age identification, Nguyen et al. [13] noted that there are different ways to classify age: age groups, life stages, and exact age. Additionally, Rangel and Rosso [16] observed that stylistic features performed better for age identification than gender identification.

## 3  Methodology

The data was collected from both Facebook and Twitter. However, for the Facebook data, corpus from the My Personality Project [21] was utilized. Then, the data underwent Feature Retrieval where the system obtained the different features included in the posts, each of which was given a corresponding value. Thereafter, it went through an iterative process of Dimension Reduction, Statistical Classification and Evaluation. The best combination of dimension reduction algorithm and statistical classifier based on the chosen evaluation metric was be used in building the final computational model.

## 3.1  Data Collection

The corpus was collected from both Twitter and Facebook. For Twitter, a website was used in gathering data from each user. Once the user authorizes access to the account, the system collects the tweets of the user. For Facebook, the data was collected from the My Personality Project [20]. A total of 250 accounts from this dataset were considered, which were chosen based on the 250 accounts from Twitter to balance the categories.

The collected posts and tweets are then cleaned by removing unnecessary elements such as email addresses and anonymized by replacing the names mentioned to "USER" to protect the privacy of the accounts.

A total of 500 accounts were used in building the corpus, 250 from Facebook and another 250 from Twitter. The dataset for Facebook and Twitter had an almost similar distribution for age and gender. Both Facebook and Twitter datasets have 133 female accounts and 117 male accounts, for a total of 250 accounts. For the age group distribution for Twitter and Facebook, both platforms have more accounts collected for the age group of 18–24 years old, while relatively equal number of accounts for the age groups of 25–34 years old and 35–44 years old. For the latter age group, 55–64 years old, only a few accounts were collected due to the few number of people from this age groups who use these social media platforms.

## 3.2  Feature Retrieval

We extracted several features from the documents organized into the following feature sets: character based, word-based, structure-based, sociolinguistics-based, bag-of-words, post time, and POS sequence patters.

**Character-Based Features.** Previous research have provided evidence that suggests the correlation between the character's frequency and an individual's age or gender [2,17]. Character-based features considered in this study include the total number of lowercase and uppercase letters, numbers, spaces, special characters (e.g. %, $, #, &), repeated alphabetical characters and punctuation marks.

**Word-Based Features.** Frequencies such as the ratio of the number of unique words and the total number of words, the length of the words, and the number of words with repeated letters etc., are good indicators of age [16]. Word-based considered in this study include word-count features, and several statistical measures which are Yule's K measure, Simpson's D measure, Sichel's S measure, Honore's R measure, and Entropy measure. Linguistic Inquiry and Word Count (LIWC) [15] features were also considered. However, since there are multiple categories in LIWC, only those under word-based were used including Summary Dimensions, Punctuation Marks, and Informal Language, which amounts to 25 features.

**Structure-Based Features.** Cheng et al. [4] used structural features to identify user gender. The structure-based features that were considered were the total number of lines, sentences, and paragraphs, average number of sentences per paragraph, average number of words per paragraph, average number of words per sentence, number of sentences beginning with uppercase and number of sentences beginning with lowercase.

**Sociolinguistics-Based Features.** Nguyen et al. [13] and Rao et al. [17] used sociolinguistic features that reflect the culture and environment that a user is in when they post. This study considered function words, disfluent and agreeing words, contextual features and emoticons or emojis.

**Bag-of-Words with Term Frequency-Inverse Document Frequency (TF-IDF).** TF-IDF was used to determine the relevant words in identifying age and gender. Nguyen et al. [13] determined that younger people used more informal words like lol, hmm, like and kinda, and Corney et al. [6] also suggests that females make more use of emotionally intensive adjectives and adverbs such as terribly, so, and awfully. Mukherjee and Liu [11] also identified that word formality affects gender identification.

**Part-of-Speech Sequence.** Mukherjee and Liu [11] introduced POS sequences that are retrieved from the POS sequence-pattern mining algorithm that captures all part of speech sequences that satisfy certain thresholds. Since the system will involve multilingual analysis, different taggers for Filipino and English will be used. For consistency in analysis, Filipino POS tags will be mapped to its equivalent English POS tags if applicable; otherwise, the tag will be retained and used.

To identify which tool to utilize, the text will undergo language detection. English tags will be used for English text and mapped Filipino-to-English tags will be used for Filipino text. If Taglish or unknown, the post will be separated into sentences and each sentence will undergo language detection for appropriate tagging.

**Post Time.** Burger and Henderson [2] determined that younger people post more often between 9PM and midnight, while those aged 24 and above post more often during the afternoon.

**Links.** Marquardt et al. [9] and Nguyen et al. [13] studied the relevance of the number of links that appear in their datasets. Websites also have categories and keywords tags that summarize their contents, and these may show biases for certain gender or age group to some contents and topics.

### 3.3   Model Production

Since there are numerous features, features that may not increase the classifier's performance were removed. Three different methods of dimension reduction techniques were utilized, namely Mutual Information, $x^2$ statistics, Truncated Singular Value Decomposition (SVD), and LASSO. After obtaining the important features, the statistical classifier will determine patterns in building the computational model in classifying gender and age. In building the model, supervised machine learning techniques were utilized because it is a classification problem. In related works, Support Vector Machines is the usual computational technique that identifies patterns from training data to apply in classifying future data [4]. However, other computational techniques such as Naive Bayes [7,11], and Decision Tree Algorithm [4,10,14] were also used and tested. Ridge Regression was used in the study of Sap et al. [18] and Schwartz et al. [19] for age identification, however since the age is defined in age brackets, and thus becomes a classification rather than a regression problem, Ridge Classifier was used instead. In this study, four statistical classifiers were considered: SVM, Naive Bayes, Decision Trees, and Ridge Classier, a classier based on Ridge Regression.

Multiple models will be made to test different combinations of different dimension reduction algorithms and statistical classifiers. In addition, because there are two outputs, age group and gender, multiple model structures will be tested, namely Age to Gender Stacked Model Structure, Gender to Age Stacked Model Structure, Parallel Model Structure, and Combined Model Structure. These were experimented on to observe if it affects the performance of the classifier.

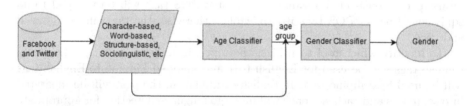

**Fig. 1.** Age to gender stacked model structure

The age to gender stacked model structure utilizes age as an additional feature for identifying gender of accounts. The two classifiers can be different algorithms (Fig. 1).

Meanwhile, the gender to age stacked model structure utilizes gender as an additional feature for identifying the gender of accounts. The two classifiers can also be different algorithms (Fig. 2).

The parallel model structure utilizes different feature sets and classifiers for age and gender identification. This is used when the classifier that performs best for gender is different from the classifier that performs best in age identification (Fig. 3).

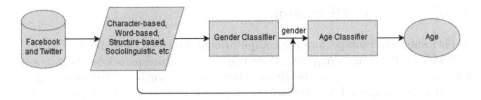

**Fig. 2.** Gender to age stacked model structure

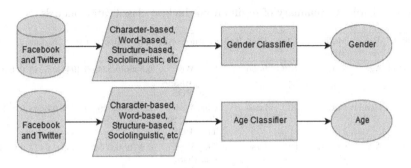

**Fig. 3.** Parallel model structure

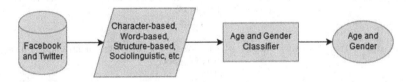

**Fig. 4.** Combined model structure

Finally, the combined model structure uses the same feature set and classifier for both age and gender (Fig. 4).

### 3.4 Evaluation

Computing accuracy and kappa may be straightforward, but the computation for recall and precision in a multiclass classification are different. In age classification, there are five age groups which means there are five classes that the classifier should learn to predict. In this case, recall and precision were computed separately for each class. The resulting values were averaged to the number of classes. The models with the highest F-measure were identified as baseline models. Since there were three data sets, three model structures, and two classification types, a total of 18 baseline models were identified. For each data set and classification type, the F-measure of the baseline models of the different model structures for each data set were compared. The model with the highest F-measure was identified as the best model.

# 4    Results and Discussion

Table 1 shows the top performing computational models for Facebook, Twitter, and Merged datasets, along with their respective feature reduction techniques, feature sets and statistical classifiers. They were primarily ranked based on f-measure then kappa, but their accuracy, precision, and recall values are also written in detail.

**Table 1.** Summary of results from the best classification models

| Dataset | Type | Mod. Struc. | Class. | Dim. Red. | Feature sets | Acc. | Rec. | Prec. | Kappa | F-meas |
|---------|------|-------------|--------|-----------|--------------|------|------|-------|-------|--------|
| Facebook | Age | Par. | SVC | lasso | Chr, Wrd, Str, Frq, Soc, Tim, POS | 0.8360 | 0.5349 | 0.4944 | 0.5243 | 0.4937 |
| Facebook | Gender | Par. | RC | lasso | Chr, Wrd, Str, Frq, Soc, Tim, POS | 0.8240 | 0.8293 | 0.8221 | 0.6443 | 0.8211 |
| Twitter | Age | Comb. | SVC | lasso | Chr, Wrd, Str, POS, Soc, Tim, POS | 0.8640 | 0.6582 | 0.6233 | 0.6427 | 0.6069 |
| Twitter | Gender | Par. | SVC | lasso | Chr, Wrd, Str, POS, Soc, Tim, POS | 0.8400 | 0.8542 | 0.8370 | 0.6775 | 0.8371 |
| Merged | Age | Par. | RC | lasso | Chr, Wrd, Str, Frq, Soc, Tim | 0.7620 | 0.3305 | 0.2273 | 0.2232 | 0.2802 |
| Merged | Gender | Par. | SVC | $x^2$ | Soc | 0.7760 | 0.7842 | 0.7777 | 0.5519 | 0.7742 |

Chr: character-based, Wrd: word-based, Str: structure-based, Frq: TF-IDF, Soc: sociolinguistic, Tim: time, POS: part-of-speech sequence

For both Facebook and Twitter, the top performing model for age utilizes SVC with lasso on a parallel model structure for Facebook while on a combined model structure for Twitter. This indicates that on top of the features used in age classification for Twitter, the model performed better with the extra features used in gender classification.

For both Facebook and Twitter, the top performing model for gender classification utilizes lasso as its dimensionality reduction technique. The classifier that gained a high f-measure for Facebook is RC, while it is SVC for Twitter, both on a parallel model structure.

Meanwhile, for the merged dataset, the top performing model for gender utilizes SVC with $x^2$ on a parallel model structure. However, this model still has a lower kappa than that of the top model for separate Facebook and Twitter datasets. There was no model that had a sufficiently high enough kappa for age prediction. The results of the merged dataset, as seen from the charts above, can be irregular and its f-measures are lower than the separate datasets. This can be attributed to the difference of the nature of Twitter tweets from Facebook posts. While tweets are usually short and expressive, filled with emoticons or emojis, posts are usually long and of the informative type. Tweets and posts will therefore have different features, and combining both datasets results in poor predictive power. It is difficult to find common features between the two of them aside from the frequency of the words present in the posts and tweets themselves.

Analyzing the features retained by Facebook and Twitter for age, both Facebook and Twitter retain several word based and socio-linguistic features, especially for Facebook where its socio-linguistic features included words like "officemate", "pamangkin" (Filipino genderless noun to mean either nephew or niece) and "nephew" (contextual features) which has been determined to be a good indicator for age. Twitter also retained more post time features which is also indicative of age. Both datasets also retained informal words. The Twitter model also retained the username_handle feature, which shows that this unique feature to Twitter is helpful in classifying age, alongside a frequency of links in tweets.

Analyzing the features of Facebook and Twitter for gender on the other hand, both datasets retained the socio-linguistic features of hobbies like "cosplay", "art" and "eating" and contextual features like "wife", "hubby" and "gf". The main difference in Facebook and Twitter for gender are that Twitter has netspeak and net abbreviation features while Facebook does not. This perhaps can be attributed to the nature of Twitter because it has a limited amount of characters per tweet. Twitter also retained emoji features while Facebook does not, but as mentioned at the start of the chapter, this can be attributed to the fact that the Facebook data used was from 2012 and emojis were not popular then. Twitter's most used emojis were the hearts ( <3 ), the smileys ( :) ) and the cloud with rain emoji.

Given the differences of the features retained by both, it is therefore better to have separate models for Facebook and Twitter because there is a considerable amount of features that are unique to each data set.

## 5   Conclusion

The corpus was composed of 250 Facebook accounts, gathered from My Personality project, and 250 Twitter accounts, gathered through a data gathering website and crawling accounts of public figures in Twitter. For each dataset, namely Facebook, Twitter, and a merge of both, 5,904 models were trained. These models differed in dimensionality reduction techniques and parameters, statistical classifiers, and model structures. From these models, a best model for age and another for gender for each dataset was determined. The results were first divided into the different model structures. Afterwards, they were sorted in descending f-measure order and compared to each other. The model with the highest f-measure is the best model.

# References

1. AMEX iSUPPORT: 10 eye-opening facts about social media in PH (2016). http://isupportworldwide.com/blog/archive/socialmediaphilippines/. Accessed 14 Feb 2017
2. Burger, J.D., Henderson, J.C.: An exploration of observable features related to blogger age. In: Spring Symposium: Computational Approaches to Analyzing Weblogs, pp. 15–20. AAAI (2006)
3. Chaffey, D.: Global social media research summary 2016 (2016). https://www.smartinsights.com/social-media-marketing/social-media-strategy/new-global-social-media-research/. Accessed 04 Sept 2017
4. Cheng, N., Chandramouli, R., Subbalakshmi, K.P.: Author gender identifcation from text. Digit. Investig. **8**(1), 78–88 (2011)
5. Choi, J.Y., Lim, G.G., Woo, M.N.: A study on the anonymity perceptions impacting on posting malicious messages in online communities. In: Proceedings of PACIS 2016 (2016)
6. Corney, M., Anderson, A., de Vel, O., Mohay, G.: Gender-preferential text mining of e-mail discourse. In: 18th Annual Computer Security Applications Conference on Proceedings, Las Vegas, pp. 282–289 (2002)
7. Hernandez, D., Guzman-Cabrera, R., Reyes, A., Rocha, M.: Semantic-based features for author profiling identification. In: Working Notes for CLEF 2013 Conference, Valencia (2013)
8. Huffaker, D.A., Calvert, S.L.: Gender, identity, and language use in teenage blogs. J. Comput. Mediat. Commun. **10**(2), 00–00 (2005)
9. Marquardt, J., Farnadi, G., Vasudevan, G., Moens, M.F., Davalos, S., Teredesai, A., De Cock, M.: Age and gender identification in social media. In: Proceedings of CLEF 2014 Evaluation Labs, pp. 1129–1136 (2014)
10. Mechti, S., Jaoua, M., Belguith, L.H., Faiz, R.: Author profiling using style-based features. In: Notebook for PAN at CLEF 2013 (2013)
11. Mukherjee, A., Liu, B.: Improving gender classification of blog authors. In: Proceedings of the 2010 conference on Empirical Methods in natural Language Processing, pp. 207–217. Association for Computational Linguistics (2010)
12. Newman, M.L., Groom, C.J., Handelman, L.D., Pennebaker, J.W.: Gender differences in language use: an analysis of 14,000 text samples. Discourse Process. **45**(3), 211–236 (2008)
13. Nguyen, D., Smith, N.A., Rose, C.P.: Author age prediction from text using linear regression. In: Proceedings of the 5th ACL-HLT Workshop on Language Technology for Cultural Heritage, Social Sciences, and Humanities, pp. 115–123. Association for Computational Linguistics (2011)
14. Patra, B.G., Banerjee, S., Das, D., Saikh, T., Bandyopadhyay, S.: Automatic author profiling based on linguistic and stylistic features. In: Notebook for PAN at CLEF (2013)
15. Pennebaker, J., Booth, R., Boyd, R., Francis, M.: Linguistic Inquiry and Word Count: LIWC2015. Pennebaker Conglomerates, Austin (2015)
16. Rangel, F., Rosso, P.: Use of language and author profiling: identification of gender and age. In: Natural Language Processing and Cognitive Science, vol. 177 (2013)
17. Rao, D., Yarowsky, D., Shreevats, A., Gupta, M.: Classifying latent user attributes in Twitter. In: Proceedings of the 2nd International Workshop on Search and Mining User-Generated Contents, pp. 37–44. ACM (2010)

18. Sap, M., Park, G., Eichstaedt, J.C., Kern, M.L., Stillwell, D., Kosinski, M., Ungar, L.H., Schwartz, H.A.: Developing age and gender predictive lexica over social media (2014)
19. Schwartz, H.A., Eichstaedt, J.C., Kern, M.L., Dziurzynski, L., Ramones, S.M., Agrawal, M., Shah, A., Kosinski, M., Stillwell, D., Seligman, M.E., Ungar, L.H.: Personality, gender, and age in the language of social media: the open-vocabulary approach. PLoS ONE **8**(9), e73791 (2013)
20. Stenzler, M.A.: How marketers are using social media to grow their businesses (2016). http://www.socialmediaexaminer.com/wp-content/uploads/2016/05/SocialMediaMarketingIndustryReport2016.pdf. Accessed 24 Aug 2017
21. Stillwell, D., Kosinski, M.: (2016) Mypersonality project, http://mypersonality.org/wiki/doku.php. Accessed 21 Feb 2017
22. Understanding Analytics: Understanding the importance of demographics in marketing (2015). http://upfrontanalytics.com/understanding-the-importance-of-demographics-in-marketing/. Accessed 02 Sept 2017

# A Wiener Causality Defined
# by Relative Entropy

Junya Chen[1], Jianfeng Feng[2], and Wenlian Lu[2(✉)]

[1] School of Mathematical Sciences, Fudan University,
Shanghai, China
junyachen15@fudan.edu.cn
[2] Institute of Science and Technology for Brain-Inspired Intelligence,
Fudan University, No. 220 Handan Road, Shanghai, China
jianfeng64@gmail.com, wenlian@fudan.edu.cn

**Abstract.** In this paper, we propose a novel definition of Wiener causality to describe the intervene between time series, based on relative entropy. In comparison to the classic Granger causality, by which the interdependence of the statistic moments beside the second moments are concerned, this definition of causality theoretically takes all statistic aspects into considerations. Furthermore under the Gaussian assumption, not only the intervenes between the co-variances but also those between the means are involved in the causality. This provides an integrated description of statistic causal intervene. Additionally, our implementation also requires minimum assumption on data, which allows one to easily combine modern predictive model with causality inference. We demonstrate that REC outperform the standard causality method on a series of simulations under various conditions.

**Keywords:** Granger causality · Time series
Relative entropy causality · Transfer entropy

## 1 Introduction

Causality, which describes the inferring interactions from data, has been of long-term interest to both statisticians and scientists in diverse fields [1]. In general, causality is defined by the models containing families of possible distributions of the variables observed and appropriate mathematical descriptions of causal structures in the data [2]. One of the most popular approaches, introduced originally by Wiener [3], has a philosophic idea as follows: Given sets of interdependent variables $x$ and $y$, it is said that $y$ causes $x$ if, in certain statistical sense, including possible distribution and model, $y$ assists in predicting the future of $x$, in comparison to the scenario that $x$ already predicts its own future.

This idea was formalised in terms of multivariate auto-regression (MVAR) by Granger, and importantly, in term of identifying a causality interaction, the measurement and statistic inference was proposed based on MVAR by comparing the variances of the residual errors with and without considering $y$ in

© Springer Nature Switzerland AG 2018
L. Cheng et al. (Eds.): ICONIP 2018, LNCS 11302, pp. 124–133, 2018.
https://doi.org/10.1007/978-3-030-04179-3_11

the prediction of $x$ [4]. Information theory provided another important quantity, transfer entropy [5], to measure directed information transfer between joint processes in complex systems [6] that is theoretically model-free. Even though transfer entropy is not identical to identifying a physically instantiated causal interaction in a system in terms of prediction but of resolution of uncertainty: the difference of the conditional entropy of $x$ with and without considering $y$, it has, however, been proved to be equivalent to Wiener-Granger causality in MVAR with Gaussian assumptions [7]. Thus, both definitions can be physically regarded the statistic aspect identified by comparing the model with and without considering the intervene of $y$ and have provided deep insights into the directed connectivity of systems in a variety of fields [8], for example, in neuroscience [9,10].

However, the differences of statistic aspects, such as the conditional covariances in Granger causality and conditional entropy in transfer entropy, are insufficient to depict the distance in statistic sense, between with and without considering intervene from $y$. Two statistic distributions may not coincide if they share the same co-variance and/or entropy. Hence, there might be the case that the measurements of causality cannot deliver the exact inferring interactions between processes. In terms of prediction, the intervene from the first moment, the mean, of $y$ should not be excluded, according to Wiener's original definition of causality, unlike what Granger did. In other words, removing the mean from the time series may cause ignorance of the causal intervenes between them.

In this paper, we propose the relative entropy causality (REC), based on Wiener's original definition of causality, including all statistic aspects. Instead of the linear and stationary assumption in Granger casusality, in REC we extend it and fit it into the nonlinear and nonstationary cases. Moreover, it turns out to be very convenient in analyzing causality combined with the nonlinear predictive models like XGBoost [11]. We show that REC outperforms standard methods like transfer entropy and Granger causality under different conditions of simulation experiments.

## 2   Notations

### 2.1   Residual Errors

The history of a time series $x$ up to $p$ lags is denoted by $x_t^p = [x_{t-1}, \cdots, x_{t-p}]$, also $x^p$ for simplicity if it is stationary.

We denote the residual errors in predicting $x$ by

$$\epsilon = x - \tilde{x} \tag{1}$$

where $\tilde{x}$ is the predictor of $x$ under certain statistic model based on $x_t^p = [x_{t-1}, \cdots, x_{t-p}]$, when as well as possibly $y_t^q = [y_{t-1}, \cdots, y_{t-q}]$.

In this paper, we will consider two predictors: $\tilde{x} = \phi(x^p, \theta)$ and $\tilde{x} = \psi(x^p, y^q, \vartheta)$ (or $\tilde{x} = \phi(x^p, z^r, \theta)$ and $\tilde{x} = \psi(x^p, y^q, z^r, \vartheta)$ when considering the conditional causality). The first one only depends on the history of $x$ itself while the second depends on the history of $y$ plus $x$'s own. The realisation of these predictors are based on the estimation of the parameters $\theta$ and $\vartheta$, which depends on the samples of $x$, denoted by $X = [X_1, \cdots, X_T]$, and the samples of $y$, denoted by $Y = [Y_1, \cdots, Y_T]$ plus $X$ respectively.

## 2.2   KL Divergence

A natural quantity to describe the distance between two distributions is the celebrated Kullback-Leibler (KL) divergence that in fact defines a relative entropy between them [12,13]. Let $P(\cdot)$ and $Q(\cdot)$ be two probability distributions on a common state space $\{\Omega, \mathcal{F}\}$ with state space $\Omega$ and $\sigma$-algebra $\mathcal{F}$. Their KL divergence from $P(\cdot)$ to $Q(\cdot)$, namely, the relative entropy of $P(\cdot)$ over $Q(\cdot)$ can be formulated as

$$D_{KL}(P \parallel Q) = \int_\Omega \ln \left( \frac{dP(x)}{dQ(x)} \right) P(dx). \tag{2}$$

An essential nature of relative entropy is: $D_{KL}(P \parallel Q) = 0$ if and only if $P = Q$. But, KL divergence is asymmetry, i.e., $D_{KL}(P \parallel Q) \neq D_{KL}(Q \parallel P)$ in general. Also, we can denote the *conditional KL divergence* between two conditional probability $P(u|v)$ and $Q(u|z)$ for random variables (vectors) $v$ and $z$ as follows [14]:

$$D_{KL}[P(\cdot|v) \parallel Q(\cdot|z)] = \int_{\Omega_{v,z}} \int_\Omega \ln \left( \frac{dP(u|v)}{dQ(u|z)} \right) P(du|v) W(dv, dz) \tag{3}$$

where $W(dv, dz)$ denotes the joint probability distribution of $(v, z)$ on their joint state space $\Omega_{v,z}$. It can be seen that $D_{KL}[P(\cdot|v) \parallel Q(\cdot|z)] = 0$ if and only if $P(\cdot|v) = Q(\cdot|z)$ with probability one in the sense of $W(dv, dz)$ when $v = z$.

## 3   REC: Relative Entropy Causality

We come back to Wiener's original definition of causality in terms of prediction by the difference of the residual errors in the statistic model with and without considering the inference of $y$.

**Definition 1.** The causality intervene from $y$ to $x$ is defined as the following conditional relative entropy:

$$REC_{y \to x} = D_{KL} \left[ P(\cdot|x^p \oplus y^q) \parallel Q(\cdot|x^p) \right]. \tag{4}$$

named *relative entropy (RE) causality*. $P(\epsilon|x^p \oplus y^q)$ is the (stationary) conditional probability distribution of the residual error for the predictor $\psi$, depending on $x$ and $y$'s history, where the symbol $\oplus$ stands for the union of two random vectors, and $Q(\epsilon|x^p)$ is that for predictor $\phi$, only depending on $x$'s history.

Take the MVAR model for example. Let $x_t$ and $y_t$ be two real random processes of one-dimension[1] in discrete time. Consider the following two MVAR models:

$$x_t = x_t^p \cdot \tilde{a} + \tilde{\epsilon}_t \tag{5}$$

$$x_t = x_t^p \cdot a + y_t^q \cdot b + \epsilon_t \tag{6}$$

with the lag orders $p$ and $q$, and $t = 1, 2, \cdots, T$, where $a, \tilde{a} \in \mathbb{R}^p$, $b \in \mathbb{R}^q$, $\tilde{\epsilon}_t, \epsilon_t \in \mathbb{R}$ and $T$ is the time duration. These two models give two predictors $\phi(x_t^p, \tilde{a}) = x_t^p \cdot \tilde{a}$ and $\psi(x_t^p, x_t^q, a, b) = x_t^p \cdot a + y_t^q \cdot b$. At this stage, we assume that all time series are stationary (w.r.t measures). We should highlight that we do not include a constant term in the formulas of MVAR, which, however, is partially absorbed in the residual terms $\epsilon$ and $\tilde{\epsilon}$. In other words, $\epsilon$ and $\tilde{\epsilon}$ may have nonzero means in these models.

To specify this definition, we consider the scenario of one dimension by further assuming that all random variables are stationary (w.r.t measures), ergodic and Gaussian, and conduct parameter estimation by some approach, for example, the least mean square (LMS) approach or the maximum likelihood approach. Consider a column random vector $u$ and a random matrix $v$ of the identical column dimension. Let $\mathbb{E}(u)$ be the expectation of $u$, $\Sigma(u) = \mathbb{E}(uu^\top) - \mathbb{E}(u)[\mathbb{E}(u)]^\top$ be the covariance of $u$, $\Sigma(u, v) = \mathbb{E}(uv^\top) - \mathbb{E}(u)\mathbb{E}(v^\top)$ be the covariance between $u$ and $v$, $\Sigma(u|v) = \Sigma(u) - \Sigma(u, v)\Sigma(v)^{-1}\Sigma(u, v)^\top$ be the partial covariance [7], and $\mu(u|v) = \mathbb{E}(u) - E(v)\Sigma(v)^{-1}\Sigma(v, u)$.

Let $X^p = [x_p^p, \cdots, x_T^p]$ and $Y^q = [y_q^q, \cdots, y_{T-1}^q]$ be the samples of the history of $x$ and $y$ picked from $X$ and $Y$ respectively. Then, by the least means square approach, suppose that the parameters $(\tilde{a}, a$ and $b)$ can be asymptotically perfectly estimated, i.e., the estimated values of $\tilde{a}$, $a$ and $b$ dependent of the samples converge to their theoretical value as the number of samples goes to infinity. Hence, we can obtain the asymptotic conditional distribution of $\tilde{\epsilon}$ and $\epsilon$ with respect to $x^p$ and $x^p \oplus y^q$ as follows:

$$\epsilon|x^p \oplus y^q \sim N_0 \left[\mu(x|x^p \oplus y^q), \Sigma(x|x^p \oplus y^q)\right]$$
$$\tilde{\epsilon}|x^p \sim N_1 \left[\mu(x|x^p), \Sigma(x|x^p)\right]. \tag{7}$$

Due to the limit of space, we omit the calculations which is not particularly difficult.

Thus, the RE causality from $y$ to $x$ in MVAR (5, 6) is:

$$REC_{y \to x} = D_{KL}(N_0 \parallel N_1)$$
$$= \frac{1}{2} \left\{ \text{tr}[\Sigma(x|x^p)^{-1}\Sigma(x|x^p \oplus y^q)] + [\mu(x|x^p \oplus y^q) - \mu(x|x^p)]^\top \right.$$
$$\Sigma(x|x^p)^{-1}[\mu(x|x^p \oplus y^q) - \mu(x|x^p)]$$
$$\left. -n - \ln \frac{\det[\Sigma(x|x^p \oplus y^q)]}{\det[\Sigma(x|x^p)]} \right\}. \tag{8}$$

---

[1] For the case of high dimensions, the similar results can be derived by the same fashion.

It can be seen that $REC_{y \to x} = 0$ if and only if $\Sigma(x/x^p) = \Sigma(x/x^p \oplus y^q)$ and $\mathbb{E}(x/x^p \oplus y^q) = \mathbb{E}(x/x^p)$. In comparison with Granger causality, in the MVAR model and under Gaussian assumption, the coefficients are estimated by the maximum likelihood approach. It can be defined as[2]

$$GC_{y \to x} = \ln\{\det[\Sigma(x|x^p)]/\det[\Sigma(x|x^p \oplus y^q)]\}.$$

Hence, $GC_{y \to x} = 0$ if and only if $\Sigma(x|x^p) = \Sigma(x|x^p \oplus y^q)$, namely, the covariance contribution of $Y^q$ can be totally replaced by $X^q$. Therefore, in this scenario, the RE causality contains the $y$'s contribution to predicting $x$ in both covariance and mean, but the Granger causality only includes its covariance contribution.

**Definition 2.** The relative entropy causality from $y$ to $x$ at time $t$ as the conditional KL divergence between two conditional probability distributions at time $t$: $REC_{y \to x}(t) = D_{KL}\left[P_t(\cdot|x_t^p \oplus y_t^q) \parallel Q_t(\cdot|x_t^p)\right]$, where $P_t(\cdot|x_t^p \oplus y_t^q)$ and $Q_t(\cdot|x_t^p)$ be the conditional distribution of $\epsilon_t$ at time $t$ with respect to $x_t^p \oplus y_t^q$ and that of $\tilde{\epsilon}_t$ at time $t$ with respect to $x_t^p$. The global RE causality as the average across the infinite time duration $[t_0, \infty)$:

$$REC_{y \to x} = \lim_{T \to \infty} \frac{1}{T} \sum_{t=t_0}^{t_0+T-1} REC_{y \to x}(t)dt, \tag{9}$$

if the limit exists and is independent of $t_0$, for example, the general Oscelec ergodicity conditions are satisfied [15].

## 4    Experiments

### 4.1    Stationary MVAR

To compare the RE causality with Granger causality in the MVAR model with Gaussian assumption (hence transfer entropy is equivalent to Granger causality), we consider one-order and one-dimensional version of (6) that generates the series $x_t$ as:

$$x_t = a \cdot x_{t-1} + b \cdot y_{t-1} + \epsilon_{1,t} \tag{10}$$

where $a = 1/2$ and $b = -1/3$, and $\epsilon_{1,t}$ is a white Gaussian noise of zero mean and variance $\sigma_1^2$ independent of all $y_t$ and all $x_t$. First, we consider the case that $y_t$ is stationary: $y_t = c \cdot y_{t-1} + \epsilon_{2,t}$ where $c = 2/3$ and $\epsilon_{2,t}$ is a white Gaussian noise of zero mean and variance $\sigma_2^2$ independent of all $y_t$ and all $x_t$.

The value of RE causality from $y$ to $x$, denoted by $\widehat{REC}_{y \to x}$ can be calculated by estimating the means and variances in (5) and (6) by maximum likelihood and employing equation (8). We use the bootstrap approach for statistic inference. Under the null hypothesis $b = 0$, we sample a series of $\tilde{x}_t$ by (10) with the estimated values $\tilde{a}$ of $a$ and the estimated noise variance $\tilde{\sigma}_1^2$. Both are estimated

---

[2] Also as $GC_{y \to x} = \ln\{\mathrm{tr}[\Sigma(x|x^p)]/\mathrm{tr}[\Sigma(x|x^p \oplus y^q)]\}.$

by the maximum likelihood. Then, the estimated values of $\widehat{REC}^k_{y \to x}$ can be obtained from $\tilde{x}_t$ and $y_t$ by Eq. (8) for the $k$-th realisation. Repeat this phase for $N = 1000$ times and get a collection of $\widehat{REC}^k_{y \to x}$, $k = 1, \cdots, N$. To avoid a zero p-value, we let $\widehat{REC}^0_{y \to x} = \widetilde{REC}_{y \to x}$. Thus, the p-value of the RE causality is calculated by $p_{REC} = \#\{k : \widehat{REC}^k_{y \to x} \geq \widetilde{REC}_{y \to x}\}/(N + 1)$. Also, we calculate the RE causality from $x$ to $y$ by the above fashion. In comparison, the Granger causality and its p-value is calculated by its definition and $F$-test for both directions. When the p-value is less than a pre-given threshold $p_{th}$, we claim that there is a causal intervene. All the above are overlapped for $M = 100$ times and the true positive (TP) rate, i.e., the ratio (among $M$ overlaps) that the directed causal intervene from $y$ to $x$ is correctly identified, and the false positive (FP) rate, i.e., the ratio (among $M$ overlaps) that the directed causal intervene from $x$ to $y$ is incorrectly identified, are used to evaluate the performance of the causality definitions (Table 1).

**Table 1.** Performance comparison on stationary MVAR between Granger causality (Transfer entropy) and RE causality.

|  | True positive | False positive |
|---|---|---|
| Granger causality | 100% | 0 |
| RE causality | 100% | 0 |

The above results show that they have similar performances to identify the intervene from $y$ to $x$ and exclude that from $x$ to $y$ for diverse p-value thresholds and model noise variances.

## 4.2  Non-stationary MVAR

To compare the RE causality with Granger causality in the nonstationary MVAR model, we define Granger causality in non-stationary case by the same token. Specified in the following model, by estimating the regressive parameters, one can define Granger causality and transfer entropy at each time (equivalent phase) and the whole Granger causality and transfer entropy can be defined as the averages across the time duration respectively.

We consider a *hidden periodic model* ([16]) to generate $y_t$ as: $y_t = d \cdot [\sin(2\pi \cdot t/T_1) + \cos(2\pi \cdot t/T_2) + \nu_t]$, where $d$ is a positive scale, $T_1 = 5$, $T_2 = 7$, and $\nu_t$ is a white Gaussian noise with zero mean $I$ and variance $\sigma^2$, independent of all $x_t$ and all $\epsilon_{1,t}$. The noise $\epsilon_{1,t}$ is white Gaussian with zero means and a static variance $\sigma_0 = 5$. $x_t$ is generated by the MVAR (10). Their sampling time series are demonstrated in Fig. 1(A). The periods and other parameters in the hidden periodic model can be estimated from the data. Here, for simplicity, we assume that the periods are precisely estimated. To calculate $REC_{y \to x}(t)$

at a specific time $t$, we have multiple recording series of $x_t$ accompanied with the same single record series of $y_t$, with different random noises, denoted by $x_t^q$ for time $t = 1, \cdots, T$ and multiple index $q = 1, \cdots, Q$. After estimating the parameters in (5) and (6), we obtain the multiple series of the residual errors, denoted by $\tilde{\epsilon}_t^q$ and $\epsilon_t^q$, respectively. Then, at each $t = t_0$, we collect $\tilde{\epsilon}_{t_0}^q$ and $\epsilon_{t_0}^q$ with $q = 1, \cdots, Q$, to calculate $\widehat{REC}_{y \to x}(t)$ by formula (8). Then, averaging $\widehat{REC}_{y \to x}(t)$ across $t = 1, \cdots, T$ leads $\widehat{REC}_{y \to x}$. Alternatively, if we do not have multiple recordings of time series and the periods are known, we can segment a long time series of $x_t$ with a period equal to the least common multiple of all periods ($T_0 = T_1 \times T_2 = 35$). After estimating the parameters, we collect the residual errors $\epsilon_l^q = \epsilon_t$ with $l = mod(t, T_0)$ and $q = \lfloor t/T_0 \rfloor$, where $mod(a, b)$ stands for the remainder of $a$ driven by $b$ and $\lfloor r \rfloor$ stands for the largest integer less than $r$. Then, by the same way, we can calculate $\widehat{REC}_{y \to x}(l)$ for each $l = 0, \cdots, T_0 - 1$ and average them to obtain $\widehat{REC}_{y \to x}$. The p-values can be estimated by the bootstrap approach in the analogical fashion mentioned above.

To calculate the Granger causality in this scenario, we perform two methods. One is the classic Granger causality approach without considering the hidden periodic property in the MVAR model. The other is analogy to the calculation process of RE causality mentioned above: using multiple recordings of time series of $x_t$ or segmenting $x_t$ with a period $T_0$, estimating the co-variances at each time $t$ (or equivalent phase), calculating the Granger causality at $t$, and averaging them across the time duration or the whole period. TP and FP are calculated by $M = 100$ overlaps.

As Fig. 1(B) and (C) show, the periodic Granger causality failed to identify any causal intervene from $y$ to $x$ when $\sigma_y$ is small, since the intervene of hidden periodic $y_t$ does not cause significant variation in the variances in the residual error in (6) in comparison to (5). However, the classic Granger causality can probe this intervene when $b$ is not very small, since the hidden periodicity in $y_t$ that leads fluctuation can cause variance in (6) different from (5). However, the RE causality is clearly more efficient than Granger causality to identify the intervene from $y$ to $x$ (larger TP) and less incorrect identification (lower FP).

## 4.3    Time-Varing Nonstationary MVAR

We consider a more complicated time-varying means with $y_t$ generated by a *hidden chaotic model*: Let

$$y_t = \sigma \cdot \zeta(t) + \nu_t \tag{11}$$

where $\zeta(t)$ is generated by the logistic iteration map, namely, $\zeta(t) = \beta \zeta(t-1)[1 - \zeta(t-a)]$ with $\beta = 3.9$ and $\zeta(1) \in (0, 1)$, which implies that $\zeta(t)$ possesses a chaotic attractor, and $\nu_t$ is an independent identical distributed process with the uniform distribution in $[0, 0.5]$, as illustrated in Fig. 2(A). As above, $x_t$ is generated by the model (10). With the knowledge of the distribution with unknown parameters, we can employ the method above to calculate the RE causality of non-stationary time series by $Q = 100$ multiple recordings of the time series up to $T = 200$.

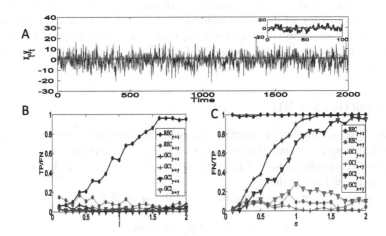

**Fig. 1.** Performance comparison among the RE causality (REC, $-*-$), classic Wiener-Granger causality (GC1, $-+-$) and time-varying Wiener-Granger causality (GC2, $-\nabla-$) by the TP rate (blue solid lines), and FN rate (red dash lines) for a causal intervene signal generated by (11). (A) the sampling hidden chaotic time series of $X_1$ (blue line) and $X_2$ (red line) as well as the inset plots for them in a small time intermal $[1, 100]$; (B) performance comparison of hidden periodic model with respect to $I$ with fixed $\sigma = 0.2$; (C) performace comparison of hidden periodic model with respect $\sigma$ with fixed $I = 2$. (Color figure online)

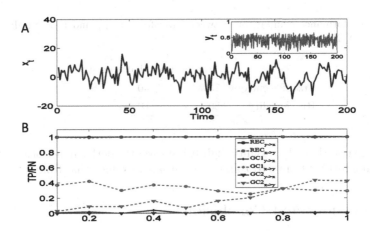

**Fig. 2.** Performance comparison among the RE causality (REC, $-*-$), classic Wiener-Granger causality (GC1, $-+-$) and time-varying Wiener-Granger causality (GC2, $-\nabla-$) by the TP rate (blue solid lines), and FN rate (red dash lines). (A) the sampling hidden chaotic time series of $X_1$ (blue line) and $X_2$ (red line); (B) the performance comparison of hidden chaotic model with respect to $\sigma$. (Color figure online)

Figure 2(B) show that the RE causality performs very well in identifying the causal intervene from $y$ to $x$ and excluding that from $x$ to $y$. In comparison the classic Granger and non-stationary Granger methods, which are deployed still with Gaussian assumption for statistic inference, RE causality is clearly better than them.

## 4.4    Predictive Models

With the help of modern predictive models, the prediction accuracy can be improved obviously compared with linear regression. As a consequence, we want to compare the performance of RE causality and Granger causality combining with some predictive model.

In this example, $x_t$ is generated by a nonlinear model:

$$x_t = a \sin(x_t - 1) + by_t - 1 + cy_{t-2} + \epsilon_{1,t} \tag{12}$$

with $y_t$ generated by: $y_t = c - dy_{t-1} + \sin(\epsilon_{2,t})$.

Applying XGBoost on the sampling time series up to $T = 1000$ to predict $x_t$ using $x$ and $y$'s history and $x$'s history respectively, we obtain the error time series $\epsilon$ and $\tilde{\epsilon}$. Assuming that the error time series follow Gaussian Mixture Model (GMM), we use EM algorithm for fitting, and thus obtain the distributions of $\epsilon$ and $\tilde{\epsilon}$. Since there is no closed form for the KL divergence between GMMs, we choose to do Monte Carlo to approximate the causalities. The last step is to calculate the TP, FP rate of RE causality and transfer entropy by $P = 100$ multiple recordings of time series.

**Table 2.** Performance comparison between Transfer entropy and RE causality combining with XGBoost.

|  | True positive | False positive |
| --- | --- | --- |
| Transfer entropy | 75% | 0 |
| RE causality | 80% | 0 |

Table 2 show that RE causality slightly improves the performance of identifying the causal intervene from $y$ to $x$ and both causality exclude that from $x$ to $y$.

## 5    Discussions and Conclusions

In conclusion, we propose a novel definition of Wiener causality by the relative entropy of the residual errors of prediction with and without considering the other time series. In comparison to the well-known Granger causality, which defines causality as the difference of the conditional co-variances, this RE causality is in the normal track to realise the Wiener's causal philosophy. For example, the intervene of the means in the time series cannot be identified by the Granger causality; however, it has contribution to prediction, in other words, span the future of $x$. We illustrate this argument by several models as intervenes with presenting the non-stationary RE causality. The fluctuation of the first-order statistics definitely causes information flow to $x$, so a causality, according to Wiener's arguments [3]. At the meantime, we may employ some predictive models, such as XGBoost, to provide high prediction accuracy, and we discover that RE causality can lead to further improvement of the reliability of causal relations inferred from data.

**Acknowledgments.** This work is jointly supported by the National Natural Sciences Foundation of China under Grant No. 61673119, the Key Program of the National Science Foundation of China No. 91630314, the Laboratory of Mathematics for Nonlinear Science, Fudan University, and the Shanghai Key Laboratory for Contemporary Applied Mathematics, Fudan University.

# References

1. Hu, S., Wang, H., Zhang, J., Kong, W., Cao, Y., Kozma, R.: Comparison analysis: granger causality and new causality and their applications to motor imagery. IEEE Trans. Neural Netw. Learn. Syst. **27**(7), 1429–1444 (2016)
2. Pearl, J.: Causality: Models, Reasoning, and Inference. Cambridge University Press, Cambridge (1999)
3. Wiener, N.: The Theory of Prediction, Modern Mathematics for Engineers. McGraw-Hill, New York (1956)
4. Granger, C.: Investigating causal relations by econometric models and cross-spectral methods. Econ. J. Econ. Soc. **37**(3), 424–438 (1969)
5. Schreiber, T.: Measuring information transfer. Phys. Rev. Lett. **85**(2), 461 (2000)
6. Liang, X.S., Kleeman, R.: Information transfer between dynamical system components. Phys. Rev. Lett. **95**(24), 244101 (2005)
7. Bernett, L., Barrett, A.B., Seth, A.K.: Granger causality and transfer entropy are equivalent for Gaussian variables. Phys. Rev. Lett. **103**(23), 238701 (2009)
8. Sobrino, A., Olivas, J.A., Puente, C.: Causality and imperfect causality from texts: a frame for causality in social sciences. In: International Conference on Fuzzy Systems, Barcelona, pp. 1–8 (2010)
9. Ding, M., Chen, Y., Bressler, S.: Handbook of Time Series Analysis: Recent Theoretical Developments and Applications. Wiley, Wienheim (2006)
10. Seth, A.K., Barrett, A.B., Barnett, L.: Granger causality analysis in neuroscience and neuroimaging. J. Neurosci. **35**(8), 3293–3297 (2015)
11. Chen, T.Q., Guestrin, C.: XGBoost: a scalable tree boosting system. arXiv:1603.02754v3 (2016)
12. Kullback, S., Leibler, R.A.: On information and sufficiency. Ann. Math. Stat. **22**(1), 79 (1951)
13. Cliff, O.M., Prokopenko, M., Fitch, R.: Minimising the Kullback-leibler divergence for model selection in distributed nonlinear systems. Entropy **20**(2), 51 (2018)
14. Cover, T.M., Thomas, J.A.: Elements of Information Theory. Wiley, New York (1991)
15. Oseledec, V.I.: A multiplicative ergodic theorem: Liapunov characteristic number for dynamical systems. Trans. Moscow Math. Soc. **19**, 197 (1968)
16. He, S.Y.: Parameter estimation of hidden periodic model in random fields. Sci. China A **42**(3), 238 (1998)

# Scene Graph Generation
# Based on Node-Relation Context Module

Xin Lin, Yonggang Li, Chunping Liu[✉], Yi Ji[✉], and Jianyu Yang

Soochow University, Su Zhou, China
{cpliu,jiyi}@suda.edu.cn

**Abstract.** For better understanding an image, the relationships between objects can provide valuable spatial information and semantic clues besides recognition of all objects. However, current scene graph generation methods don't effectively exploit the latent visual information in relationships. To dig a better relationship hidden in visual content, we design a node-relation context module for scene graph generation. Firstly, GRU hidden states of the nodes and the edges are used to guide the attention of subject and object regions. Then, together with the hidden states, the attended visual features are fed into a fusion function, which can obtain the final relationship context. Experimental results manifest that our method is competitive with the current methods on Visual Genome dataset.

**Keywords:** Scene graph · Relationship detection · Visual information
Visual attention

## 1 Introduction

Nowadays, with deeper understanding of images, classifying and locating objects is not enough for some tasks, such as Visual Question Answering problems [1, 2] and Image Caption [3, 4]. It is important to understand the relationships between the object pairs. Through understanding the relationships we can understand not only the spatial structure but also the semantic relationships. As a result, understanding the relationships can help more precise image retrieval [5], object detection [6] and image understanding problems [7].

With the development of neural networks, there are some fast and accurate object detection models [8–10]. They concentrate to recognize a wide variety of objects and regress their bounding boxes. Fast R-CNN [11] is a classic object detection algorithm. Our paper is also based on it. However, only object categories in the image cannot fully represent the complicated real world.

Besides the categories of multiple objects, relationships in the image can provide rich and semantic information. Lu *et al.* propose to use triple structural language to represent relationships between object pairs, such as <object1-predicate-object2> [12]. Relationship is detected using language priors. However, this work concentrates to detect pair-wise relationship rather than understanding the whole image. Xu *et al.* propose a scene graph generation task based on iterative message passing [13]. They train a model to generate scene graph from an image automatically to represent objects

© Springer Nature Switzerland AG 2018
L. Cheng et al. (Eds.): ICONIP 2018, LNCS 11302, pp. 134–145, 2018.
https://doi.org/10.1007/978-3-030-04179-3_12

in the image and the complicated topological relationships between them. However, latent visual information is ignored in their message pooling stage. To overcome this problem, we propose a new edge context message pooling method, in which we reenter visual features and we use node GRUs' (Gated Recurrent Unit) [14] hidden state to guide the attention to attend to the more important regions in the corresponding relations. In this way, latent visual information in relationships can be obtained.

On the other hand, the contribution of subject and object in sentence comprehension is not balanced in linguistics [22]. For Example, in the phrase "person holding cup", subject "person" is more important. When do predicate classification, if the subject is "table", "under" or "above" became essential. Based on this assumption, we do importance measure after visual attention when computing edge context message.

In summary, based on Xu *et al.* s' remarkable work [13], we design a node-relation context module. The method consists of two parts: node states guided relation attention module and a better fusion function. In the node states guided relation attention module, we use node GRUs' hidden state guided attention to better utilize the ignored latent visual information. In the fusion function, we measure the contribution of the subject and objects' visual information as well as the hidden states in both edge message pooling and node message pooling to obtain better context messages.

## 2 Related Work

### 2.1 Baseline Scene Graph Generation

Xu *et al.* [13] pass context message through node GRUs and edge GRUs iteratively so that the prediction of objects and relationships can benefit from its neighboring context. In the scene graph topology, a relationship triple consists of two node and one edge. For a node, there are inbound edges and outbound edges. For an edge GRU, it receives messages from its neighboring node GRUs. And for a node GRU, it receives message pooling from its inbound and outbound edge GRUs. The module to generate node messages and edge messages is called message pooling. The overview frame work is shown in Fig. 1.

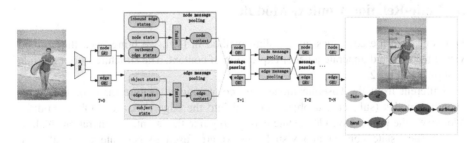

**Fig. 1.** The architecture of scene graph generation by iterative message passing [13].

Given an image, convolution feature maps are first extracted. Afterwards, the Region Proposal Network [11] generates the region proposals. Using the region proposals, node and relationship feature maps are extracted using ROI-pooling method and fed into corresponding GRUs. Afterwards, the hidden states are fed into message pooling module to generate context message. After a few message passing iterations, a scene graph is generated consisting of object categories, relationships and bounding boxes. The training process is a multi-task learning [15].

## 2.2 Attention Models

Since Xu *et al.* [13] focused on hidden states formed context, initial visual information is ignored during edge message pooling stage. Instead of treating all feature maps equally, attention mechanism tries to discover different weights of image regions according to their value. Soft attention has been widely used in machine translation [16] and image captioning [3]. In the soft attention mechanism, the hidden state of the LSTM(Long Short-Term Memory) [17] is used to guide attention. In our work, we use the processed node GRU hidden states to guide the attention of the corresponding feature maps to obtain better expression of edge context. To the best of our knowledge, it's the first time to apply attention module in scene graph generation.

## 2.3 Relationship Referring

Natural language referring expression is presented and let the model tag the objects referring [18, 19]. Position information, color, object classes and etc. are needed. Recently, structural language is used to referring the object pairs engaged in the relationship expression [12, 20, 21]. Using structural language can reduce the cost of understanding the natural language so that visual understanding task can be focused on. Attention shift algorithm is applied to model relationship in this work. Relationship referring task is a reverse task of scene graph generation. Scene graph generation is a more complicated task since we need to understand all the relationships existing in the image.

## 3 Node-Relation Context Module

Different from the edge message pooling method in [13], we bring back the ignored visual features and generate a better expression of context. The structure of our module is shown in Fig. 2.

Different from the baseline model, the convolutional feature maps of the corresponding nodes (subject and object) are fed into the edge message pooling module. Moreover, the processed hidden state is used to guide the attention generation. At last, both hidden state context and visual context are fused to generate the final edge context. The node-relation context module consists of two parts: node state guided relation attention module and fusion function. We will present our method from these two aspects.

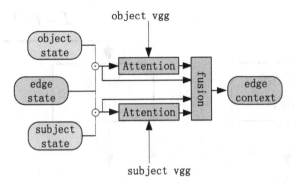

**Fig. 2.** Illustration of the node-relation context module

### 3.1 Attention Module for Relationship Prediction

Considering different regions of the object pairs have different contribution to relationship classification, the hidden states of object and subject are used to guide the attention. The concrete framework is shown in Fig. 3 where, $h_{i,t}$ represents the hidden state of $node_i$ at time step t and $h_{i+j,t}$ represents the hidden state of edge GRU relating $node_i$ and $node_j$ at time step t. When $node_i$ and $node_j$ have different characters in the relationship, $h_{i+j,t}$ has different meanings, as expressed in Eq. 1.

$$h_{i+j,t} = \begin{cases} h_{i \to j,t}, sub(i) = True \\ h_{j \to i,t}, else \end{cases} \tag{1}$$

where the decision function $sub(i)$ decides if $node_i$ is subject, if true then true, otherwise false.

We first element-wise multiply the hidden state of node GRU and edge GRU as the guidance of the attention, shown in Eq. 2. Moreover, $s_{i,t}$ is also taken as components of hidden state context to embed the final context expressed in Sect. 3.2.

$$s_{i,t} = h_{i,t} \odot h_{i+j,t} \tag{2}$$

Here, the same as the baseline model, images are fed into a VGG-16 [4] convNet. The convolution feature maps from the last conv layer (i.e., conv5_3) are used. Roi shifts are fed into the model to obtain the feature maps of $node_i$ through roi pooling function, denoted as $V_i$. $V_i$ is composed of k small regions, $V_i = (v_1, v_2, \ldots, v_k)$.

Then, $s_{i,t}$ indicates how much attention the module is placing on different regions. A softmax function is used to gain attention distribution.

$$\alpha_t = \text{softmax}(s_{i,t} \odot V_i) \tag{3}$$

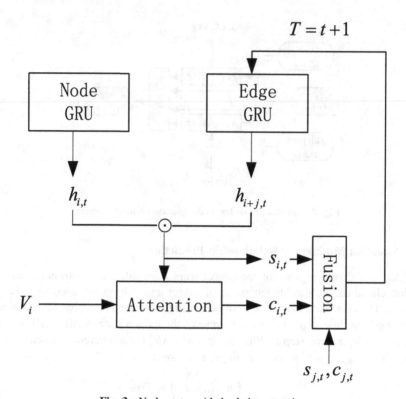

**Fig. 3.** Node state guided relation attention

Based on the attention distribution, the weighted node feature can be gained by:

$$c_{i,t} = \sum_{l=1}^{k} \alpha_{tl} v_{tl} \qquad (4)$$

Due to different context, even the same $node_i$ has different attention distribution in different relationship.

### 3.2 Fusion Function

After $s_{i,t}$ and $c_{i,t}$ obtained, we will fuse the context vectors. First, contributions of the subject and object in the corresponding relation are computed.

$$sub\_score = \tanh(W_s s_{i,t} + b_s) \qquad (5)$$

$$obj\_score = \tanh(W_o s_{j,t} + b_o) \qquad (6)$$

Specially, tanh function is also used in node message pooling stage instead. The advantage of this function will be proved in ablation study (Sect. 4.2).

Then, we fuse the subject and object components, shown in Eqs. 7 and 8:

$$v\_c_{i+j,t} = sub\_score \bullet c_{i,t} + obj\_score \bullet c_{j,t} \tag{7}$$

$$g\_c_{i+j,t} = sub\_score \bullet s_{i,t} + obj\_score \bullet s_{j,t} \tag{8}$$

After weight fusion, the hidden state context $g\_c_{i+j,t}$ and the visual context $v\_c_{i+j,t}$ are concatenated and fed into a fully connected layer, as the final context of this relationship.

## 4  Experiments

Applying our module to the baseline model [13], we can generate better context message and obtain more accurate scene graph. We conduct experiments on large-scale benchmark: Visual Genome [20]. This dataset is a human-annotated scene graph dataset, containing 108,077 images. Each image involves 25 objects and 22 relationships on average. In this section, we analyze our model in four parts: evaluation metrics, ablation study, comparison with existing works and qualitative analysis. The ablation study includes **Baseline** model [13], **Baseline + V**(visual-context), **Baseline + V**(visual-context) + **H**(hidden-states-context, not sharing the weights), **Baseline + V**(visual-context) + **H**(hidden-states-context) + **SW** (sharing weights, our proposed final model, short as ours).

### 4.1  Evaluation Metrics

Given an image, the scene graph generation task includes locating the objects, predicting their categories and figuring out the relationship between each object pairs. According to the metrics in [12], the evaluation is divided into three levels.

**PREDCLS(predicate classification):**   Given the locations and the categories of the objects, relationships between object pairs are to be predicted.

**SGCLs(scene graph classification):**   Given the bounding boxes of the objects, categories of the objects and the relationships between them are to be predicted.

**SGGen(scene graph generation):**   Locations, categories of the objects and the relationships between the objects all need to be predicted

The difficulty level of these three tasks is from easy to difficult. Image-wise recall evaluation metrics, R@50 and R@100, are adopted to evaluate the three tasks. R@x is abbreviation of Recall@x, meaning the fraction of the ground truth relationships prediction among top x predictions in an image. Obviously, the bigger fraction of the ground truth relationships prediction among top x predictions, the better the model performs. The reight prediction means to classify the triple structure.

## 4.2  Ablation Study

According the improvement on model message passing [13], we design module analysis experiments to prove the superiority of our node-relation context module. In this subsection, we perform ablative studies to analyze the contribution of each improvement to our module, shown in Table 1.

**Table 1.** Evaluation results of the contribution of each improvement to our module.

| Methods | PREDCLs | | SGCls | | SGGen | |
|---|---|---|---|---|---|---|
| | R@50 | R@100 | R@50 | R@100 | R@50 | R@100 |
| Baseline [13] | 44.75 | 53.08 | 21.72 | 24.38 | 3.44 | 4.24 |
| Baseline + V | 43.09 | 50.96 | 22.12 | 24.69 | 4.36 | 5.53 |
| Baseline + H | **45.83** | **53.93** | 22.25 | 24.96 | **4.51** | **5.61** |
| Baseline + V + H | 45.37 | 53.61 | 22.37 | 25.05 | 4.18 | 5.39 |
| Baseline + V + H + SW(ours) | 45.55 | 53.66 | **23.37** | **26.29** | 4.33 | 5.57 |

As we can see, though only visual context model does not perform well in task PREDCLs, it performs well in the other two tasks. Only H-states context model performs best in task PREDCLs and SGGen. Combining these two context, we obtain a model performing best in SGCls. Though a bit of poorer than H-states context model in task PREDCLs and SGGen, considering relatively large exceeding in task SGCls, we make it our final model. As shown in Table 1, sharing the weights between the visual context and hidden states context performs better. In addition, one benefit is achieved that we can save calculation amount obviously. Our model performs best in SGCls task and performs second best in other two tasks. Since there is a larger improvement in SGCls task, we use Baseline + V + H + SW as our final model

To better prove the advantage of tanh function to weigh the contribution of subject and object components, we design a comparison of some common activation functions, including sigmoid function, ReLU (Rectified Linear Unit) and tanh, on both the baseline model and our model. As shown in Table 2, tanh function performs best in both the baseline models and our improved models. Due to the improvements proposed, our model using tanh function performs better.

## 4.3  Comparison with Existing Works

Table 3 shows the performance of our model against two existing ones. We can see that in each task, our proposed model has exceeded the existing two models. Specially, there is a relatively large increase in task SGCls. Categories of the objects are not offered in task SGCls. Thanks to re-entering the visual features, more valuable information is gained to classify the objects and the relationships.

**Table 2.** Evaluation results of different fusion functions, including tanh, sigmoid and relu function to calculate weights of subject component and object component.

| Methods | Fusion function | PREDCLs | | SGCls | | SGGen | |
|---|---|---|---|---|---|---|---|
| | | R@50 | R@100 | R@50 | R@100 | R@50 | R@100 |
| Baseline [13] | Sigmoid | 44.75 | 53.08 | 21.72 | 24.38 | 3.44 | 4.24 |
| Baseline + r | ReLU | 34.03 | 43.32 | 18.34 | 21.37 | 3.29 | 4.28 |
| Baseline + t | tanh | 45.54 | **53.96** | 21.89 | 24.62 | 4.36 | 5.43 |
| Ours + s | Sigmoid | 44.37 | 51.97 | 21.97 | 24.58 | 4.36 | 5.47 |
| Ours + r | ReLU | 44.22 | 52.19 | 22.21 | 24.85 | **4.38** | 5.48 |
| Ours | tanh | **45.55** | 53.66 | **23.37** | **26.29** | 4.33 | **5.57** |

**Table 3.** Comparison with two existing models.

| Methods | PREDCLs | | SGCls | | SGGen | |
|---|---|---|---|---|---|---|
| | R@50 | R@100 | R@50 | R@100 | R@50 | R@100 |
| Language priors [12] | 27.88 | 35.04 | 11.79 | 14.11 | 0.47 | 0.32 |
| Message passing [13] | 44.75 | 53.08 | 21.72 | 24.38 | 3.44 | 4.24 |
| Ours | **45.55** | **53.66** | **23.37** | **26.29** | **4.33** | **5.57** |

## 4.4 Qualitative Analysis

In this section, part of the experimental results are shown below.

Figure 4 shows qualitative results using human annotated bounding boxes. The results show that the baseline model confuses about the categories of the objects. For example, it predicts the head of a cute owl <sheep-of-bird>, because the head looks fury. Our model predicts <head-of-bird> instead. Since we obtain better representation of context, it's easier for model to distinguish the categories having similar appearance. What is more interesting, our model predicts the man play tennis wearing "short" and the baseline model predicts "pant". As we know, male tennis players usually wear short and our model can infer this from the context.

Figure 5 shows the results using the bounding box produced by RPN. Our model is par with the baseline model. Using the region proposals, the models produce more reasonable answers than using human annotated even if there are less objects. That's because the RPN network and the generation model share the same feature maps and understand the images from similar angle. Compared with the results generated by human annotated bounding boxes, this model tends to predict uncertain predicates such as "has".

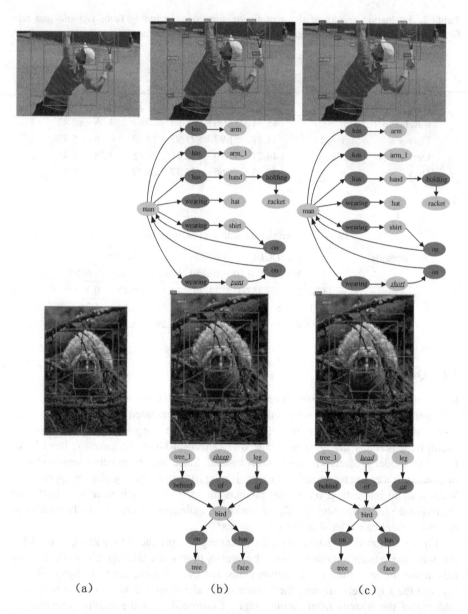

**Fig. 4.** Part of results from model message passing [13] and our final model, where the (a) is human annotated bounding boxes, (b) are the results of the original model and (c) are the results of our model. The 1st and the 3rd lines are the visualized results over the images and the 2nd and the 4th lines are the final scene graph. The baseline model is retrained by us using the default parameters.

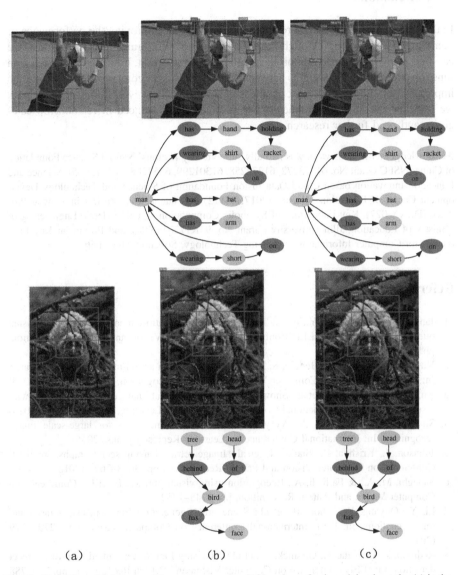

(a)                    (b)                    (c)

**Fig. 5.** Part of results from model message passing [13] and our final model, where the (a) is the bounding boxes generated by RPN, (b) are the results of the original model and (c) are the results of our model. The 1st and the 3rd lines are the visualized results over the images and the 2nd and the 4th lines are the final scene graph.

# 5   Conclusions

In this paper, we propose node-relation context module, improving the performance of scene graph generation. Through introducing the attention guided visual features, we can find latent visual information in relationship. In addition, by using a better fusion function and sharing the weights between subject and object components, we obtain the importance of subject and object context components. However, during the research, we find that the Visual Genome dataset is severely unbalanced in relationship categories and need further research.

**Acknowledgement.** This work was partially supported by National Natural Science Foundation of China (NSFC Grant No. 61773272, 61272258, 61301299, 61572085, 61272005), Science and Education Innovation based Cloud Data fusion Foundation of Science and Technology Development Center of Education Ministry (2017B03112), Six talent peaks Project in Jiangsu Province (DZXX-027), Key Laboratory of Symbolic Computation and Knowledge Engineering of Ministry of Education, Jilin University (Grant No. 93K172016K08), and Provincial Key Laboratory for Computer Information Processing Technology, Soochow University.

# References

1. Jang, Y., Song, Y., Yu, Y., et al.: TGIF-QA: Toward spatio-temporal reasoning in visual question answering. In: IEEE Conference on Computer Vision and Pattern Recognition, pp. 1359–1367 (2017)
2. Vinyals, O., Toshev, A., Bengio, S., et al.: Show and tell: a neural image caption generator. In: IEEE Conference on Computer Vision and Pattern Recognition, pp. 3156–3164 (2015)
3. Xu, K., Ba, J., Kiros, R., et al.: Show, Attend and tell: neural image caption generation with visual attention. In: International Conference on Machine Learning, pp. 2048–2057 (2015)
4. Simonyan K., Zisserman A.: Very deep convolutional networks for large-scale image recognition. In: International Conference on Learning Representations (2015)
5. Johnson, J., Krishna, R., Stark, M., et al.: Image retrieval using scene graphs. In: IEEE Conference on Computer Vision and Pattern Recognition, pp. 3668–3678 (2015)
6. Sadeghi, M. A., & Farhadi, A.: Recognition using visual phrases. In: IEEE Conference on Computer Vision and Pattern Recognition, pp. 1745–1752
7. Li, Y., Ouyang, W., Zhou, B., et al.: Scene graph generation from objects, phrases and region captions. In: IEEE International Conference on Computer Vision, pp. 1270–1279 (2017)
8. Redmon, J., Divvala, S., Girshick, R., et al.: You only look once: unified, real-time object detection. In: IEEE Conference on Computer Vision and Pattern Recognition, pp. 779–788 (2016)
9. Wang, Z., Chen, T., Li, G., et al.: Multi-label image recognition by recurrently discovering attentional regions. In: IEEE International Conference on Computer Vision, pp. 464–472 (2017)
10. Wang, J., Yang, Y., Mao, J., Huang, Z., Huang, C., Xu, W.: CNN-RNN: a unified framework for multi-label image classification. In: IEEE Conference on Computer Vision and Pattern Recognition, pp. 2285–2294 (2016)
11. Girshick, R. B.: Fast R-CNN. In: International Conference on Computer Vision, pp. 1440–1448 (2015)

12. Lu, C., Krishna, R., Bernstein, M.S., et al.: Visual relationship detection with language priors. In: European Conference on Computer Vision, pp. 852–869 (2016)
13. Xu, D., Zhu, Y., Choy, C.B., et al.: Scene graph generation by iterative message passing. In: IEEE Conference on Computer Vision and Pattern Recognition, pp. 3097–3106 (2017)
14. Dey, R., Salemt, F.M.: Gate-variants of gated recurrent unit (GRU) neural networks. In: International Midwest Symposium on Circuits and Systems, pp. 1597–1600 (2017)
15. Xue, Y., Liao, X., Carin, L., et al.: Multi-task learning for classification with Dirichlet process priors, **8**(1), 35–63 (2007)
16. Bahdanau, D., Cho, K., Bengio, Y.: Neural machine translation by jointly learning to align and translate. In: International Conference on Learning Representations (2015)
17. Hochreiter, S., Schmidhuber, J.: Long short-term memory. Neural Comput. **9**(8), 1735–1780 (1997)
18. Hu, R., Rohrbach, M., Andreas, J., et al.: Modeling relationships in referential expressions with compositional modular networks. In: IEEE Conference on Computer Vision and Pattern Recognition, pp. 1115–1124 (2016)
19. Yu, L., Lin, Z., Shen, X., et al.: MAttNet: modular attention network for referring expression comprehension. arXivpreprint: 1801.08186 (2018)
20. Krishna, R., Zhu, Y., Groth, O., et al.: Visual genome: connecting language and vision using crowdsourced dense image an-notations. Int. J. Comput. Vision **123**(1), 32–73 (2016)
21. Krishna, R., Chami, I., Bernstein, M., et al. Referring relationship. arXivpreprint: 1803. 10362 (2018)
22. Kibrik, A.E.: Beyond subject and object: toward a comprehensive relational typology. Linguist. Typology **1**(3), 279–346 (1997)

# Cross-Layer Convolutional Siamese Network for Visual Tracking

Yanyin Chen[1,2], Xing Chen[1,2], Huibin Tan[1,2], Xiang Zhang[2,3(✉)],
Long Lan[2,3(✉)], Xuhui Huang[4], and Zhigang Luo[1,2(✉)]

[1] Science and Technology on Parallel and Distributed Laboratory,
NUDT, Changsha 410073, Hunan, People's Republic of China
{chenyanyin16,chenxing16,zgluo}@nudt.edu.cn
[2] College of Computer, NUDT, Changsha 410073, Hunan, People's Republic of China
{zhangxiang08,long.lan}@nudt.edu.cn
[3] HPCL, NUDT, Changsha 410073, Hunan, People's Republic of China
[4] Department of Computer Science and Technology, College of Computer,
NUDT, Changsha 410073, Hunan, People's Republic of China

**Abstract.** In most trackers for visual tracking, Siamese network based
trackers construct a pair of twin structures to learn a similarity metric between tracked object and search region to predict the position of
the object in the coming frame. They have achieved impressive performance in both speed and accuracy. However, semantic features from
different layers are not fully explored in most current Siamese network
based tracker. To this, we propose a cross-layer convolutional Siamese
network tracker (Siam-CC) which attempts to explore more semantic
features of different layers from two aspects. Firstly, we combine the
shallow-to-deep cross-layer convolutional response maps to capture various semantic-aware features and meanwhile enforce Siam-CC to only
focus on the most interesting location, because much more semantic
information is able to reduce negative effect of background. Secondly, to
further boost the discrimination of responses, an adaptive contrastive loss
is additionally developed together with traditional logistical loss, which,
to some extent, assists in filtering out some noisy responses. Experiments
on a large-scale benchmark dataset show the effectiveness of Siam-CC as
compared to the state-of-the-art trackers.

**Keywords:** Visual tracking · Cross-layer convolutional
Contrastive loss

## 1  Introduction

Visual tracking is still a challenging topic in computer vision with broad applications such as visual surveillance, etc. It attends to track a specific object
in a video sequence with the ground truth location of the first frame known.
Although some existing trackers have achieved impressive tracking performance,

© Springer Nature Switzerland AG 2018
L. Cheng et al. (Eds.): ICONIP 2018, LNCS 11302, pp. 146–156, 2018.
https://doi.org/10.1007/978-3-030-04179-3_13

there remain various unsolved challenges like clutter environments and target occlusion. Then it is non-trivial to devise a real-time and effective tracker.

Recently, discriminative trackers have become a main focus in visual tracking community. Among them, convolutional neural networks (CNNs, [13]) based trackers have gained the state-of-the-art tracking performance on popular benchmarks. Their success can be attributed to discriminative features and effective classifiers which strengthen discriminant ability and generality of trackers. For instance, CNN-SVM [12] uses SVM to classify features learned from R-CNN [5], and locates the object by a generative model. MDNet [16] considers the network with a shared layer and a domain-specific layer. The former learns a generic representation for all the objects while the latter learns an exclusive representation for each object. FCNT [22] explores feature representations of different layers to greatly boost tracking performance. However, they are very expensive in time.

A feasible solution to this problem is offline tracking. Held et al. proposed GOTURN [10] to take as inputs two successive image frames and predict the relative position of the object in the coming frame. Without online fine-tuning, it is very fast. As a competitive counterpart, Siam-FC [2] has achieved the state-of-the-art performance in both speed and accuracy. It locates the object position by estimating the response map via a fully convolutional operation. However, they cannot fully explored diverse semantic features from different layers.

To address this issue, we develop a cross-layer convolution Siamese network tracker with adaptive loss (Siam-CC) to leverage semantic features of different layers to enhance the location accuracy of the tracked object. Specifically, we exploit the outputs of different layers to generate the so-called cross-layer response maps via multiple efficient cross-layer convolution operations. The resultant maps can capture various levels of semantic features across different scales of receptive fields to suppress noisy responses. This can reduce negative effect of background information of either object region or search region. Further, to boost discrimination of responses, we develop an adaptive contrastive loss together with logistical loss to consider the pair-wise relationships among different entries of response maps without tuning the margin parameter. Experiments on a benchmark dataset including OTB-2013 [23], OTB-2015 [24] and OTB-50 show the efficacy of Siam-CC as compared to the state-of-the-art trackers.

## 2   Related Work

With the development of deep networks in computer vision, CNNs has shown outstanding performance in visual tracking. Nowadays, CNNs-based trackers are mainly divided into three categories: (1) direct CNNs features, (2) online fine-tuning CNNs, and (3) only offline training. The first category is to directly use features of pre-trained CNNs-based classification network on large-scale image classification datasets. Among them, FCNT [22] explored features of different layers based on VGG-net [19] to locate the tracked object. CF2 [15] integrated features of different layers and multiple correlation filters to predict the position of the object in a hierachical manner. They are based on the assumption where deeper layer extracts more semantic features while the early layers learn spatial details of the object. This assumption also motivates us to incorporate diverse semantic features to generate multiple response maps.

Different from the first kind of methods, the second want to track the object whilst to fine-tune the designed networks in online learning fashion [16,20]. For instance, Nam et al. constructed MDNet [16] by treating visual tracking as multi-domain learning problem, with fully-connected layers online updating. Besides, Teng et al. proposed TSN [20] to extract spatial-temporal features by collecting key historical samples to predict the candidate locations. In contrast to them, Siam-CC is very efficient in tracking.

The last kind of methods offline learn the tracker on large-scale videos. Siam-FC is a typical representative, which estimates the region-wise feature similarity through fully convolutional operation. Without fine-tuning, they are very efficient. There are numerous trackers [4,6,8,21] based on Siam-FC. Our tracker is a variant of Siam-FC, but differs from existing counterparts because we first explore diverse semantic features to generate multiple response maps for object location via cross-layer convolution.

## 3   Method

We briefly introduce the network framework of our tracker with Siamese networks, as in Fig. 1. Our tracker still shares the novel idea with Siam-FC [2] and has two branches: exemplar image branch (Branch A) and search region branch (Branch B); but we add another convolutional layer (Conv6) to Conv5 of Branch B in order to allow it to extract more semantic information. The most difference from Siam-FC lies in that four cross-layer convolutional response maps are introduced here. Because of them, we call our tracker the cross-layer convolutional Siamese network tracker (Siam-CC). Three of four response maps are learned through the logistic loss, while the rest is trained by using an adaptive contrastive loss. The network components are detailed in Table 1.

**Table 1.** Network components of Siam-CC

| Branch A | Branch B | Layers | Filter size | Channels | Stride |
|----------|----------|--------|-------------|----------|--------|
| √ | √ | Conv1 | 11×11 | 96 | 2 |
| √ | √ | Pool1 | 3×3 | - | 2 |
| √ | √ | Conv2 | 5×5 | 256 | 1 |
| √ | √ | Pool2 | 3×3 | - | 2 |
| √ | √ | Conv3 | 3×3 | 384 | 1 |
| √ | √ | Conv4 | 3×3 | 256 | 1 |
| √ | √ | Conv5 | 3×3 | 256 | 1 |
| × | √ | Conv6 | 3×3 | 256 | 1 |

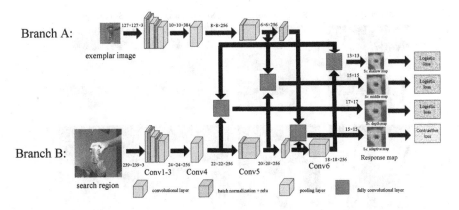

**Fig. 1.** The architecture of Siam-CC: exemplar image branch (*Branch A*) and search region branch (*Branch B*). There are four cross-layer convolutional operations to calculate four scales of response maps. The logistic loss and an adaptive contrastive loss are used to train the network.

### 3.1 Cross-Layer Convolutional Response Maps

As stated in [15], diverse layers can extract different semantic information. Generally, shallow layers of CNNs contain fine-grained spatial details of the object; deep layers of CNNs embrace semantic regions of the tracked object, which are robust to object deformation and rotation. Thus, different layers have distinctive effects on the response maps. In terms of this insight, we consider cross-layer convolutional operations to generate response maps. For efficiency, we adopt the outputs of three convolutional layers, i.e., Conv4, Conv5 and Conv6, of Branch B and that of the Conv5 layer of Branch A to calculate three scale response maps. We intuitively display the response maps of different layers on two frames of two datasets to show the motivation of Siam-CC. As in Fig. 2, the cross-layer convolutional response maps display different location regions. In *carDark*, the response map of deep layers has good location; inversely, the response map of shallow layers obtains accurate location on *basketball*. Thus, it is possible to derive the reliable location from the response maps of three cross layers.

Note that Conv5 has larger object scale than that of Conv6. It appears like dilated convolution to enlarge the receptive fields of traditional convolution. In some cases, the larger the dilated size of dilated convolutional operation, the more information loses. This situation cannot occur in our cross-layer convolution. This is because our convolutional filters have no rigid dilations. That is why we choose the cross-layer convolution to learn response maps. We leave the forth response map of Fig. 1 to the next subsection due to its distinctness. In Siam-CC, we employ the logistic loss to connect three cross-layer convolutional response maps, respectively. Although three response maps have different sizes, to avoid information loss, we do not rescale them into identical size maps. Subsequently, we need to construct different sizes of groundtruth labels for these three response maps. To guide more discriminative maps, we adopt a piece-wise label function

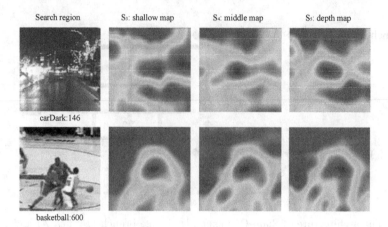

**Fig. 2.** Three response maps in the 146-th frame of *carDark* and the 600-th frame of *basketball*.

to guide different regions of each response map. In our empirical studies, this modification can boost the location accuracy. Concretely, our piece-wise label function is defined as:

$$Y_{lg}^i[j] = \begin{cases} 1 & l \leq 0.2R \\ 0.6 & 0.2R < l \leq 0.5R \\ 0.3 & 0.5R < l \leq R \\ -1 & l > R \end{cases}, \tag{1}$$

where $Y_{lg}^i[j]$ is the logistic label of the $j$-th pixel of the $i$-th response map, and $l$ is the distance between $j$ and target position. $R$ is a hyper-parameter which is set to 2 in our experiment. We locate the object's position in search region by finding maximum response point of response map. By our label function, the maximum value point of the response map might approach the center of object.

## 3.2   Adaptive contrastive loss

Contrastive loss [7] is often used to measure the similarity of two samples, i.e.,

$$L = \frac{1}{2N} \sum_{n=1}^{N} Y[n]d[n]^2 + (1 - Y[n]) \max(m - d[n], 0)^2, \tag{2}$$

where $d[n]=\|a[n] - b[n]\|_2$ is the Euclidean distance between two sample features $a[n]$ and $b[n]$, $y[n] \in \{0, 1\}$ denotes whether the two samples are matched, $m$ is a margin value, and $N$ is the number of input sample pairs. It encourages positive pairs to be closer and puts negative pairs away from each other.

Inspired by this idea, we regard the response map as discriminative embedding learning problem. Intuitively, each entry of the response map can be viewed

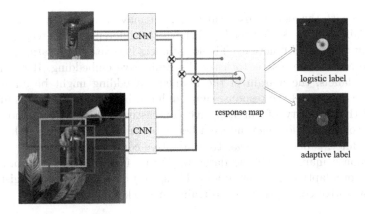

**Fig. 3.** The difference between logistic label and adaptive label. There are different colour boxes in search region since it has different values in logistic label. The blue box in the figure indicates background, the yellow point and the red point are candidate points of the adaptive label, but red point has a larger value in the logistic label because it is closer to object position. (Color figure online)

as the similarity between exemplar image and a sliding window of search region. Thus, we can adopt contrastive loss to measure the gap between the response maps and the ground truth labels. Then we replace the Euclidean distance with the values of the response map. Note that the Euclidean distance reflects the dissimilarity while the response map indicates the similarity. That is, the larger the values of the response map, the more similar two objects appear to be. Therefore, the contrastive loss can be formulated as follows:

$$L_{cl} = \frac{1}{2N} \sum_{j=1}^{N} (1 - Y_{al}[j]) S_{al}[j] + Y_{al}[j] \max(m - S_{al}[j], 0), \qquad (3)$$

where $S_{al}[j]$ denotes the $j$-th element value of the response map $S_{al}$, and $Y_{cl}[j]$ is the contrastive loss label in location $j$. To avoid tuning the margin parameter $m$, we intend to adaptively learn it with a simple regularization term. This regularization term avoids the trivial value. Thus, we rewrite Eq. (3) as:

$$L_{al} = \frac{1}{2N} \sum_{j=1}^{N} (1 - Y_{al}[j]) S_{al}[j] + Y_{al}[j] \cdot max(m - S_{al}[j], 0) + \frac{\lambda}{m}, \qquad (4)$$

where $\lambda$ is a positive parameter, which we set to 1 in our experiments, and the label function $Y_{al}$ of our adaptive contrastive loss is defined as:

$$Y_{al}[j] = \begin{cases} 1 & l \le R \\ -1 & l > R \end{cases}, \qquad (5)$$

wherein $R$ is the same as that of Eq. (2). Note that our adaptive contrastive loss is connected to the fourth response map while the other three response maps are trained with the logistic loss. Naturally, one may question why we define two different label functions. To this, Fig. 3 shows the difference of the response maps induced by both label functions. The logistic loss is treated as location function, while the contrastive loss is to learn discriminative embedding. If we use the piece-wise labels, the margin of the learned embedding might be ambiguous. Finally, we display all the response maps induced by two loss functions in Fig. 4 to verify the necessity of our adaptive contrastive loss. As Fig. 4 shows, our adaptive contrastive loss can induce a better location than the other three ones on *Bike*. But this does not mean that adaptive contrastive loss always induces good response maps on all the datasets. Figure 4 only implies the promising potentials of adaptive contrastive loss. Thus, we jointly employ the logistic loss and the adaptive contrastive loss to train our tracker.

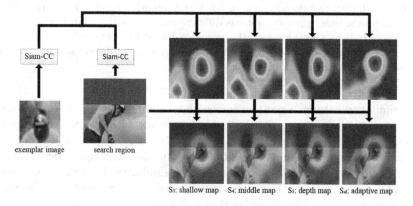

**Fig. 4.** Different response maps on the 76-th frame of *Biker*. The fourth response map has a better performance than the other three ones on this dataset.

## 4   Implementation Details

**Training:** We use Stochastic Gradient Descent (SGD) to train our tracker by minimizing both logistic loss and adaptive contrastive loss. The training dataset ILSVRC-2015 [18] is large enough to train Siam-CC. Through improved Xavier method [9], we can initialize the weights of convolutional layers. The training batch size is 8. After 50 training epoch, the decay weights reduce from $10^{-2}$ to $10^{-5}$. Then we can get our Siam-CC model.

**Tracking:** During tracking, Siam-CC does not fine-tune the network weights. As shown in Fig. 1, Siam-CC outputs four response maps per frame. Then, these maps are normalized and resized to $239 \times 239$. We reweight the four response maps to get the final one. The location of the highest point is the target's location. To get the scale variations, the input of search region has three scales $1.0375^{\{-1,0,1\}}$. Then there are three final response maps and the maximum value of the corresponding map to determine whether to change the scale.

# 5 Experiments

We choose eight state-of-art trackers used in experiments to verify the effectiveness of Siam-CC on OTB benchmark. This benchmark has three datasets including OTB-2013, OTB-2015 and OTB-50, which respectively contain 51 videos, 100 videos and 50 videos. Eight compared state-of-art trackers include KCF [11], Staple [1], CNN-SVM [12], MEEM [25], SAMF [14], DSST [3], Siam-FC3s [2] and HDT [17]. The area under curve (AUC) is utilized to measure the overlap ratio of bounding box between ground-truth and tracker evaluation. There are eleven attributes in OTB videos.

As in Fig. 5, Siam-CC gets the highest average AUC on three datasets. Besides, we make quantitative comparison in Table 2. Figure 6 shows that Siam-CC has the highest AUC in most attributions.

To qualitatively compare the trackers, we select four challenging videos covering eleven attributions to perform tracking task. As shown in Fig. 7, Siam-CC

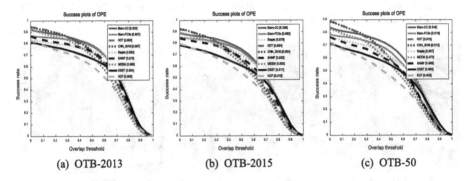

(a) OTB-2013      (b) OTB-2015      (c) OTB-50

**Fig. 5.** The Success plots for one pass evaluation (OPE) on OTB-2013, OTB-100 and OTB-50, respectively.

**Table 2.** Average AUC scores and two overlap thresholds (0.4 and 0.7) on three datasets. The best is in bold font. The results are in the form of OTB-2013/OTB-2015/OTB-50.

| Tracker | Average AUC | Overlap = 0.4 | Overlap = 0.7 |
|---|---|---|---|
| Siam-CC(ours) | **0.623/0.598/0.548** | **0.828/0.801/0.740** | **0.540**/0.507/**0.448** |
| Siam-FC3s | 0.607/0.582/0.516 | 0.818/0.781/0.701 | 0.420/0.503/0.415 |
| HDT | 0.603/0.554/0.515 | 0.799/0.739/0.680 | 0.356/0.417/0.345 |
| CNN-SVM | 0.597/0.578/0.512 | 0.810/0.742/0.690 | 0.336/0.388/0.331 |
| Staple | 0.593/0.564/0.507 | 0.771/0.752/0.664 | 0.447/**0.516**/0.439 |
| MEEM | 0.566/0.530/0.473 | 0.774/0.711/0.638 | 0.303/0.355/0.292 |
| SAMF | 0.579/0.553/0.469 | 0.769/0.730/0.617 | 0.359/0.455/0.356 |
| DSST | 0.554/0.517/0.460 | 0.712/0.653/0.589 | 0.393/0.461/0.392 |
| KCF | 0.505/0.473/0.403 | 0.670/0.613/0.523 | 0.278/0.347/0.274 |

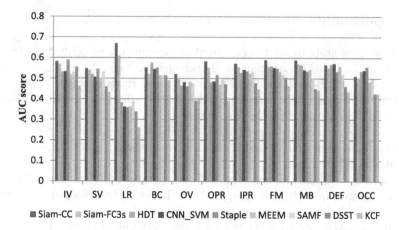

**Fig. 6.** The AUC of eleven attributes including Illumination Variation (IV), Scale Variation (SV), Low Resolution (LR), Background Clutters (BC), Out-of-View (OV), Out-of-Plane Rotation (OPR), In-Plane Rotation (IPR), Fast Motion (FM), Motion Blur (MB), Deformation (DEF) and Occlusion (OCC) on OTB-2015.

**Fig. 7.** The tracking results of Siam-CC and eight state-of-art trackers on *Bike, Motor-Rolling, Skiing* and *Walking2*.

can effectively track the objects of these four video sequences and it achieves a more accurate and tighter bounding box. In short, both qualitative and quantitative results validate the effectiveness of Siam-CC.

# 6    Conclusion

This paper developed a cross-layer convolutional Siamese network tracker (Siam-CC) to capture diverse semantic features from different layers by learning the cross-layer convolutional response maps. By incorporating a adaptive contrastive loss and the logistical loss, Siam-CC can learn discriminative response maps and further filter out some noisy points in the response map. Both strategies can make the final response map more reliable. Experiments on OTB database show that Siam-CC performs favorably against the state-of-the-art trackers.

**Acknowledgment.** This work was supported by the National Natural Science Foundation of China [61806213], the National Natural Science Foundation of China [U1435222] and the National High-tech R&D Program [2015AA020108].

# References

1. Bertinetto, L., Valmadre, J., Golodetz, S., Miksik, O., Torr, P.H.S.: Staple: complementary learners for real-time tracking. In: IEEE Conference on Computer Vision and Pattern Recognition, pp. 1401–1409 (2016)
2. Bertinetto, L., Valmadre, J., Henriques, J.F., Vedaldi, A., Torr, P.H.S.: Fully-convolutional siamese networks for object tracking. In: Hua, G., Jégou, H. (eds.) ECCV 2016. LNCS, vol. 9914, pp. 850–865. Springer, Cham (2016). https://doi.org/10.1007/978-3-319-48881-3_56
3. Danelljan, M., Häger, G., Khan, F.S.: Accurate scale estimation for robust visual tracking. In: British Machine Vision Conference, pp. 65.1–65.11 (2014)
4. Deng, C., Huang, G., Xu, J., Tang, J.: Extreme learning machines: new trends and applications. Sci. China Inf. Sci. **58**(2), 1–16 (2015)
5. Girshick, R., Donahue, J., Darrell, T., Malik, J.: Rich feature hierarchies for accurate object detection and semantic segmentation. In: IEEE Conference on Computer Vision and Pattern Recognition, pp. 580–587 (2014)
6. Guo, Q., Feng, W., Zhou, C., Huang, R., Wan, L., Wang, S.: Learning dynamic Siamese network for visual object tracking. In: IEEE International Conference on Computer Vision, pp. 1–9 (2017)
7. Hadsell, R., Chopra, S., Lecun, Y.: Dimensionality reduction by learning an invariant mapping. In: IEEE Conference on Computer Vision and Pattern Recognition, pp. 1735–1742 (2006)
8. He, A., Luo, C., Tian, X., Zeng, W.: A twofold Siamese network for real-time object tracking. arXiv:1802.08817 (2018)
9. He, K., Zhang, X., Ren, S., Sun, J.: Delving deep into rectifiers: surpassing human-level performance on imagenet classification. In: IEEE International Conference on Computer Vision, pp. 1026–1034 (2015)
10. Held, D., Thrun, S., Savarese, S.: Learning to track at 100 FPS with deep regression networks. In: Leibe, B., Matas, J., Sebe, N., Welling, M. (eds.) ECCV 2016. LNCS, vol. 9905, pp. 749–765. Springer, Cham (2016). https://doi.org/10.1007/978-3-319-46448-0_45
11. Henriques, J.F., Rui, C., Martins, P., Batista, J.: High-speed tracking with kernelized correlation filters. IEEE Trans. Pattern Anal. Mach. Intell. **37**(3), 583–596 (2015)
12. Hong, S., You, T., Kwak, S., Han, B.: Online tracking by learning discriminative saliency map with convolutional neural network. In: International Conference on Machine Learning, pp. 597–606 (2015)
13. Krizhevsky, A., Sutskever, I., Hinton, G.E.: ImageNet classification with deep convolutional neural networks. In: Advances in Neural Information Processing Systems, pp. 1097–1105 (2012)
14. Li, Y., Zhu, J.: A scale adaptive kernel correlation filter tracker with feature integration. In: Agapito, L., Bronstein, M.M., Rother, C. (eds.) ECCV 2014. LNCS, vol. 8926, pp. 254–265. Springer, Cham (2015). https://doi.org/10.1007/978-3-319-16181-5_18

15. Ma, C., Huang, J.B., Yang, X., Yang, M.H.: Hierarchical convolutional features for visual tracking. In: IEEE International Conference on Computer Vision, pp. 3074–3082 (2015)
16. Nam, H., Han, B.: Learning multi-domain convolutional neural networks for visual tracking. In: IEEE Conference on Computer Vision and Pattern Recognition, pp. 4293–4302. IEEE (2016)
17. Qi, Y., et al.: Hedged deep tracking. In: IEEE Conference on Computer Vision and Pattern Recognition, pp. 4303–4311 (2016)
18. Russakovsky, O., et al.: ImageNet large scale visual recognition challenge. Int. J. Comput. Vis. **115**(3), 211–252 (2015)
19. Simonyan, K., Zisserman, A.: Very deep convolutional networks for large-scale image recognition. arXiv preprint arXiv:1409.1556 (2014)
20. Teng, Z., Xing, J., Wang, Q., Lang, C., Feng, S., Jin, Y.: Robust object tracking based on temporal and spatial deep networks. In: IEEE Conference on Computer Vision and Pattern Recognition, pp. 1144–1153 (2017)
21. Valmadre, J., Bertinetto, L., Henriques, J., Vedaldi, A., Torr, P.H.: End-to-end representation learning for correlation filter based tracking. In: IEEE Conference on Computer Vision and Pattern Recognition, pp. 5000–5008. IEEE (2017)
22. Wang, L., Ouyang, W., Wang, X.: Visual tracking with fully convolutional networks. In: IEEE International Conference on Computer Vision, pp. 3119–3127 (2015)
23. Wu, Y., Lim, J., Yang, M.H.: Online object tracking: A benchmark. In: IEEE Conference on Computer Vision and Pattern Recognition, pp. 2411–2418. IEEE (2013)
24. Wu, Y., Lim, J., Yang, M.H.: Object tracking benchmark. IEEE Trans. Pattern Anal. Mach. Intell. **37**(9), 1834–1848 (2015)
25. Zhang, J., Ma, S., Sclaroff, S.: MEEM: robust tracking via multiple experts using entropy minimization. In: Fleet, D., Pajdla, T., Schiele, B., Tuytelaars, T. (eds.) ECCV 2014. LNCS, vol. 8694, pp. 188–203. Springer, Cham (2014). https://doi.org/10.1007/978-3-319-10599-4_13

# A Data Augmentation Model Based on Variational Approach

Lei Xia, Jiancheng Lv$^{(\boxtimes)}$, and Yong Xu

MILab, Computer Science College, SiChuan University, Chengdu, China
lvjiancheng@scu.edu.cn

**Abstract.** The labeled training data are very rare in actual environment. Generating new data based on given label is one of the most commonly approaches in data augmentation. This paper proposes a new data augmentation model that can extract the deformation features between the given deformation image and the original image. The model generates similar images to the given deformation images according to the deformation feature. The model can keep the new generation images have the same probability distribution as the given deformation images. Experiments on MNIST and CIFAR-10 prove that the new deformation images can get a similar classification accuracy with the given deformation images, which proves that the new sample is effective.

**Keywords:** Generation · Deformation features · Distribution
Data augmentation

## 1 Introduction

Deep learning requires a large number of labeled data to train the network converge to a better state. However, due to the lack of labeled data in experiments and the existence of different number of training data with different catalog, the training of deep learning model can not learn features well. In this paper, a data augmentation model for image deformation is proposed. By extracting the deformation features between the original images and the given labeled deformation images, model learns the distribution of the deformation features and sample from it to generate new deformation images. The advantage of this model is that the generated images can have the same distribution with the given labeled deformation images, so it reserves the deformation invariance in original samples.

At present, the common approaches be used of data augmentation in image processing are rotating images, cutting images, changing image chromatic aberration, distorting image features, changing image size, enhancing image noise (usually using Gauss noise, salt pepper noise), and so on. The classification performance to the augmented data through random deformation approaches to raw

This work was supported by the National Science Foundation of China (Grant No. 61625204), partially supported by the State Key Program of National Science Foundation of China (Grant No. 61432012 and 61432014).

© Springer Nature Switzerland AG 2018
L. Cheng et al. (Eds.): ICONIP 2018, LNCS 11302, pp. 157–168, 2018.
https://doi.org/10.1007/978-3-030-04179-3_14

images can be improved by extracting invariant features [1,2]. This approach has achieved good results in many tasks [3]. But it mostly relies on a manual processing [4]. This processing is mainly embodied in the invariance of the image by a series of simple linear deformation operations designed by mankind experts, such as flip, rotation, or zoom, for a specified task of image classification. The essence of this process is actually a feature engineering. This makes data become a second role, but experts become the dominant one. This operation costs a lot human resources and time [4].

If we realized the data augmentation by extracting the deformation feature $T^\theta$ among the same catalog images, then the given sample "$x$" with class label "$y$" will generate a new sample "$x \bullet T^\theta$" which should also have the same class label with given sample "$x$". This idea lies at the heart of Grenanders Pattern Theory [5], which encodes invariances by describing the data itself as transformations acting on reference objects.

To count this, A widely-used approach is to synthesize new observations by applying the known $T^\theta$ to the training data, and then train a classifier on the augmented data set [6]. Eigen et al. [7] consider multiple invariances including length scaling, rotation, translation, horizontal flip and color scaling in a deep multiscale network for single-image depth estimation. In work on speech recognition, Jaitly and Hinton [8] apply Vocal Tract Normalization (VTLN) as a way to artificially transform utterances of one speaker to the voice of another. The Infinite MNIST data set [9] was generated by considering horizontal and vertical translations, rotations, horizontal and vertical scalings, hyperbolic transformations, and random Gaussian perturbations.

All the above methods assume that the deformation feature $T^\theta$ between samples has already obtained. But in fact, the prerequisite is hard to satisfy. The purpose of this paper is to find the prior deformation feature. In order to make sure the generated samples to have the identical distribution with given deformation samples, we need to get another deformation features which should be similar to the prior deformation feature. This is the objective of the current manuscript.

The remaining part of this article is organized as follows: the second section introduces the related work of this model; the third section details the structure and detailed inference of this model; the fourth section evaluates the performance on data augmentation and the classification accuracy; the fifth section is the summary and prospect of the work.

## 2   Motivation

To generate large amounts of new samples, decrease human intervention, and meanwhile, ensure that the new samples are distributed identically with the given deformation samples, we consider an approach for learning image deformations and generate new samples according to the similar deformations. Assume the given image $x$ and its deformation $\hat{x}$ have the deformation feature $T^\theta$ with parameter $\theta$. If there is a feature $\hat{T}^\theta$ has the same distribution to the given deformation feature $T^\theta$, then another deformation image $\tilde{x}$ will be obtained under the parameter $\hat{T}^{\theta}$. If the classification result of $\tilde{x}$ is consistent with the result of a given deformation image $\hat{x}$, then we consider the $\tilde{x}$ will be an effective generative sample of $\hat{x}$.

The strategy of this paper is to achieve the data augmentation by retrieving deformation feature $T^{\theta}$ and generating new $\hat{T}^{\theta}$. The identical distribution of two deformation features means that they will result to a similar outputs. Through the deep learning network, the deformation feature between the given image pairs can be obtained. Generate the new deformation feature to obey the prior distribution $T^{\theta}$, then the data augmentation for the given deformation image is completed. During this process, only a small amount of deformation target samples will be provided by user. The proposed model will greatly decrease human intervention, generate augmented data and maintain identically distribution with given deformation data.

## 3  Related Work

### 3.1  Factored Gated Restricted Boltzmann Machine

Gated Restricted Boltzmann Machine (GRBM) is a gated conditional restricted boltzmann machine [10] proposed by Memisevic and Hinton, et al. [11]. By placing multiplicative interactions between the input sample pair $(x, y)$ and its hidden layer $h$, the model is used to reveal the deformation features between the input sample pairs and to encode the feature in the hidden layer. However, as the weight matrix $W$ of the model is a three-dimensional tensor, the number of parameters in $W$ increases exponentially while the input neural unit number increase, which makes the model more difficult to train. So Hinton et al. improved it, and put forward factored gated restricted boltzmann machine (FGRBM) [12]. The FGRBM introduces a factor $F$ into the GRBM, which reduces the weight matrix from three dimensions to two dimensions. It results the number of parameters of the model to decrease and enhanced the training efficiency of the model. The energy function of the model is defined in Eq. (1)

$$E(y, h; x) = -\sum_{f}(\sum_{i} W_{if}^{x} x_i)(\sum_{j} W_{jf}^{y} y_j)(\sum_{k} W_{kf}^{h} h_k) - \sum_{j} b_j y_j - \sum_{k} c_k h_k$$

(1)

where the $(x, y)$ is the input sample pair. $h$ is the hidden layer. $i, j, k$ are the subindex of $x, y, h$ respectively. $W_{if}, W_{jf}, W_{kf}$ stand for the weight matrix of $x, y, h$ which connect to factor $F$ who there are $f$ units. The $b, c$ are the bias terms of $y$ and $h$. The target of FGRBM is to find the parameter configuration to maximize the joint probability of $p(y, h)$ under given $x$. So the corresponding joint probability function is defined in Eq. (2).

$$p(y, h; x) = \frac{exp(-E(y, h; x))}{z}$$

(2)

where the $z = \sum_{y,h} exp(-E(y,h;x))$ is partition function According to the joint probability function the training target is equivalent to maximize the log-likelihood $logp(y,h;x)$ in Eq. (3).

$$logp(y,h;x) = -E(y,h;x) - logz \qquad (3)$$

## 3.2 Variational Autoencoder

If it can find the distribution of a given sample $x$, the model can directly sample and generate new samples from the distribution. But the distribution is always intractable. Variational autoencoder (VAE) provides a way to transform the unknown distribution of original data to a specific distribution. New samples are obtained by sampled from this specific distribution. The model uses $p(z)$ to express the prior distribution of $z$ in hidden layer. In the case of given data sample $x$, the distribution of $z$ is notated by $p_\theta(z|x)$, which is called a posteriori distribution. Because the distribution is usually intractable, we usually take $q_\phi(z|x)$ as an approximation to $p_\theta(z|x)$. Because $z$ is the latent representation of input $x$, the $q_\phi(z|x)$ can be seen as an encoder, and the $p_\theta(x|z)$ is a corresponding decoder. The observed samples are obey to the unknown distribution $q(x)$. It is intractable to find the configuration of parameter $\theta$ of $p_\theta(x)$ to maximize the log-likelihood estimation [13]. Therefore, VAE [14,15] tries to obtain the maximum lower bounds for $p_\theta(x)$ by maximizing the variational expression $\ell_{VAE}(\theta,\phi)$, as shown in Eq. (4), in which the $(\theta,\phi)$ is a related parameter.

$$\ell_{VAE}(\theta,\phi) = \mathbb{E}_{q_\phi(x,z)} log \frac{p_\theta(x,z)}{q_\phi(z|x)} = \mathbb{E}_{q(x)}[logp_\theta(x) - \text{KL}(q_\phi(z|x)||p_\theta(z|x))] \qquad (4)$$

As the items $\mathbb{E}_{q_\phi(x,z)}$ and $\mathbb{E}_{q(x)}$ are intractable, the model has to do the approximate computation by sampling. It is equally to maximize the $\ell_{VAE}(\theta,\phi)$ to calculate the lower bound of $\frac{1}{N}\sum_{i=1}^{N} logp_\theta(x_i)$. Therefore, the training to VAE equals to get the configuration of parameter $\theta$ to make the $KL(q(x)||p_\theta(x))$ minimum.

# 4 Variational Factored Gated Restricted Boltzmann Machine

## 4.1 Model Description

In order to achieve the data augmentation under given deformation samples, it is a key processing to obtain the deformation features between the given deformation sample and the original sample, and form the deformation code according to the feature. It is a key problem to restrict the distribution of the deformation code to a specific distribution and sampling from it to get a new deformation code. The problem can be discussed from two aspects: i. how to learn the distribution of the deformation code when encoded the deformation features from given images; ii. How to sample from the learned distribution to generate new deformation codes and further more to generate new deformation samples. Based on the above problems, the structure of the proposed model is shown in Fig. 1. Model takes the sample pair $\{X, Y\}$ as input. Sample $X$ is the given original input and $Y$ is the target deformation sample of given $X$. The $X, Y$ and $H$ connect to factor $(F)$ by $W_{if}^X, W_{jf}^Y, W_{kf}^H$ respectively, where $i, j, k, f$ are the subindex of $X, Y, H, F$. From the [12] we can consider that when the model learns the joint distribution of $Y$ and $H$ under given samples $X$, marked $p(Y, H; X)$, then we can take the $H$ as deformation codes between $X$ and $Y$. Model maps the distribution of $H$ into a multi-dimensional Gaussian distribution, marked $H_{enc}$. It can calculate the mean vector$(\mu)$ and standard deviation vector$(\sigma)$ of $H_{enc}$ to make them identically to the target distribution, such as standard normal distribution $\mathcal{N}(0, 1)$. Sampling from the $H_{enc}$ will get a $Z$ which can be decoded to a new deformation code $H_{dec}$. Finally, the new deformation sample $Y_{new}$ achieved through $H_{dec}$ and $X$.

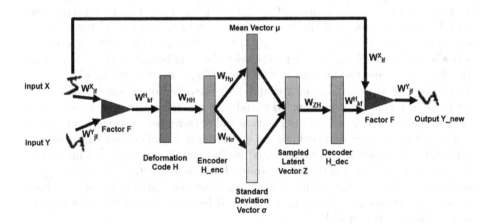

**Fig. 1.** Model structure of VFGRBM

This model hopes to make the joint probability distribution of $(Y, H)$, say $p(Y, H; X)$, to achieve the maximum value, and meanwhile find the true distribution of $p_\theta(Z|H)$. But as the $p_\theta(Z|H)$ can't be calculated directly, we have to approximate the $p_\theta(Z|H)$ by the variational distribution called $q_\phi(Z|H)$, this requires the $KL(q_\phi(Z|H)||p_\theta(Z|H))$ to be minimum. After a similar deduction to VAE on the $KL(q_\phi(Z|H)||p_\theta(Z|H))$ part, the goal of this model equivalents to maximize the value of Eq. (5).

$$Target = logp(Y, H; X) + \mathcal{L}(\theta, \phi; H)$$
$$= log\frac{exp(-E(y, h; x))}{z} + E_{q_\phi(Z|H)}[logp_\theta(H|Z)] - KL[q_\phi(Z|H)||p_\theta(Z)]$$
$$(5)$$

where the formula of $E(y, h; x)$ is shown in Eq. (1), and $z = \sum_{y,h} exp(-E(y, h; x))$.

To ensure samples $Y$ and $Y_{new}$ are in the same distribution, the proposed model shares the parameters which connected to factor $F$. For deformation coding $H$, it is like a black box to find another deformation coding $H_{dec}$ that obeys the same distribution. During the processing of changing the dimension of $H$, we constrain the $H_{enc}$ to the multidimensional Gaussian distribution with zero mean and unit invariance. According to FGRBM, it is equivalent to minimize $E(y, h; x)$ and maximize the $p(Y, H; X)$. Different with FGRBM, the proposed model does not sample on the original dimension of $H$ but the $H_{enc}$. The advantage of sampling on $H_{enc}$ will not result to the generation invalid for it's huge deviation. Therefore, we introduce the sharing parameters mechanism into the model to make the generation samples have the same distribution to the given deformation samples.

To maximize the $E_{q_\phi(Z|H)}[logp_\theta(H|Z)]$ and minimize the $KL[q_\phi(Z|H)||p_\theta(Z|H)]$ under given $H$ in the second and third part in Eq. (5), it means to make $q_\phi(Z|H)$ as close as possible to the real distribution $p_\theta(Z|H)$. We assume that the real distribution of $p_\theta(Z|H)$ is the standard normal distribution $\mathcal{N}(0,1)$ for convenience of discussion. The model encodes the $H$ through the formula $H_{enc} = Relu(W_{HH} * H + b_{H_{enc}})$ to another dimensionality. The $b_{H_{enc}}$ is the bias term of $H_{enc}$. In order to restrict the distribution of $H_{enc}$ to normal distribution, we use the formula $\mu = H_{enc} * W_{H\mu} + b_\mu$, $\sigma = H_{enc} * W_{H\sigma} + b_\sigma$ to calculate the expectation and variance of its distribution, where the $b_\sigma$ and $b_\mu$ are the bias terms corresponding to $\sigma$ and $\mu$, respectively. In order to make the sampling of $Z$ to obey normal distribution and perform back propagation, model first samples from the standard normal distribution $\mathcal{N}(0,1)$ to get $\epsilon$ just like the VAE did. Therefore, $Z$ can be obtained by formula $Z = \sigma * \epsilon + \mu$. Finally, the $H_{dec}$ is obtained from the decoding process by $H_{dec} = Relu(W_{ZH} * Z + b_{H_{dec}})$. In essential, $H_{dec}$ is the result of sampling from the distribution of $H$, which is the reconstruction of $H$. $H_{dec}$ and $X$ will be used as conditional inputs to achieve the generation of $Y_{new}$.

## 4.2   Inference

The model chooses stochastic gradient descent algorithm to update its parameters. The general gradient derivation formula shown in Eq. (6).

$$\frac{\partial Target}{\partial \Theta} = \frac{\partial(-E(y,h;x))}{\partial \Theta} - \frac{\partial \sum_{y,h}(-E(y,h;x))}{\partial \Theta} + \frac{\partial E_{q_\phi(Z|H)}[logp_\theta(H|Z)]}{\partial \Theta} - \frac{\partial KL[q_\phi(Z|H)||P_\theta(Z)]}{\partial \Theta} \tag{6}$$

where the $\Theta = \{W_{if}^X, W_{jf}^Y, W_{kf}^H, W_{HH}, W_{H\mu}, W_{H\sigma}, W_{ZH}\}$.

During the processing of solving the first part of Eq. (6), we should obtain the latent representation $H$ under given $X$ and $Y$ according to the Eq. (3). The formula of $H$ is shown in Eq. (7).

$$H \sim p(h_k = 1|X, Y) = \sigma(\sum_f (\sum_i W_{if}^x x_i)(\sum_j W_{jf}^y y_j) W_{kf}^h + c_k) \tag{7}$$

where the $\sigma(\bullet)$ is the sigmoid function, $c$ is the bias term of $H$, $k$ is the subindex of $H$. Symbol "$\sim$" stands for sampling. During the calculate the second term of Eq. (6), the model needs to take all the possible configuration of $Y$ and $H$ into account. It is obviously intractable. According to the restricted boltzmann machine(RBM), the model chooses the contrastive divergence(CD) algorithm prompted by Hinton et al. [16] to approximate computing the gradient of $Z$ in Eq. (6). Thus, the approximation of $Y$ under given $X$ and $H\_dec$ is shown in Eq. (8)

$$Y \sim p(y_j = 1|X, H\_{dec}) = \sigma(\sum_f(\sum_i W_{if}^x x_i)(\sum_k W_{kf}^h h\_{dec_k})W_{jf}^y + b_j) \quad (8)$$

where $H\_{dec}$ represents the sampling from the distribution of $H$. The third term in Eq. (5) is to make sure that the $H\_{dec}$ and $H$ should be as close as possible with each other. In this way, the $Y\_{new}$ can be generated with given different $H\_{dec}$. Model takes the least square procedure as the training algorithm. The fourth term in Eq. (5) is the KullbackLeibler divergence between $q_\phi(Z|H)\|p_\theta(Z)$. The purpose of the fourth term is to get the sampling distribution $q_\phi(Z|H)$ of $Z$ under given $H$ to as close to its true distribution $p_\theta(Z)$. According to the formula of KL divergence of multidimensional distribution, we can have the formula of $KL[q_\phi(Z|H)\|p_\theta(Z)]$ shown in Eq. (9).

$$KL[q_\phi(Z|H)\|p_\theta(Z)] = \frac{1}{2}\sum_L^{l=1}(1 + log(\sigma_l^2) - \mu_l^2 - \sigma_l^2) \quad (9)$$

## 5   Experiment

We evaluate the learned augmentation scheme on the MNIST dataset and CIFAR10. We first transform the original training dataset respectively to the target deformation dataset. Finally, we evaluate whether the generated samples are effective or not according to the classification results between the original samples and the target deformation samples.

### 5.1   MNIST

The original MNIST dataset consists of 60,000 training images and 10,000 test images. We hold out 10,000 of the training images to form a validation set. The augmentations are, thus, based on only 40,000 training images

First, we rotate 40000 training samples of the MNIST clockwise 90°, clockwise 180°, anticlockwise 90°, shift 3 pixels upward, horizontal flip and vertical flip respectively, and get six new datasets. We group these new datasets with corresponding samples of original dataset together to be input pairs to train various deformations. Finally, we obtain the generated datasets named get_MC90, gen_MC180, gen_MU90, gen_MUP3, gen_MFliplr, gen_MFlipud.

We evaluate our model on MNIST dataset with 100 units on factor and 2000 units on hidden layer. Figure 2 shows the classification accuracy matrix by

1-Layer NN with training for 300 epoches. Observe from it, the classification result is almost equal to the state-of-art performance. Figure 3 shows some generated samples by the proposed model. It shows each number in four lines. Each line shows a kind of deformations. The first column is the original sample and the second column is the deformation target samples, while from the third to sixth columns are generated samples. Observe from the Fig. 3, we consider that however the data augmentation samples are more similar to the given deformation samples, but still there are some slight differences between them. For example, there are obvious differences on pixel brightness between samples in $3^{rd}$ columns and $5^{th}$ columns of $3^{rd}$ rows. It means the generation hold the feature of original sample while there are some additional noise or features in them.

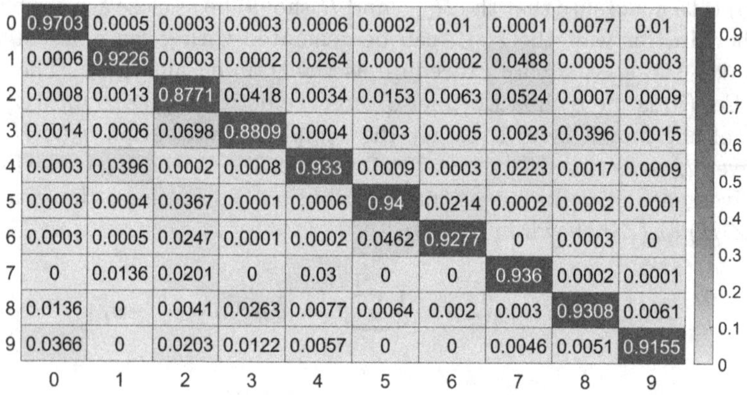

| | 0 | 1 | 2 | 3 | 4 | 5 | 6 | 7 | 8 | 9 |
|---|---|---|---|---|---|---|---|---|---|---|
| 0 | 0.9703 | 0.0005 | 0.0003 | 0.0003 | 0.0006 | 0.0002 | 0.01 | 0.0001 | 0.0077 | 0.01 |
| 1 | 0.0006 | 0.9226 | 0.0003 | 0.0002 | 0.0264 | 0.0001 | 0.0002 | 0.0488 | 0.0005 | 0.0003 |
| 2 | 0.0008 | 0.0013 | 0.8771 | 0.0418 | 0.0034 | 0.0153 | 0.0063 | 0.0524 | 0.0007 | 0.0009 |
| 3 | 0.0014 | 0.0006 | 0.0698 | 0.8809 | 0.0004 | 0.003 | 0.0005 | 0.0023 | 0.0396 | 0.0015 |
| 4 | 0.0003 | 0.0396 | 0.0002 | 0.0008 | 0.933 | 0.0009 | 0.0003 | 0.0223 | 0.0017 | 0.0009 |
| 5 | 0.0003 | 0.0004 | 0.0367 | 0.0001 | 0.0006 | 0.94 | 0.0214 | 0.0002 | 0.0002 | 0.0001 |
| 6 | 0.0003 | 0.0005 | 0.0247 | 0.0001 | 0.0002 | 0.0462 | 0.9277 | 0 | 0.0003 | 0 |
| 7 | 0 | 0.0136 | 0.0201 | 0 | 0.03 | 0 | 0 | 0.936 | 0.0002 | 0.0001 |
| 8 | 0.0136 | 0 | 0.0041 | 0.0263 | 0.0077 | 0.0064 | 0.002 | 0.003 | 0.9308 | 0.0061 |
| 9 | 0.0366 | 0 | 0.0203 | 0.0122 | 0.0057 | 0 | 0 | 0.0046 | 0.0051 | 0.9155 |

**Fig. 2.** The classification accuracy heat map on MNIST by 1-Layer NN.

In order to prove the availability of the sample generated by this model, we use several classification models announced by official, such as Linear Classifier (1-layer NN), K-Nearest-Neighbors (KNN), SVM DEG 4 polynomial (SVM), hidden layer neurons 300 (2-layer NN), convolutional neural network (CNN)and unsupervised sparse features + SVM(USF+SVM) to compare the classification error rate between generations and original samples. Table 1 shows the classification error rate between generation and original samples. The results of the classification of original samples in that table are all derived from the publication of official [17]. As we can see from the Table 1, the classification error rate of the original sample is lower than the generation samples, but the classification error rate of the model is almost close to the original error rate of given sample. This indicates that the augmented data is available and effective.

A standard approach for one-shot learning is to learn an appropriate distance between representation which can be classified by a nearest neighbour algorithm under some metric. We choose the Matching Networks [18] to learn a representation space to produce a good classifier from it. Because the Matching Networks can only learn to match based on individual samples, we train the VFGRBM on

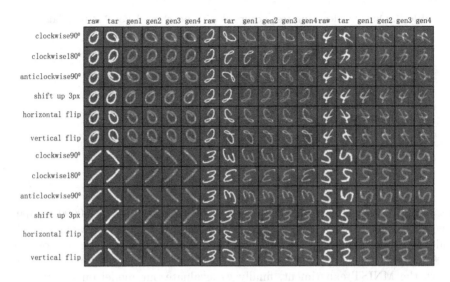

**Fig. 3.** Generations of six kind deformations on MNIST.

**Table 1.** Comparison of the training classification error rate between generation and original samples on MNIST dataset.

| Model | Original | gen_MC90 | gen_MC180 | gen_MU90 | gen_MUP3 | gen_MFliplr | gen_MFlipud |
|---|---|---|---|---|---|---|---|
| 1-Layer NN | 8.4% | 9.69% | 9.74% | 9.69% | 9.3% | 9.63% | 9.67% |
| 2-Layer NN | 1.6% | 1.85% | 1.67% | 1.68% | 1.9% | 1.66% | 1.68% |
| KNN | 2.4% | 2.9% | 2.7% | 2.8% | 2.6% | 2.7% | 2.9% |
| SVM | 1.1% | 1.9% | 1.5% | 1.3% | 1.4% | 1.4% | 1.5% |
| CNN | 1.1% | 1.52% | 1.62% | 1.44% | 1.4% | 1.53% | 1.61% |
| USF+SVM | 0.59% | 0.92% | 0.85% | 0.74% | 0.77% | 0.86% | 0.79% |

the source domain and then so do the Matching Networks. To make the process be differentiable, we train the previous two models with a $\epsilon$ to select the best sample to create the augmentation data. Besides the Matching Networks, Prototypical Networks [19] and Neural Statistician [20] are another benchmarks. All of these two models are designed for one-shot or few-shot learning. We combine the Pixel Distance with proposed model and classified by a vanilla classifier. To simulate the lack of training data, we chose 50 original samples for each class in MNIST and augment them to fullfil the comparison. The result is shown in Table 2. From the table we can find that, the proposed model promotes the accuracy from 30.24% to 53.78% after the data augmentation and to the Matching Networks and Prototypical Networks, there are almost 1% enhancement.

## 5.2  CIFAR-10

Dataset CIFAR10 is composed by 60000 three channel color images with 32*32 pixels. It is divided into 10 cataloges: airplane, automobile, bird, cat, deer,

**Table 2.** One-shot learning test accuracy with benchmarks on MNIST dataset.

| Model name | Test accuracy |
|---|---|
| Pixcel Distance | 0.3024 |
| Pixcel Distance+VFGRBM | **0.5378** |
| Matching Networks | 0.9044 |
| Matching Networks+VFGRBM | **0.9135** |
| Progotypical Networks | 0.9301 |
| Progotypical Networks+VFGRBM | **0.941** |

dog, frog, horse, ship, truck. There are 40000 training images, 10000 validate images and 10000 test images. The data set is divided into 4 training blocks and 1 test blocks, each block contains 10000 images. The training set contains 4000 images per class, and the test set contains 1000 images per class. Similar to the MNIST experiment, finally we evaluate our model on the six new generation datasets: gen_CC90, gen_CC180, gen_CU90, gen_CUP3, gen_CFliplr, gen_CFlipud based on the given deformation datasets deformed from the original dataset. To decrease the burden on the computation, we just use the first channel of each image to be the training input. Similar to the previous section, Fig. 4 shows some generation samples. For each subplot, the first column is part of the original dataset, the second column is the target deformation samples and from third to six columns are generated data.

Table 3 shows the classification error rate on CIFAR-10 dataset. Observed from the Table 3 we can find that, the classification error rate on the last deformation is relative higher than the others. The reason probably is the shift-up

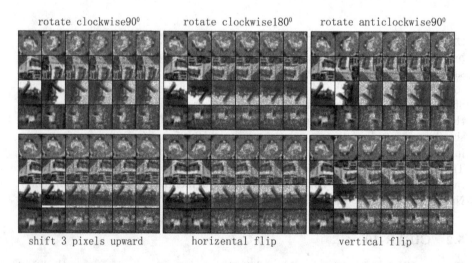

**Fig. 4.** Generation of six deformations on CIFAR-10.

**Table 3.** Comparison of the training classification error rate between generation and original samples on CIFAR-10 dataset.

| Model | Original | gen_CC90 | gen_CC180 | gen_CU90 | gen_CUP3 | gen_CFliplr | gen_CFlipud |
|---|---|---|---|---|---|---|---|
| 1-Layer NN | 11.25% | 11.68% | 12.22% | 11.55% | 13.06% | 11.76% | 11.82% |
| 2-Layer NN [21] | 7.51% | 8.86% | 8.33% | 8.72% | 8.85% | 8.41% | 8.63% |
| KNN | 9.89% | 12.07% | 12.73% | 11.24% | 10.37% | 12.37% | 12.46% |
| SVM | 10.67% | 13.9% | 13.02% | 14.74% | 11.42% | 13.22% | 13.03% |
| CNN [22] | 3.47% | 5.89% | 5.76% | 6.29% | 4.54% | 5.77% | 5.81% |
| USF+SVM | 7.67% | 9.98% | 11.65% | 10.82% | 9.59% | 10.93% | 11.14% |

deformation changed the feature in the original dataset and break the distribution of the original samples, as shown at the third row of the left bottom image in Fig. 4, while the others are only rotated with different angles and keep the distribution of the original samples well.

## 6   Conclusion

This paper presents a data augmentation model for deformation images. The model can transform the distribution of the feature to a specified distribution by extracting the deformation features between the original images and the deformation images. By sampling the distribution, new deformation images are generated. This model can solve the problem of lacking labeled data. Through data augmentation, we can provide more effective samples for feature extraction. Next research will focus on the acquisition and integration expression of multiple deformation features. In this paper, we train a model for each deformation dataset. Following work will discuss how to improve the model's representation ability to retrieve multi-deformation features from input image pairs.

## References

1. Baird, H.S.: Document image defect models. In: Baird, H.S., Bunke, H., Yamamoto, K. (eds.) Structured Document Image Analysis. Springer, Heidelberg (1992). https://doi.org/10.1007/978-3-642-77281-8_26
2. Simard, P., Victorri, B., Cun, Y. L., Denker, J.: Tangent Prop: a formalism for specifying selected invariances in an adaptive network. In: International Conference on Neural Information Processing Systems, pp. 895–903 (1991)
3. Simard, P.Y., Steinkraus, D., Platt, J.C.: Best practices for convolutional neural networks applied to visual document analysis. In: International Conference on Document Analysis and Recognition, p. 958 (2003)
4. Hauberg, S., Freifeld, O., Larsen, A.B.L., Fisher Iii, J.W., Hansen, L.K.: Dreaming more data: class-dependent distributions over diffeomorphisms for learned data augmentation. In: Computer Science (2015)
5. Grenander, U.: General Pattern Theory: A Mathematical Study of Regular Structures. Clarendon Press, New York (1993)

6. Krizhevsky, A., Sutskever, I., Hinton, G.E.: Imagenet classification with deep convolutional neural networks. In: International Conference on Neural Information Processing Systems, pp. 1097–1105 (2012)
7. Eigen, D., Puhrsch, C., Fergus, R.: Depth map prediction from a single image using a multi-scale deep network, pp. 2366–2374 (2014)
8. Jaitly, N., Hinton, G.E.: Vocal tract length perturbation (VTLP) improves speech recognition. In: ICML Workshop on Deep Learning for Audio, Speech and Language, p. 958 (2013)
9. Loosli, G., Canu, S., Bottou, L.: Invariant SVM using selective sampling training invariant support vector machines using selective sampling, pp. 301–320 (2014)
10. Taylor: Modeling human motion using binary latent variables. In: International Conference on Neural Information Processing Systems, pp. 1345–1352 (2006)
11. Memisevic, R., Hinton, G.: Unsupervised learning of image transformations. In: IEEE Conference on Computer Vision and Pattern Recognition, CVPR 2007, pp. 1–8 (2007)
12. Memisevic, R., Hinton, G.E.: Learning to represent spatial transformations with factored higher-order Boltzmann machines. Neural Comput. **22**(6), 1473–1492 (2010)
13. Pu, Y., et al.: Adversarial symmetric variational autoencoder (2017)
14. Rezende, D.J., Mohamed, S., Wierstra, D.: Stochastic backpropagation and approximate inference in deep generative models, pp. 1278–1286 (2014)
15. Kingma, D.P., Welling, M.: Auto-encoding variational Bayes (2013)
16. Hinton, G.E., Osindero, S., Teh, Y.W.: A fast learning algorithm for deep belief nets. Neural Comput. **18**(7), 1527–1554 (2014)
17. Lecun, Y., Bottou, L., Bengio, Y., Haffner, P.: Gradient-based learning applied to document recognition. Proc. IEEE **86**(11), 2278–2324 (1998)
18. Vinyals, O., Blundell, C., Lillicrap, T., Kavukcuoglu, K., Wierstra, D.: Matching networks for one shot learning (2016)
19. Snell, J., Swersky, K., Zemel, R.S.: Prototypical networks for few-shot learning (2017)
20. Edwards, H., Storkey, A.: Towards a neural statistician (2016)
21. Agostinelli, F., Hoffman, M., Sadowski, P., Baldi, P.: Learning activation functions to improve deep neural networks. In: Computer Science (2015)
22. Graham, B.: Fractional max-pooling. Eprint Arxiv (2014)

# Asynchronous Value Iteration Network

Zhiyuan Pan, Zongzhang Zhang[(⊠)], and Zixuan Chen

School of Computer Science and Technology, Soochow University,
Suzhou, People's Republic of China
zzzhang@suda.edu.cn

**Abstract.** Value iteration network (VIN) improves the generalization of
a policy-based neural network by embedding a planning module. However, this module performs value iteration on the entire state space of
a Markov decision process and all states in the space are updated by
sweeping the state space systematically, regardless of their significance.
This paper introduces an improved version of VIN with a novel planning module, called asynchronous value iteration network (AVIN), performing value updates on some states more frequently than other states
asynchronously, depending on their significance/urgency to improve a
policy. The new planning module utilizes the urgency of the states to
prioritize updates at important states. We measure the urgency in a way
of enhancing the global awareness, leading to an improvement of the
generalization ability of policies. AVIN with the new module makes the
value updates more efficient and effective, thus significantly demonstrating better generalization on unknown environments.

**Keywords:** Markov decision process · Value iteration
Asynchronous update

## 1 Introduction

Markov decision processes (MDPs) [2] provide a good mathematical framework
for sequential decision problems. In unknown MDP environments, the agent
attempts to find an optimal policy derived from the value function with two different basic approaches: imitation learning (IL) [12] and reinforcement learning
(RL) [15]. In IL, the learning process of value function is conducted by imitating
the behavior of experts in a supervised way. In RL, the goal of the agent is to
maximize the cumulative reward signals from the interaction of the environment.
In general, the cumulative reward to the goal is also defined as a value function,
which is used to search an optimal policy. The similar part — learning an optimal policy, in these two approaches can be formalized by minimizing the costs

---

Z. Pan, Z. Zhang and Z. Chen—Contributed equally to this work.

This work is in part supported by the National Natural Science Foundation of China
under Grant Nos. 61876119 and 61502323.

© Springer Nature Switzerland AG 2018
L. Cheng et al. (Eds.): ICONIP 2018, LNCS 11302, pp. 169–180, 2018.
https://doi.org/10.1007/978-3-030-04179-3_15

of deep neural networks, and the costs indicate the difference between expert's behavior and current behavior in IL [11] and the difference between the optimal value function and the current value function in RL [7], respectively.

Value iteration network (VIN) [17] is a fully differentiable network and can be trained end-to-end by using back propagation in IL or RL methods. The key planning module is embedded into the network by representing the value iteration as the form of a convolutional neural network (CNN), and thus a policy can be computed using the planning information. The policies equipped with the planning information can generalize better than reactive policies, such as fully convolutional network [6], on new, unseen domains. The main drawback of the value iteration is that the entire state space has to be updated at each iteration and planning has the risk of falling into some meaningless sweeps. Meanwhile, the number of value iterations in the planning module will increase as the problem size increases. For example, on navigation domains the shortest path from a starting state to the goal is $L$ steps, then the planning module at least requires $L$ iterations to propagate the reward information of the goal to the starting state. Any action prediction obtained with less than $L$ iterations at the starting state may be unaware of the goal location, and therefore undesirable.

It is well known that asynchronous planning algorithms, such as prioritized sweeping (PS) [8], can be more efficient by updating in a reasonable order. Prioritized sweeping updates a state sampled from a priority queue and states with high urgency can be added to the queue. It is nature to use the temporal-difference (TD) error as the urgency of states because it measures the difference between the current value function and the optimal value function. Prioritized experience replay (PER) [13] also utilizes the TD error as the criterion of the priority and guarantees the probability of being sampled is monotonically increasing according to the priority.

In this paper, we focus on the planning module of VIN and combine the asynchronous update with the VI module. We investigate a heuristic definition of the number of states for asynchronous update, which uses the local connectivity structure to contain the information of the entire state space, and a measure of priority for asynchronous update, which can make the policies more aware of global concept to improve its generalization. In the experimental part, we show that AVIN has better generalization than VIN for problems with different sizes, and explain how to apply the urgency to select important states to make subsequent value iteration more accurate. Finally, we demonstrate that in a more complex navigation environment, AVIN can still find a successful trajectory.

## 2   Background

### 2.1   Markov Decision Process

In sequential decision problems, the general MDP framework can be described as a tuple $(S, A, T, R, \gamma)$, where $S$ represents the state space, $A$ represents the action space, $T$ represents the transition function $T(s'|s, a)$ that indicates the probability distribution over the next state $s'$ after executing action $a$ in state

$s$, $R$ represents the reward function $R(s, a)$ that indicates the numerical reward when executing action $a$ in state $s$, and $\gamma \in [0, 1)$ represents a discount factor. A policy $\pi$ in an MDP environment is a mapping from the state space $S$ to the action space $A$ and the aim of the learning methods is to find an optimal policy $\pi^*(a|s)$ that maximizes the cumulative reward $V^\pi(s)$ obtained in state $s$. The cumulative reward is also defined as follows, called the value function:

$$V^\pi(s) = \mathbb{E}^\pi[\sum_{t=0}^{\infty} \gamma^t R(s_t, a_t)|s_0 = s] \tag{1}$$

In MDPs, learning methods can be divided into model-based methods and model-free methods. The model-free methods learn policies directly through the interaction with the environment, whereas the model-based methods try to learn how environment will respond to actions of the agent and do planning according to the learned model of environment. Dyna-Q [14] unifies model-free methods and model-based methods in one framework.

## 2.2   Learning and Planning

VIN is a fully differentiable neural network trained in a model-free way. The planning module embedded in it is based on the implicit model which is represented by network parameters. The key module performs value iteration (VI) with $K$ recurrence to generate planning information. Generalized value iteration network (GVIN) [9] utilizes novel differentiable kernels as graph convolution operators, which can plan on the irregular spatial graphs. Another related work inspired by the Dyna-Q is value prediction network (VPN) [10]. It replaces the VI process with the Monte-Carlo tree search [3] to predict the future values on semi-MDPs [16], meanwhile it eliminates the limitation that the state space is a fixed topology thus is more applicable. QMDP-Net [5] is a deep recurrent policy network for partially observable Markov decision processes (POMDPs), which embeds their model priors for planning in partial observability.

## 2.3   Asynchronous Methods

Because VI has to sweep entire state spaces, it requires prohibitively expensive time to update if the state space is very large. For example, VI is difficult to be implemented in the game of Backgammon. The asynchronous methods provide a reasonable solution to this problem. Since the efficiency of asynchronous methods is related to the order of updates, how to determine the urgency of states to be updated becomes very important. Real-time dynamic programming [1] focuses on states that related to the optimal policy, and its order of update is determined by the sequence in which the states encountered in the actual or simulated trajectory. PS maintains a priority queue in which the TD errors of states are larger than a small threshold. When the top state in the queue updates, the urgency of its predecessor states will be updated and those states with urgency greater than the threshold will be appended to the queue with the new priority.

PER uses the same criterion to define a prioritized sampling method for a large replay memory. The probability of being sampled is monotonically increasing according to the priority of transition samples measured by the TD error.

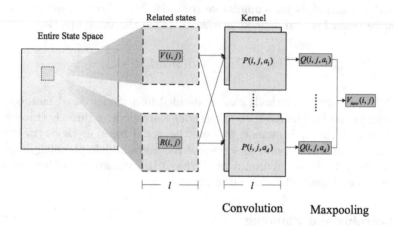

**Fig. 1.** A value update through CNN

# 3   Asynchronous Value Iteration Network

In this section, we will focus on the VI module, a core component in VIN. Then we discuss how many states to be updated and how to choose these states in the asynchronous update. At last, we combine the asynchronous update with the VI module.

## 3.1   VIN

The VIN framework embeds a planning module that uses a model represented by the network parameters to predict the values of states, and an attention mechanism that helps to reduce the number of network parameters for learning due to the local connectivity structure, in which the optimal decision on the current state only depends on a subset of values that related to the state.

If the MDP environment is known, VI can be used to conduct planning. The update operation combines the policy improvement and the policy evaluation:

$$V_{i+1}(s) = \max_a Q_i(s, a) \tag{2}$$

$$= \max_a [R(s, a) + \gamma \sum_{s'} P(s'|s, a)V_i(s')] \tag{3}$$

where $i$ represents the $i$th iteration. Focusing on Eq. 3, the update of the VI module can be approximated as a CNN with a convolution layer and a maxpooling layer. The truncated policy evaluation part corresponds to the convolution layer to calculate the state-action value for each action. The channels and the kernels of the convolution correspond to different actions and transition functions, respectively. The policy improvement part corresponds to the maxpooling layer which selects the best state-action value of the current state. The output of the planning module includes all the information that can achieve the optimal planning on the entire state space. Figure 1 shows the value update at a certain state $s = (i, j)$ represented by red color in one iteration. Each dashed blue square includes the states that are related with the convolution, whose size is the same as the convolution kernel size, and their values and rewards are convolved with the convolution kernel weights to get a new scalar value. Each round of VI is to perform value update on each state of the state space systematically.

There are two ways of viewing the VIN framework. The forward view is a model-free process to generate a policy that maps states with model-based planning information to a specific action, and the backward view is to train the MDP model using the standard back propagation. In this paper, we focus on the planning module in the forward view where the implicit MDP model represented by network parameters is fixed.

## 3.2    Asynchronous VI Module

The original VI needs to repeat the sweep of the entire state space simultaneously, whereas asynchronous update can update some important states more frequently based on their urgency. The efficiency of asynchronous update is closely related to which states to be updated and how many states to be updated in each iteration. Next we will explain in one iteration, (1) how many states are required for asynchronous update; (2) how to select these states according to their urgency; and (3) how to combine the asynchronous update with the VI module.

First, we need to decide the number of states to be updated. In PS, after the value of a state-action pair changes, the state-action pair at the top of the priority queue is selected preferentially for planning, and each planning only predicts the actions of its predecessor states. In VIN, if a state with the highest priority is updated preferentially like PS, its value will merely affect the values of its related states in the next update. If the state space size is large, then updating one state asynchronously will be meaningless for the next iteration. Therefore, before the next VI begins, we hope to select some states that could contain the information (value function) of the entire state space to perform asynchronous update, so that the values of these updated states can have enough influence on the next iteration. Meanwhile, these states should be selected as few as possible to improve the efficiency of the asynchronous update.

In most MDP environments, the local connectivity structure plays an important role, for example, it is used in attention mechanism to reduce the number of network parameters to improve the learning performance. Specifically, the local connectivity structure represents a property, i.e., a one-step decision of the state $s$ depends only on the next states $s'$ satisfying $P(s'|s, a) > 0$. The local connectivity structure implicitly specifies which parts of information of states are required for the update, and the states of the asynchronous update should involve the information of the entire state space to make the update meaningful. Therefore, in order to select as few states as possible to contain the information of the entire state space, we can utilize the local connectivity structure to heuristically define the number of states for asynchronous update. According to the local connectivity structure and the transition probabilities, the topology of convolution kernel implicitly specifies the states that required for the update. If the length of the two-dimensional convolution kernel is $l$, then the value of a certain state will require values of $l^2$ states to perform the convolution for the value update. Assuming that the states to be updated asynchronously can be evenly distributed over the entire state space, then we need at least $\lceil |S|/l^2 \rceil$ states to ensure that the states required for their updates could utilize the information of the entire state space, where $|S|$ represents the entire state space size. At last, we consider the current state in the attention mechanism that needs to be predicted, so an asynchronous update will update $\lceil |S|/l^2 \rceil + 1$ states preferentially. It should be noted that $\lceil |S|/l^2 \rceil + 1$ is a heuristic lower bound because the states that required for updates will be overlapped with each other due to the location of the states to be updated and the shape of the state space, which makes the information of these states not enough to cover the entire state space.

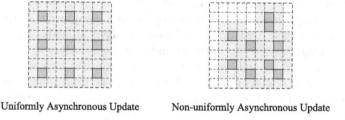

Uniformly Asynchronous Update          Non-uniformly Asynchronous Update

**Fig. 2.** Uniformly vs. Non-uniformly asynchronous update (Color figure online)

Figure 2 shows the states to be updated in one asynchronous update and their local states required for the convolution, in which the size of convolution kernel is $3 \times 3$, indicating that a state can reach the adjacent states and itself after one step. The left figure represents a special case where the states to be updated are evenly distributed over the entire state space, and the required local states will cover the entire state space. The figure on the right represents the general case that the states to be updated (red grids) and its local states will not cover the entire state space.

Second, we describe a way of selecting states to be updated. The urgency of state is defined as the quantity that can be learned from the current state. The TD error in PS is used as the measure of urgency to determine whether a state-action pair can be inserted into the priority queue. However, in VI, still using TD error as a measure becomes inconvenient. So we replace the TD error with the Bellman error as a measure, which will be very suitable for the VI module to select $\lceil |S|/l^2 \rceil + 1$ states for asynchronous update.

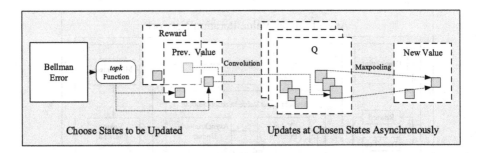

**Fig. 3.** Asynchronous update process

There are two situations that will cause the Bellman error changes. One is the change of model. However, in VIN, since the model is fixed in forward propagation, we do not need to consider the effect of the model change on the Bellman error. The other one is the update of values. In VI, the Bellman error of the entire state space can be easily obtained by calculating the difference between the value function before and after the iteration. The update of the Bellman error $E(s)$ in the $i$th iteration is shown as below:

$$E_i(s) = |V_i(s) - \max_a Q_i(s,a)| = |V_i^{be}(s) - V_i^{af}(s)| \qquad (4)$$

where $V_i^{be}(s)$ and $V_i^{af}(s)$, respectively, represent the value functions before the $i$th iteration and after the $i$th iteration. Intuitively, the larger the Bellman error of a state, the greater the amount of information it can provide. So we choose to update states with largest absolute Bellman errors to learn more from them:

$$S_i^{BE} = \arg \underset{s \in S}{topk}(E_i(s), k = N) \qquad (5)$$

where $S_i^{BE}$ represents the set of states to be updated according to the Bellman error in the $i$th iteration, $topk$ is a function that returns the $k$ largest elements of the given input, and $N = \lceil |S|/l^2 \rceil + 1$ is the number of states to be updated in one asynchronous update. The states that are updated preferentially will make their Bellman errors decrease due to their values are closer to the optimal value. Thus, more accurate values of these states will be very helpful to the subsequent iterations. As we know, VIN can focus prediction accuracy on more important parts of a trajectory to obtain the generalization ability in an unknown environment, while the Bellman error is used as the measure to select important states for asynchronous update, making the agent adaptively focus on important parts of the state space in different iteration periods to obtain a more reasonable global value function. Figure 3 shows the process of asynchronous update. The dashed line just maintains the shape of value function and reward function on the entire state space, the states represented by the solid line will be updated in each iteration. Note that the input of the $topk$ function is the Bellman error of the entire state space.

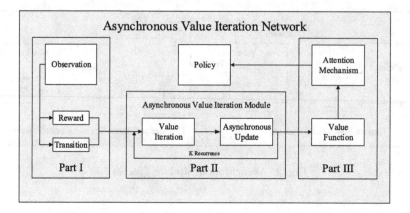

**Fig. 4.** Asynchronous value iteration network

Third, we describe a way of combining asynchronous update with the VI module. Since updated states in each asynchronous update are only parts of the state space, if the total number of updated states remains the same, the number of updates for asynchronous update will be much more than the number of recurrences for VI, which causes the speed of back propagation to become very slow. Therefore only using asynchronous update for planning is impractical. However, the asynchronous update of important states will be very helpful for subsequent iterations, so we try to append the asynchronous update to the VI module, which contributes to evaluate the value function more accurately to improve the generalization ability.

The new asynchronous VI module consists of VI and asynchronous update function sequentially. In other words, in one iteration, an asynchronous update is appended after VI. The asynchronous update function in the new module has two advantages. First, in each round of iterations it updates some important states, and more accurate values of these important states will be propagated in the next VI to make the subsequent VI more efficient. Second, it may speed up the propagation of valuable information. For example, VI requires at least $L$ iterations to deliver the information of the goal to the starting point if the shortest path has $L$ steps. The asynchronous update function might have a greater chance to accelerate the propagation of valuable information since important states are likely to be the states on the shortest path, so the new module can achieve better performance with fewer iterations.

Finally, we embed the new asynchronous VI module into the original VIN framework and call it AVIN. The entire process requires the VIN design to specify the input and output for the reward function, the transition, the new planning module and the attention mechanism, as shown in Fig. 4. The first part of VIN design maps the observation to the reward function and the transition. The second part uses the reward function and the transition as input and outputs the value function of the entire state space through the new planning module. In the

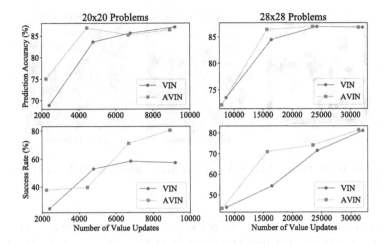

**Fig. 5.** Comparisons of VIN and AVIN in terms of prediction accuracy and success rate on navigation problems with different sizes.

third part, the values related to the current state are obtained by the attention mechanism, and are added to the reactive policy as the planning information. To illustrate that the performance improvement comes from asynchronous update rather than coming from other aspects, we remain the same VIN design as in original VIN. In the next section, we show that compared to VIN, AVIN can generalize better on large unseen navigation problems.

## 4   Experiments

In this section, we will compare the performance of AVIN and VIN on a navigation domain with different sizes and explain how the additional new asynchronous update works in each iteration.

The navigation map is discretized as a gridworld, in which each grid represented as state can be accessible region or obstacle. The agent has five actions: up, down, left, right and stay, which deterministically lead one-step move in the respective direction on the grid. Expert trajectories from 3 random starting states to the goal are generated by the shortest path function. The number and the size of obstacles vary according to the problem sizes. Different with VIN, the number of expert trajectories in the training set was changed from the original 7 to 3 to increase the difficulty of training. So, here the accuracy and success rate of VIN appear lower than the ones in [17].

We evaluate the prediction accuracy and the success rate of original VIN and AVIN on two navigation problems of $16 \times 16$ and $28 \times 28$ sizes. We mainly evaluate the performance in terms of success rate which measures how well policies can generalize on unseen domains, whereas the accuracy is used as a secondary criterion because it only evaluates behaviors along expert trajectories rather than over the entire state space. For each unknown map, the complete trajectory from

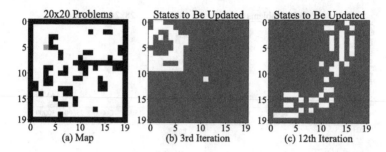

**Fig. 6.** States to be updated at different iterations using the Bellman error (Color figure online)

a random initial state is generated by iteratively predicting the next states by the trained network. The entire trajectory is succeed only if it reaches the goal without encountering any obstacle. The success rate can reflect the behavior of policies in the entire state space, including the expert trajectories, therefore the success rate better indicates the generalization ability of policies. Figure 5 shows the performance of prediction accuracy and success rate of two networks with different rounds of iterations. We considered the total number of the value updates. In order to highlight the performance of AVIN, the total number of the value updates in VIN is always greater than that of AVIN. On the problems with 20 × 20 size, AVIN respectively has performed 2230, 4460, 6690, 8920 value updates at the end of the 5th, 10th, 15th, 20th iterations, while VIN respectively has performed 2400, 4800, 6800, 9200 value updates at the end of the 6th, 12th, 17th, 23rd iterations. On the problems with 28 × 28 size, AVIN respectively has performed 7857, 15714, 23571, 31428 value updates at the end of the 9th, 18th, 27th, 36th iterations, while VIN respectively has performed 8624, 16464, 24304, 32144 value updates at the end of the 11th, 21st, 31st, 41st iterations. Although AVIN has lower success rate than VIN at the end of the 10th iteration on the problems with 20 × 20 size, might caused by the difference in iteration rounds between AVIN and VIN (10th in AVIN vs. 12th in VIN), the success rate of AVIN increases significantly while VIN increases slowly after performing more iterations. After a sufficient number of iterations, the accuracy of VIN is close to AVIN, but the success rate of AVIN is still much higher than VIN, which means that AVIN has better generalization in unknown environments.

In different iterations of the new asynchronous VI module, states that the agent pays attention to are also different: In early iterations, since the reward signal of the goal is larger than the ones at other positions, VI will deliver the information of the goal to the adjacent states. The value of these states will have large difference before and after VI and therefore the agent selects these states for asynchronous update to speed up the delivery of information. As the number of iterations increases, the Bellman errors of these states that near the goal will gradually decrease. However, values of obstacles in the reward function is different from those of accessible grids, resulting in a relative large differ-ence between the values before and after the iteration at these states, then the

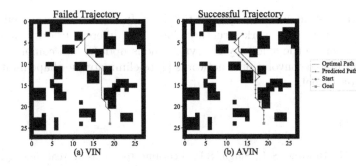

**Fig. 7.** Trajectories generated by VIN and AVIN in the same environment (Color figure online)

important states for asynchronous update will gradually move closer to obstacles. Figure 6(a) represents a map of 20 × 20 size in the training set, where yellow color indicates goal, black color indicates obstacles and the remaining white grids indicate non-goal places without obstacles. Figure 6(b–c) show the states selected by the Bellman error to be updated asynchronously at different iterations. The states to be updated (white grids) at the 3rd iteration are concentrated around the goal, while the states to be updated at the 12th iteration are closer to the obstacles distributed over the entire state space. Thus, updating these states preferentially will help the agent get more accurate values for subsequent iterations. So, the asynchronous VI module will obtain a more accurate value function on the entire state space.

Figure 7 shows the predicted trajectories generated by VIN and AVIN in the same test environment. When the predicted trajectory represented by the green line hits an obstacle, it terminates immediately. In Fig. 7(a), the prediction of VIN at the state $(11, 5)$ causes the agent to end with an obstacle. In Fig. 7(b), the predicted trajectory generated by AVIN successfully reached the goal, indicating that AVIN has better generalization on the unseen environment.

## 5   Conclusion and Future Work

In this paper, we propose an improved VIN architecture with asynchronous update, which performs the value updates more frequently on some important states according to their urgency. We heuristically define the number of states for asynchronous update and find a way to measure the urgency of states for selecting important states, then we combine the asynchronous update with the VI module to make the value updates more efficient and effective. Empirically, policies of AVIN can generalize better on unseen problems than VIN.

There are some drawbacks in the AVIN framework. In AVIN, the number of the states to be updated asynchronously is computed based on the assumption that the selected states evenly distribute over the entire state space. We leave the possibility of relaxing the assumption as a research topic. At the same time, the convolutional kernel is limited by the local connectivity structure of the problem and must maintain a specific topology. In the future, we would like to design some neural network structures based on different priority sampling methods to improve the efficiency of asynchronous update, and combine AVIN with the deformable convolutional networks [4] to eliminate the shape limitation of convolution kernel.

# References

1. Barto, A.G., Bradtke, S.J., Singh, S.P.: Learning to act using real-time dynamic programming. Artif. Intell. **72**(1–2), 81–138 (1995)
2. Bertsekas, D.P.: Dynamic Programming and Optimal Control, 3rd edn. Athena Scientific, Belmont (2005)
3. Browne, C., et al.: A survey of Monte Carlo tree search methods. IEEE Trans. Comput. Intell. AI Games **4**(1), 1–43 (2012)
4. Dai, J., et al.: Deformable convolutional networks. CoRR abs/1703.06211 (2017). http://arxiv.org/abs/1703.06211
5. Karkus, P., Hsu, D., Lee, W.S.: QMDP-Net: deep learning for planning under partial observability. In: NIPS, pp. 4697–4707 (2017)
6. Long, J., Shelhamer, E., Darrell, T.: Fully convolutional networks for semantic segmentation. In: CVPR, pp. 3431–3440 (2015)
7. Mnih, V., et al.: Human-level control through deep reinforcement learning. Nature **518**(7540), 529–533 (2015)
8. Moore, A.W., Atkeson, C.G.: Prioritized sweeping: reinforcement learning with less data and less time. Mach. Learn. **13**, 103–130 (1993)
9. Niu, S., Chen, S., Guo, H., Targonski, C., Smith, M.C., Kovacevic, J.: Generalized value iteration networks: life beyond lattices. In: AAAI, pp. 6246–6253 (2018)
10. Oh, J., Singh, S., Lee, H.: Value prediction network. In: NIPS, pp. 6120–6130 (2017)
11. Ross, S., Gordon, G.J., Bagnell, D.: A reduction of imitation learning and structured prediction to no-regret online learning. In: AISTATS, pp. 627–635 (2011)
12. Schaal, S.: Is imitation learning the route to humanoid robots? Trends Cogn. Sci. **3**(6), 233–242 (1999)
13. Schaul, T., Quan, J., Antonoglou, I., Silver, D.: Prioritized experience replay. CoRR abs/1511.05952 (2016). http://arxiv.org/abs/1511.05952
14. Sutton, R.S.: Integrated architectures for learning, planning, and reacting based on approximating dynamic programming. In: ICML, pp. 216–224 (1990)
15. Sutton, R.S., Barto, A.G.: Reinforcement Learning: An Introduction. Adaptive Computation and Machine Learning. MIT Press, Cambridge (1998)
16. Sutton, R.S., Precup, D., Singh, S.P.: Between MDPs and Semi-MDPs: a framework for temporal abstraction in reinforcement learning. Artif. Intell. **112**(1–2), 181–211 (1999)
17. Tamar, A., Levine, S., Abbeel, P., Wu, Y., Thomas, G.: Value iteration networks. In: NIPS, pp. 2146–2154 (2016)

# FVR-SGD: A New Flexible Variance-Reduction Method for SGD on Large-Scale Datasets

Mingxing Tang[1], Zhen Huang[1(✉)], Linbo Qiao[1], Shuyang Du[2], Yuxing Peng[1], and Changjian Wang[1]

[1] Science and Technology on Parallel and Distributed Laboratory, National University of Defense Technology, Changsha 410073, China
tomingxing@gmail.com, {huangzhen,qiao.linbo}@nudt.edu.cn,
pengyuxing@aliyun.com, c_j_wang@yeah.net
[2] Tongji University, Shanghai, China
dushuyang@126.com

**Abstract.** Stochastic gradient descent (SGD) is a popular optimization method widely-used in machine learning, while the variance of gradient estimation leads to slow convergence. To accelerate the speed, many variance reduction methods have been proposed. However, most of these methods require additional memory cost or computational burden on full gradient, which results in low efficiency or even unavailable while applied to real-world applications with large-scale datasets. To handle this issue, we propose a new flexible variance reduction method for SGD, named FVR-SGD, which can reduce memory overhead and speedup the convergence using flexible subset size without extra operation. The details of convergence property are presented, the convergence of variance reduction method using flexible subset size can be guaranteed. Several numerical experiments are conducted on a genre of real-world large-scale datasets. The experimental results demonstrated that FVR-SGD outperforms contemporary SVRG algorithm. Specifically, the proposed method can achieve up to 40% reduction in the training time to solve the optimization problem of logistic regression.

**Keywords:** Machine learning · Optimization
Stochastic gradient descent · Variance reduction
Distributed optimization

## 1 Introduction

In machine learning, the issues we need to solve are often abstracted as an empirical risk minimization problem, given as

$$\min_{w} F(w) = \frac{1}{n} \sum_{i=1}^{n} f_i(w), \tag{1}$$

© Springer Nature Switzerland AG 2018
L. Cheng et al. (Eds.): ICONIP 2018, LNCS 11302, pp. 181–193, 2018.
https://doi.org/10.1007/978-3-030-04179-3_16

where $w$ is the parameter to learn, $n$ is the number of instances, $f_i(\cdot) : \mathbb{R}^d \to \mathbb{R}$ is a loss function defined on the instance indexed $i$. SGD is widely used to solve problem (1) due to its efficacy for solving large-scale problems. Specifically, SGD method only needs to compute one derivative per iteration, and the update rule of SGD can be formally described as

$$w_k = w_{k-1} - \eta \nabla f_{i_k}(w_{k-1}), \tag{2}$$

where $\eta$ is learning rate, $i_k$ is the instance index at the $k$-th iteration. Since SGD uses a noisy estimation of true gradient, a diminishing step size is proposed to ensure the convergence. However, this may slow down the convergence.

Variance reduction [1] can speedup the convergence rate through reducing the variance of gradient estimation at each iteration. And a linear convergence speed can be achieved theoretically. The update of SGD with variance reduction can be formulated as follows

$$v_k = \nabla f_{i_k}(w_{k-1}) - \nabla f_{i_k}(\tilde{w}_{i_k}) + \tilde{u}, \qquad w_k = w_{k-1} - \lambda v_k, \tag{3}$$

where $\tilde{w}_{i_k}$ is a snapshot of $w$ at the latest iteration for instance indexed $i_k$ and $\tilde{u}$ is an unbiased estimate of the true gradient. So $v_k$ is an unbiased estimate of $\nabla F(w_{k-1})$, while the estimation error will be much smaller than SGD.

In recent years, several works on variance reduction have been proposed [2]. One of the research ways is to store the snapshot of gradients of all instances to achieve a fast linear convergence, such as SAG [3], SAGA [4] and CentralVR [5]. However, the huge memory cost makes them impractical to train on large-scale datasets. To handle the problem, SVRG [6] periodically makes an unbiased estimation with full gradients without storing the stalled gradients. However, SVRG becomes computational costly as the increasing scale of training data. Moreover, the synchronous computational operations cause the poor scalability in a distributed manner. Furthermore, CentralVR and SVRG divide the training process into multiple epochs, while each epoch requires to go through all the datasets. When the data is too large to be loaded in memory, disk I/O operations will be performed, which greatly increase the time consumption. Experimental results in literatures demonstrated that variance reduction can be achieved depending on the history and partial current data, when applied to applications with large-scale dataset.

So we propose an efficient variance reduction algorithm adopted flexible subset size, without extra computational or memory consumption. Specifically, the algorithm divides the whole dataset into several subsets flexibly and uses them to train the model separately in each epoch, which makes it to be adaptively loaded into memory or distributed environments for large-scale datasets, with the convergence property guaranteed in the training process. The major contributions of this paper include:

(1) A novel variance reduction method is proposed. Different from the traditional method, FVR-SGD achieves variance reduction by using the history

and partial currently data, which makes it of good parallelism and can handle large-scale datasets efficiently. We analyze the convergence of the method theoretically and the details of the convergence analysis are presented.

(2) A distributed algorithm of the method FVR-SGD is proposed and implemented. Based on the parameter server model, the distributed algorithm of FVR-SGD can flexibly adjust memory overhead according to the subset size, which supports a large number of model replicas.

(3) Several experiments are carried out to validate the performance of the algorithm. The experimental results show that, whether in stand-alone or distributed occasion, FVR-SGD converges faster than SGD and other variance reduction methods when training large scale datasets.

## 2  Background

### 2.1  Variance Reduction

The variance reduction methods require two parameter values at each iteration: full gradient $\tilde{u}$ and the latest snapshot of $w$. SVRG obtains these values through extra computation, which divides the training process into multiple epochs. In each epoch, $w$ can be initialized as $\tilde{w}$ and the full gradient of $\tilde{w}$ be computed as $\tilde{u}$, while the $k$-th iteration can be given as

$$w_k = w_{k-1} - \eta(\nabla f_{i_k}(w_{k-1}) - \nabla f_{i_k}(\tilde{w}) + \tilde{u}). \tag{4}$$

SAGA obtains these values by storing the latest snapshot of $w$ and $\nabla f(w)$ for each instance, which can be saved as $\tilde{w}_i$ and $\nabla f(\tilde{w}_i)$. And the update rule is given as

$$w_k = w_{k-1} - \eta(\nabla f_{i_k}(w_{k-1}) - \nabla f_{i_k}(\tilde{w}_{i_k}) + \frac{1}{n}\sum_{j=1}^{n}\nabla f_j(\tilde{w}_j)). \tag{5}$$

Since the value of update item approaches to 0, variance reduction methods can use a large step size to accelerate the convergence. However, most current variance reduction methods aim to obtain a fast convergence rate and they are mainly trained on the small datasets. When the training datasets become large, heavy computational burden or memory consumption may cause them be impractical to real world applications.

### 2.2  Distributed SGD

In recent years, datasets used in machine learning community become bigger and bigger, which brings great challenges to traditional optimization methods. Parallel or distributed optimization algorithms become more and more popular, such as $HOGWILD!$ [7] and $DownpourSGD$ [8]. $HOGWILD!$ allows multiple processors access to shared memory without lock, which achieves a nearly optimal rate of convergence among gradient based methods. $DownpourSGD$ is an

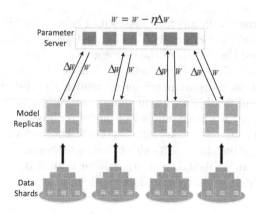

**Fig. 1.** Parameter server model.

asynchronous SGD procedure supporting a large number of model replicas, for which the basic model is depicted as Fig. 1. *Downpour SGD* divides the dataset into multiple subsets and trains each to obtain the parameters of replicas. These parameters are communicated to update through a centralized server, which keeps the latest state of all parameters for the model.

## 3   FVR-SGD Algorithm

With the growth of datasets, current variance reduction methods become ineffi-cient caused by the additional overhead on memory or computation. We observe that variance reduction can be achieved using history and partial current data while training large-scale datasets. Moreover, history data is easy to access and using partial current data has better memory efficiency and may achieve higher parallelism.

So we propose FVR-SGD, a method of variance reduction through accumu-lated parameter values and flexible subset. FVR-SGD divides the dataset into $J$ subsets uniformly and each has $K$ instances that can be loaded into memory flexibly, where $K$ could be adaptively set according to the amount of memory. The training scheme can be divided into a series of epochs, in each of which one subset will be processed randomly and $M$ stochastic updates will be performed, where we set $M = 2K$ in this paper according to other variance reduction meth-ods [6].

Let $w_m^k$ denotes the parameters at the $k$-th iteration of the $m$-th epoch and $i_k$ denotes the index of the instance sampled at the $k$-th iteration. Let $\tilde{w}^j$ and $\tilde{g}^j$ denote the latest stored parameter and gradient for subset $j$, supposed that the $j$-th subset is selected at the $m+1$-th epoch. Let $\tilde{w}_m$ and $\tilde{g}_m$ denote the average of all latest stored parameters and gradients at the $m$-th epoch. The update rule of FVR-SGD can be given as

$$w_{m+1}^{k+1} = w_{m+1}^k - \eta v_{m+1}^k, \tag{6}$$

where, $v_{m+1}^k$ is the corrected gradient of the $k$-th iteration in the $m$-th epoch, which can be formulated as

$$v_{m+1}^k = \nabla f_{i_k}(w_{m+1}^k) - \nabla f_{i_k}(\tilde{w}_m) + \tilde{g}_m. \tag{7}$$

During the iterations, $w_{m+1}^{k+1}$ and $\nabla f_{i_k}(w_{m+1}^k)$ are accumulated. In the end of the $m+1$-th epoch, parameters $\tilde{w}^j$ and $\tilde{g}^j$ for the selected subset can be updated as follows:

$$\tilde{w}^j = \frac{1}{2K} \sum_{k=1}^{2K} w_{m+1}^k, \tilde{g}^j = \frac{1}{2K} \sum_{k=1}^{2K} \nabla f_{i_k}(w_{m+1}^k). \tag{8}$$

Since the dataset is divided into several subsets, FVR-SGD only needs to store the latest average $\tilde{w}^j$ and $\tilde{g}^j$ for each subset, which indicates that the memeory requirement can be adjusted flexibly. And the parameters $\tilde{w}_{m+1}$ and $\tilde{g}_{m+1}$ can be updated as follows:

$$\tilde{w}_{m+1} = \frac{1}{J} \sum_{j=1}^{J} \tilde{w}^j, \tilde{g}_{m+1} = \frac{1}{J} \sum_{j=1}^{J} \tilde{g}^j. \tag{9}$$

---

**Algorithm 1.** FVR-SGD.

---

**Parameters :** subset number $J$, subset size $K$, learning rate $\eta$
**Initialize :** obtain $w_0^{2K}, \tilde{w}^j$ and $\tilde{g}^j$ using SGD
1: $\tilde{w}_0 = \frac{1}{J} \sum_{j=1}^{J} \tilde{w}^j, \tilde{g}_0 = \frac{1}{J} \sum_{k=1}^{J} \tilde{g}^j$
2: **for** $m = \{1, 2, ...\}$ **do**
3:     Select a subset index $j$ randomly
4:     $w_m^0 = w_{m-1}^{2K}, w_a = 0, g_a = 0$
5:     **for** $k$ *in* $\{1, ..., 2K\}$ **do**
6:         Sample instance indexed $i_k$ from subset indexed $j$
7:         $w_m^k = w_m^{k-1} - \eta(\nabla f_{i_k}(w_m^{k-1}) - \nabla f_{i_k}(\tilde{w}_{m-1}) + \tilde{g}_{m-1})$
8:         $w_a += w_m^k, g_a += \nabla f_{i_k}(w_m^{k-1})$
9:     **end for**
10:    $\tilde{w}^j = \frac{1}{2K} w_a, \tilde{g}^j = \frac{1}{2K} g_a$
11:    $\tilde{w}_m = \frac{1}{J} \sum_{j=1}^{J} \tilde{w}^j, \tilde{g}_m = \frac{1}{J} \sum_{j=1}^{J} \tilde{g}^j$
12: **end for**

---

FVR-SGD uses the average of parameters $\tilde{w}^j$ and gradients $\tilde{g}^j$ to correct gradient without extra overhead on computation of full gradient. The initialization of algorithm uses SGD going through instances of each subset to obtain the original $\tilde{w}^j$ and $\tilde{g}^j$. The details of FVR-SGD algorithm are described in Algorithm 1, which can adjust memory cost by the subset size $K$ that is robust and efficient on updates in memory. Also subsets can be distributed to multiple machines, which brings the algorithm to enjoy the ability to be scalable.

## 4  Convergence Analysis

In this paper, we assume that the cost function $f_i$ is $L$-smooth, which means

$$f_i(x) \leq f_i(y) + \nabla f_i(y)^T(x-y) + \frac{L}{2} \parallel x-y \parallel^2 \ or \ \parallel \nabla f_i(x) - \nabla f_i(y) \parallel \leq L \parallel x-y \parallel,$$
$$(10)$$

and the cost function $f_i$ is $\mu$ convex, which means

$$f_i(x) \geq f_i(y) + \nabla f_i(y)^T(x-y) + \frac{\mu}{2} \parallel x-y \parallel^2 \ or \ \parallel \nabla f_i(x) - \nabla f_i(y) \parallel \geq \mu \parallel x-y \parallel.$$
$$(11)$$

We can prove that the algorithm converges even though only one subset is processed per epoch. Inequality (12) has been proved in [6] and inequality (13) can be obtained from the continuity and convexity of $F$.

$$\mathbb{E} \parallel \nabla f_i(w) - \nabla f_i(w_*) \parallel^2 \leq 2L(F(w) - F(w_*)),$$
$$(12)$$

$$\parallel \nabla f_i(w) - \nabla f_i(w_*) \parallel^2 \leq L^2 \parallel w - w_* \parallel^2 \leq \frac{2L^2}{\mu}(F(w) - F(w_*)).$$
$$(13)$$

Suppose that $w_*$ is optimum and $\nabla F(w_*) = 0$, $\phi_l$ denotes the latest accumulated $w$ and $\tilde{G}_m = \sum_{l=1}^{2n} F(\phi_l)/2n$. Now we can take expectation of $v_{m+1}^k$ with respect to $i_k$ and obtain

$$\mathbb{E} \parallel v_{m+1}^k \parallel^2 = \mathbb{E} \parallel \nabla f_{i_k}(w_{m+1}^k) - \nabla f_{i_k}(\tilde{w}_m) + \tilde{g}_m \parallel^2$$

$$\leq 2\mathbb{E} \parallel \nabla f_{i_k}(w_{m+1}^k) - \nabla f_{i_k}(w_*) \parallel^2 + 2\mathbb{E} \parallel \nabla f_{i_k}(\tilde{w}_m) - \nabla f_{i_k}(w_*) - \tilde{g}_m \parallel^2$$

$$\leq 2\mathbb{E} \parallel \nabla f_{i_k}(w_{m+1}^k) - \nabla f_{i_k}(w_*) \parallel^2 + 4\mathbb{E} \parallel \nabla f_{i_k}(\tilde{w}_m) - \nabla f_{i_k}(w_*) \parallel^2$$

$$+ 8 \parallel \frac{1}{2n} \sum_{l=1}^{2n} [\nabla f_{i_l}(\phi_l) - \nabla f_{i_l}(w_*)] \parallel^2 + 8 \parallel \frac{1}{2n} \sum_{l=1}^{2n} \nabla f_{i_l}(w_*) \parallel^2$$

$$\leq 4L(F(w_{m+1}^k) - F(w_*)) + 8L(F(\tilde{w}_m) - F(w_*)) + \frac{4}{n} \sum_{l=1}^{2n} \parallel \nabla f_{i_l}(\phi_l) - \nabla f_{i_l}(w_*) \parallel^2$$

$$\leq 4L(F(w_{m+1}^k) - F(w_*)) + (8L + \frac{8L^2}{\mu})(\tilde{G}_m - F(w_*)).$$
$$(14)$$

The first and second inequalities use $\parallel a+b \parallel^2 \leq 2 \parallel a \parallel^2 + 2 \parallel b \parallel^2$. The third inequality uses inequality (12) and $\parallel \sum_{i=1}^n a_n \parallel^2 \leq n \sum_{i=1}^n \parallel a_i \parallel^2$. The forth inequality uses inequality (13) and $F(\tilde{w}_m) = F(\frac{1}{2n} \sum_{l=1}^{2n} \phi_l) \leq \frac{1}{2n} \sum_{l=1}^{2n} F(\phi_l)$. Noticing that $\mathbb{E}[v_{m+1}^k] = \nabla F(w_{m+1}^k) - \nabla F(\tilde{w}_m) + \tilde{g}_m$, we can obtain the bound:

$$\mathbb{E} \parallel w_{m+1}^{2K+1} - w_* \parallel^2 = \mathbb{E} \parallel w_{m+1}^{2K} - w_* \parallel^2 - 2\eta(w_{m+1}^{2K} - w*)\mathbb{E}[v_{m+1}^{2K}] + \eta^2 \mathbb{E} \parallel v_{m+1}^{2K} \parallel^2$$

$$\leq \mathbb{E} \parallel w_{m+1}^{2K} - w_* \parallel^2 - 2\eta\mathbb{E}(F(w_{m+1}^{2K}) - F(w_*)) - 2\eta\mathbb{E}(w_{m+1}^{2K} - w_*)^{\mathrm{T}}(\tilde{g}_m - \nabla F(\tilde{w}_m))$$
$$+ \eta^2 \mathbb{E} \parallel v_{m+1}^{2K} \parallel^2$$

$$\leq \mathbb{E} \parallel w_{m+1}^{2K} - w_* \parallel^2 - 2\eta\mathbb{E}(F(w_{m+1}^{2K}) - F(w_*)) - 2\eta\mathbb{E}(w_{m+1}^{2K} - w_*)^{\mathrm{T}}(\tilde{g}_m - \nabla F(\tilde{w}_m))$$
$$+ 4\eta^2 L\mathbb{E}(F(w_{m+1}^{2K}) - F(w_*)) + \eta^2(8L + \frac{8L^2}{\mu})\mathbb{E}(\tilde{G}_m - F(w_*))$$

$$= \mathbb{E} \parallel w_{m+1}^{2K} - w_* \parallel^2 - (2\eta - 4\eta^2 L)\mathbb{E}(F(w_{m+1}^{2K}) - F(w_*))$$
$$- 2\eta\mathbb{E}(w_{m+1}^{2K} - w_*)^{\mathrm{T}}(\tilde{g}_m - \nabla F(\tilde{w}_m)) + \eta^2(8L + \frac{8L^2}{\mu})\mathbb{E}(\tilde{G}_m - F(w_*))$$

$$\leq \mathbb{E} \parallel w_{m+1}^0 - w_* \parallel^2 - (2\eta - 4\eta^2 L)\sum_{i=1}^{2K}\mathbb{E}(F(w_{m+1}^i) - F(w_*))$$
$$- 4K\eta\mathbb{E}(\tilde{w}_{m+1} - w_*)^{\mathrm{T}}(\tilde{g}_m - \nabla F(\tilde{w}_m)) + 16K\eta^2(L + \frac{L^2}{\mu})\mathbb{E}(\tilde{G}_m - F(w_*))$$

$$\leq \mathbb{E} \parallel w_{m+1}^0 - w_* \parallel^2 - 4K\eta\mathbb{E}(\tilde{w}_{m+1} - w_*)^{\mathrm{T}}(\tilde{g}_m - \nabla F(\tilde{w}_m)) \qquad (15)$$
$$+ 16K\eta^2(L + \frac{L^2}{\mu})\mathbb{E}(\tilde{G}_m - F(w_*)) - 4n(\eta - 2\eta^2 L)\mathbb{E}(\tilde{G}_{m+1} - F(w_*))$$
$$+ 4n(\eta - 2\eta^2 L)\mathbb{E}(\tilde{G}_m - F(w_*)).$$

The first inequality uses the convexity of $F$. The second inequality uses inequality (14). The third inequality is summing the previous inequality over $2K$ at $m + 1$ epoch. Let $\tilde{G}_{m+1}^{-j}$ denotes the $\tilde{G}_{m+1}$ removing the selected subset $j$ at $(m + 1)$-epoch, which means $\tilde{G}_{m+1}^{-j} = \frac{1}{2n-2K}\sum_{l=1}^{2n-2K} F(\phi_l)$. The forth inequality uses

$$- \sum_{i=1}^{2K}\mathbb{E}(F(w_{m+1}^i) - F(w_*))$$

$$\leq -\mathbb{E}[\sum_{i=1}^{2K} F(w_{m+1}^i) + \tilde{G}^{-j} - 2nF(w_*) - (\tilde{G}^{-j} - (2n - 2K)F(w_*))]$$

$$\leq -2n\mathbb{E}(\tilde{G}_{m+1} - F(w_*)) + 2n\mathbb{E}(\tilde{G}_m - F(w_*)).$$

And the inner product term can be bounded as follow:

$$- 2(\tilde{w}_{m+1} - w_*)^{\mathrm{T}}(\tilde{g}_m - \nabla F(\tilde{w}_m)) \leq \parallel \tilde{w}_{m+1} - w_* \parallel^2 + \parallel \tilde{g}_m - \nabla F(\tilde{w}_m) \parallel^2$$

$$\leq \frac{2}{\mu}(F(\tilde{w}_{m+1}) - F(w_*)) + 2 \parallel \tilde{g}_m - \nabla F(w_*) \parallel^2 + 2 \parallel \nabla F(\tilde{w}_m) - \nabla F(w_*) \parallel^2$$

$$\leq \frac{2}{\mu}(F(\tilde{w}_{m+1}) - F(w_*)) + \frac{4L^2}{\mu}(\tilde{G}_m - F(w_*)) + 4L(F(\tilde{w}_m) - F(w_*))$$

$$\leq \frac{2}{\mu}(\tilde{G}_{m+1} - F(w_*)) + (\frac{4L^2}{\mu} + 4L)(\tilde{G}_m - F(w_*)). \qquad (16)$$

The first and second inequalities use from Cauchy-Schwartz inequality and inequality (11), the third inequality uses inequality (13) and inequality (12), the forth inequality uses $F(\tilde{w}_m) = F(\frac{1}{2n}\sum_{l=1}^{2n}\phi_l) \leq \frac{1}{2n}\sum_{l=1}^{2n}F(\phi_l)$. And we also have

$$\mathbb{E}\parallel w_{m+1}^0 - w_* \parallel^2 \leq \frac{1}{2n}\sum_{l=1}^{2n}\mathbb{E}\parallel \phi_l - w_* \parallel^2 \leq \frac{1}{2n}\sum_{l=1}^{2n}\frac{2}{\mu}\mathbb{E}(F(\phi_l) - F(w_*))$$

$$\leq \frac{2}{\mu}\mathbb{E}(\tilde{G}_m - F(w_*)). \tag{17}$$

From (15), (16), (17), we can get the bound:

$$\mathbb{E}(\tilde{G}_{m+1} - F(w_*)) \leq \frac{\mu}{4\eta(n\mu - 2n\mu\eta L - k)}[\frac{2}{\mu}\mathbb{E}(\tilde{G}_m - F(w_*))$$

$$+ 8K\eta(\frac{L^2}{\mu} + L)(\tilde{G}_m - F(w_*))$$

$$+ 16K\eta^2(L + \frac{L^2}{\mu})\mathbb{E}(\tilde{G}_m - F(w_*)) + 4n(\eta - 2\eta^2 L)\mathbb{E}(\tilde{G}_m - F(w_*))]$$

$$= \frac{1 + 4\eta KL^2 + 8\eta^2 KL^2 + 4\eta\mu KL + 8\eta^2\mu KL + 2\eta\mu n - 4\eta^2\mu nL}{2\eta(n\mu - 2n\mu\eta L - K)}\mathbb{E}(\tilde{G}_m - F(w_*)).$$

It indicates that $\mathbb{E}(\tilde{G}_{m+1} - F(w_*)) \leq \alpha^{m+1}\mathbb{E}(\tilde{G}_0 - F(w_*))$. When $n$ is large enough and $\eta$ is small enough, $\alpha$ can be located in the range of (0,1), which means the convergence of algorithm can be guaranteed.

## 5   Distributed Scheme

In this section, we propose an efficient distributed scheme for FVR-SGD. Since the algorithm logically divides the dataset into multiple subsets, which can be distributed to multiple computing nodes, a distributed algorithm of FVR-SGD can be designed based on the parameter server model. We distribute the dataset to workers uniformly, each worker can only access its own data, which can also be divided into subsets. The latest parameters and gradient information are stored in the server. At the beginning of each epoch, each worker pulls them from server and then randomly selects a subset to train the model. The updated information is pushed to server at the end of each epoch.

The details of algorithm running on server are illustrated in Algorithm 3, while the details of algorithm running on worker are illustrated in Algorithm 2. The communication frequency between workers and server can be controlled by adjusting the subset size $K$. As compared to previous work, it is practical to adapt the network bandwidth. Since there is no interaction between workers, the algorithm is allowed to be high stable.

---

**Algorithm 2.** Distributed FVR-SGD Worker.

---

**Parameters :** instances number $N$, subset number $J$, subset size $K$, learning rate $\eta$
**Initialize :** obtain $w$, $\tilde{w}^j$ and $\tilde{g}^j$ using SGD
1:  $\triangle\tilde{g} = v_a = \frac{1}{J}\sum_{k=1}^{J}\tilde{g}^j$
2:  Send $\triangle\tilde{g}$, $v_a$ to server
3:  Receive $\tilde{g}$, $w$ from server
4:  **for** $m = \{1, 2, ...\}$ **do**
5:      Select a subset index $j$ randomly
6:      $w_a = 0$, $g_a = 0$, $v_a = 0$
7:      **for** $k$ in $\{1, ..., 2K\}$ **do**
8:          Sample instance index $i_k$ from subset indexed $j$
9:          $v = \nabla f_{i_k}(w) - \nabla f_{i_k}(\tilde{w}^j) + \tilde{g}$
10:         $w = w - \eta v$
11:         $w_a += w$, $g_a += \nabla f_{i_k}(w)$, $v_a += v$
12:     **end for**
13:     $\triangle g = \frac{1}{K}g_a - \tilde{g}^j$, $\tilde{g}^j = \frac{1}{K}g_a$, $\tilde{w}^j = \frac{1}{K}w_a$
14:     Send $\triangle\tilde{g}$, $v_a$ to server
15:     Receive $\tilde{g}$, $w$ from server
16: **end for**

---

**Algorithm 3.** Distributed FVR-SGD Sever.

---

**Parameters :** total subset number $J$, worker number $W$, learning rate $\eta$
**Initialize :** $w$
1:  **while** *true* **do**
2:      Receive $\triangle\tilde{g}$, $v_a$ from worker $i$
3:      $\tilde{g} += \frac{1}{J}\triangle\tilde{g}$
4:      $w -= \frac{1}{W}\eta v_a$
5:      Send $\tilde{g}$, $w$ to worker $i$
6:  **end while**

---

## 6  Experiments and Evaluations

In this section, we select a classical machine learning algorithm to evaluate the performance of our FVR-SGD as compared to the contemporary methods SGD and SVRG. Specifically, the experiments are conducted on logistic regression with $\ell_2$-norm regularization, which can be formulated as follows:

$$\min_{w} F(w) = \frac{1}{n}\sum_{i=1}^{n} \log[1 + \exp(-y_i x_i^T w)] + \frac{\lambda}{2} \parallel w \parallel_2^2,$$

where we set the $\ell_2$ regularization parameter to be $\lambda = 10^{-6}$. Then we carry out several experiments both in the single and distributed mode to validate the efficacy of our algorithm.

### 6.1  Single Machine

In this subsection, we carry out experiments on a single machine using two real-world large-scale datasets, of which the statistic details are illustrated in Table 1

[9]. These two datasets contain large number of instances and each of them contains large number of features, which will lead to huge memory consumption to load them into the RAM. Since SAGA is impractical on the memory consumption, we focus on the comparisons among FVR-SGD, SGD and SVRG. The learning rate $\eta$ of FVR-SGD, SGD, SVRG are set to be 0.005, 0.005, 0.001 respectively.

**Table 1.** Datasets for single machine.

| Name | Source | Instance number | Feature number |
|------|--------|-----------------|----------------|
| Real-sim | LIBSVM | 72,309 | 20,958 |
| rcv1.binary | LIBSVM | 20,242 | 47,236 |

Firstly we choose the computational overhead as the evaluation criteria, which can be computed as the number of computations on gradient. The experimental results are depicted in the Fig. 2. In order to validate the flexibility of our method, we divide the datasets into 1, 2, 20 subsets and label them into FVR-1, FVR-2 and FVR-20 respectively.

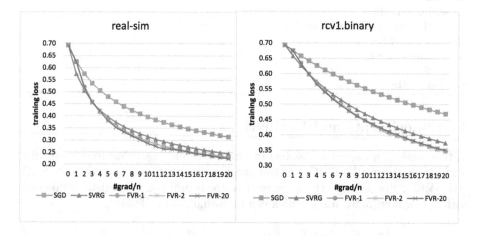

**Fig. 2.** Computational overhead on real-sim and rcv1.binary. The $x$-axis is the number of gradient computations divided by $n$, where $n$ is the number of samples, the $y$-axis is the training loss.

We can observe that different subset sizes have a similar computational overhead in the decreasing process of loss, which indicates that the variance reduction can be achieved based on flexible partial data. Moreover, we can observe that the computational overhead of them can be about 60% lower than SGD at the same loss and also about 10% lower than SVRG, which indicates that our FVR-SGD method can reach the lowest computational overhead in the two large-scale datasets.

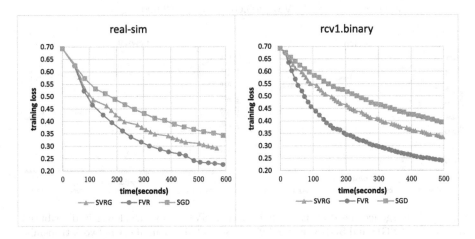

**Fig. 3.** Convergence speed on real-sim and rcv1.binary. When datasets can not fit into RAM, FVR-SGD divides them into subsets, improve the efficiency while guaranteeing convergence, and the convergence speed is increased.

Then we choose the convergence speed as the evaluation criteria, which is measured as the training time. The experimental results are depicted in the Fig. 3. Actually the matrix of datasets is too huge to be loaded into the RAM at the same time. Supposed that half of the data can be loaded in, we can divide the datasets into two subsets. However, SVRG must use the matrix of datasets multiple times in per epoch, which will cause frequent swap I/O between disk and memory. We can observe that the training time of SVRG on large-scale dataset is high in our experiments. The experimental results show that FVR-SGD can achieve 40% faster than SVRG and 60% faster than SGD, which indicates that FVR-SGD can reach the fastest convergence speed in the two large-scale datasets.

## 6.2   Distributed Mode

In this subsection, we carry out experiments on several machines using two standard large-scale datasets, of which the details are illustrated in Table 2 [10] [11]. We construct a distributed system with 21 nodes, of which one is parameter server and the other 20 are workers.

In the distributed system, we mainly carry out experiments on the comparisons of FVR-SGD, SVRG and *DownpourSGD* (DSGD), for which the results are depicted in the Fig. 4. The learning rate $\eta$ of FVR-SGD, SVRG, DSGD are set to be 0.005, 0.005, 0.001 respectively.

**Table 2.** Datasets for distributed mode.

| Name | Source | Instance number | Feature number |
|------|--------|-----------------|----------------|
| news20.binary | LIBSVM | 19,996 | 1,355,191 |
| Epsilon | LIBSVM | 400,000 | 2,000 |

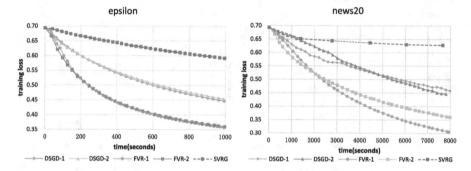

**Fig. 4.** Convergence speed in distributed system. SVRG is much slower in distributed system, FVR-SGD and SGD use two different subset size, can affect network transmission.

Because of the huge computational overhead and synchronization waiting, we can observe that SVRG converges even much slower than DSGD. We set the subset size in FVR-SGD the same as DSGD, where FVR-SGD and DSGD with the number of $i$ subsets are labeled as FVR-$i$ and DSGD-$i$ respectively, in which $i$ can be 1 or 2 in our experiments. And we can see that the convergence speed of FVR-SGD is 50% faster than that of DSGD. Also we can observe that subset size have effect on network transmission, $news20$ dataset has large feature number, so a relative larger subset size can reduce network traffic and have a faster convergence speed.

## 7    Conclusion

In this paper, we propose FVR-SGD, a variance reduction method adopting flexible subset size, which can flexibly adjust the memory costs without extra computations. We at first prove the convergence of FVR-SGD theoretically and then propose a distributed algorithm of FVR-SGD based on the parameter server, which enjoys well scalability, flexibility and stability. Several experiments on the real-world large-scale datasets in the single and distributed environments demonstrated that FVR-SGD outperforms existing methods in computational overhead and convergence speed while processing large-scale datasets.

**Acknowledgments.** This work is supported by the National Basic Research Program (973) of China (No. 2014CB340303).

# References

1. Roux, N.L., Schmidt, M., Bach, F.: A stochastic gradient method with an exponential convergence rate for finite training sets. Adv. Neural Inf. Process. Syst. **4**, 2663–2671 (2012)
2. Reddi, S.J., Hefny, A., Sra, S., et al.: On variance reduction in stochastic gradient descent and its asynchronous variants. Evid. Rep./Technol. Assess. **31**(183), 2647–2655 (2015)
3. Schmidt, M., Le Roux, N., Bach, F.: Minimizing finite sums with the stochastic average gradient. Math. Programm. **162**(1–2), 83–112 (2017)
4. Defazio, A., Bach, F., Lacoste-Julien, S.: SAGA: a fast incremental gradient method with support for non-strongly convex composite objectives. In: Advances in Neural Information Processing Systems, pp. 1646–1654 (2014)
5. De, S., Goldstein, T.: Efficient Distributed SGD with Variance Reduction. In: Mathematics (2015)
6. Johnson, R., Zhang, T.: Accelerating stochastic gradient descent using predictive variance reduction. In: International Conference on Neural Information Processing Systems, pp. 315–323. Curran Associates Inc. (2013)
7. Niu, F., Recht, B., Re, C., et al.: HOGWILD!: a lock-free approach to parallelizing stochastic gradient descent. Adv. Neural Inf. Process. Syst. **24**, 693–701 (2011)
8. Dean, J., Corrado, G., Monga, R., et al.: Large scale distributed deep networks. In: Advances in Neural Information Processing Systems, pp. 1223–1231 (2012)
9. Lewis, D.D., Yang, Y., Rose, T.G., et al.: RCV1: a new benchmark collection for text categorization research. J. Mach. Learn. Res. **5**(2), 361–397 (2004)
10. Lang, K.: NewsWeeder: learning to filter netnews. In: Twelfth International Conference on International Conference on Machine Learning, pp. 331–339. Morgan Kaufmann Publishers Inc. (1995)
11. West, M., Blanchette, C., Dressman, H., et al.: Predicting the clinical status of human breast cancer by using gene expression profiles. Proc. Nat. Acad. Sci. U.S.A. **98**(20), 11462 (2001)

# A Neural Network Model for Gating Task-Relevant Information by Rhythmic Oscillations

Ryo Tani and Yoshiki Kashimori[✉]

Department of Engineering Science, University of Electro-Communications,
Chofu, Tokyo 182-8585, Japan
r.tani@uec.ac.jp, kashi@pc.uec.ac.jp

**Abstract.** Visual system processes simple object features in early visual areas, and visual features become more complex as visual information is sending to downstream areas. In addition to the feedforward pathway, visual system has abundant feedback connections, whose number is even larger than feedforward ones. This suggests that top-down signal from higher visual areas may strongly affect sensory representation of early visual areas. Also, visual processing along the feedforward and feedback pathways is coordinated by brain rhythms. However, little is known about how the bidirectional visual processing is related with brain rhythms. To address this issue, we focus on an experimental study using two tasks in a visual perception. We develop a model of visual system which consists of a V1 and a V2 network. Using the model, we show that tuning modulations of V1 neurons are caused by a top-down influence mediated by the change in long-range connections of V1 neurons. We also show that top-down signal reflecting a slower oscillation in V2 neurons, coupled with a fast oscillation of V1 neurons, enables the efficient gating of task-relevant information encoded by V1 neurons.

**Keywords:** Rhythmic oscillation · Top-down influence · Neural model

## 1 Introduction

We can recognize rapidly and effortlessly complex visual scenes. Such amazing ability in visual recognition needs the effective processing of visual information along the multiple stages of visual pathways. Neurophysiological experiments have provided evidence for a "simple-to-complex" processing model based on a hierarchy of increasing complex image features, performed along the feedforward pathway of the ventral visual system [1]. On the other hand, visual system has abundant feedback connections, whose number is even larger than the feedforward ones [2]. Li et al. [3] showed that top-down signals allowed neurons of the primary visual cortex (V1) to engage stimulus components that were relevant to a perceptional task and to discard influences from components that were irrelevant to the task. They also showed that V1 neurons exhibited characteristic tuning patterns depending on the array of stimulus

© Springer Nature Switzerland AG 2018
L. Cheng et al. (Eds.): ICONIP 2018, LNCS 11302, pp. 194–202, 2018.
https://doi.org/10.1007/978-3-030-04179-3_17

components. Ramalingam et al. [4] further examined dynamic aspects of V1 neurons in the tasks used by Li et al., and revealed the difference in the dynamic correlations between V1 responses evoked by two tasks. Using a V1 model, we also proposed the neural mechanism of the tuning modulations by top-down signal [5].

Visual information processing along the feedforward and feedback pathways is coordinated by brain rhythms. Different brain rhythms are known and involved in gating of information flow [6–8]. Fast oscillations such as gamma rhythms are involved in sensory coding and feature binding in local circuits, while slower oscillations such as alpha and beta rhythms are evoked in higher brain areas and may contribute to the coupling of distinct brain areas. However, it is poorly understood how sensory information relevant to task context is coordinated by these brain oscillations.

To address this issue, we present a model of visual system which consists of networks of V1 and V2. We consider two types of perceptual tasks used by Li et al., bisection task and vernier one. V2 neurons receive the top-down signal reflecting task behaviors, as well as the feedforward inputs from V1 neurons, and feed their outputs back to V1 neurons. We show that the tuning modulations of V1 neurons are caused by a top-down influence mediated by the change in long-range connections of V1 neurons. We also show that top-down signal reflecting a slower oscillation in V2 neurons, coupled with a fast oscillation of V1 neurons, enables the efficient gating of task-relevant information encoded by V1 neurons. This study provides a useful insight to understanding how rhythmic oscillations in distinct brain areas are coupled to gate task-relevant information encoded in early sensory areas.

## 2 Model

### 2.1 Visual Tasks

We consider two visual tasks conducted by Li et al. [3]. The stimuli used consist of five bars, as shown in Fig. 1. The center bar is fixed, and two side-flanking bars and two end-flanking bars are located at different positions. The stimuli used are 25 in all. Monkeys were trained by two visual tasks: bisection and vernier tasks. In bisection task, the monkeys have to decide which side bar is closer to the central bar. In vernier task, they have to determine to which side the central bar is displaced relative to a collinear line of end bars. We consider here the bisection task, as a first step to study the mechanism underlying both tasks and the roles of brain rhythms in the task performance.

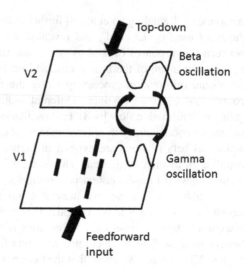

**Fig. 1.** Five-bar stimulus. The stimulus consists of a central bar, two side-flanking bars, and two end-flanking bars.

**Fig. 2.** Network model of visual system. Feedforward input to V1 neurons evokes a subthreshold membrane oscillation with a gamma frequency. Top-down signal to V2 neurons elicits a subthreshold membrane oscillation with a beta frequency.

## 2.2   Overview of Our Model

Figure 2 shows a model of visual system that performs bisection and vernier tasks. The model consists of two network models, the models of V1 and V2. V1 and V2 neurons are connected with each other in each network and between both networks. Neurons in the V1 network receive inputs from thalamus, and also receive feedback input from V2 neurons, which reflects task-relevant behavior. Feedforward and top-down inputs evoke the membrane oscillations of V1 and V2 neurons in their subthreshold regions.

## 2.3   The Model of V1

The model of V1 has two-dimensional array of V1 neurons that are tuned to a small vertical line, as shown in Fig. 3a. Neurons encoding the features of five-bar stimuli are depicted in gray scale. The unit model of V1 neuron is based on the leaky integrate-and-fire (LIF) model. Then, the membrane potential of the $(i, j)$th neuron, $V_{ij}^{V1}$, is determined by

$$\tau_{V1}\frac{dV_{ij}^{V1}}{dt} = -V_{ij}^{V1} + \sum_{kl} w_{ij,kl}^{V1} X_{kl}^{V1}(t) + \sum_{mn} w_{ij,mn}^{FB} X_{mn}^{V2}(t) + I_{ij}^{FF} + I_{osci}^{V1}(t) + \xi_{ij}^{V1}(t),$$

(1)

(a)                                         (b)

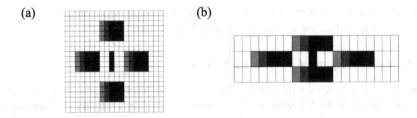

**Fig. 3.** (a) The network model of V1. A bar component of a five-bar stimulus is encoded by four V1 neurons. Side-flanking pairs and end-flanking pairs are depicted in gray scale. (b) The network model of V2. V2 neuron encoding each bar component is depicted in the similar way to the panel **a**.

$$\tau_X \frac{dX_{kl}^{V1}(t)}{dt} = -X_{kl}^{V1}(t) + \lambda\delta(t - t_i), \tag{2}$$

where $w_{ij,kl}^{V1}$ is the synaptic weight of the connection from the $(k, l)$th V1 neuron to the $(i, j)$th one, and $w_{ij,mn}^{FB}$ is the synaptic weight of the feedback connection from the $(m, n)$th V2 neuron to the $(i, j)$th V1 neuron. $\tau_{V1}$ is the time constant of $V_{ij}^{V1}$. $X_{kl}^{V1}(t)$ is the output of the $(k, l)$th V1 neuron and determined by Eq. (2), and $\tau_X$ is the time constant of $X_{kl}^{V1}(t)$ and $\delta(t - t_i)$ is the Dirac's delta function. $X_{mn}^{V2}$ is the output of the $(m, n)$th V2 neuron. The feedforward input from thalamus, $I_{ij}^{FF}$, was set to be a constant value, and $I_{osci}^{V1}$ is an input to induce a subthreshold membrane oscillation of V1 neuron, which is given by $I_0 \sin(2\pi ft)$, where $f = 50$ Hz within the gamma-range frequencies. $\xi_{ij}^{V1}(t)$ is noise input.

## 2.4 The Model of V2

The model of V2 is constructed with two-dimensional array of V2 neurons, as shown in Fig. 3b. Each neuron integrates the outputs of four V1 neurons encoding the feature of a bar element. The membrane potential of the $(i, j)$th V2 neuron, $V_{ij}^{V2}$, is given by

$$\tau_{V2}\frac{dV_{ij}^{V2}}{dt} = -V_{ij}^{V2} + \sum_{kl} w_{ij,kl}^{V2}X_{kl}^{V2}(t) + \sum_{mn} w_{ij,mn}^{FF}X_{mn}^{V1}(t) + I_{ij}^{Top} + \xi_{ij}^{V2}(t), \tag{3}$$

where $w_{ij,kl}^{V2}$ is the synaptic weight of the connection from the $(k, l)$th V2 neuron to the $(i, j)$th one, and $w_{ij,mn}^{FF}$ the synaptic weight of the feedforward connection from the $(m, n)$th V1 neuron to the $(i, j)$th V2 neuron. The output of V2 neuron, $X_{kl}^{V2}(t)$, is determined by the similar equation to Eq. (2). $I_{ij}^{Top}$ is a top-down signal from higher recognition areas, reflecting perceptual strategy of task context. In the bisection task, the top-down signal reflects the strategy of perceptual grouping, and thereby would elicit synchronous spiking of V2 neurons encoding the central bar and a side-flanking bar. Thus, $I_{ij}^{Top}$ is given by $I_0^{Top}\sin(2\pi ft + \phi)$, where $f = 20$ Hz within beta-range frequencies and

$\phi = 0$. This indicates that the top-down signal evokes in-phase membrane oscillation of V2 neuron encoding the central and a side-flanking bar.

## 2.5  Synaptic Weights of V1 and V2 Neurons

V1 neurons are connected with an excitatory and an inhibitory synapse and form long-range connections in the V1 network. The inhibitory synapse is actually mediated by an inhibitory interneuron, but, for simplicity, our model does not include the interneuron. The synaptic connection of a V1 neuron pair has a balanced weights of excitatory and inhibitory synapses, $w^E$ and $w^I$. We assumed that the balanced weight is $w^E > |w^I|$ for the connection between V1 neurons encoding the central bar and a side-flanking bar and that the weight is $|w^I| > w^E$ for the connection between V1 neurons encoding the central bar and an end-flanking bar. V2 neurons have the balance weights similar to V1 neurons.

# 3  Results

## 3.1  Tuning Modulations of V1 Neurons by Top-Down Influence

Figure 4 illustrates the tuning curves of V1 responses to five-bar stimuli in the bisection task. The tuning curves exhibited different modulations depending on the position of side-flanking bars. Our model reproduced the four types of tuning curves shown by the experimental study by Li et al. [3]. Our model also accounts for the mechanism by which the tuning modulations are generated. The key points of the mechanism are the top-down activation of task-relevant V1 neurons and the balanced weights of V1 neurons described in Sect. 2.5. The bar stimulus with offset of 0 does not generate top-down influence because of the equidistance of the left and right side-flanking bars from the central bar. The V-shaped curve, shown in Fig. 4a, is generated by the activity of the neurons encoding the central bar. Top-down signal elevates the activity of V2 neurons encoding the central bar and a side- flanking bar, resulting in the increased activity of V1 neurons encoding the same bars via the net excitatory connection of $w^E > |w^I|$. As a result, the activity of V1 neurons increases except for no-offset case. The caret-shaped curve, shown in Fig. 4b, comes from the activity of V1 neurons encoding an end-flanking bar with the offset of 0. The increased activity of V2 neurons encoding the central bar, elicited by top-down signal, increases the activity of V1 neurons encoding the central bar, leading to the reduced activity of V1 neurons encoding the end-flanking bar. This is due to the net inhibitory connection of $|w^I| > w^E$. The asymmetric curves, shown in Fig. 4c and d, are generated by the activity of V1 neurons encoding the side-flanking bars with the offsets of $+1$ and $-1$ [5].

### 3.2    A Role of Rhythmic Oscillations in Gating Task-Relevant Information

Rhythmic oscillations widely appear in the brain areas [6–8]. These oscillations are known to be involved in encoding of object's features within a brain area and in coordinating activity between brain areas. Figure 5a shows the spike patterns of a V1 neuron encoding the central bar in the bisection task and control. Top-down signal elicited by bisection task allows the V1 neuron to evoke a rhythmic burst spike with a beta frequency (20 Hz). The intraburst interval of the burst spikes is about 20 ms, indicating the spike interval of a gamma frequency (50 Hz). The spike pattern of a V2 neuron exhibits the tendency similar to that of the V1 neuron, as shown in Fig. 5b. Thus, the features of a bar is encoded in to a spike pattern of V1 neuron which has a fast (gamma) frequency, and information relevant to the bisection task can be extracted by synchronous spike patterns of V1 and V2 neurons, with a slower (beta) frequency.

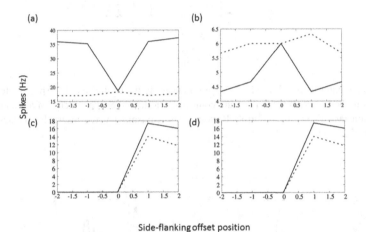

**Fig. 4.** Modulations of tuning curves in bisection task. (a) V-shaped curve, (b) Caret-shaped curve, (c), (d) Asymmetric curves. The tuning curves for bisection task and control are denoted by the solid and dashed lines, respectively.

Figure 6 shows the cross-correlations of the spikes of a V1 neuron pair, each encoding the central bar and a side-flanking bar. Top-down signal elicited by the bisection task strengthens the characteristic of a beta oscillation. Figure 7 shows the local field potential (LFP) coherence of a pair of V1 neuron groups, each encoding the central bar and a side-flanking bar. The LFP coherence exhibits the characteristic peaks of a gamma (50 Hz) and a beta (20 Hz) oscillation. The results shown in Figs. 6 and 7 indicate that the gamma oscillation is involved in encoding of bar features and the beta oscillation, coupled with V2 responses, is responsible for the grouping of bar features relevant to the bisection task.

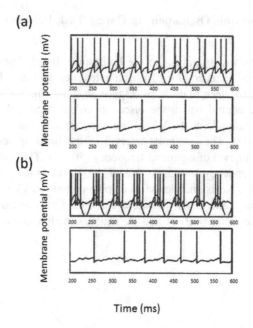

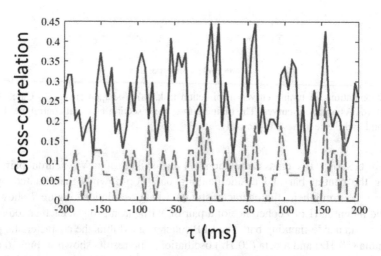

**Fig. 5.** Spike patterns of a V1 and a V2 neuron encoding the central bar. (a) Spike patterns of a V1 neuron in the bisection task (Top) and control (Bottom). (b) Spike patterns of a V2 neuron in the bisection task (Top) and control (Bottom). The sinusoidal wave inserted indicates a beta oscillation of 20 Hz.

**Fig. 6.** Cross-correlations of spikes of a V1 neuron pair. The solid and dashed lines indicate the cross-correlations in the bisection task and control, respectively.

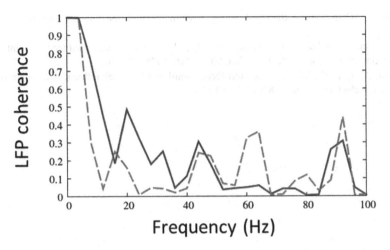

**Fig. 7.** LFP coherence of a pair of V1 neuron groups. Each group encodes the central bar and a side-flanking bar. The LFP coherence in the bisection task and control are indicated by the solid and dashed lines, respectively.

## 4  Conclusion

We have shown how top-down signal modulates the tuning curves of V1 neurons under the bisection task, and how brain rhythms evoked in V1 and V2 are involved in gating task-relevant information. The modulations of the tuning curves are caused by a top-down influence on V1 responses through the change in long-range connections of V1 neurons. A slower oscillation of V2 neurons, evoked by top-down signal, gates task-relevant information by cross-coupling of a fast oscillation of V1 neurons. These results would provide new insights to understanding the roles of top-down signal and brain rhythms in visual recognition.

## References

1. Rolls, E.T., Deco, G.: Computational Neuroscience of Vision. Oxford University Press Inc., New York (2002)
2. Budd, J.M.: Extrastriate feedback to primary visual cortex in primates: a quantitative analysis of connectivity. Proc. Biol. Sci. **265**, 1037–1044 (1998)
3. Li, W., Piëch, V., Gilbert, C.D.: Perceptual learning and top-down influences in primary visual cortex. Nat. Neurosci. **7**, 651–657 (2004)
4. Ramalingam, N., McManus, J.N.J., Li, W., Gilbert, C.D.: To-down modulation of lateral interaction in visual cortex. J. Neurosci. **33**, 1773–1789 (2013)
5. Kamiyama, A., Fujita, K., Kashimori, Y.: A neural mechanism of dynamic gating of task-relevant information by top-down influence in primary visual cortex. BioSystems **50**, 138–148 (2016)

6. Zheng, C., Colgin, L.L.: Beta and gamma rhythms go with the flow. Neuron **85**, 236–237 (2015)
7. von Stein, S., Chiang, C., König, P.: Top-down processing mediated by interareal synchronization. Proc. Natl. Acad. Sci. **97**, 14748–14753 (2000)
8. Bastos, A.M., et al.: Visual areas exert feedforward and feedback influences through distinct frequency channels. Neuron **85**, 390–401 (2015)

# A Hybrid Model Based on the Rating Bias and Textual Bias for Recommender Systems

Jiao Dai[1,2], Mingming Li[1,2(✉)], Songlin Hu[1,3], and Jizhong Han[1]

[1] Institute of Information Engineering, Chinese Academy of Sciences, Beijing, China
{daijiao,limingming,husonglin,hanjizhong}@iie.ac.cn
[2] School of Cyber Security, University of Chinese Academy of Sciences,
Beijing, China
[3] University of Chinese Academy of Sciences, Beijing, China

**Abstract.** Matrix Factorization is a useful approach in recommender systems. However, it only considers the user-item matrix, which will result in the data sparsity problem. To remit this issue, most researchers focus on using the item side-information to improve the performance and make a great success such as CDL, ConvMF. But these models all ignore the effect of specific item bias which is important because the same word represented different semantic for the different item. For example, the word "long" is a good description of the battery renewal time but opposite for the logistics of an item. In our work, we present a hybrid model that integrates the textual bias and rating bias to the PMF framework simultaneously. This model can exploit and modified the item specific word representation by CNN and obtain more precise side-information. Experiments on the three real-world datasets show that our model outperforms state-of-the-art method.

**Keywords:** Side-information · Textual bias · Rating bias
Recommender system · Matrix factorization

## 1 Introduction

In recent years, with the growing number of products and services, recommender system is becoming more and more significant and it has made great breakthroughs in multiple domains such as news, movie, music and so on. One successful approach is matrix factorization (MF) [5,7,17], which could learn the latent factors of user and item from the rating matrix. Concretely, WNMF [17] increases a non-negativity constraint for the latent vector to make the model easier to be interpreted and understood intuitively. BiasMF [5] takes the user and item biases into account. Mnih et al. [7] proposes Probabilistic Matrix Factorization (PMF) which increase the interpretability of model by the view of probability. But the performance will be discounted since the user-item matrix

© Springer Nature Switzerland AG 2018
L. Cheng et al. (Eds.): ICONIP 2018, LNCS 11302, pp. 203–214, 2018.
https://doi.org/10.1007/978-3-030-04179-3_18

is sparsity in the real-world. Besides, when new user or item comes, the prediction may be inaccurate due to lack of historical rating.

To alleviate the above problems, some researchers try to incorporate auxiliary information to PMF (or BiasMF). Wang et al. [12] proposes a collaborative topic regression model (CTR) using the item content as side-information by utilizing latent dirichlet allocation (LDA) [1]. CTR is the first attempt to integrate topic model into recommender system. It provides a new thought to remit the sparsity problem which achieves a great success. There are many further works by incorporating social network or other auxiliary information to CTR to improve performance [8,13,14] whereas its core components LDA is based on the assumption of words independence, thus there are more opportunities for optimizing by the document representation learning. Wang et al. presents CDL [15] that integrates the stacked denoising autoencode [11] to the PMF firstly. What's more, there are many extended models biased on CDL [2,6,16]. Compared to CTR, CDL further improves the performance by benefiting from the representation learning of depth network. Recently, Kim et al. proposes the ConvMF model [3] which uses the convolutional neural network (CNN) [4,18] to handle the additional information and makes a better prediction. However the above approaches ignore the effect of document bias for each item, it is significant that the same word may express different meaning in reviews context of different users or items which has been validated in sentiment analysis [9,10,16]. For instance, the review of "long" is a good description of the battery, but opposite for the logistics of an item.

Thus, in this paper, to handle the rating bias and textual bias simultaneously, we present a hybrid model that integrates the modified side-information and rating bias to the PMF framework. This model could obtain a specific word representation for each item review, and get an accurate latent factor for the item. To demonstrate the effectiveness of the model, we design some experiments comparing with the classical model on three real-world datasets. And results show that the Recall@M has improved obviously, and the RMSE is also better than state-of-the-art method.

The rest of this paper is organized as follows. Section 2 introduces the preliminary of recommendation system. Section 3 elaborates the proposed model. Section 4 gives the experimental results and analysis. Section 5 concludes the paper.

## 2   Preliminary

In this section, we will define some notations, and explain the classical models such as PMF, ConvMF in detail.

**Notation.** Given $N$ users, $M$ items and an user-item rating matrix $R \in \mathbb{R}^{N \times M}$, $R_{i,j}$ denotes the rating of user $i$ for item $j$. $U \in \mathbb{R}^{N \times D}$ and $V \in \mathbb{R}^{M \times D}$ denotes latent user and item feature matrices. $U_i$ and $V_j$ denotes user-specific and item-specific latent feature vector respectively. We note the side-information as $X \in \mathbb{R}^{M \times L \times K}$, where $L$ is the number of words for each item side-information, and

$K$ is the word dimension. Given $X_j = \{x_1, x_2, \ldots x_l, \ldots, x_L\}, j = \{1, 2, \ldots, M\}$, $x_l$ is the word representation vector.

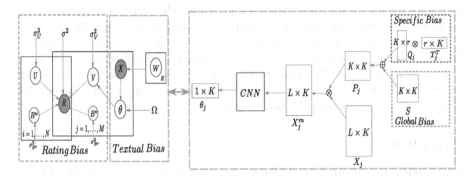

**Fig. 1.** The full architecture of proposed model, where the left is the graphical model which consists of two parts: rating bais module (red dotted rectangle) and textual bias module (blue dotted rectangle). The right shows the specific process of handling the bias of text. ⊕ denotes element-wise addition and ⊗ denotes the operation of matrix multiplication. (Color figure online)

**PMF.** Mnih et al. present the Probabilistic Matrix Factorization (PMF) model [7] which models the user preference matrix as a product of two lower-rank user and item matrices. Meanwhile, PMF supposes that the latent factors of user and item, predicted deviations are adopt the a prior of Gaussian distribution as $U \sim \mathcal{N}(0, \sigma_U^2)$, $V \sim \mathcal{N}(0, \sigma_V^2)$, $R \sim \mathcal{N}(UV^T, \sigma^2)$. Then, we could calculate the latent factors for $U$ and $V$, and use the $UV^T$ to predict. This model make a great success on the Netflix datasets. But it only use the user-item matrix which will result in the sparsity problem.

**ConvMF.** Convolutional Matrix Factorization [3] is mixed by PMF and CNN. CNN is a variant of feed-forward neural networks, and it has been widely applied in information retrieval and NLP such as search query retrieval, sentence modeling and classification and other traditional NLP tasks because of the advantages of mining the local features. The biggest difference between SDAE and CNN is that the former is based on unsupervised learning while the latter is supervised learning which could obtain more accurate representation vector. However, ConvMF ignores the effect of rating bias and textual bias. In other words, the same rating represents the different preference for different users and the same word in the reviews context of different items will have different semantics. Model proposed in this paper helps alleviate the above issues by adjusting ratings and word representation for the specific item.

## 3  Proposed Model

We propose a hybrid model which could handle the rating bias and textual bias for specific item simultaneously. The complete architecture of the model is presented in Fig. 1. Our model consists of two parts: rating bias module and textual bias module. Rating bias module is to update the latent vector for both user and item, and to make a prediction for unknown ratings. Textual bias module is to handle the item reviews and reduce the error of V. Specifically, firstly, we represent item reviews by word vector utilizing the pre-training model of Glove[1] which could map the word to a dense vector space. Secondly, we modify the word vector by using the bias of specific item in the embedding layer. And then we feed the embedding layer to the CNN module and get the item latent vector. Finally, we ingrate these two kinds of information to PMF model. All parameters could be learned by back propagation algorithms.

## Rating Bias

Rating bias is a very common phenomenon in recommender systems. Due to different evaluation criteria of different users, the same rating has different meanings. For example, a critical user usually gives lower ratings while a lenient user favors giving higher ratings. Thus, we use the bias of user rating and item rating to adjust the original observation rating according to [5]. Define the bias of user and item rating as $B^u \in \mathbb{R}^{1 \times N}$ and $B^v \in \mathbb{R}^{1 \times M}$ respectively. The modified rating could be amended as:

$$R^* = R - (B^u)^T \mathbf{1}_M - (B^v)^T \mathbf{1}_N \tag{1}$$

where $\mathbf{1}_M$ and $\mathbf{1}_N$ are all-one matrices with shape $1 \times M$ and $1 \times N$ respectively.

## Textual Bias

Given the side-information (text) $X$ and each word vector $x_l$ coming from the pre-train model of Glove which ignores the effect of the bias of item comments, we consider modifying the word vector for each specific item. Suppose that the number of vocabulary is $n$, we could construct a weight matrix with $n \times K$ for each item, but there are too many parameters need to be learned. In order to simplify the problem, we use an item-word-bias matrix to represent the weight matrix, denoting as $P \in \mathbb{R}^{M \times K \times K}$, $K \ll n$. $P_j$ is the bias of $X_j$, so we could get the modified side-information as $X_j^m = X_j P_j$. There is also a problem that $P_j$ is of high dimension and hard to be updated. To solve this problem, we reduce its dimension by matrix factorization as follows:

$$P_j = Q_j T_j^T + diag(S) \tag{2}$$

where $Q_j \in \mathbb{R}^{K \times r}, T_j \in \mathbb{R}^{K \times r}$, the value of $r$ is small; $S$ is the deviation matrix of all items and $S \in \mathbb{R}^{K \times K}$.

The CNN [4] model contains three significant components: convolutional layer, pooling layer and a full connect layer. We use $X_j^m$ obtaining by embedding layer as the input data of CNN. A convolution operation involves a filter

---

[1] https://nlp.stanford.edu/projects/glove/.

$W \in \mathbb{R}^{h \times K}$, which is applied to a window of $h$ words to produce a new feature. For example, a feature $c_i$ is generated from a window of words $x_i : x_{i+h-1}$ by

$$c_i = f(W \cdot x_i : x_{i+h-1} + b) \tag{3}$$

We can get a map features as $c = [c_1, c_2, \ldots, c_{L-h+1}]$. Then we apply a pooling operation on the $c$ and take the maximum value $\hat{c} = \max\{c\}$ as the feature corresponding to this particular filter. This idea is to capture the most important feature—one with the highest value—for each feature map. Finally, we use the full connect to transform the features to the specific dense vector space as $\theta = f(w\hat{c} + b)$. Above all, for each raw input $X_j$, we can get a item latent vector about the side-information as

$$\theta_j = \text{CNN}(W, X_j, Q_j, T_j, S) \tag{4}$$

## Integrating the Bias of Rating and Text

Now we combine above information to the PMF model. Suppose that $U$, $\varepsilon$, $B = \{B^u, B^v\}$ and predicted deviations all adopt a probabilistic linear model with Gaussian distribution as $U \sim \mathcal{N}(0, \sigma_U^2)$, $\varepsilon \sim \mathcal{N}(0, \sigma_\varepsilon^2)$, $B \sim \mathcal{N}(0, \sigma_B^2)$, $R^* \sim \mathcal{N}(UV^T, \sigma^2)$. We set item vector to be $V = \varepsilon + \theta$, $V \sim \mathcal{N}(\theta, \sigma_V^2)$. So the conditional distribution over the observed ratings could be defined as

$$P(R^*|U, V, B, \sigma^2) = \prod_{i=1}^{N} \prod_{j=1}^{M} \left[ \mathcal{N}\left((R_{i,j} - B_i^u - B_j^v)|U_iV_j^T, \sigma^2\right) \right]^{I_{i,j}} \tag{5}$$

where $I_{i,j}$ is the indicator function that is equal to 1 if user $i$ rated item $j$ and equal to 0 otherwise. Thus, the problem becomes to maximize the posterior probability:

$$\max_{U,V,W,Q,T,S,B} P(U, V, Q, T, S, B|R^*, \sigma^2, \sigma_U^2, \sigma_V^2,)$$

$$\propto \max_{U,V,W,Q,S,T,B} P(R^*|U, V, B, \sigma^2, \sigma_B^2)P(V|W, Q, S, T, \sigma_V^2, \sigma_W^2, \sigma_Q^2, \sigma_T^2, \sigma_S^2)$$

$$P(U|\sigma_U^2)P(W|\sigma_W^2)P(Q|\sigma_Q^2)P(T|\sigma_T^2)P(S|\sigma_S^2)P(B|\sigma_B^2) \tag{6}$$

Define $\frac{\sigma^2}{\sigma_U^2} = \frac{\sigma^2}{\sigma_{B^u}^2} = \lambda_U$, $\frac{\sigma^2}{\sigma_V^2} = \frac{\sigma^2}{\sigma_{B^v}^2} = \lambda_V$, $\frac{\sigma^2}{\sigma_W^2} = \frac{\sigma^2}{\sigma_S^2} = \lambda_W$, $\frac{\sigma^2}{\sigma_Q^2} = \frac{\sigma^2}{\sigma_T^2} = \lambda_Q$, $||\cdot||_F^2$ denotes the Frobenius norm, and then use the function of log to simplify the formula (6) as:

$$\min \mathcal{L} = \frac{1}{2}\sum_{i=1}^{N}\sum_{j=1}^{M} I_{i,j}(R_{i,j} - B_i^u - B_j^v - U_iV_j^T)^2 + \frac{\lambda_V}{2}\sum_{j=1}^{M}\left(||V_j - \theta_j||_2^2 + ||B_j^v||_2^2\right)$$

$$+ \frac{\lambda_U}{2}\sum_{i=1}^{N}\left(||U_i||_2^2 + ||B_i^u||_2^2\right) + \frac{\lambda_W}{2}(||W||_2^2 + ||S||_F^2) + \frac{\lambda_Q}{2}\sum_{j=1}^{M}(||Q_j||_F^2 + ||T_j||_F^2)$$

$$\tag{7}$$

## Optimization

For $U_i$ and $V_j$, we can minimization the $\mathcal{L}$ follows in a similar fashion as for probabilistic matrix factorization. Given the current estimate of $\theta_j$, taking the gradient of $\mathcal{L}$ with respect to $U_i$ and $V_j$ and setting it to zero leads to (recall the matrix definition $U = (U_i)_{i=1}^N$, $V = (V_j)_{j=1}^M$, $R_i = (R_{i,j})_{j=1}^M$)

$$U_i \leftarrow (R_i - B_i^u 1_M - B^v) I_i V (V I_i V^T + \lambda_U E_D)^{-1}$$
$$V_j \leftarrow \left( (R_j - B^u - B_j^v 1_N) I_j U + \lambda_V \theta_j \right) (U I_j U^T + \lambda_V E_D)^{-1}$$
$$B_i^u \leftarrow \frac{\sum_j^M I_{i,j}(R_{i,j} - U_i V_j^T - B_j^v)}{||I_i||_1 + \lambda_U}$$
$$B_j^v \leftarrow \frac{\sum_i^N I_{i,j}(R_{i,j} - U_i V_j^T - B_i^u)}{||I_j||_1 + \lambda_V}$$

where $I$ is a diagonal matrix with $I_{i,j}$, for item j, $I_j$ and $R_j$ are similarly defined; $E_D$ is the identity matrix with rank of $D$.

For $\theta$ we can fix the $U$ and $V$, and then use the back-propagation learning algorithm. The second term of object function (6) can be seen as a supervised learning which target is $V_j$ and predicted output is $\theta_j$. So the gradient with respect of the $\Omega = \{W, S, Q_j, T_j\}$ is as follows:

$$\nabla_W \mathcal{L} = \lambda_W W - \lambda_V \sum (V_j - \theta_j) \nabla_W \theta_j$$
$$\nabla_S \mathcal{L} = \lambda_W S - \lambda_V \sum (V_j - \theta_j) \nabla_S \theta_j$$
$$\nabla_{Q_j} \mathcal{L} = \lambda_Q Q_j - \lambda_V (V_j - \theta_j) \nabla_{Q_j} \theta_j$$
$$\nabla_{T_j} \mathcal{L} = \lambda_Q T_j - \lambda_V (V_j - \theta_j) \nabla_{T_j} \theta_j$$

Thus, update the parameters $\Omega$ in each iterations, $\omega \leftarrow \omega - \eta \nabla_\omega \mathcal{L}$ where $\omega \in \Omega$, $\eta$ is the learning rate.

For prediction, we have obtain the latent actors for user and item, so unknown rating could be predicted by $\hat{R}_{i,j} \approx U_i V_j^T + B_i^u + B_j^v = U_i(\epsilon_j + \theta_j)^T + B_i^u + B_j^v$, and then a list of ranked items is generated for each user.

## 4    Experiments

In order to verify the availability and high efficiency of our model, we design some comparison experiments, and evaluate the experimental results. What is more, we will describe the parameters setting in detail.

### 4.1    Datasets

We experiment on three public real-world datasets: Movie-Lens (1M,10M)[2] and Amazon[3]. These datasets all consist of explicit ratings on a scale of 1 to 5.

---

[2] https://grouplens.org/datasets/movielens/.
[3] http://jmcauley.ucsd.edu/data/amazon/.

**Table 1.** Statistics of three datasets

| Datasets | #Users | #Items | #Ratings | #Density |
|---|---|---|---|---|
| Movie-Lens1M | 6,040 | 3,544 | 993,482 | 4.641% |
| Movie-Lens10M | 69,878 | 10,073 | 9,945,875 | 1.413% |
| AIV | 29,757 | 15,149 | 135,188 | 0.030% |

The item reviews of Amazon Instant Video (AIV) are provided by itself. But Movie-Lens does not contain reviews information, so we replace them with the description information which could be crawled from IMDB[4]. In addition, we remove users who have rated less than 3 times for AIV to improve the performance. Table 1 summarizes the statistics on the three datasets.

### 4.2 Experimental Setup

We design some comparison tests with classical models (i.e., PMF, WNMF, BiasMF, CTR, and CDL) and the state-of-the-art model of ConvMF which we have introduced in Sect. 2. In this paper, we use the explicit rating, so we define the rating of CTR and CDL to 1 if $R_{i,j}$ is observed and 0 otherwise. Several parameters play an important rule in the performance, and better value of $\lambda_V$ and $\lambda_U$ are taken from [3] which are shown precision. For our model, we set $\lambda_U = 1$, $\lambda_V = 190$, $\lambda_W = 1000$, $\lambda_Q = 10000$ for MovieLens, and $\lambda_U = 1$, $\lambda_V = 100$, $\lambda_W = 1000$, $\lambda_Q = 10000$ for AIV. The length of side-information $L$ is 300 because of the limitations of compute resources. The dimension of the latent factors $D$ is 50; the low rank $r$ of $Q_j$ and $T_j$ is 3, and the learning rating $\eta$ is $10^{-3}$.

**Table 2.** Average RMSE of compared models on three datasets

| Model | Datasets | | |
|---|---|---|---|
| | Movie-Lens1M | Movie-Lens10M | AIV |
| PMF [7] | 0.8971 | 0.8311 | 1.2889 |
| WNMF [17] | 0.9296 | 0.8834 | 1.2062 |
| BiasMF [5] | 0.8834 | 0.8287 | 1.2062 |
| CTR [12] | 0.8969 | 0.8275 | 1.5496 |
| CDL [15] | 0.8879 | 0.8186 | 1.3594 |
| ConvMF [3] | 0.8531 | 0.7958 | 1.1337 |
| Ours | **0.8407** | **0.7815** | **1.1060** |

---

[4] http://www.imdb.com/.

### 4.3    Evaluation Metrics

In this paper, we split the original dataset into train, validation and test sets with a 80:10:10 split. To improve the effectiveness of the measurement, we guarantee that every part contains all the users with one rating as least. We employ the root mean squared error (RMSE) as one of the evaluation metrics as:

$$\text{RMSE} = \sqrt{\frac{1}{|T|} \sum_{R_{i,j} \in T} (\hat{R}_{i,j} - R_{i,j})^2} \tag{8}$$

where $T$ is the test sets, $|T|$ is the total number of rating.

The other evaluation metrics is Recall@M:

$$\text{Recall@M} = \frac{\text{number of items the user likes in top M}}{\text{total number of items the user likes}} \tag{9}$$

We think the user likes a item if the rating is more than four for Movie-Lens, and two for AIV. Finally, we can obtain the result by computing the average of all users.

### 4.4    Experimental Results

As the same datasets and the same parameters, we directly compare our results with the experimental results in [3]. Experimental results of RMSE and Recall@M is shown in Tables 2 and 3 respectively. As we can see that our proposed model outperforms the others. On the datasets Movie-Lens10M, our model has a better improvement than others, which are 1.43% than state-of-the-art method of ConvMF, and 3.71% than CDL. On the AIV datasets, Our model has a large promotion than the ConvMF. We can find more observations from the results: there are huge differences of the result between Movie-Lens and AIV, because we use the item description information to replace the item specific reviews for Movie-Lens, but the real-world item reviews for AIV, and the description information has minor differences for each item, but the reviews inversely. This result also proves our assumption that there is a huge reviews basis for the item in real-world. Thus, that is necessary to consider the bias of text.

What is more, we compute the average Recall@M on three datasets. Table 3 shows that the Recall@3 of ours is about 4% higher than ConvMF on the Movie-Lens dataset, and 5% on the AIV dataset. On the AIV dataset, the Recall@M will not increase as M when $M >= 30$, because the total number of items the user likes for the test is fixed, in other words, the ratings of the item that we predicted is lower two. The smaller of the $M$ we use, the higher the accuracy we will get, thus, our model predicts the unknown rating more accurately than ConvMF. The essence of this result is we can obtain a precise item latent vector by considering the basis of reviews.

To investigate the ability of handling the sparse problems, we design some test over various sparseness train data on Movie-Lens1M dataset, and the result

**Table 3.** Average Recall@M on three datasets

| Recall@M | Datasets | | | | | |
|---|---|---|---|---|---|---|
| | Movie-Lens1M | | Movie-Lens10M | | AIV | |
| | ConvMF | **Ours** | ConvMF | **Ours** | ConvMF | **Ours** |
| 3 | 0.2058 | **0.2428** | 0.2041 | **0.2492** | 0.2707 | **0.3167** |
| 5 | 0.2520 | **0.2950** | 0.2392 | **0.2907** | 0.2725 | **0.3186** |
| 10 | 0.2986 | **0.3463** | 0.2721 | **0.3279** | 0.2731 | **0.3191** |
| 20 | 0.3208 | **0.3689** | 0.2868 | **0.3438** | 0.2731 | **0.3192** |
| 30 | 0.3253 | **0.3736** | 0.2897 | **0.3468** | 0.2732 | **0.3192** |
| 40 | 0.3264 | **0.3746** | 0.2905 | **0.3477** | 0.2732 | **0.3192** |
| 50 | 0.3268 | **0.3751** | 0.2908 | **0.3480** | 0.2732 | **0.3192** |

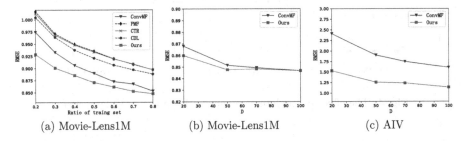

(a) Movie-Lens1M          (b) Movie-Lens1M          (c) AIV

**Fig. 2.** (a) shows the effect of ratio of train set. (b) and (c) shows the effect of latent dimension on the Movie-Lens1M and AIV respectively.

is shown in Fig. 2(a). Our model has a better performance than others obviously. Specifically, on the sparse environment such as ratio = 0.2, the average RMSE of our model is about 4% lower than ConvMF, and 8% lower than CDL. We can also find that ConvMF boosts effects more than CDL because of the advantages of CNN on text representation learning.

In addition, the dimension of latent factors plays an important rule in Matrix Factorization. So we evaluate the average RMSE and Recall@M by setting the dimension $D$ to 20, 50, 70, 100, respectively. Figure 2(b) is the result of RMSE on the Movie-Lens1M dataset. We can find that our model has a better performance than ConvMF when $D = 20$, but the RMSE is also higher. And when $D >= 70$, RMSE will not change as $D$, which indicate that we have obtained a better latent factor. The result on the AIV dataset is shown in Fig. 2(c), it is obvious that our model boost about 90% than ConvMF when $D = 20$, this can explain that our model could learn a useful and precise latent factor although the latent space is lower. On the whole, the RMSE will decline with increasing of $D$. But we have to consider the computational efficiency, because it is fairly difficult for us to MF when $D >= 100$. And in general, we set $D = 50$ to take into account both accuracy and efficiency.

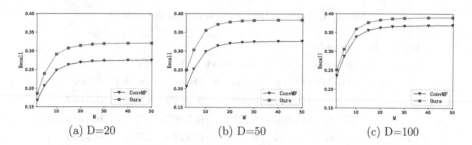

**Fig. 3.** The effect of the latent dimension D on the Movie-Lens1M dataset.

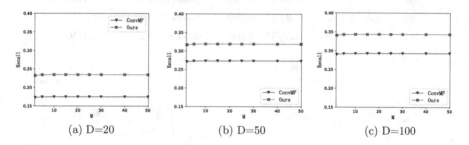

**Fig. 4.** The effect of the latent dimension D on the AIV dataset.

The result of Recall@M on the Movie-Lens and AIV dataset is shown in Figs. 3 and 4, respectively. The Recall@M of ours are all higher than ConvMF on different datasets. When $D = 100$, our model has little advantage on Movie-Lens1M, but inversely on AIV dataset, and the reason is that the difference of data source we mentioned before.

## 5  Conclusion

In this paper, we focus on the effect of item reviews bias inspired by the bias of user rating, and present a hybrid model which handles the textual bias and rating bias to the PMF framework simultaneously. Proposed model could obtain a specific word representation for each item, and get a more accurate latent factors for the item. Experiments shows that, the RMSE of our model is better than the stat-of-the-art method, and the recall@M is also better obviously. There is still much work to do, for example, we could consider dealing with the selective bias of observations simultaneously.

**Acknowledgments.** This work was supported in part by National Key R&D Program of China under Grant 2017YFB101000.

# References

1. Blei, D.M., Ng, A.Y., Jordan, M.I.: Latent dirichlet allocation. J. Mach. Learn. Res. **3**(Jan), 993–1022 (2003)
2. Dong, X., Yu, L., Wu, Z., Sun, Y., Yuan, L., Zhang, F.: A hybrid collaborative filtering model with deep structure for recommender systems. In: AAAI, pp. 1309–1315 (2017)
3. Kim, D., Park, C., Oh, J., Lee, S., Yu, H.: Convolutional matrix factorization for document context-aware recommendation. In: Proceedings of the 10th ACM Conference on Recommender Systems, pp. 233–240. ACM (2016)
4. Kim, Y.: Convolutional neural networks for sentence classification. arXiv preprint arXiv:1408.5882 (2014)
5. Koren, Y., Bell, R., Volinsky, C.: Matrix factorization techniques for recommender systems. Computer **42**(8), 30–37 (2009)
6. Li, S., Kawale, J., Fu, Y.: Deep collaborative filtering via marginalized denoising auto-encoder. In: Proceedings of the 24th ACM International on Conference on Information and Knowledge Management, pp. 811–820. ACM (2015)
7. Mnih, A., Salakhutdinov, R.R.: Probabilistic matrix factorization. In: Advances in neural information processing systems, pp. 1257–1264 (2008)
8. Purushotham, S., Liu, Y., Kuo, C.C.J.: Collaborative topic regression with social matrix factorization for recommendation systems. arXiv preprint arXiv:1206.4684 (2012)
9. Tan, J., Wan, X., Xiao, J.: A neural network approach to quote recommendation in writings. In: Proceedings of the 25th ACM International on Conference on Information and Knowledge Management, pp. 65–74. ACM (2016)
10. Tang, D., Qin, B., Liu, T.: Learning semantic representations of users and products for document level sentiment classification. In: Proceedings of the 53rd Annual Meeting of the Association for Computational Linguistics and the 7th International Joint Conference on Natural Language Processing (Volume 1: Long Papers), vol. 1, pp. 1014–1023 (2015)
11. Vincent, P., Larochelle, H., Lajoie, I., Bengio, Y., Manzagol, P.A.: Stacked denoising autoencoders: learning useful representations in a deep network with a local denoising criterion. J. Mach. Learn. Res. **11**(Dec), 3371–3408 (2010)
12. Wang, C., Blei, D.M.: Collaborative topic modeling for recommending scientific articles. In: Proceedings of the 17th ACM SIGKDD International Conference on Knowledge Discovery and Data Mining, pp. 448–456. ACM (2011)
13. Wang, H., Chen, B., Li, W.J.: Collaborative topic regression with social regularization for tag recommendation. In: IJCAI, pp. 2719–2725 (2013)
14. Wang, H., Li, W.J.: Relational collaborative topic regression for recommender systems. IEEE Trans. Knowl. Data Eng. **27**(5), 1343–1355 (2015)
15. Wang, H., Wang, N., Yeung, D.Y.: Collaborative deep learning for recommender systems. In: Proceedings of the 21th ACM SIGKDD International Conference on Knowledge Discovery and Data Mining, pp. 1235–1244. ACM (2015)

16. Zhang, F., Yuan, N.J., Lian, D., Xie, X., Ma, W.Y.: Collaborative knowledge base embedding for recommender systems. In: Proceedings of the 22nd ACM SIGKDD International Conference on Knowledge Discovery and Data Mining, pp. 353–362. ACM (2016)
17. Zhang, S., Wang, W., Ford, J., Makedon, F.: Learning from incomplete ratings using non-negative matrix factorization. In: Proceedings of the 2006 SIAM International Conference on Data Mining, pp. 549–553. SIAM (2006)
18. Zhang, Y., Wallace, B.: A sensitivity analysis of (and practitioners' guide to) convolutional neural networks for sentence classification. arXiv preprint arXiv:1510.03820 (2015)

# Phase and Amplitude Modulation in a Neural Oscillatory Model of the Orientation Map

Bhadra S. Kumar, Avinash Kori, Sundari Elango,
and V. Srinivasa Chakravarthy[✉]

Computational Neuroscience Lab, Bhupat and Jyothi Mehta School of Biosciences,
Indian Insitite of Technology Madras, Chennai, India
schakra@iitm.ac.in

**Abstract.** The traditional approach to characterization of orientation maps as they were expounded by Hubel and Wiesel treats them as static representations. Only the magnitude of a neuron's firing response to orientation is considered and the neuron with the highest response is said to be "tuned" to that response. But the neuronal response to orientation is a time-varying spike train and, if the response of an entire cortical area that potentially responds to orientations in a given part of the visual field is considered, the response must be considered as a spatio-temporal wave. We propose a neural field model consisting of FitzHugh-Nagumo neurons, that generates such a wave. Reflecting the dynamics of a single FitzHugh-Nagumo neuron, the neural field also exhibits excitatory and oscillatory regimes as an offset parameter is increased. We consider the question of the manner in which the input orientation is coded in the response of the neural field and discovered that two different codes − Amplitude Modulation and Phase Modulation − are present. Whereas for smaller offset values, when the model is in excitatory regime the orientation is coded in terms of amplitude, for larger offset values when the model is in the oscillatory regime, the orientation is coded in terms of phase.

**Keywords:** NFM · Oscillator · Amplitude · Phase · FN

## 1 Introduction

In their seminal studies on mammalian visual cortex Hubel and Wiesel found neurons that respond to oriented bars [14]. This discovery was significant in that neurons at lower levels of the visual hierarchy − retinal layers, Lateral Geniculate Nucleus etc − showed responses to simple dot-like patterns, while the visual cortical neurons showed no response to dots, but responded to linear oriented patterns. Furthermore, neurons are tuned to bars of specific orientation, $\theta$. The mapping of the orientation, on the two dimensional surface of the primary visual

© Springer Nature Switzerland AG 2018
L. Cheng et al. (Eds.): ICONIP 2018, LNCS 11302, pp. 215–226, 2018.
https://doi.org/10.1007/978-3-030-04179-3_19

cortex, $\theta(x,y)$ is known as the orientation map. Among neurons that exhibit orientation tuning, there are the simple cells that respond to an oriented line present in the center of their Receptive Field (RF) and complex cells that respond to an oriented line present almost anywhere in their RF. Orientation tuning changes continuously as a function of cortical location except at isolated points: these points, mathematically speaking, are singularities and are commonly referred to as "pinwheels" [2]. An extensive amount of computational modelling effort has been invested in understanding these intriguing features of orientation maps (see Chap. 5 of Miikkulainen et al. [22] for an extensive review of both biological and computational aspects of orientation maps). There are spectral models that show how to compute maps that resemble visual orientation maps without invoking neurons or learning [13, 26]. These models are based on the insight that orientation map like patterns can be produced simply by filtering random dot images. Then there are correlation-based learning models in which there are neurons with lateral and afferent connections. In these models, the afferent connections are trained while the lateral connections are fixed [23, 32, 36]. A particularly elegant map model of this category was the one proposed by Obermayer et al. [27]. This model, which was based on the Self-organizing Map (SOM) model proposed by Kohonen [18], was able to account for several key features of orientation maps like pinwheels and fractures. Another line of models of orientation maps that involved trainable lateral connections was the Laterally Interconnected Synergistically Self-Organizing Map (LISSOM) proposed by Miikkulainen and colleagues [21]. The LISSOM was later used to model a wide variety of topographical maps in the visual cortex.

But the visual cortical "maps" denote static representations of visual stimuli. They embody a static notion of the kind of information processing performed by the visual cortex. There is an entire perspective of cortical processing that posits that the fundamental building block of cortical computation is not a single neuron but a neuronal ensemble [5, 11, 19, 28] and the collective activity of such an ensemble is not a spike train but a Local Field Potential (LFP) [15, 25]. Unlike spike trains, the LFP is usually resolved into a small number of oscillatory components described in terms of amplitude, frequency and phase. Information transfer between two such ensembles can be achieved by mechanisms like entrainment and synchronization [8, 16, 29, 31]. There is a growing body of experimental literature that describes cortical responses to stimuli and learning in terms of LFP oscillations [7, 12, 20, 33, 34]. Theta oscillations in visual cortical LFPs show adaptation to stimulus and the ability to predict time of reward prediction [37]. Klimesh et al. [17] proposed that the visual alpha waves play a key role in encoding the visual stimulus in the early post stimulus interval (<150 ms). The varied roles of all the Electroencephalogram (EEG) waves not just in visual perception but in cognitive processes in general have been extensively reviewed [1]. In view of the extensive literature on the roles of oscillatory neural activity in cortical processing, we propose that the notion of cortical "maps" is slightly outmoded. There must be a more expanded notion of cortical representations in terms of oscillatory activity. With such a motivation we set out to construct an oscillatory neural model of orientation sensitivity. The model consists of a two-dimensional array – a "neural field" – of FitzHugh-Nagumo neurons with afferent and lateral connections. The network is trained to respond to oriented bar stimuli. An analysis of the network responses shows that orientation is coded in two forms: Amplitude Modulated (AM) form and Phase Modulated (PM) form. The network is operated in two modes: (1) excitatory mode and (2) oscillatory mode. Whereas the AM code dominates in the excitatory mode, PM is predominantly found in the oscillatory mode. The paper is outlined as follows. Section 2 presents the model architecture and learning equations. Section 3 describes the simulation results and the emergence of AM and FM codes of orientation sensitivity. The modeling results and future work is discussed in the last section

## 2   Methods

The architecture of the proposed model of oscillatory orientation maps consists of a cascade of two components: the Self-organizing Map (SOM) and a Neural Field Model (NFM) consisting of the FitzHugh-Nagumo (FN) neurons. The SOM represents a traditional and one of the simplest models of orientation selectivity. The output of the SOM is presented to the NFM which encodes the input oriented bar in terms of its oscillations. The architecture is shown in Fig. 1.

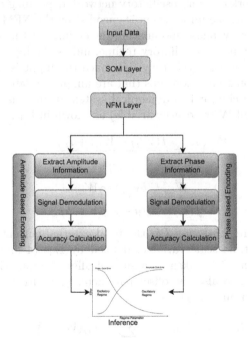

**Fig. 1.** The above figure shows flow of the data, modulation and demodulation strategies followed in our method

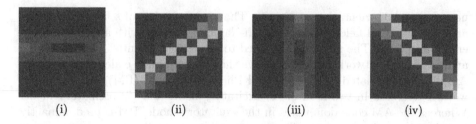

(i)          (ii)          (iii)          (iv)

**Fig. 2.** Gaussian bars at different orientations which are used as input stimulus for the experiment

## 2.1   The Network Architecture

The SOM layer is trained using orientation bars shown in Fig. 2. The output of the SOM network is given as a one to one projection to the oscillatory neural network called Neural Field Model (NFM). Each element of the NFM is a Fitzhugh Nagumo (FN) oscillator. Each FN oscillator is connected among the neighbors by means of a fixed Mexican Hat connectivity (M) to form the NFM. The strength of the lateral connections can be made strong or weak depending on the variable $\beta$. The input influences the neuron both in an additive and multiplicative manner. The NFM layer can exist in two states and behaves as an excitatory network or as an oscillatory network depending on the parameter $C$. It is observed that the information is encoded by the NFM either by amplitude modulation or by phase modulation depending on the state in which it exists. For example, in the oscillatory regime, unless the input is presented, all of the neurons oscillate in synchrony; but once the input is presented, neural oscillations attain phase relationships that are uniquely related to the input orientation. The FN oscillator is a dynamical system defined using two variables, V (fast variable) and W (recovery variable) as shown in Eqs. (1–3).

$$\dot{V}_{i,j} = \frac{f(V_{i,j}) - W_{i,j} + I_{i,j}}{\varepsilon_{i,j}} \tag{1}$$

$$\dot{W}_{i,j} = bV_{i,j} - \gamma W_{i,j} \tag{2}$$

$$f(V) = V(a - V)(V - 1) \tag{3}$$

where $b$, $\gamma$ and $a$ are constants. $I$ is the total external current felt by the neuron, and $\varepsilon$ is the time scale variable, defined by Eqs. (4 and 6) respectively. The external current, $I$ is taken as a sum of afferent input ($A$), current from lateral connections and also the offset parameter ($C$) with appropriate scaling parameters as shown in Eq. (4).

$$I_{i,j} = C + A_{i,j} + \beta \sum_{k,l} M_{kl,ij}(V_{k,l} - V_{i,j}) \tag{4}$$

where, $M$ denotes the fixed mexican hat weight connection among the neighbouring neurons, $\beta$ is a scaling parameter and $A$ is the afferent input given by:

$$A_{i,j} = g \times SOM \tag{5}$$

where SOM is the output of the self orientation map layer and $g$ is a scaling parameter. The factor $g$ ensures that the input does not override the lateral connections.

The time scale variable $\varepsilon$ is given by the Eq. (6)

$$\varepsilon_{i,j} = 1 - SOM_{i,j} + p \tag{6}$$

where $p$ is a constant. The value of $p$ is chosen to ensure the stability of the oscillator such that a stable limit cycle exists for all values of SOM outputs ranging between 0 and 1. Considering the facts that the value of $\varepsilon$ is ensured to lie between $p(p \rightarrow 0^+)$ and 1, the parameters $b$ and $\gamma$ are constants and the afferent and lateral inputs are bounded (absolute normalization of lateral weights), the model complies with the classic Fitzhugh Nagumo model and is hence stable at both excitatory and inhibitory regimes. The values of the parameters used are given in Table 1.

**Table 1.** Table of parameters

| Parameter | Value chosen |
| --- | --- |
| $a$ | 0.5 |
| $b$ | 0.1 |
| $\gamma$ | 0.1 |
| $\beta$ | 0.6 |
| $g$ | 0.05 |
| $p$ | 0.01 |

## 2.2 Oscillatory Neural Response Demodulation

The response of each oscillatory neuron is observed for 50 time units. Two different methods of data processing, phase demodulation and amplitude demodulation is done to extract the phase and amplitude information respectively from the response waveforms.

## 2.3 Phase Demodulation

In order to extract the phase information from the NFM response, the relative phase difference (RP) of each neural response is calculated by taking the phase of first neuron - the neuron at the location (1,1) of the NFM as reference. Instantaneous phase of the output of a neuron $V_{i,j}$ is calculated using Hilbert transform

of a signal and Mean of the Relative Phases (MRP) is used to train the logistic
regression model.

$$RP_{i,j}(t) = \angle H\{V_{i,j}(t)\} - \angle H\{V_{1,1}(t)\} \qquad (7)$$

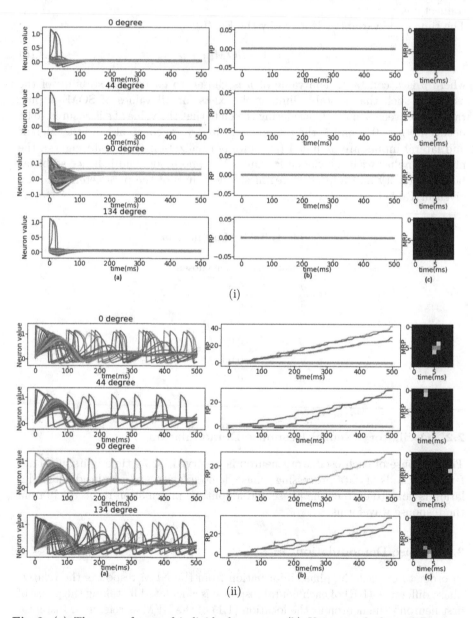

**Fig. 3.** (a) The wave forms of individual neurons, (b) Unwrapped phase difference
across time, (c) Mean relative phase, during (i) excitatory regime and (ii) oscillatory
regime.

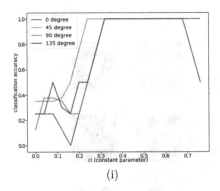
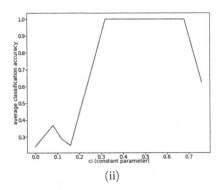

(i)                              (ii)

**Fig. 4.** (i) Classification accuracy on new test data, and figure (ii) Average classification accuracy for all the test angles (obtained by Phase demodulation).

where $H\{X\}$ defines hilbert transform of $X$ and $\angle H\{X\}$ calculates the instantaneous angle of $H\{X\}$

$$MRP_{i,j} = \frac{1}{T}(\sum_{t=0}^{T} RP_{i,j}(t))$$  (8)

Using the data generated from 20 different trials for each orientation, and 4 different Orientations (Fig. 2), a multi-nominal logistic regression model was trained with the $n^2-1$ dimensional input data with corresponding $K$ dimensional one hot vectors as labels, where $K$ is the number of classes (In this case, $K = 4$). The classification accuracy was calculated on held out test data.

### 2.4   Amplitude Demodulation

The amplitude demodulation was done by simply averaging the output, $V_{i,j}$, of the NFM across time. The averaged NFM response ($A$) is given by.

$$A_{i,j} = \frac{1}{T}(\sum_{t=0}^{T} V_{i,j}(t))$$  (9)

Similar to Phase demodulation, the data was generated from 20 different trials for each orientation and 4 different orientations. The data was used to train a multi-nominal logistic regression model with $n^2$ dimensional input data with corresponding $K$ dimensional one hot vectors as labels. The classification accuracy was calculated on held out test data.

## 3   Results and Discussion

The response of the neural sheet is observed for 50 time units and the waveform is analyzed for each network characteristics. By varying the value of offset

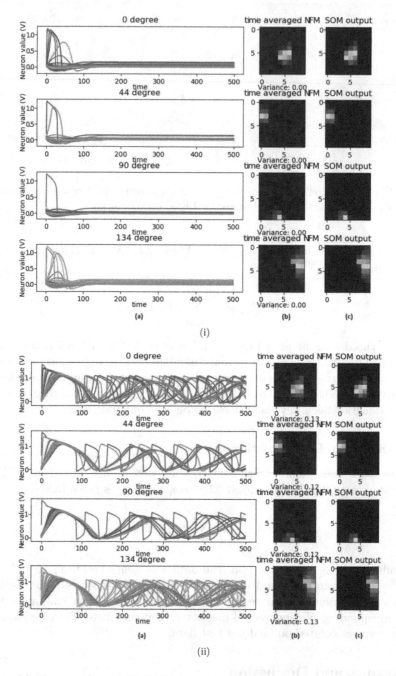

**Fig. 5.** (a) The wave forms of individual neurons, (b) average responses of NFM across time, (c) SOM output, during (i) excitatory regime and (ii) oscillatory regime.

parameter, $C$, between 0 to 0.8, the NFM shifts from excitatory regime to oscillatory regime. The response of the network in both regimes, while presented with a SOM input is shown in plot (a) in Figs. 3 and 5.

In Fig. 3(i) and (ii), plot(b) shows the relative phase of each neuron with respect to the phase first neuron and plot (c) shows the mean relative phase vector of all neurons across time($MRP$). As can be seen, the phase difference during excitatory regime (Fig. 3: i.b and i.c) is almost zero everywhere, carrying no information, while the phase difference plot in oscillatory regime (Fig. 3: ii.b and ii.c) shows a unique pattern. As seen in Fig. 3: ii.a, in the oscillatory regime, most of the background neurons oscillate in phase and the neurons encoding the input oscillate at a higher frequency. This indicates the encoding of information in phase during oscillatory regime. This is further verified by calculating classification accuracy using the MRP vectors by means of logistic regression. The results are shown in Fig. 4 where the classification accuracy increases once the network enters the oscillatory regime.

In the Fig. 5(i.b and ii.b), the average sheet response across time is observed for both excitatory regime and oscillatory regime. Though the average sheet resembles the SOM response in both regimes, the variance (noise) in the average sheet responses is zero in the excitatory regime Fig. 3(i.b) and it increases once the neuron enters the oscillatory regime as shown in Fig. 3(ii.b). This suggests a stronger amplitude coding in the excitatory regime than in the oscillatory regime which is further confirmed by computing average classification accuracy using logistic regression as shown in Fig. 6 where the classification accuracy decreases once the network enters the oscillatory regime.

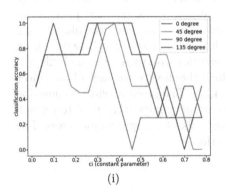
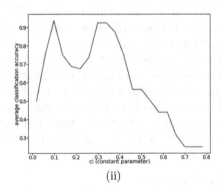

(i)                                        (ii)

**Fig. 6.** (i) Classification accuracy on new test data, and figure (ii) Average classification accuracy for all the test angles (obtained by Amplitude demodulation).

## 4    Conclusion

We proposed a neural field model ($n \times n$) of an orientation map in which the orientation information is encoded in the spatio-temporal response of the neural field. The model consists of a laterally connected 2D layer of FN neurons. Oriented bar patterns are used to train a SOM (n × n), whose output is presented to the NFM in a one-to-one fashion. The neural field model was shown to exist in two regimes: in the excitatory regime for lower values of the offset parameter, and in the oscillatory regime for higher values of the offset parameter. The classical orientation map which presents a static picture of orientation responses of the visual cortical neurons, can be reinterpreted in terms of neural oscillations. The proposed NFM model encoded the input orientation information primarily using amplitude modulation while in excitatory regime and using phase modulation while in the oscillatory regime. The classical depictions of orientation maps, or for that matter any topographic map of the brain, present the impression that information is encoded in static forms in the cortical surface. However the brain is a dynamical system and cortical responses to stimuli are more ideally considered as spatio-temporal responses. A primary instance of such a characterization of neural responses to sensory stimuli, is the study of the responses of the olfactory bulb by Freeman and colleagues [9]. In these studies Electroencephalogram (EEG) recordings were made by grid electrodes from the rabbit's olfactory bulb. The responses of the bulb to odors showed stable and repeatable spatio-temporal patterns.

But the aforementioned recordings in the olfactory bulb are Local Field Potentials (LFPs) and not spike patterns. A prominent perspective of neural computations posits that the fundamental unit of computations in the brain is not a single neuron but a neural ensemble. The activity of a neural ensemble is captured by a LFP, often described in terms of a small number of oscillatory components. Neural models often fall under two broad categories: rate coded or spiking neuron models. But a large number of cerebral phenomena can be best described in terms of oscillatory activity [4]. For example, gamma rhythm is thought to have a key role in information transfer between brain regions [3,6,10,35]. Synchronized gamma oscillations are thought to underlie feature binding, a temporal mechanism by which the diverse features of an object are integrated in a unified representation [24,30]. The growing effort to describe brain function in terms of dynamics and oscillations points to the need to rethink static notions of neural information processing like the topographic maps. The proposed model is a small step in that direction.

# References

1. Başar, E., Başar-Eroglu, C., Karakaş, S., Schürmann, M.: Gamma, alpha, delta, and theta oscillations govern cognitive processes. Int. J. Psychophysiol. **39**(2–3), 241–248 (2001)
2. Bonhoeffer, T., Grinvald, A.: Iso-orientation domains in cat visual cortex are arranged in pinwheel-like patterns. Nature **353**(6343), 429 (1991)
3. Bosman, C.A., et al.: Attentional stimulus selection through selective synchronization between monkey visual areas. Neuron **75**(5), 875–888 (2012)
4. Buzsaki, G.: Rhythms of the Brain. Oxford University Press, New York (2006)

5. Christopher deCharms, R., Merzenich, M.M.: Primary cortical representation of sounds by the coordination of action-potential timing. Nature **381**(6583), 610 (1996)

6. Colgin, L.L., et al.: Frequency of gamma oscillations routes flow of information in the hippocampus. Nature **462**(7271), 353 (2009)

7. Einevoll, G.T., Kayser, C., Logothetis, N.K., Panzeri, S.: Modelling and analysis of local field potentials for studying the function of cortical circuits. Nat. Rev. Neurosci. **14**(11), 770 (2013)

8. Engel, A.K., König, P., Singer, W.: Temporal coding in the visual cortex: new vistas on integration in the nervous system. Trends Neurosci. **15**(6), 218–226 (1992)

9. Freeman, W.J., Schneider, W.: Changes in spatial patterns of rabbit olfactory eeg with conditioning to odors. Psychophysiology **19**(1), 44–56 (1982)

10. Fries, P.: Neuronal gamma-band synchronization as a fundamental process in cortical computation. Annu. Rev. Neurosci. **32**, 209–224 (2009)

11. Georgopoulos, A.P., Schwartz, A.B., Kettner, R.E.: Neuronal population coding of movement direction. Science **233**(4771), 1416–1419 (1986)

12. Gray, C.M., Singer, W.: Stimulus-specific neuronal oscillations in orientation columns of cat visual cortex. Proc. Natl. Acad. Sci. **86**(5), 1698–1702 (1989)

13. Grossberg, S., Olsen, S.J.: Rules for the cortical map of ocular dominance and orientation columns. Technical report. Boston University Center for Adaptive Systems and Department of Cognitive and Neural Systems (1994)

14. Hubel, D.H., Wiesel, T.N.: Receptive fields of single neurones in the cat's striate cortex. J. Physiol. **148**(3), 574–591 (1959)

15. Katzner, S., et al.: Local origin of field potentials in visual cortex. Neuron **61**(1), 35–41 (2009)

16. Klimesch, W.: Memory processes, brain oscillations and eeg synchronization. Int. J. Psychophysiol. **24**(1–2), 61–100 (1996)

17. Klimesch, W., Fellinger, R., Freunberger, R.: Alpha oscillations and early stages of visual encoding. Front. Psychol. **2**, 118 (2011)

18. Kohonen, T.: Self-organized formation of topologically correct feature maps. Biol. Cybern. **43**(1), 59–69 (1982)

19. Lee, C., Rohrer, W.H., Sparks, D.L.: Population coding of saccadic eye movements by neurons in the superior colliculus. Nature **332**(6162), 357 (1988)

20. Liu, J., Newsome, W.T.: Local field potential in cortical area mt: stimulus tuning and behavioral correlations. J. Neurosci. **26**(30), 7779–7790 (2006)

21. Miikkulainen, R., Bednar, J.A., Choe, Y., Sirosh, J.: Self-organization, plasticity, and low-level visual phenomena in a laterally connected map model of the primary visual cortex. In: Psychology of Learning and Motivation. volume 36: Perceptual Learning, pp. 257–308. Academic Press, San Diego CA (1997)

22. Miikkulainen, R., Bednar, J.A., Choe, Y., Sirosh, J.: Computational maps in the visual cortex. Springer, New York (2006). https://doi.org/10.1007/0-387-28806-6

23. Miller, K.D., Keller, J.B., Stryker, M.P.: Ocular dominance column development: analysis and simulation. Science **245**(4918), 605–615 (1989)

24. Milner, P.M.: A model for visual shape recognition. Psychol. Rev. **81**(6), 521 (1974)

25. Mitzdorf, U.: Current source-density method and application in cat cerebral cortex: investigation of evoked potentials and eeg phenomena. Physiol. Rev. **65**(1), 37–100 (1985)

26. Niebur, E., Wörgötter, F.: Orientation columns from first principles. In: Eeckman, F.H., Bower, J.M. (eds.) Computation and Neural Systems, pp. 409–413. Springer, Boston (1993). https://doi.org/10.1007/978-1-4615-3254-5_62

27. Obermayer, K., Ritter, H., Schulten, K.: A principle for the formation of the spatial structure of cortical feature maps. Proc. Natl. Acad. Sci. **87**(21), 8345–8349 (1990)
28. Pasupathy, A., Connor, C.E.: Population coding of shape in area v4. Nat. Neurosci. **5**(12), 1332 (2002)
29. Singer, W.: Synchronization of cortical activity and its putative role in information processing and learning. Annu. Rev. Physiol. **55**(1), 349–374 (1993)
30. Singer, W., Gray, C.M.: Visual feature integration and the temporal correlation hypothesis. Annu. Rev. Neurosci. **18**(1), 555–586 (1995)
31. Sirota, A., Montgomery, S., Fujisawa, S., Isomura, Y., Zugaro, M., Buzsáki, G.: Entrainment of neocortical neurons and gamma oscillations by the hippocampal theta rhythm. Neuron **60**(4), 683–697 (2008)
32. Tanaka, S.: Theory of self-organization of cortical maps: mathematical framework. Neural Netw. **3**(6), 625–640 (1990)
33. Tort, A.B., Komorowski, R.W., Manns, J.R., Kopell, N.J., Eichenbaum, H.: Theta-gamma coupling increases during the learning of item-context associations. Proc. Natl. Acad. Sci. **106**(49), 20942–20947 (2009)
34. Van Der Meer, M.A., Redish, A.D.: Low and high gamma oscillations in rat ventral striatum have distinct relationships to behavior, reward, and spiking activity on a learned spatial decision task. Front. Integr. Neurosci. **3**, 9 (2009)
35. Womelsdorf, T., Fries, P., Mitra, P.P., Desimone, R.: Gamma-band synchronization in visual cortex predicts speed of change detection. Nature **439**(7077), 733 (2006)
36. Yuille, A., Kammen, D., Cohen, D.: Quadrature and the development of orientation selective cortical cells by hebb rules. Biol. Cybern. **61**(3), 183–194 (1989)
37. Zold, C.L., Shuler, M.G.H.: Theta oscillations in visual cortex emerge with experience to convey expected reward time and experienced reward rate. J. Neurosci. **35**(26), 9603–9614 (2015)

# Neural Networks Models for Analyzing Magic: The Gathering Cards

Felipe Zilio, Marcelo Prates[(✉)], and Luis Lamb

Federal University of Rio Grande do Sul, Institute of Informatics, Av Bento Goncalves, Porto Alegre 9500, Brazil
{fzmorais,morprates,lamb}@inf.ufrgs.br,
http://www.inf.ufrgs.br

**Abstract.** Historically, games of all kinds have often been the subject of study in scientific works of Computer Science, including the field of machine learning. By using machine learning techniques and applying them to a game with defined rules or a structured dataset, it's possible to learn and improve on the already existing techniques and methods to tackle new challenges and solve problems that are out of the ordinary. The already existing work on card games tends to focus on gameplay and card mechanics. This work aims to apply neural networks models, including Convolutional Neural Networks and Recurrent Neural Networks, in order to analyze Magic: the Gathering cards, both in terms of card text and illustrations; the card images and texts are used to train the networks in order to be able to classify them into multiple categories. The ultimate goal was to develop a methodology that could generate card text matching it to an input image, which was attained by relating the prediction values of the images and generated text across the different categories.

## 1 Introduction

Games and similar forms of entertainment, due to their constrained nature and usually tight set of rules, lend themselves especially well to be subjects of study in scientific works in the field of machine learning.

There have been several studies applying machine learning techniques to learn and analyze several different games to great success, from traditional board games such as Go [15] to video games including DotA 2 [9] and Starcraft II [16].

More recently, card games have also become a source of interest in machine learning research. In a typical card game, players are allowed to collect (or buy, or trade) several different copies of cards, with different in-game effects, which are then used to construct a 'deck', to play against other players. These types of games present several different angles of analysis, from the cards themselves, to deck construction, to the gameplay and its rules.

Among the different card games, a good number of studies have focused on the bigger and most enfranchised games, such as Hearthstone [18] and Magic: the Gathering [2]. This work will focus on the latter.

© Springer Nature Switzerland AG 2018
L. Cheng et al. (Eds.): ICONIP 2018, LNCS 11302, pp. 227–239, 2018.
https://doi.org/10.1007/978-3-030-04179-3_20

Magic: the Gathering (MTG) is a fantasy-themed trading card game (TCG) created by Wizards of the Coast. Since the release of its first edition in 1993, the game has grown to be one of the biggest card games in the world in terms of number of players, with an estimated 20 million players worldwide [3], and multiple international MTG tournaments being run every year across the world. As of the time of this writing, there are over 16 thousand unique cards with around 30 thousand images currently in the game, with more being released every few months in new editions, or sets (each set containing up to a few hundred cards, between new cards and 'reprints' of older cards, which may include new illustrations).

Although the vast number of cards in the game - each with its own different effect and mechanics - makes it an extremely complex game, the core gameplay concepts behind it are actually quite simple. The focus of this work, however, will be on analyzing the cards themselves, and not on the gameplay.

Each individual MTG card usually contains the following fields:

- A name, which uniquely identifies the card.
- A mana cost, which is the number and colors of 'mana' (the game's main resource system) required to play the card.
- An illustration.
- A number of types and subtypes, that define which type categories the card is a part of.
- The set symbol, which represents the edition in which the card was released.
- Rules text defining a card's effect.
- Flavor text.
- Power/Toughness (only on creature-type cards), values that determine the strength of a creature card.

The 'cost' and 'rules text' of a card will also determine what is called the card's 'color identity'. In Magic: the Gathering, a card's color identity refers to the set of all the color symbols that appear anywhere on the card,. A card's color identity can be any of the game's five colors - 'White', 'Blue', 'Black', 'Red' or 'Green'. It can also be two, three, four or five of those colors, or even none of them ('Colorless'). Alternately, a card's color identity can be thought of as a point $x \in \{0,1\}^5$ in five-dimensional space. As such, in this work, a card's 'color' will refer to its color identity as defined by the mechanics and rules of the game only.

## 1.1   Color

In Magic: the Gathering, the five-color system is one of the biggest design principles of the game. Not only does each color have its own set of effects and mechanics for its cards, but those are also related to the ideas and concepts each color represents.

Investigating the classification of card images by color is an intuitive choice since, visually, it's easy to notice a certain correlation between the images of cards that are the same color - the tonality of an image tends to be related to the color of its card (that is, 'Red' cards do tend to have images with a lot of red tones in it, and so on), and the images of the cards themselves are usually created on purpose to look similar to other cards of the same color, to create a sort of 'visual identity' for each color through the card images. The similarities between cards of a certain color also extend to the cards' rules text - since there is a clear mechanical and gameplay separation for each different card type, the words used and even the way the card's effect is structured will usually be different for each card type. This distinction also extends to the cards' color identity, albeit in a more subtle manner; just as each color has a different visual theme for its images, the gampeplay mechanics and effects of each color also tends to differ - that is, there is a mechanical, or gameplay, identity for each color that tends to appear in the cards' text. As such, since this definition will remain relevant to the rest of this work, this section will provide a brief description of each of the game's colors, both in terms of card art and game mechanics, taken directly from a series of articles on the subject by one of the game's developers [11] (Table 1).

**Table 1.** Conceptual attributes and common elements associated with each card color.

| Color | Attributes | Common elements in illustrations |
|-------|-----------|----------------------------------|
| White | Order, Structure, Law | Soldiers, Angels, Clerics, Light, The Sun |
| Blue | Intellect, Manipulation, Cold, Water, Air | Books, Wizards, Clouds, The Sky, Bodies of Water, Birds, Sea Creatures |
| Black | Ambition, Amorality, Sacrifice, Death | Zombies, Demons, Vampires, Darkness |
| Green | Nature, Life, Growth | Animals, Plants, Elves |
| Red | Impulse, Chaos, Earth, Fire, Lightning | Fire, Aggression, Lightning, Dragons, Goblins |

## 1.2 Card Type

Each card in Magic: the Gathering belongs to one or more card types. The main card types - those being **Land**, **Creature**, **Instant**, **Sorcery**, **Artifact**, **Enchantment** and **Planeswalker** – are already very descriptive: Land cards are places or landscapes, Creature cards are beings like beasts or humanoids, Instant and Sorcery cards represent magical spells, Artifacts are inanimate objects, Enchantments are long lasting changes to the battlefield or other enhancements, Planeswalker cards represent specific named wizards that are part of the game's story. More secondary card types exist, but any card in the game belongs to at least one of the previously mentioned card types.

## 2   Our Goals

Magic: the Gathering cards are comprised of two elements: the rules text (including the textbox as well as its name, cost, and other relevant information printed on the card), and its illustration. Previous projects on the subject of this card game have mostly focused on a card's text and its in-game mechanics, such as the card-text generating neural network RoboRosewater, which posts the results of its generated cards on Twitter [8], and Google DeepMind's project for generating programming code to implement cards through the use of latent prediction networks [6].

In this work, the focus will be on the cards' illustrations, and how they relate to elements of the game such as the different categories of cards, and the card text itself. A process will be described to classify card images according to different types of card characteristics, which will then be used to relate input images to randomly generated card texts according to those parameters, learned from training with a dataset containing all of the game's card images.

## 3   Methods

Card data was pulled directly from Wizards of the Coast's database [17], containing all the current cards in the game. All card images were downloaded in JPG format, and all card text information in CSV format. All the images were converted to binary format, and encoded in batch files along with the relevant label for that dataset. Two datasets were created, one for card type classification, one for color identity classification, of encoded images and labels separated in batches. These datasets were then fed to a convolutional neural network, configured slightly differently for each classification, the specifics of which will be detailed in later sections of this work. Another neural network was used to generate random cards from the dataset of existing card text already in the game. These cards were used to form a database of 20 thousand generated cards, to which input images could then be related according to their characteristics.

All card images experimented with in this study are suppressed from the manuscript due to copyright restraints. Nevertheless, they can be obtained by typing the card names in the search engine http://gatherer.wizards.com.

### 3.1   Environment

All experiments were conducted in a machine with the specifications: i5-4690k quad-core CPU @ 3.5 GHz, NVIDIA GTX 970 graphics card with 4 GB VRAM, 16 GB of RAM, running the Ubuntu 10.4 OS.

# 4    Classifying MTG Card Illustrations by Color

Out of the 32.000 images in the dataset, about 26.000 images were used during the training. The rest was randomly selected to form the test batch of images, for evaluation purposes. This ensures a good mix of different image styles from all different eras and editions of Magic: the Gathering history, for both training and testing. A set amount of images was also distorted at random through cropping and displacing, a method which has been shown to increase the performance of the network in the past [14] and was effective here as well, offering a slight gain in performance when compared to a training session that was run with just the regular images as inputs.

The convolutional neural network architecture that was used for this task was adapted from a CNN model [5] for image classification. The network was created through the use of consecutive convolution, pooling, normalization, and softmax layers, connected to form the network structure.

For the network parameters, a small initial learning rate of 0.01 was used - higher initial values usually resulted in the network diverging.

Our neural network is composed of 8 convolution, pooling, normalization, fully connected and softmax layers arranged in the following way: **CONV →  POOL → NORM → CONV → NORM → POOL → FULLY → SOFT-MAX**.

Table 2 shows the distribution of color identity across the 32.017 cards in the dataset. There are 3824 multicolored cards, or 11.94%. It would be extremely hard to classify card images across all color combinations using single labels: not only are there $2^5 = 32$ possible combinations, from 'Colorless' to 'All colors', but there is a big imbalance across each category in terms of numbers of cards. In fact, while there are 5.068 Green cards in total, there are only two cards currently in the game in the White/Blue/Red/Green color identity, for example.

While imbalanced categories are common problems with image datasets, and can be usually addressed with simple solutions such as re-weighting the training cost function or resampling the training images in order to balance the dataset [10], in this case, the imbalance present in the dataset was just too big for these solutions to be considered feasible. Sticking to 6 labels, then - one for each color, plus colorless - was the system that was used for initially classifying the dataset.

We can visually verify that multicolored cards do usually contain traces of all colors in their color identity in their illustration picture – a card that was both Red and Blue, for example, would usually present characteristic exhibited by both Red and Blue cards in its picture. As such, we have decided to include a copy of each multicolored card in the dataset of each of its respective colors.

The final accuracy of the network, when trained with the merged dataset, was 0.595, or 59.5%. This value, while not necessarily low, might possibly be explained at least in part by the huge variance in image styles that are present in the dataset, as well as the fact that different eras of Magic: the Gathering cards present different ideas and sometimes completely different styles of cards when it comes to color, since the color system is always in flux as the game develops [12].

**Table 2.** Color distribution in the dataset - number of images

| Color | Images | Color | Images |
|-------|--------|-------|--------|
| G     | 5068   | UBR   | 99     |
| B     | 5021   | WRG   | 86     |
| R     | 4973   | WUG   | 79     |
| W     | 4878   | BRG   | 79     |
| U     | 4828   | WUB   | 75     |
| Cl    | 3425   | WUBRG | 66     |
| WG    | 378    | WBR   | 49     |
| BR    | 376    | WBG   | 41     |
| RG    | 372    | UBG   | 38     |
| UB    | 354    | WUR   | 37     |
| WU    | 346    | URG   | 37     |
| BG    | 288    | UBRG  | 2      |
| WB    | 266    | WUBR  | 2      |
| UR    | 260    | WUBG  | 2      |
| WR    | 249    | WBRG  | 2      |
| UG    | 238    | WURG  | 2      |

Consequently, it needs to be taken into account, then, that at least some of the prediction 'misses' will be the result of images from early in the game's history, some of which are unfitting for their color identity as per today's Magic: the Gathering card design standards. Table 3 shows an example of prediction values obtained for a multicolored card.

**Table 3.** Prediction values for card color

| Type      | Prediction value |
|-----------|------------------|
| White     | 27,49%           |
| Green     | 27,15%           |
| Black     | 27,14%           |
| Blue      | 9,73%            |
| Red       | 8,49%            |
| Colorless | 0,00%            |

# 5  Classifying MTG Card Illustrations by Type

Similarly to color classification, classifying cards by type is also an intuitive notion. Since the cards are already pre-labeled with their type, and each card type does share a few similarities and characteristics between its images, it stands to reason that using that criteria for classification would be an effective choice.

Card type as a method of classification also presents a much clearer correlation to images outside of the context of the game, when compared to the color classification, since the characteristics of the images of each type are much more well defined ('creatures' are living things, 'artifacts' are inanimate objects, 'lands' are landscapes or places, etc). Table 4 shows the distribution of cards among all types.

This classification system suffers from the exact same problem as the color system; that is, there are cards that belong to more than one type category(for example, Enchantment Creature or Artifact Creature). The solution described before for the case of multicolor cards was also proposed and utilized here, since, in this case as well, images of cards that have multiple types are usually representative of both types. We have also decided to merge 'Instant' and 'Sorcery' types into a single category, given the poor differentiation between both types in the philosophy of the game.

The implemented CNN was trained with this dataset, and it was able to obtain an accuracy of 0.685, or 68.5%, in its predictions - a marked increase in performance when compared to the color classification.

Table 5 shows an example of prediction values obtained from evaluating the card art in the 2011 core set Ornitopher card with the trained type classification CNN.

**Table 4.** Type distribution in the dataset - number of images

| Type | # Cards |
| --- | --- |
| Creature | 14081 |
| Instant | 3962 |
| Sorcery | 3638 |
| Enchantment | 3519 |
| Land | 3297 |
| Artifact | 2259 |
| Creature/Artifact | 920 |
| Planeswalker | 179 |
| Creature/Enchantment | 98 |
| Land/Artifact | 14 |
| Instant/Sorcery | 13 |
| Artifact/Enchantment | 8 |
| Creature/Land | 2 |

**Table 5.** Prediction values for card type

| Type | Prediction value |
|---|---|
| Creature | 44,06% |
| Artifact | 43,82% |
| Enchantment | 7,95% |
| Instant/Sorcery | 4,17% |
| Land | 0,00% |

## 6   Text Classification

Another common problem studied in the field of machine learning is that of text classification. Previous works on the subject of text analysis have studied methods used to classify text according to anything from sentiment analysis [7], to even more practical applications such as spam filtering [13]. Convolutional Neural Networks have been used in the past to solve natural language processing tasks of various sorts [1].

The method used in this work embeds each piece of text into a sparse vector representations [4] which can then be used as inputs to be passed to the CNN on the training or evaluation steps. The data is organized much like on the experiments described in the previous section, with the same categories for color and card type. The complete dataset was split into a training dataset and an evaluation dataset, with a ratio of 5/6 for training to 1/6 for evaluation. The overall method utilized for the task of text classification in this work differs slightly from the reference model in [4], but the network structure is otherwise very similar. After the preliminary embedding layer, the data is passed through the usual convolution, pooling and fully connected layers, providing the output through the result of a softmax function. Although the details of this architecture differ heavily from the network used previously for the task of image classification, the general idea of each layer remains mostly the same and as such will not be repeated in detail in this section.

Given the relative simplicity of card text when compared to card images, the overall better results obtained from this classification experiment, when compared to the image classification, are not surprising. The CNN used to classify the color dataset was able to predict categories with 0.78 accuracy, and the second dataset, used for type classification, yielded a prediction accuracy of 0.91. Tables 6 and 7 shows an example of text classification done through the network, for card type and color, on the text of the "Journey into Nyx" expansion "Kruphix, God of Horizons" card.

**Table 6.** Prediction values for card type

| Type | Prediction value |
| --- | --- |
| Enchantment | 37,65% |
| Creature | 36,44% |
| Artifact | 18,44% |
| Instant/Sorcery | 7,48% |
| Land | 0,00% |

**Table 7.** Prediction values for card color

| Type | Prediction value |
| --- | --- |
| Green | 27,34% |
| Blue | 26,37% |
| White | 17,79% |
| Black | 16,34% |
| Red | 12,16% |
| Colorless | 0,00% |

# 7   Matching Randomly Generated Card Text to Input Images

As a part of this research endeavor, we used the prediction values for card color and type obtained from the CNNs for image and text classification described in Sects. 4 and 5 in order to develop a method that can relate an input image to a randomly generated card text that better fits it – effectively automatically "creating" a Magic: the Gathering card for any input image. A few of these examples are described above.

All randomly generated card text used in this work was entirely generated through the method described in [8], which contains a set of methods to encode the dataset of MTG cards in JSON format to be used as input. Then, it utilizes a Recurrent Neural Network model to generate a set number of examples from the training data. The card generating RNN was used to generate a database of 30.000 card text data. This card text was then processed through the text classifying CNN developed previously on Sect. 6, to classify the generated text according to both type and color.

An input image will first be run through two CNNs – the first one, described in Sect. 4, will classify the image according to its potential color identity, and return the prediction values obtained for each possible color label. The same process will be repeated for the second CNN, on Sect. 5, to obtain prediction values for each card type label.

After a normalization process these values can be directly compared to the prediction values previously obtained and recorded for each generated card. This

(a)                              (b)

**Fig. 1.** Generated card examples 1 and 2

comparison is a simple sum of the differences between the value of each label for the two categories, in order to compute the total color and type 'distance' $C_d$ and $T_d$: $C_d = \sum_{c_i \in C} |I_{c_i} - T_{c_i}|$, where $c_i$ represents the possible color labels in the space of color labels $C$, $I_{c_i}$ is the prediction value of the color label $ci$, obtained by the CNN for the image $I$, and $T_{c_i}$ is the prediction value of the color label $c_i$ on the generated card text $T$. Likewise, for obtaining the type distance $T_d$, the same formula can be applied, by utilizing the type labels instead of the color ones.

Finding the best possible match between the predicted color and type values for the input image and the generated card texts becomes a simple matter of finding the card text that minimizes both distance values $C_d$ and $T_d$ - this selected card text will then will be the sole output of the method. Some examples of our results are described above.

### 7.1 Card Generation Examples

Figure 1a shows a herd of elephants. The prediction values for type and color, on Table 8, show a strong preference for the Creature type, and Green color, closely followed by Red. This decision is consistent with the philosophy of the Green color, which is home to most animals and beasts. The generated card text ( *"Cumulative upkeep 1. At the beginning of your upkeep, you may pay 3G. If you do, put a +1/+1 counter on it."*) both fits the description of a Green Creature as well as being a coherent card effect.

Figure 1b shows an owl mid-flight. The values predicted by the CNN for this image are reported on Table 9. Once again the NN easily recognizes it as Creature type card. Color-wise, strong values of Blue and White are predicted. The fact that in MTG birds of all kinds are very common on both colors probably explains the Blue classification even though the image lacks Blue hues.

The card text generated ( *"Flying. When {this card} enters the battlefield, detain target creature an opponent controls."*), again, is correct by the rules of the game, and features the already existing 'Detain' ability, which shows up exclusively in Blue/White cards. The card having the ability 'Flying' is

**Table 8.** Color and type prediction values for examples 1

| Color | Color values | Type | Type values |
|-------|-------------|------|-------------|
| Green | 27,22% | Creature | 39,73% |
| Red | 26,02% | Enchantment | 27,89% |
| White | 17,19% | Artifact | 22,97% |
| Black | 17,17% | Instant/Sorcery | 9,41% |
| Colorless | 12,39% | Land | 0,00% |
| Blue | 0,00% | - | - |

**Table 9.** Color and type prediction values for example 2

| Color | Color values | Type | Type values |
|-------|-------------|------|-------------|
| Blue | 29,49% | Creature | 55,11% |
| White | 29,03% | Artifact | 26,66% |
| Black | 15,80% | Instant/Sorcery | 12,87% |
| Red | 14,65% | Enchantment | −5,36% |
| Green | 11,04% | Land | 0,00% |
| Colorless | 0,00% | - | - |

interesting, since it's not possible to identify that it is an aerial creature only by the 'Blue/White Creature' description, which is all the network has.

# 8 Discussion and Future Work

The main objective of this work was to assess whether state-of-the-art convolutional neural network architectures could tackle a classification problem that remained up to now untouched – that is, the dataset of a major trading card game. The image classification problem presented an interesting challenge in regards to the usual image classification problems: instead of labelling images based on elements that were or not present, a more subjective classification system was employed, based on game rules and mechanics not necessarily related to the images themselves.

The relative success of the classification phase presented an opportunity to test a hypothesis – that it would possible to relate the values of predictions obtained on the images for the different categories, with the values obtained from a classification operation on the card text, in a way that would enable the approximation of a given image to a text that most matches it. Thus, the second CNN model was utilized, to classify existing card text in order to train a network to be able to classify newly generated card text. This classification yielded even better results. A model for generating random card text, based on a RNN model, was then utilized, and each of its cards were classified with the already trained text classification CNN. This allowed for a direct comparison to the prediction values obtained from the input image to be made.

The method developed was tested with real-world images, from outside the game. The generality of the categories used, color identity and card type, allowed images that had nothing to do with the game to be analyzed and classified according to the same criteria used for card images in the game – this stems from the intuitive notion that, for example, if the Creature card type is comprised of animals and creatures, then an image of a real animal, if recognized, should fit into the same classification. The testing process, while not extensive, already yielded a few significant insights and conclusions.

Overall, the results of this work present an exciting possibility for future research – since there are so many independent components to this work, from the classifying CNNs to the generating RNN, there are multiple areas that could be improved upon to make the task of generating cards to match input images yield better results. The obvious two candidates are the image classification and text generation phases, which can be improved on multiple fronts. Also, there still lacks a formal, extensive review of the effectiveness of the methodology utilized – one possible idea is to conduct a quantitative research, asking Magic: the Gathering players to evaluate the quality of the generated cards on a certain scale.

# References

1. Collobert, R., et al.: Natural language processing (almost) from scratch. J. Mach. Learn. Res. **12**(Aug), 2493–2537 (2011)
2. Cowling, P.I., Ward, C.D., Powley, E.J.: Ensemble determinization in monte carlo tree search for the imperfect information card game magic: The gathering. IEEE Trans. Comput. Intell. AI Games **4**(4), 241–257 (2012)
3. Duffy, O.: How magic: the gathering became a pop-culture hit - and where it goes next (2015). https://www.theguardian.com/technology/2015/jul/10/magic-the-gathering-pop-culture-hit-where-next. Accessed 20 Dec 2017
4. Kim, Y.: Convolutional neural networks for sentence classification. arXiv preprint arXiv:1408.5882 (2014)
5. Krizhevsky, A.: Cuda-convnet: High-performance C++/cuda implementation of convolutional neural networks (2011). https://code.google.com/archive/p/cuda-convnet/. Accessed 25 Dec 2017
6. Ling, W., Grefenstette, E., Hermann, K.M., Kočiský, T., Senior, A., Wang, F., Blunsom, P.: Latent predictor networks for code generation. arXiv preprint arXiv:1603.06744 (2016)
7. Maas, A.L., Daly, R.E., Pham, P.T., Huang, D., Ng, A.Y., Potts, C.: Learning word vectors for sentiment analysis. In: Proceedings of the 49th Annual Meeting of the Association for Computational Linguistics: Human Language Technologies-Volume 1, pp. 142–150. Association for Computational Linguistics (2011)
8. Milewicz, R.M.: Generating magic cards using deep, recurrent neural networks (2015). http://www.mtgsalvation.com/forums/magic-fundamentals/custom-card-creation/612057-generating-magic-cards-using-deep-recurrent-neural. Accessed 25 Dec 2017

9. OpenAI: More on dota 2 (2017). https://blog.openai.com/more-on-dota-2/. Accessed on 20 Dec 2017
10. Oquab, M., Bottou, L., Laptev, I., Sivic, J.: Learning and transferring mid-level image representations using convolutional neural networks. In: Proceedings of the IEEE Conference on Computer Vision and Pattern Recognition, pp. 1717–1724 (2014)
11. Rosewater, M.: Seeing red (2004). https://magic.wizards.com/en/articles/archive/making-magic/seeing-red-2004-07-19. Accessed 29 Dec 2017
12. Rosewater, M.: Mechanical color pie (2017). https://magic.wizards.com/en/articles/archive/making-magic/mechanical-color-pie-2017-2017-06-05. Accessed on 25 Dec 2017
13. Sahami, M., Dumais, S., Heckerman, D., Horvitz, E.: A Bayesian approach to filtering junk e-mail. In: Learning for Text Categorization: Papers from the 1998 Workshop, vol. 62, pp. 98–105 (1998)
14. Simard, P.Y., Steinkraus, D., Platt, J.C., et al.: Best practices for convolutional neural networks applied to visual document analysis. ICDAR **3**, 958–962 (2003)
15. Singh, S., Okun, A., Jackson, A.: Artificial intelligence: learning to play go from scratch. Nature **550**(7676), 336 (2017)
16. Vinyals, O., et al.: Starcraft ii: a new challenge for reinforcement learning. arXiv preprint arXiv:1708.04782 (2017)
17. WOTC: Gatherer - magic: The gathering (1995). http://gatherer.wizards.com/Pages/Default.aspx. Accessed 25 Dec 2017
18. Zhang, S., Buro, M.: Improving hearthstone AI by learning high-level rollout policies and bucketing chance node events. In: 2017 IEEE Conference on Computational Intelligence and Games (CIG), pp. 309–316. IEEE (2017)

# MulAttenRec: A Multi-level Attention-Based Model for Recommendation

Zhipeng Lin[1], Wenjing Yang[1(✉)], Yongjun Zhang[2], Haotian Wang[1], and Yuhua Tang[1]

[1] State Key Laboratory of High Performance Computing, College of Computer, National University of Defense Technology, Changsha, China
{linzhipeng13,wenjing.yang,wanghaotian13,yhtang}@nudt.edu.cn
[2] National Innovation Institute of Defense Technology, Beijing, China
yjzhang@nudt.edu.cn

**Abstract.** It is common nowadays for online buyers to rate shopping items and write review text. This review text information has been proven to be very useful in understanding user preferences and item properties, and thus enhances the capability of Recommender Systems (RS). However, the usefulness of reviews and the significance of words in each review are varied. In this paper, we introduce a multi-level attention mechanism to explore the usefulness of reviews and the significance of words and propose a Multi-level Attention-based Model (MulAttRec) for the recommendation. In addition, we introduce a hybrid prediction layer that model the non-linear interaction between users and items by coupling Factorization Machine (FM) to Deep Neural Network (DNN), which emphasizes both low-order and high-order feature interaction. Extensive experiments show that our approach is able to provide more accurate recommendations than the state-of-the-art recommendation approaches including PMF, NMF, LDA, DeepCoNN, and NARRE. Furthermore, the visualization and analysis of keyword and useful reviews validate the reasonability of our multi-level attention mechanism.

**Keywords:** Recommender systems · Attention-based mechanism
Convolutional neural network · Factorization machine

## 1 Introduction

In the era of big data, Recommender Systems (RS) are the most common solutions to alleviate information overload and have been adopted by a wide range of websites and applications. One of the most popular approaches for RS is Collaborative Filtering (CF), which focus on using historical data like clicking and rating to model user preferences and item features.

Z. Lin—This work was funded through National Science Foundation of China (No. 91648204).

© Springer Nature Switzerland AG 2018
L. Cheng et al. (Eds.): ICONIP 2018, LNCS 11302, pp. 240–252, 2018.
https://doi.org/10.1007/978-3-030-04179-3_21

Although CF has made a considerable progress, it has its own limitations. First, the exponential growth in the number of users and items increases the sparseness of the rating matrix and pose a challenge in making the prediction for items and users with few rating records (the famous cold-start problem). In addition, it is not easy for CF techniques to capture the underlying non-linear coupling between users and items.

One of the approaches to address the cold-start problem is using the review text. Review text usually contains rich information about the item features and user preferences. Recently, some studies [11] have shown that using the abundant information in review text can alleviate cold–start problem.

However, it is a challenging and important issue to get most informative features from an enormous number of related reviews [2]. In real life, people tend to be attracted to both keywords in a review text and the most useful review in all reviews. Inspired by this phenomenon, we design a multi-level attention-based mechanism to extract most useful information from reviews. The significance of a word in a review text and the helpfulness of a review in all reviews are respectively defined according to the informativeness and the usefulness of themselves.

To learn the weight of words and reviews, we propose a Multi-level Attention-based Model (MulAttenRec) in this paper. We apply the attention mechanism on both word level and full-text level in the convolutional neural network (CNN), which extract embedded representations of users and items from their corresponding review text in a joint manner. The word attention layer is a local window of words to measure the informativeness of words. The full-text attention layer is a multi-layer neural network to capture the characteristics. In the prediction layer, we combine the Factorization Machine (FM) with Deep Neural Network to capture the non-linear user-item interactions. We evaluate MulAttenRec extensively on three popular datasets. Experimental results show that our model consistently outperforms the state-of-the-art methods including PMF, NMF, LDA, DeepCoNN, and NARRE.

The main contributions of this paper are:

- To the best of our knowledge, we are the first to introduce a novel attention mechanism that selects both keywords and highly-helpful reviews automatically in RS, which helps to raise the accuracy and give us the ability to explaining what the RS is doing.
- In prediction layer, we integrate the architectures of FM and deep neural networks (DNN) to model the non-linear interactions between users and items, which not only models low-order feature interactions like FM but also high-order feature interactions like DNN.
- Experimental results on benchmark datasets show that our MulAttenRec model obtains better prediction results than the state-of-the-art models and the average relative improvement over PMF is up to 30.06%.

## 2  Related Work

### 2.1  Attention-Based Mechanism in RS

The idea of attention in neural networks is inspired from the visual attention mechanism found in humans. Attention mechanism has been widely used in many tasks, such as information retrieval [13], recommendation [12] and machine translation [5]. In the field of recommendation systems, He *et al.* [3] introduce an attention mechanism in CF to learn to assign attentive weights for a set of features for the multimedia recommendation. When comes for review text, Chen *et al.* [2] adopts attention mechanism to learn the weight of each review for predicting item ratings. In this paper, we propose a multi-level attention mechanism to better model users and items for predicting item ratings.

### 2.2  Combination of FM and Deep Neural Networks

Deep neural networks have been shown effective in RS for learning sophisticated feature interactions [14]. Some models extend CNN for click-through rate (CTR) prediction like FNN [15], Wide and Deep [4] DeepFM [6]. Among these hybrid models, DeepFM shows some impressed performance. It is an end-to-end model without any pre-train or feature engineering, and makes the "FM part" and "deep part" share the same input. However, DeepFM mainly focuses on CTR prediction. Inspired form DeepFM, we design a hybrid prediction layer for rating predictons.

## 3  Our Approach

### 3.1  Overview

In order to select both useful words and reviews, we utilize the attention mechanism to automatically assign weights to reviews and words when modeling users and items. Figure 1 illustrates the architecture of *MulAttenRec*. The model contains two similar attention-based neural networks for users ($N_{et_u}$) and items ($N_{et_i}$) relatively, after which is the vector concatenation layer and the hybrid prediction layer. Note that the hybrid prediction layer consisted of FM and MP to capture the feature of concatenated vector more efficiently. Reviews written by the same users or to the same items are fed into to $N_{et_u}$ and $N_{et_i}$. The corresponding rating is calculated as the sum of FM and MP (we can also use other methods to combine the output of FM and MP). In the following subsections, due to the similarity between $N_{et_u}$ and $N_{et_i}$, we only give the details $N_{et_i}$. The same analysis is also practicable for $N_{et_u}$.

In the first stage of $N_{et_i}$, the word-attention CNN Processor is applied to process the textual reviews of item $i$. In this processor, the model process each review of item $i$ respectively. Specifically, each review is first transformed into a matrix of word vectors from the word-embedding, which we denoted as $V_{i1}, V_{i2}, \cdots, V_{ik}$. Then, these matrices are sent to word attention CNN and the feature vectors of them can be obtained from the output. These feature vectors are noted as $O_{i1}$, $O_{i2}, \cdots, O_{ik}$.

In the full-text-attention layer, we calculate the contribution of each review and aggregate these vectors to get the representation of item $i$:

$$O_i = \sum_{l=1}^{k} a_{il} O_{il} \qquad (1)$$

In the Eq. (1), the $a_{il}$ is the weight of each review for a specific item, which is learned by the full-text attention layer.

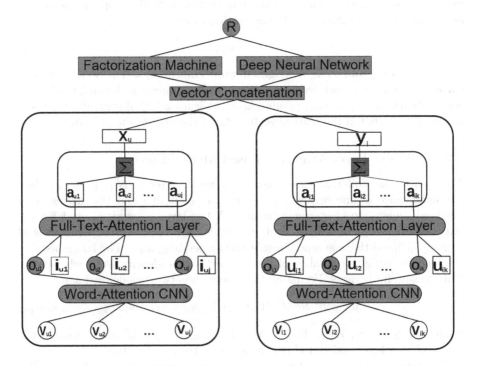

**Fig. 1.** The architecture of MulAttenRec

## 3.2  Word-Level Attention-Based Mechanism

Inspired from the human visual attention, we further develop a word attention mechanism for reviews. When we read text or see images, we focus on certain part of the input to understand or recognize them more efficiently. Generally, people tend to be attracted by the significant word in a local part. Here, we will introduce a local Logistics Regression to measure the significance of one word.

Suppose $D_u$ is a local word embeddings with $w$ words $(x_1, x_2, \cdots, x_w)$. Here, $x_i$ is the center word and $w$ is the width of local window. The significance for each each word in window is given by:

$$X_{att,i} = (x_{i+\frac{-w+1}{2}}, x_{i+\frac{-w+3}{2}}, \cdots, x_i, \cdots, x_{i+\frac{w-1}{2}})^T \tag{2}$$

$$s_i = g(X_{att,i} * W^1_{att} + b^1_{att}) \tag{3}$$

where $s_i$ is the score showing how much the $i$th word is informative and $W^1_{att} \in R^{w \times d}$ is a parameter matrix and $b^1_{att}$ is a bias. The score can be directly used as a weight for $i$th word embedding or we can apply a threshold to remove "trivial" words and only consider informative attention words. In this work, we use scores as weights. We use sigmoid for the activation function $g$.

$$\widetilde{x}_t = s_t x_t \tag{4}$$

where $t \in [1, T]$ and $\widetilde{x}_t$ is a weighted word embeddings. The weighted word embedding matrix is then fed into textual CNN module, which consists of a 1-D convolutional layer and max pooling layer, to learn a global semantic representation. The output is the feature vector $O_{ik}$ of the $k$th review in item $i$.

### 3.3   Full-Text Level Attention-Based Mechanism

The goal of the Full-Text Level Attention-based Layer in $N_{et_i}$ is to calculate the significance of one review for the features of item $i$ and the aggregate of weighted reviews to characterize item $i$. A two-layer network is used for the attention score $a_{il}$. The input contains the feature vector of the $l$th review of item $i$ ($O_{il}$) and the user who wrote it (ID embedding, $u_{il}$). The ID embedding is added to model the quality of users, which helps identify users who always write less-useful reviews. Formally, the attention network is defined as:

$$a^*_{il} = h^T ReLU(W_O O_{il} + W_u u_{il} + b_1) + b_2 \tag{5}$$

where $W_O \in R^{t \times k_1}$, $W_u \in R^{t \times k_2}$, $b_1 \in R^t$, $h \in R^t$, $b_2 \in R^1$ and $t$ denotes the size of hidden layer and ReLU is a nonlinear activation function.

The final weight of reviews are predicted by the softmax function to normalize the above attention scores. The contribution of the $l$-th review to the final feature vector of item $i$ is given by:

$$a_{il} = \frac{exp(a^*_{il})}{\sum_{l=0}^{k} exp(a^*_{il})} \tag{6}$$

After we obtain the attention weight of each review, the feature vector of item $i$ is calculated as the Eq. (1).

Then $O_i$ is sent to a fully connected layer with weight matrix $W_0 \in R^{n \times k_1}$ and bias $b_0 \in R^n$. The final representation of item $i$ is given by:

$$Y_i = W_0 O_i + b_0 \tag{7}$$

### 3.4  Hybrid Prediction Layer

We aim to capture both linear and non-linear interactions between users and items. We propose the hybrid prediction layer. As depicted in Fig. 1, the prediction layer consists of two components, *FM component* and *Deep component* sharing the same input. First, let us concatenate $x_u$ and $y_i$ into a single feature vector $\tilde{z}_i = (x_u, y_i)$. For the feature vector $z_i$, the predicted rating $\tilde{R}$ is given by:

$$\tilde{R} = sigmoid(R_{FM} + R_{NN}) \tag{8}$$

where $R_{FM}$ and $R_{NN}$ are respectively the output of FM component and deep component.

The FM component is a factorization machine [9]. Not only modeling a first-order linear interactions among features, FM also use a dot product between vectors to model a second-order pairwise feature interactions. The output of FM is the sum of a bias and the two kinds of interactions. The $R_{FM}$ is given by:

$$R_{FM} = w_0 + \sum_{i=1}^{|z_i|} <w_i, z_i> + \sum_{i=1}^{|z_i|} \sum_{j=i+1}^{|z_i|} <v_i, v_j> z_i z_j \tag{9}$$

where $w_0$ is the global bias, $w_i$ measures the impact of the $i$th variable in $z_i$ and $<v_i, v_j>$ models the second order interactions.

The deep component is a deep neural network for learning the high-order feature interactions. The architecture of network is given by:

$$a^{(l+1)} = \sigma(W^{(l)} a^{(l)} + b^{(l)}) \tag{10}$$

where $l$ is the depth of neural network and $\sigma$ is an activation function. $a^{(l)}$, $W^{(l)}$, $b^{(l)}$ are the input, weight matrix, and bias of the $l$-th layer.

The prediction layer for $R_{NN}$ in deep component is given by:

$$R_{NN} = \sigma(W^{H+1}) a^H + b^{H+1} \tag{11}$$

where $H$ is the number of hidden layers and $\sigma$ is the sigmoid function.

On the basis of the study on DeepFM, all parameters, including $w_i$, $v_i$, and the network parameters ($W^{(l)}$, $b^{(l)}$) are trained jointly for the combined prediction model.

### 3.5  Learning

Since the task we focus in this paper is rating prediction, which actually is a regression problem. For regression, a popular objective function is the squared loss. The objective function is given by:

$$J = \sum_{u,i \in \Omega} (\tilde{R}_{u,i} - R_{u,i})^2 \tag{12}$$

where $\Omega$ denotes the set of instances for training, and $R_{u,i}$ is the rating assigned by the user $u$ to the item $i$. To optimize the objective function, we adopt the Adaptive Moment Estimation (Adam) as the optimizer. It could adjust the learning rate during the training phase, which avoid the process of choosing an efficient learning rate and result in the faster convergence than the vanilla SGD.

To alleviate overfitting, we consider dropout [10], a widely used method in deep learning models. Dropout stop working during testing and we use the whole network for prediction. Trough dropout, we can prevent complex coadaptations of neurons on training data. Moreover, dropout may potentially improve the performance of the whole neural network due to the side effect of performing model averaging with smaller neural networks.

Specially, we make use of dropout on the Full-Text Level Attention Layer. After obtaining the $k_1$-dimensional vector of latent factors $O_i$ which is a, we randomly drop $\rho$ percent of latent factors, where $\rho$ is termed as the dropout ratio. Moreover, we also apply dropout after obtaining $h_0$ at the same way to prevent overfitting.

## 4 Experiment

### 4.1 Experimental Settings

We test our method on three popular datasets, which not only provide rating information but also user reviews. The first dataset is from **Yelp**[1], which consists of $5,200,000$ reviews. The other two datasets is from **Amazon**[2] product data with 5-core: **Books** ($8,898,041$ reviews and $22,507,155$ ratings) and **Electronics** ($1,689,188$ reviews and $7,824,482$ ratings). We preprocessed datasets to ensure that all users and items have at least five ratings records.

To show the efficiency and potential of our method, we implemented five baselines including PMF [8], NMF [7], LDA [1], DeepCoNN [16] and NARRE [2]. The first two methods only utilize ratings at the training stage, while the latter three are representative review utilizing methods for rating prediction.

The dataset is randomly split into training set (80%), validation set (10%), and test set (10%) sets. The parameters for the algorithms mentioned above were initialized and precisely tuned according to the corresponding papers to obtain optimal performance. The key parameters for baselines are shown as follows. The number of latent factors for PMF and NMF $k$ is 10 and the number of topics for LDA $K$ is 10. For deep models, we set the number of neurons $m = 100$ for the convolutional layer, and the window size $t = 3$. Moreover, Google News was imported as a pre-trained word embeddings. We use w = 5 window size for the word attention layer. The size of word embedding is 300 and the dropout ratio $\rho$ is 0.5. The number of hidden layers in the hybrid prediction layer is 2.

---

[1] https://www.yelp.com/dataset/challenge.
[2] http://jmcauley.ucsd.edu/data/amazon/.

## 4.2   Hyper-Parameter Study

Based on the Yelp dataset, we analyze the influence of different hyper-parameters of the deep models including DeepCoNN, NARRE, and MulAttRec. The order is: (1) dropout rate; (2) number of hidden layers.

We set the dropout to be $0.1, 0.3, 0.5, 0.9$. As shown in Fig. 2(a), all the models are able to reach the peak of performance when the dropout is set as 0.5. The result prove that some reasonable randomness could enhance the robustness of model.

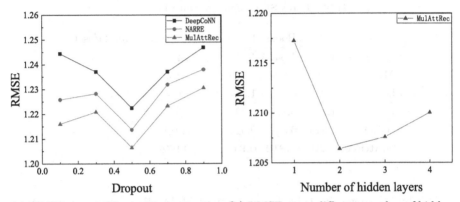

(a) RMSE w.r.t. different dropout ratios  (b) RMSE w.r.t. different number of hidden layers

**Fig. 2.** Hyper-Parameter Study

Because the DeepCoNN and NARRE have no hidden layers in prediction layer, we only carry out the second parameter study on MulAttRec. As presented in Fig. 2(b), increasing number of hidden layers in the hybrid prediction layer could raise the accuracy of the models at the beginning. However, if we keep increasing the number of hidden layers, their performance is degraded, as a result of overfitting.

## 4.3   Performance Evaluation

The performance of MulAttRec and the baselines are reported in terms of RMSE in Table 1. The best performance is shown in bold and the averages on three datasets are reported. From the results, several observations can be made:

To begin with, we can find that both traditional matrix factorization methods like PMF and NMF do not obtain comparable performance to that of those methods which utilize the use reviews. The gap in performance between MF methods and content-based methods validates our hypothesis that review text could supply additional information, and considering reviews in models can further improve the accuracy of rating prediction.

Secondly, although the simple employment of LDA to learn topic characters from item reviews can improve the performance of recommendation system, deep models can better capture the feature of review when compared with LDA. In this experiments, LDA models reviews without the feedback from users ratings. Thus, the learned unsupervised features from LDA may be not so efficient as the expectation. Therefore, by modeling ratings and reviews together and using supervised learning for regression tasks, DeepCoNN, NARRE, and MulAttRec obtain additional improvements.

**Table 1.** RMSE comparison with baselines

|           | Yelp      | Books     | Electronics | Average on all datasets |
|-----------|-----------|-----------|-------------|-------------------------|
| PMF       | 1.401     | 1.041     | 1.373       | 1.272                   |
| NMF       | 1.356     | 0.947     | 1.092       | 1.132                   |
| LDA       | 1.327     | 0.898     | 1.012       | 1.078                   |
| DeepCoNN  | 1.222     | 0.827     | 0.933       | 0.994                   |
| NARRE     | 1.214     | 0.817     | 0.921       | 0.984                   |
| MulAttRec | **1.206** | **0.812** | **0.915**   | **0.978**               |

Thirdly, as shown in Table 1, our method MulAttRec consistently outperforms all the baseline methods. Although review information is useful in recommendation, the performance can vary depending on how the review information is utilized. In our model, we propose a multi-level attention mechanics for extracting both word-level and full-text-level information. This allows a review to be modeled with a finer granularity, which can lead to a better performance according to the results. Compared to PMF, our approach gains 30.06% improvement on average.

## 4.4   Effect of Multi-level Attention

We now focus on analyzing the effect of the multi-level attention mechanism by a controlled trial. Note that when we do not use full-text-level attention mechanics, a normalized constant weight will be assigned to each review and when we do not consider the word-level attention, the Word-Attention degenerates to a normal textual CNN module. The average RMSE on three datasets are shown in Fig. 3(a).

From the figure, we can see that when the attention mechanism is applied, the performance of rating prediction is improved significantly as compared with the non-attention approach. Furthermore, the multi-level attention approach makes the most precise prediction, which validates the usefulness of our approach. From the better performance of the full-text level approaches compared the word-level one, we find that the full-text-level attention has a more significant contribution to the improvement.

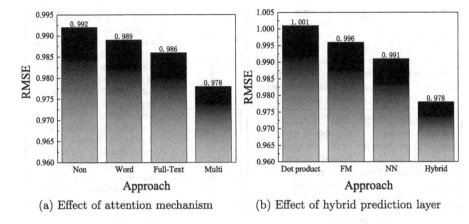

**Fig. 3.** Comparing variants of the proposed model

### 4.5   Effect of Hybrid Prediction Layer

In this section, our controlled experiment on the effect of hybrid prediction layer is presented. There are three new layers. The first one is the traditional dot-product prediction layer: $\widetilde{R} = X_u Y_i$. The last two approaches use FM or NN solely as prediction layer. We reuse the hyper-parameter settings in the last two approaches. The average RMSE on three datasets are shown in Fig. 3(b).

Compared to the simple dot-product prediction, the other three approaches apparently achieve better performance. This comparison justifies the popular assumption that traditional latent factor model may miss the information about the non-linear interaction between users and items and increase the error of predicted rating. Furthermore, we can see that the hybrid prediction layer makes the most precise prediction, which validates the usefulness of our hybrid prediction layer.

### 4.6   Keyword and Review Analysis

In Fig. 4(a), to confirm our design on the word-level attention module, we highlight keywords that with high weights in the attention module. Colored words are considered as informative words, and green words have higher attention scores than those of blue words. We select the same review from Yelp but highlighted differently by the user network and the item network.

We can make two key observation. First, the two networks choose different attention words, because the two networks are trained with the different sets of reviews and the network decided the keywords in reviews. Second, all of these keywords are likely words that describe properties of the item or some more personalized words.

As for the analysis of full-text-level attention module, we select some reviews according to their final attention weights in Fig. 4(b) to illustrate the results

| category:Yelp(user) |
| --- |
| It seems as though good sausage is a way to my heart...or so my doctor tells me. But enough about me, this is all about a fantastic Phoenix gem that I would have never known of had it not popped up on my Yelp app one Sunday afternoon; this is about Caffe Sarajevo. I was excited when we decided on the Cevapi. |

| category:Yelp(item) |
| --- |
| It seems as though good sausage is a way to my heart...or so my doctor tells me. But enough about me, this is all about a fantastic Phoenix gem that I would have never known of had it not popped up on my Yelp app one Sunday afternoon; this is about Caffe Sarajevo. I was excited when we decided on the Cevapi. |

| | | |
| --- | --- | --- |
| Item1 | $a(a_{ij}= 0.1793)$ | The service is very friendly and eager to please. If you're in the area looking for a great lunch, the Asian Island is a solid place to go. |
| | $b(a_{ij}= 0.0138)$ | Just had dinner there w/a big group (4 adults & 6 kids). Our bill came out to $84 for 2 plates of sushi & 6 large size entrees |
| Item2 | $a(a_{ij}= 0.2046)$ | Seriously though - the food is really delicious. If you like flavors - it's all really flavorful. Great hummus too. Try the pizza puff. |
| | $b(a_{ij}= 0.0252)$ | I am a Greek Salad lover so get them often. Unfortunately I have to say that the George's on McKellips does a much better Greek Salad. |

(a) keyword                                    (b) review

**Fig. 4.** Keyword and review analysis

on review usefulness identification. In the figure, examples of the high-weight and low-weight reviews are selected by our model and $a_{ij}$ means the weight of attention.

Generally, the reviews with high attention weight contain more information about the item. For example, the buyers can easily get the feature of each item from Review 1a and 2a, which is highly instructive for making purchasing decisions. In contrast, the low attention reviews only contain the authors' general opinions, but give fewer details to help make a decision.

## 5    Conclusion and Future Work

We have presented a multi-level attention-based CNN model (MulAttenRec) that combines multi-level review information and ratings for recommendation. It selects both useful words and reviews automatically to provide word-level and full-text-level explanations and make a more precise prediction. In addition, it models the non-linear interaction between user and item in a hybrid prediction layer, which couple the factorization machine to a deep neural network. Extensive experiments have been made on three real-life datasets from Amazon and Yelp. The visualization and analysis of keyword and useful reviews validate the reasonability of our multi-level attention mechanism. In terms of recommendation performance, the proposed MulAttenRec consistently outperforms the state-of-the-art recommendation models based on matrix factorization and deep learning in rating prediction. We believe this work offers a new approach to capture the context of recommendation systems.

In the future, we plan to explore the knowledge transfer between the latent item vector and reviews information by low-rank subspace transfer kearning. Moreover, we are interested in the exploration of more advanced neural network e.g., Long Short-Term Memory (LSTM) network, which using sequence learning, to handle sequence and sentiment analysis in the review texts.

# References

1. Blei, D.M., Ng, A.Y., Jordan, M.I.: Latent dirichlet allocation. J. Mach. Learn. Res. **3**(Jan), 993–1022 (2003)
2. Chen, C., Zhang, M., Liu, Y., Ma, S.: Neural attentional rating regression with review-level explanations. In: Proceedings of the 2018 World Wide Web Conference on World Wide Web, pp. 1583–1592. International World Wide Web Conferences Steering Committee (2018)
3. Chen, J., Zhang, H., He, X., Nie, L., Liu, W., Chua, T.S.: Attentive collaborative filtering: multimedia recommendation with item-and component-level attention. In: Proceedings of the 40th International ACM SIGIR Conference on Research and Development in Information Retrieval, pp. 335–344. ACM (2017)
4. Cheng, H.T., et al.: Wide & deep learning for recommender systems. In: Proceedings of the 1st Workshop on Deep Learning for Recommender Systems, pp. 7–10. ACM (2016)
5. Diao, Q., Qiu, M., Wu, C.Y., Smola, A.J., Jiang, J., Wang, C.: Jointly modeling aspects, ratings and sentiments for movie recommendation (jmars). In: Proceedings of the 20th ACM SIGKDD International Conference on Knowledge Discovery and Data Mining, pp. 193–202. ACM (2014)
6. Guo, H., Tang, R., Ye, Y., Li, Z., He, X.: DeepFM: a factorization-machine based neural network for CTR prediction. In: Proceedings of the 26th International Joint Conference on Artificial Intelligence, pp. 1725–1731. AAAI Press (2017)
7. Lee, D.D., Seung, H.S.: Algorithms for non-negative matrix factorization. In: Advances in Neural Information Processing Systems, pp. 556–562 (2001)
8. Mnih, A., Salakhutdinov, R.R.: Probabilistic matrix factorization. In: Advances in Neural Information Processing Systems, pp. 1257–1264 (2008)
9. Rendle, S.: Factorization machines. In: 2010 IEEE 10th International Conference on Data Mining (ICDM), pp. 995–1000. IEEE (2010)
10. Srivastava, N., Hinton, G., Krizhevsky, A., Sutskever, I., Salakhutdinov, R.: Dropout: a simple way to prevent neural networks from overfitting. J. Mach. Learn. Res. **15**(1), 1929–1958 (2014)
11. Tang, D., Qin, B., Liu, T.: Learning semantic representations of users and products for document level sentiment classification. In: Proceedings of the 53rd Annual Meeting of the Association for Computational Linguistics and the 7th International Joint Conference on Natural Language Processing (Volume 1: Long Papers), vol. 1, pp. 1014–1023 (2015)
12. Xiao, J., Ye, H., He, X., Zhang, H., Wu, F., Chua, T.S.: Attentional factorization machines: learning the weight of feature interactions via attention networks. In: Proceedings of the 26th International Joint Conference on Artificial Intelligence, pp. 3119–3125. AAAI Press (2017)
13. Xiong, C., Callan, J., Liu, T.Y.: Learning to attend and to rank with word-entity duets. In: Proceedings of the Annual International ACM SIGIR Conference on Research and Development in Information Retrieval, vol. 763, p. 772 (2017)
14. Zhang, S., Yao, L., Sun, A.: Deep learning based recommender system: a survey and new perspectives. ACM J. Comput. Cult. Herit. **1**(1) (2017). Article 35, 35 p

15. Zhang, W., Du, T., Wang, J.: Deep learning over multi-field categorical data. In: Ferro, N., et al. (eds.) ECIR 2016. LNCS, vol. 9626, pp. 45–57. Springer, Cham (2016). https://doi.org/10.1007/978-3-319-30671-1_4

16. Zheng, L., Noroozi, V., Yu, P.S.: Joint deep modeling of users and items using reviews for recommendation. In: Proceedings of the Tenth ACM International Conference on Web Search and Data Mining, pp. 425–434. ACM (2017)

# SSteGAN: Self-learning Steganography Based on Generative Adversarial Networks

Zihan Wang[1,2,3,4], Neng Gao[2,3,4], Xin Wang[2(✉)], Xuexin Qu[1,2,3,4], and Linghui Li[5]

[1] School of Cyber Security, University of Chinese Academy of Sciences, Beijing, China

[2] Institute of Information Engineering, Chinese Academy of Sciences, Beijing, China
{wangzihan,gaoneng,wangxin,quxuexin}@iie.ac.cn

[3] State Key Laboratory of Information Security, Chinese Academy of Sciences, Beijing, China

[4] Data Assurance and Communications Security Research Center, Chinese Academy of Sciences, Beijing, China

[5] Institute of Computing Technology, Chinese Academy of Sciences, Beijing, China
lilinghui@ict.ac.cn

**Abstract.** Steganography is designed to conceal a secret message within public media. Traditional steganography needs a lot of expert knowledge and complex artificial rules. To solve this problem, we propose a novel self-learning steganographic algorithm based on the generative adversarial network, which we called SSteGAN. This method learns the steganographic algorithm in an unsupervised manner without expert knowledge and directly generates the stego image from the secret message without the cover image. We define a game with four parts: Alice, Bob, Dev and Eve. Alice and Bob attempt to communicate securely. Eve eavesdrops on their conversation and wants to distinguish whether the secret message is embedded in the image. Dev attempts to determine real images from generated images. Experiment results demonstrate that Alice can produce vivid stego images and Bob can successfully decode the secret message with 98.8% accuracy.

**Keywords:** Steganography · Generative adversarial network
Unsupervised training · Self-learning · Decoding accuracy

## 1 Introduction

Steganography is the practice of unobtrusively concealing a secret message within public digital media, for example, images, audios or videos. It is a great challenge to design a good steganographic algorithm because embedding a message will modify the appearance and statistical features of the carrier [2]. The degree of alteration depends on three factors: First, the steganographic algorithm which

© Springer Nature Switzerland AG 2018
L. Cheng et al. (Eds.): ICONIP 2018, LNCS 11302, pp. 253–264, 2018.
https://doi.org/10.1007/978-3-030-04179-3_22

we use. Different steganographic algorithms treat the cover image differently, therefore lead to varying degrees of change of cover image. Second, the length of the message we embed in the cover image. We usually measure message length with bits-per-pixel (bpp). In the vast majority of steganography research, we set it to 0.4 bpp or lower. The longer the message is, the larger the bpp is, and therefore the more the cover image is altered. Third, the degree of alteration depends on the cover image itself. Traditional steganographic algorithms embed information in noise regions or complex textures of the cover images. If one image has more complex regions than another, the more messages can be embedded in it.

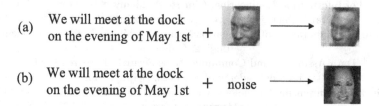

**Fig. 1.** Examples of stego image generated by SteGAN [6] (a) and our SSteGAN (b) with the embedding rate of 0.4 bpp. (a) is generated by using secret information and cover. (b) is generated by using secret information and n bit random noise.

Most of the traditional steganographic algorithms need expertise and complex artificial rules. In order to improve this situation, we attempt to introduce deep learning into the steganography and hide information without expertise. Hayes and Danezis [6] first propose SteGAN method to automatically learn a steganographic algorithm in the technology of adversarial training. However, as shown in Fig. 1, the stego image generated by this method alters lots of pixels of the cover image, which can be easily detected by human eyes or steganalyzer.

To solve these problems, in this paper, we propose a novel SSteGAN model based on the generative adversarial network, which can be learned in an unsupervised manner and without expertise. Instead of embedding the message in the cover image, our model uses secret message and noise as input to generate the stego images By this way, we can not only hide information in pixel values but also in the image contents and features. We define a game with four parts, Alice, Bob, Dev and Eve. Alice and Bob attempt to communicate a secret message contained in the image. Alice generates a stego image by inputting noise and the secret message, Bob receives this stego image and can accurately decode the secret message. Alice and Dev are essentially a pair of generator and discriminator. Alice attempts to approximate the distribution of the real images and synthesize as realistic images as possible, while Dev tries to distinguish the real image from the generated image. Eve is a steganalyzer which is used to detect the presence of hidden information in images. We iteratively train these four parts. Finally, Alice learns to generate realistic stego images by simultaneously deceiving the discriminator Dev and the steganalyzer Eve. In addition, the decoder network Bob which has been simultaneously trained with the Alice learns to decode the secret message accurately.

In summary, the contributions of our work can be concluded as follows: (1) To the best of our knowledge, this is the first steganographic algorithm which directly generates the stego image from secret message and noise instead of embedding the message into the cover image. (2) Our SSteGAN model automatically learns the steganographic method through the generative adversarial network, which is entirely trained in an unsupervised manner and without expert knowledge. (3) Experiment results demonstrate that our model can generate realistic images and successfully deceive the steganalyzer on the database of CelebA.

# 2   Related Work

## 2.1   Generative Adversarial Networks

Generative Adversarial Networks (GANs) is a generative model proposed by Goodfellow et al. [4]. The GAN is composed of a generator and a discriminator. This framework can be described as a minimax two-player game. The generator is trained to generate samples by capturing the real data distribution, thus fooling the discriminator. The discriminator is optimized to distinguish between real data samples and fake samples produced by the generator. Formally, the game between the generator G and the discriminator D is the minimax objective, where $P_{data}$ is the data distribution and $P_z$ is the noise distribution:

$$\min_{G}\max_{D}V(D,G) = \mathbb{E}_{x \sim P_{data}(x)}[logD(x)] + \mathbb{E}_{z \sim P_z(z)}[log(1 - D(G(z)))] . \quad (1)$$

Researchers have studied GAN vigorously, and several pioneering techniques have been proposed to stabilize the training and generate more realistic images. Deep Convolutional GAN (DCGAN) [13] is the first to introduce convolutional architecture which improves visual quality. Wasserstein GAN with Gradient Penalty (WGAN-GP) [5] is proposed to stably train GANs by optimizing the Earth-Mover (Wasserstein distance). Boundary Equilibrium GAN (BEGAN) [3] uses an auto-encoder as a discriminator and attempts to match auto-encoder loss distributions instead of matching the distribution of the samples directly. This method has a more comfortable training procedure and uses a simpler network architecture compared to typical GANs.

## 2.2   Steganography

Steganography research can be split into two subfields: steganographic algorithm and steganalysis. A steganographic algorithm refers to concealing a message, image, or video within another message, image or video while minimizing the perturbations to the carrier. On the contrary, steganalysis refers to the process of determining whether there is embedded secret information in the carrier.

The most common steganographic algorithm is the Least significant bit (LSB) [10], which hides the secret by replacing the least significant bits of cover images. Though these alters are not visually observable, the statistical features of the image have been destroyed. Thus, it is easy to be detected by the steganalyzer. Advanced methods attempt to preserve the statistical characteristics of images by using a distortion function to select the embedding localization (noisy regions or complex textural regions) of the cover image. The most representative sophisticated steganographic schemes are HUGO [11], WOW [7], and S-UNIWARD [8]. HUGO defines a distortion function domain by assigning costs to pixels based on the effect of embedding some information within a pixel, and it uses a weighted norm function to represent the feature space. WOW is another advanced steganographic algorithm. If a region of an image is more texturally complex than another, more pixel values within that region will be modified. In particular, S-UNIWARD proposes a universal distortion function which is unrelated to the embedding domain.

There are few relative attempts to incorporate adversarial training into the steganography [1,6,12,14,15]. To assure security, Abadi and Andersen [1] adversarially train two neural networks to encrypt and decrypt the short message. Steganographic GAN (SGAN) [15] additionally introduces a steganalyzer into the standard DCGAN model, which should be adversarially trained with the generator. Finally, the generator learns to generate a secure cover image that is more suitable to hide information. Secure Steganography GAN (SSGAN) [14] is the follow-up work by using WGAN loss and a new steganalyzer model to improve the effect of SGAN. Different from the two studies above, Hayes and Danezis [6] input the cover image, and automatically learn steganographic method through Adversarial training.

## 3    Self-learning Steganography Based on Generative Adversarial Networks

This section describes our SSteGAN model. It elaborates the proposed model architecture and the function of each participant in this scheme. We also present experiments to support our claims in the next section.

### 3.1    Model Design and Objective Function

Our Steganographic model has four parts: Alice, Bob, Dev and Eve. As a classic scenario in security, Alice and Bob wish to communicate securely, and Eve wishes to eavesdrop on the communication. Ultimately, Alice learns to generate a realistic stego image. Bob can decode the secret message from the stego image accurately. Meanwhile, Eve can't do anything except randomly guess whether this image is a stego image.

The full model is depicted in Fig. 2. Alice produces a stego image $C'$ with the input of noise $Z$ and secret message $M$. The length of secret information $M$ depends on the size of stego images and the steganographic embedding rate.

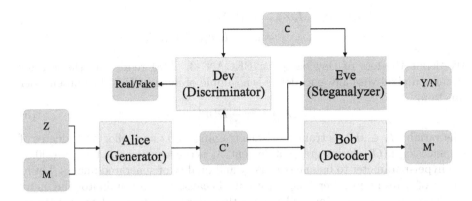

**Fig. 2.** Self-learning steganography based on generative adversarial networks: Alice, Bob, Dev and Eve.

We input noise due to ensure the diversity and safety of the generated images. For example, when we enter the same secret information twice, random noise can guarantee that generated images will not be identical. Bob attempts to decode secret message $M'$ from $C'$. If $M = M'$, we can say that Bob performs perfectly. Both Eve and Dev receive real image $C$ and stego image $C'$. Eve learns to distinguish stego images from cover images. The role of Dev is to distinguish between real images and generated images. At the beginning of training, a people can easily distinguish the difference between the cover image and the stego image. At the moment, the image is blurry because Alice has not learned how to generate a realistic image from the noise and the secret message. Much as in the definitions of GANs, we would like Alice and Bob to defeat the best possible version of Eve and Dev, so we adversarially train each part rather than a fixed pre-training Eve or Dev. As training continues, based on the loss of Eve and Dev, Alice updates its parameters and produces the realistic stego image which can deceive both Eve and Dev. The four parts are essentially like a generator (Alice), a decoder (Bob), a steganalyzer (Eve) and a discriminator (Dev).

In our model, Alice, Bob, Dev and Eve are all neural networks. We describe their detailed structures in Sect. 3.2. We let $\theta_A$, $\theta_B$, $\theta_D$ and $\theta_E$ respectively denote the parameters of Alice, Bob, Dev and Eve. We write $A(\theta_A, Z, M)$ for Alice's output on input $Z$ and $M$, write $B(\theta_B, C')$ for Bob's output on input $C'$, write $D(\theta_D, C, C')$ for Dev's output on input $C$ and $C'$, and $E(\theta_E, C, C')$ for Eve's output on input $C$ and $C'$. $L_A, L_B, L_D, L_E$ respectively represent the loss of Alice, Bob, Dev and Eve. Obviously, we can get the following relations:

$$C' = A(\theta_A, Z, M) , \tag{2}$$

$$M' = B(\theta_B, A(\theta_A, Z, M)) . \tag{3}$$

We introduce a distance funcion $d$ on $M$ and $M'$, which is the secret message reconstruction loss. Concretely, we take the L2 distance $d(M, M') = \sum_{i=1}^{N}(M_i - M_i')^2$, where $N$ is the length of secret message. We define the loss function for Bob:

$$L_B(\theta_A, \theta_B, M, Z) = d(M, M') \qquad (4)$$
$$= d(M, B(\theta_B, A(\theta_A, Z, M))) .$$

We define the loss of Dev based on BEGAN [3]. $L(v)$ represents the loss for training a pixel-wise autoencoder. $v$ is a sample of images. $D$ is the autoencoder function.

$$L(v) = |v - D(v)| , \qquad (5)$$

The variable $k_t \in [0, 1]$ is trained at each training step $t$ and reflects the degree of emphasis on $L(C')$ during gradient descent. $\lambda_k$ is the learning rate for $k$. $\gamma \in [0, 1]$ is a hyper-parameter to balance diversity and quality of generated images. Lower values of $\gamma$ result in lower image diversity because the discriminator put more emphasis on auto-encoding real images than discriminate real from generated images. We use $k_0 = 0$, $\lambda_k = 0.001$ and $\gamma = 0.5$ in our experiments.

$$L_D(C, C') = L(C) - k_t \cdot L(C') , \qquad (6)$$

$$k_{t+1} = k_t + \lambda_k(\gamma L(C) - L(C')) . \qquad (7)$$

$$\gamma = \frac{\mathbb{E}[L(C')]}{\mathbb{E}[L(C)]} , \qquad (8)$$

Same as the traditional GAN discriminator implementations, we calculate Eve's loss use sigmoid cross entropy loss:

$$L_E(\theta_E, C, C') = -log(E(\theta_E, C)) - log(1 - E(\theta_E, C')) . \qquad (9)$$

Alice's loss is given as a weighted sum of Bob's loss, Eve's loss and Dev's loss:

$$L_A(\theta_A, C, C') = \lambda_B L_B(\theta_B, C, C') + \lambda_D L_D(\theta_D, C') + \lambda_E L_E(\theta_E, C') . \qquad (10)$$

where the $\lambda_B, \lambda_D, \lambda_E \in \mathbb{R}$ define the weight of each loss term. The size of these weights controls the trade-off among the image quality, the ability of anti-steganalysis and the decoding accuracy. In different datasets or embedding rate, we should perform a grid search to find the optimal weight value.

## 3.2 Structure Details

**Alice(Generator).** Alice accepts the noise $Z$ and the random n-bit binary secret message $M$ as input. First, we concatenate the noise $Z$ and the secret message $M$ together, which is then fed into a fully connected layer. After that, we reshape it to $8 \times 8 \times N_h$, where $N_h$ is usually set as 128. We denote the structure block of $3 \times 3$ convolution with exponential linear units (ELUs) activation function as Conv2D-ELUs. Each of the next three layers is a sequence of Conv2D-ELUs-Conv2D-ELUs and a nearest neighbor upsampling, except for the final layer does not contain the upsampling layer (Fig. 3).

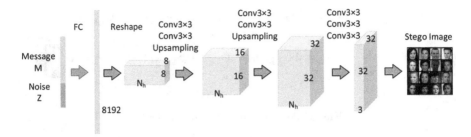

**Fig. 3.** The network structure of Alice.

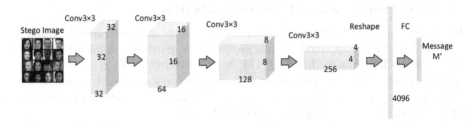

**Fig. 4.** The network structure of Bob.

**Bob(Decoder).** Bob receives the stego image $C'$ as input. It is made up of four convolutional layers and one final fully connected layer. Batch normalization and LeakRelu is used in all but except the final layer where we use tanh as the activation function. The output layer has $n$ neuron corresponding to the n bit secret message $M$ (Fig. 4).

**Dev(Discriminator).** Dev accepts both real images $C$ and stego images $C'$ as input. It is a convolutional deep neural network architecture as an auto-encoder. The structure of the decoder is the same as Alice. The encoder has three layers Conv2D-ELUs-Conv2D-ELUs-downsampling, where the down-sampling is a convolutional layer with stride 2. In the meantime, the convolutional filters are increased linearly through each down-sampling layer. At the boundary between the encoder and the decoder, the tensor is mapped via fully connected layers to $N_e$ dimension tensor, and not followed by any activation functions (Fig. 5).

**Eve(Steganalyzer).** Eve accepts both real images $C$ and stego images $C'$ as input. Eve has a similar network structure to Bob but uses the sigmoid activation function at the final fully connected layer to output probabilities.

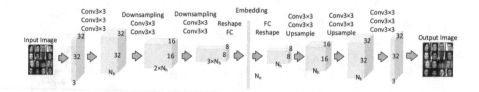

**Fig. 5.** The network structure of Dev.

## 4    Experiment

### 4.1    Data Preparation and Parameters Setting

To validate our method, we conduct some experiments on CelebFaces Attributes Dataset (CelebA) [9]. CelebA is a large-scale face attributes dataset with 202, 599 celebrity face images from 10, 177 unique identities. We randomly select 190 K pictures as the training set, and the rest pictures are used to validate the security of our model. We first pre-process these pictures by centering the images on the facial region and cropping them to $32 \times 32$ pixels.

Then, we concatenate n-bit random secret message $M$ and 100-bit random noise $Z$ as the input of Alice. We change the length $n$ of the message to test the effect on image quality and the decoding accuracy. As almost steganographic researches, we test the embedding rate from 0.1 bpp to 0.4 bpp. That is, we vary n from 100 to 400, which is equivalent to successfully hiding a 12 to 50 character ASCII message.

Now we have prepared all the inputs. Then we train our SSteGAN model by using Adam optimization algorithm with the initial learning rate of $8^{-5}$. The learning rate will decay by a factor of 2 when the measure of convergence stalls. According to our experience, we use $N_h = 128$, $N_e = 64$ in our experiment with this dataset. We alternately train Alice, Bob, Dev and Eve at each batch.

(a) Stego images of 0.1bpp.

(b) Stego images of 0.4bpp.

**Fig. 6.** Stego images generated by Alice at the embedding rate of 0.1 bpp and 0.4 bpp.

## 4.2   Experimental Results

As shown in Fig. 6(a) and (b), we first demonstrate that our model can successfully produce stego images at the embedding rate of 0.1 bpp and 0.4 bpp, which are almost indistinguishable from the real images. Moreover, there is no noticeable image quality decrease from 0.1 bpp to 0.4 bpp.

In Fig. 7(a), pictures of each row are generated with the same noise and different message. On the contrary, pictures of each row in Fig. 7(b) are generated with the same message and different noise. Although these pictures are generated by same noise or message, people in these pictures may have different race, gender, age, hair style, and so on. The result indicates that the stego images generated by our SSteGAN model have good randomness and diversity. Even if we repeatedly send the same secret message, our model will not produce the same picture.

(a) Fixed noise (0.4bpp).                    (b) Fixed message (0.4bpp).

**Fig. 7.** Stego images generated by Alice with fixed noise or fixed message.

We find that if we increase the length of noise and secret information continuously, although Bob can successfully decode the message, the quality of the stego images will decrease. By this time, Alice can not balance realistic image creation capacity with message decoding accuracy. In general, we can deem images as a distribution in the low-dimensional manifold space. The generator aims to learn a mapping from the hidden space to the data space. The search space will be larger as the hidden spatial dimension (the length of noise and message) increased, which leads to more saddle points, lower learning speed and failed convergence.

Figure 8(a) and (b) show the average error rate of the decoder Bob in the training process. In the first few rounds of training, the visual quality of stego images are low, and Bob can not do better than random guessing. Then, Bob gradually learns to decode the correct secret message. After 5 K training steps, Bob can decode the secret message with the accuracy of 100%. Then, Alice strives to improve the quality and security of images around the premise that the decoding accuracy is ensured.

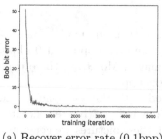
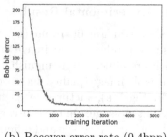

(a) Recover error rate (0.1bpp).    (b) Recover error rate (0.4bpp).

**Fig. 8.** The recover error rate of Bob during training process.

### 4.3    Comparison with Related Work

In order to evaluate the performance of our method, we compare our model against other steganographic algorithms (HUGO, WOW, S-UNIWARD, Ste-GAN). Additionally, We train an independent steganalyzer EVE which we called $S^*$. First, we randomly select 10,000 cover images and respectively generate the corresponding 10,000 stego images with different steganographic algorithms. Then, we choose 5000 stego images of them and the corresponding 5000 cover images as the training set to train $S^*$. Finally, for each steganographic algorithm, we calculate the accuracy of the $S^*$ on the rest of 5000 stego images.

Results are provided in Table 1. The accuracy of our model is 59.16%. It is clear that our model performs well against other steganographic methods. It demonstrate that Alice tries to learn an steganographic method such that $S^*$ can not do better than random guessing ($p_{accuracy} = 0.5$). We do not train Alice to minimize the prediction accuracy of Eve, because Eve could be completely right by simply flipping her decision. Our unsupervised learning method is without expert domain knowledge and cover images, but it is as effective as the traditional method in steganographic tasks.

**Table 1.** Accuracy of distinguishing test images for the steganalyzer S* at embedding rate of 0.4 bpp.

| Steganographic algorithm | HUGO | WOW | S-UNIWARD | SteGAN | SSteGAN |
|---|---|---|---|---|---|
| Accuracy | 78.28% | 77.28% | 76.36% | 60.34% | 59.16% |

### 4.4    Decryption Security Evaluation

As shown in Table 2, to demonstrate the robustness and security of the model, we train two SSteGAN models which is called $model_1$ and $model_2$ at the same time. First, we use $Alice_1$ (Alice of $model_1$) to randomly generate 20 K stego images,

**Table 2.** We use $Alice_1$ to generate $20\,K$ stego images randomly, and calculate the recover accuracy of $Bob_1$ and $Bob_2$.

| Different Bob | $Bob_1$ | $Bob_2$ |
| --- | --- | --- |
| Accuracy | 98.8% | 50.04% |

and decode the secret message by $Bob_1$. Then we calculate that the recover accuracy is 98.8%. If we use a tolerate error encryption algorithm to encrypt the message before embedding within an image, the receiver can successfully recover and decrypt the information. Because of that, our model does not require perfect decoding accuracy.

However, when we use $Bob_2$ to decode the same $20\,K$ stego images generated by $Alice_1$, we can only get a recover accuracy of 50.04%. $Bob_2$ can not do better than random guessing. These two experiments prove that only when the sender and the receiver use the Alice and Bob from the simultaneously trained process, that they can successfully decode the secret message. Although the adversary uses the same method to train a decoder, they can not recover the secret message.

## 5 Conclusion

In this paper, we propose a model called SSteGAN which directly generates realistic and secure steganographic images without using cover images. Our model consists of four parts: Alice, Bob, Dev and Eve. The generator Alice learns to generate eligible stego images by fooling the discriminator Dev and the steganalyzer Eve. In the meantime, we get a decoder Bob which can accurately recover the secret information. Our method is entirely unsupervised learning and does not require expert domain knowledge to design network structure. We demonstrate the effectiveness of our model on CelebA dataset. Meanwhile, we evaluate the performance of our model by comparison with related work and analyzing the security of decryption. In total, our study acquires expected performance and opens a new avenue for incorporating generative adversarial networks into the steganography.

**Acknowledgments.** This research was supported by the National Key Research and Development Program of China (No. 2016YFB0800504), the National Natural Science Foundation of China (No. U163620068) and the Strategy Cooperation Project (AQ-18-01).

## References

1. Abadi, M., Andersen, D.G.: Learning to protect communications with adversarial neural cryptography. arXiv preprint arXiv:1610.06918 (2016)
2. Baluja, S.: Hiding images in plain sight: deep steganography. In: Advances in Neural Information Processing Systems, pp. 2066–2076 (2017)

3. Berthelot, D., Schumm, T., Metz, L.: BEGAN: boundary equilibrium generative adversarial networks. arXiv preprint arXiv:1703.10717 (2017)
4. Goodfellow, I.J., et al.: Generative adversarial networks. Adv. Neural Inf. Process. Syst. **3**, 2672–2680 (2014)
5. Gulrajani, I., Ahmed, F., Arjovsky, M., Dumoulin, V., Courville, A.C.: Improved training of wasserstein GANs. In: Advances in Neural Information Processing Systems, pp. 5769–5779 (2017)
6. Hayes, J., Danezis, G.: Generating steganographic images via adversarial training. In: Advances in Neural Information Processing Systems, pp. 1951–1960 (2017)
7. Holub, V., Fridrich, J.: Designing steganographic distortion using directional filters. In: 2012 IEEE International Workshop on Information Forensics and Security (WIFS), pp. 234–239. IEEE (2012)
8. Holub, V., Fridrich, J., Denemark, T.: Universal distortion function for steganography in an arbitrary domain. EURASIP J. Inf. Secur. **2014**(1), 1 (2014)
9. Liu, Z., Luo, P., Wang, X., Tang, X.: Deep learning face attributes in the wild. In: Proceedings of the IEEE International Conference on Computer Vision, pp. 3730–3738 (2015)
10. Mielikainen, J.: LSB matching revisited. IEEE Signal Process. Lett. **13**(5), 285–287 (2006)
11. Pevný, T., Filler, T., Bas, P.: Using high-dimensional image models to perform highly undetectable steganography. In: Böhme, R., Fong, P.W.L., Safavi-Naini, R. (eds.) IH 2010. LNCS, vol. 6387, pp. 161–177. Springer, Heidelberg (2010). https://doi.org/10.1007/978-3-642-16435-4_13
12. Qian, Y., Dong, J., Wang, W., Tan, T.: Deep learning for steganalysis via convolutional neural networks. In: Media Watermarking, Security, and Forensics 2015, vol. 9409, p. 94090J. International Society for Optics and Photonics (2015)
13. Radford, A., Metz, L., Chintala, S.: Unsupervised representation learning with deep convolutional generative adversarial networks. arXiv preprint arXiv:1511.06434 (2015)
14. Shi, H., Dong, J., Wang, W., Qian, Y., Zhang, X.: SSGAN: secure steganography based on generative adversarial networks. In: Zeng, B., Huang, Q., El Saddik, A., Li, H., Jiang, S., Fan, X. (eds.) PCM 2017. LNCS, vol. 10735, pp. 534–544. Springer, Cham (2018). https://doi.org/10.1007/978-3-319-77380-3_51
15. Volkhonskiy, D., Nazarov, I., Borisenko, B., Burnaev, E.: Steganographic generative adversarial networks. arXiv preprint arXiv:1703.05502 (2017)

# A Multidimensional Interaction-Focused Model for Ad-Hoc Retrieval

Qiang Sun$^{(\boxtimes)}$, JiaLiang Wu, and Yue Wu

Department of Computer Engineering and Science, Shanghai University,
Shanghai, China
{sun1,ywu}@shu.edu.cn, JohnWu@gwu.edu

**Abstract.** Ad-hoc Retrieval based on deep learning model often suffers from the limitation of embedding semantic abuse problem. Inspired by the success of convolutional neural network based models in image processing, where a series of hidden layers extracts increasingly abstract features from a image, we propose a multidimensional interaction-focused model to solve the above problem in a image processing way. Firstly, we construct the query-document similarity matrix as a 3d tensor which means a word similarity value becomes a vector. Then we apply a CNN layer to capture complicated interaction patterns on every similarity chanel and a Bi-LSTM layer will map the output of CNN to a vector of fixed dimensionality. Finally a feed forward network will calculate a matching score. Experiments on the question-answer task with dataset WikiQA have achieved the state-of-the-art results compared to traditional statistical methods and deep neural network methods.

**Keywords:** Ad-hoc retrieval · Matching tensor · CNN · Bi-LSTM

## 1 Introduction

Deep Neural Network models are extremely powerful that can achieve excellent performance on many hard problems, such as question answering (QA) [1] and Machine Translation (MT) [2]. A striking example of DNN's power is its ability to map input sequences to a fixed-sized vector while their lengths are unknown. Furthermore, DNN models can be trained with a large number of parameters which can contain rich information about the source text data. Thus, if there exists a well labeled training set, back propagation will find the appropriate parameters to solve the problems [3].

Different from using one hot word representation in the hand-craft features such as BM25 [10] and language model, recently most deep information retrieval models adopt pretrained word embeddings as the word representation, except the letter-trigrams used in DSSM [11] and CDSSM [12]. The advantage of adopting word embedding as the word representation is to investigate semantic matching information into the model. However, on the flip side, semantic matching brings too much noise matching signals, which covers up the exact matching signals and dominates the final matching score. For example, given the following three texts:

© Springer Nature Switzerland AG 2018
L. Cheng et al. (Eds.): ICONIP 2018, LNCS 11302, pp. 265–274, 2018.
https://doi.org/10.1007/978-3-030-04179-3_23

*Q*: *How old was sue lyon when she made Lolita.*

$D_{right}$: *The actress who played Lolita, Sue Lyon, was fourteen at the time of filming.*

$D_{wrong}$: *The man is aged aged aged aged aged aged aged aged aged.*

The first two sentences are query text and right matching text, they all come from dataset WikiQA, the third one is a fake and obviously wrong matching sentence we create. The given query term "old" has a high similarity with so many same related words "aged" in Wrong document. Thus the sum of matching signals in the wrong document is higher than that in the right document, even the wrong document dose not has any exact matching signal with query term "old".

The embedding semantic abuse problem [19] poses a challenge for current DNN IR models because we need to focus on the exact matching signals rather than how many document key words similar to some query key words. In this paper, we introduce the multidimensional interaction-focused model to solve the problem. The idea is to take 3d matching tensor intead of matching matrix to obtain better representation of matching signal, and then use a CNN layer capture complicated interaction patterns on every similarity chanel, finnaly use a Bi-LSTM to joint different level patterns for final matching socre.

The main result of this work is the below. On the WikiQA question answering task, we obtained a MAP score of 0.655. This is by far the best score in the dataset. For comparison, the MAP score of a statistical based baseline BM25 on this dataset is 0.550. In short, the major contributions of this paper include:

1. We take 3d matching tensor intead of matching matrix to obtain better representation of matching signal.
2. We use a CNN layer to capture complicated interaction patterns on every similarity chanel.
3. We apply a Bi-LSTM layer to map the output of CNN output to joint different level patterns for final matching socre.

## 2    The Model

In the field of information retrieval, specially for deep models, it's very common for making a shared similarity matrix which we called matching matrix for further modeling. Given a query sequence $q$ : $\{q_1, q_2, ..., q_m\}$ and a document sequence $d$ : $\{d_1, d_2, ..., d_n\}$, both $q_i$ and $d_j$ represent a word. Firstly mapping every words into a fixed-length vector $v_q$ : $\{v_{q_1}, v_{q_2}, ..., v_{q_m}\}$ and $v_d$ : $\{v_{d_1}, v_{d_2}, ..., v_{d_n}\}$, then calculate the similarity between word $v_{q_i}$ and $v_{d_j}$. Generally we use $M_{i,j} = v_{q_i}^T \cdot v_{d_j}$ as the similarity. It's easy to treat it as a 2D pixel grid image and dig more interaction information from it [6]. But this method is too rough to find out delicate parts of text matching.

**Multidimensional Matching Vector.** In MIF we represent the interaction of query and document word as a vector, which means a 3d tensor will replace the

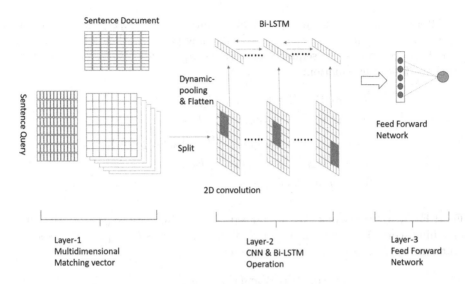

**Fig. 1.** The architecture of MIF. Given input query sentence and document sentence, Layer-1 maps them into multidimensional matching tensor. Then Layer-2 use a CNN layer to capture matching patterns in each matching matrix and then a BiLSTM layer will compress these patterns. Finally, a MLP layer produce the final ranking score.

matching matrix. We construct the 3d tensor by a simple concat of $v_{q_i}$ and $v_{d_j}$ and then map it to a specified dimension:

$$M_{i,j} = \sigma_1([v_{q_i}; v_{d_j}]U^T) \tag{1}$$

where $[v_{q_i}; v_{d_j}] \in \mathbb{R}^{|v_{q_i}|+|v_{d_j}|}$ and $U^T \in \mathbb{R}^{(|v_{q_i}|+|v_{d_j}|) \times c}$, if we set the dimension as $c$, the matching tensor can be represented as: $M_{i,j} = [z_{i,j,1}, ..., z_{i,j,x}, ..., z_{i,j,c}]$. $z_{i,j,x}$ is a matching matrix as the comman use of query and document interaction. In other words, we apply many matching matrix to denote the interaction.

**CNN and LSTM Extract Layer.** The CNN & LSTM Layer is to learn different levels of matching patterns and compress them to a fix vector for the final Feed Forward Network. In the layer of CNN, the $k$-th kernel $w^{(1;k)}$ maps over the whole matching matrix $z^{(0,x)} = M[:, :, x]$ to generate a feature map $z^{(1,k,x)}$:

$$z_{i,j}^{(1,k,x)} = \sigma\left(\sum_{s=0}^{r_k-1} \sum_{t=0}^{r_k-1} W_{s,t}^{(1,k,x)} z_{i+s,j+t}^{(0,x)} + b^{(1,k,x)}\right) \tag{2}$$

The dynamic pooling strategy [17] is used to deal with the text length variability. After applying dynamic pooling, we will get fixed-size feature maps:

$$z_{i,j}^{(2,k,x)} = max_{0 \le s < w_k} max_{0 \le t < l_k} z_{i \cdot w_k + s, j \cdot l_k + t}^{(1,k,x)} \tag{3}$$

where $w_k$ and $l_k$ denote the width and length of the corresponding pooling kernel, which are determined by the text lengths $n$ and $m$, and output feature map size $n'$ $m'$, i.e. $w_k = \lceil n/n' \rceil$; $l_k = \lfloor m/m' \rfloor$.

After CNN layer, we get a sequence of outport $z^{(0)}, z^{(1)}, ..., z^{(x)}, ..., z^{(c)}$, which represents each dimension's compressed matching patterns, then a Bidirectional LSTM [18] computes two sequence of outputs $[h_1^\rightarrow, ..., h_c^\rightarrow], [h_1^\leftarrow, ..., h_c^\leftarrow]$ by iterating the following equation:

$$
\begin{aligned}
i_t &= \sigma(W_{zi}z^{(t)} + W_{hi}h_{t-1} + b_i), \\
f_t &= \sigma(W_{zf}z^{(t)} + W_{hf}h_{t-1} + b_f), \\
c_t &= f_t c_{t-1} + i_t \tanh(W_{zc}z^{(t)} + W_{hc}h_{t-1} + b_c), \\
o_t &= \sigma(W_{zo}z^{(t)} + W_{ho}h_{t-1} + b_o), \\
h_t &= o_t \tanh(c_t)
\end{aligned}
$$

**Feed Forward Network.** We use a Feed Forward Network to produce the final matching score. Take the concat of two direction outputs $h_x = [h_x^\rightarrow, h_x^\leftarrow]^T$ as the input, we apply a two layer fully connected layer:

$$
s(q, d) = W_2 \sigma_2(W_1 h + b_1) + b_2 \tag{4}
$$

where $q$ and $d$ are the query and document sentence of the first input.

## 3　Experiment Methodology

This section describes the dataset, our baseline models and the evaluation method.

### 3.1　Dataset

*WikiQA* The WikiQA [8] corpus is collected and annotated for research on opendomain question answering. Each question comes from Bing query log and linked to a Wikipedia page that potentially has the answer and it use the summary section sentence as the candidate answers. With the help of crowdsourcing, it includes 3,047 questions and 29,258 sentences, where 1,473 sentences were labeled as answer sentences to their corresponding questions.

### 3.2　Baseline Methods

We adopt statistical baseline BM25 and deep model baseline including K-NRM et al. for comparison.

**BM25:** The BM25 [10] is a ranking function used by search engines based on the probabilistic retrieval framework.

**CDSSM:** CDSSM [12] is the convolutional version of DSSM [11] based on a convolutional neural network (CNN) to learn lowdimensional semantic vectors for search queries and Web documents (Table 1).

**Table 1.** Training and testing dataset characteristics.

| | WikiQA | |
|---|---|---|
| | Training | Testing |
| Fields | Question answering | |
| Queries | 2,118 | 236 |
| Query average length | 3.7 | 3.2 |
| Docs per query | 9.6 | 9.9 |
| Document average length | 15.7 | 15.5 |
| Vocabulary size | 18243 | 7623 |

**ARC-I, ARC-II:** ARC-I [13] is a general representation-focused deep matching model that finds the representation of each sentence, and then compares the representation for the two sentences with a multi-layer perceptron. ARC-II [13] was proposed to fix the drawbacks of the model ARC-I, it focuses on learning hierarchical matching patterns from local interactions using a CNN.

**MatchPyramid:** MatchPyramid [6] is another state-of-the-art interaction-focused deep matching model that uses convolutional neural network to capture rich matching patterns in a layer-by-layer way.

**DRMM:** DRMM [4] performs histogram pooling on the embedding based translation matrix and uses the binned soft-TF as the input to a ranking neural network.

**KNRM:** KNRM [14] is a kernel based neural model for document ranking. It used a new kernel-pooling technique to softly count word matches at different similarity levels and provide soft-TF ranking features.

**MVLSTM:** MVLSTM [16] matches two sentences with multiple positional sentence representations and each positional sentence representation generated by a bidirectional long short term memory (Bi-LSTM).

### 3.3 Evaluation Method

The measurement of a ranking result mainly includes: $P@k$ (Precision at $k$), $R@k$ (Recall at $k$), $MAP$ (Mean Average Precision) and $nDCG$ (normalized Discounted Cumulative Gain) [5]. Suppose there are $G_k$ matching documents in the $k$ real ranking documents, while $Y_k$ matching documents in the k predict ranking documents. The $P@k$ and $R@k$ are defined as: $P@k = \frac{Y_k}{k}$, $R@k = \frac{Y_k}{G_k}$. For the $MAP$, we assume that the position of real matching document in predict ranking documents are $k_1, k_2, ..., k_r$, the index $r$ represents the total number of matching documents, so we define $MAP$ as: $MAP = \frac{\sum_{i=1}^{r} P@k_i}{r}$. For realistic situation, we introduce the matching level. Given the optimal ranking, the matching score at every position are $\hat{rel}_1, \hat{rel}_2, ..., \hat{rel}_3$, while the matching scores are set as $rel_1, rel_2, ..., rel_n$ in predict ranking, so the $nDCG$ is defined as: $nDCG = \frac{DCG_p}{IDCG}$, where $DCG = rel_1 + \sum_{i=2}^{n} \frac{rel_i}{log_2 i}$, $IDCG = \hat{rel}_1 + \sum_{i=2}^{n} \frac{\hat{rel}_i}{log_2(i)}$.

# 4 Experiments Details

## 4.1 Training Details

Firstly, we construct a 3d matching tensor to obtain the representation of matching signal. Then we use a CNN layer capture complicated interaction patterns on every similarity chanel, and we use a Bi-LSTM to joint different level patterns. Finally we apply a feed forward network to obtain the final matching socre. Earlier studies [17, 18] show that BM25 is also a strong baseline on these datasets, which is even better than some deep models such as DSSM and CDSSM. So we choose it as the compare baseline of the experiments. The complete training details are given below:

- We choose max query length as 20 and max doc length as 100 at dataset WikiQA.
- We used GloVe as pre-trained word embedding to embed the query and document sentences with the embed size of 300 and the vocab size of 18678.
- We then use 12 as the dimension of multidimensional matching tensor.
- In the CNN layer we use kernel size of $3 \times 3$, the hidden representation dimensions of Bi-LSTM are also set to 16.
- We use two MLP layers with the parameters initialized with the uniform distribution between $-0.08$ and $0.08$.

**Table 2.** Ranking performances of MIF and baseline models. Relative performance compared with BM25 are in percentages. Significant improvement or degradation with respect to BM25 is indicated $(+/-)$ $(p\text{-}value <= 0.05)$.

| Method | NDCG@1 | | NDCG@3 | | NDCG@5 | | NDCG@10 | | MAP | |
|---|---|---|---|---|---|---|---|---|---|---|
| BM25 | 0.376 | – | 0.522 | – | 0.608 | – | 0.643 | – | 0.550 | – |
| CDSSM | 0.198 | −47% | 0.356 | −32% | 0.445 | −27% | 0.515 | −20% | 0.401 | −27% |
| ARCII | 0.367 | −2% | 0.547 | −5% | 0.602 | −1% | 0.649 | +1% | 0.555 | +1% |
| K-NRM | 0.412 | +10% | 0.557 | +7% | 0.624 | +3% | 0.669 | +4% | 0.577 | +5% |
| ARCI | 0.439 | +17% | 0.571 | +9% | 0.637 | +5% | 0.676 | +5% | 0.593 | +8% |
| MATCHPYRAMID | 0.426 | +13% | 0.604 | +16% | 0.663 | +9% | 0.695 | +8% | 0.607 | +10% |
| MVLSTM | **0.502** | +34% | 0.612 | +17% | 0.669 | +10% | 0.716 | +11% | 0.633 | +15% |
| DRMM | 0.485 | +29% | **0.643** | +23% | 0.681 | +12% | **0.724** | +13% | 0.647 | +18% |
| MIF-LSTM | 0.446 | +18% | 0.607 | +16% | 0.611 | +1% | 0.655 | +2% | 0.634 | +15% |
| MIF-BiLSTM | 0.456 | +21% | 0.622 | +19% | **0.690** | +13% | 0.722 | +12% | **0.656** | +19% |

## 4.2   Experiment Results

Table 2 shows the ranking performance of MIF and our baselines over WikiQA. As we can see, MIF-BiLSTM outperformed almost all the statistical baselines and almost all the DNN based baselines in MAP. The closest baseline on WikiQA is DRMM, another interaction based models built upon the embedding translation matrix. MIF performed better in larger ranking positions: it's NDCG@5 score have achieved nine point promotion compared to the BM25. Meanwhile we can see two directional LSTM have a better information extraction ability of the matching patterns.

# 5   Analysis of MIF

## 5.1   Matching Tensor Dimension

One of the attractive features of our model is the usage of muldimension of matching tenso. Figure 2 visualizes the impact of different dimension. It seems that the proper dimension is subject to specific circumstance. In our model dimension of 12 is the best choice for MAP evluation. The larger dimension performance bad and time consuming because of the CNN&LSTM layer.

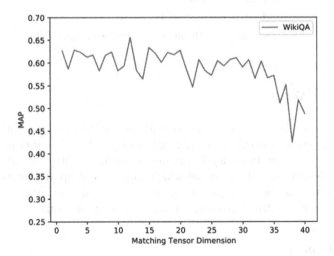

**Fig. 2.** The impact of different dimension.

## 5.2   Error Analysis

The Table 3 presents several examples of short sentence pair and long sentence pair. Our model works well in most long sentence matching pair because of adequate diversity of matching patterns for CNN&BiLSTM layer. However, to be honest, it's trapped by short matching pair sometimes as the Table 3 shows. The main reason is that short sentences lack of multiple levels of semantic expression. It can be solved by combining with traditional statistic model like tf-idf in practice.

**Table 3.** A few examples of best results produced by our model alongside the ground truth matching pairs. The reader can verify that our model is sensitive to the length of text.

| Type | Sentence |
|------|----------|
| Question | What states allow same sex marriage As of May 2013 |
| Right answer | Ten states Connecticut, Iowa, Maine, Maryland, Massachusetts, New Hampshire, New York, Rhode Island, Vermont, and Washington as well as the District of Columbia and three Native American tribes have legalized same-sex marriage, representing 16.1 % of the U.S. population |
| Our model | Ten states Connecticut, Iowa, Maine, Maryland, Massachusetts, New Hampshire, New York, Rhode Island, Vermont, and Washington as well as the District of Columbia and three Native American tribes have legalized same-sex marriage, representing 16.1 % of the U.S. population |
| Question | What are land parcels |
| Right answer | Land lot, a piece of land |
| Our model | A package, sent through the mail or package delivery |

## 6   Related Work

There have been a lot of related works to address the embedding semantic abuse problem with neural network. Our approach is inspired by Xiong et al. [14] who used kernels to extract multi-level soft match features. Pang [4] introduced a novel deep relevance matching model which employs a deep architecture at the query term level for relevance matching. Wan [16] used another way to match two sentences with multiple positional sentence representations.

## 7   Conclusion

This paper presents MIF, a multidimensional interaction-focused model, for Ad-hoc Retrieval. The model applies multidimensional interactions on query and document text and ranks documents using a feed forward neural network. The key of our model is the multidimensional matching tensor which uses a series of matching matrix to grab a far more nuanced interaction of a query and document pair. Then, we use CNN layer to capture matching patterns in each matching matrix and then a BiLSTM layer will compress these patterns as inputs of a final feed forward network.

Our experiments on WikiQA question answer task demonstrate the advantage of multidimensional matching tensor. It overcomes the embedding semantic abuse problem for the multidimensional matching tensor enrich the interaction of word pairs.

Most importantly, we show that an multidimensional matching tensor can be a replacement of matching matrix in tasks contain text matching or similarity. The results suggest that our approach will do well on other similar tasks.

# References

1. Yu, L., Hermann, K.M., Blunsom12, P., Pulman, S.: Deep Learning for Answer Sentence Selection. arXiv preprint arXiv:1412.1632 (2014)
2. Bahdanau, D., Cho, K., Bengio, Y.: Neural machine translation by jointly learning to align and translate. arXiv preprint arXiv:1409.0473 (2014)
3. Sutskever, I., Vinyals, O., Le, Q.V.: Sequence to sequence learning with neural networks. In: Advances in Neural Information Processing Systems, pp. 3104–3112 (2014)
4. Guo, J., Fan, Y., Ai, Q., Croft, W.B.: A deep relevance matching model for ad-hoc retrieval. In: Proceedings of the 25th ACM International on Conference on Information and Knowledge Management, pp. 55–64. ACM (2016)
5. Reed, G.F., Lynn, F., Meade, B.D.: Use of coefficient of variation in assessing variability of quantitative assays. Clin. Diagn. Lab. Immunol. **9**, 1235–1239 (2002)
6. Pang, L., Lan, Y., Guo, J., Xu, J., Wan, S., Cheng, X.: Text Matching as Image Recognition. In: AAAI, pp. 2793–2799 (2016)
7. Liu, T.-Y.: Learning to rank for information retrieval. INR **3**, 225–331 (2009)
8. Yang, Y., Yih, W., Meek, C.: WikiQA: a challenge dataset for open-domain question answering. In: Proceedings of the 2015 Conference on Empirical Methods in Natural Language Processing, pp. 2013–2018 (2015)
9. Qin, T., Liu, T.-Y.: Introducing LETOR 4.0 datasets. arXiv preprint arXiv:1306.2597 (2013)
10. Zhai, C., Lafferty, J.: A study of smoothing methods for language models applied to ad hoc information retrieval. In: ACM SIGIR Forum, pp. 268–276. ACM (2017)
11. Huang, P.-S., He, X., Gao, J., Deng, L., Acero, A., Heck, L.: Learning deep structured semantic models for web search using clickthrough data. In: Proceedings of the 22nd ACM International Conference on Conference on Information & Knowledge Management, pp. 2333–2338. ACM (2013)
12. Shen, Y., He, X., Gao, J., Deng, L., Mesnil, G.: Learning semantic representations using convolutional neural networks for web search. In: Proceedings of the 23rd International Conference on World Wide Web, pp. 373–374. ACM (2014)
13. Hu, B., Lu, Z., Li, H., Chen, Q.: Convolutional neural network architectures for matching natural language sentences. In: Advances in Neural Information Processing Systems, pp. 2042–2050 (2014)
14. Xiong, C., Dai, Z., Callan, J., Liu, Z., Power, R.: End-to-end neural ad-hoc ranking with kernel pooling. In: Proceedings of the 40th International ACM SIGIR Conference on Research and Development in Information Retrieval, pp. 55–64. ACM (2017)
15. Järvelin, K., Kekäläinen, J.: Cumulated gain-based evaluation of IR techniques. ACM Trans. Inf. Syst. (TOIS) **20**, 422–446 (2002)

16. Wan, S., Lan, Y., Guo, J., Xu, J., Pang, L., Cheng, X.: A deep architecture for semantic matching with multiple positional sentence representations. In: AAAI, pp. 2835–2841 (2016)
17. Socher, R., Huang, E.H., Pennin, J., Manning, C.D., Ng, A.Y.: Dynamic pooling and unfolding recursive autoencoders for paraphrase detection. In: Advances in Neural Information Processing Systems, pp. 801–809 (2011)
18. Pang, L., Lan, Y., Guo, J., Xu, J., Xu, J., Cheng, X.: DeepRank: a new deep architecture for relevance ranking in information retrieval. In: Proceedings of the 2017 ACM on Conference on Information and Knowledge Management, pp. 257–266. ACM (2017)
19. Pang, L., Lan, Y., Guo, J., Xu, J., Cheng, X.: A deep investigation of deep ir models. arXiv preprint arXiv:1707.07700 (2017)

# Accounting Results Modelling with Neural Networks: The Case of an International Oil and Gas Company

Yang Duan[✉], Chung-Hsing Yeh, and David L. Dowe

Faculty of Information Technology, Monash University, Clayton,
VIC 3800, Australia
yang.duan@monash.edu

**Abstract.** Accounting results are crucial information closely monitored by managers, investors and government agencies for decision making. Understanding various endogenous and exogenous business factors affecting accounting results is an essential step in managing them. However, how to model the relationship between accounting results and their business factor antecedents remains an unresolved issue. To address this issue, this paper develops neural network (NN) models for modelling complex interactions between the business factors and accounting results. Based on empirical data from an international leading oil and gas company, 15 original data points, 8 inputs and 6 outputs are used, and 4 NN architectures in 2 training settings are tested. The experiments conducted show satisfactory results. Comparisons of various training settings suggest that a recurrent NN architecture with multiple outputs is best suited for accounting results modelling. The relative contribution factor analysis with the best-performing NN model provides new insights in understanding crucial business factors for the case company and accounting professionals to manage accounting results. As a pilot study, this paper contributes to business, accounting and finance research by providing a promising approach for accounting results modelling.

**Keywords:** Neural network modelling · Accounting results · Business factors
Oil and gas company

## 1 Introduction

Accounting results, namely accounting information presented on the statements of the balance sheet, the profit and loss, and the cash flow, are critical information to managers, investors and the government agencies. For public traded companies, published accounting results have significant impacts on their share prices and influences on their ability in funding future operations. Many studies in management, finance and economics use accounting results for performance measurements, asset pricing, bankruptcy prediction and tax fraud detection.

The modern business operates in a complex environment that exogenous factors would have the economic consequences to its accounting results. These exogenous factors have been neglected in traditional accounting research and their complex

© Springer Nature Switzerland AG 2018
L. Cheng et al. (Eds.): ICONIP 2018, LNCS 11302, pp. 275–285, 2018.
https://doi.org/10.1007/978-3-030-04179-3_24

non-linear relationships with the accounting results are also overlooked due to the use of traditional statistical models [1]. Numerous endogenous influential factors, such as operational settings, managerial preferences and financial settings, have been well investigated. However, the interactive effects among these endogenous factors and the effects caused by exogenous factors from the corresponding industry, financial market and macro economy are not considered. Complex and fast-changing business environments make accounting results projection an unresolved issue in business, accounting and finance research.

As a pilot study to address this challenging issue, this paper develops a neural network (NN) approach by considering both endogenous and exogenous factors relevant to the selected accounting results for modelling. With empirical data from a leading multinational integrated oil and gas company, we experiment with 4 different NN architectures and training techniques to demonstrate the feasibility of NN modelling of accounting results. The results of the experiments conducted provide new insights in managing accounting results for the case company and general accounting professionals.

In subsequent sections, we first briefly review the accounting result related studies and NN modelling in Sect. 2. We next present our NN modelling approach with application to the case company in Sect. 3. We then discuss the results and implications of this study in Sect. 4. Finally, we conclude and summarize this study in Sect. 5.

## 2 Related Works

Accounting is a systematic and comprehensive approach to measuring, processing, recording and communicating companies' operation in financial information. Accounting results are crucial information and widely used in business decision making and academic research, such as contracting, performance evaluation, investment decision making, share price prediction, bankruptcy prediction and accounting information quality [1–4].

Accounting results are categorized and reorganized financial information of the business operation. The relationships between accounting results and business operation have been investigated [2, 5]. Companies have the discretion in choosing suitable accounting process for its business operation to achieve better quality accounting results.

Current accounting research examines real-world accounting practice and connects accounting events with economic consequences, which is known as positive accounting research. Mixed views have been developed. Existing studies have concluded that accounting results are the true reflection of the business operation, with empirical data of 260,000 observations over 50 years [6]. Similar conclusions are reached by other studies using accounting results for research and development (R&D) projects and initial public offering (IPO) respectively [7, 8]. By contrast, another view believes that accounting results are driven by the financial demands for meeting or beating analysts' expectations and senior management personal benefits [1, 3, 9, 10]. For instance, accounting choices determine how business operations to be recorded in financial information. As discretionary decisions, they have often been used to purposefully affect accounting results to smooth or boost accounting results for a variety of reasons, such as personal benefits [11–13], draw funding [14, 15] and tax saving [1, 5, 16, 17]. For instance, senior management would influence the accounting results through the deliberate selection of accounting choice to serve their personal needs, such as bonus and tenure. Therefore, it is believed that we only have limited understanding of the accounting results and the process producing them [18]. Further, conflicting theories cannot effectively support business decision making.

In modelling complex problems, such as business and financial decision making, neural networks (NNs) have been proven to be effective and promising [4]. It requires less theoretical understanding of the problem and high accuracy in performance [19]. Difference NN architectures suit different problems [20]. Selection of relevant features, sufficient data, appropriate NN architectures and limiting data fitting problems are essential to successful NN applications [21, 22].

Despite their wide applications, NNs have not been widely applied in the context of accounting yet. One study that uses NNs in forecasting quarterly accounting earning has yielded less satisfactory results in comparison with linear time series models [23]. The study uses a one-layer feedforward NN model to predict the current period based on the previous 4 periods of quarterly earnings.

## 3   Neural Network Modelling of Accounting Results

### 3.1   Accounting Results of the Case Company

The case company, Eni S. p. A., is a leading multinational integrated oil and gas company with 50 billion euros of market capital. It has a complex capital structure that funds multi-section operations, including activities in upstream and mid-downstream. Debt, corporate bond and shares are the main sources of its funding. Its operation spreads among oil and natural gas exploration, field development and production, supplying, trading and shipping of natural gas, liquid natural gas (LNG), electricity, fuels and chemical products worldwide. In addition, the business model, business strategy and corporate governance policies, produced by executives, guide its daily operation, decision making, performance control and improvement.

As Eni operates globally, it is also exposed to the influences of macroeconomic factors, such as oil commodity price, currency exchange rates and interest rates. Those factors are identified as risks and can cause significant fluctuations on the accounting results. For instance, sale prices of crude oil and refinery products are considerably affected by oil future prices, as the mechanism of the oil market is highly integrated with the future market. Thereby, operation revenue and profit are to be affected. Additionally, the effects of currency exchange rates and interest rates would impact the revenue and expense when involving foreign currencies and debts.

Thus, Eni's accounting results are the results of its operation under the influences of several endogenous and exogenous factors, such as capital structure, production, sales, management preference and financial market impacts.

## 3.2 Selection of Accounting Results as NN Outputs

To determine the most crucial accounting items for modelling, we review a number of Eni's key performance indicators (KPIs) and financial ratios. We choose one set of commonly used financial ratios that are regarded as the basis for comprehending the business performance and evaluating the value and the profitability of a company, in practice. These financial ratios are return on asset (ROA), return on equity (ROE) and financial leverage. The items used for calculating the ratio are thus selected, which are asset, liability, equity and net profit. Moreover, net sale from the operation and net profit attributed to shareholders are also included since sales made by the operation is the essential part that accounting is recording, and the amount of profit attributed to shareholders determines the level of appeal in investing the company. Each accounting item corresponds to a category of item accumulated, as shown in the following:

- Total asset (R1): the resources owned or controlled that are able to provide future economic benefits.
- Total liability (R2): the financial obligations that the company owes to external entities.
- Total equity (R3): the resources brought by the ownership of the company. It also can be defined as the value of the business after its liabilities.
- Net sale from operation (R4): the value of sales after returns, damaging, missing and discounts from its normal operation.
- Operation profit (R5): the profits generated from its normal business operation.
- Net profit attributed to shareholders (R6): the net profit that has been determined to be distributed to shareholders.

## 3.3 Selection of Business Factors as NN Inputs

Relevant business factors are to be identified for the selected accounting items. Existing accounting literature, Eni's annual reports and industry reports are reviewed and analyzed. Based on the relationships between Eni's operation and accounting items, we classify the identified business factors into four groups, namely, resource-related factors, production-related factors, financial factors, and accounting factors.

First, the resource-related factor is proved oil and gas reserves (B1). Three types of reserves are accounted for, which are liquids (O11), natural gas (O12) and hydrocarbons (O13). They are the most important asset as the source generates future economic benefits. The reserves that can be recovered with a reasonable level of certainty are proved reserves, which might or might not be able to be recovered economically. Thus, not all found proved reserves are assets. Hence, determining whether reserves are to be recorded as asset depends on the project and the conditions of the operation including technology applied, operation and financial efficiency, and managerial preference of control and risk.

Second, the production-related factors are the production volume (B2) and company size (B3). The production volumes of liquids (O21), natural gas (O22) and hydrocarbons (O23) are the values created by the operation. Eni uses its resources to create values for revenue. The number of employee (O31) is considered to approximate the capacity of the production with given other conditions.

Third, the financial factors include the effects brought by company shares (B4), oil commodity (B5), interest rates (B6) and exchange rate (B7). The oil and gas industry is a capital-intensive industry. The share price (O41) and traded share volume (O42) mostly affect the operation funding by investment. The oil future prices (O51) and traded volume (O52) in two world major future markets, the Light Sweet Crude Oil in the ICE West Texas Intermediate future market and the Brent future market, significantly influences the prices of crude oil and refinery products. Eni is located in Europe and uses EURO as the base currency, whereas USD is the primary currency in the international oil and gas market. The interest rates of USD (O61) and EURO (O62) and exchange rate of USD/EURO (O71) can fluctuate their accounting results significantly through USD debts and revenue.

Finally, the accounting factor is the discretionary decision that is related to the accounting process, namely, accounting choice (B8). Despite accounting standards and laws existing to regulate accounting practice, within the accounting process, accounting choice can be made by management discretion. It makes the accounting process more suitable to the circumstance of the company. Eni complies with the International Financial Reporting Standard (IFRS). It has changed one of the critical accounting choices, exploration costing (O81), from Full Cost (FC) to Successful Effort (SE) in January 2016. Its effects on the accounting results are substantial by changing how to record the cost that is related to searching for, acquiring and developing the reserves. FC defers unsuccessful exploration and development costs to be charged to the company as an entity in the future, whereas SE requires unsuccessfully projects to be expensed immediately on a field-by-field basis. Although the accounting results would be the same in the long term, assuming the identical operation result, the FC would report more net income and profit in the early stage of new projects than SE. The effect reverses at a later stage from the information recording perspective without considering other factors.

Therefore, we select 15 business factors to produce 8 NN inputs, shown in Table 1.

**Table 1.** Business factors for NN inputs

| Business factors | NN inputs |
|---|---|
| Reserves volume: liquids (O11), natural gas (O12), hydrocarbons (O13) | B1: Reserve |
| Production volume: liquids (O21), natural gas (O22), hydrocarbons (O23) | B2: Production |
| Number of employees (O31) | B3: Company size |
| Share price (O41), traded volume (O42) | B4: Company share effect |
| Future price (O51), traded volume (O52) | B5: Oil commodity effect |
| Interest rate for USD (O61) and EURO (O62) | B6: Interest rates |
| Exchange rate for USD/EURO (O71) | B7: Exchange rate |
| Exploration costing method FC/SE (O81) | B8: Accounting choice |

### 3.4    Data Collection and Pre-processing

The dataset used for NN training consists of 45 sets of 15 quarterly data points from the 4th quarter of 2005 to the 4th quarter of 2016. They are collected and reorganized from two types of sources. Company relevant data, including operational data and accounting data, are sourced from publicly available Eni's annual reports, Eni's Factbooks and U.S. Securities and Exchange Commission (SEC) 20-F filings. Data for exogenous factors, including shares, features and interest rates are sourced from yahoo finance and the economic research data repository of Federal Reserve Bank of St. Louis (FRED).

Three scaling methods are applied to examine the performance of the NN models constructed as the business factors scale differently. These three methods are (a) standard score normalization, (b) feature scaling normalization and (c) decimal scaling, given respectively as

$$X' = (X - \mu)/\sigma \tag{1}$$

$$X' = (X - X_{min})/(X_{max} - X_{min}) \tag{2}$$

$$X' = X/10^i \tag{3}$$

where $X'$ and $X$ are the normalize data and original data respectively. $\mu$, $\sigma$, $X_{max}$, $X_{min}$ and $10^i$ denote the mean, standard deviation, the maximum of the $X$, minimum of the $X$ and 10 to the $i$th power to bring $X'$ into the range $(-10, 10)$ respectively. Decimal scaling has achieved better results. The following experiments and results are reported based on the decimal scaled data.

### 3.5    NN Architectures and Training

We choose 4 different NN architectures for testing in two settings. First, we construct the NN model with all the inputs and multiple outputs (MO) shown in Fig. 1. Second, we test the NN model with the same inputs but with a single activated output (SO) neuron in training. The 4 NN architectures are

- Single hidden layer feedforward NN with standard backpropagation;
- Two hidden layers feedforward NN with standard backpropagation;
- Recurrent NN with feedback connection from output neurons to input neurons, suiting better to time series data and often being used in predicting financial data; and
- Ward nets with multiple activation functions to detect different features in data.

As shown in Fig. 1, business factors are used to compose the 8 NN inputs for the 6 NN outputs. When training in SO settings, only one neuron is active. To model all 6 selected accounting items, we train 6 separate NN models, each with a single output.

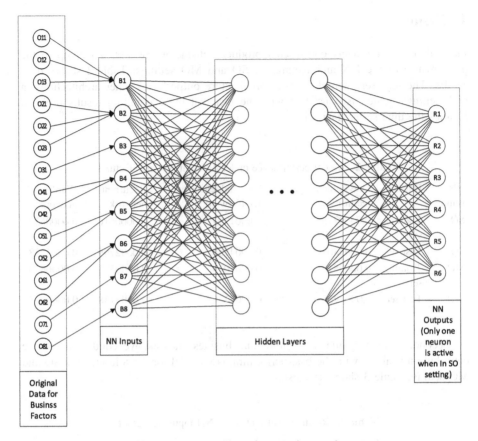

**Fig. 1.** NN modelling of accounting results

With a 10-fold cross-validation procedure, the test dataset is extracted randomly which consists of 20% instances of the total dataset. The performance of the NN models is measured with the minimum value of the mean squared prediction error (MSPE) which is the averaged of the mean squared error between model prediction and actual output over all of training or testing patterns. Different parameters are tested, including the number of neurons for each hidden layer, initial weights, learning rate and updating momentum. With the best performing parameters, we conduct the experiments for all 4 NN architectures in MO and SO settings with 70,000 learning epochs.

## 4   Results

First, all the NN models constructed produce satisfactory results. To compare the performance of the NN architectures in SO and MO settings, Table 2 shows their MAEs. The recurrent neural network architecture outperforms other architectures in both settings, which is consistent with the fact that the accounting results problem features the time series data set.

**Table 2.** Model performance under different NN architectures

| NN setting | Dataset | One hidden layer feedforward | Two hidden layers feedforward | Recurrent network | Ward network |
|---|---|---|---|---|---|
| SO | Training set | 0.0000775 | 0.0001397 | 0.0000616 | 0.0000627 |
| SO | Test set | 0.0003808 | 0.0004755 | 0.0002799 | 0.0003238 |
| MO | Training set | 0.0000815 | 0.0001018 | 0.0000595 | 0.0000630 |
| MO | Test set | 0.0003935 | 0.0004246 | 0.0003138 | 0.0003295 |

Second, to identify the most influential business factors, we conduct a relative contribution analysis with the best-performing NN model for the 8 inputs in a SO and MO setting. Table 3 shows the result.

**Table 3.** Relative contribution of NN inputs to outputs

| NN output | NN setting | NN Input | | | | | | | |
|---|---|---|---|---|---|---|---|---|---|
| | | B1 | B2 | B3 | B4 | B5 | B6 | B7 | B8 |
| R1 | SO | 0.114 | 0.083 | 0.128 | 0.129 | 0.087 | 0.109 | 0.115 | 0.134 |
| | MO | 0.042 | 0.035 | 0.068 | 0.043 | 0.083 | 0.103 | 0.04 | 0.041 |
| R2 | SO | 0.075 | 0.061 | 0.146 | 0.08 | 0.199 | 0.243 | 0.073 | 0.077 |
| | MO | 0.048 | 0.03 | 0.081 | 0.042 | 0.082 | 0.118 | 0.042 | 0.039 |
| R3 | SO | 0.097 | 0.076 | 0.164 | 0.111 | 0.089 | 0.161 | 0.096 | 0.117 |
| | MO | 0.048 | 0.025 | 0.07 | 0.045 | 0.08 | 0.102 | 0.041 | 0.036 |
| R4 | SO | 0.134 | 0.084 | 0.122 | 0.124 | 0.086 | 0.079 | 0.114 | 0.127 |
| | MO | 0.042 | 0.028 | 0.059 | 0.048 | 0.094 | 0.064 | 0.04 | 0.046 |
| R5 | SO | 0.107 | 0.056 | 0.215 | 0.094 | 0.09 | 0.178 | 0.092 | 0.086 |
| | MO | 0.045 | 0.022 | 0.071 | 0.047 | 0.082 | 0.095 | 0.043 | 0.035 |
| R6 | SO | 0.123 | 0.06 | 0.182 | 0.108 | 0.081 | 0.139 | 0.11 | 0.094 |
| | MO | 0.054 | 0.022 | 0.063 | 0.042 | 0.081 | 0.108 | 0.043 | 0.041 |

Table 2 shows mixed results in the relative contribution of the inputs made to individual output. Thus, we consolidate the rankings produced in 6 NNs in SO settings to compare with the ranking in MO settings for further analysis. Table 4 shows the result.

By analyzing individual business factors' contribution to individual accounting items and overall accounting results as shown in Tables 3 and 4, the results produced by MO settings are more consistent with the perceptions of the oil and gas industry and of Eni's analysts and accounting professionals [24]. For instance, as a matured company, Eni's operating profit (R5) is mainly influenced by the total revenue less its variable cost. In addition, for total asset (R1), as Eni explores and develops reserves outside Europe while using EURO as the base currency, the contribution of the exchange rate should surpass company size given Eni's operational settings.

**Table 4.** Consolidated ranking of contributing factors to overall outputs

| NN Setting | Ranking |
|---|---|
| SO | B3 > B6 > B8 > B1 > B4 > B5 > B7 > B2 |
| MO | B6 > B5 > B3 > B1 > B4 > B7 > B8 > B2 |

The performance of NN models in SO and MO settings suggests that SO settings, in more cases, produce better accuracy results with the same NN architecture. However, in SO settings, the contributions of the inputs that are to other inactive output neurons may be accounted for the active output in this case. Thus, the results of relative contribution analysis in SO settings are not consistent with the current understanding of the Eni business and accounting results. Whereas, having similar MSPE, MO settings better reflect the accounting information processing. Hence, we conclude that the MO setting is better suited for modelling accounting results in this case study.

Additionally, based on the best-performing NN model in MO settings, the quantitative relative contribution values shown in Table 5 provide new insights into understanding the relative importance of business factors for managing accounting results. The effects of oil futures (B5) and interest rates (B6) can be almost twice as influential as other NN inputs to the accounting results. Specifically, the oil future price (O51), traded volume (O52) and interest rates of USD (O61) and EURO (O62) are among the most important business factors. This result is consistent with the qualitative conclusions made by Eni and industry reports [24], although the trade volume (O52) was not considered in the reports. Moreover, production (B2) is the least contributing factor, whereas company size (B3) seems to have an unexpected high impact on the accounting results. This result offers new evidence to the inconclusive theories regarding the accounting results, production and company size [1, 25]. Lastly, the accounting choices (B8) is as significant as reserves (B1), company shares (B4) and exchange rate (B7) in terms of its impact on accounting results, which has not been investigated quantitatively in previous accounting studies.

**Table 5.** Relative contribution of NN inputs to overall accounting results

| NN Input | B1 | B2 | B3 | B4 | B5 | B6 | B7 | B8 |
|---|---|---|---|---|---|---|---|---|
| Relative Contribution | 0.28 | 0.16 | 0.41 | 0.27 | 0.50 | 0.59 | 0.25 | 0.24 |

## 5 Conclusion

Accounting results are crucial, and the projection of the accounting results is a challenging issue. To address this issue, we have proposed an NN modelling approach by considering complex interactions between the business factors and accounting results. We have experimented with 4 NN architectures in SO and MO settings and achieved satisfactory performance. The mixed results by SO and MO settings imply interacting effects among accounting items, although NN models with multiple outputs perform better. The experiments conducted show that a recurrent NN model with multiple outputs is the most appropriate NN architecture in terms of its performance and consistency in relevant contribution factor analysis. In addition, this study provides new insights into understanding the business factors that have high impact on the accounting results for the case company. This study offers a promising approach in modelling accounting results which can be used to predict accounting results based on business factors for business, accounting and finance research. The NN modelling approach proposed in this study has general application for other companies with different accounting settings. In future research, we will investigate the correlations between accounting items and apply the NN modelling approach in alternative accounting problems that are significant for decision making for other companies and industries.

## References

1. Fields, T.D., Lys, T.Z., Vincent, L.: Empirical research on accounting choice. J. Account. Econ. **31**, 255–307 (2001)
2. Watts, R.L., Zimmerman, J.L.: Positive accounting theory: a ten year perspective. Account. Rev. **65**, 131–156 (1990)
3. Kim, J.-B., Zhang, L.: Accounting conservatism and stock price crash risk: firm-level evidence. Contemp. Account. Res. **33**, 412–441 (2016)
4. Tkáč, M., Verner, R.: Artificial neural networks in business: two decades of research. Appl. Soft Comput. **38**, 788–804 (2016)
5. Groot, T.d.: Accounting choices of controllers: an insight into controller deliberations. vol. Doctor of Philosophy. Tilburg University. CentER, Center for Economic Research, Tilburg (2015)
6. Dichev, I.D., Li, F.: Growth and accounting choice. Aust. J. Manag. **38**, 221–252 (2013)
7. Aharony, J., Lin, C.-J., Loeb, M.P.: Initial public offerings, accounting choices, and earnings management. Contemp. Account. Res. **10**, 61–81 (1993)
8. Cazavan-Jeny, A., Jeanjean, T., Joos, P.: Accounting choice and future performance: the case of R&D accounting in France. J. Account. Public Policy **30**, 145–165 (2011)

9. Huang, T.-L., Wang, T., Seng, J.-L.: Voluntary accounting changes and analyst following. Int. J. Account. Inf. Manag. **23**, 2–15 (2015)
10. Gietzmann, M., Ireland, J.: Cost of capital, strategic disclosures and accounting choice. J. Bus. Financ. Account. **32**, 599–634 (2005)
11. Watts, R.L., Zimmerman, J.L.: Towards a positive theory of the determination of accounting standards. Account. Rev. **53**, 112–134 (1978)
12. Zhang, W.: CEO Tenure and Aggressive Accounting, vol. 3421490, p. 63. The University of Texas at Dallas, Ann Arbor (2010)
13. Balsam, S.: Discretionary accounting choices and CEO compensation. Contemp. Account. Res. **15**, 229–252 (1998)
14. Oler, M.: Determinants of the length of time a firm's book-to-market ratio is greater than one. Rev. Quant. Financ. Acc. **45**, 509–539 (2015)
15. Friedlan, J.M.: Accounting choices of issuers of initial public offerings. Contemp. Account. Res. **11**, 1–31 (1994)
16. Lennox, C., Lisowsky, P., Pittman, J.: Tax aggressiveness and accounting fraud. J. Account. Res. **51**, 739–778 (2013)
17. Hanlon, M., Slemrod, J.: What does tax aggressiveness signal? Evidence from stock price reactions to news about tax shelter involvement. J. Public Econ. **93**, 126–141 (2009)
18. Hirshleifer, D., Kewei, H., Teoh, S.H., Yinglei, Z.: Do investors overvalue firms with bloated balance sheets? J. Account. Econ. **38**, 297–331 (2004)
19. Gardner, M.W., Dorling, S.: Artificial neural networks (the multilayer perceptron)—a review of applications in the atmospheric sciences. Atmos. Environ. **32**, 2627–2636 (1998)
20. Russell, S., Norvig, P.: Artificial Intelligence: A Modern Approach. Prentice-Hall, Egnlewood Cliffs (2003)
21. Hagan, M.T., Demuth, H.B., Beale, M.H.: Neural Network Design. PWS Pub, Boston (1996)
22. Coakley, J.R., Brown, C.E.: Artificial neural networks in accounting and finance: modeling issues. Int. J. Intell. Syst. Account. Financ. Manag. **9**, 119–144 (2000)
23. Callen, J.L., Kwan, C.C.Y., Yip, P.C.Y., Yuan, Y.: Neural network forecasting of quarterly accounting earnings. Int. J. Forecast. **12**, 475–482 (1996)
24. PwC: Financial reporting in the oil and gas industry - International Financial Reporting Standards (2017)
25. Dechow, P.M., Hutton, A.P., Sloan, R.G.: Economic consequences of accounting for stock-based compensation. J. Account. Res. **34**, 1–20 (1996)

# Attention Based Dialogue Context Selection Model

Weidi Xu[1], Yong Ren[2], and Ying Tan[1(✉)]

[1] Key Laboratory of Machine Perception (Ministry of Education)
and Department of Machine Intelligence,
School of Electronics Engineering and Computer Science, Peking University,
Beijing 100871, People's Republic of China
wead_hsu@pku.edu.cn, ytan@pku.edu.cn
[2] Complex Engineered System Lab (CESL), Department of Electronic Engineering,
Tsinghua University, Beijing 100084, People's Republic of China
reny@tsinghua.edu.cn

**Abstract.** The particular phenomena of Information Overload and Conversational Dependency in multi-turn dialogues have brought massive noise for feature learning in existing deep learning models. To solve the problem, the Attention Based Dialogue Context Selection Model (ABDCS) is proposed in this paper. This model uses attention mechanism to extract the relationship between current response utterance and previous utterances. Qualitative and quantitative analysis show that ABDCS is able to choose the semantically related utterances in its dialogue history as context and be robust against the noise.

## 1 Introduction

Dialogue System [1] is computer system that interacts with humans through natural language. It involves many frontier research directions such as natural language understanding, information retrieval, logical reasoning, text generation and speech recognition. In recent years, thanks to the breakthrough in the NLP field, data-driven dialogue system has become the research hotspot.

Dialogue system, according to the way of interaction, can be divided into single-turn and multi-turn dialogue system. Multi-turn dialogue, which are usually seen in the open-domain scenario, is more liberal and usually consists of interwoven interrogation questions. The multi-turn dialogue systems should take long-term historical information into consideration, which requires sophisticating logical reasoning.

This paper considers two major problems in multi-turn dialogue, i.e., the information overload and conversational dependency. The information overload is very common in the conversations among a group of people, e.g., the barrage of Twitch TV [10]. The users have limited capability of processing information in the conversations. Therefore, when the responses are produced quickly, the users will be overwhelmed by numerous utterances and have to perceive the information selectively. Figure 1 illustrates the curve about the rate of response

© Springer Nature Switzerland AG 2018
L. Cheng et al. (Eds.): ICONIP 2018, LNCS 11302, pp. 286–295, 2018.
https://doi.org/10.1007/978-3-030-04179-3_25

and information bandwidth. When the rate is low (on the left), the information can be easily processed and the replies are generated quickly. On the other side, when the information is too heavy to be received by individuals, the rate will be decreased. Due to this phenomenon, the agent should better process the sentences selectively in the multi-turn dialogue.

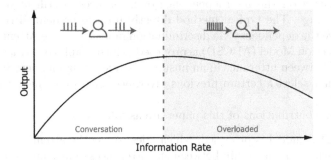

**Fig. 1.** The illustration of information overload [10].

In terms of conversational dependency, [6] and [5] demonstrates that each utterance has a corresponding previous utterance that is most semantically related. The dependencies of the utterances can be used to parse the dialog structure, as shown in Fig. 2.

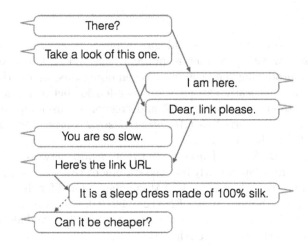

**Fig. 2.** The dependency of utterances [5]

The study of the dependency of utterance is of great significance. The interdependence between utterance helps us to analyze the structure of the dialogue and can be applied to other tasks, e.g., dialogue generation, topic analysis. Especially in conversational tasks, the information in previous utterances plays a crucial role in the generation of statements [14]. By learning the dependencies between the sentences, the relevant sentences in the dialogue can be selected more accurately while ignoring irrelevant information and noise, in order to improve the quality of the dialogue.

The research on the interdependence of utterances is still at a relatively preliminary stage. The typical method directly use the predefined rule [13], or simply derive the dependency via discrimination models [5]. The Attention Based Context Selection Model (ABCSD) is proposed, which is able to extract the interdependency between utterance in an unsupervised learning manner. Specifically, the model first select a certain previous utterance, based on which a new reply is generated.

The main contributions of this paper are as follows:

1. The ABDCS model is proposed, which tries to use the attention mechanism to establish the relationship between the utterances. The comparative experiments on the Chinese-English dialogue dataset show that compared with the existing models, this method outperforms.
2. It is the first time to use unsupervised generative model to obtain the dependencies of the conversational statements. Qualitative analysis shows that the dialog dependencies obtained by this method can effectively extract the key information of dialogues and the parallel structure of the dialogue.

## 2    Model

### 2.1    Framework

When generating statements in a dialogue, a previous utterance that is most informative for generating current response is selected and it is used as the main context in producing the utterance. When doing this, we use the attention mechanism to automatically extract the dependencies between statements. In this work, the "statement", "utterance" and "response" are used exchangeable. The proposed model, referred as (*Attention Based Dialogue Context Selection Model (ABDCS)*) is shown in Fig. 3.

Briefly, the *ABDCS* model uses a recurrent neural network to encode the dialogue statements independently into the statement vector $D$ ($F_{BiLSTM}$ in the figure), and simultaneously use the LSTM language model for dialog generation ($F_{LM}$ in the figure). Similar to *HRED* [12], *ABDCS* also uses the dialog-level recurrent neural network ($F_{RNN}$ in the figure) to describe the dialogue state $R$. The entire model is shown by the Algorithm 1.

**Sentence Encoding via Bidirectional Recurrent Neural Networks.** We use bidirectional LSTM network [3,4,11] to encode the raw input sentences, including a forward LSTM networks $F_{LSTM}^{fw}$ and a backward LSTM networks $F_{LSTM}^{bw}$.

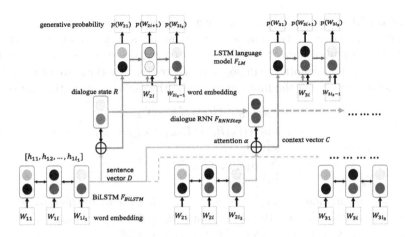

**Fig. 3.** The framework of the *ABDCS*.

---

**Algorithm 1.** *ABDCS* Model

---

**Require:** Dialgoue set $D = \;< S_1, S_2, \cdots, S_{l_D} >$

**Ensure:** Loss $J$

1: Initializes $J = 0$, $t = 1$, and dialog state $R = \mathbf{0}$, $D_0 = \mathbf{0}$;;

2: Encoding   dialog   into   vectors   $< \quad D_0, D_1, D_2, \cdots, D_{l_D} \quad > \xleftarrow{\text{word vector}}_{\text{RNN}} <$
   $S_1, S_2, \cdots, S_{l_D} >$;

3: **repeat**

4:      Calculating attention $\alpha \xleftarrow{\text{Attention mechanism}} (R, < D_0, D_1, \cdots, D_{t-1} >)$;

5:      Calculates the above vector $C \xleftarrow{\text{weighted sum}} (\alpha, < D_0, D_1, \cdots, D_{t-1} >)$;

6:      Calculates the loss of $S_t$ $J'_{S_t} \xleftarrow[\text{language model}]{\text{LSTM}} (C, S_t)$;

7:      Update loss $J = J + J'_{S_t}$;

8:      Updates the dialog state $R \xleftarrow[\text{status update}]{\text{RNN}} (R, C, D_t), t = t + 1$;

9: **until** $t > l_D$

---

Let $W_{i,j}$ denote $j$-th word in the sentence $S_i$, $h_{i,j}^{fw}$ and $h_{i,j}^{bw}$ be the state of forward LSTM and backward LSTM respectively, the vector $D_i$ presenting the entire sentence can be formulated by:

$$[h_{i,1}^{fw}, h_{i,2}^{fw}, \cdots, h_{i,l_{S_i}}^{fw}] = F_{LSTM}^{fw}(S_i), \; h_{i,j}^{fw} \in \mathbb{R}^{U_{rnn}} \tag{1}$$

$$[h_{i,1}^{bw}, h_{i,2}^{bw}, \cdots, h_{i,l_{S_i}}^{bw}] = F_{LSTM}^{bw}(S_i), \; h_{i,j}^{bw} \in \mathbb{R}^{U_{rnn}} \tag{2}$$

$$D_i = [h_{i,l_{S_i}}^{fw}, h_{i,1}^{bw}], \; V_i \in \mathbb{R}^{U_{rnn} \times 2} \tag{3}$$

After the $S_{<t}$ is processed with the bidirectional LSTM, the vectors of the preceding statements are represented as $< D_1, D_2, \cdots, D_{t-1} >$.

**Attention Mechanism.** After encoding each sentence into a vector, the dialogue can be processed in the sentence-level semantic space. Denote the attention weight of each sentence in $S_{<t}$ as the vector $\alpha_t = (\alpha_{t,1}, \alpha_{t,2}, \cdots, \alpha_{t,t-1})$, where $\sum_{k=1}^{t-1} \alpha_{t,k} = 1$, and denote the dialog state $R_t$ to describe the state of the entire dialog before generating the first $t$ statements, the dialogue state transforms as:

$$R_t = F_{RNNStep}(R_{t-1}, D_{t-1}, C_{t-1})$$
$$= \tanh(W_h[R_{t-1}, D_{t-1}, C_{t-1}] + b_h) \tag{4}$$

$$C_t = \sum_{k=1}^{t-1} \alpha_k D_k, \quad C_t \in \mathbb{R}^{U_{rnn} \times 2}, \tag{5}$$

where $C_t$ is the previous vector (*Context Vector*) that the model used to generate the $t$-th statement. We use the dialog state $R_t$ and the preceding statement's $< D_1, D_2, \cdots, D_{t-1} >$ to calculate the attention weight:

$$\phi_{t,i} = F_{att}(R_t, D_i) \tag{6}$$

$$\alpha_{t,i} = \frac{\exp(\phi_{t,i})}{\sum_{k=1}^{t-1} \exp(\phi_{t,i})} \tag{7}$$

$$F_{att}(R_t, D_i) = \begin{cases} V^T \tanh(W[R_t, D_i] + b) & concat \\ R_t^T M D_i & bilinear, \end{cases} \tag{8}$$

where $\phi_{t,i}$ is the attention score and $F_{att}$ is the attention score function.

**Dialogue Generation.** This article uses the LSTM network to generate the dialog statement $S_t$. The LSTM uses the vector $C_t$ as the condition input as well as the preceding $j - 1$ words $w_{t,1}, w_{t,2}, \cdots, w_{t,j-1}$ to predict the probability of generating $j$-th word $w_{t,j}$. Let the implicit state of the LSTM language model be $g_{t,j-1}$, and the entire process is given by the following conditional probabilities:

$$p(w_{t,j}) = p(w_{t,j}|C_t, w_{t,1}, w_{t,2}, \cdots, w_{t,j-1})$$
$$= p(w_{t,j}|C_t, W_{t,j-1}, g_{t,j-1})$$
$$= p(w_{t,j}|g_{t,j}) = softmax(W_V g_{t,j} + b_v), \tag{9}$$

where $W_V \in \mathbb{R}^{|V| \times U_{rnn}}$ and $b_v \in \mathbb{R}^{|V|}$ and $|V|$ represents the vocabulary size.

**Loss Function.** The loss function of the LSTM language model is the log likelihood function of the statement. In the dialog generation task, we optimize the model by minimizing the negative logarithm generation probability of each statement in the conversation.

$$J(D) = \sum_{i=1}^{l_D} -\log p(S_i)$$

$$= \sum_{i=1}^{l_D} \sum_{j=1}^{l_{S_i}} -\log p(w_{i,j}|C_i, W_{i,j-1}, G_{i,j-1}) \tag{10}$$

As every function in the *ABDCS* model is differentiable, the model can be trained by the end-to-end stochastic gradient descent method (*SGD*).

## 3   Experiments

Deep model requires a large corpus. The experiments in this paper are mainly based on the following data sets:

*Ubuntu-Chat*    [9]. The Ubuntu-Chat dataset is a multi-round dialogue data set which focuses on Ubuntu system-related issues. This article uses the NLTK natural language processing package [2] to preprocess the original corpora in the data set. The training data consists of 1.5 million conversations, the size of validation and test sets is 10,000. We use *Ubuntu* to refer to this dataset in the following.

*Ubuntu-Chat-200K.* In order to compare the effects of amount of training samples, 200K conversations were randomly selected from the *Ubuntu* data set as training sets. The preprocessing methods were identical to the *Ubuntu* data sets. Similarly, we use *Ubuntu-200K* to refer to this dataset.

*Baidu Tieba Data.* This article uses crawlers to randomly grab online text from Baidu Tieba and creates an open field Chinese multi-round dialogue data set. We use *Tieba* to refer to this data set. Finally, training data with a data volume of 2 million conversations and validation/test sets with a size of 10,000 were obtained.

**Table 1.** Statistical summary of *Ubuntu* and *Tieba*

| Data sets | Fields | Training sets | Verification/ Test sets | Means rounds | Mean words | Total sentences | Total word counts | Lexword size |
|---|---|---|---|---|---|---|---|---|
| *Ubuntu* | Computers | 1.5 Million | 1 Million | 6.9 | About 70 | 103.780 Million | 1.05 Billion | 21157 |
| *Tieba* | Open | 2 Million | 1 Million | 5.3 | About 45 | 1063.3 Million | 9047.8 Million | 28777 |

The detailed statistics of the data set are summarized in the Table 1.

### 3.1   Evaluation Metrics and Results

**Perplexity.** Perplexity (*Perplexity, PPL*) is a common index for evaluating natural language generation models. Its basic definition is:

$$PPL(S) = \exp\left[\frac{1}{n_S} \cdot \sum_{w_i \in S} -\log p(w_i)\right], \tag{11}$$

where sentence $S$ has $n_S$ words. In the dialogue generation task, in order to describe the generation of the dialogue from different perspectives, we slightly adjust the calculation of the confusion degree and define it in the form of (12) and (13).

$$PPL(D) = \exp\left[\frac{1}{n_D} \cdot \sum_{S_i \in D} \sum_{w_{i,j} \in S_i} -\log p(w_{i,j})\right] \qquad (12)$$

$$PPL@L(D) = \exp\left[\frac{1}{n_{S_{|D|}}} \cdot \sum_{w_j \in S_{|D|}} -\log p(W_j)\right], \qquad (13)$$

where $C$ represents a complete conversation containing a total of $n_D$ words, and $S_{|D|}$ represents the last statement in conversation $D$, containing $n_{S_{|D|}}$ words. The metric (12) is used to calculate the perplexity of the entire dialogue, reflecting the overall situation of the dialogue; The formula 13 calculates the perplexity of the last reply. The results are summarized in Tables 2, 3 and 4. ABDCS-WA is the ABDCS model with the word-level attention mechanism.

**Table 2.** *Ubuntu* dataset perplexity (*PPL*) results on validation sets/test sets

| Ubuntu | PPL | PPL@L |
|---|---|---|
| LSTM | 43.30/44.97 | 43.15/44.66 |
| HRED | 43.82/44.08 | 44.02/44.81 |
| ABDCS | 42.53/43.29 | 42.74/43.51 |
| ABDCS-WA | **41.79/42.14** | **41.92/42.27** |

**Table 3.** *Ubuntu-200K* dataset perplexity (*PPL*) results on validation sets/test sets

| Ubuntu-200K | PPL | PPL@L |
|---|---|---|
| LSTM | 49.89/51.35 | 49.12/50.88 |
| HRED | 48.53/49.92 | 48.91/50.57 |
| ABDCS | 47.83/49.15 | 47.85/49.45 |
| ABDCS-WA | **46.40/47.56** | **46.59/47.82** |

It should be noted that this article uses the perplexity *PPL* as the only criteria for evaluating the optimal model. We used the early stop strategy [7] in the training to avoid overfitting.

**Table 4.** *Tieba* dataset perplexity (*PPL*) results on validation set/test set

| Model | PPL | PPL@L |
|---|---|---|
| LSTM | 123.3/123.7 | 123.8/124.0 |
| HRED | 125.1/126.0 | 126.7/127.2 |
| ABDCS | 120.5/120.8 | 121.2/121.3 |
| ABDCS-WA | **116.6/117.1** | **116.9/117.4** |

**Table 5.** The word error rate (*WER*) of the *Ubuntu* data set on the validation set/test set

| Ubuntu | WER % | WER@L % |
|---|---|---|
| LSTM | 67.85/68.02 | 67.57/67.91 |
| HRED | 66.45/66.79 | 66.32/66.65 |
| ABDCS | 66.04/66.49 | 65.97/66.31 |
| ABDCS-WA | **65.25/65.71** | **65.22/65.58** |

**Word Error Rate.** The word error rate (*WER*) [8] is also one of the commonly used indicators for evaluating natural language generation models. Its basic definition is as follows:

$$WER(S) = 1 - \frac{n_D}{n_S}, \tag{14}$$

where $S$ is a sentence containing $n_S$ words, and $n_D$ represents the number of correctly predicted words. The WER of entire dialogue and the WER of last reply is defined as (15) and (16).

$$WER(D) = \frac{\sum_{S_i \in D} \sum_{w_{i,j} \in S_i} \#[w_{i,j} \neq \arg max(p(w|w_{i,<j}))]}{n_D} \tag{15}$$

$$WER@L(D) = \frac{\sum_{w_j \in S_{|D|}} \#[w_j \neq \arg max(p(w|w_{<j}))]}{n_{S_{|D|}}} \tag{16}$$

The word error rate results are summarized in Tables 5, 6 and 7.

**Table 6.** The word error rate (*WER*) of the *Ubuntu-200K* data set on the validation set/test set

| Ubuntu-200K | WER % | WER@L % |
|---|---|---|
| LSTM | 69.58/70.05 | 69.36/69.87 |
| HRED | 68.12/68.22 | 68.03/68.16 |
| ABDCS | 67.21/67.65 | 67.29/67.83 |
| ABDCS-WA | **66.87/67.24** | **66.95/67.34** |

**Table 7.** *Tieba* dataset word error rate (*WER*) results on validation set/test set

| Model | WER % | WER@L % |
|---|---|---|
| LSTM | 74.89/75.11 | 75.01/75.59 |
| HRED | 75.68/76.42 | 75.86/76.78 |
| ABDCS | 73.54/73.65 | 73.67/73.83 |
| ABDCS-WA | **72.26/72.61** | **72.42/72.90** |

## 4    Conclusion

Overall, this article presents the attention-based models *ABDCS* compared to the baseline models *LSTM* and *HRED*. There is a clear improvement in the results of perplexity and word error rate. Experimental results prove that the attention mechanism is very effective in dialog generation.

**Acknowledgment.** This work was supported by the Natural Science Foundation of China (NSFC) under grant no. 61673025 and 61375119 and Supported by Beijing Natural Science Foundation (4162029), and partially supported by National Key Basic Research Development Plan (973 Plan) Project of China under grant no. 2015CB352302.

## References

1. Arora, S., Batra, K., Singh, S.: Dialogue system: a brief review. CoRR abs/1306.4134 (2013). http://arxiv.org/abs/1306.4134
2. Bird, S.: NLTK: the natural language toolkit. In: Proceedings of the ACL Workshop on Effective Tools and Methodologies for Teaching Natural Language Processing and Computational Linguistics. Association for Computational Linguistics, Philadelphia (2002)
3. Cheng, L., et al.: Recurrent neural network for non-smooth convex optimization problems with application to the identification of genetic regulatory networks. IEEE Trans. Neural Netw. **22**(5), 714–726 (2011)
4. Czajkowski, K., Fitzgerald, S., Foster, I., Kesselman, C.: Grid information services for distributed resource sharing. In: 10th IEEE International Symposium on High Performance Distributed Computing. Proceedings, pp. 181–194. IEEE (2001)

5. Du, W., Poupart, P., Xu, W.: Discovering conversational dependencies between messages in dialogs. CoRR abs/1612.02801 (2016). http://arxiv.org/abs/1612.02801

6. Elsner, M., Charniak, E.: Disentangling chat with local coherence models. In: The 49th Annual Meeting of the Association for Computational Linguistics: Human Language Technologies, Proceedings of the Conference, Portland, Oregon, USA, 19–24 June 2011, pp. 1179–1189 (2011). http://www.aclweb.org/anthology/P11-1118

7. Hansen, L.K., Larsen, J., Fog, T.: Early stop criterion from the bootstrap ensemble. In: 1997 IEEE International Conference on Acoustics, Speech, and Signal Processing, ICASSP 1997, Munich, Germany, 21–24 April 1997, pp. 3205–3208 (1997). https://doi.org/10.1109/ICASSP.1997.595474

8. Klakow, D., Peters, J.: Testing the correlation of word error rate and perplexity. Speech Commun. **38**(1–2), 19–28 (2002). https://doi.org/10.1016/S0167-6393(01)00041-3

9. Lowe, R., Pow, N., Serban, I., Pineau, J.: The Ubuntu dialogue corpus: a large dataset for research in unstructured multi-turn dialogue systems. CoRR abs/1506.08909 (2015). http://arxiv.org/abs/1506.08909

10. Nematzadeh, A., Ciampaglia, G.L., Ahn, Y., Flammini, A.: Information overload in group communication: from conversation to cacophony in the twitch chat. CoRR abs/1610.06497 (2016). http://arxiv.org/abs/1610.06497

11. Schuster, M., Paliwal, K.K.: Bidirectional recurrent neural networks. IEEE Trans. Signal Process. **45**(11), 2673–2681 (1997). https://doi.org/10.1109/78.650093

12. Serban, I.V., Sordoni, A., Bengio, Y., Courville, A.C., Pineau, J.: Building end-to-end dialogue systems using generative hierarchical neural network models. In: Proceedings of the Thirtieth AAAI Conference on Artificial Intelligence, Phoenix, Arizona, USA, 12–17 February 2016, pp. 3776–3784 (2016). http://www.aaai.org/ocs/index.php/AAAI/AAAI16/paper/view/11957

13. Shen, D., Yang, Q., Sun, J., Chen, Z.: Thread detection in dynamic text message streams. In: SIGIR 2006: Proceedings of the 29th Annual International ACM SIGIR Conference on Research and Development in Information Retrieval, Seattle, Washington, USA, 6–11 August 2006, pp. 35–42 (2006). http://doi.acm.org/10.1145/1148170.1148180

14. Sordoni, A., et al.: A neural network approach to context-sensitive generation of conversational responses. In: NAACL HLT 2015, The 2015 Conference of the North American Chapter of the Association for Computational Linguistics: Human Language Technologies, Denver, Colorado, USA, May 31–June 5 2015, pp. 196–205 (2015). http://aclweb.org/anthology/N/N15/N15-1020.pdf

# Memory-Based Model with Multiple Attentions for Multi-turn Response Selection

Xingwu Lu[1], Man Lan[1,2(✉)], and Yuanbin Wu[1,2(✉)]

[1] School of Computer Science and Software Engineering,
East China Normal University, Shanghai 200062, People's Republic of China
51174506023@stu.ecnu.edu.cn, {mlan,ybwu}@cs.ecnu.edu.cn
[2] Shanghai Key Laboratory of Multidimensional Information Processing,
Shanghai, China

**Abstract.** In this paper, we study the task of multi-turn response selection in retrieval-based dialogue systems. Previous approaches focus on matching response with utterances in the context to distill important matching information, and modeling sequential relationship among utterances. This kind of approaches do not take into account the position relationship and inner semantic relevance between utterances and query (i.e., the last utterance). We propose a memory-based network (MBN) to build the effective memory integrating position relationship and inner semantic relevance between utterances and query. Then we adopt multiple attentions on the memory to learn representations of context with multiple levels, which is similar to the behavior of human that repetitively think before response. Experimental results on a public data set for multi-turn response selection show the effectiveness of our MBN model.

**Keywords:** Multi-turn conversation · Response selection
Neural networks · Memory network

## 1 Introduction

Building an open domain dialogue system has always been a challenge in natural language processing (NLP). There are two ways to build a dialog system, including the methods of generation-based [13–15,17] and retrieval-based [4,9,11,20,22]. The retrieval-based methods have been successfully applied in industry because they can ensure that the selected responses are fluent. Herein, we study the retrieval-based dialogue systems.

In existing work, the dialogue systems have the scenarios of single-turn and multi-turn. Single-turn conversation only needs to take the query and response into account. In contrast, multi-turn dialogue further needs to consider previous utterances in the context. The task of multi-turn dialogue has two major challenges: (1) how to identify the important semantic information in the context

© Springer Nature Switzerland AG 2018
L. Cheng et al. (Eds.): ICONIP 2018, LNCS 11302, pp. 296–307, 2018.
https://doi.org/10.1007/978-3-030-04179-3_26

that is useful for response selection; (2) how should we model the relationship among historical utterances.

Recently, there have been some stages of progress in addressing these two challenges. Different researchers have proposed different solutions including dual LSTM [9], Multi-View LSTM [22], and Sequential Matching Network (SMN) [20]. Among them, SMN uses the convolutional neural network (CNN) [6] to match each utterance in historical conversation with the response on two levels (i.e., word level and sentence level) and uses a recurrent neural network to model sequential relationship among historical conversation. SMN considers the solutions of the above two challenges and seems to be a great success in multi-turn response selection, but it still has insufficiency of coherence and relevance.

On the one hand, previous work only considers the sequential relationship between utterances in the context. They ignore the position relationship between utterances and the query, as well as the inner semantic relevance. Hence, we build the effective memory which considers the position relationship and inner semantic relevance between utterances and query. On the other hand, inspired by the behavior of human that think about before response, we apply the multiple attentions on the memory, which allows the neural network to learn representations of context with multiple levels of abstraction. It also has the advantage of using global contextual information to capture implied context features. In other words, it allows neural network to review context information to deepen understanding.

In this work, we propose a new network structure called memory-based network (MBN). Specifically, our framework first obtains matching vectors based on [20]. Then we adopt matching vectors to build the memory weighted by position relationship and semantic relevance between utterances and query. Next, we obtain the presentation of context read from the memory by multiple attentions, and feed the results of attention into a Long Short-term Memory (LSTM) [3] networks. At last, we apply softmax function on the output of the LSTM to predict the final matching score.

The contributions of this paper are reflected in the following three points: (1) building an effective memory integrating position relationship and inner semantic relevance between utterances and query; (2) inspired by the behavior of humans that think about before response, we adopt multiple attentions on the memory to learn representations of context with multiple levels; (3) we validate the effectiveness of the proposed memory-based network model on a public corpus provided by NLPCC 2018.

## 2   Related Work

Together with the rapid growth of social conversation data on Internet, building a chatbot on open domain conversation with data-driven approaches has drawn significant attention. Researchers have proposed both generative-based methods [13–15,17] and retrieval-based methods [4,9,11,20,22] for open domain chatbots. Generative-based chatbots synthesize a response with natural language generation techniques and retrieval-based chatbots select a response from a pre-built conversation repository. Retrieval-based methods have advantages of selecting fluent and informative responses, and have been successfully applied to many real products such as the XiaoIce [16] from Microsoft and AliMe [7] from Alibaba Group.

Research on retrieval-based dialogue systems has the scenarios of single-turn and multi-turn. Most early studies focus on response selection for single-turn conversation. [11] propose a matching model with Deep Neural Network (DNN) for response selection. [4] improve the performance using CNN. [18] further extract dependency tree matching patterns as inputs of a DNN for context-response matching. However, those models built for single-turn response selection only consider the last utterance for matching response, which makes it difficult to be extended to the tasks of multi-turn response selection. The multi-turn response selection takes not only the query but also the utterances in previous turns into consideration to select a response that is relevant and coherent to the entire context.

In recent years, researchers have begun to focus on multi-turn conversation. [9] match a response with the concatenation of context utterances. [22] propose a multi-view model from the view of sentence and word to process this task. [20] further extract important matching information on word and segment levels with a CNN and model the sequential relationship among utterances with gated recurrent unit (GRU) [2].

Although the previous work has considered the sequential relationship among utterances, they ignore the position relationship and inner semantic relevance between utterances and query. We propose a memory-based network (MBN) for multi-turn conversation in retrieval-based chatbots. MBN builds effective memory integrating position relationship and inner semantic relevance between utterances and query. Furthermore, inspired by the behavior of human that repetitively think before response, we apply multiple attentions to learn representations of context with multiple levels of abstraction. The effectiveness of multiple-attention mechanism has been proven in many tasks such as dependency parsing [21], sentiment analysis [1] and coherence modeling [8].

Our one most recent work published in NLPCC [10] is closely related to this work. The difference between [10] and this work lies in three: (1) We use single model instead of the ensemble of multiple module. (2) Instead of using human-designed features, we encourage our model to learn the matching information between response and utterances. (3) We design more effective memory integrating the position relationship and inner semantic relevance between utterances and query.

## 3    The Approach

### 3.1    Model Overview

Figure 1 gives an overview of our model, which generally can be divided into three layers. The first layer is matching vectors layer, which follows the work of [20]. We design this layer to capture important semantic structures among the utterances in context. To build effective memory considering position relationship and inner semantic relevance between utterances and query, we design the memory building layer. In order to capture the representations of context with multiple levels of abstraction, we design the multiple attentions layer. Finally, we apply softmax on the output of the multiple attentions layer to predict the matching score for each response candidate.

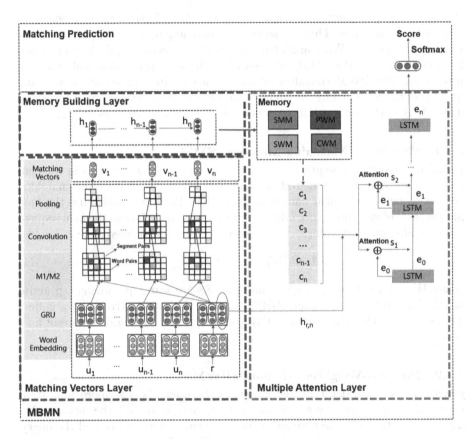

**Fig. 1.** Overview of our system architecture. The dotted lines indicate a memory is alternative.

## 3.2   Matching Vectors Layer

This layer is designed to capture and distill the important semantic structures among the utterances. We follow [20] and design the matching vectors layer. Firstly, we input the word embedding vectors of utterances and response to the GRU, named utterance GRU and response GRU, to get the hidden vectors. Then, we use word embedding vectors and hidden vectors to match response candidates with utterances in the context on different granularities. Finally, we leverage convolution with max-pooling operations to capture the important semantic features. Output of this layer is matching vector $[v_1, \ldots, v_n]$. The matching vectors are used directly to build memory in [10], however it ignores the sequential relationship among the utterances in the context.

### 3.3   Memory Building Layer

We design three different methods to explore the effectiveness of memory in memory building layer. They are sequential matching memory (SMM), position-weighted memory (PWM) and similarity-weighted memory (SWM). Their differences lie in that the SMM only considers the sequential relationship among utterances. The PWM considers position relationship between utterances and query and the SWM considers inner semantic relevance between utterances and query. It is worth noting that we denote all the memory slices as $[c_1, \ldots, c_n]$.

#### 3.3.1   Sequential Matching Memory (SMM)

In order to model the sequential relationship among the utterances in the context, we use a recurrent neural network with GRU to process the matching vectors sequentially, which named as final GRU. Besides, we also use the final hidden state of each utterance GRU to build the memory. Then, the final sequential matching memory is $[c_1, \ldots, c_n]$ where $c_i = t_i$, $t_i$ is defined as:

$$t_i = tanh(W_{1,1}h_{u_i,n} + W_{1,2}h_i + b_i), \tag{1}$$

where $W_{1,1} \in \mathbb{R}^{q \times p}$, $W_{1,2} \in \mathbb{R}^{q \times q}$ and $b_1 \in \mathbb{R}^q$ are learnt parameters, p and q are the hidden size of utterance GRU and final GRU respectively. $h_{u_i,n}$ is final hidden state of the $i$-th utterance GRU, and $h_i$ is the $i$-th hidden state of final GRU.

#### 3.3.2   Position-Weighted Memory (PWM)

Intuitively, we consider that the closer to the query is, the more important its utterance is. Hence we adopt an intuitive method to edit the memory. We define the distance as the number of turn between the utterance and the query. Precisely, the weight for the utterance at position $i$ is calculated as:

$$w_{p_i} = 1 - \frac{|i - N|}{N}, \tag{2}$$

where $N$ is truncation length of the dialogue utterances. Then, the final position-weighted memory is $[c_1, \ldots, c_n]$ where $c_i = (w_{p_i} \cdot t_i)$, $t_i$ is the vector in sequential matching memory.

#### 3.3.3   Similarity-Weighted Memory (SWM)

Usually the utterances in the conversation are short, so the relevant information for response is not fully contained in the query, but appears in the context utterances. Context utterances related to query can be useful, and irrelevant utterances can lead to more information about noise. Hence, we compute the semantic similarity between the context and query by the cosine measure:

$$\delta_{s_i} = sim(h_{u_i,n}, h_{u_n,n}) = \frac{h_{u_i,n} \cdot h_{u_n,n}}{\| h_{u_i,n} \| \cdot \| h_{u_n,n} \|}, \tag{3}$$

where $h_{u_i,n}$ is the final hidden state of the $i$-th utterance GRU. Then, we normalize the semantic similarity and obtain the weight with softmax function:

$$w_{s_i} = \frac{exp(\delta_{s_i})}{\sum_{j=0}^{n} exp(\delta_{s_i})}, \tag{4}$$

the final similarity-weighted memory is $[c_1, \ldots, c_n]$ where $c_i = (w_{s_i} \cdot t_i)$, $t_i$ is the vector in sequential matching memory.

### 3.3.4 Combined-Weighted Memory (CWM)

As shown in Sects. 3.3.2 and 3.3.3, these two types of memory respectively consider the position relationship and inner semantic relevance between utterances and query. Hence, we propose to explicitly combine the position weight and similarity weight by simple scalar multiplication and renormalization. The updated weight is

$$w_{b_i} = \frac{w_{p_i} \times w_{s_i}}{\sum_{j=0}^{n} w_{p_j} \times w_{s_j}}, \tag{5}$$

the final combined-weighted memory is $[c_1, \ldots, c_n]$ where $c_i = (w_{b_i} \cdot t_i)$, $t_i$ is the vector in sequential matching memory.

### 3.4 Multiple Attentions Layer

The memory in second layer respectively considers the sequential relationship among utterances, position relationship and inner semantic relevance between utterances and query. Inspired by the behavior of human that repetitively think before response, we design multiple attentions layer to learn representations of context with multiple levels of abstraction.

On the one hand, we employ multiple attentions to capture important context information from the memory with global context information. On the other hand, we use LSTM to nonlinearly combine the attention results. The process of updating the hidden state $e_t$ of LSTM is shown as follows:

$$i_t = \sigma(W_i s_t + U_i e_{t-1}) \tag{6}$$

$$f_t = \sigma(W_f s_t + U_f e_{t-1}) \tag{7}$$

$$o_t = \sigma(W_o s_t + U_o e_{t-1}) \tag{8}$$

$$g = tanh(W_c s_t + U_c e_{t-1}) \tag{9}$$

$$C_t = f \odot C_{t-1} + i \odot g \tag{10}$$

$$h_t = o_t \odot tanh(C_t) \tag{11}$$

where $e_{t-1}$ denotes the hidden state at time $t-1$ in LSTM and $s_t$ is the current representation of attention results, $W_i, W_f, W_o, W_c \in \mathbb{R}^{h \times c}$, $U_i, U_f, U_o, U_c \in \mathbb{R}^{h \times h}$ are learned in training, h is the hidden size of LSTM and c is the size of memory vector $c_i$, $e_0$ is a zero vector.

In order to get the attended result $s_t$ at time $t$, we first compute the attention weights of each memory slice $c_i$ by:

$$g_i^t = v^T tanh(W_c c_i + W_e e_{t-1} + W_r h_{r,n} + b_{attn}), \tag{12}$$

where $c_i (1 \le i \le N)$ is the memory slice, $e_{t-1}$ is the hidden state at time $t-1$ and $h_{r,n}$ is the final hidden state of the response GRU in the first layer. The parameters of $W_c$, $W_e$, $W_r$ and $b_{attn}$ are learned in training.

Then we use softmax function to normalize the attention score of each memory slice as:

$$\alpha_i^t = \frac{exp(g_i^t)}{\sum_{j=1}^T exp(g_j^t)} \tag{13}$$

Finally, we feed the attended result $s_t$ and the previous hidden state $e_{t-1}$ into LSTM at time t. The content $s_t$ is obtained by dynamically selection and linearly combination of memory slices:

$$s_t = \sum_{i=1}^N \alpha_i^t c_i \tag{14}$$

### 3.5   Matching Prediction

For each memory, we respectively apply multiple attentions to obtain the final representation $e_n$ of context with LSTM. Given a dialogue data set $D$, we denote $u$ as utterances in a dialogue context and $r$ as a response candidate. $y \in \{0, 1\}$ is a binary label, indicating whether $r$ is a proper response for $u$. Our goal is to learn a matching model $g(u, r)$ with $D$, which can measure the relevance between the utterances u and candidate response $r$. We finally compute matching score $g(u, r)$ as follows:

$$g(u, r) = softmax(W_2 e_n + b_2), \tag{15}$$

where $W_2$ and $b_2$ are learnt parameters, and softmax function gives the probability if the candidate response $r$ is proper. The loss function is cross entropy, formulated as:

$$-\sum_{i=1}^{|D|} [y_i log(g(u_i, r_i)) + (1 - y_i)(1 - log(g(u_i, r_i)))] \tag{16}$$

# 4    Experiments

## 4.1    Datasets

We test our model on the public multi-turn response selection dataset provided by NLPCC 2018 Task 5[1]. This dataset is crawled from Sina Weibo[2] which is a Chinese social networking on open-domain topics. The training set contains 5 million conversation sessions, and each session has context, query and one positive reply. The testing set has 10k sessions, where each session is provided with context, query, one positive reply and nine negative replies.

We split the training set as training/validation by 4960,000/40,000 randomly. Then we randomly sample negative replies from the 5 million data with 1:1 positive-negative ratios in training set and 1:9 positive-negative ratios in validation set. We use precision at top 1 among 10 candidataes as evaluation metric.

## 4.2    Model Training

Our model is implemented with Tensorflow. We set the maximum utterances number as 10 and use zero-pad to handle the variable-sized input. We use the same vocabularies, word embedding sizes for all models. Word embeddings are pre-trained using training sets via word2vec [12] and the size is 200. The parameters in matching vectors layer is set as same as the SMN. The hidden size of final GRU in memory building and LSTM in the multiple attentions layer is 200. We report our results with 5 attention cycles because it gains the best scores on validation. We tune our model with Adam optimizer [5] to minimize loss function and the initial learning rate is 0.001. The batch-size is 256. We select the best models on validation set and report their performances on test set.

## 4.3    Experimental Results

### 4.3.1    Comparison with Different Models

In order to verify the effects of the memory-based model, we conduct the corresponding experiments. Table 1 shows the performance of different models on test set. We observe the following findings.

The SMN-last performs best among the three baseline in [20] on the datasets provided by NLPCC 2018. Our models outperform all baselines, which verifies the effects of the memory-based model we proposed.

Among our models, MBN-PWM and MBN-SWM both outperform MBN-SMM, and MBMN-CWM performs the best. This result indicates that one cannot neglect the position relationship and inner semantic relevance between utterances and query. The memory weighted by the position relationship and semantic relevance between utterances and query is helpful for the neural network to learn the representations of context with emphasis.

---

[1] http://tcci.ccf.org.cn/conference/2018/dldoc/taskgline05.pdf.
[2] http://weibo.com/.

Our MBN-SMM model outperforms the SMN-dynamic model, which only employs single attention mechanism to combine the vectors in sequential matching memory (SMM). This indicates that the multiple attentions can learn representations of context with multiple levels of abstraction, which is similar to the behavior of human that repetitively think before response.

**Table 1.** Performance of different models on test set.

| Model | $P@1(\%)$ |
|---|---|
| SMN-last [20] | 61.04 |
| SMN-static [20] | 60.99 |
| SMN-dynamic [20] | 60.10 |
| MBN-SMM | 61.13 |
| MBN-PWM | 62.06 |
| MBN-SWM | 61.85 |
| MBN-CWM | **62.18** |

### 4.3.2 Comparison with Other Results on NLPCC

Table 2 shows the comparison with results of other participants on NLPCC 2018 Task 5. Compared to the top-ranked systems ECNU [10], on the one hand, our model does not use multiple module ensemble, on the other hand, we do not use human-designed features. We design more effective memory structure integrating the position relationship and inner semantic relevance between utterances and query, thus encouraging the model to learn better representations of context through the multiple attentions layer. The performance of model MBN-CWM is comparable to [10], which is promising.

**Table 2.** Comparisons with top systems on test data of NLPCC 2018 Task 5. † mark means the results are announced by NLPCC 2018 task organizer.

| Team ID | $P@1(\%)$ |
|---|---|
| ECNU [10] | 62.61 |
| wyl_buaa [19] | 59.03 |
| YiwiseDS† | 26.68 |
| laiye_rocket† | 18.13 |
| ELCU_NLP† | 10.54 |
| MBN-CWM | 62.18 |

### 4.4   Case Study

We qualitatively analyze our model with some examples from the test data given in Table 3. As shown in example 1, our MBN-CWM model can capture cue context features separated by a long distance, such as the word *"boyfriends"* and *"introduction"*. It indicates that besides position relationship between utterances and query, our model also considers the inner semantic relevance between utterances and query. As shown in example 2, MBN-CWM model can also synthesize cue features in complex context, such as the context with *"a male model in woman's dress"* and *"female model"*. This benefits from the multiple attentions mechanism, which learns representations of context with multiple levels of abstraction.

**Table 3.** Examples of MBN-CWM model compared with SMN-last model.

| Example 1 | **Context** |
|---|---|
| Context 1 | 男朋友也是不错的Boyfriends are good, too. |
| Context 2 | 求介绍Ask for introduction. |
| Context 3 | 讲真啊Tell me the truth. |
| Query | 讲真讲真。I'm telling you the truth. |
| | **Response Candidates** |
| MBN-CWM | 要啥类型的What style would you like? ✓ |
| SMN-last | 愚人节。April fool 's day. ✗ |
| Example 2 | **Context** |
| Context 1 | 所以是男模拍了女装? So it's a photo taken by a male model in woman's dress? |
| Context 2 | 女模啊This is a female model! |
| Context 3 | 这是个女人? Is this a woman? |
| Context 4 | 对啊That's right |
| Query | 听说那天拍的不错! 期待啊I heard it was taken well that day! Looking forward to! |
| | **Response Candidates** |
| MBN-CWM | 带了假发其实是圆寸女孩Wearing a wig, actually a girl with round brush cut. ✓ |
| SMN-last | 不完全是Not exactly ✗ |

## 5   Conclusion

We propose a memory-based model for response selection in multi-turn conversation. We leverage three effective methods to build memory which integrates sequential relationship among utterances, and considers position relationship and inner semantic relevance between utterances and query. We also apply multiple attentions to learn the representations of context with multiple levels of abstraction. The performance of our models exceeds the baselines. In future work, we consider extending the memory structure to generation-based model.

We believe it will help construct a better representation of context and improve the performance.

**Acknowledgements.** This work is supported by the Science and Technology Commission of Shanghai Municipality Grant (No. 15ZR1410700) and the open project of Shanghai Key Laboratory of Trustworthy Computing (No. 07dz22304201604).

# References

1. Chen, P., Sun, Z., Bing, L., Yang, W.: Recurrent attention network on memory for aspect sentiment analysis. In: EMNLP, pp. 452–461 (2017)
2. Chung, J., Gulcehre, C., Cho, K., Bengio, Y.: Empirical evaluation of gated recurrent neural networks on sequence modeling. arXiv preprint arXiv:1412.3555 (2014)
3. Hochreiter, S., Schmidhuber, J.: Long short-term memory. Neural Comput. **9**(8), 1735–1780 (1997)
4. Hu, B., Lu, Z., Li, H., Chen, Q.: Convolutional neural network architectures for matching natural language sentences. In: NIPS, pp. 2042–2050 (2014)
5. Kingma, D.P., Ba, J.: Adam: a method for stochastic optimization. arXiv:1412.6980 (2014)
6. LeCun, Y., et al.: Backpropagation applied to handwritten zip code recognition. Neural Comput. **1**(4), 541–551 (1989)
7. Li, F.L., et al.: AliMe assist: an intelligent assistant for creating an innovative e-commerce experience. In: Proceedings of the 2017 ACM on Conference on Information and Knowledge Management, pp. 2495–2498. ACM (2017)
8. Logeswaran, L., Lee, H., Radev, D.: Sentence ordering and coherence modeling using recurrent neural networks. arXiv:1611.02654 (2018)
9. Lowe, R., Pow, N., Serban, I., Pineau, J.: The Ubuntu dialogue corpus: a large dataset for research in unstructured multi-turn dialogue systems. arXiv:1506.08909 (2015)
10. Lu, X., Lan, M., Wu, Y.: Memory-based matching models for multi-turn response selection in retrieval-based chatbots. In: NLPCC, pp. 294–303 (2018)
11. Lu, Z., Li, H.: A deep architecture for matching short texts. In: NIPS, pp. 1367–1375 (2013)
12. Mikolov, T., Sutskever, I., Chen, K., Corrado, G.S., Dean, J.: Distributed representations of words and phrases and their compositionality. In: NIPS, pp. 3111–3119 (2013)
13. Serban, I.V., Sordoni, A., Bengio, Y., Courville, A.C., Pineau, J.: Building end-to-end dialogue systems using generative hierarchical neural network models. In: AAAI, vol. 16, pp. 3776–3784 (2016)
14. Serban, I.V., et al.: A hierarchical latent variable encoder-decoder model for generating dialogues. In: AAAI, pp. 3295–3301 (2017)
15. Shang, L., Lu, Z., Li, H.: Neural responding machine for short-text conversation. In: ACL, vol. 1, pp. 1577–1586 (2015)
16. Shum, H.Y., He, X., Li, D.: From Eliza to Xiaoice: challenges and opportunities with social chatbots. arXiv preprint arXiv:1801.01957 (2018)
17. Sordoni, A., et al.: A neural network approach to context-sensitive generation of conversational responses. In: NAACL, pp. 196–205 (2015)
18. Wang, M., Lu, Z., Li, H., Liu, Q.: Syntax-based deep matching of short texts. arXiv:1503.02427 (2015)

19. Wang, Y., Yan, Z., Li, Z., Chao, W.: Response selection of multi-turn conversation with deep neural networks. In: NLPCC, pp. 110–119 (2018)
20. Wu, Y., Wu, W., Xing, C., Zhou, M., Li, Z.: Sequential matching network: a new architecture for multi-turn response selection in retrieval-based chatbots. In: ACL, vol. 1, pp. 496–505 (2017)
21. Zhang, Z., Liu, S., Li, M., Zhou, M., Chen, E.: Stack-based multi-layer attention for transition-based dependency parsing. In: EMNLP, pp. 1677–1682 (2017)
22. Zhou, X., et al.: Multi-view response selection for human-computer conversation. In: EMNLP, pp. 372–381 (2016)

# A Robust LPNN Technique for Target Localization Under Hybrid TOA/AOA Measurements

Muideen Adegoke[1]([☒]), Andrew Chi Sing Leung[1], and John Sum[2]

[1] Department of Electronic Engineering, City University of Hong Kong,
Hong Kong, Hong Kong
maadegoke2-c@my.cityu.edu.hk, eeleungc@cityu.edu.hk
[2] Institute of Technology Management, National Chung Hsing University,
Taichung, Taiwan
pfsum@dragon.nchu.edu.tw

**Abstract.** This paper presents an approach based on the Lagrange programming neural network (LPNN) framework for target localization under the outlier situation. The problem is formulated as a minimization problem of the mixture of $l_1$ and $l_2$ norms of the measurement errors with time of arrival (TOA) and angle of arrival (AOA) measurements. In our approach, we introduce an approximation function for the $l_1$ norm to avoid non-differentiable points of the $l_1$ norm. Moreover, we study the network stability of the proposed approach. We carry out simulations with various settings to validate the performance of the presented approach. In addition, the proposed method is compared with the existing linear and nonlinear target localization techniques. Based on the mean square error (MSE) values obtained under different settings, the proposed approach is superior to the other comparison algorithms.

**Keywords:** Target localization · Time-of-arrival · Angle-of-arrival
LPNN · Outliers

## 1 Introduction

Locating a target is an important problem in many areas, such as radar [1], communication [2], sonar [3] and wireless sensor network [4]. It has a lot of applications, including intelligent transports and rescue systems [5,6]. Many techniques have been proposed to solve this problem based on different measurement methods, such as time of arrival (TOA) [7], time difference of arrival (TDOA) [2], angle of arrival(AOA) [8], and frequency difference of arrival (FDOA) [9]. More details on the different aforementioned measurement methods and target localization techniques can be found in a recent survey [10] and the reference therein.

It is worth knowing that combining different types of measurement methods can improve the accuracy of target location estimate, especially in the presence of noise. In practical, the noise may contain outliers or heavily tailed errors

© Springer Nature Switzerland AG 2018
L. Cheng et al. (Eds.): ICONIP 2018, LNCS 11302, pp. 308–320, 2018.
https://doi.org/10.1007/978-3-030-04179-3_27

which can render the performance of one(single) measurement technique useless. Recently, the compressed sensing approach has been proposed [11] by utilizing joint TDOAs and FDOAs measurements. In similar manner, the hybrid approach which involves a combination of TDOAs and FDOAs using multidimensional scaling analysis was presented in [12].

This paper explores a combined approach based on hybrid TOA and AOA measurements. Also, we formulate the problem as an optimization problem with the mixture of $l_1$ and $l_2$ norm errors. We then use the Lagrange programming neural network (LPNN) [13] framework to solve the problem. In order to avoid non-differentiable points of the $l_1$ norm, we use an approximate function for the $l_1$ norm. Generally speaking, using the hybrid measurement method, that considers TOA and AOA measurements, can improve the target position estimation. Furthermore, in compressed sensing problems, using the mixture of $l_1$ and $l_2$ norm errors has been shown to be better than using the $l_1$ norm only [14]. Hence, this paper considers the hybrid approach for the challenge of target localization.

The main contribution of the paper is to develop a robust localization technique that uses hybrid TOA/AOA measurements to handle complicated noise scenarios, including non-line of sight situations. Besides, we utilize the LPNN framework to handle the mixture of $l_1$ and $l_2$ norms based objective function. However, the LPNN framework cannot handle the $l_1$ norm term. Then we devise a distinct approach, namely smooth approximation function to handle it. In addition, we carry out stability analysis for the proposed approach, and perform simulations to validate the proposed approach. From the simulation results, the proposed approach performs well, even under complicated noise situations.

This article is arranged as follows. Section 2 gives the backgrounds on TOA, AOA, and LPNN. Furthermore, Sect. 3 presents the proposed approach and the network stability analysis of the proposed method. Section 4 presents the simulation result. Finally, Sect. 5 presents the concluded remark.

## 2   General Background

This section gives brief backgrounds on TOA, AOA, and LPNN.

### 2.1   Time of Arrival Model

In the TOA model formulation, there are $p$ sensors with known positions and a target with an unknown location in the 2D/3D space. Here we consider the 2D case. It should be noticed that our approach can be easily extended to the 3D scenario. Let $\boldsymbol{u} = [u_1, u_2]^{\mathrm{T}}$ be the unknown coordinates of a mobile source (a target), and let $\boldsymbol{s}_i = [s_{i1}, s_{i2}]^{\mathrm{T}}, \forall\, i = 1, \cdots, p$, be the known coordinates of the $p$ sensors. Let $g_i$ be the distance between the $i$th sensor and the target, given by

$$g_i = \|\boldsymbol{u} - \boldsymbol{s}_i\|_2, \quad i = 1, \cdots, p. \tag{1}$$

The collection of the distances between the target and the sensors is denoted as $\boldsymbol{g} = [g_1, \cdots g_p]^{\mathrm{T}}$.

Consider that the source generates a signal at time $\tau_o$ and the $i$th sensor at the location $s_i$ receives the emitted signal from the source at time $\tau_i$. Hence the signal propagating time between the sensor and the target is given by $\tau_i - \tau_o$. From this propagating time, a measurement distance between the sensor at location $s_i$ and the target location at the unknown position $u$ can be obtained as $d_i = (\tau_i - \tau_o) \times c$, where $c$ is the light speed.

Under the noiseless situation, we have $d_i = g_i = \|u - s_i\|_2, \quad \forall i = 1, \cdots, p$. The concept of the TOA based target position estimation is to utilize the measurement distances $d = [d_i, \cdots, d_p]^T$ to estimate the source(target) position. Given the measurement distances $d_i$'s, for $i = 1, \cdots, p$, with $p \geq 3$, the target coordinates can be easily calculated under the noiseless situation.

The measurement distances $d = [d_i, \cdots, d_p]^T$ usually contain some noise. Therefore, $d$ is not equal to the true distance vector $g$. The relationship between the measurement distances and the true distances can be described as:

$$d = g + \eta, \tag{2}$$

where $\eta = [\eta_1, \cdots, \eta_p]^T$ represents the measurement noise vector. The main task here is to estimate the target position $u$ based on the measurement distance $d$. In order to solve this problem, we usually assume that the noise component $\{\eta_i\}, i = 1, \cdots p$, follows the Gaussian distribution and its mean is equal to zero. Therefore, to mitigate the effect of Gaussian noise, the TOA model objective function is set as:

$$\min_{u} \frac{1}{2} \sum_{i}^{p} (d_i - \|u - s_i\|_2)^2. \tag{3}$$

The traditional gradient based algorithms cannot handle the problem stated in (3) directly due to the occurrence of some term "$(1/\|u - s_i\|_2)$" in the gradient vector. It is obvious that if the target gets closer to any of the sensors positions, then the term $(1/\|u - s_i\|_2)$'s will blow-up. Hence the algorithm becomes unstable. This ill-posed case is addressed in many algorithms such as in [15,16]. The problem is then formulated as a constrained optimization problem, given by

$$\min_{u,g} \frac{1}{2} \|d - g\|_2^2 \quad \text{s.t. } g_i^2 = \|u - s_i\|_2^2 \text{ and } g_i \geq 0. \tag{4}$$

The objective function in (4) is the $l_2$ norm formulation of the error between the $d$ and $g$. It cannot handle outliers and/or heavily tailed errors. Hence, those algorithms in [15,16] are not capable to handle outliers/heavily tailed errors.

## 2.2   Angle of Arrival (AOA) Model

In the AOA case, angles of arrival of signal received from the unknown source co-ordinate is the key measurements. Figure 1 describes the AOA measurement scenario. The $i$th sensor at location $s_i$ received the signal emitted from a source at the unknown location $u$. Let $\phi_i$ be the measured angle of the direction of the signal received by the $i$th sensor, and let $\theta_i$ be the true angle. Practically, the measured angle by the sensor contains some noise, either due to sensors inaccuracy or some disturbances in the non-line of sight environment. The measured angles can be modeled as

$$\boldsymbol{\phi} = \boldsymbol{\theta} + \boldsymbol{\zeta}, \tag{5}$$

where $\boldsymbol{\phi} = [\phi_i, \cdots, \phi_p]^T$, $\boldsymbol{\theta} = [\theta_i, \cdots, \theta_p]^T$, $\boldsymbol{\zeta} = [\zeta_i, \cdots, \zeta_p]^T$, and the vector $\boldsymbol{\zeta}$ is regarded as a vector of noise. It should be noticed that from the geometry shown in Fig. 1, the true angle $\theta_i$ is given by

$$\tan(\theta_i) = \frac{s_{i2} - u_2}{s_{i1} - u_1} \quad \forall\, i = 1, \cdots, p. \tag{6}$$

Under the noiseless situation, using some algebraic manipulation, we have

$$b_i = \boldsymbol{a}_i^T \boldsymbol{u} \quad \forall\, i = 1, \cdots, p, \tag{7}$$

where $b_i = s_{i1} \tan(\theta_i) - s_{i2}$ and $\boldsymbol{a}_i^T = [\tan(\theta_i), -1]$ with $\boldsymbol{u} = [u_1, u_2]$ denotes unknown position of the target.

Under the noisy situation, the angle in $b_i$ and $\boldsymbol{a}_i$ is replaced by the measured angle $\phi_i$, i.e.,

$$b_i = s_{i1} \tan(\phi_i) - s_{i2}, \text{ and } \boldsymbol{a}_i^T = [\tan(\phi_i), -1]. \tag{8}$$

The target position $\boldsymbol{u}$ could be obtained by solving the following optimization problem, given by

$$\min_{\boldsymbol{u}} \quad \frac{1}{2} \sum_{i=1}^{p} (b_i - \boldsymbol{a}_i^T \boldsymbol{u})^2. \tag{9}$$

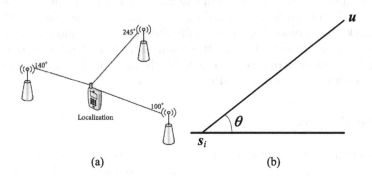

**Fig. 1.** The structure of angle of arrival measurement.

## 2.3  Lagrange Programming Neural Network (LPNN)

The LPNN framework [13,16] is an analog neural network model that is capable of handling general nonlinear optimization problems. The LPNN model aims at solving a constrained optimization problem with equality constraints, given by

$$\min_{\boldsymbol{y}} \quad f(\boldsymbol{y}) \quad \text{s.t} \quad \boldsymbol{h}(\boldsymbol{y}) = \boldsymbol{0}, \tag{10}$$

where $\boldsymbol{y} = [y_1, \cdots, y_{p_y}]^{\mathrm{T}}$ denotes decision variables vector, $f(\boldsymbol{y}) : \mathbb{R}^{p_y} \rightarrow \mathbb{R}$ represents the fitness(objective) function, and $\boldsymbol{h}(\boldsymbol{y}) : \mathbb{R}^{p_y} \rightarrow \mathbb{R}^{p_h}$ is a vector-valued function with known $p_y$ equality constraints. Also, in the LPNN formulation, $p_y \leq p_h$. To solve the problem stated in (10), we first consider the Lagrangian function, given by

$$L(\boldsymbol{y}, \boldsymbol{\lambda}) = f(\boldsymbol{y}) + \boldsymbol{\lambda}^{\mathrm{T}} \boldsymbol{h}(\boldsymbol{y}), \tag{11}$$

where $\boldsymbol{\lambda} = [\lambda_i, \cdots, \lambda_{p_h}]^{\mathrm{T}}$ represents the Lagrange multiplier vector. In the LPNN, there are two types of neurons to hold available variables. There are $p_y$ variable neurons to hold decision variables and $p_h$ Lagrange neurons to hold Lagrange variables. From (11), we can construct the neuron dynamics as

$$\frac{d\boldsymbol{y}}{dt} = -\frac{1}{\upsilon} \frac{\partial L(\boldsymbol{y}, \boldsymbol{\lambda})}{\partial \boldsymbol{y}} \tag{12a}$$

$$\frac{d\boldsymbol{\lambda}}{dt} = \frac{1}{\upsilon} \frac{\partial L(\boldsymbol{y}, \boldsymbol{\lambda})}{\partial \boldsymbol{\lambda}}, \tag{12b}$$

where $\upsilon$ denotes the characteristic time constant which is a function of capacitance and resistance of the neural circuit. Without loss of generality we set $\upsilon = 1$. Equation (12a) is designed to seek for the decision variables (vector $\boldsymbol{y}$). Also, (12b) is the derivative of Lagrangian function with respective Lagrange multiplier which prevents the constraints from being violated. In sum, under mild conditions [13], the network will reach an equilibrium point. At this point, one can obtain the solution by recording or measuring the neuron outputs when the network settles down (i.e. at an equilibrium point). For more details on the LPNN theory one can check [13]. **One vital condition that can make the dynamic of the network to be definable is that both the constraints $h$ and the objective $f$ must be differentiable.**

## 3  The Proposed Approach

In this section, we first introduce the hybrid TOA/AOA localization problem as a constrained optimization problem. Afterwards, we formulate the LPNN model to solve the problem. That is, we introduce an approximation function for the $l_1$ norm to avoid non-differentiable points of the $l_1$ norm. Furthermore, we present the stability analysis.

## 3.1    Problem Formulation

Most existing algorithms for the target localization problem, stated in (4) and (9), assume that the noise in the TOA or AOA measurements are governed by Gaussian distribution. Hence, they formulate the optimization problem as a minimization of the $l_2$ norm of errors between the measured target distance and the estimated ones. However, if the measurements contain outliers and or heavily tailed errors, the $l_2$ norm performance is poor. To mitigate this, we consider the mixture of $l_2$ and $l_1$, we referred to this as $l_{1-2}$ norm. Moreover, our formulation considers hybrid TOA/AOA measurement. Note that in [15, 16], only one kind of measurements is considered. Using this modification we formulate our optimization problem as:

$$\min_{u,c,g} \quad \mathcal{L}_2(u) + \mathcal{L}_1(c,g) \tag{13a}$$

$$\text{s.t} \quad c_i = a_i^T u, \quad i = 1, \cdots, p \tag{13b}$$

$$g_i^2 = \|u - s_i\|_2^2 \text{ and } g_i \geq 0, \tag{13c}$$

where

$$\mathcal{L}_2(u) = 0.5\alpha \sum_{i=1}^{p} (b_i - a_i^T u)^2 + 0.5\beta \sum_{i=1}^{p} (d_i^2 - \|u - s_i\|_2^2)^2, \tag{14}$$

$$\mathcal{L}_1(c,g) = \sum_{i=1}^{p} |b_i - c_i| + \sum_{i=1}^{p} |d_i - g_i|, \tag{15}$$

where $b_i$ and $a_i$ are defined in (9). In addition, $\alpha$ and $\beta$ are positive constant. In this formulation, $\mathcal{L}_1$ is designed to mitigate the effect of outliers/heavily tailed error that can occur such as in non-line of sight situations and $\mathcal{L}_2$ is introduced to stabilize the LPNN.

In order to employ the LPNN framework to solve the aforementioned formulation stated in (13), two issues need to be addressed. First, the formulation in (13c) has inequality constraints but the original LPNN model can accommodate equality constraints only. Hence we need to handle these inequality constraints. In addition, the objective function in (13a) contains the $l_1$ norm which is non-differentiable. That means, the LPNN model cannot be applied directly to solve the formulated problem stated in (13). The first problem can be solved based on Theorem 1, which shows that the inequality constraints in (13c) can be deleted without affecting the LPNN model to reach a solution.

**Theorem 1.** *The problem formulated in (13) is equivalent to:*

$$\min_{u,c,g} \quad \mathcal{L}_2(u) + \mathcal{L}_1(c,g) \quad \text{s.t } c_i = a_i^T u, \ g_i^2 = \|u - s_i\|_2^2 \tag{16}$$

*where $i = 1, \cdots, p$, and $\mathcal{L}_2(u)$ and $\mathcal{L}_1(c,g)$ are as defined before.*

*Proof.* Here, we will justify that if $(u^*, c^*, g^*)$ is an optimal solution of (13), then it is also an optimal solution for (16). That means that the inequality in (16) can be removed without affecting the optimality solution.

Firstly, the objective function can be rewritten as

$$\mathcal{L}_2(\boldsymbol{u}) + \mathcal{L}_1(\boldsymbol{c}, \boldsymbol{g}) = \mathcal{L}_2(\boldsymbol{u}) + \sum_{i=1}^{p} |b_i - c_i| + \sum_{i=1}^{p} |d_i - g_i|. \tag{17}$$

Assume that the optimal solution of the formulation (16) is $(\boldsymbol{u}^*, \boldsymbol{c}^*, \boldsymbol{g}^*)$. Note that if $(\boldsymbol{u}^*, \boldsymbol{c}^*, \boldsymbol{g}^*)$ is the optimal solution, then $(\boldsymbol{u}^*, \text{abs}(\boldsymbol{c}^*), \text{abs}(\boldsymbol{g}^*))$ is a feasible solution (see the constraints in (13c)), where $\text{abs}(\cdot)$ is the absolute operator.

Utilizing the basic algebra, we obtain

$$\sum_{i=1}^{p} |d_i - g_i^*| \geq \sum_{i=1}^{p} |\text{abs}(d_i) - \text{abs}(g_i^*)|. \tag{18}$$

Since the measured distances $d_i$'s are greater than zero, we have

$$\sum_{i=1}^{p} |d_i - g_i^*| \geq \sum_{i=1}^{p} |\text{abs}(d_i) - \text{abs}(g_i^*)| = \sum_{i=1}^{p} |d_i - \text{abs}(g_i^*)| \tag{19}$$

From (19), we have

$$\mathcal{L}_2(\boldsymbol{u}^*) + \mathcal{L}_1(\boldsymbol{c}^*, \boldsymbol{g}^*) \geq \mathcal{L}_2(\boldsymbol{u}^*) + \sum_{i=1}^{p} |b_i - c_i^*| + \sum_{i=1}^{p} |d_i - \text{abs}(g_i^*)|. \tag{20}$$

Inequality (23) means that the optimal objective value "$\mathcal{L}_2(\boldsymbol{u}^*) + \mathcal{L}_1(\boldsymbol{c}^*, \boldsymbol{g}^*)$" is greater than or equal to the objective value "$\mathcal{L}_2(\boldsymbol{u}^*) + \mathcal{L}_1(\boldsymbol{c}^*, \text{abs}(\boldsymbol{g}^*))$" achieved by the feasible solution $(\boldsymbol{u}*, \boldsymbol{c}^*, \text{abs}(\boldsymbol{g}^*))$.[1] Hence the equality in (23) must hold if and only if $g_i^* = \text{abs}(g_i^*)$ for all $i = 1, \cdots, p$. Therefore, the optimal solution should be $g_i \geq 0$ for all $i$. Hence, the problem stated in (16) is equivalent to the formulation stated in (13). This is the end of the proof. ∎

## 3.2   Approximation of $l_1$ Norm

Since the $l_1$ norm term is non-differentiable, the LPNN method cannot be used to handle the corresponding analog circuit directly. In order to resolve this issue, in the objective function (16), we introduce a differentiable approximation function given by (21). Afterwards, we utilize the LPNN model to solve the problem. The approximation of the $l_1$ norm is

$$|z| \approx \rho \, \log(\cosh(z/\rho)), \tag{21}$$

where $\rho$ is a small positive number. It should be noticed that a smaller value of $\rho$ leads to a better approximation to the $l_1$ norm. **One may argue that the approximation is quite complicated and is difficult to implement. In**

---

[1] Note that from the constraints in (13), if $(\boldsymbol{u}*, \boldsymbol{c}^*, \boldsymbol{g}^*)$ is the optimal solution, $(\boldsymbol{u}*, \boldsymbol{c}^*, \text{abs}(\boldsymbol{g}^*))$ is a feasible solution.

fact, the approximation is very interesting, in the sense that we only need to implement the derivative of the approximation function. Its derivative is equal to $(\tanh(z/\rho))$ that is a commonly used activation function in artificial neural networks.

Now, with the introduced approximation function for the $l_1$ norm, we can recast the problem stated in (16) as follows:

$$\min_{u,c,g} \mathcal{L}_2(u) + \sum_{i=1}^p \rho \log(\cosh((b_i - c_i)/\rho)) + \sum_{i=1}^p \rho \log(\cosh((d_i - g_i)/\rho))$$

$$\text{s.t} \qquad c_i = a_i^T u, \quad g_i^2 = \|u - s_i\|_2^2, \forall i = 1, \cdots, p. \tag{22}$$

With this formulation, we define Lagrangian function, given by

$$L(u, g, c, \lambda, \Gamma) = \mathcal{L}_2(u) + \sum_{i=1}^p \rho \log(\cosh((d_i - g_i)/\rho)) + \sum_{i=1}^p \rho \log(\cosh((b_i - c_i)/\rho))$$

$$+ \sum_{i=1}^p \lambda_i(c_i - a_i^T u) + \sum_{i=1}^p \Gamma_i(g_i^2 - \|u - s_i\|_2^2). \tag{23}$$

where $\Gamma = [\Gamma_1, \cdots, \Gamma_p]^T$ and $\lambda = [\lambda_1, \cdots, \lambda_p]^T$ are the Lagrange multipliers, and $\mathcal{L}_2(u) = 0.5\alpha \sum_{i=1}^p (b_i - a_i^T u)^2 + 0.5\beta \sum_{i=1}^p (d_i^2 - \|u - s_i\|_2^2)^2$. In addition, $\alpha$ and $\beta$ are positive constant greater than zero. In this function, the $\mathcal{L}_2$ term serves to enhance the stability and convexity of the dynamics while the approximated $l_1$ norm handles any available outliers or heavily tailed error in the TOA/AOA measurements. Based on (12a), (12b) and (23), the dynamics of the network ar given by

$$\frac{du}{dt} = -\frac{\partial L}{\partial u_{xy}} = \alpha \sum_{i=1}^p (a_i^T(b_i - a_i^T u_{xy}^T) + 2\beta \sum_{i=1}^p (u - s_i)(d_i^2 - \|u - s_i\|_2^2)$$

$$+ \sum_{i=1}^p \lambda_i a_i^T + 2 \sum_{i=1}^p \Gamma_i(u - s_i), \tag{24a}$$

$$\frac{dg_i}{dt} = -\frac{\partial L}{\partial g_i} = \tanh((d_i - g_i)/\rho) - 2\Gamma_i g_i \tag{24b}$$

$$\frac{dc_i}{dt} = -\frac{\partial L}{\partial c_i} = \tanh((b_i - c_i)/\rho) - \lambda_i, \tag{24c}$$

$$\frac{d\lambda_i}{dt} = \frac{\partial L)}{\partial \lambda_i} = c_i - a_i^T u \tag{24d}$$

$$\frac{d\Gamma_i}{dt} = \frac{\partial L}{\partial \Gamma_i} = g_i^2 - \|u_{xy} - s_i\|_2^2, \tag{24e}$$

where $i = 1, \cdots, p$. The computational complexity of (24) reflects certain difficulty in the implementation of (24). From (24), it can be seen that the complexity is $\mathcal{O}(p)$.

## 3.3  Network Stability Study

It is important to investigate local stability of the LPNN network dynamic. In essence for local stability, it means that a minimum point should be stable. In other words, if the minimum point is not stable, then the network will not converges to the optimal value. Denote $\boldsymbol{x} = [\boldsymbol{u}^{\mathrm{T}}, \boldsymbol{g}^{\mathrm{T}}, \boldsymbol{c}^{\mathrm{T}}]^{\mathrm{T}}$ as the collection of all the decision variables and $\{\boldsymbol{x}^*, \boldsymbol{\lambda}^*, \boldsymbol{\Gamma}^*\}$ as the optimal point of the dynamic network of (24). Based on Theorem 1 stated in [13], there are two sufficient conditions for a local stability. They are

- The Hessian matrix $\left( \frac{\partial^2 L(\boldsymbol{x}^*, \boldsymbol{\lambda}^*, \boldsymbol{\Gamma}^*)}{\partial \boldsymbol{x}^2} \right)$ of the dynamic system at $\{\boldsymbol{x}^*, \boldsymbol{\lambda}^*, \boldsymbol{\Gamma}^*\}$ should be positive definite
- The gradient vector of the constraints must be linearly independent at the minimum point.

The first one is achieved via the use of the $l_2$ norm combine with the Lagrange multiplier term. In order words, if $\alpha$ and $\beta$ in the Lagrangian function (23) are large enough, then at the equilibrium point the Hessian matrix is positive definite under mild conditions [17]. For the second case, the gradient vectors of the constraints with respect to $\boldsymbol{x}$ must be linearly independent. Hence, denote $h_i$ as the $i$th constraint, then the $2p$ constraints can be expressed as

$$h_i^1 = c_i - \boldsymbol{a}_i^{\mathrm{T}} \boldsymbol{u}, \tag{25a}$$
$$h_i^2 = g_i^2 - \|\boldsymbol{u} - \boldsymbol{s}_i\|_2^2 \tag{25b}$$

for $i = 1, \cdots p$. Note that $\boldsymbol{a}_i = [\tan(\phi_i), -1]^{\mathrm{T}}$. The gradient vectors at the minimum point are given by

$$\frac{\partial h}{\partial \boldsymbol{x}} = \left\{ \frac{\partial h_i^1}{\partial \boldsymbol{x}}, \frac{\partial h_i^2}{\partial \boldsymbol{x}} \right\} \Big|_{\boldsymbol{x} = \boldsymbol{x}^*} \quad \text{for} \quad i = 1, \cdots p \tag{26a}$$

$$= \left\{ \begin{array}{cccc|cccc} -\tan(\phi_1) & -\tan(\phi_2) & \ldots & -\tan(\phi_p) & r_{11}^* & r_{21}^* & \ldots & r_{p1}^* \\ 1 & 1 & \ldots & 1 & r_{12}^* & r_{22}^* & \ldots, & r_{p2}^* \\ 0 & 0 & \ldots & 0 & 2g_1^* & 0 & \ldots & 0 \\ 0 & 0 & \ldots & 0 & 0 & 2g_2^* & \ldots & 0 \\ \vdots & \vdots & \ddots & \vdots & \vdots & \vdots & \ddots & \vdots \\ 0 & 0 & \ldots & 0 & 0 & 0 & \ldots & 2g_p^* \\ 1 & 0 & \ldots & 0 & 0 & 0 & \ldots & 0 \\ 0 & 1 & \ldots & 0 & 0 & 0 & \ldots & 0 \\ \vdots & \vdots & \ddots & 0 & 0 & 0 & \ddots & 0 \\ 0 & 0 & \ldots & 1 & 0 & 0 & \ldots & 0 \end{array} \right\}, \tag{26b}$$

where the $r_{i1}^* = -2(u_1^* - s_{i1})$ and $r_{i2}^* = -2(u_2^* - s_{i2})$, for $i = 1, \cdots, p$. It is clear that the gradient vectors stated in (26) is linearly independent, if $g_i \neq 0$ for $i = 1, \cdots, p$. Hence, this is interpreted to means, if the position of any of the sensors and that of the estimated position of the target are not the same, then the gradient vectors satisfied the linear independence conditions.

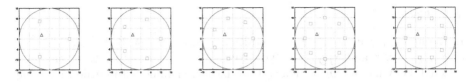

**Fig. 2.** The positions of sensors.

# 4 Simulation

In this section, we present simulation studies on our proposed approach for target localization. We compare our method with the existing techniques, including the linear approaches namely LLS algorithm (LSS) [18], two step weighted least square (TSWLS), [15], and the nonlinear approach namely NRML [19], which utilizes a maximum likelihood non-linear objective function based on the Newton-Raphson algorithm [20].

## 4.1 Setting

For our simulations, we considers different number of sensors which are set as $\{3, 5, 7, 10\}$, the sensors are uniformly distributed on the circumference of a circle, which is centered at the origin $(0, 0)$ and has a radius equals 10. The positions of the those settings are shown in Fig. 2. Furthermore, in all our experiments, we assume that the sensors TOA and AOA measurements are corrupted with uncorrelated Gaussian noise with variance equal to $\sigma_{TOA}^2$ and $\sigma_{AOA}^2$ for TOAs and AOAs measurements, respectively.

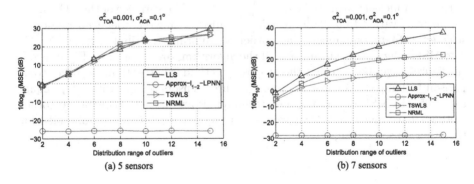

(a) 5 sensors          (b) 7 sensors

**Fig. 3.** MSE performance under different settings with fixed target position at $(-4,2)$ over 500 runs. Both $\sigma_{TOA}^2$ and $\sigma_{AOA}^2$ are fixed while the outliers are distributed from 2 to 15.

## 4.2   Case 1

In the first experiment, the target position is fixed at $(-4, 2)$. We fix the value of $\sigma^2_{AOA} = 0.1^o$ and $\sigma^2_{TOA} = 0.001$. We inject a single outlier to a randomly selected TOA measurement. For a TOA measured value with outlier, its value is given by

$$d_i = g_i + n_i + \zeta_i, \qquad (27)$$

where $n_i$ is the Gaussian noise and $\zeta_i$ is the outlier noise. The outlier noise is follow uniformly distributed with a width equal to 2. In our setting, in order to control the intensity of the outlier, we vary the mean value of the outlier noise from 1 to 15. For instance, if we set the mean value to 4, the outlier is uniformly distributed between 3 to 5. We measure the mean square error (MSE) between the true target position and the estimated ones over 500 runs. In addition, we consider the 5 and 7 sensor settings in this part. The result is shown in Fig. 3. It is obvious from the figure that the performance of the proposed method is much better than that of the other three algorithms. For instance, we could see that for the 5 sensors case, when the outlier intensity is 4, the MSE value of the proposed method is around $-25.89$ dB, while the other three comparison methods have much higher MSE values (around 5.61 dB). In addition, for our proposed method, the MSE values do no sensitive to the outlier intensity. For instance, for the 5 sensors case, the MSE values are around $-25.8$ dB even though the intensity of the outlier changes to 2 to 15.

## 4.3   Case 2

In the second case, we consider the performance of the algorithms under a complicated noise situation. The target is fixed at position $(-3, 8)$. The noise variance of the AOA measurement is fixed at $\sigma^2_{AOA} = 1.5^o$. The variance of the TOA measurement is varied from 0.001 to 0.5. We inject a single outlier to a randomly selected TOA measurement. The outlier is uniformly distributed from 2 to 11. The MSE values over 500 runs for all the algorithms are reported in Fig. 4. From the figure, it can be seen that the proposed method is better than the other comparison techniques. For instance, for the three sensors setting with $\sigma^2_{TOA}$ equal to $=0.05$, the MSE of the proposed technique is around $-15$ dB while the other methods have much high MSE values. The similar results are also obtained for the other sensor settings.

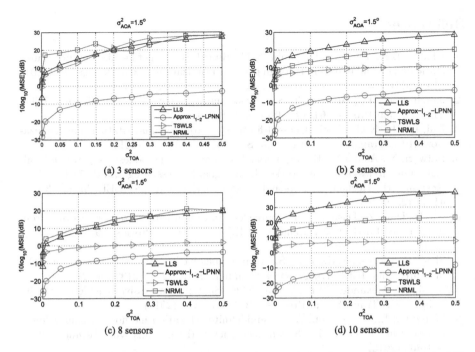

**Fig. 4.** MSE performance under different settings with fixed target position at $(-3,8)$ over 500 runs. The outlier is uniformly distributed from 2 to 11. We randomly pick one TOA measurement and add one outlier to it. The noise variance of the AOA measurement is fixed at $\sigma^2_{AOA} = 1.5°$. The variance of the TOA measurement is varied from 0.001 to 0.5.

## 5  Conclusion

We proposed an approach for robust target localization using hybrid TOA and AOA measurements. We formulated the target localization as an optimization problem, that is, minimization of hybrid $l_1$ and $l_2$ norm of the measurement errors. In this contribution we investigated the robustness of the proposed technique under the impact of outliers situations, besides we showed that the dynamics of the method presented are locally stable. The simulation results showed that the proposed approach has better resistant to outliers than the other state of art linear and nonlinear techniques for source or target localization.

**Acknowledgment.** The work was supported by a research grant from City University of Hong Kong (7004842).

# References

1. Chalise, B.K., Zhang, Y.D., Amin, M.G., Himed, B.: Target localization in a multi-static passive radar system through convex optimization. Signal Process. **102**, 207–215 (2014)
2. Jamali-Rad, H., Leus, G.: Sparsity-aware multi-source TDOA localization. IEEE Trans. Signal Process. **61**(19), 4874–4887 (2013)
3. Rui, L., Ho, K.C.: Efficient closed-form estimators for multi-static sonar localization. IEEE Trans. Aerosp. Electron. Syst. **51**(1), 600–614 (2015)
4. Patwari, N., Ash, J.N., Kyperountas, J.N.: Locating the nodes: cooperative localization in wireless sensor networks. IEEE Signal Process. Mag. **22**(4), 54–69 (2005)
5. Koshima, H., Hoshen, J.: Personal locator services emerge. IEEE Spectr. **37**(2), 41–48 (2000)
6. Xu, E., Ding, Z., Dasqupta, S.: Source localization in wireless sensor networks from signal time-of-arrival measurements. IEEE Trans. Signal Process. **59**(6), 2887–2897 (2011)
7. Shen, J., Andreas, F.M., Jussi, S.: Accurate passive location estimation using TOA measurements. IEEE Trans. Wirel. Commun. **11**(6), 2182–2192 (2012)
8. Rossi, M., Alexander, M.H., Yonina, C.E.: Spatial compressive sensing for MIMO radar. IEEE Trans. Signal Process. **62**(2), 419–430 (2014)
9. Wang, G., Li, Y., Nirwan, A.: A semidefinite relaxation method for source localization using TDOA and FDOA measurements. IEEE Trans. Veh. Technol. **62**(2), 853–862 (2013)
10. Kumar, S., Hegde, R.M.: A Review of Localization and Tracking Algorithms in Wireless Sensor Networks. arXiv preprint arXiv:1701.02080 (2017)
11. Einemo, M., So, H.C.: Weighted least squares algorithm for target localization in distributed MIMO radar. Signal Process. **115**, 144–150 (2015)
12. Wei, H.W., et al.: Multidimensional scaling analysis for passive moving target localization with TDOA and FDOA measurements. IEEE Trans. Signal Process. **58**(3), 1677–1688 (2010)
13. Zhang, S., Constantinides, A.G.: Lagrange programming neural networks. IEEE Trans. Circ. Syst. II Analog Digit. Signal Process. **39**(7), 441–452 (1992)
14. Yin, P., et al.: Minimization of $l_{1-2}$ for compressed sensing. SIAM J. Sci. Comput. **37**(1), 536–563 (2015)
15. Chan, Y.T., Ho, K.C.: A simple and efficient estimator for hyperbolic location. IEEE Trans. Signal Process. **42**(8), 1905–1915 (1994)
16. Leung, C.S., et al.: Lagrange programming neural networks for time-of-arrival-based source localization. Neural Comput. Appl. **24**(1), 109–116 (2014)
17. Zhu, X., Zhang, S., Constantinides, A.G.: Lagrange neural networks for linear programming. J. Parallel Distrib. Comput. **14**(3), 354–360 (1992)
18. Chen, J.C., Ralph, E.H., Kung, Y.: Maximum-likelihood source localization and unknown sensor location estimation for wideband signals in the near-field. IEEE Trans. Signal Process. **50**(8), 1843–1854 (2002)
19. So, H.C.: Source localization: algorithms and analysis. In: Handbook of Position Location: Theory, Practice, and Advances, pp. 25–66 (2011)
20. Ypma, T.J.: Historical development of the Newton? Raphson method. SIAM Rev. **37**(4), 531–551 (1995)

# Central Pattern Generator Based on Interstitial Cell Models Made from Bursting Neuron Models

Takahiro Toizumi[1] and Katsutoshi Saeki[2]([⊠])

[1] Graduate School of Science and Technology, Nihon University,
1-8-14 Kandasurugadai, Chiyodaku, Tokyo 101-8308, Japan
[2] College of Science and Technology, Nihon University, 7-24-1 Narashinodai,
Funabashi-shi, Chiba 274-8501, Japan
saeki.katsutoshi@nihon-u.ac.jp

**Abstract.** It is well understood that basic locomotion in living organisms is controlled by a central pattern generator (CPG). We previously studied a hardware CPG model based on interstitial cell models. The interstitial cell models can output low-frequency oscillations (approximately 2 to 5 Hz) with a 900-fF capacitor. However, in order to obtain these low frequencies, the signal must cascade through dozens of stages of neuron models, which dramatically increases the mounting area. In this paper, the mounting area was reduced by constructing an interstitial cell model using bursting neuron models area. By using the bursting neuron models, the mounting area was reduced by approximately 73% while obtaining the same frequency. Moreover, the proposed CPG model can generate transitions between quadruped locomotion patterns.

**Keywords:** Bursting neuron model · Analog circuit · Central pattern generator Interstitial cell model

## 1 Introduction

Central pattern generators (CPGs), which are neural networks that exist in spinal nerves, generate and control rhythmic locomotion patterns. Numerous researchers have focused on CPG models in order to develop control adaptations for robots [1–6]. Matsuoka [1] proposed a mathematical CPG model, moreover analyzing oscillating conditions in connection with nervous oscillators. Taga [2] proposed the bipedal locomotion CPG model by manipulating a musculoskeletal system using Matsuoka's model. Ito et al. [3] studied the consumption of energy of quadruped locomotion. Fukuoka and Kimura [4] and Takemura et al. [5] developed quadruped locomotion robots capable of walking on even ground by using mathematical CPG models and sense feedback from each limb. In contrast, researchers have also developed hardware CPG models that generate locomotion patterns by changing connecting structures [6–8]. Even though these researchers utilized fine VLSI technology, using control boards, circuits with large capacitance value, and a mixed-signal circuit [9–12].

© Springer Nature Switzerland AG 2018
L. Cheng et al. (Eds.): ICONIP 2018, LNCS 11302, pp. 321–329, 2018.
https://doi.org/10.1007/978-3-030-04179-3_28

VLSI technology is essential for the construction of adaptation robots based on CPG models. Until now, we proposed a CPG model using analog electronic circuits (by a common name: pulse-type hardware neuron models (P-HNM)) [13, 14]. Based on the P-HNM, we proposed a CPG model generated 1–10-Hz patterns through a 0.18-μm CMOS process [15].

In this paper, we suggest a new CPG model to reduce the area of an interstitial cell model constructed from bursting neuron models (BNMs).

## 2   Interstitial Cell Model

Figure 1 shows a schematic diagram of an interstitial cell model made from BNMs. The interstitial cell model consists of a self-excited neuron model (*Self*) and BNMs ($B_n$: $n = 1,...,N$). The attached open and solid circles respectively indicate excitatory and inhibitory connections of the synaptic model. The operating principle of this model is similar to our already proposed model [15], but BNMs ($B_1 - B_N$, $B_Y$, and *Self*) produce low-frequency bursting pulses.

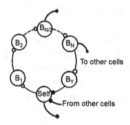

**Fig. 1.**   Schematic diagram of an interstitial cell model using BNMs.

Figure 2 shows a circuit diagram of a P-HNM (*Self* in Fig. 1). The P-HNM consists of a cell body model and a synaptic model. The cell body model is based on a simple analog circuit that is composed of a voltage source ($V_{dd}$ = 1.8 V), negative resistance parts ($M_C$ = 0.06, $M_D$ = 0.06, $M_N$ = 10, $M_P$ = 10, and $C_g$ = 16 fF), a membrane capacitor ($C_m$ = 16 fF), and a membrane leak resistance part utilizing the MOS resistance ($M_R$ = 0.06). The cell body model enables the generation of a continuous action potential ($V_{out}$) through both self- and separately excited oscillations. The synaptic model consists of a circuit composed of a secondary voltage source ($V_{int2}$ = 0.2 V), integration components ($M_{s1}$ = 1, $M_{s2}$ = 1, $M_{s3}$ = 5, $M_{s4}$ = 0.06, and $C_{s2}$ = 200 fF), excitatory output current components ($M_{k1}$ = 2.5, $M_{k2}$ = 1, and $M_{k3}$ = 2), and inhibitory output current components ($M_{y1}$ = 10 and $M_{y2}$ = 10). The synaptic model configures each input voltage as a cell body model output ($V_{out}$) by using an integration system consisting of a current mirror circuit ($M_{s1}$, $M_{s2}$, and $M_{s3}$), capacitance ($C_{s2}$), and a leak resistance part ($M_{s4}$). Figure 3 shows the output waveform of a P-HNM. The output waveform looks like the active potential of a biological neuron. The result shown in Fig. 3 was simulated using HSPICE for a 0.18-μm CMOS process.

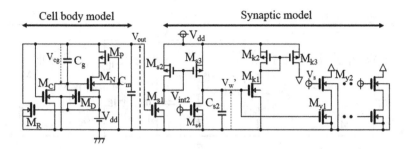

**Fig. 2.** Circuit diagram of a P-HNM.

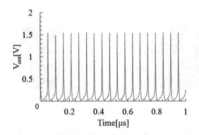

**Fig. 3.** Output waveform of a P-HNM.

Figure 4 shows the circuit diagram of a BNM ($B_1 - B_N$ and $B_Y$ in Fig. 1) with the following circuit parameters: $C_s = 16$ fF, $M_C = 0.06$, $M_{cg} = 10$, $M_{cm} = 1$, $M_D = 0.17$, $M_N = 10$, $M_{k1} = 2.5$, $M_{k2} = 1$, $M_{k3} = 2$, $M_{n1} = 1$, $M_{n2} = 1$, $M_{n3} = 0.17$, $M_P = 10$, $M_R = 0.06$, $M_{s1} = 1$, $M_{s2} = 1$, $M_{s3} = 5$, $M_{s4} = 0.06$, $M_{sw} = 2.5$, $M_{y1} = 10$, $V_{int} = 1.2$ V, and $V_{int2} = 0.2$ V. The BNM consists of the burst-generating unit, a cell body model, and a synaptic model. The cell body model and synaptic model are almost equivalent to those in the P-HNM. However, the cell body model is instead controlled by a threshold voltage via $M_{sw}$. Moreover, this model has fewer capacitors. The burst-generating unit includes a low-frequency oscillator. The output voltage ($V_w$) of this unit can be controlled through the threshold of the cell body model. The process starts with $V_w = 0$ V. If a stimulus input current (I) is applied, $V_w$ becomes large. Furthermore, if $M_{sw}$ changes to OFF, the threshold voltage of the cell body model falls. These interactions generate a beating pulse. As this pulse passes through the integration circuit, a low-frequency square pulse is generated, and electric charge is extracted from $V_w$ by switching $M_{n1}$ at regular intervals; in this circumstance, $V_w$ is low, and $M_{sw}$ is ON, and the threshold voltage of the cell body model rises and suspends oscillation. Namely, the output voltage of the cell body model through feedback with the low-frequency oscillator can be controlled by $M_{n1}$ to have a low frequency and a low duty ratio. As a result, this circuit fires the desired bursts.

Figure 5 shows the output waveform of a BNM. The output waveform looks like the burst firing of a biological neuron. The result shown in Fig. 4 was simulated using HSPICE for a 0.18-μm CMOS process.

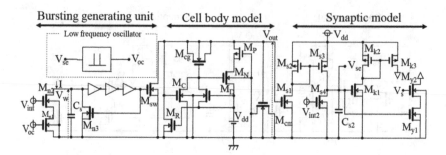

**Fig. 4.** Circuit diagram of the BNM.

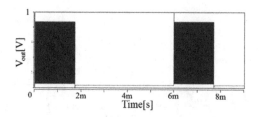

**Fig. 5.** Output waveform of a BNM.

Figure 6 compares a traditional low-frequency oscillation circuit with the proposed circuit. In the traditional circuit, the frequency of $V_w$' depends on the time constant derived from $C_s$ and $I_s$. In contrast, the proposed circuit can generate low-frequency oscillation using low capacitance because the switching voltage ($V_{oc}$) reduces the charge that must be stored on $C_s$, as it is instead dissipated through the MOSFETs.

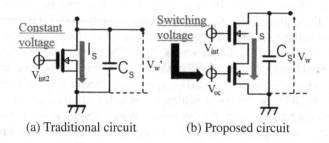

(a) Traditional circuit          (b) Proposed circuit

**Fig. 6.** Low-frequency oscillation circuits.

Figure 7 shows the output waveform of *Self* and six stages ($N = 6$) of BNMs in the interstitial cell model. This figure shows that low-frequency oscillation is achieved, and the duty cycle increases with each stage of BNM. The interstitial cell model oscillates at approximately 5 Hz over six stages of BNM, which is suitable for locomotion control of a robot.

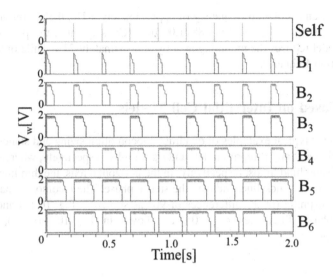

**Fig. 7.** Output waveform of *Self* and each BNM in the interstitial cell model.

Figure 8 shows the layout design of a BNM for the ROHM 0.18-µm process. The horizontal length is 45.60 µm, and the vertical length is 28.80 µm. Therefore, the mount area is 1313.28 µm². Figure 9 displays the layout design of an interstitial cell model with six stages of BNMs ($N = 6$). In this figure, the horizontal length is 187.65 µm, and the vertical length is 71.04 µm. Therefore, the mount area is 13330.66 µm².

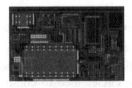

**Fig. 8.** Layout design of a BNM ($N = 1$).

**Fig. 9.** Layout design of an interstitial cell model which is the six stages of BNMs ($N = 6$).

To achieve an oscillation frequency of approximately 5 Hz, the mount area for the traditional interstitial cell model is 48811.00 $\mu m^2$. Consequently, the proposed interstitial cell model reduces the mounting area by approximately 73% while obtaining the same oscillation frequency.

## 3    CPG Based on Interstitial Cell Models

Figure 10 shows typical examples of quadruped locomotion swing and stance patterns. In the vertical axis L and R represent left and right limbs, respectively, whereas F and H represent fore and hind limbs, respectively. In the horizontal axis, the thin lines indicate the swing phase, the thick lines indicate the stance phase. The locomotion patterns can be divided four-phase (phase differences of 90° as shown in Fig. 10 (d) and (e)) and two-phase (phase differences of 180°, as shown in Fig. 10 (a), (b), and (c)) synchronization.

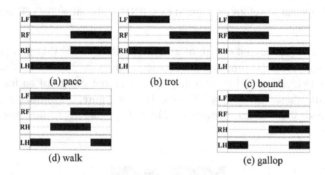

**Fig. 10.** Typical examples of quadruped locomotion patterns.

Figure 11 shows a schematic diagram of CPG models for generating quadruped locomotion patterns. The white circles used to construct the CPG model represent interstitial cell models. The limb notation is the same as that used in Fig. 10. The solid lines shows connections from $B_N$ to *Self*. Additionally, the dashed and dot-dashed lines indicate connections from $B_Y$ to *Self* and from $B_{N/2}$ to *Self*, respectively. Each interstitial cell model causes oscillation according to the order and coupling direction by suppressing each other interstitial cell model. As a result, it is possible to generate each locomotion pattern.

Figure 12 shows the output waveforms of the $B_N$ in each interstitial cell model of the CPG model for pacing. This model successfully produces the two-phase synchronization pattern. In contrast, Fig. 13 shows the output waveforms of the $B_N$ in each interstitial cell model of the CPG model for walking. This model successfully produces the four-phase synchronization pattern.

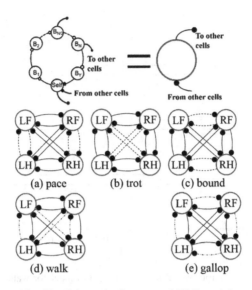

**Fig. 11.** Schematic diagrams of CPG models.

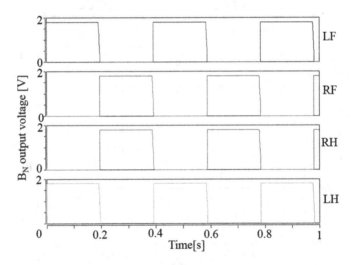

**Fig. 12.** Output waveforms of the CPG model for pacing

Figure 14 shows the output waveforms of each interstitial cell model in a CPG model demonstrating a transition from walking to pacing. In this figure, the CPG model for transition between locomotion patterns is achieved by changing the network connection by switching the state of $V_s$ (i.e., High or Low) after 1 s of walking, the timing of which is shown by the dashed line. These waveforms demonstrate that the CPG model successfully transitions from a two-phase synchronization pattern to a four-phase synchronization pattern, as represented by the transition from walking to pacing.

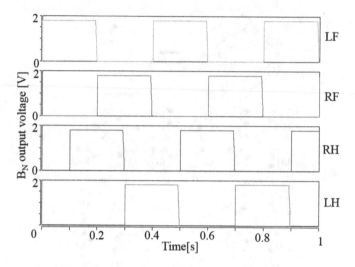

**Fig. 13.** Output waveforms of the CPG model for walking.

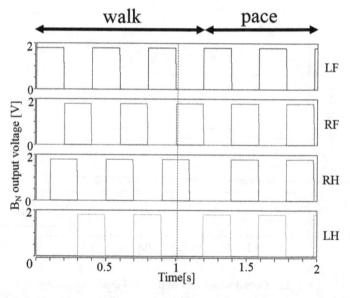

**Fig. 14.** Output waveforms of a CPG model with transition from walking to pacing.

## 4 Conclusion

In this paper, we presented a CPG model using interstitial cell models constructed from BNMs. By using the BNMs, the mounting area was reduced by approximately 73% while obtaining the same frequency. Moreover, the CPG model can straightforwardly generate transitions between quadruped locomotion patterns.

In future work, we will propose a CPG model considering flexor and extensor muscles using the BNM. Furthermore, we will construct an IC chip through the VLSI Design and Education Center (VDEC) in Japan.

# References

1. Matsuoka, K.: Sustained oscillations generated by mutually inhibiting neurons with adaptation. Biol. Cybern. **52**(6), 367–376 (1985)
2. Taga, G.: A model of the neuro-musculo-skeletal system for human locomotion. Biol. Cybern. **73**(2), 97–111 (1995)
3. Ito, S., Yuasa, H., Ito, K.: Oscillator-mechanical model of the pattern transition on quadrupedal locomotion based on energy expenditure. Trans. Soc. Instrum. Control Eng. **32**(11), 1535–1543 (1996)
4. Fukuoka, Y., Kimura, H.: Biologically inspired adaptive dynamic walking of a quadruped on irregular terrain-realization of walking in outdoor environment using a self-contained robot: "Tekken2". J. Robot. Soc. Jpn. **25**(1), 138–154 (2007)
5. Takemura, H., Ueda, J., Matsumoto, Y., Ogasawara, T.: Three dimensional adaptive walking of quadruped robot using sideways sway motion and posture reflex via neural oscillators. J. Robot. Soc. Jpn. **22**(4), 528–534 (2004)
6. Tenore, F., Vogelstein, R.J., Cummings, R.E.: Sensor-based dynamic control of the central pattern generator for locomotion. In: IEEE International Symposium on Circuits and Systems, pp. 613–616 (2007)
7. Nakada, K., Asai, T., Hirose, T., Amemiya, Y.: Analog current-mode CMOS implementation of central pattern generator for robot locomotion. In: Proceedings of International Joint Conference on Neural Networks, vol. 1, pp. 639–644. IEEE Press, New York (2005)
8. Nakada, K., Asai, T., Amemiya, Y.: Analog CMOS implementation of a CNN-based locomotion controller with floating-gate devices. IEEE Trans. Circuits Syst. **52**(6), 1095–1103 (2005)
9. Hasan, S.M.R., Xu, W.L.: Low-voltage analog current-mode central pattern generator circuit for robotic chewing locomotion using 130nanometer CMOS technology. In: International Conference on Microelectronics, pp. 155–160 (2007)
10. Arena, P., Fortuna, L., Frasca, M., Patane, L., Pollino, M.: An autonomous mini-hexapod robot controlled through a CNN-based CPG VLSI chip. In: International Workshop on Cellular Neural Networks and Their Applications, pp. 1–6 (2006)
11. Kier, R.J., Ames, J.C., Beer, R.D., Harrison, R.R.: Design and implementation of multi-pattern generators in analog VLSI. IEEE Trans. Neural Netw. **17**(4), 1025–1038 (2006)
12. Yang, Z., Cameron, K., Lewinger, W., Webb, B., Murray, A.: Neuromorphic control of stepping pattern generation: a dynamic model with analog circuit implementation. IEEE Trans. Neural Netw. Learn. Syst. **23**(3), 373–384 (2012)
13. Saeki, K., Sekine, Y.: CMOS implementation of a pulse-type hardware neuron model and its application. IEICE Trans. Fundam. **15**(1), 27–38 (2008)
14. Ono, K., Saeki, K., Sekine, Y.: A study on a pulse-type hardware neuron model using CMOS. In: IEEJ Technical Meeting on Electronic Circuits, vol. ECT-5, no. 79, pp. 41–44 (2005)
15. Saeki, K., Nihei, D., Tatebe, T., Sekine, Y.: IC implementation of interstitial cell-based CPG model. Int. J. Analog Integr. Circuits Signal Process. **81**(3), 551–559 (2014)

# Active Feedback Framework
# with Scan-Path Clustering
# for Deep Affective Models

Li-Ming Zhao[1], Xin-Wei Li[1], Wei-Long Zheng[1], and Bao-Liang Lu[1,2,3]($\boxtimes$)

[1] Center for Brain-like Computing and Machine Intelligence,
Department of Computer Science and Engineering, Shanghai Jiao Tong University,
Shanghai, China
[2] Key Laboratory of Shanghai Education Commission for Intelligent Interaction
and Cognitive Engineering, Shanghai Jiao Tong University, Shanghai, China
[3] Brain Science and Technology Research Center, Shanghai Jiao Tong University,
Shanghai, China
{lm_zhao,college_lxw,weilong,bllu}@sjtu.edu.cn

**Abstract.** The attention of subjects to EEG-based emotion recognition experiments could seriously affect their emotion induction level and annotation quality of EEG data. Therefore, it is important to evaluate the raw EEG data before training the classification model. In this paper, we propose a framework to filter out low quality EEG data from participants with low attention using eye tracking data and boost the performance of deep affective models with CNN and LSTM. We introduce a novel attention-deprived experiment with dual tasks, in which the dominant task is auditory continuous performance test, identical pairs version (CPT-IP) and the subtask is emotion eliciting experiment. Motivated by the idea that subjects with attention share similar scan-path patterns under the same clips, we adopt the cosine distance based spatial-temporal scan-path analysis with eye tracking data to cluster these similar scan-paths. The average accuracy of emotion recognition using the selected EEG data with attention is about 3% higher than that of original training dataset without filtering. We also found that with the increasing distance of scan-paths between outliers and cluster center, the performance of corresponding EEG data tends to decrease.

**Keywords:** Eye tracking · Scan-path · Attention evaluation
EEG data filtering

# 1 Introduction

Recently, multimodal emotion recognition based on EEG and eye movement data has attracted increasing attention. Combining eye movements and EEG can considerably improve the performance of emotion recognition systems because eye movements and EEG are complementary to emotion recognition [8]. Eye movement data has the advantages that the device is wearable and the data is

© Springer Nature Switzerland AG 2018
L. Cheng et al. (Eds.): ICONIP 2018, LNCS 11302, pp. 330–340, 2018.
https://doi.org/10.1007/978-3-030-04179-3_29

easy to handle. Meanwhile, the eye movement data can be used as a multi-modal data to complement the EEG data in emotion recognition, and can also be used as a measure of whether the subject is seriously involved in the experiment.

In emotion recognition from EEG signals, using emotional film clips as stimuli to elicit emotions is one of the most popular and effective methods [12,14,16]. However, it is hard to know whether the participants have been elicited corresponding emotions through watching these clips. Traditionally, questionnaires were sent to participants for self-assessment [7], which is rather subjective. In our previous work [13], we proposed that eye tracking data could be an effective reference for evaluating EEG data quality in emotion recognition experiments. However, this work failed to explain the following points: (a) What is the relationship among the participants' attention, the scan-path pattern and the emotion recognition performance. (b) How to guide the participants to watch each clip without attention. (c) How to better measure the trend variation of different scan-paths. In this study, we propose a modified framework which is capable of evaluating the quality of data and filtering EEG data for emotion recognition.

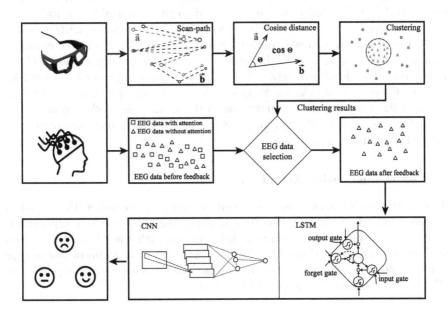

**Fig. 1.** The proposed framework for attention evaluation and feedback.

The flow chart of our framework is shown in Fig. 1. The EEG data and eye tracking data are collected when the participants watch the emotion film clips. The distance matrix is calculated based on cosine distances between any two scan-paths and is clustered with DBSCAN method [5]. In the feedback step, EEG data with high attention is selected by using the scan-path clustering results. The training dataset of the classifiers consists of the selected high-quality data after feedback.

In order to evaluate our framework, we design a novel attention-deprived experiment. The dominant task named as Auditory CPT-IP, derived from CPT-IP [2] which is used to activate particular regions of brain. We conduct both the attention-deprived experiment as well as the normal emotion eliciting experiment, and the data go through the process shown in Fig. 1. We compare the selected training dataset to original full data, and test the affective models on the same test dataset. The models include SVM, CNN and LSTM, which are popular in emotion recognition areas [15]. Finally, we analyze the relationship between the dissimilarity of scan-path and the performance of corresponding EEG data.

## 2    Experiment Settings

Previous studies have explored the reliability of using movie clips to induce emotions [12]. In our work, 15 Chinese movie clips with highly emotional contents are selected to induce three corresponding emotional states, i.e., positive, negative and neutral [16]. The subjects are instructed to sit comfortably, facing a large screen and move as least as possible to reduce the interference of artifact. There is a 5 s hint for starting, a 60 s rest for subjects to recover from elicited emotions in each trial. During these experiments both eye movement signal and EEG signal are collected simultaneously. Eye movement signals are recorded using SMI 30 Hz ETG eye tracking glasses. EEG signals are recorded by an ESI NeuroScan System with sampling rate 1000 Hz from a 62-channel electrode cap according to the international 10–20 system. A set of control experiments are performed under these common conditions.

### 2.1    Experiment with Attention

Sixteen subjects (7 males) aged between 20 and 24 years old, with normal or corrected-to-normal vision and normal hearing, participate in this experiment. All the participants are required to watch the clips with intently and elicit their own corresponding emotions. One loudspeakers plays the audio of movie clips and the volume is adjusted to the appropriate size. Both eye tracking data and EEG data are collected as shown in Fig. 2(a).

### 2.2    Attention-Deprived Experiment

In this experiment, according to the theory of cognitive psychology, attention-deprived tasks are added to video-based emotion experiments. Study shows that multitasking will cause switching costs and mixing costs [10], which could seriously affect the task-handle ability of the brain. In the video-based emotional stimulation experiment, participants have to understand both the auditory and visual information. Therefore, we design an auditory CPT-IP experiment to fight for the attention of the subjects in video task.

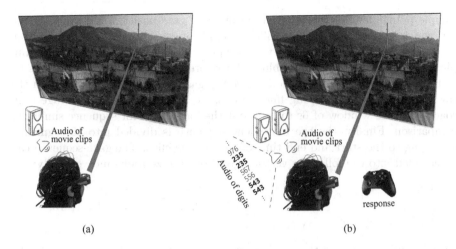

(a)                                           (b)

**Fig. 2.** Setting of experimental scene. (a) In the experiment with whole attention, the participants watched the movie clips normally. (b) In the attention-deprived experiment, two loudspeakers are used to play the audio of movies and digits. Gamepad is used to response and the response key needs to be pressed once a number was repeated.

As shown in Fig. 2(b), in the CPT-IP experiment, participants work through several conditions of a continuous performance task with the task to identify identical pairs of 3-digit numbers. Participants are presented a continuous stream of 3-digit numbers per second. The response key needs to be pressed once a number was repeated (Go trials). For non-repeating stimuli, participants are instructed to wait for the next repetition (NoGo trials).

Two separate loudspeakers are used in this attention-deprived experiment, one for the audio of movie clips and the other for CPT-IP auditory digits. Both speakers are set in the same place, with the same volume. Participants are required to complete the dominant auditory CPT-IP task as well as possible while watching the movie clips at the same time. We use the same equipment to collect the eye tracking data and the EEG data. Fourteen subjects, 7 males and 7 females, aged from 19 to 24, with self-reported normal or corrected-to-normal vision and normal hearing, participate in this experiment.

## 3   Method

### 3.1   Attention Evaluation

**Generating Gaze Sequence.** When subjects watching videos, their gaze position in each video frame would form a scan-path. Therefore, the eye movements can be regarded as a spatial-temporal sequence. The gaze sequence generation is a manual process using the BeGaze software from SMI. Since SMI ETG has a sampling rate of 30 Hz, we can acquire 30 raw sample points per second. Each

sample point includes the information of time, gaze position and pupil diame-
ter, which is labeled with three different kinds of event type, including fixation,
saccade and blink. In particular, the gaze position will be recorded as 0 for sample points whose event type is blink. Therefore, we fix these gaze positions by
using linear interpolation method. Since the gaze sequence we extract from the
raw data usually have strong fluctuations, we apply the moving average approach with the window of 6s to filter out the local jitter for sequence similarity
comparison. Finally, the eye movement sequence is divided into 15 segments,
according to the start and end time of each movie clips. The gaze sequence can
be encoded into the following vector space both horizontally and vertically:

$$
\begin{cases}
\boldsymbol{S}_{i_c}^{x} = [x_{c_1}^{i}, \ x_{c_2}^{i}, \ \ldots, \ x_{c_m}^{i}] \\
\boldsymbol{S}_{i_c}^{y} = [y_{c_1}^{i}, \ y_{c_2}^{i}, \ \ldots, \ y_{c_m}^{i}]
\end{cases}
\qquad i = 1, \ 2, \ldots, \ 30; c = 1, \ 2, \ldots, \ 15, \qquad (1)
$$

where $i$ and $c$ represent the participant number and clip number, respectively,
$m$ is the dimension of the gaze vector associated with the length of each movie
clip, and $x$ and $y$ stand for the horizontal gaze position and the vertical gaze
position, respectively.

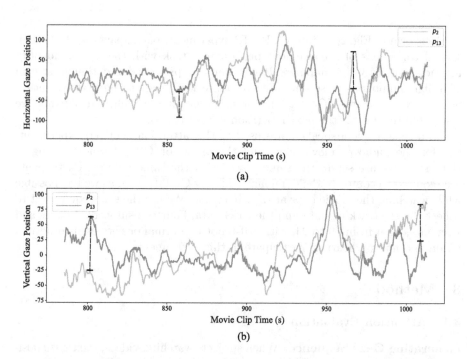

**Fig. 3.** The scan-path of two subjects with attention in movie clip 3. The black dotted
line indicates a large difference in amplitude. (a) The map of horizontal gaze vector.
(b) The map of vertical gaze vector.

**Similarity Measures.** The gaze sequence, as described above, should share high similarity among data collected from subjects with high attention, whereas there will be large differences between the subjects involved in the attention-deprived experiment. Before clustering, we must determine how to measure the similarity between different gaze sequence. Moreover, choosing an appropriate similarity measure is also crucial for cluster analysis. A wide variety of similarity measures can be used for clustering, such as Euclidean distance and cosine similarity which are very popular similarity measures for clustering [6]. The Euclidean distance of two vectors is defined as:

$$D_E(a, b) = (\sum_{t=1}^{m} |a_t - b_t|^2)^{1/2}. \tag{2}$$

Cosine similarity is one of the most popular similarity measures applied to clustering. Given two vectors $a$ and $b$, their cosine similarity is

$$D_C(a, b) = \frac{a \cdot b}{|a| \times |b|}. \tag{3}$$

When gaze sequence is represented as gaze vectors in two directions, the similarity of two sequences corresponds to the correlation between the vectors. In general, the gaze vector has more than 7000 dimensions. Euclidean distance will not be a good metric in such a high dimensional space. Figure 3 depicts the horizontal and vertical gaze vector of movie clip 3 from two subjects with high attention. From Fig. 3, we can see that although the red and blue line have similar trend, they can sometimes be very different in amplitude, as illustrated in black dotted line. This difference is caused by head movements, which is hard to eliminate. In other words, although two participants looked at the same place, when they moved their heads, the SMI ETG would move along with it, thus the recorded gaze position would also be different. In this case, Euclidean measurement calculates the absolute distance between the two vectors, which is not suitable for our data set.

Compared with Euclidean distance, cosine distance is good at capturing the similarity of patterns of feature changes, at the same time disregarding the absolute amplitude of the compared feature vectors [11]. Therefore, we choose cosine distance as the similarity measurement method finally. The average cosine distance between two gaze sequences $S_i$ and $S_j$ can be calculated as

$$Dis_C(S_i, S_j) = \frac{1}{15} \sum_{c=1}^{15} \left( \frac{1}{2} \left( D_C(S_{i_c}^x, S_{j_c}^x) + D_C(S_{i_c}^y, S_{j_c}^y) \right) \right)$$

$$= \frac{1}{30} \sum_{c=1}^{15} \left( \frac{S_{i_c}^x \cdot S_{j_c}^x}{|S_{i_c}^x| \times |S_{j_c}^x|} + \frac{S_{i_c}^y \cdot S_{j_c}^y}{|S_{i_c}^y| \times |S_{j_c}^y|} \right), \tag{4}$$

where $S_{i_c}^x$ and $S_{i_c}^y$ are defined in Eq. 1.

**Clustering.** Through clustering the scan-path by unsupervised clustering algorithm, we can distinguish whether the subjects are involved in the video. Based on the cosine distance matrix calculated by the previous steps, we perform a density-based clustering algorithm called DBSCAN [5] to fulfil our task. We chose DBSCAN because it is density-based and allows using precalculated distance matrix as the input. The DBSCAN algorithm views clusters as areas of high density separated by areas of low density. Meanwhile, clusters found by DBSCAN can be any shape, as opposed to k-means [4] which assumes that clusters are convex shaped. There are two parameters to the algorithm, *min_samples* and *eps*, which define formally what we mean when we say dense. Higher *min_samples* or lower *eps* indicate higher density necessary to form a cluster. By performing this algorithm, the subjects are classified into outliers and non-outliers, which are considered as subjects without attention and subjects with attention, respectively. Metric Multidimensional scaling (MDS) [1] is used for visualizing the clustering results. Both DBSCAN and MDS algorithms used here are built from the scikit-learn toolkit [9].

## 3.2 EEG Data Filtering for Emotion Recognition

The clustering results provide us with an indicator for filtering EEG data, which means if one subject is labeled with outliers, his or her corresponding EEG data will be removed from the training data set.

**EEG Feature Extraction and Smoothing.** Considering the effectiveness of differential entropy (DE) in EEG-based emotion recognition [3], we choose DE as the EEG feature. The DE features are extracted in five frequency bands : $\delta(1 - 3\ Hz)$, $\theta(4 - 7\ Hz)$, $\alpha(8 - 13\ Hz)$, $\beta(14 - 30\ Hz)$ and $\gamma(31 - 50\ Hz)$. A 256-point Short-time Fourier transform (STFT) with 1 s non-overlapping Hanning window is used to calculate the average DE features of each channel on these bands. Since 62-channel EEG signals are collected, we obtain 310 dimensional features for each sample. A linear dynamic system (LDS) approach is used to eliminate the rapid changes of DE features, which makes the features more reliable.

**Classification.** We test the clustering result on a series of classification models including Linear SVM, LSTM and CNN. To explore the influence of feedback for EEG-based emotion recognition on these models, we use all 30 subjects' EEG data recorded while they watching the first 9 movie clips, including 3 positive, 3 negative and 3 neutral clips, as the training set before feedback and remove the EEG data who is labeled with outliers from the training set when doing the feedback session. To ensure fairness, the rest six sessions, including 2 positive, 2 negative and 2 neutral clips from 16 subjects in video-base emotion experiment are used as the testing data set which keep the same before and after feedback.

# 4   Experimental Results

## 4.1   Results of Attention Evaluation

Firstly, we count the performance of the 14 subjects in the CPT-IP task. The accuracies of 12 subjects are higher than 93%, while the accuracies of the rest 2 subjects are below 85%. We randomly select 6 subjects for visualization, 3 subjects with attention and 3 subjects from attention-deprived experiment, denoted by $p$ and $n$ respectively. Figure 4 illustrate the horizontal and vertical gaze position along with time in one movie clip. The subjects with attention share a similar scan-path, while the subjects in attention-deprived experiment have different scan-paths. In particular, the scan-path of $n_1$ whose CPT-IP accuracy is 85%, is very similar to the average eye movement trajectory of the 16 subjects with attention.

The two parameters of DBSCAN, *eps* and *min_samples*, are set to 0.53 and 2, respectively. Finally, the subjects with similar scan-path are clustered into a cluster, while other subjects are scattered around them, among which 3 subjects, i.e. $p_{16}$, $n_1$ and $n_5$ do not match the experimental category to which they belong.

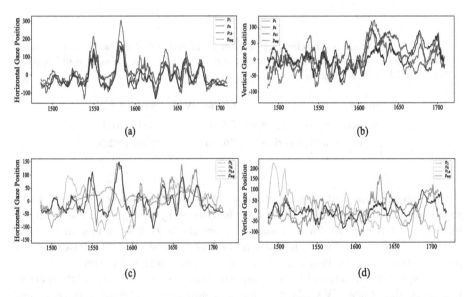

**Fig. 4.** Similarity comparison of scan-paths between different subject. The black dot line in each figure stands for the average scan-paths of all subjects with attention. (a), (b) The subjects with attention share a similar scan-path in horizontal gaze position and vertical gaze position. (c), (d) The subjects in the attention-deprived experiments have different eye movement trajectories.

## 4.2   Results of Feedback

According to the clustering results, we filtered out the EEG data of the subjects corresponding to the outliers. In order to verify the validity of the feedback, we test on different classifiers. The hyper-parameters and their corresponding range of these models are shown in Table 1.

**Table 1.** The hyper-parameters and their corresponding range of different models.

| Model | Linear SVM | LSTM | CNN |
|---|---|---|---|
| c | $2^{-10} \sim 2^{10}$ | - | - |
| Learning rate | - | $10^{-6} \sim 10^{-3}$ | $10^{-5} \sim 10^{-3}$ |
| Hidden layer | - | 2 | 1 conv |
| Hidden size | - | $128 \sim 512$ | $32 \sim 64$ |
| Time step | - | $5 \sim 30$ | - |
| Epoch | - | 500 | 300 |

**Table 2.** Classification accuracies (%) of different models with and without EEG data filtering. The size of training data set is 60390 before EEG data filtering, and becomes 34221 after EEG data filtering.

| Model | Linear SVM | LSTM | CNN |
|---|---|---|---|
| Accuracy without filtering | 72.39 | 75.81 | 81.1 |
| Accuracy with filtering | **76.30** | **80.39** | **82.3** |

The recognition performance of these four models are shown in Table 2. As shown above, although the training data is reduced by nearly 50%, the accuracy in testing set is improved after feedback when the testing set keeps unchanged. We believe the reason for the increase in accuracy is that the EEG data removed by our attention evaluation algorithm contain less emotional patterns.

To explore the relationship between attention and emotion recognition accuracy, we train a linear SVM classifier for each subject's EEG data. Firstly, we use the EEG data for the first nine clips as the training set and the EEG data for the remaining six clips as the test data, then we sort the classification accuracy of each subject from high to low. There is a correlation between EEG accuracy and the average cosine distance, as shown in Fig. 5, when the accuracy rate of emotion recognition decreases, the divergence of scan-path between subjects tends to increase.

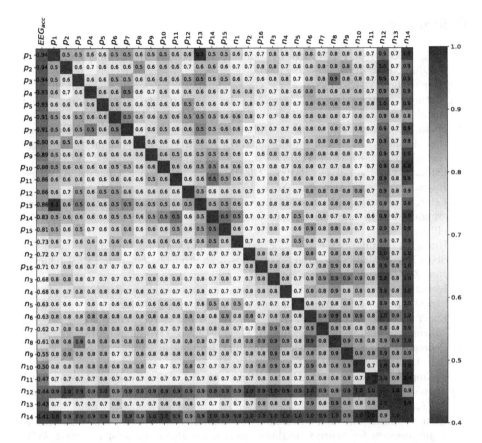

**Fig. 5.** The matrix shows the relationship between scan-path similarity and the emotion recognition accuracy. The first column is the EEG accuracy. The rest columns form a 30*30 matrix, in which each element is the average cosine distance between two scan-path.

## 5 Conclusion

In this paper, we have proposed a modified framework to evaluate the quality of EEG data for emotion recognition by using spacial-temporal scan-path analysis. The performance of emotion recognition using the selected EEG data with high engagement is better than that of original training dataset without filtering. The experimental results have demonstrated the effectiveness of our proposed framework and have indicated that the scan-path is related to the quality of data as well as the attendance level of participants.

**Acknowledgments.** This work was supported in part by the grants from the National Key Research and Development Program of China (Grant No. 2017YFB1002501), the National Natural Science Foundation of China (Grant No. 61673266), and the Fundamental Research Funds for the Central Universities.

# References

1. Borg, I., Groenen, P.: Modern multidimensional scaling: theory and applications. J. Educ. Meas. **40**(3), 277–280 (2003)
2. Cornblatt, B.A., Risch, N.J., Faris, G., Friedman, D., Erlenmeyer-Kimling, L.: The continuous performance test, identical pairs version (CPT-IP): new findings about sustained attention in normal families. Psychiatr. Res. **26**(2), 223–238 (1988)
3. Duan, R.N., Zhu, J.Y., Lu, B.L.: Differential entropy feature for EEG-based emotion classification. In: 6th International IEEE/EMBS Conference on Neural Engineering, pp. 81–84. IEEE (2013)
4. Duda, R.O., Hart, P.E.: Pattern Classification and Scene Analysis. A Wiley-Interscience Publication, New York (1973)
5. Ester, M., Kriegel, H.P., Sander, J., Xu, X.: A density-based algorithm for discovering clusters in large spatial databases with noise. In: KDD, vol. 96, pp. 226–231 (1996)
6. Huang, A.: Similarity measures for text document clustering. In: Proceedings of the Sixth New Zealand Computer Science Research Student Conference, Christchurch, New Zealand, pp. 49–56 (2008)
7. Koelstra, S., et al.: Patras: DEAP: a database for emotion analysis; using physiological signals. IEEE Trans. Affect. Comput. **3**(1), 18–31 (2012)
8. Lu, Y., Zheng, W.L., Li, B., Lu, B.L.: Combining eye movements and EEG to enhance emotion recognition. In: IJCAI, vol. 15, pp. 1170–1176 (2015)
9. Pedregosa, F., et al.: Scikit-learn: machine learning in Python. J. Mach. Learn. Res. **12**, 2825–2830 (2011)
10. Philipp, A.M., Kalinich, C., Koch, I., Schubotz, R.I.: Mixing costs and switch costs when switching stimulus dimensions in serial predictions. Psychol. Res. **72**(4), 405–414 (2008)
11. Qian, G., Sural, S., Gu, Y., Pramanik, S.: Similarity between Euclidean and cosine angle distance for nearest neighbor queries. In: Proceedings of the 2004 ACM Symposium on Applied Computing, pp. 1232–1237. ACM (2004)
12. Schaefer, A., Nils, F., Sanchez, X., Philippot, P.: Assessing the effectiveness of a large database of emotion-eliciting films: a new tool for emotion researchers. Cognit. Emot. **24**(7), 1153–1172 (2010)
13. Shi, Z.F., Zhou, C., Zheng, W.L., Lu, B.L.: Attention evaluation with eye tracking glasses for EEG-based emotion recognition. In: 8th International IEEE/EMBS Conference on Neural Engineering, pp. 86–89. IEEE (2017)
14. Wang, X.W., Nie, D., Lu, B.L.: Emotional state classification from EEG data using machine learning approach. Neurocomputing **129**, 94–106 (2014)
15. Yan, X., Zheng, W.L., Liu, W., Lu, B.L.: Investigating gender differences of brain areas in emotion recognition using LSTM neural network. In: Liu, D., Xie, S., Li, Y., Zhao, D., El-Alfy, E.S. (eds.) ICONIP 2017. LNCS, vol. 10637, pp. 820–829. Springer, Cham (2017). https://doi.org/10.1007/978-3-319-70093-9_87
16. Zheng, W.L., Lu, B.L.: Investigating critical frequency bands and channels for EEG-based emotion recognition with deep neural networks. IEEE Trans. Auton. Ment. Dev. **7**(3), 162–175 (2015)

# Stability Analysis

# A New Robust Stability Result
# for Delayed Neural Networks

Ozlem Faydasicok[1], Cemal Cicek[1], and Sabri Arik[2(✉)]

[1] Department of Mathematics, Istanbul University, Istanbul, Turkey
{kozlem,cicekc}@istanbul.edu.tr
[2] Department of Computer Engineering,
Istanbul University-Cerrahpasa, Istanbul, Turkey
ariks@istanbul.edu.tr

**Abstract.** This work proposes a further improved global robust stability condition for neural networks involving intervalized network parameters and including single time delay. For the sake of obtaining a new robust stability condition, a new upper bound for the norm of the intervalized interconnection matrices is established. The homeomorphism mapping and Lyapunov stability theorems are employed to derive the proposed stability condition by making use of this upper bound norm. The obtained result is applicable to all nondecreasing slope-bounded activation functions and imposes constraints on parameters of neural network without involving time delay.

**Keywords:** Delayed systems · Neural networks · Robust stability
Lyapunov theorems

## 1 Introduction

Analysis of various dynamics of neural networks has recently become an interesting research topic due to their qualitative properties that are employed to solve various practical real-word problems related to combinatorial optimization, image processing and control systems. When solving these types of problems by using neural networks, one needs to establish a neural system possessing a unique and globally asymptotically stable equilibrium point. Thus, one needs to deal with stability of neural networks. The fact that neurons implemented by amplifiers usually have finite switching speeds will result in time delays, which may have undesired affects on the dynamics of neural networks. Another problem is that the parameters of neural systems may involve some uncertainties, which can also have an affect on the equilibria of neural networks. Because of these reasons, for a proper stability analysis, the time delay in the states and uncertainties in the network parameters need to be included in the mathematical model of neural networks. That is to say, the key requirement would be the establishment of robust stability of neural systems which also involve time delay. When reviewing past literature, it can be realized that many researchers published useful

© Springer Nature Switzerland AG 2018
L. Cheng et al. (Eds.): ICONIP 2018, LNCS 11302, pp. 343–352, 2018.
https://doi.org/10.1007/978-3-030-04179-3_30

robust stability criteria for delayed neural systems, (see references [1–19]). This paper uses Lyapunov and Homeomorphic mapping theorems to derive a novel condition for global robust asymptotic stability.

Notations: Let $z = (z_1, z_2, ..., z_n)^T$. We will use $|z| = (|z_1|, |z_2|, ..., |z_n|)^T$. For a given matrix $E = (e_{ij})_{n \times n}$, we will use $|E| = (|e_{ij}|)_{n \times n}$, and $\lambda_m(E)$ will represent the minimum eigenvalue of $E$. If of $E = E^T$, $E > 0$ will show that $E$ is positive definite. $E = (e_{ij})_{n \times n}$ is nonnegative matrix if $e_{ij} \geq 0, \forall i, j$. Assume that $E = (e_{ij})_{n \times n}$ and $F = (f_{ij})_{n \times n}$ are nonnegative matrices. In this case, $E \preceq F$ will denote that $e_{ij} \leq f_{ij}, \forall i, j$. For the vector $z$, we will use the norm $||z||_2^2 = \sum_{i=1}^{n} z_i^2$, and for $E$, we use $||E||_2 = [\lambda_{\max}(E^T E)]^{1/2}$.

## 2   Preliminaries

Consider the neural network of the mathematical form

$$\frac{dx_i(t)}{dt} = -c_i x_i(t) + \sum_{j=1}^{n} a_{ij} f_j(x_j(t)) + \sum_{j=1}^{n} b_{ij} f_j(x_j(t - \tau)) + u_i, \forall i \quad (1)$$

in this above equation, $a_{ij}$ and $b_{ij}$ are interconnection parameters, $c_i$ are the neurons charging rates, $x_i(t)$ represent state of neuron $i$, the functions $f_i(\cdot)$ are the nonlinear activation functions, $\tau$ represents the time delay, $u_i$ are the inputs.

Neural system (1) can be put into an equivalent system governed by the differential equation:

$$\dot{x}(t) = -Cx(t) + Af(x(t)) + Bf(x(t - \tau)) + u \quad (2)$$

where $C = diag(c_i)$, $A = (a_{ij})_{n \times n}$, $B = (b_{ij})_{n \times n}$, $x(t) = (x_1(t), x_2(t), ..., x_n(t))^T$, $u = (u_1, u_2, ..., u_n)^T$, $f(x(\cdot)) = (f_1(x_1(\cdot)), f_2(x_2(\cdot)), ..., f_n(x_n(\cdot)))^T$.

The functions $f_i$ possess the following properties:

$$0 \leq \frac{f_i(x) - f_i(\tilde{x})}{x - \tilde{x}} \leq k_i, \quad \forall i, \quad \forall x, \tilde{x} \in R, x \neq \tilde{x}$$

with $k_i$ being positive constants. The functions satisfying the above conditions are denoted by $f \in \mathcal{K}$.

The matrices $A = (a_{ij})$, $B = (b_{ij})$ and $C = diag(c_i > 0)$ in (1) are stated by the following intervals:

$$C_I := \{C : 0 \preceq \underline{C} \preceq C \preceq \overline{C}, \; i.e., \; 0 < \underline{c}_i \leq c_i \leq \overline{c}_i\}$$
$$A_I := \{A : \underline{A} \preceq A \preceq \overline{A}, \; i.e., \; \underline{a}_{ij} \leq a_{ij} \leq \overline{a}_{ij}\} \quad (3)$$
$$B_I := \{B : \underline{B} \preceq B \preceq \overline{B}, \; i.e., \; \underline{b}_{ij} \leq b_{ij} \leq \overline{b}_{ij}\}$$

We now introduce the following lemma which is of great importance to obtaining our main result:

**Lemma 1:** Let $D$ be a positive diagonal matrix with $n$ diagonal entries, $x$ be any real vector having $n$ elements, and consider any real $n \times n$ dimensional matrix $A = (a_{ij})$ with being intervalized as $\underline{A} \preceq A \preceq \overline{A}$. In this case, the following inequality is satisfied:

$$x^T A^T D A x \leq |x^T| [|A^{*T} D A^*| + |A^{*T}| D A_* + A_*^T D |A^*| + A_*^T D A_*] |x|$$

in which $A^* = \frac{1}{2}(\overline{A} + \underline{A})$ and $A_* = \frac{1}{2}(\overline{A} - \underline{A})$.

**Proof:** If $A \in A_I$, then, $a_{ij}$ can be written as

$$a_{ij} = \frac{1}{2}(\overline{a}_{ij} + \underline{a}_{ij}) + \frac{1}{2}\sigma_{ij}(\overline{a}_{ij} - \underline{a}_{ij}), \quad -1 \leq \sigma_{ij} \leq 1, \forall i, j.$$

Assume that $\tilde{A} = (\tilde{a}_{ij})_{n \times n}$ is a real constant matrix and whose elements are defined as $\tilde{a}_{ij} = \frac{1}{2}\sigma_{ij}(\overline{a}_{ij} - \underline{a}_{ij})$. Then, $A$ can be written as

$$A = \frac{1}{2}(\overline{A} + \underline{A}) + \tilde{A} = A^* + \tilde{A}$$

We can now express the following:

$$
\begin{aligned}
x^T A^T D A x &= x^T (A^* + \tilde{A})^T D (A^* + \tilde{A}) x \\
&= x^T (A^{*T} D A^* + A^{*T} D \tilde{A} + \tilde{A}^T D A^* + \tilde{A}^T D \tilde{A}) x \\
&\leq |x^T| |A^{*T} D A^* + A^{*T} D \tilde{A} + \tilde{A}^T D A^* + \tilde{A}^T D \tilde{A}| |x| \\
&\leq |x^T| |A^{*T} D A^*| |x| + |x^T| |A^{*T}| D |\tilde{A}| |x| \\
&\quad + |x^T| |\tilde{A}^T| D |A^*| |x| + |x^T| |\tilde{A}^T| D |\tilde{A}| |x|
\end{aligned}
$$

Since $|\tilde{a}_{ij}| \leq \frac{1}{2}(\overline{a}_{ij} - \underline{a}_{ij}), \forall i, j$, it follows that $|\tilde{A}| \preceq A_*$. Then, we obtain

$$
\begin{aligned}
x^T A^T D A x &\leq |x^T| |A^{*T} D A^*| |x| + |x^T| |A^{*T}| D |A_*| |x| \\
&\quad + |x^T| |A_*^T| D |A^*| |x| + |x^T| |A_*^T| D |A_*| |x| \\
&= |x^T| (|A^{*T} D A^*| + |A^{*T}| D A_* + A_*^T D |A^*| + A_*^T D A_*) |x|
\end{aligned}
$$

Below are two lemmas and a fact that will be needed in the proofs:

**Lemma 2 [1]:** Let $D$ be a positive diagonal matrix with $n$ diagonal entries, $x$ be any real vector having $n$ elements, and consider any real $n \times n$ dimensional matrix $A = (a_{ij})$ with being intervalized as $\underline{A} \preceq A \preceq \overline{A}$. In this case, the following inequality is satisfied:

$$x^T (DA + A^T D) x \leq |x^T| |S| |x|$$

where $S = (s_{ij})$ such that $s_{ii} = 2d_i\bar{a}_{ii}$ and $s_{ij} = max(|d_i\bar{a}_{ij} + d_j\bar{a}_{ji}|, |d_i\underline{a}_{ij} + d_j\underline{a}_{ji}|)$ for $i \neq j$.

**Lemma 3** [2]: Let the map $H(y) \in C^0$ posses two properties: $H(y) \neq H(z)$, $\forall y \neq z$ and $||H(y)|| \rightarrow \infty$ as $||y|| \rightarrow \infty$ with $y \in R^n$ and $z \in R^n$. Then, $H(y)$ is said to be homeomorphism of $R^n$.

**Fact 1:** If $A = (a_{ij})$ and $B = (b_{ij})$ satisfy (3), then, $A$ and $B$ have bounded norms, i.e., we can find some positive real constants $\epsilon$ and $\varepsilon$ satisfying the following

$$||A||_2 \leq \epsilon \text{ and } ||B||_2 \leq \varepsilon$$

## 3    Existence and Uniqueness Analysis

The following theorem presents the criterion which ensures that system (1) possesses a unique equilibrium point for each constant input:

**Theorem 1:** Let neuron activation functions belong $\mathcal{K}$, and assume that the uncertain network elements $A$, $B$ and $C$ are defined by (3). Then, delayed neural network described by (1) possesses only one equilibrium point, if one can find a matrix $D = diag(d_i > 0)$ satisfying the following condition

$$\Theta = 2\underline{C}DK^{-1} - D - S - Q > 0$$

where $K = diag(k_i > 0)$, $Q = B_*^T D|B^*| + B_*^T DB_* + |B^{*T} DB^*| + |B^{*T}|DB_*$, $S = (s_{ij})$ is the matrix whose diagonal elements are defined by $s_{ii} = 2d_i\bar{a}_{ii}$ and off-diagonal elements are defined by $s_{ij} = max(|d_i\bar{a}_{ij} + d_j\bar{a}_{ji}|, |d_i\underline{a}_{ij} + d_j\underline{a}_{ji}|)$, the matrix $B^*$ included in $Q$ is defined as $B^* = \frac{1}{2}(\bar{B} + \underline{B})$ and the nonnegative matrix $B_*$ included in $Q$ is defined as $B_* = \frac{1}{2}(\bar{B} - \underline{B})$.

**Proof:** Consider the associated map $H(x)$ representing neural network (2)

$$H(x) = -Cx + Af(x) + Bf(x) + u \tag{4}$$

For every equilibrium point $x^*$ of (2), by definition of equilibrium equation, the following must be satisfied

$$-Cx^* + Af(x^*) + Bf(x^*) + u = 0$$

Apparently, when a vector $x$ satisfies $H(x) = 0$, it results in the fact that $H(x) = 0$ also corresponds to the equilibrium points representing the solutions of (2). Thus, by the virtue of Lemma 3, one can conclude that neural model (2) possesses only one equilibrium point for the constant $u$ if $H(x)$ fulfills conditions in Lemma 3. For any randomly selected vectors $x \neq y$, using (4), we express

$$H(x) - H(y) = -C(x - y) + A(f(x) - f(y)) + B(f(x) - f(y))$$

Let $H(x,y) = H(x) - H(y)$ and $f(x,y) = f(x) - f(y)$. Then, the previous equation can be put in the form:

$$H(x,y) = -C(x-y) + Af(x,y) + Bf(x,y) \qquad (5)$$

Since $f \in \mathcal{K}$, if $x \neq y$ and $f(x) = f(y)$, (5) yields

$$H(x,y) = -C(x-y)$$

in which $C = diag(c_i > 0)$. Therefore, $x - y \neq 0$ will ensure the condition that $H(x,y) \neq 0$. For $x - y \neq 0$, $f(x,y) \neq 0$, and $D = diag(d_i > 0)$, multiplying (5) by the nonzero vector $2f^T(x,y)D$ leads to

$$2f^T(x,y)DH(x,y) = -2f^T(x,y)DC(x-y)$$
$$+ 2f^T(x,y)DAf(x,y)$$
$$+ 2f^T(x,y)DBf(x,y)$$

The following can be written

$$2f^T(x,y)DAf(x,y) = f^T(x,y)(DA + A^TD)f(x,y)$$

Thus, one would obtain

$$2f^T(x,y)DH(x,y) = -2f^T(x,y)DC(x-y)$$
$$+ f^T(x,y)(DA + A^TD)f(x,y)$$
$$+ f^T(x,y)DBf(x,y) \qquad (6)$$

For activation functions in $\mathcal{K}$, the following can be derived

$$-2f^T(x,y)DC(x-y) = -\sum_{i=1}^{n} 2d_i c_i (f_i(x_i) - f_i(y_i))(x_i - y_i)$$
$$\leq -2\sum_{i=1}^{n} \frac{d_i c_i}{k_i}(f_i(x_i) - f_i(y_i))^2$$
$$= -2f^T(x,y)\underline{C}DK^{-1}f(x,y) \qquad (7)$$

Lemma 2 leads to

$$f^T(x,y)(DA + A^TD)f(x,y) \leq |f^T(x,y)|S|f(x,y)| \qquad (8)$$

It is worth noting that

$$2f^T(x,y)DBf(x,y) \leq f^T(x,y)Df(x,y) + f^T(x,y)B^TDBf(x,y)$$

By using Lemma 1, one would get

$$f^T(x,y)B^TDBf(x,y) \leq |f^T(x,y)|Q|f(x,y)|$$

Thus

$$2f^T(x,y)DBf(x,y) \leq f^T(x,y)Df(x,y) + |f^T(x,y)|Q|f(x,y)| \qquad (9)$$

Using (7)–(9) in (6) will give the following

$$\begin{aligned}
2f^T(x,y)DH(x,y) &\leq -2|f^T(x,y)||CDK^{-1}|f(x,y)| \\
&\quad + |f^T(x,y)|(S+D+Q)|f(x,y)| \\
&= -|f(x,y)|^T\Theta|f(x,y)|
\end{aligned}$$

Since $\Theta > 0$, one can observe that

$$2f^T(x,y)DH(x,y) \leq -\lambda_m(\Theta)||f(x,y)||_2^2 \qquad (10)$$

Obviously, $f(x,y) \neq 0$ with $\Theta$ being positive definite, that is $\Theta > 0$, (10) leads to

$$2f^T(x,y)DH(x,y) < 0$$

where $f(x,y) \neq 0$ guarantees condition that $H(x) \neq H(y)$ for all $x \neq y$.
Choosing $y = 0$, (10) will directly result in

$$2(f(x) - f(0))^T D(H(x) - H(0)) \leq -\lambda_m(\Theta)||f(x) - f(0)||_2^2$$

It follows from the above inequality that

$$|2(f(x) - f(0))^T D(H(x) - H(0))| \geq \lambda_m(\Theta)||f(x) - f(0)||_2^2$$

yielding

$$||H(x) - H(0)||_1 > \frac{\lambda_m(\Theta)||f(x) - f(0)||_2^2}{2||D||_\infty||f(x) - f(0)||_\infty}$$

Using some basic properties of the vector norms, we can state

$$||H(x)||_1 > \frac{\lambda_m(\Theta)(||f(x)||_2 - ||f(0)||_2 - 2||D||_\infty||H(0)||_1)}{2||D||_\infty}$$

Knowing that $||H(0)||_1$, $||D||_\infty$, and $||f(0)||_2$ have limited upper bounds will enable us to conclude that $||H(x)|| \to \infty$ if $||x|| \to \infty$. Q.E.D.

## 4    Stability Analysis

Stability of neural network (1) will be studied in this section. If $x^*$ is defined to denote an equilibrium point of (1), then, by means of $z_i(\cdot) = x_i(\cdot) - x_i^*$, neural system (1) is replaced by a new model whose dynamics is governed by:

$$\dot{z}_i(t) = -c_i z_i(t) + \sum_{j=1}^n a_{ij}g_j(z_j(t)) + \sum_{j=1}^n b_{ij}g_j(z_j(t-\tau)) \qquad (11)$$

Note that $g_i(z_i(\cdot)) = f_i(z_i(\cdot) + x_i^*) - f_i(x_i^*)$. We can easily observe that the functions $g_i$ belong to the class $\mathcal{K}$, that is, $g \in \mathcal{K}$ satisfying $g_i(0) = 0$.

The vector-matrix form of neural system (11) is

$$\dot{z}(t) = -Cz(t) + Ag(z(t)) + Bg(z(t-\tau)) \tag{12}$$

In this new system, $z(t) = (z_1(t), z_2(t), ..., z_n(t))^T$, and the new nonlinear output functions state vector is $g(z(\cdot)) = (g_1(z_1(\cdot)), g_2(z_2(\cdot)), ..., g_n(z_n(\cdot)))^T$.

The stability result is given as follows:

**Theorem 2:** Let neuron activation functions belong $\mathcal{K}$, and assume that the uncertain network elements $A$, $B$ and $C$ are given by (3). Then, the origin of delayed neural system described by (11) is globally asymptotically stable, if one can find an appropriate matrix $D = diag(d_i > 0)$ satisfying the following condition

$$\Theta = 2\underline{C}DK^{-1} - D - S - Q > 0$$

where $K = diag(k_i > 0)$, $Q = B_*^T D|B^*| + B_*^T DB_* + |B^{*T} DB^*| + |B^{*T}|DB_*$, $S = (s_{ij})$ is the matrix whose diagonal elements are defined by $s_{ii} = 2d_i\bar{a}_{ii}$ and off-diagonal elements are defined by $s_{ij} = max(|d_i\bar{a}_{ij} + d_j\bar{a}_{ji}|, |d_i\underline{a}_{ij} + d_j\underline{a}_{ji}|)$, the matrix $B^*$ included in $Q$ is defined as $B^* = \frac{1}{2}(\overline{B} + \underline{B})$ and the nonnegative matrix $B_*$ included in $Q$ is defined as $B_* = \frac{1}{2}(\overline{B} - \underline{B})$.

**Proof:** The Lyapunov functional to be exploited for the proof of this theorem is chosen as:

$$V(z(t)) = \sum_{i=1}^{n}(z_i^2(t) + 2\gamma \int_0^{z_i(t)} d_i g_i(s)ds)$$

$$+ \gamma \int_{t-\tau}^{t} |g^T(z(\zeta))|Q|g(z(\zeta))|d\zeta + \xi \int_{t-\tau}^{t} \|g(z(\zeta))\|_2^2 d\zeta$$

where the $d_i$, $\gamma$ and $\xi$ are constants. $\dot{V}(z(t))$ is determined to be as

$$\begin{aligned}
\dot{V}(z(t)) = &-2z^T(t)Cz(t) + 2z^T(t)Ag(z(t)) + 2z^T(t)Bg(z(t-\tau)) \\
&- 2\gamma g^T(z(t))DCz(t) + 2\gamma g^T(z(t))DAg(z(t)) \\
&+ 2\gamma g^T(z(t))DBg(z(t-\tau)) \\
&+ \gamma|g^T(z(t))|Q|g(z(t))| - \gamma|g^T(z(t-\tau))|Q|g(z(t-\tau))| \\
&+ \xi\|g(z(t))\|_2^2 - \xi\|g(z(t-\tau))\|_2^2
\end{aligned} \tag{13}$$

Let $\alpha = \|A\|_2^2\|C^{-1}\|_2$ and $\beta = \|A\|_2^2\|C^{-1}\|_2$. We now observe the inequalities:

$$2z^T(t)Ag(z(t)) - z^T(t)Cz(t) \le \alpha\|g(z(t))\|_2^2 \tag{14}$$

$$2z^T(t)Bg(z(t-\tau)) - z^T(t)Cz(t) \le g^T(z(t-\tau))B^T C^{-1} Bg(z(t-\tau))$$
$$\le \beta\|g(z(t-\tau))\|_2^2 \tag{15}$$

$$2\gamma g^T(z(t))DAg(z(t)) = \gamma g^T(z(t))(DA + A^T D)g(z(t))$$
$$\leq \gamma |g^T(z(t))|S|g(z(t))| \tag{16}$$

$$2\gamma g^T(z(t))DBg(z(t-\tau)) \leq \gamma g^T(z(t))Dg(z(t))$$
$$+ \gamma g^T(z(t-\tau))B^T DBg(z(t-\tau))$$
$$\leq \gamma g^T(z(t))Dg(z(t))$$
$$+ \gamma |g^T(z(t-\tau))|Q|g(z(t-\tau))| \tag{17}$$

$$- 2\gamma g^T(z(t))DCz(t) \leq -2\gamma g^T(z(t))DCK^{-1}g(z(t)) \tag{18}$$

Inserting (14)–(18) into (13) yields

$$\dot{V}(z(t)) \leq \alpha\|g(z(t))\|_2^2 + \beta g^T(z(t))g(z(t))$$
$$- 2\gamma g^T(z(t))P\underline{C}K^{-1}g(z(t)) + \gamma |g^T(z(t))|S|g(z(t))|$$
$$+ \gamma g^T(z(t))Pg(z(t)) + \gamma |g^T(z(t-\tau))|Q|g(z(t-\tau))|$$
$$+ \gamma |g^T(z(t))|Q|g(z(t))| - \gamma |g^T(z(t-\tau))|Q|g(z(t-\tau))|$$
$$+ \xi\|g(z(t))\|_2^2 - \xi\|g(z(t-\tau))\|_2^2$$

Taking $\xi = \beta$ leads to

$$\dot{V}(z(t)) \leq (\beta + \alpha)\|g(z(t))\|_2^2 - 2\gamma g^T(z(t))D\underline{C}K^{-1}g(z(t))$$
$$+ \gamma |g^T(z(t))|S|g(z(t))| + \gamma g^T(z(t))Dg(z(t))$$
$$+ \gamma |g^T(z(t))|Q|g(z(t))|$$
$$= (\beta + \alpha)\|g(z(t))\|_2^2 - \gamma |g^T(z(t))|(D\underline{C}K^{-1} - D - S - Q)|g(z(t))|$$
$$= (\beta + \alpha)\|g(z(t))\|_2^2 - \gamma |g^T(z(t))|\Theta|g(z(t))| \tag{19}$$

Since $\Theta > 0$, (19) gives

$$\dot{V}(z(t)) \leq - (\gamma\lambda_m(\Theta) - (\beta + \alpha)\|)g(z(t))\|_2^2$$

Thus

$$\gamma > \frac{(\alpha + \beta)}{\lambda_m(\Theta)}$$

guarantees that $\dot{V}(z(t))$ will have negative values $\forall g(z(t)) \neq 0$, or equivalently $\dot{V}(z(t)) < 0$ when $z(t) \neq 0$.

Let $g(z(t)) = 0$. Taking $z(t) \neq 0$ leads to

$$\dot{V}(z(t)) = -2z^T(t)Cz(t) + 2z^T(t)Bg(z(t-\tau))$$
$$- \gamma |g^T(z(t-\tau))|Q|g(z(t-\tau))| - \xi\|g(z(t-\tau))\|_2^2$$

Then

$$-z^T(t)Cz(t) + 2z^T(t)Bg(z(t-\tau)) \leq \xi\|g(z(t-\tau))\|_2^2$$

leads to
$$\dot{V}(z(t)) \le -z^T(t)Cz(t)$$
in which $\dot{V}(z(t)) < 0 \ \forall z(t) \neq 0$. Finally, $g(z(t)) = z(t) = 0$ leads to
$$\dot{V}(z(t)) \le -\xi \|g(z(t-\tau))\|_2^2$$
Apparently, $\dot{V}(z(t)) < 0 \ \forall g(z(t-\tau)) \neq 0$. Hence, $\dot{V}(z(t)) = 0$ iff $z(t) = g(z(t)) = g(z(t-\tau)) = 0$. This also means $\dot{V}(z(t)) < 0$ when states and delayed states are not equal to zero. The radially unboundedness the Lyapunov functional is to be easily checked by proving $V(z(t)) \to \infty$ when $\|z(t)\| \to \infty$. Q.E.D.

## 5    Conclusions

This work proposed a further improved condition for the robustness of neural networks involving intervalized network parameters and including single time delay. For the sake of obtaining a new robust stability condition, a new upper bound for the norm of the intervalized interconnection matrices has been established. The homeomorphism mapping and Lyapunov stability theorems are employed to derive the proposed stability condition by making use of this upper bound norm. The obtained result is applicable to all nondecreasing slope-bounded activation functions and imposes constraints on parameters of neural network without involving time delay.

## References

1. Ensari, T., Arik, S.: New results for robust stability of dynamical neural networks with discrete time delays. Expert Syst. Appl. **37**, 5925–5930 (2010)
2. Faydasicok, O., Arik, S.: Robust stability analysis of a class of neural networks with discrete time delays. Neural Netw. **29–30**, 52–59 (2012)
3. Faydasicok, O., Arik, S.: A new upper bound for the norm of interval matrices with application to robust stability analysis of delayed neural networks. Neural Netw. **44**, 64–71 (2013)
4. Shao, J.L., Huang, T.Z., Zhou, S.: Some improved criteria for global robust exponential stability of neural networks with time-varying delays. Commun. Nonlinear Sci. Numer. Simul. **15**, 3782–3794 (2010)
5. Zhu, Q., Cao, J.: Robust exponential stability of markovian jump impulsive stochastic Cohen-Grossberg neural networks with mixed time delays. IEEE Trans. Neural Netw. **21**, 1314–1325 (2010)
6. Huang, T., Li, C., Duan, S., Starzyk, J.A.: Robust exponential stability of uncertain delayed neural networks with stochastic perturbation and impulse effects. IEEE Trans. Neural Netw. Learn. Syst. **23**, 866–875 (2012)
7. Zhou, J., Xu, S., Zhang, B., Zou, Y., Shen, H.: Robust exponential stability of uncertain stochastic neural networks with distributed delays and reaction-diffusions. IEEE Trans. Neural Netw. Learn. Syst. **23**, 1407–1416 (2012)
8. Shen, Y., Wang, J.: Robustness analysis of global exponential stability of recurrent neural networks in the presence of time delays and random disturbances. IEEE Trans. Neural Netw. Learn. Syst. **23**, 87–96 (2012)

9. Song, Q., Yu, Q., Zhao, Z., Liu, Y., Alsaadi, F.E.: Boundedness and global robust stability analysis of delayed complex-valued neural networks with interval parameter uncertainties. Neural Netw. **103**, 55–62 (2018)
10. Li, X., et al.: Extended robust global exponential stability for uncertain switched memristor-based neural networks with time-varying delays. Appl. Math. Comput. **325**, 271–290 (2018)
11. Ali, M.S., Gunasekaran, N., Rani, M.E.: Robust stability of hopfield delayed neural networks via an augmented L-K functional. Neurocomputing **234**, 198–204 (2017)
12. Wang, Z., Wang, J., Wu, Y.: State estimation for recurrent neural networks with unknown delays: a robust analysis approach. Neurocomputing **227**, 20–36 (2017)
13. Wan, Y., Cao, J., Wen, G., Yu, W.: Robust fixed-time synchronization of delayed Cohen-Grossberg neural networks. Neural Netw. **73**, 86–94 (2016)
14. Muthukumar, P., Subramanian, K., Lakshmanan, S.: Robust finite time stabilization analysis for uncertain neural networks with leakage delay and probabilistic time-varying delays. J. Franklin Inst. **353**, 4091–4113 (2016)
15. Gong, W., Liang, J., Kan, X., Nie, X.: Robust state estimation for delayed complex-valued neural networks. Neural Process. Lett. **46**, 1009–1029 (2017)
16. Dong, Y., Liang, S., Guo, L.: Robustly exponential stability analysis for discrete-time stochastic neural networks with interval time-varying delays. Neural Process. Lett. **46**, 135–158 (2017)
17. Arik, S.: New criteria for global robust stability of delayed neural networks with norm-bounded uncertainties. IEEE Trans. Neural Netw. Learn. Syst. **25**, 1045–1052 (2014)
18. Wang, J., Zhang, H., Wang, Z., Shan, Q.: Local synchronization criteria of markovian nonlinearly coupled neural networks with uncertain and partially unknown transition rates. IEEE Trans. Syst. Man Cybern. Syst. **47**, 1953–1964 (2017)
19. Ding, Z., Zeng, Z., Wang, L.: Robust finite-time stabilization of fractional-order neural networks with discontinuous and continuous activation functions under uncertainty. IEEE Trans. Neural Netw. Learn. Syst. **29**, 1477–1490 (2018)

# A Novel Criterion for Global Asymptotic Stability of Neutral-Type Neural Networks with Discrete Time Delays

Ozlem Faydasicok[1] and Sabri Arik[2]([⊠])

[1] Department of Mathematics, Istanbul University, Istanbul, Turkey
kozlem@istanbul.edu.tr
[2] Department of Computer Engineering,
Istanbul University-Cerrahpasa, Istanbul, Turkey
ariks@istanbul.edu.tr

**Abstract.** The main target in this work is to propose a new condition for global asymptotic stability for neutral-type neural networks including constant delay parameters. This newly obtained criterion is derived by making the use a properly modified Lyapunov functional with the employment of Lipschitz activation functions, which establishes a new set of relationships among the constant system parameters of this class of neural networks. The obtained condition is expressed independently of delay parameters and can be easily approved by simply checking the validity of some algebraic equations.

**Keywords:** Discrete time delays · Stability theory
Neutral-type neural systems · Lyapunov theorems

## 1 Introduction

Recently, stability problem for different types of neural network systems including Hopfield, Fuzzy, Markovian, Cohen-Grossberg and cellular neural networks has been extensively studied as these systems capable of solving various real world engineering problems regarding image processing, system control, pattern recognition and data compression. In such applications, the key property to be established is the stability of these neural networks. Besides, when neural networks are electronically implemented, some transmissions delays will occur because of unavoidable finite switch speed of neurons. Thus, we need to consider the affect of these delay parameters in the dynamical analysis of neutral networks. In this case, one needs to include these delays into mathematical models describing neural networks. One of usual approaches to addressing this question is involving these delay parameters in the state variables governing the dynamics of neutral neural systems. Besides, we know that the time derivatives of state variables are also changing in time, time derivatives of state variables must also include these delay parameters, which will enable us to carry out a more complete

© Springer Nature Switzerland AG 2018
L. Cheng et al. (Eds.): ICONIP 2018, LNCS 11302, pp. 353–360, 2018.
https://doi.org/10.1007/978-3-030-04179-3_31

stability analysis of neural systems. Such more generalized class of neutral-type systems possessing delay parameters in states and as well as in derivatives of states are known as delayed neutral-type neural networks.

Recently, numerous research papers have dealt with stability of some classes of neutral-type systems involving delay parameters, where the authors of these papers have derived various valuable results on global asymptotic stability of these types of systems [1–15]. The majority of existing stability criterion for this neural network model are mainly stated in forms of linear matrix inequalities. However, stability conditions in linear matrix inequality forms require to solve the high dimensional matrix equations to derive the constraint conditions imposed on system parameters of neutral network model. In this research work, with the help of a adequate Lyapunov functional, we submit an algebraic criterion for stability of neutral neural systems that include discrete delay parameters. This newly proposed condition establishes a simple and easily verifiable relationship among the network parameters of the system.

Consider a neutral-type neural network governed by differential equations given below:

$$\dot{x}_i(t) = -c_i x_i(t) + \sum_{j=1}^{n} a_{ij} f_j(x_j(t)) + \sum_{j=1}^{n} b_{ij} f_j(x_j(t - \tau_j))$$

$$+ \sum_{j=1}^{n} e_{ij} \dot{x}_j(t - \tau_j) + u_i, \; i = 1, 2, \cdots, n. \tag{1}$$

in this above equation, $a_{ij}$, $b_{ij}$ and $e_{ij}$ are constant system parameters, $c_i$ are the neurons charging rates, $x_i(t)$ represent state of neuron $i$, the functions $f_i(\cdot)$ are the nonlinear activation functions, $\tau_j$ represents the time delay, $u_i$ are the inputs.

The functions $f_i$ possess the property:

$$|f_i(x) - f_i(\hat{x})| \le \ell_i |x - \hat{x}|, \; \forall x \ne \hat{x}$$

with $k\ell_i$ being positive constants. The functions satisfying the above conditions are denoted by $f \in \mathcal{L}$.

## 2    Global Stability Analysis

Stability of neural network (1) will be studied in this section. If $x^*$ is defined to denote an equilibrium point of (1), then, by means of $z_i(\cdot) = x_i(\cdot) - x_i^*$, neural system (1) is replaced by a new model whose dynamics is governed by:

$$\dot{z}_i(t) = -c_i z_i(t) + \sum_{j=1}^{n} a_{ij} g_j(z_j(t)) + \sum_{j=1}^{n} b_{ij} g_j(z_j(t - \tau_j)) + \sum_{j=1}^{n} e_{ij} \dot{z}_j(t - \tau_j) \tag{2}$$

The vector-matrix form of neural system (2) is

$$\dot{z}(t) = -Cz(t) + Ag(z(t)) + Bg(z(t - \tau)) + E\dot{z}(t - \tau) \tag{3}$$

in which $C = diag(c_i > 0)$, $A = (a_{ij})_{n \times n}$, $B = (b_{ij})_{n \times n}$, $E = (e_{ij})_{n \times n}$, $z(t) = [z_1(t), z_2(t), \cdots, z_n(t)]^T$ is the vector representing neurons states, $g(z(t)) = [g_1(z_1(t)), g_2(z_2(t)), \cdots, g_n(z_n(t))]^T$ is the vector representing nonlinear neurons outputs, $g(z(t - \tau)) = [g_1(z_1(t - \tau_1)), g_2(z_2(t - \tau_2)), \cdots, g_n(z_n(t - \tau_n))]^T$ is the vector representing delayed nonlinear neurons outputs, and $\dot{z}(t - \tau) = [\dot{z}_1(t - \tau_1), \dot{z}_2(t - \tau_2), \cdots, \dot{z}_n(t - \tau_n)]^T$ is the vector representing time derivatives of the neurons states.

After this transformation, the activation functions $g_i(z_i(t))$ of this new system satisfy the following conditions:

$$g_i(z_i(t)) = f_i(z_i(t) + x_i^*) - f_i(x_i^*)$$

from which it follows that

$$|g_i(z_i(\cdot))| \le \ell_i |z_i(\cdot)|$$

Note that $f \in \mathcal{L}$ implies that $g \in \mathcal{L}$.

The stability result is given as follows:

**Theorem 1:** Let neuron activation functions belong to $\mathcal{L}$. Then, the origin of delayed neutral-type neural network system described by (2) is globally asymptotically stable if it is possible to find some appropriate positive real numbers $\alpha$ and $\beta$ satisfying the conditions:

$$\delta_i = \frac{c_i^2}{\ell_i^2} - \left(\frac{2\alpha + 1}{\alpha}\right) \sum_{j=1}^{n} |\sum_{k=1}^{n} a_{ki} a_{kj}| - \left(\frac{2\beta + 1}{\beta}\right) \sum_{j=1}^{n} |\sum_{k=1}^{n} b_{ki} b_{kj}| > 0, \forall i$$

$$\Theta = -E^T E - \alpha E^T E - \beta E^T E + I > 0$$

**Proof:** The Lyapunov functional to be exploited for the proof of this theorem can be chosen as formulated in following:

$$V(z(t)) = \sum_{i=1}^{n} (c_i z_i^2(t) + \int_{t-\tau_i}^{t} \dot{z}_i^2(s) ds + \int_{t-\tau_i}^{t} \gamma_i g_i^2(z_i(s)) ds)$$

In this Lyapunov functional, the $\gamma_i$ are some positive constants. $\dot{V}(z(t))$ is determined to be

$$\dot{V}(z(t)) = \sum_{i=1}^{n} (2c_i z_i(t) \dot{z}_i(t) + \dot{z}_i^2(t) - \dot{z}_i^2(t - \tau_i))$$

$$+ \sum_{i=1}^{n} \gamma_i g_i^2(z_i(t)) - \sum_{i=1}^{n} \gamma_i g_i^2(z_i(t - \tau_i)) \qquad (4)$$

The matrix-vector form of (4) is:

$$\dot{V}(z(t)) = 2z^T(t)C\dot{z}(t) + \dot{z}^T(t)\dot{z}(t) - \dot{z}^T(t-\tau)\dot{z}(t-\tau)$$
$$+ \sum_{i=1}^{n}\gamma_i g_i^2(z_i(t)) - \sum_{i=1}^{n}\gamma_i g_i^2(z_i(t-\tau_i))$$
$$= [2Cz(t) + \dot{z}(t)]^T\dot{z}(t) - \dot{z}^T(t-\tau)\dot{z}(t-\tau)$$
$$+ \sum_{i=1}^{n}\gamma_i g_i^2(z_i(t)) - \sum_{i=1}^{n}\gamma_i g_i^2(z_i(t-\tau_i))$$
$$= [Cz(t) + Ag(z(t)) + Bg(z(t-\tau)) + E\dot{z}(t-\tau)]^T$$
$$\times [-Cz(t) + Ag(z(t)) + Bg(z(t-\tau)) + E\dot{z}(t-\tau)]$$
$$- \dot{z}^T(t-\tau)\dot{z}(t-\tau)$$
$$+ \sum_{i=1}^{n}\gamma_i g_i^2(z_i(t)) - \sum_{i=1}^{n}\gamma_i g_i^2(z_i(t-\tau_i)) \tag{5}$$

One can write the equality

$$[Cz(t) + Ag(z(t)) + Bg(z(t-\tau)) + E\dot{z}(t-\tau)]^T$$
$$\times [-Cz(t) + Ag(z(t)) + Bg(z(t-\tau)) + E\dot{z}(t-\tau)]$$
$$= z^T(t)C^2 z(t) + g^T(z(t))A^T Ag(z(t)) + 2g^T(z(t))A^T Bg(z(t-\tau))$$
$$+ 2g^T(z(t))A^T E\dot{z}(t-\tau) + g^T(z(t-\tau))B^T Bg(z(t-\tau))$$
$$+ 2g^T(z(t-\tau))B^T E\dot{z}(t-\tau) + \dot{z}^T(t-\tau)E^T E\dot{z}(t-\tau) \tag{6}$$

Using (6) in (5) yields

$$\dot{V}(z(t)) = -z^T(t)C^2 z(t) + g^T(z(t))A^T Ag(z(t)) + 2g^T(z(t))A^T Bg(z(t-\tau))$$
$$+ 2g^T(z(t))A^T E\dot{z}(t-\tau) + g^T(z(t-\tau))B^T Bg(z(t-\tau))$$
$$+ 2g^T(z(t-\tau))B^T E\dot{z}(t-\tau) + \dot{z}^T(t-\tau)E^T E\dot{z}(t-\tau)$$
$$- \dot{z}^T(t-\tau)\dot{z}(t-\tau) + \sum_{i=1}^{n}\gamma_i g_i^2(z_i(t)) - \sum_{i=1}^{n}\gamma_i g_i^2(z_i(t-\tau_i)) \tag{7}$$

Now note that:

$$2g^T(z(t))A^T Bg(z(t-\tau)) \le g^T(z(t))A^T Ag(z(t)) + g^T(z(t-\tau))B^T Bg(z(t-\tau)) \tag{8}$$

$$2g^T(z(t))A^T E\dot{z}(t-\tau) \le \frac{1}{\alpha}g^T(z(t))A^T Ag(z(t)) + \alpha\dot{z}^T(t-\tau)E^T E\dot{z}(t-\tau) \tag{9}$$

$$2g^T(z(t-\tau))B^T E\dot{z}(t-\tau) \le \frac{1}{\beta}g^T(z(t-\tau))B^T Bg(z(t-\tau))$$
$$+ \beta\dot{z}^T(t-\tau)E^T E\dot{z}(t-\tau) \tag{10}$$

with some particularly selected positive constants $\alpha$ and $\beta$. Inserting (8)–(10) into (7) leads to

$$\dot{V}(z(t)) = -z^T(t)C^2 z(t) + (\frac{2\alpha + 1}{\alpha})g^T(z(t))A^T Ag(z(t))$$

$$+ (\frac{2\beta + 1}{\beta})g^T(z(t - \tau))B^T Bg(z(t - \tau))$$

$$+ (\alpha + \beta + 1)\dot{z}^T(t - \tau)E^T E\dot{z}(t - \tau) - \dot{z}^T(t - \tau)\dot{z}(t - \tau)$$

$$+ \sum_{i=1}^{n} \gamma_i g_i^2(z_i(t)) - \sum_{i=1}^{n} \gamma_i g_i^2(z_i(t - \tau_i)) \tag{11}$$

We know state a well-known matrix property. If $A$ and $B$ are real matrices, then, the following conditions hold

$$g^T(z(t))A^T Ag(z(t)) \leq \sum_{i=1}^{n}\sum_{j=1}^{n}|\sum_{k=1}^{n} a_{ki}a_{kj}|g_i^2(z_i(t)) \tag{12}$$

and

$$g^T(z(t - \tau))B^T Bg(z(t - \tau)) \leq \sum_{i=1}^{n}\sum_{j=1}^{n}|\sum_{k=1}^{n} b_{ki}b_{kj}|g_i^2(z_i(t - \tau_i)) \tag{13}$$

Combining (12) and (13) with (11) leads to the following inequality

$$\dot{V}(z(t)) \leq -\sum_{i=1}^{n} c_i^2 z_i^2(t) + (\frac{2\alpha + 1}{\alpha})\sum_{i=1}^{n}\sum_{j=1}^{n}|\sum_{k=1}^{n} a_{ki}a_{kj}|g_i^2(z_i(t))$$

$$+ (\frac{2\beta + 1}{\beta})\sum_{i=1}^{n}\sum_{j=1}^{n}|\sum_{k=1}^{n} b_{ki}b_{kj}|g_i^2(z_i(t - \tau_i)) - \dot{z}^T(t - \tau)\dot{z}(t - \tau)$$

$$+ \sum_{i=1}^{n} \gamma_i g_i^2(z_i(t)) - \sum_{i=1}^{n} \gamma_i g_i^2(z_i(t - \tau_i))$$

$$+ (1 + \alpha + \beta)\dot{z}^T(t - \tau)E^T E\dot{z}(t - \tau) \tag{14}$$

Letting

$$\gamma_i = (\frac{2\beta + 1}{\beta})\sum_{j=1}^{n}|\sum_{k=1}^{n} b_{ki}b_{kj}|$$

will make (14) to lead to

$$\dot{V}(z(t)) \leq -\sum_{i=1}^{n} c_i^2 z_i^2(t) + (\frac{2\alpha + 1}{\alpha})\sum_{i=1}^{n}\sum_{j=1}^{n}|\sum_{k=1}^{n} a_{ki}a_{kj}|g_i^2(z_i(t))$$

$$+ (\frac{2\beta + 1}{\beta})\sum_{i=1}^{n}\sum_{j=1}^{n}|\sum_{k=1}^{n} b_{ki}b_{kj}|g_i^2(z_i(t)) - \dot{z}^T(t - \tau)\dot{z}(t - \tau)$$

$$+ (\alpha + \beta + 1)\dot{z}^T(t - \tau)E^T E\dot{z}(t - \tau) \tag{15}$$

F $f \in \mathcal{L}$ will directly satisfy the inequality

$$\frac{g_i^2(z_i(t))}{\ell_i^2} \leq z_i^2(t) \forall i$$

Thus, (15) yields

$$\dot{V}(z(t)) \leq -\sum_{i=1}^{n} \frac{c_i^2}{\ell_i^2} g_i^2(z_i(t)) + \left(\frac{2\alpha+1}{\alpha}\right) \sum_{i=1}^{n} \sum_{j=1}^{n} \left| \sum_{k=1}^{n} a_{ki}a_{kj}\right| g_i^2(z_i(t))$$

$$+ \left(\frac{2\beta+1}{\beta}\right) \sum_{i=1}^{n} \sum_{j=1}^{n} \left| \sum_{k=1}^{n} b_{ki}b_{kj}\right| g_i^2(z_i(t))$$

$$+ (1+\alpha+\beta)\dot{z}^T(t-\tau)E^T E\dot{z}(t-\tau) - \dot{z}^T(t-\tau)\dot{z}(t-\tau)$$

$$= -\sum_{i=1}^{n} \left[\frac{c_i^2}{\ell_i^2} - \left(2+\frac{1}{\alpha}\right) \sum_{j=1}^{n} \left| \sum_{k=1}^{n} a_{ki}a_{kj}\right| - \left(2+\frac{1}{\beta}\right) \sum_{j=1}^{n} \left| \sum_{k=1}^{n} b_{ki}b_{kj}\right| \right] g_i^2(z_i(t))$$

$$- \dot{z}^T(t-\tau)(I - (1+\alpha+\beta)E^T E)\dot{z}(t-\tau)$$

$$= -\sum_{i=1}^{n} \delta_i g_i^2(z_i(t)) - \dot{z}^T(t-\tau)\Theta\dot{z}(t-\tau) \tag{16}$$

Let $\delta_m = min\{\delta_i\}$. Then, since $\Theta$ is positive definite, from (16), we get

$$\dot{V}(z(t)) \leq -\sum_{i=1}^{n} \delta_i g_i^2(z_i(t)) \leq -\sum_{i=1}^{n} \delta_m g_i^2(z_i(t)) = -\delta_m g^T(z(t))g(z(t)) \tag{17}$$

directly implying the condition of $\dot{V}(z(t))$ taking real negative values if $g(z(t)) \neq 0$. When taking the nonlinear output function vector $g(z(t)) = 0$ together with taking the state vector $z(t) \neq 0$, the inequality expressed by (15) leads to

$$\dot{V}(z(t)) \leq -\sum_{i=1}^{n} c_i^2 z_i^2(t) - \dot{z}^T(t-\tau)\Theta\dot{z}(t-\tau)$$

$$\leq -c_m z^T(t)z(t)$$

where $c_m = min\{c_i\}$. One can observe the result that $V(z(t))$ will have real negative values when $z(t) \neq 0$ together with the fact that $c_m$ is a positive constant. When taking the nonlinear output function vector $g(z(t)) = 0$ together with taking the state vector $z(t) = 0$, the inequality expressed by (15) leads to

$$\dot{V}(z(t)) \leq -\dot{z}^T(t-\tau)\Theta\dot{z}(t-\tau)$$

in the above expression, the positive definiteness of $\Theta$ would directly lead to the fact of $\dot{V}(z(t)) < 0$ for every $\dot{z}(t-\tau) \neq 0$. As for the last case of the analysis of $\dot{V}(z(t))$, one can observe that, if $g(z(\cdot)) = z(\cdot) = \dot{z}(\cdot) = 0$, then, one would easily get the expression

$$\dot{V}(z(t)) \leq -\gamma_m \|g(z(t-\tau))\|_2^2$$

noting that $\gamma_m = min\{\gamma_i\}$. Note that $\dot{V}(z(t)) < 0$ for all nonlinear delayed output vector $g(z - \tau) \neq 0$ as $\gamma_m$ is a positive constant. Hence, $\dot{V}(z(t)) = 0$ iff $g(z(\cdot)) = z(\cdot) = \dot{z}(\cdot) = 0$, leading the fact that $\dot{V}(z(t)) < 0$ in all other cases. Hence, in the light of Lyapunov stability theorems, we can observe the fact that the origin $z(t) = 0$ is asymptotically stable. This, we proves that neutral-type neural system (1) is asymptotically stable. In addition, $V(z(t))$ exploited for stability analysis can be proved to be radially unbounded, that is, the condition of $V(z) \to \infty$ as $||z|| \to \infty$ is satisfied. Q.E.D.

## 3   Conclusions

This work submitted a novel sufficient condition for global asymptotic stability for neutral-type neural networks including constant discrete an constant time delays. This newly obtained criterion was derived by making the use a properly modified Lyapunov functional with the employment of Lipschitz activation functions, which established a new set of relationships among the constant system parameters of these classes of neural networks. The obtained condition was expressed independently of delay parameters and it is easy to be verified simply checking the validity of some algebraic equations.

## References

1. Samidurai, S., Marshal, A., Balachandran, R.K.: Global exponential stability of neutral-type impulsive neural networks with discrete and distributed delays. Nonlinear Anal. Hybrid Syst. **4**, 103–112 (2010)
2. Shi, K., Zhu, H., Zhong, S., Zeng, Y., Zhang, Y.: New stability analysis for neutral type neural networks with discrete and distributed delays using a multiple integral approach. J. Franklin Inst. **352**, 155–176 (2015)
3. Liao, X., Liu, Y., Wang, H., Huang, T.: Exponential estimates and exponential stability for neutral-type neural networks with multiple delays. Neurocomputing **149**, 868–883 (2015)
4. Arik, S.: An analysis of stability of neutral-type neural systems with constant time delays. J. Franklin Inst. **351**, 4949–4959 (2014)
5. Samli, R., Arik, S.: New results for global stability of a class of neutral-type neural systems with time delays. Appl. Math. Comput. **210**, 564–570 (2009)
6. Orman, Z., Arik, S.: An analysis of stability of a class of neutral-type neural networks with discrete time delays. Abstr. Appl. Anal. (2013). Article ID 143585
7. Liu, P.L.: Further improvement on delay-dependent robust stability criteria for neutral-type recurrent neural networks with time-varying delays. ISA Trans. **55**, 92–99 (2015)
8. Zhang, Z., Liu, K., Yang, Y.: New LMI-based condition on global asymptotic stability concerning BAM neural networks of neutral type. Neurocomputing **81**, 24–32 (2012)
9. Zhang, G., Wang, T., Li, T., Fei, S.: Multiple integral Lyapunov approach to mixed-delay-dependent stability of neutral neural networks. Neurocomputing **275**, 1782–1792 (2018)

10. Shi, K., Zhong, S., Zhu, H., Liu, X., Zeng, Y.: New delay-dependent stability criteria for neutral-type neural networks with mixed random time-varying delays. Neurocomputing **168**, 896–90730 (2015)
11. Wang, B., Liu, X., Zhong, S.: New stability analysis for uncertain neutral system with time-varying delay. Appl. Math. Comput. **197**, 457–465 (2008)
12. Zhang, W.A., Yu, L.: Delay-dependent robust stability of neutral systems with mixed delays and nonlinear perturbations. Acta Automat. Sinica **33**, 863–866 (2007)
13. Lakshmanan, S., Park, J.H., Jung, H.Y., Kwon, O.M., Rakkiyappan, R.: A delay partitioning approach to delay-dependent stability analysis for neutral type neural networks with discrete and distributed delays. Neurocomputing **111**, 81–89 (2013)
14. Lien, C.H., Yu, K.W., Lin, Y.F., Chung, Y.J., Chung, L.Y.: Global exponential stability for uncertain delayed neural networks of neutral type with mixed time delays. IEEE Trans. Syst. Man Cybern. B Cybern. **38**, 709–720 (2008)
15. Akça, H., Covachev, V., Covacheva, Z.: Global asymptotic stability of Cohen-Grossberg neural networks of neutral type. J. Math. Sci. **205**, 719–732 (2015)

# Anti-synchronization of Neural Networks with Mixed Delays

Dan Liu and Dan Ye[✉]

College of Information Science and Engineering, Northeastern University,
Shenyang 110189, Liaoning, People's Republic of China
yedan@ise.neu.edu.cn

**Abstract.** This paper focuses on the anti-synchronization of neural networks (NNs) with mixed delays. Under state feedback control, by establishing suitable Lyapunov functional and applying inequality techniques, corresponding algebraic conditions are given to ascertain anti-synchronization of drive-response NNs with time-varying delays and unbounded distributed delays. The availability of the proposed result is verified by an example.

**Keywords:** Neural networks · Anti-synchronization · Mixed delays

## 1 Introduction

As we all know, synchronization of NNs is a very significant dynamical performance in secure communication. In recent years, many different synchronization control methods have been proposed, including state or output feedback control, intermittent control, adaptive control and impulsive control [1–9]. As the extension of synchronization, anti-synchronization, complete synchronization, pinning synchronization, and projective synchronization are also proposed and widely investigated by researchers [10–14]. Among of these extensions, anti-synchronization is a noteworthy dynamical performance, that is the state variables of anti-synchronous system possess same absolute values, but their symbols are inverse. Thus, by anti-synchronization control method to relate systems can continuously produce digital signals between synchronization and anti-synchronization. This process is more valuable for secure communication.

On the other hand, time delay is ineluctable in the hardware implementation of neural networks. As result of this, several types of delay are considered in NNs, such as constant delay, time-varying delays, distributed delays. It is should be pointed that the existent delays in real dynamical systems are mixed rather than single type of delays. In order to accurately depict the impacts of delays in real neuron systems, mixed delays (time-varying delays and distributed delays) are introduced in addressed systems. In the past few decades, many classical achievements on synchronization of NNs with time-varying delays and bounded distributed delays have been reported [15–20]. However, there are few works

© Springer Nature Switzerland AG 2018
L. Cheng et al. (Eds.): ICONIP 2018, LNCS 11302, pp. 361–370, 2018.
https://doi.org/10.1007/978-3-030-04179-3_32

about anti-synchronization of NNs with time-varying delays and unbounded distributed delays, though it is very important.

In this paper, we process the anti-synchronization of NNs with time-varying delays and unbounded distributed delays. Under the state feedback control, several algebraic criteria are obtained to realize the anti-synchronization of NNs on the basis of drive-response concept. The given conditions can be easy to directly verify by parameters of systems. Finally, the availability of the proposed algebraic criteria is verified by an example.

## 2    Preliminaries

We consider NNs with mixed delays as follows:

$$\dot{p}_i(t) = -d_i p_i(t) + \sum_{j=1}^{n} a_{ij} f_j(p_j(t)) + \sum_{j=1}^{n} b_{ij} g_j\left(p_j(t - \pi(t))\right)$$

$$+ \sum_{j=1}^{n} c_{ij} \int_{-\infty}^{t} \mathcal{K}_{ij}(t-s) h_j(p_j(s)) ds, \quad t > 0, \quad i = 1, 2, \cdots, n, \quad (1)$$

where $p_i(t)$ is state variable; $d_i$ is a positive constant; $a_{ij}$, $b_{ij}$, $c_{ij}$ are the connection weights; $f_j(p_j(t))$, $h_j(p_j(t))$, $g_j(p_j(t-\pi(t)))$ are activation functions; $\pi(t)$ is time-varying delay, which satisfies $0 \leq \pi(t) \leq \tau$, $\dot{\pi}(t) \leq \pi < 1$ ($\tau$ and $\pi$ are positive constants); $\mathcal{K}_{ij}$ represents delay kernels, which is piecewise continuous function form $\mathbb{R}_+$ to $\mathbb{R}_+$, and satisfies $\int_0^\infty \mathcal{K}_{ij}(t)dt = 1$, $\int_0^\infty \mathcal{K}_{ij}(t)e^{\beta t}dt < \infty$ for some constants $\beta$. The initial condition of (1) is $p_i(s) = \phi_i(s)$, $s \in [-\tau, 0]$. For $\tau > 0$, $\mathfrak{C}([-\tau, 0]; \mathbb{R}^n)$ stands for the Banach space of continuous functions mapping form $[-\tau, 0]$ to $\mathbb{R}^n$. In addition, the norm is defined as $\|\chi\| = \sup\limits_{-\tau \leq t \leq 0} \left[\sum\limits_{i=1}^{n} |\chi_i(t)|^\sigma\right]^{\frac{1}{\sigma}}$, where $\sigma > 1$.

**Assumption 1.** *For any $p$, $q \in \mathbb{R}$, $x \neq y$, the activation functions $f_i$, $g_i$ and $h_i$ are odd functions, and there exist constants $l_i > 0$, $w_i > 0$, and $v_i > 0$ such that*

$$|f_i(p) - f_i(q)| \leq l_i |p - q|,$$
$$|g_i(p) - g_i(q)| \leq w_i |p - q|,$$
$$|h_i(p) - h_i(q)| \leq v_i |p - q|.$$

According to the drive-response concept, system (1) as considered as drive system, then, we given the following response system

$$\dot{q}_i(t) = -d_i q_i(t) + \sum_{j=1}^{n} a_{ij} f_j(q_j(t)) + \sum_{j=1}^{n} b_{ij} g_j\left(q_j(t - \pi(t))\right)$$

$$+ \sum_{j=1}^{n} c_{ij} \int_{-\infty}^{t} \mathcal{K}_{ij}(t-s) h_j(q_j(s)) ds + \xi_i(t), \quad t \geq 0, \quad (2)$$

where $\xi_i(t)$ is appropriate control input to realize anti-synchronization. The initial condition of (2) is $q_i(s) = \varphi_i(s)$, $s \in [-\tau, 0]$.

Define $r(t) = q(t) + p(t)$ as anti-synchronization error, then error system is

$$\dot{r}_i(t) = -d_i r_i(t) + \sum_{j=1}^{n} a_{ij} \mathcal{F}_j(r_j(t)) + \sum_{j=1}^{n} b_{ij} \mathcal{G}_j(r_j(t - \pi(t)))$$

$$+ \sum_{j=1}^{n} c_{ij} \int_{-\infty}^{t} \mathcal{K}_{ij}(t - s) \mathcal{H}_j(r_j(s)) ds + \xi_i(t), \quad t \geq 0, \tag{3}$$

where $\mathcal{F}_j(r_j(t)) = f_j(q_j(t)) + f_j(p_j(t))$, $\mathcal{G}_j(r_j(t - \pi(t))) = g_j(q_j(t - \pi(t))) + g_j(p_j(t - \pi(t)))$, $\mathcal{H}_j(r_j(t)) = h_j(q_j(t)) + h_j(p_j(t))$, the initial condition of (3) is $r_i(s) = \psi_i(s) = \phi_i(s) + \varphi_i(s)$ $s \in [-\tau, 0]$.

The control input $\xi_i(t)$ in system (3) is given as

$$\xi_i(t) = -\kappa_i (q_i(t) + p_i(t)), \quad i = 1, 2, \cdots, n. \tag{4}$$

*Remark 1.* Due to the activation functions $f_j$, $g_j$, $h_j$ are odd functions, under Assumption 1, for $r_j(t) \neq 0$, we know $\mathcal{F}_j(r_j(t))$, $\mathcal{G}_j(r_j(t - \pi(t)))$, $\mathcal{H}_j(r_j(t))$ have the following properties

$$|\mathcal{F}_j(r_j(t))| \leq l_j |r_j(t)|,$$
$$|\mathcal{G}_j(r_j(t - \pi(t)))| \leq w_j |r_j(t - \pi(t))|,$$
$$|\mathcal{H}_j(r_j(t))| \leq v_j |r_j(t)|,$$

and

$$|\mathcal{F}_j(0)| = f_j(p_j(t)) + f_j(-p_j(t)) = 0,$$
$$|\mathcal{G}_j(0)| = g_j(p_j(t - \pi(t))) + g_j(-p_j(t - \pi(t))) = 0,$$
$$|\mathcal{H}_j(0)| = h_j(p_j(t)) + h_j(-p_j(t)) = 0.$$

So error system (3) admits a zero solution $r_j(t) \equiv 0$.

**Definition 1.** *Drive system (1) globally anti-synchronize with response system (2) with controller (4), if there exist constants $\lambda \geq 1$ and $\epsilon > 0$ such that*

$$\left[ \sum_{i=1}^{n} |p_i(t) + q_i(t)|^\sigma \right]^{\frac{1}{\sigma}} \leq \lambda e^{-\epsilon t} \|\psi + \phi\|, \tag{5}$$

*for $t \geq 0$, where $\epsilon$ is globally exponential anti-synchronization rate.*

# 3   Main Results

**Theorem 1.** *Under Assumption 1, if there exist constants $0 < \epsilon \leq \beta$, $\rho_i > 0$, and $\sigma > 1$ such that the following condition*

$$\sigma \rho_i \left(\epsilon - d_i - \kappa_i\right) + \sum_{j=1}^{n} \left[ \rho_i \left(\sigma - 1\right) \left(|a_{ij}| + |b_{ij}| + |c_{ij}|\right) + \rho_i |a_{ji}| l_i^\sigma \right.$$

$$\left. + \frac{\rho_j |b_{ji}| e^{\sigma \epsilon \tau} w_i^\sigma}{1 - \pi} + v_i^\sigma \rho_i |c_{ji}| \int_0^\infty \mathcal{K}_{ji}\left(s\right) e^{\sigma \epsilon s} ds \right] < 0, \qquad (6)$$

*then, system (1) and system (2) are said to be exponentially anti-synchronized under the control input (4).*

*Proof.* First, it is should be noticed that $\int_0^{+\infty} \mathcal{K}_{ij}(t) e^{\beta t} dt < \infty$, hence, we choose $\epsilon$ $(0 < \epsilon \leq \beta)$ that satisfies $\int_0^{+\infty} \mathcal{K}_{ij}(t) e^{\epsilon t} dt < \infty$.

Then, establishing Lyapunov functional

$$V(t) = \sum_{i=1}^{n} \rho_i \left[ |r_i(t)|^\sigma e^{\sigma \epsilon t} + \frac{1}{1 - \pi} \sum_{j=1}^{n} |b_{ij}| \int_{t-\pi(t)}^{t} |\mathcal{G}_j(r_j(s))|^\sigma e^{\sigma \epsilon (s + \tau)} ds \right.$$

$$\left. + \sum_{j=1}^{n} |c_{ij}| \int_0^\infty \mathcal{K}(\theta) e^{\sigma \epsilon \theta} \left( \int_{t-\theta}^{t} e^{\sigma \epsilon s} |\mathcal{H}_j(r_j(s))|^\sigma ds \right) d\theta \right]. \qquad (7)$$

According to (3), the derivative of $V(t)$ is

$$D^+ V(t) = e^{\sigma \epsilon t} \sum_{i=1}^{n} \sigma \epsilon \rho_i |r_i(t)|^\sigma + e^{\sigma \epsilon t} \sum_{i=1}^{n} \sigma \rho_i |r_i(t)|^{\sigma - 1} sgn\left(r_i(t)\right)$$

$$\times \left[ -d_i r_i(t) + \sum_{j=1}^{n} a_{ij} \mathcal{F}_{ij}\left(r_j(t)\right) + \sum_{j=1}^{n} b_{ij} \mathcal{G}_j\left(r_j(t - \pi(t))\right) \right.$$

$$\left. + \sum_{j=1}^{n} c_{ij} \int_{-\infty}^{t} \mathcal{K}_{ij}(t - s) \mathcal{H}_j\left(r_j(s)\right) ds - \kappa_i r_i(t) \right]$$

$$+ \sum_{i=1}^{n} \sum_{j=1}^{n} \frac{\rho_i |b_{ij}| e^{\sigma \epsilon t}}{1 - \pi} \left[ |\mathcal{G}_j\left(r_j(t)\right)|^\sigma e^{\sigma \epsilon \tau} - (1 - \dot{\pi}(t)) \right.$$

$$\left. \times |\mathcal{G}_j\left(r_j(t - \pi(t))\right)|^\sigma e^{\sigma \epsilon (\tau - \pi(t))} \right] + \sum_{i=1}^{n} \sum_{j=1}^{n} \rho_i |c_{ij}| \int_0^\infty \mathcal{K}_{ij}(\theta)$$

$$\times \left[ e^{\sigma \epsilon \theta} e^{\sigma \epsilon t} |\mathcal{H}_j\left(r_j(t)\right)|^\sigma - e^{\sigma \epsilon t} |\mathcal{H}_j\left(r_j(t - \theta)\right)|^\sigma \right] d\theta$$

$$\leq e^{\sigma \epsilon t} \sum_{i=1}^{n} \sigma \epsilon \rho_i |r_i(t)|^{\sigma} + e^{\sigma \epsilon t} \sum_{i=1}^{n} \sigma \rho_i |r_i(t)|^{\sigma-1}$$

$$\times \left[ -d_i |r_i(t)| + \sum_{j=1}^{n} |a_{ij}||\mathcal{F}_j(r_j(t))| + \sum_{j=1}^{n} |b_{ij}||\mathcal{G}_j(r_j(t - \pi(t)))| \right.$$

$$\left. + \sum_{j=1}^{n} |c_{ij}| \int_{-\infty}^{t} \mathcal{K}_{ij}(t-s)|\mathcal{H}_j(r_j(s))|\,ds - \kappa_i |r_i(t)| \right]$$

$$+ e^{\sigma \epsilon t} \sum_{i=1}^{n} \sum_{j=1}^{n} \frac{\rho_i |b_{ij}| e^{\sigma \epsilon \tau}}{1 - \pi} |\mathcal{G}_j(r_j(t))|^{\sigma}$$

$$- e^{\sigma \epsilon t} \sum_{i=1}^{n} \sum_{j=1}^{n} \rho_i |b_{ij}||\mathcal{G}_j(r_j(t - \pi(t)))|^{\sigma}$$

$$+ e^{\sigma \epsilon t} \sum_{i=1}^{n} \sum_{j=1}^{n} \rho_i |c_{ij}| \int_{0}^{\infty} \mathcal{K}_{ij}(\theta) e^{\sigma \epsilon \theta} |\mathcal{H}_j(r_j(t))|^{\sigma}\,d\theta$$

$$- e^{\sigma \epsilon t} \sum_{i=1}^{n} \sum_{j=1}^{n} \rho_i |c_{ij}| \int_{0}^{\infty} \mathcal{K}_{ij}(\theta) |\mathcal{H}_j(r_j(t - \theta))|^{\sigma}\,d\theta$$

$$= e^{\sigma \epsilon t} \sum_{i=1}^{n} \sigma \epsilon \rho_i |r_i(t)|^{\sigma} - e^{\sigma \epsilon t} \sum_{i=1}^{n} d_i \sigma \rho_i |r_i(t)|^{\sigma}$$

$$+ e^{\sigma \epsilon t} \sum_{i=1}^{n} \sum_{j=1}^{n} \sigma \rho_i |a_{ij}||r_i(t)|^{\sigma-1} |\mathcal{F}_j(r_j(t))|$$

$$+ e^{\sigma \epsilon t} \sum_{i=1}^{n} \sum_{j=1}^{n} \sigma \rho_i |b_{ij}||r_i(t)|^{\sigma-1} |\mathcal{G}_j(r_j(t - \pi(t)))|$$

$$+ e^{\sigma \epsilon t} \sum_{i=1}^{n} \sum_{j=1}^{n} \sigma \rho_i |c_{ij}||r_i(t)|^{\sigma-1} \int_{-\infty}^{t} \mathcal{K}_{ij}(t-s)|\mathcal{H}_j(r_j(s))|\,ds$$

$$- e^{\sigma \epsilon t} \sum_{i=1}^{n} \sigma \rho_i \kappa_i |r_i(t)|^{\sigma} + e^{\sigma \epsilon t} \sum_{i=1}^{n} \sum_{j=1}^{n} \frac{\rho_i |b_{ij}| e^{\sigma \epsilon \tau}}{1 - \pi} |\mathcal{G}_j(r_j(t))|^{\sigma}$$

$$- e^{\sigma \epsilon t} \sum_{i=1}^{n} \sum_{j=1}^{n} \rho_i |b_{ij}||\mathcal{G}_j(r_j(t - \pi(t)))|^{\sigma}$$

$$+ e^{\sigma \epsilon t} \sum_{i=1}^{n} \sum_{j=1}^{n} \rho_i |c_{ij}| \int_{0}^{\infty} \mathcal{K}_{ij}(\theta) e^{\sigma \epsilon \theta} |\mathcal{H}_j(r_j(t))|^{\sigma}\,d\theta$$

$$- e^{\sigma \epsilon t} \sum_{i=1}^{n} \sum_{j=1}^{n} \rho_i |c_{ij}| \int_{0}^{\infty} \mathcal{K}_{ij}(\theta) |\mathcal{H}_j(r_j(t - \theta))|^{\sigma}\,d\theta. \tag{8}$$

In light of Young inequality, we get

$$|r_i(t)|^{\sigma-1}|\mathcal{F}_j(r_j(t))| \le \frac{\sigma-1}{\sigma}\left(|r_i(t)|^{\sigma-1}\right)^{\frac{\sigma}{\sigma-1}} + \frac{1}{\sigma}|\mathcal{F}_j(r_j(t))|^{\sigma}, \tag{9}$$

$$|r_i(t)|^{\sigma-1}|\mathcal{G}_j(r_j(t-\pi(t)))|$$
$$\le \frac{\sigma-1}{\sigma}\left(|r_i(t)|^{\sigma-1}\right)^{\frac{\sigma}{\sigma-1}} + \frac{1}{\sigma}|\mathcal{G}_j(r_j(t-\pi(t)))|^{\sigma}, \tag{10}$$

$$|r_i(t)|^{\sigma-1}\int_{-\infty}^{t}\mathcal{K}_{ij}(t-s)|\mathcal{H}_j(r_j(s))|ds$$
$$\le \int_0^{\infty}\mathcal{K}_{ij}(s)\left[\frac{\sigma-1}{\sigma}\left(|r_i(t)|^{\sigma-1}\right)^{\frac{\sigma}{\sigma-1}} + \frac{1}{\sigma}|\mathcal{H}_j(r_j(t-s))|^{\sigma}\right]. \tag{11}$$

Substituting (9)–(11) into (8), we have

$$D^+V(t) \le e^{\sigma\epsilon t}\sum_{i=1}^{n}\sigma\epsilon\rho_i|r_i(t)|^{\sigma} - e^{\sigma\epsilon t}\sum_{i=1}^{n}d_i\sigma\rho_i|r_i(t)|^{\sigma} - e^{\sigma\epsilon t}\sum_{i=1}^{n}\sigma\rho_i\kappa_i|r_i(t)|^{\sigma}$$

$$+ e^{\sigma\epsilon t}\sum_{i=1}^{n}\sum_{j=1}^{n}\sigma\rho_i|a_{ij}|\left[\frac{\sigma-1}{\sigma}\left(|r_i(t)|^{\sigma-1}\right)^{\frac{\sigma}{\sigma-1}} + \frac{1}{\sigma}|\mathcal{F}_j(r_j(t))|^{\sigma}\right]$$

$$+ e^{\sigma\epsilon t}\sum_{i=1}^{n}\sum_{j=1}^{n}\sigma\rho_i|b_{ij}|\left[\frac{\sigma-1}{\sigma}\left(|r_i(t)|^{\sigma-1}\right)^{\frac{\sigma}{\sigma-1}} + \frac{1}{\sigma}|\mathcal{G}_j(r_j(t-\pi(t)))|^{\sigma}\right]$$

$$+ e^{\sigma\epsilon t}\sum_{i=1}^{n}\sum_{j=1}^{n}\sigma\rho_i|c_{ij}|\int_0^{\infty}\mathcal{K}_{ij}(s)\left[\frac{\sigma-1}{\sigma}\left(|r_i(t)|^{\sigma-1}\right)^{\frac{\sigma}{\sigma-1}}\right.$$

$$\left. + \frac{1}{\sigma}|\mathcal{H}_j(r_j(t-s))|^{\sigma}\right]ds + e^{\sigma\epsilon t}\sum_{i=1}^{n}\sum_{j=1}^{n}\frac{\rho_i|b_{ij}|e^{\sigma\epsilon\tau}}{1-\pi}|\mathcal{G}_j(r_j(t))|^{\sigma}$$

$$- e^{\sigma\epsilon t}\sum_{i=1}^{n}\sum_{j=1}^{n}\rho_i|b_{ij}||\mathcal{G}_j(r_j(t-\pi(t)))|^{\sigma}$$

$$+ e^{\sigma\epsilon t}\sum_{i=1}^{n}\sum_{j=1}^{n}\rho_i|c_{ij}|\int_0^{\infty}\mathcal{K}_{ij}(\theta)e^{\sigma\epsilon\theta}|\mathcal{H}_j(r_j(t))|^{\sigma}d\theta$$

$$- e^{\sigma\epsilon t}\sum_{i=1}^{n}\sum_{j=1}^{n}\rho_i|c_{ij}|\int_0^{\infty}\mathcal{K}_{ij}(\theta)|\mathcal{H}_j(r_j(t-\theta))|^{\sigma}d\theta$$

$$\le e^{\sigma\epsilon t}\sum_{i=1}^{n}\sigma\epsilon\rho_i|r_i(t)|^{\sigma} - e^{\sigma\epsilon t}\sum_{i=1}^{n}d_i\sigma\rho_i|r_i(t)|^{\sigma} - e^{\sigma\epsilon t}\sum_{i=1}^{n}\sigma\rho_i\kappa_i|r_i(t)|^{\sigma}$$

$$+ e^{\sigma\epsilon t}\sum_{i=1}^{n}\sum_{j=1}^{n}\rho_i|a_{ij}|(\sigma-1)|r_i(t)|^{\sigma} + e^{\sigma\epsilon t}\sum_{i=1}^{n}\sum_{j=1}^{n}\rho_i|a_{ij}|l_j^{\sigma}|r_j(t)|^{\sigma}$$

$$+ e^{\sigma\epsilon t}\sum_{i=1}^{n}\sum_{j=1}^{n}\rho_i|b_{ij}|(\sigma-1)|r_i(t)|^{\sigma} + e^{\sigma\epsilon t}\sum_{i=1}^{n}\sum_{j=1}^{n}\rho_i|c_{ij}|(\sigma-1)|r_i(t)|^{\sigma}$$

$$+ e^{\sigma \epsilon t} \sum_{i=1}^{n} \sum_{j=1}^{n} \frac{\rho_i |b_{ij}| e^{\sigma \epsilon \tau} w_j^{\sigma}}{1 - \pi} |r_j(t)|^{\sigma}$$

$$+ e^{\sigma \epsilon t} \sum_{i=1}^{n} \sum_{j=1}^{n} \rho_i v_j^{\sigma} |c_{ij}| \int_0^{\infty} \mathcal{K}_{ij}(s) e^{\sigma \epsilon s} ds |r_j(t)|^{\sigma}$$

$$= e^{\sigma \epsilon t} \sum_{i=1}^{n} \left\{ \sigma \rho_i (\epsilon - d_i - \kappa_i) + \sum_{j=1}^{n} \left[ \rho_i (\sigma - 1) |a_{ij}| \right. \right.$$

$$+ \rho_j |a_{ji}| l_i^{\sigma} + \rho_i |b_{ij}| (\sigma - 1) + \rho_i |c_{ij}| (\sigma - 1)$$

$$+ \frac{\rho_j |b_{ji}| e^{\sigma \epsilon \tau} w_i^{\sigma}}{1 - \pi} + v_i^{\sigma} \rho_i |c_{ji}| \int_0^{\infty} \mathcal{K}_{ji}(s) e^{\sigma \epsilon s} ds \left. \right] \right\} |r_i(t)|^{\sigma}$$

$$< 0. \tag{12}$$

It follows from (6) that, for $t > 0$,

$$V(t) \leq V(0). \tag{13}$$

On the one hand

$$V(0) = \sum_{i=1}^{n} \rho_i \left[ |r_i(0)|^{\sigma} + \frac{1}{1 - \pi} \sum_{j=1}^{n} |b_{ij}| \int_{-\pi(0)}^{0} |\mathcal{G}_j(r_j(s))|^{\sigma} e^{\sigma \epsilon (s + \tau)} ds \right.$$

$$\left. + \sum_{j=1}^{n} |c_{ij}| \int_0^{\infty} \mathcal{K}(\theta) e^{\sigma \epsilon \theta} \left( \int_{-\theta}^{0} e^{\sigma \epsilon s} |\mathcal{H}_j(r_j(s))|^{\sigma} ds \right) d\theta \right]$$

$$\leq \left[ \max_{1 \leq k \leq n} \rho_k + \frac{\tau e^{\sigma \epsilon \tau} \max_{1 \leq k \leq n} w_k^{\sigma}}{1 - \pi} \sum_{i=1}^{n} \sum_{j=1}^{n} \rho_i |b_{ij}| \right.$$

$$\left. + \tau \max_{1 \leq k \leq n} v_k^{\sigma} \sum_{i=1}^{n} \sum_{j=1}^{n} \rho_i |c_{ij}| \int_0^{\infty} \mathcal{K}_{ij}(s) e^{\sigma \epsilon s} ds \right]$$

$$\times \sup_{-\tau \leq t \leq 0} \sum_{i=1}^{n} |\psi_i(t) + \phi_i(t)|^{\sigma}. \tag{14}$$

On the other hand

$$V(t) \geq e^{\sigma \epsilon t} \sum_{i=1}^{n} \rho_i |r_i(t)|^{\sigma} \geq e^{\sigma \epsilon t} \min_{1 \leq k \leq n} (\rho_k) \sum_{i=1}^{n} |r_i(t)|^{\sigma}. \tag{15}$$

Thus, we have

$$\left[ \sum_{i=1}^{n} |p_i(t, \psi) + q_i(t, \phi)|^{\sigma} \right]^{\frac{1}{\sigma}} \leq W e^{-\epsilon t} \sup_{-\tau \leq t \leq 0} \left[ \sum_{i=1}^{n} |\psi_i(t) + \phi_i(t)|^{\sigma} \right]^{\frac{1}{\sigma}}, \tag{16}$$

that is

$$\|p(t, \psi) + q(t, \phi)\| \leq W e^{-\epsilon t} \|\psi + \phi\|, \tag{17}$$

where

$$W = \left\{ \frac{1}{\min\limits_{1\leq k\leq n} \rho_k} \times \left[ \max\limits_{1\leq k\leq n} \rho_k + \frac{\tau e^{\sigma\epsilon\tau} \max\limits_{1\leq k\leq n} w_k^\sigma}{1-\pi} \sum_{i=1}^{n}\sum_{j=1}^{n} \rho_i|b_{ij}| \right. \right.$$

$$\left. \left. +\tau \max\limits_{1\leq k\leq n} v_k^\sigma \sum_{i=1}^{n}\sum_{j=1}^{n} \rho_i|c_{ij}| \int_0^\infty \mathcal{K}_{ij}(s)e^{\sigma\epsilon s}ds \right] \right\}^{\frac{1}{\sigma}}. \tag{18}$$

**Corollary 1.** *Under the Assumption 1, system (1) anti-synchronize with system (2) with controller (4), if*

$$2\left(\epsilon - d_i - \kappa_i\right) + \sum_{j=1}^{n}\left[|a_{ij}| + |b_{ij}| + |c_{ij}| + |a_{ji}|l_i^2\right.$$

$$\left. + \frac{|b_{ji}|e^{2\epsilon\tau}w_i^2}{1-\pi} + v_i^2|c_{ji}| \int_0^\infty \mathcal{K}_{ji}(s)\,e^{2\epsilon s}ds \right] < 0. \tag{19}$$

*Proof.* The proof is similar to Theorem 1 by letting $\sigma = 2$, $\rho_i = 1 (i = 1, 2, \cdots, n)$.

## 4   Numerical Example

*Example 1.* Consider drive system (1) with

$$A = (a_{ij})_{2\times 2} = \begin{bmatrix} 1.01 & -2.09 \\ -4.8 & 3.6 \end{bmatrix}, \quad B = (b_{ij})_{2\times 2} = \begin{bmatrix} -3.2 & 2.09 \\ 2.07 & -2.5 \end{bmatrix},$$

$$C = (c_{ij})_{2\times 2} = \begin{bmatrix} -1.1. & 0.7 \\ 0.5 & 1.2 \end{bmatrix}, \quad D = diag(d_1, d_2) = \begin{bmatrix} 2 & 0 \\ 0 & 4 \end{bmatrix}.$$

and $\pi(t) = \frac{e^t}{1+e^t}$, $\tau = 1$, $\dot\pi(t) \leq \pi = 0.25 < 1$, $f(p) = g(p) = h(p) = \tanh(p)$, $\mathcal{K}_{ij} = e^{-t}$. Taking $\sigma = 2$, $\rho_i = 1$, $\epsilon = 0.4$, $l_j = w_j = v_j = 1$ $(j = 1, 2)$, and choosing feedback control gains as $\kappa_1 = \kappa_2 = 16$. It is easy to find that

$$2\left(\epsilon - d_1 - \kappa_1\right) + \sum_{j=1}^{2}\left[|a_{1j}| + |b_{1j}| + |c_{1j}| + |a_{j1}|l_1^2\right.$$

$$\left. + \frac{|b_{j1}|e^{2\epsilon\tau}w_1^2}{1-\pi} + v_1^2|c_{j1}| \int_0^\infty \mathcal{K}_{1j}(s)\,e^{2\epsilon s}ds \right] = -1.3719 < 0, (20)$$

$$2\left(\epsilon - d_2 - \kappa_2\right) + \sum_{j=1}^{2}\left[|a_{2j}| + |b_{2j}| + |c_{2j}| + |a_{j2}|l_2^2\right.$$

$$\left. + \frac{|b_{j1}|e^{2\epsilon\tau}w_2^2}{1-\pi} + v_2^2|c_{j2}| \int_0^\infty \mathcal{K}_{2j}(s)\,e^{2\epsilon s}ds \right] = -1.4097 < 0, (21)$$

According to Corollary 1, we can say system (1) anti-synchronize with corresponding response system (2) under control input (4) (see Fig. 1). Figure 2 depicts the anti-synchronization between $p_1(t)$, $q_1(t)$ and $p_2(t)$, $q_2(t)$. The initial condition is chosen as $e(s) = (4, -1.35)^T$, $s \in [-1, 0)$.

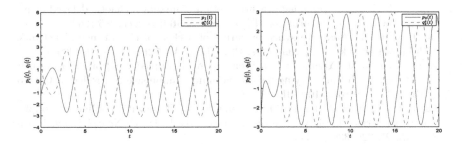

**Fig. 1.** The evolution of state variables.

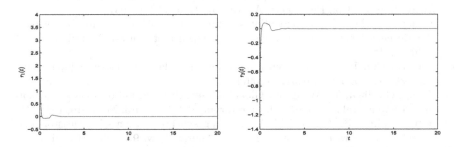

**Fig. 2.** The anti-synchronization errors of variables $p_1(t)$, $q_1(t)$ $p_2(t)$, $q_2(t)$.

## 5    Conclusions

This paper considered discrete and unbounded distributed delays in NNs to deal with their anti-synchronization under the state feedback control. Several new algebraic criteria were reported to realize anti-synchronization of addressed system via using Lyapunov functional and inequality techniques. The availability of the proposed result was verified by an example.

**Acknowledgments.** This work was supported by the Funds of National Natural Science of China under Grant no. 61773097, the Fundamental Research Funds for the Central Universities under Grant no. N160402004, and the Liaoning BaiQianWan Talents Program (201517).

# References

1. Gao, X., Zhong, S., Gao, F.: Exponential synchronization of neural networks with time-varying delays. Nonlinear Anal. **71**(5), 2003–2011 (2009)
2. Cao, J., Lu, J.: Adaptive synchronization of neural networks with or without time-varying delay. Chaos **16**, 013133 (2006)
3. Sun, Y., Cao, J.: Adaptive lag synchronization of unknown chaotic delayed neural networks with noise perturbation. Phys. Lett. A **364**(3), 277–285 (2007)
4. He, H., Cao, J.: Adaptive synchronization of a class of chaotic neural networks with known or unknown parameters. Phys. Lett. A **372**(4), 408–416 (2008)
5. Yang, X., Cao, J., Yang, Z.: Synchronization of coupled reaction-diffusion neural networks with time-varying delays via pinning-impulsive controller. J. Control Optim. **51**(5), 3486–3510 (2013)
6. Lu, J., Ho, D., Cao, J., Kurths, J.: Exponential synchronization of linearly coupled neural networks with impulsive disturbances. IEEE Trans. Neural Netw. Learn. Syst. **22**(2), 329–336 (2011)
7. Hu, C., Yu, J., Jiang, H., Teng, Z.: Exponential stabilization and synchronization of neural networks with time-varying delays via periodically intermittent control. Nonlinearity **23**(10), 2369–2391 (2010)
8. Yang, X., Cao, J.: Stochastic synchronization of coupled neural networks with intermittent control. Phys. Lett. A **373**(36), 3259–3272 (2009)
9. Zhang, G., Shen, Y.: Exponential synchronization of delayed memristor-based chaotic neural networks via periodically intermittent control. Neural Netw. **55**, 1–10 (2014)
10. Zhu, H., Cui, B.: The anti-synchronization of a class of chaotic delayed neural networks. Chaos **17**(4), 043122 (2007)
11. Zhang, G., Shen, Y., Wang, L.: Global anti-synchronization of a class of chaotic memristive neural networks with time-varying delays. Neural Netw. **46**, 1–8 (2013)
12. Yang, Z., Luo, B., Liu, D., Li, Y.: Pinning synchronization of memristor-based neural networks with time-varying delays. Neural Netw. **93**, 143–151 (2017)
13. Li, Y., Li, C.: Complete synchronization of delayed chaotic neural networks by intermittent control with two switches in a control period. Neurocomputing **173**, 1341–1347 (2016)
14. Abdurahman, A., Jiang, H., Teng, Z.: Function projective synchronization of impulsive neural networks with mixed time-varying delays. Nonlinear Dyn. **78**(4), 2627–2638 (2014)
15. Li, T., Fei, S., Zhu, Q., Cong, S.: Exponential synchronization of chaotic neural networks with mixed delays. Neurocomputing **71**, 3005–3019 (2008)
16. Li, X., Rakkiyappan, R.: Impulse controller design for exponential synchronization of chaotic neural networks with mixed delays. Commun. Nonlinear Sci. Numer. simul. **18**(6), 1515–1523 (2013)
17. Gan, Q., Xu, R., Kang, X.: Synchronization of chaotic neural networks with mixed time delays. Commun. Nonlinear Sci. Numer. Simul. **16**, 966–974 (2011)
18. Liu, Z., Zhang, H., Yang, D.: Adaptive synchronization of a class of chaotic neural networks with mixed time-varying delays. J. Northeast. Univ. **30**(4), 475–478 (2009)
19. Song, Q.: Design of controller on synchronization of chaotic neural networks with mixed time-varying delays. Neurocomputing **72**(13–15), 3288–3295 (2009)
20. Zheng, C., Liang, W., Wang, Z.: Anti-synchronization of Markovian jumping stochastic chaotic neural networks with mixed time delays. Circ. Syst. Sig. Process. **33**, 2761–2792 (2014)

# Incremental Stability of Neural Networks with Switched Parameters and Time Delays via Contraction Theory of Multiple Norms

Hao Qiang$^{(\boxtimes)}$ and Wenlian Lu

School of Mathematical Sciences, Fudan University, Shanghai 200433, China
hqiang14@fudan.edu.cn

**Abstract.** In this paper, we propose a new approach to investigate the incrementally exponentially asymptotically stability and contraction property of the switched recurrently connected neural networks with time-varying delays. This method of contraction theory extends the current result from ordinary differential systems to delayed differential systems. Thus, we derive sufficient conditions of incremental stability of a class of neural networks with time-varying parameters and time-delays and its asymptotical periodicity as a consequence when the time-variation is periodic. Numerical examples are presented to illustrate the power of theoretical results.

**Keywords:** Incrementally exponentially asymptotically stability
Multiple norms · Delayed neural networks
Contraction · Asymptotical periodicity

## 1 Introduction

It is particularly important to study the incremental stability of nonlinear systems since it can be applied to many application areas, such as synchronization and consensus problems in network control. [19] introduced contraction theory be highlighted as a promising approach to research incremental exponential asymptotic stability of nonlinear systems [2,14,22]. Also, the existing researches of contraction theory to switched system depend on the appliance of a unique matrix measure. Recently, more related to the present paper, [23] studied the incremental exponential asymptotic stability of a class of Carathéodory nonlinear systems.

Recurrent neural network is an important class of nonlinear systems [1] that shows power approximation capability [7,9,13,16] and widely been used to imaging process, signal analysis and formation pattern. Dynamical analysis is an

This work is jointly supported by the National Natural Sciences Foundation of China under Grant No. 61673119.

© Springer Nature Switzerland AG 2018
L. Cheng et al. (Eds.): ICONIP 2018, LNCS 11302, pp. 371–380, 2018.
https://doi.org/10.1007/978-3-030-04179-3_33

important step towards theoretical analysis and designment of neural networks in real-world application. In practice, time delay is inevitable in application due to the limit of transmission speed and so has attracted a lot of attention.

It is important to study the stability theory of recurrent neural networks. Some stability criteria related to global asymptotic stability in dependent of delays were obtained by the appliance of the Lyapunov function method [3,20] to guarantee global exponential stability of delayed neural networks [5,8]. [24] utilized novel Lyapunov function to analyze the global convergence of a class of neural network models with time delays and derived a new sufficient condition which guarantees the existence, uniqueness and global exponential stability of the equilibrium point. Asymptotical periodicity of delayed neural network systems is also an important aspect to research. [21,29,30] derived the criteria of global exponential stability, existence and asymptotically periodicity of its periodical solution.

In the present paper, we study the switched recurrently connected neural networks with time-varying delays and present sufficient conditions for incrementally exponentially asymptotically stability ($\delta$EAS) of the neural network systems under multiple norms. As a consequence, we present the sufficient conditions for asymptotic periodicity of the neural network systems in case of periodic switching signal. The theoretical results are illustrated by numerical examples. These works motivate the development of artificial intelligence.

## 2 Preliminaries

Recurrently connected neural networks with switched parameters and delays are described by the following dynamical systems

$$
\begin{aligned}
\frac{du_i}{dt} = &- d_i(r(t))u_i(t) + \sum_{j=1}^{n} a_{ij}(r(t))g_j(u_j(t)) \\
&+ \sum_{j=1}^{n} b_{ij}(r(t))f_j(u_j(t - \tau_{ij}(t))) + I_i(r(t)) \quad (i = 1, 2, \ldots, n),
\end{aligned}
\tag{1}
$$

where $u = (u_1, \ldots, u_n) \in \mathbb{R}^n$. We assume that the switching signal $r(t)$ is a real-valued staircase function: there exist countable discontinuous points $0 = t_0 < t_1 < \cdots < t_i < \cdots$ such that $r(t_i\pm)$ exist and $r(t_i) = r(t_i+)$, $r(t) = \eta_i$, $\eta_i \in \mathbb{R}$, for $t_i \leq t < t_{i+1}$. And the interconnection weights $a_{ij}$, $b_{ij}$, self-inhibition $d_i$ which are positive and inputs $I_i$ depend on the switching signal $r(t)$. Here we denote by $N(t,s) = \#\{i : s < t_i \leq t\}$, the number of time instants $t_i$ in the time interval $(s,t]$. And $N(t) = N(t,0)$.

We define $u(t; \phi(\theta), r_t)$ is the solution of system (1) with initial state $\phi(\theta)$, $\theta \in [-\tau, 0]$. And $r_t$ denotes the trajectory of $r(t)$ up to t, i.e., $r_t = \{r(s)\}_{0 \leq s \leq t}$. $\|\cdot\|_{\chi(t)}$ denotes the multiple norms. In the present paper, we simply define that $\chi(t) = r(t)$.

**Definition 1.** *If a continuous function $f : [0, a) \to [0, \infty)$ satisfies (i) it is strictly increasing; (ii) $f(0) = 0$, we say that it belongs to class $\mathcal{K}$.*

*If a continuous function $f(s, t) : [0, a) \times [0, \infty) \to [0, \infty)$ satisfies (i) for each fixed $t$, $f(s, t)$ belongs to class $\mathcal{K}$; (ii) for each fixed $s$, $f(s, t)$ is decreasing with respect to $t$ and $\lim_{t \to \infty} f(s, t) = 0$, we say that it belongs to class $\mathcal{KL}$.*

*In addition, if $f(s, t)$ of class $\mathcal{KL}$ converges to $0$ exponentially as $t \to \infty$, we say that $f(s, t)$ belongs to class $\varepsilon\mathcal{KL}$.*

**Definition 2.** *System (1) is said to be incrementally asymptotically stable ($\delta AS$ for short) with $r(t)$ in the region $C \subset \mathbb{R}^n$ if there exists a function $f(s, t)$ of class $\mathcal{KL}$ such that for any initial data $\phi(0), \varphi(0) \in C$, the following property holds*

$$|u(t; \phi(\theta), r_t) - u(t; \varphi(\theta), r_t)| \leq f(|\phi(\theta) - \varphi(\theta)|, t) \tag{2}$$

*for some norm $|\cdot|$.*

*Furthermore, if $f(s, t)$ belongs to class $\varepsilon\mathcal{KL}$, then system (1) is said to be incrementally exponentially asymptotically stable (i.e. $\delta EAS$).*

## 3   Main Results

We can give the incremental stability of system (1) under the case of multiple norms.

**Theorem 1.** *Suppose that there exist a sequence of positive constants $\xi_{1k}$, $\xi_{2k}$, $\cdots$, $\xi_{nk}$, $\alpha_k$, nonnegative constants $\beta_k, k = 1, 2, \cdots$, such that the following conditions hold*

$$- \xi_{ik}(d_i(r(t)) - \alpha_k) + \sum_{j=1}^{n} \xi_{jk} G_j |a_{ij}(r(t))|$$

$$+ \sum_{j=1}^{n} \xi_{jk} F_j e^{\alpha_k \tau_{ij}(t)} |b_{ij}(r(t))| \leq 0, \quad \forall t \in [t_{k-1}, t_k), i = 1, \cdots, n, \tag{3}$$

$$|g_i(x + h) - g_i(x)| \leq G_i |h|, |f_i(x + h) - f_i(x)| \leq F_i |h|, \quad i = 1, \cdots, n, \tag{4}$$

$$\| \cdot \|_{r(t_k)} \leq \beta_k \| \cdot \|_{r(t_k-)}, \tag{5}$$

*and for all $T > 0$, there exist a positive constant $c$, such that*

$$\frac{1}{T}\left(- \int_0^T \alpha_{N(s)+1} ds + \sum_{i=1}^{N(T)} \log \beta_i \right) < -c. \tag{6}$$

*Then the system (1) is incrementally exponentially asymptotically stable.*

*Proof.* For two different initial states $x_0(\theta)$ and $y_0(\theta)$, define $\varphi(\lambda) = (1 - \lambda)x_0 + \lambda y_0$, $\lambda \in [0, 1]$, such that $\varphi(0) = x_0$ and $\varphi(1) = y_0$, $\varphi'(\lambda) = y_0 - x_0$.

Let $\psi(t, \lambda) = u(t; \varphi(\lambda), r_t)$ be the solution of system (1) with the initial value $\psi(0, \lambda) = \varphi(\lambda)$. Define $\omega = \dfrac{\partial \psi}{\partial \lambda}$, which is the solution of

$$
\begin{cases}
\dfrac{d\omega_i}{dt} = -d_i(r(t))\omega_i(t) + \sum_{j=1}^{n} a_{ij}(r(t))g_j'(\psi_j(t))\omega_j(t) \\
\qquad + \sum_{j=1}^{n} b_{ij}(r(t))f_j'(\psi_j(t - \tau_{ij}(t)))\omega_j(t - \tau_{ij}(t)), \\
\omega(0, \lambda) = \varphi'(\lambda), \quad i = 1, \cdots, n.
\end{cases}
\tag{7}
$$

If $t \in [t_{k-1}, t_k)$, let $v = \omega \cdot e^{\alpha_k t}$. And we construct multiple norms $\|v\|_{r(t)} = \|v\|_{\xi_k} = \max_i \xi_{ik}^{-1}|v_i(t)|$. First, we consider the situation that $t \in [0, t_1)$. There exist $i^*$ such that $\|v\|_{\xi_1} = \xi_{i^*1}^{-1}|v_{i^*}(t)|$. Assuming that $\|v\|_{\xi_1}$ is increasing in $[0, t]$, we have

$$
\frac{d|v_{i^*}(t)|}{dt} = sign(v_{i^*}(t))\Big((-d_{i^*}(r(t)) + \alpha_1)v_{i^*}(t) + \sum_{j=1}^{n} a_{i^*j}(r(t))g_j'(\psi_j(t))v_j(t)
$$

$$
+ \sum_{j=1}^{n} b_{i^*j}(r(t))f_j'(\psi_j(t - \tau_{i^*j}(t)))v_j(t - \tau_{i^*j}(t)) \cdot e^{\alpha_1 \tau_{i^*j}(t)}\Big)
$$

$$
\leq (-d_{i^*}(r(t)) + \alpha_1)\xi_{i^*1}\xi_{i^*1}^{-1}|v_{i^*}(t)| + \sum_{j=1}^{n} |a_{i^*j}(r(t))|G_j\xi_{j1}\xi_{j1}^{-1}|v_j(t)|
$$

$$
+ \sum_{j=1}^{n} |b_{i^*j}(r(t))|F_j\xi_{j1}\xi_{j1}^{-1}|v_j(t - \tau_{i^*j}(t))| \cdot e^{\alpha_1 \tau_{i^*j}(t)}
$$

$$
\leq \Big((-d_{i^*}(r(t)) + \alpha_1)\xi_{i^*1} + \sum_{j=1}^{n} |a_{i^*j}(r(t))|G_j\xi_{j1}
$$

$$
+ \sum_{j=1}^{n} |b_{i^*j}(r(t))|F_j\xi_{j1}e^{\alpha_1 \tau_{i^*j}(t)}\Big)\|v(t)\|_{\xi_1} \leq 0.
$$

It is contradictory to the assumption above which implies that $\|v\|_{\xi_1}$ is nonincreasing. That is, $\|v(t_1)\|_{r(t_1-)} = \|v(t_1)\|_{\xi_1} \leq \|v(0)\|_{\xi_1}$, i.e.,

$$
\|\omega(t_1, \lambda)\|_{\xi_1} \leq \|\omega(0, \lambda)\|_{\xi_1} \cdot e^{-\alpha_1 t_1}.
\tag{8}
$$

By the same approach used above, we can prove that

$$
\|\omega(t_k, \lambda)\|_{\xi_k} \leq \|\omega(t_{k-1}, \lambda)\|_{\xi_k} \cdot e^{-\alpha_k(t_k - t_{k-1})}.
\tag{9}
$$

Noting that $t_{N(t)} \leq t < t_{N(t)+1}$, then we can conclude for any $t$,

$$\|\omega(t,\lambda)\|_{r(t)} \leq e^{-\alpha_{N(t)+1}(t-t_{N(t)})}\|\omega(t_{N(t)},\lambda)\|_{r(t_{N(t)})}$$

$$\leq e^{-\alpha_{N(t)+1}(t-t_{N(t)})}\beta_{N(t)}\|\omega(t_{N(t)},\lambda)\|_{r(t_{N(t)}-)}$$

$$\leq e^{-\alpha_{N(t)+1}(t-t_{N(t)})-\alpha_{N(t)}(t_{N(t)}-t_{N(t)-1})}\beta_{N(t)}\|\omega(t_{N(t)-1},\lambda)\|_{r(t_{N(t)-1})}$$

$$\leq e^{-\int_0^t \alpha_{N(s)+1}\mathrm{d}s+\sum_{i=1}^{N(t)}\log\beta_i}\|\omega(0,\lambda)\|_{r(0)} < e^{-ct}\|\varphi'(\lambda)\|_{r(0)}$$

$$(10)$$

Since $\|\cdot\|_{r(t)}$ is uniformly equivalent, there exist $D > 0$ such that $\|\cdot\|_{r(t)} \leq D\|\cdot\|_{r(t')}$ for any $t, t' > 0$. Then by (10), we have

$$\|u(t;y_0(\theta),r_t) - u(t;x_0(\theta),r_t)\|_{r(0)} \leq D\|u(t;y_0(\theta),r_t) - u(t;x_0(\theta),r_t)\|_{r(t)}$$

$$=D\|\int_0^1 \frac{\partial\psi(t,\lambda)}{\partial\lambda}\mathrm{d}\lambda\|_{r(t)} \leq D\int_0^1 \|\omega(t,\lambda)\|_{r(t)}\mathrm{d}\lambda$$

$$\leq D\int_0^1 e^{-ct}\|\varphi'(\lambda)\|_{r(0)}\mathrm{d}\lambda \leq De^{-ct}\|y_0(\theta) - x_0(\theta)\|_{r(0)}$$

which proves that the system (1) is $\delta$EAS.

If $r(t)$ is an $\omega$-periodic staircase function which means $r(t+\omega) = r(t)$, and it has $m$ switching time instants $\{\tilde{t}_j\}_{j=1}^n (0 < \tilde{t}_1 < \cdots < \tilde{t}_m = \omega)$ in the time interval $(0,\omega]$, we have the following theorem.

**Theorem 2.** *If*

$$|g_j(s)| \leq G_j|s| + C_j, |f_j(s)| \leq F_j|s| + D_j, \quad j = 1, \cdots, n, \quad (11)$$

*where $G_j > 0$, $F_j > 0$, $C_j$ and $D_j$ are constants, and there exist positive constants $\xi_{1k}, \cdots, \xi_{nk}, (k = 1, \cdots, m)$ such that for $\forall t$, if $t \in [\tilde{t}_{k-1}, \tilde{t}_k)$*

$$-\xi_{ik}d_i(r(t)) + \sum_{j=1}^n \xi_{jk}G_j|a_{ij}(r(t))| + \sum_{j=1}^n \xi_{jk}F_j|b_{ij}(r(t))| < \eta < 0, \quad i = 1, \cdots, n,$$

$$(12)$$

*then the system (1) has an $\omega$-periodic solution $x(t)$ at least. Further more, if conditions (3), (4), (5) and (6) in Theorem 1 are satisfied, then for any solution $u(t)$ of system (1), if $r(t) = r(\tilde{t}_{k-1})$,*

$$\|u(t) - x(t)\|_{r(t)} = O(e^{-ct}), \quad (13)$$

*i.e., system (1) is asymptotically periodic.*

*Proof.* Let us choose a constant $M$ satisfying $M > \dfrac{J}{\eta}$. Here,

$$J = \max_{i,t}\left(\sum_{j=1}^n |a_{ij}(r(t))|C_j + \sum_{j=1}^n |b_{ij}(r(t))|D_j + |I_i(r(t))|\right). \quad (14)$$

Let $C = C([-\tau, 0], \mathbb{R}^n)$, we know that $\phi(\theta) \in C$. Define the norm of space $C$ as $\|\phi\| = \sup_{-\tau \leq \theta \leq 0} \|\phi(\theta)\|_{r(0)}$. So, $C$ is an Banach space. Here $\| \cdot \|_{r(t)}$ is the multiple norms.

Denote

$$\Omega = \{x(\theta) : \|x(\theta)\| \leq M, \|\dot{x}(\theta)\| \leq N\}$$

where

$$N = (a + b + c)M + d, \quad a = \max_{i,k} \sup_t |d_i(r(t))|\xi_{ik}^{-1},$$

$$b = \max_{i,j,k} \sup_t |a_{ij}(r(t))|G_j\xi_{ik}^{-1}, \quad c = \max_{i,j,k} \sup_t |b_{ij}(r(t))|F_j\xi_{ik}^{-1},$$

$$d = \max_{i,k} \sup_t |I_i(r(t))|\xi_{ik}^{-1}.$$

Obviously, $\Omega$ is a convex compact set. Next, let us define a map $T$ from $\Omega$ to $\Omega$ by

$$T : \phi(\theta) \mapsto x(\theta + \omega, \phi)$$

where $x(t, \phi)$ is the solution of system (1) with the initial state $\phi(\theta)$.

First, we should prove that this map is well defined which means that we need to prove

$$\|x\| = \sup_{-\tau \leq \theta \leq 0} \|x(\theta + \omega)\|_{r(\omega)} \leq M. \tag{15}$$

Since the switching signal $r(t)$ is $\omega$-periodic, we have $\| \cdot \|_{r(\omega)} = \| \cdot \|_{r(0)}$

Assume that $t_0 \in [0, \omega]$ is the earliest time such that

$$\|x(t_0)\|_{r(\omega)} = \|x(t_0)\|_{r(0)} = \xi_{i_0 1}^{-1}|x_{i_0}(t_0)| = M. \tag{16}$$

Then direct calculation gives

$$\frac{d|x_{i_0}(t)|}{dt}\bigg|_{t=t_0} = sign(x_{i_0}(t_0))\bigg( - d_{i_0}(r(t_0))x_{i_0}(t_0) + \sum_{j=1}^n a_{i_0 j}(r(t_0))g_j(x_j(t_0))$$

$$+ \sum_{j=1}^n b_{i_0 j}(r(t_0))f_j(x_j(t_0 - \tau_{i_0 j}(t_0))) + I_{i_0}(r(t_0))\bigg)$$

$$\leq - d_{i_0}(r(t_0))|x_{i_0}(t_0)| + \sum_{j=1}^n |a_{i_0 j}(r(t_0))|G_j|x_j(t_0)|$$

$$+ \sum_{j=1}^n |b_{i_0 j}(r(t_0))|F_j|x_j(t_0 - \tau_{i_0 j}(t_0))| + J$$

$$\leq ( - d_{i_0}(r(t_0))\xi_{i_0 1} + \sum_{j=1}^n |a_{i_0 j}(r(t_0))|G_j\xi_{j1})\|x(t_0)\|_{r(0)}$$

$$+ \sum_{j=1}^n |b_{i_0 j}(r(t_0))|F_j\xi_{j1}\|x(t_0 - \tau_{i_0 j}(t_0))\|_{r(0)} + J$$

$$\leq ( - d_{i_0}(r(t_0))\xi_{i_0 1} + \sum_{j=1}^n |a_{i_0 j}(r(t_0))|G_j\xi_{j1}$$

$$+ \sum_{j=1}^n |b_{i_0 j}(r(t_0))|F_j\xi_{j1})M + J \leq -\eta M + J < 0$$

$$\tag{17}$$

which means $\|x(\theta + \omega, \phi)\|_{r(\omega)}$ can never exceed $M$. Obviously, $\|\dot{x}\| \leq N$. So $T$ is well defined.

We pick two different functions $\phi, \varphi \in \Omega$. By Theorem 1, we have

$$\|x(\theta + \omega, \phi) - x(\theta + \omega, \varphi)\|_{r(\omega)} \leq e^{-c\omega}\|\phi(\theta) - \varphi(\theta)\|_{r(0)} \qquad (18)$$

Thus $T$ is a contraction map. According to Contraction Mapping Theorem, there exist one and only one $\phi^*$ such that $T\phi^* = \phi^*$. Therefore, $x(t, T\phi^*) = x(t, \phi^*)$, i.e., $x(t + \omega, \phi^*) = x(t, \phi^*)$, which is an $\omega$-periodic solution to system (1).

For $t \in [\tilde{t}_{k-1}, \tilde{t}_k)$, let $z(t) = (u(t) - x(t))e^{\alpha_k t}$ and $i^*$ be the index such that $\|z(t)\|_{r(t)} = \xi_{i^*k}^{-1}|z_{i^*}(t)|$. We have

$$\frac{\mathrm{d}|z_{i^*}(t)|}{\mathrm{d}t} \leq -(d_{i^*}(r(t)) - \alpha_k)|z_{i^*}(t)| + \sum_{j=1}^{n}|a_{i^*j}(r(t))||G_j||z_j(t)|$$

$$+ \sum_{j=1}^{n}|b_{i^*j}(r(t))||F_j||z_j(t - \tau_{i^*j}(t))|e^{\alpha_k \tau_{i^*j}(t)}$$

$$\leq \Big( -(d_{i^*}(r(t)) - \alpha_k)\xi_{i^*k} + \sum_{j=1}^{n}|a_{i^*j}(r(t))||G_j\xi_{jk}\Big)\|z(t)\|_{r(t)}$$

$$+ \sum_{j=1}^{n}|b_{i^*j}(r(t))||F_j\xi_{jk}e^{\alpha_k \tau_{i^*j}(t)}\|z(t - \tau_{i^*j}(t))\|_{r(t)} \qquad (19)$$

$$\leq \Big(-(d_{i^*}(r(t)) - \alpha_k)\xi_{i^*k} + \sum_{j=1}^{n}|a_{i^*j}(r(t))||G_j\xi_{jk}$$

$$+ \sum_{j=1}^{n}|b_{i^*j}(r(t))||F_j\xi_{jk}e^{\alpha_k \tau_{i^*j}(t)}\Big)\|z(t)\| \leq 0,$$

which means that $\|z(t)\|_{r(t)}$ is nonincreasing and bounded. By the same approach used in Theorem 1, we have $\|u(t) - x(t)\|_{r(t)} \leq e^{-ct}\|u(\theta) - x(\theta)\|_{r(0)}$. Therefore, $\|u(t) - x(t)\|_{r(t)} = O(e^{-ct})$. Then Theorem 2 is proved.

## 4    Numerical Example

We choose $\xi_{11} = 3$, $\xi_{21} = 1$, $\xi_{12} = 2$, $\xi_{22} = 1$, $\alpha_1 = \frac{1}{8}$, $\alpha_2 = \frac{4}{9}$, $\omega = 2$. So, we have

$$\|x(t)\|_1 = \max_i \xi_{i1}^{-1}|x_i(t)|, \quad \|x(t)\|_2 = \max_i \xi_{i2}^{-1}|x_i(t)|.$$

It can be seen $\|x(t)\|_2 \leq 1.5\|x(t)\|_1$.

Then we give an example as follows. For $t \in [0,1)$,

$$\begin{cases} \dfrac{du_1}{dt} = -3u_1(t) + \tanh(u_1(t)) + \tanh(u_2(t)) \\ \qquad + \arctan(u_1(t-1)) + \arctan(u_2(t-1)) + 1 \\ \dfrac{du_2}{dt} = -5u_2(t) - \dfrac{1}{2}\tanh(u_1(t)) - \dfrac{1}{2}\tanh(u_2(t)) \\ \qquad + \dfrac{1}{2}\arctan(u_1(t-1)) + \dfrac{1}{2}\arctan(u_2(t-1)) + 2 \end{cases} \qquad (20)$$

and for $t \in [1,2)$,

$$\begin{cases} \dfrac{du_1}{dt} = -9u_1(t) - \tanh(u_1(t)) - \tanh(u_2(t)) \\ \qquad + \arctan(u_1(t-1)) + \arctan(u_2(t-1)) - 1 \\ \dfrac{du_2}{dt} = -6u_2(t) + \dfrac{1}{4}\tanh(u_1(t)) + \dfrac{1}{4}\tanh(u_2(t)) \\ \qquad + \dfrac{1}{4}\arctan(u_1(t-1)) + \dfrac{1}{4}\arctan(u_2(t-1)) + 2 \end{cases} \qquad (21)$$

We can see that all the conditions in Theorem 2 are satisfied.

Figure 1 shows that each $u_i(t)$ converges to a periodic solution respectively.

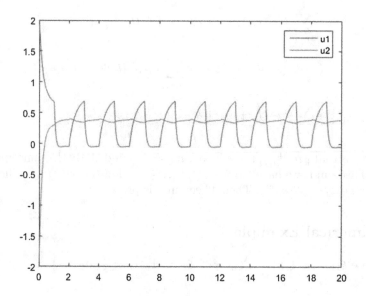

**Fig. 1.** Asymptotically periodic dynamics.

## 5    Conclusions

In conclusion, based on the variant of Halanay inequality, we presented an extension of contraction analysis in [23] to delayed differential systems with switched parameters and derives sufficient conditions for δEAS and furthermore asymptotic periodicity as a consequence of the neural network systems. Numerical example showed efficiency of the theoretical results. This methodology can be generalized to develop contraction analysis approach towards incremental stability of delayed differential systems. This is the orient of our future research.

# References

1. Aleksander, I., Morton, H.: An Introduction to Neural Computing. Chapman and Hall, London (1990)
2. Angeli, D.: A lyapunov approach to incremental stability properties. IEEE Trans. Autom. Control **47**(3), 410–421 (2002)
3. Arik, S., Tavsanoglu, V.: On the global asymptotic stability of delayed cellular neural networks. IEEE Trans. Circuits Syst. I: Fundam. Theory Appl. **47**(4), 571–574 (2000)
4. di Bernardo, M., Liuzza, D., Russo, G.: Contraction analysis for a class of non-differentiable systems with applications to stability and network synchronization. SIAM J. Control Optim. **52**(5), 3203–3227 (2014)
5. Chen, T., Amari, S.I.: Exponential convergence of delayed dynamical systems. Neural Comput. **13**(3), 621–635 (2001)
6. Chen, T., Lu, W., Chen, G.: Dynamical behaviors of a large class of general delayed neural networks. Neural Comput. **17**(4), 949–968 (2005)
7. Chen, T., Chen, H., Liu, R.: A constructive proof and an extension of Cybenko's approximation theorem. In: Page, C., LePage, R. (eds.) Computing Science and Statistics. Springer, New York (1992). https://doi.org/10.1007/978-1-4612-2856-1_21
8. Chen, T.: Global exponential stability of delayed hopfield neural networks. Neural Netw. **14**(8), 977–980 (2001)
9. Chen, T., Chen, H.: Universal approximation to nonlinear operators by neural networks with arbitrary activation functions and its application to dynamical systems. IEEE Trans. Neural Netw. **6**(4), 911–917 (1995)
10. Chen, T., Lu, W.: Stability analysis of dynamical neural networks. In: Proceedings of the 2003 International Conference on Neural Networks and Signal Processing, vol. 1, pp. 112–116 (Dec 2003)
11. Chua, L.O.: Cellular neural network: applications. IEEE Trans. Circ. Syst. **35**(10), 1273–1290 (1988)
12. Chua, L.O.: Cellular neural network: theory. IEEE Trans. Circ. Syst. **35**, 1257–1272 (1988)
13. Cybenko, G.: Approximation by superpositions of a sigmoidal function. Math. Control Signals Syst. **2**(4), 303–314 (1989)
14. Forni, F., Sepulchre, R.: A differential lyapunov framework for contraction analysis. IEEE Trans. Autom. Control **59**(3), 614–628 (2014)
15. Gopalsamy, K., He, X.Z.: Stability in asymmetric hopfield nets with transmission delays. Physica D **76**(4), 344–358 (1994)
16. Hecht-Nielsen, R.: Neurocomputing. Nature **359**(6394) (1990)
17. Hopfield, J.J.: Neurons with graded response have collective computational properties like those of two-state neurons. Proc. Nat. Acad. Sci. U.S.A. **81**(10), 3088–3092 (1984)

18. Hopfield, J.J., Tank, D.W.: Computing with neural circuits: a model. Science **233**(4764), 625–633 (1986)
19. Lewis, D.C.: Metric properties of differential equations. Am. J. Math. **71**(2), 294–312 (1949)
20. Liao, T.L., Wang, F.C.: Global stability for cellular neural networks with time delay. IEEE Trans. Neural Netw. **11**(6), 1481–1484 (2000)
21. Liu, Z., Liao, L.: Existence and global exponential stability of periodic solution of cellular neural networks with time-varying delays. J. Math. Anal. Appl. **290**(1), 247–262 (2004)
22. Lohmiller, W., Slotine, J.J.E.: On contraction analysis for non-linear systems. Automatica **34**(6), 683–696 (1998)
23. Lu, W., di Bernardo, M.: Contraction and incremental stability of switched carathéodory systems using multiple norms. Automatica **70**, 1–8 (2016)
24. Lu, W., Rong, L., Chen, T.: Global convergence of delayed neural network systems. Int. J. Neural Syst. **13**(03), 193–204 (2003). pMID: 12884452
25. Russo, G., di Bernardo, M.: Contraction theory and master stability function: linking two approaches to study synchronization of complex networks. IEEE Trans. Circuits Syst. II Express Briefs **56**(2), 177–181 (2009)
26. Russo, G., di Bernardo, M., Slotine, J.J.E.: A graphical approach to prove contraction of nonlinear circuits and systems. IEEE Trans. Circuits Syst. I Regul. Pap. **58**(2), 336–348 (2011)
27. Wenlian, L.U., Chen, T.: On periodic dynamical systems. Chin. Ann. Math. **25**(4), 455–462 (2004)
28. Xu, Z.B., Kwong, C.P.: Global convergence and asymptotic stability of asymmetric hopfield neural networks. J. Math. Anal. Appl. **191**(3), 405–427 (1995)
29. Zheng, Y., Chen, T.: Global exponential stability of delayed periodic dynamical systems. Phys. Lett. A **322**(5), 344–355 (2004)
30. Zhou, J., Liu, Z., Chen, G.: Dynamics of periodic delayed neural networks. Neural Netw. **17**(1), 87–101 (2004)

# Lag Synchronization of Complex-Valued Neural Networks with Time Delays

Jiarong Li, Haijun Jiang$^{(\boxtimes)}$, Cheng Hu, and Juan Yu

College of Mathematics and System Sciences, Xinjiang University, Urumqi 830046, People's Republic of China
jianghaijunxju@163.com

**Abstract.** In this paper, the global exponential lag synchronization has been investigated for complex-valued neural networks with time delays. By using Lyapunov functionals and matrix inequality techniques, designing feedback control and adaptive feedback control, some new sufficient conditions for checking the global exponentially lag synchronization of the addressed complex-valued neural networks are established. Finally, numerical examples are given to demonstrate the effectiveness of theoretical results.

**Keywords:** Complex-valued neural networks · Lag synchronization
Matrix inequality technique · Adaptive control · Feedback control

## 1 Introduction

In recent years, there has been increasing research interest in analyzing the dynamic behaviors of real-valued neural networks [1]. Compared with real-valued recurrent neural networks, research for complex-valued ones has achieved slow and little progress as there are more complicated properties inside such systems. Complex-valued neural networks with complex-valued state, activation function, connection weight and output become strongly desired because of their practical applications in electromagnetic, quantum waves and other areas [2,3]. It is an interesting and challenging issue to make complex-valued systems achieving synchronization.

In the past decades, there were some researches on the stability and synchronization of complex-valued neural networks. In [4], author discussed complex-valued system based on Halanay inequalities and the results obtain here were more accurate under less restrictive conditions. In [5], using homeomorphism theory and Lyapunov-Krasovskii functional, the stability of delayed complex-valued Cohen-Grossberg BAM neural networks was considered. To the best of our knowledge, the previous results for checking the stability and synchronization of the addressed complex-valued neural networks were somewhat conservative. In addition, there are few results for the synchronization of complex-valued neural networks [6,7]. Hence, the study of exponential synchronization for complex-valued neural networks is an interesting and significant problem in recent years.

© Springer Nature Switzerland AG 2018
L. Cheng et al. (Eds.): ICONIP 2018, LNCS 11302, pp. 381–392, 2018.
https://doi.org/10.1007/978-3-030-04179-3_34

Motivated by the above discussions, the main purpose of this paper is to study the exponential lag synchronization for complex-valued neural networks with time delays. The main contributions of this paper are reflected as follows: 1. By using Lyapunov functional and matrix inequality techniques, some simple and verifiable synchronization criteria are obtained through designing feedback controllers. 2. Lag synchronization criteria derived in this paper can also generalize to the globally exponentially lag synchronization. 3. Adaptive control schemes are designed to reduce the conservativeness of the derived synchronization criteria. The paper is organized as follows. The system description, assumptions and lemmas are provided in Sect. 2. Some new sufficient conditions are derived in Sect. 3. Numerical examples are given to show the effectiveness of our main results in Sect. 4. Finally, the conclusion of this paper is given in Sect. 5.

*Notations*: $C^n$ denotes the $n(n \geq 1)$ dimensional complex vector space. $A^T$ and $A^{-1}$ stand for the transposition and inverse of $A$, respectively. Let $\bar{z}$ and $z^*$ be the conjugate and conjugate transpose of $z = (z_1, ..., z_n)^T \in C^n$, respectively. And $|z| = (|z_1|, ..., |z_n|)^T$, where $|z_i| = \sqrt{z_i \bar{z}_i}, i = 1, ..., n, ||z|| = \sqrt{z^* z}$. $\lambda_{\min}(A)$ and $\lambda_{\max}(A)$ are the minimum and the maximum eigenvalue of matrix $A$. $P > 0$ means that $P$ is symmetric and positive. $z = a + ib$, where $i = \sqrt{-1}$ denotes the imaginary unit.

## 2    Preliminaries

This paper considers a class of complex-valued neural networks with time delays as the driving system:

$$\dot{z}(t) = -Dz(t) + Wf(z(t)) + Nf(z(t - \tau)) + J, \quad t \geq 0, \tag{1}$$

where $z(t) = (z_1(t), ..., z_n(t))^T \in C^n$ is the state vector, $D = \text{diag}\{d_1, ..., d_n\} \in R^{n \times n}$ is the self-feedback connection weight matrix, where $d_i > 0(i = 1, ..., n)$, $\tau(\tau \geq 0)$ corresponds to the transmission delay, $W = (w_{ij})_{n \times n} \in C^{n \times n}, N = (n_{ij})_{n \times n} \in C^{n \times n}$ are the connection weight matrix and the delayed connection weight matrix. $f(z(t)) = (f_1(z_1(t)), ..., f_n(z_n(t))) : C^n \to C^n$ and $f(z(t - \tau)) = (f_1(z_1(t - \tau)), ..., f_n(z_n(t - \tau))) : C^n \to C^n$ are the complex-valued vector-valued activation functions at time $t$ and time $t - \tau$, $J \in C^n$ is the external bias.

Consider the following response system:

$$\dot{\tilde{z}}(t) = -D\tilde{z}(t) + Wf(\tilde{z}(t)) + Nf(\tilde{z}(t - \tau)) + J + U(t), \tag{2}$$

where $U(t)$ are the controllers and $U(t) \in C^n$.

The initial conditions associated with (1) and (2) are given by:

$$\begin{cases} z_j(s) = \phi_j(s), \\ \tilde{z}_j(s) = \tilde{\phi}_j(s), \end{cases}$$

where $s \in [-\tau, 0]$, $\phi_j(s), \tilde{\phi}_j(s) \in C([-\tau, 0], C^n), j = 1, ..., n$, $C([-\tau, 0], C^n)$ represents the set of all $n$-dimensional complex-valued continuous functions defined on the interval $[-\tau, 0]$.

In this paper, we consider the complex-valued activation functions satisfying the following assumptions:

**Assumption 1.** For any $j = 1, ..., n$, the activation function $f_j$ satisfies the Lipschitz continuity condition in the complex domain, that is, for any $z_j^1, z_j^2 \in C$, there exists a real number $l_j > 0$ such that

$$|f_j(z_j^1) - f_j(z_j^2)| \leq l_j|z^1 - z^2|, \quad j = 1, ..., n,$$

where $l_j$ is called Lipschitz constant. Or equivalently, there exists a definite matrix $L = \text{diag}(l_1, ..., l_n)$ satisfying

$$|f(z^1) - f(z^2)| \leq L|z^1 - z^2|$$

for all $z^1, z^2 \in C^n$.

## 3 Main Results

In this section, we will give some novel sufficient conditions to ensure the lag synchronization of the considered systems. The lag synchronization error can be characterized as $e(t) = \tilde{z}(t) - z(t - \xi)$ for some constant lag time $\xi > 0$, where $e(t) \in C^n$, then, we can obtain the error system as

$$\dot{e}(t) = -De(t) + W\Phi(e(t)) + N\Phi(e(t - \tau)) + U(t), \tag{3}$$

where

$$\begin{aligned}
\Phi(e(t)) &= (\Phi_1(e_1(t)), \cdots, \Phi_n(e_n(t)))^T \\
&= f(e(t) + z(t - \xi)) - f(z(t - \xi)), \\
\Phi(e(t - \tau)) &= (\Phi_1(e_1(t - \tau)), \cdots, \Phi_n(e_n(t - \tau))^T) \\
&= f(e(t - \tau) + z(t - \tau - \xi)) - f(z(t - \tau - \xi)).
\end{aligned}$$

And the controller is designed as follows:

$$U(t) = Ke(t), \tag{4}$$

where $K = \text{diag}(k_1, k_2, \cdots, k_n) \in C^{n \times n}$ is a gain matrix to be designed in order to achieve lag synchronization of systems (1) and (2).

**Theorem 1.** Based on Assumption 1, systems (1) and (2) are lag synchronized under the controller (4) if the following LMI holds:

$$\begin{pmatrix} 2D - 2L^T L - (K^* + K) & B \\ B^* & I_{2n} \end{pmatrix} > 0 \tag{5}$$

where $B = [W, N]$.

**Proof.** Based on the error dynamical system (3), we define the following Lyapunov functional:

$$V(t) = e^*(t)e(t) + \sum_{i=1}^{n} \int_{t-\tau}^{t} \Phi_i(e_i(s))\bar{\Phi}_i(e_i(s))ds.$$

In the following,taking the derivative of $V(t)$ with respect to time $t$ along with the solutions of error system (3) yields

$$\dot{V}(t) \leq -e^*(t)(2D - WW^* - NN^* - (K + K^*) - 2L^T L)e(t).$$

In view of (5) and Schur complement [8], we can acquire

$$2D - WW^* - NN^* - (K + K^*) - 2L^T L > 0,$$

so,

$$\dot{V}(t) \leq -\lambda e^*(t)e(t),$$

where $\lambda = \lambda_{\min}[2D - WW^* - NN^* - (K + K^*) - 2L^T L] > 0$.
For $t \geq 0$, integrating both sides of inequality (10) over $[0, t]$, we get

$$V(t) - V(0) \leq -\int_0^t \lambda e^*(s)e(s)ds. \tag{6}$$

It follows from (6) that

$$\int_0^\infty e^*(t)e(t)dt < +\infty.$$

It is easy to see that $e^*(t)e(t)$, $(e^*(t)e(t))'$ is bounded on $(0, +\infty)$. Hence, $(e^*(t)e(t))^2$ and $[(e^*(t)e(t))^2]'$ is bounded on $(0, +\infty)$, that is $||e(t)||^2$ are uniformly continuous on $(0, +\infty)$. From Barbalat Lemma, it follows that

$$\lim_{t \to +\infty} ||e(t)|| = 0,$$

which implies that systems (1) and (2) are lag synchronized under the controller (4). This completes the proof.

**Corollary 1.** Assume that the conditions in Theorem 1 hold, then systems (1) and (2) are globally exponentially lag synchronized.

**Proof.** Let $V(t)$ defined in Theorem 1 is a positive definite Lyapunov functional. Using Theorem 1, we have

$$\dot{V}(t) \leq -\lambda e^*(t)e(t).$$

And if we choose a positive number $\epsilon > 0$, such that

$$\epsilon - \lambda + (e^{\epsilon\tau} - 1)||L^2|| < 0. \tag{7}$$

Then, we can acquire

$$\frac{d}{dt}(e^{\epsilon t}V(t)) \leq (\epsilon - \lambda)e^{\epsilon t}e^*(t)e(t) + \epsilon e^{\epsilon t}\sum_{i=1}^{n}\int_{t-\tau}^{t}\Phi_i(e_i(s))\bar{\Phi}_i(e_i(s))ds, \quad (8)$$

And then

$$\int_0^s d(e^{\epsilon t}V(t)) \leq (\epsilon - \lambda)\int_0^s e^{\epsilon t}e^*(t)e(t)dt$$
$$+ \epsilon\int_0^s e^{\epsilon t}\sum_{i=1}^{n}\int_{t-\tau}^{t}\Phi_i(e_i(\zeta))\bar{\Phi}_i(e_i(\zeta))d\zeta dt \quad (9)$$

Furthermore, based on Assumption 1, we have the following inequality

$$\epsilon\int_0^s e^{\epsilon t}\sum_{i=1}^{n}\int_{t-\tau}^{t}\Phi_i(e_i(\zeta))\bar{\Phi}_i(e_i(\zeta))d\zeta dt$$
$$\leq (e^{\epsilon\tau} - 1)\|L^2\|(\int_{-\tau}^{0}e^{\epsilon\zeta}e^T(\zeta)e(\zeta)d\zeta + \int_0^s e^{\epsilon\zeta}e^T(\zeta)e(\zeta)d\zeta). \quad (10)$$

By combining (7), (9) and (10), we can obtain

$$e^{\epsilon s}V(s) - V(0) \leq (e^{\epsilon\tau} - 1)\|L^2\|\int_{-\tau}^{0}e^{\epsilon t}e^T(t)e(t)dt$$
$$= M_1\|\Psi\|^2, \quad (11)$$

where $M_1 = (e^{\epsilon\tau} - 1)\|L^2\|\int_{-\tau}^{0}e^{\epsilon t}dt$, $\|\Psi\|^2 = \max_{-\tau \leq \xi \leq 0}\{e^T(\xi)e(\xi)\}$.
Thus,

$$V(t) \leq (V(0) + M_1\|\Psi\|^2)e^{-\epsilon t}, \quad \forall t > 0, \quad (12)$$

where

$$V(0) = e^*(0)e(0) + \sum_{i=1}^{n}\int_{-\tau}^{0}\Phi_i(e_i(s))\bar{\Phi}_i(e_i(s))ds$$
$$\leq \|\Psi\|^2 + \tau\|L^2\|\|\Psi\|^2$$
$$\equiv M_2\|\Psi\|^2,$$

where $M_2 = 1 + \tau\|L^2\|$.
By (12),

$$e^*(t)e(t) \leq V(t) \leq (M_1 + M_2)\|\Psi\|^2 e^{-\epsilon t}, \quad \forall t > 0,$$

which shows that
$$\|e(t)\| \leq \sqrt{M_1 + M_2}\|\Psi\|e^{-\frac{\epsilon}{2}t}.$$

Hence systems (1) and (2) are globally exponentially lag synchronized.

Next, we design an adaptive feedback controller to further investigate the synchronization of complex-valued neural networks (1) and (2).

$$U(t) = \varepsilon(t)e(t), \tag{13}$$

the feedback strength $\varepsilon(t) = \text{diag}(\varepsilon_1(t), \ldots, \varepsilon_n(t)) \in R^{n \times n}$ is adapted by the following law:

$$\dot{\varepsilon}_i(t) = -\alpha_i \bar{e}_i(t) e_i(t), \tag{14}$$

where $\alpha_i > 0, \alpha = \text{diag}(\alpha_1, \ldots, \alpha_n) \in R^{n \times n}$.

Then, the error system (3) is transformed into:

$$\dot{e}(t) = -De(t) + W\Phi(e(t)) + N\Phi(e(t - \tau)) + \varepsilon(t)e(t), \tag{15}$$

**Theorem 2.** Assume that Assumption 1 holds, the feedback strength $\varepsilon(t)$ is adapted by (14); Then the drive system (1) and response system (2) are lag synchronized under adaptive feedback controller (13).

**Proof.** In order to establish the result of Theorem 2, we define the following Lyapunov functional:

$$V(t) = e^*(t)e(t) + \sum_{i=1}^{n} \int_{t-\tau}^{t} \Phi_i(e_i(s))\bar{\Phi}_i(e_i(s))ds + \sum_{i=1}^{n} \frac{1}{\alpha_i}(\varepsilon_i(t) + h_i)^2,$$

where $H = \text{diag}(h_1, \ldots, h_n)$ is a constant matrix.

In the following, we calculate the derivative of $V(t)$ along the solution of error system (15).

$$\dot{V}(t) \leq -e^*(t)(2D - 2L^T L - WW^* - NN^* + 2H)e(t).$$

we choose

$$H = \frac{1}{2}\{\lambda_{\max}(-2D + 2L^T L + WW^* + VV^*) + 1\}I_n.$$

Then, we finally obtain

$$\dot{V}(t) \leq -e^*(t)e(t).$$

Similarly to the proof of Theorem 1, we have

$$\lim_{t \to +\infty} ||e(t)|| = 0,$$

which implies that the systems (1) and (2) are lag synchronization under the controller (13). This completes the proof.

In the following, we will decompose the complex-valued differential equations to real part and imaginary part, and then recast it into an equivalent real-valued differential system.

Let $e(t) = e^R(t) + ie^I(t), W = W^R + iW^I, N = N^R + iN^I, U(t) = U^R(t) + iU^I(t), \Phi(e(t)) = \Phi^R(e^R, e^I) + i\Phi^I(e^R, e^I), \Phi(e^\tau) = \Phi^R(e^R_\tau, e^I_\tau) + i\Phi^I(e^R_\tau, e^I_\tau)$, where $e^R = e^R(t), e^I = e^I(t), e^R_\tau = e^R(t - \tau), e^I_\tau = e^I(t - \tau)$. Therefore, error system can be separated into real and imaginary parts as:

$$\dot{e}^R(t) = -De^R(t) + W^R\Phi^R(e^R, e^I) - W^I\Phi^I(e^R, e^I) + N^R\Phi^R(e^R_\tau, e^I_\tau)$$
$$- N^I\Phi^I(e^R_\tau, e^I_\tau) + U^R(t),$$
$$\dot{e}^I(t) = -De^I(t) + W^R\Phi^I(e^R, e^I) + W^I\Phi^R(e^R, e^I) + N^R\Phi^I(e^R_\tau, e^I_\tau)$$
$$+ N^I\Phi^R(e^R_\tau, e^I_\tau) + U^I(t).$$

Let

$$\hat{e}(t) = \begin{bmatrix} e^R(t) \\ e^I(t) \end{bmatrix}, \quad \hat{D} = \begin{bmatrix} D & 0 \\ 0 & D \end{bmatrix}, \quad \hat{W} = \begin{bmatrix} W^R & -W^I \\ W^I & W^R \end{bmatrix}, \quad \hat{N} = \begin{pmatrix} N^R & -N^I \\ N^I & N^R \end{pmatrix},$$

$$\hat{U}(t) = \begin{bmatrix} U^R(t) \\ U^I(t) \end{bmatrix}, \quad \hat{\Phi}(\hat{e}(t)) = \begin{bmatrix} \Phi^R(e^R, e^I) \\ \Phi^I(e^R, e^I) \end{bmatrix}, \quad \hat{\Phi}(\hat{e}(t - \tau)) = \begin{bmatrix} \Phi^R(e^R_\tau, e^I_\tau) \\ \Phi^I(e^R_\tau, e^I_\tau) \end{bmatrix}.$$

Then, error system (3) is equivalent to

$$\dot{\hat{e}}(t) = -\hat{D}\hat{e}(t) + \hat{W}\hat{\Phi}(\hat{e}(t)) + \hat{N}\hat{\Phi}(\hat{e}(t - \tau)) + \hat{U}(t). \tag{16}$$

Obviously, $\|\hat{e}(t)\| \to 0$ as $t \to +\infty$ means the drive system (1) and the response system (2) are synchronized under the controller $\hat{U}(t)$.

From controller (4), we get

$$\hat{U}(t) = \hat{K}\hat{e}(t), \tag{17}$$

where $\hat{K} = \begin{bmatrix} K^R & -K^I \\ K^I & K^R \end{bmatrix}$.

It is clear from Assumption 1 and definition of $\Phi(e(t))$, we have

$$\hat{\Phi}^T(\hat{e}(t))\hat{\Phi}(\hat{e}(t)) \le \hat{e}^T(t)\hat{L}\hat{e}(t),$$

where $\hat{L} = \begin{bmatrix} L^T L & 0 \\ 0 & L^T L \end{bmatrix}$.

Then, we can obtain the following Theorem.

**Theorem 3.** Under Assumption 1, systems (1) and (2) are lag synchronized under the controller (17), if the following LMI holds:

$$\begin{pmatrix} 2\hat{D} - 2\hat{L} - (\hat{K}^T + \hat{K}) & Q \\ Q^T & I_{4n} \end{pmatrix} > 0$$

where $Q = [\hat{W}, \hat{N}]$.

**Corollary 2.** Assume that the conditions in Theorem 3 hold, then systems (1) and (2) are globally exponentially lag synchronized under the controller (17).

Next, we design the following adaptive feedback controller.

$$\begin{cases} U^R(t) = \varepsilon(t)e^R(t), \\ U^I(t) = \rho(t)e^I(t), \end{cases} \tag{18}$$

where $\varepsilon(t) = \text{diag}(\varepsilon_1(t), \cdots, \varepsilon_n(t)) \in R^{n \times n}$, $\rho(t) = \text{diag}(\rho_1(t), \cdots, \rho_n(t)) \in R^{n \times n}$ are the coupling strength matrices. The coupling strength matrices $\varepsilon(t)$ and $\rho(t)$ are adapted by the following laws:

$$\begin{cases} \dot{\varepsilon}_i(t) = -\alpha_i(e_i^R(t))^2, \\ \dot{\rho}_i(t) = -\beta_i(e_i^I(t))^2, \end{cases} \tag{19}$$

where $\alpha_i > 0, \alpha = \text{diag}(\alpha_1, \cdots, \alpha_n)$, $\beta_i > 0, \beta = \text{diag}(\beta_1, \cdots, \beta_n)$.

Then, we can obtain the following theorem.

**Theorem 4.** Suppose that Assumption 1 satisfies, the feedback strength $\varepsilon(t)$ and $\rho(t)$ are adapted by (19); Then systems (1) and (2) are lag synchronized under adaptive feedback controller (18).

**Proof.** Based on the error system (16), we construct the following Lyapunov functional:

$$V(t) = \hat{e}^T(t)\hat{e}(t) + \sum_{i=1}^{2n} \int_{t-\tau}^{t} \hat{\Phi}_i^2(\hat{e}_i(s))ds + \sum_{i=1}^{n} \frac{1}{\alpha_i}(\varepsilon_i(t) + h_{1i})^2$$

$$+ \sum_{i=1}^{n} \frac{1}{\beta_i}(\rho_i(t) + h_{2i})^2.$$

where $H_1 = \text{diag}(h_{11}, h_{12}, \cdots, h_{1n}) \in R^{n \times n}$ and $H_2 = \text{diag}(h_{21}, \cdots, h_{2n}) \in R^{n \times n}$ are constant matrices.

From (18), we let

$$\hat{U}(t) = \begin{pmatrix} U^R(t) \\ U^I(t) \end{pmatrix} = \hat{F}(t)\hat{e}(t),$$

where $\hat{F}(t) = \begin{pmatrix} \varepsilon(t) & 0 \\ 0 & \rho(t) \end{pmatrix}$.

Based on Assumption 1, we can calculate the derivative of $V(t)$ along the solution of error system (16).

$$\dot{V}(t) \le -\hat{e}^T(t)(2\hat{D} - 2\hat{L} - \hat{W}\hat{W}^T - \hat{N}\hat{N}^T + 2\hat{H})\hat{e}(t),$$

where $\hat{H} = \begin{pmatrix} H_1 & 0 \\ 0 & H_2 \end{pmatrix}$.

We choose

$$\hat{H} = \frac{1}{2}\{\lambda_{\max}(-2\hat{D} + 2\hat{L} + \hat{W}\hat{W}^T + \hat{N}\hat{N}^T) + 1\}I_{2n}.$$

And we finally obtain

$$\dot{V}(t) < -\hat{e}^T(t)\hat{e}(t),$$

which implies that systems (1) and (2) are lag synchronization under adaptive feedback controller (18). This completes the proof.

## 4  Numerical Example

In this section, we give some numerical simulations to demonstrate the above results.

Consider a two-neuron complex-valued neural networks as the driving system:

$$\dot{z}(t) = -Dz(t) + Wf(z(t)) + Nf(z(t-\tau)) + J(t), \quad t \geq 0, \qquad (20)$$

where $\tau = 1, z(t) = (z_1(t), z_2(t))^T$, $z_j = a_j + ib_j$, $j = 1, 2$, $f_j(z_j) = 2(|x_j| + i|j_j|)$, $j = 1, 2$. $J_1 = 1 + 2i$, $J_2 = i$ and

$$D = \begin{bmatrix} 10 & 0 \\ 0 & 6 \end{bmatrix}, W = \begin{bmatrix} -3 & -2-3i \\ 1 & 1 \end{bmatrix}, N = \begin{bmatrix} -1 & 6i \\ -3 & 0.5 \end{bmatrix},$$

Under the initial condition $z_1(s) = -0.5-3i, z_2(s) = 1.8-0.3i$, for $s \in [-1, 0]$, the state trajectories of real parts and imaginary parts of system (20) are given in Fig. 1.

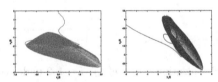

**Fig. 1.** The chaotic behaviors of real parts and imaginary parts of system (20).

Consider (20) as the drive system, the response system is given by:

$$\dot{\tilde{z}}(t) = -D\tilde{z}(t) + Wf(\tilde{z}(t)) + Nf(\tilde{z}(t-\tau)) + J(t) + U(t), \quad t \geq 0, \qquad (21)$$

where $\tilde{z}(t) = (\tilde{z}_1(t), \tilde{z}_2(t))^T$, $\tilde{z}_j = \tilde{a}_j + i\tilde{b}_j$, $j = 1, 2$.

For any $z^1 = (z_1^1, z_2^1)^T \in C^2$ and $z^2 = (z_1^2, z_2^2)^T \in C^2$, from Assumption 1, we have

$$|f(z^1) - f(z^2)| \leq L|z^1 - z^2|$$

where $L = \begin{bmatrix} 2 & 0 \\ 0 & 2 \end{bmatrix}$.

Now, in order to realize the synchronization of systems (20) and (21), we choose

$$K = \begin{bmatrix} -25 + 2i & 0 \\ 0 & -5 + i \end{bmatrix}.$$

It can be checked that

$$\begin{bmatrix} 2D - 2L^T L - (K^* + K) & B \\ B^* & I_{2n} \end{bmatrix}$$

is a positive-definite matrix. Thus, the conditions of Theorem 1 hold.

Under the initial condition $z_1(s) = -0.5 - 3i$, $z_2(s) = 1.8 - 0.3i$, $\tilde{z}_1(s) = -0.5 + 2.6i$, $\tilde{z}_2(s) = -1.4 + 3i$ for $s \in [-1, 0]$, then the synchronization errors between $\tilde{x}_i$ and $x_i$ ($i = 1, 2$) are shown in Fig. 2. Figure 3 shows the synchronization errors between $\tilde{y}_i$ and $y_i$ ($i = 1, 2$). It clear that systems (20) and (21) are synchronized under the feedback control (4).

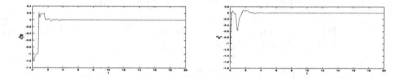

**Fig. 2.** Synchronization errors $e_i^R(t)$, $i = 1, 2$ under control (4).

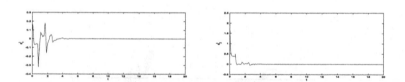

**Fig. 3.** Synchronization errors $e_i^I(t)$, $i = 1, 2$ under control (4).

*Example 1.* In the following, one exchanges the feedback controller to adaptive controller (18) and choose $\alpha_1 = 0.2$, $\alpha_2 = 0.3$, $\beta_1 = 0.3$, $\beta_2 = 0.2$. The dynamical coupling strength $\alpha_i(t)$, $\beta_i(t)$, $i = 1, 2$ of the controller (18) are shown in Fig. 4.

Under the initial conditions $z_1(s) = -0.5 - 3i$, $z_2(s) = 1.8 - 0.3i$, for $s \in [-1, 0]$, and $\varepsilon_i(0) = \rho_i(0) = 0$, $i = 1, 2$, the synchronization errors between $\tilde{x}_i$ and $x_i$ ($i = 1, 2$) are shown in Fig. 5, the synchronization errors between $\tilde{y}_i$ and $y_i$ ($i = 1, 2$) are showing in Fig. 6. It's clear that systems (20) and (21) are synchronized under the control (18). The simulation results are consistent with the theory obtained above, which shows the effectiveness of our results.

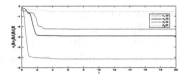

**Fig. 4.** The dynamical coupling strengths of the controller (18).

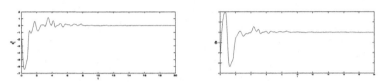

**Fig. 5.** Synchronization errors $e_i^R(t)$, $i = 1, 2$ under control (18).

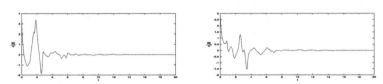

**Fig. 6.** Synchronization errors $e_i^I(t)$, $i = 1, 2$ under control (18).

## 5   Conclusion

In this paper, the problem of exponential lag synchronization for complex-valued neural networks with time delay is investigated. Our results obtained in this paper not only extend and improve some previous works on the synchronization of real-valued neural networks, but also easy to apply for determining the exponential lag synchronization of the complex-valued neural networks.

## References

1. Aram, Z., Jafari, S., Ma, J.: Using chaotic artificial neural networks to model memory in the brain. Commun. Nonlinear Sci. Numer. Simul. **44**, 449–459 (2017)
2. Chistyakov, Y.S., Kholodova, E.V., Minin, A.S.: Modeling of electric power transformer using complex-valued neural networks. Energy Procedia **12**, 638–647 (2011)
3. Chen, X., Zhao, Z., Song, Q.: Multistability of complex-valued neural networks with time-varying delays. Appl. Math. Comput. **294**, 18–35 (2017)
4. Liu, B., Lu, W., Chen, T.: Generalized Halanay inequalities and their applications to neural networks with unbounded time-varying delays. IEEE Trans. Neural Netw. **22**, 1508–1513 (2011)
5. Subramanian, K., Muthukumar, P.: Existence, uniqueness, and global asymptotic stability analysis for delayed complex-valued Cohen Grossberg BAM neural networks. Neural Comput. Appl. **29**, 1–20 (2016)
6. Nitta, T.: Solving the XOR problem and the detection of symmetry using a single complex-valued neuron. Neural Netw. **16**, 1101–1105 (2003)

7. Aizenberg, I.: Solving the XOR and parity N problems using a single universal binary neuron. Soft Comput. **12**, 215–222 (2008)
8. Zhang, Z., Yu, S.: Global asymptotic stability for a class of complex-valued Cohen Grossberg neural networks with time delays. Neurocomputing **171**, 1158–1166 (2016)

# Continuous Attractors of Nonlinear Neural Networks with Asymmetric Connection Weights

Jiali Yu[1]($\boxtimes$), Zhang Yi[2], Chunxiao Wang[3], Yong Liao[4], and Zhixin Pang[1]

[1] School of Mathematical Sciences, University of Electronic Science and Technology of China, Chengdu 611731, China
{yujiali,Liaoyong}@uestc.edu.cn, zhangyi@scu.edu.cn, xiao2166@126.com
[2] College of Computer Science, Sichuan University, Chengdu 610065, China
[3] School of Science, Shandong Jianzhu University, Jinan 250101, China
[4] School of Information and Software Engineering, University of Electronic Science and Technology of China, Chengdu 610054, China

**Abstract.** Continuous attractor neural networks with symmetric connection weights have been studied widely. However, there is short of results on continuous attractor neural networks with asymmetric connection weights. This paper studies the networks with asymmetric connection weights. To overcome the difficulties caused by asymmetric connection weights, an interesting new norm is proposed in a general vector space which is not Euclidean space. Then the new distance and attractivity are defined. Finally, the explicit expressions of continuous attractors of Cellular Neural Networks with asymmetric connection weights are obtained.

**Keywords:** Cellular neural networks
Asymmetric connection weights · Continuous attractors · Norm

## 1 Introduction

Recently, there has been increasing interest in multistability analysis for neural networks. Multistability is significantly different from mono-stability. In mono-stability analysis, the network has unique equilibrium point, and all the trajectories of the networks converge to it [1,2]. Whereas in multistability analysis, the networks are allowed to have multiple equilibrium points. Based on the distribution of the equilibrium points, there are two kinds of attractors: discrete attractors [3] and continuous attractors [4–7]. A neural network is said to have discrete attractors if its equilibrium points are discretely distributed in the state space. Discrete attractors computation has been founded in many practical

This work is supported by National Natural Science Foundation of China under Grant 61572112, 61103041, 61432012, 61803228, the Fundamental Research Funds for the Central Universities under Grant ZYGX2016J136.

© Springer Nature Switzerland AG 2018
L. Cheng et al. (Eds.): ICONIP 2018, LNCS 11302, pp. 393–402, 2018.
https://doi.org/10.1007/978-3-030-04179-3_35

applications in some conventional neural networks, such as the Hopfield recurrent neural networks [8,9]. However, discrete attractors may not be appropriate for continuous stimuli. Continuous attractors have been used to describe the encoding of continuous stimuli such as the eye position [4], head direction [10], the moving direction [11,12], path integrator [13–15], cognitive map [16] and population decoding [17,18]. Continuous attractors have been widely studied by many authors [19–21].

Nowadays most of the continuous attractor neural networks have symmetric connection weights [22]. It was reported in [23] that continuous attractors can be designed by tuning the external inputs to networks that have a connectivity matrix with Toeplitz symmetry. When the reciprocal connections between two neurons are not always equal, the synaptic weight is asymmetric. It is more difficult to study the continuous attractor neural networks with asymmetric connection weights than symmetric ones. A continuous attractor is a low-dimensional manifold embedded in the high-dimensional state space. Form the mathematical view, it is a subspace [24]. Euclidean space is a special space which is used in the subspace research commonly, we may study continuous attractors in a more general space but not in Euclidean space. Based on this reason, an interesting new norm is proposed in more general space. Using this new norm, we have a new distance definition and attractivity definition. Finally, Explicit expression of the continuous attractor can be obtained. The expression could be looked as a solution of the network, it can give complete description to the continuous attractor.

This paper is organized as follows. Preliminaries are given in Sect. 2. The main results on the continuous attractors of Cellular Neural Networks are given in Sect. 3. Simulations are given in Sect. 4 to illustrate the theory. Finally, conclusions are drawn in Sect. 5.

## 2  Preliminaries

Consider the following Cellular Neural Networks (CNNs)

$$\dot{r}(t) + r(t) = f\left(Wr(t) + h\right) \tag{1}$$

for $t \geq 0$, where $r = (r_1, \cdots, r_n)^T \in R^n$ is the state vector, $W = (W_{ij})_{n \times n}$ is the connection weight which can be asymmetric, $h = (h_1, \cdots, h_n)^T$ denotes the external input, the transfer function $f(\cdot)$ is a nonlinear function defined by

$$f(s) = \frac{|s + 1| - |s - 1|}{2}, \quad s \in R.$$

Equilibrium points of (1) are some vectors $r^* \in R^n$ which satisfy

$$r^* = f\left(Wr^* + h\right).$$

Let $b_j (j = 1, \cdots, n)$ be a set of independent unit vectors of $R^n$. Then, given each $r(t) \in R^n$, there exist constants $x_j(t)(j = 1, \cdots, n)$ such that $r$ can be uniquely represented by

$$r(t) = \sum_{j=1}^{n} x_j(t) \cdot b_j.$$

**Lemma 1.** *Denote*

$$\|r\|_{\mathcal{S}} = \sqrt{\sum_{j=1}^{n} x_j^2}, \quad r \in R^n.$$

*Then,* $\| \cdot \|_{\mathcal{S}}$ *defines a norm in the space of* $R^n$.

*Proof:* Given any $r \in R^n$, clearly, it holds that

$$\|r\|_{\mathcal{S}} \geq 0$$

and $\|r\|_{\mathcal{S}} = 0$ if and only if $r = 0$.

Given any constant $c \in R$, it is easy to see that

$$\|c \cdot r\|_{\mathcal{S}} = |c| \cdot \|r\|_{\mathcal{S}}.$$

Next, we prove that

$$\|r + u\|_{\mathcal{S}} \leq \|r\|_{\mathcal{S}} + \|u\|_{\mathcal{S}}$$

for $r, u \in R^n$. Suppose that

$$\begin{cases} \|r\|_{\mathcal{S}} = \sqrt{\sum_{j=1}^{n} x_j^2}, & r \in R^n, \\[2ex] \|u\|_{\mathcal{S}} = \sqrt{\sum_{j=1}^{n} u_j^2}, & u \in R^n. \end{cases}$$

By the well known Cauchy inequality, it holds that

$$\sum_{j=1}^{n} x_j \cdot u_j \leq \sqrt{\sum_{j=1}^{n} x_j^2} \cdot \sqrt{\sum_{j=1}^{n} u_j^2}.$$

Then, it follows that

$$\|r + u\|_{\mathcal{S}} = \sqrt{\sum_{j=1}^{n} (x_j + u_j)^2}$$

$$= \sqrt{\sum_{j=1}^{n} x_j^2 + 2 \sum_{j=1}^{n} x_j u_j + \sum_{j=1}^{n} u_j^2}$$

$$\leq \sqrt{\sum_{j=1}^{n} x_j^2 + 2 \sqrt{\sum_{j=1}^{n} x_j^2} \cdot \sqrt{\sum_{j=1}^{n} u_j^2} + \sum_{j=1}^{n} u_j^2}$$

$$= \sqrt{\sum_{j=1}^{n} x_j^2} + \sqrt{\sum_{j=1}^{n} u_j^2}$$

$$= \|r\|_S + \|u\|_S.$$

Clearly, $\| \cdot \|_S$ is a norm in $R^n$. The proof is complete.

Let $\Omega$ be an nonempty set, denote the distance from a point $r \in R^n$ to the set $\Omega$ by

$$dist\,(r, \Omega) = \inf_{r^\dagger \in \Omega} \|r - r^\dagger\|_S.$$

A set of equilibrium points $C$ is said to be stable, if given any constant $\epsilon > 0$, there exists a constant $\delta > 0$ such that

$$dist\,(r(0), C) \leq \delta$$

implies that

$$dist\,(r(t), C) \leq \epsilon$$

for all $t \geq 0$.

**Definition 1.** *A set of equilibrium points $C$ is called a continuous attractor if it is connected and stable.*

The above definition are based on general vector space but not Euclidean space.

## 3    Representations of Continuous Attractors

Let's assume that $W$ has $n$ unit linearly independent eigenvectors $b_1, b_2 \cdots b_n$, and $\lambda_1 \geq \lambda_2 \cdots \geq \lambda_n$ are the corresponding eigenvalues. Since $b_1, b_2 \cdots b_n$ is a basis for $R^n$, Let the multiplicity of $\lambda_1$ is $k$ and denote by $V_{\lambda_1}$ the eigensubspace associated with the eigenvalue $\lambda_1$. Suppose that

$$h = \sum_{j=1}^{n} \tilde{h}_j b_j. \tag{2}$$

**Theorem 1.** *The set*

$$G = \{r \,|\, |r_j| \leq 1, (j = 1, \cdots, n)\}$$

*is an invariant set of the network (1), i.e., each trajectory starting in $G$ remains in $G$ for ever.*

*Proof:* Given any $r(0) \in G$, we will prove that $r(t) \in G$ for all $t \geq 0$. We have

$$r_j(t) = r_j(0)e^{-t} + \int_0^t e^{-(t-s)} f\left(\sum_{j=1}^{n} w_{ij} r_j(s) + h_j\right) ds \tag{3}$$

for $t \geq 0$ and $i = 1, \cdots, n$. Then, from (3),

$$|r_j(t)| \leq |r_j(0)| e^{-t} + \int_0^t e^{-(t-s)} \left| f\left( \sum_{j=1}^n w_{ij} r_j(s) + h_j \right) \right| ds$$

$$\leq e^{-t} + \int_0^t e^{-(t-s)} ds$$

$$= 1$$

for $t \geq 0$. The result now follows and the proof is completed.

Given a constant $\epsilon > 0$ such that

$$\left| x_j - \frac{|\tilde{h}_j|}{1 - \lambda_j} \right| \leq \epsilon, \tag{4}$$

define a neighborhood $D_\epsilon$ by

$$D_\epsilon = \left\{ r \in R^n \,\middle|\, r = \sum_{j=1}^n x_j b_j \right\}. \tag{5}$$

**Theorem 2.** *Suppose $\lambda_1 = 1$ is the largest eigenvalue of $W$ with multiplicity $k$ and $\tilde{h}_1 = \cdots = \tilde{h}_k = 0$. If $r(0) \in D_\epsilon$, then, CNN (1) possesses a continuous attractor*

$$CA = \left\{ r \,\middle|\, r = \sum_{j=1}^k x_j(0) b_j + \sum_{j=k+1}^n \frac{\tilde{h}_j}{1 - \lambda_j} b_j, \right.$$

$$\left. \sqrt{ \sum_{j=1}^k x_j(0)^2 + \sum_{j=k+1}^n \left( \frac{|\tilde{h}_j|}{1 - \lambda_j} \right)^2 } \leq 1, 0 \leq x_j(0) \leq 1 \right\}.$$

*Proof:* Given any $r^* \in CA$,

$$r^* = \sum_{j=1}^k c_j^* b_j + \sum_{j=k+1}^n \frac{\tilde{h}_j}{1 - \lambda_j} b_j.$$

where

$$\sqrt{ \sum_{j=1}^k c_j^{*2} + \sum_{j=k+1}^n \left( \frac{|\tilde{h}_j|}{1 - \lambda_j} \right)^2 } \leq 1.$$

Then

$$\|r^*\|_S = \sqrt{ \sum_{j=1}^k c_j^{*2} + \sum_{j=k+1}^n \left( \frac{|\tilde{h}_j|}{1 - \lambda_j} \right)^2 } \leq 1.$$

and

$$r^* \in G.$$

So $CA \subset G$.

Next, for any $r^* \in CA$,

$$
f(Wr^* + h)
$$

$$
= f\left(\sum_{j=1}^{k} c_j^* Wb_j + \sum_{j=k+1}^{n} \frac{\tilde{h}_j}{1-\lambda_j} Wb_j + \sum_{j=1}^{n} \tilde{h}_j \cdot b_j\right)
$$

$$
= f\left(\sum_{j=1}^{k} c_j^* b_j + \sum_{j=k+1}^{n} \frac{\tilde{h}_j}{1-\lambda_j} \lambda_j b_j + \sum_{j=k+1}^{n} \tilde{h}_j \cdot b_j\right)
$$

$$
= f\left(\sum_{j=1}^{k} c_j^* b_j + \sum_{j=k+1}^{n} \frac{\tilde{h}_j}{1-\lambda_j} b_j\right)
$$

$$
= f(r^*)
$$

$$
= r^*.
$$

So $r^*$ is an equilibrium point of (1).
Next, we prove that $CA$ is stable.
From Theorem 1,

$$
r(t) = \sum_{j=1}^{n} x_j(t) b_j \in G, t \geq 0.
$$

Then we have

$$
\sum_{j=1}^{n} x_j(t)^2 \leq 1.
$$

Moreover, we have

$$
Wr(t) = \sum_{j=1}^{n} x_j(t) Wb_j = \sum_{j=1}^{n} x_j(t) \lambda_j b_j,
$$

for $t \geq 0$.
Since $\tilde{h}_1 = \cdots = \tilde{h}_k = 0$, if $Wr(t) + h \in G$ for all $t \geq 0$, it follows from (1) that

$$
\dot{x}_j(t) = 0, \quad (j = 1, \cdots, k) \tag{6}
$$

and

$$
\dot{x}_j(t) = (\lambda_j - 1) \cdot x_j(t) + \tilde{h}_j, \quad (j = k+1, \cdots, n) \tag{7}
$$

for $t \geq 0$. Solving Eqs. (6) and (7), we have

$$
x_j(t) = x_j(0), \quad (j = 1, \cdots, k)
$$

and

$$
x_j(t) = \left(x_j(0) - \frac{\tilde{h}_j}{1-\lambda_j}\right) e^{(\lambda_j - 1)t} + \frac{\tilde{h}_j}{1-\lambda_j},
$$

for $j = k+1, \cdots, n$.

Then, it follows that

$$r(t) = \sum_{j=1}^{k} x_j(0)b_j + \sum_{j=k+1}^{n} \left( x_j(0) - \frac{\tilde{h}_j}{1-\lambda_j} \right) b_j e^{(\lambda_j-1)t} + \sum_{j=k+1}^{n} \frac{\tilde{h}_j}{1-\lambda_j} b_j$$

for $t \geq 0$.

Now we prove that $Wr(t) + h \in G$.

$$Wr(t) + h$$
$$= \sum_{j=1}^{k} x_j(0)b_j + \sum_{j=k+1}^{n} \lambda_j \left( x_j(0) - \frac{\tilde{h}_j}{1-\lambda_j} \right) b_j e^{(\lambda_j-1)t} + \sum_{j=k+1}^{n} \frac{\lambda_j \tilde{h}_j}{1-\lambda_j} b_j + \sum_{j=k+1}^{n} \tilde{h}_j b_j$$
$$= \sum_{j=1}^{k} x_j(0)b_j + \sum_{j=k+1}^{n} \lambda_j x_j(0) e^{(\lambda_j-1)t} b_j - \sum_{j=k+1}^{n} \lambda_j \frac{\tilde{h}_j}{1-\lambda_j} e^{(\lambda_j-1)t} b_j + \sum_{j=k+1}^{n} \frac{\tilde{h}_j}{1-\lambda_j} b_j$$
$$= \sum_{j=1}^{k} x_j(0)b_j + \sum_{j=k+1}^{n} \left( \frac{\tilde{h}_j}{1-\lambda_j} + \lambda_j e^{(\lambda_j-1)t} (x_j(0) - \frac{\tilde{h}_j}{1-\lambda_j}) \right) b_j$$

If $r(0) \in B$, then we have

$$\| Wr(t) + h \| = \sqrt{ \sum_{j=1}^{k} x_j(0)^2 + \sum_{j=k+1}^{n} \left( \frac{|\tilde{h}_j|}{1-\lambda_j} \right)^2 } \leq 1.$$

So $Wr(t) + h \in G$.

Given any $\epsilon > 0$, choose a constant $\delta = \epsilon$, if

$$dist\,(r(0), CA)$$
$$= \inf_{r^* \in C} \{ \| r(0) - r^* \|_S \}$$
$$= \inf_{x_j(0) \in R(1 \leq j \leq k)} \left\{ \left\| \sum_{j=1}^{k} (x_j(0) - x_j(0)) \cdot b_j + \sum_{j=k+1}^{n} \left( x_j(0) - \frac{\tilde{h}_j}{1-\lambda_j} \right) \cdot b_j \right\|_S \right\}$$
$$= \inf_{x_j(0) \in R(1 \leq j \leq k)} \left\{ \sqrt{ \sum_{j=k+1}^{n} \left( x_j(0) - \frac{\tilde{h}_j}{1-\lambda_j} \right)^2 } \right\}$$
$$\leq \delta,$$

then,

$$dist\,(r(t), CA)$$

$$= \inf_{r^* \in C} \{\|r(t) - r^*\|s\}$$

$$= \inf_{x_j(0) \in R(1 \le j \le k)} \left\{ \left\| \sum_{j=1}^{k} (x_j(0) - x_j(0)) \cdot b_j + \sum_{j=k+1}^{n} \left( x_j(0) - \frac{\tilde{h}_j}{1 - \lambda_j} \right) \cdot b_j e^{(\lambda_j - 1)t} \right\|_s \right\}$$

$$= \inf_{x_j(0) \in R(1 \le j \le k)} \left\{ \sqrt{ \sum_{j=k+1}^{n} \left( x_j(0) - \frac{\tilde{h}_j}{1 - \lambda_j} \right)^2 e^{2(\lambda_j - 1)t} } \right\}$$

$$\le \inf_{x_j(0) \in R(1 \le j \le k)} \left\{ \sqrt{ \sum_{j=k+1}^{n} \left( x_j(0) - \frac{\tilde{h}_j}{1 - \lambda_j} \right)^2 } \right\}$$

$$\le \delta = \epsilon$$

for all $t \ge 0$.

By Definition 1, the set $CA$ is a continuous attractor of the network (1).

Geometrically, if the multiplicity of $\lambda_1 = 1$ is $k$, then the asymptotical continuous attractor $CA$ is the $k$ dimensional manifold in $R^n$.

## 4   Simulations

First consider a two dimensional CNN with an asymmetric connection weight first

$$\dot{r}(t) + r(t) = f \left( \begin{bmatrix} -1 & 8 \\ 1/2 & -1 \end{bmatrix} r(t) + \begin{bmatrix} -0.4 \\ 0.1 \end{bmatrix} \right) \tag{8}$$

for $t \ge 0$.

It can be checked that $W$ has two eigenvalues $\lambda_1 = 1$ and $\lambda_2 = -3$. Moreover, $\tilde{h}_1 = 0$. By Theorem 2, the network (8) possess a continuous attractor which is one-dimensional. In Fig. 1, The open circles are randomly selected initial points, the line attractor is the red line in the two dimensional square.

The next example is a three dimensional CNN:

$$\dot{r}(t) + r(t) = f \left( \begin{bmatrix} 2 & 1 & 0 \\ -1 & 0 & 0 \\ 0 & 0 & 0 \end{bmatrix} r(t) + \begin{bmatrix} 0 \\ 0 \\ 0.3 \end{bmatrix} \right) \tag{9}$$

for $t \ge 0$.

The multiplicity of the largest eigenvalue $\lambda_1 = 1$ of $W$ is 2, and $\tilde{h}_1 = \tilde{h}_2 = 0$, so the continuous attractor of network (9) is in two dimensional. It is called a plane attractor (Fig. 2).

## 5   Conclusions

In this paper, we have investigated the representation of continuous attractors of CNNs with asymmetric connection weights. The connection matrix can be

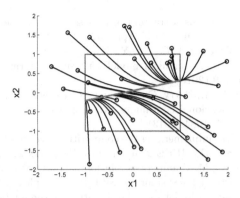

**Fig. 1.** Line attractor of the network (8). (Color figure online)

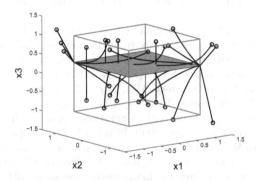

**Fig. 2.** Plane attractor of the model (9). The gray plane in the three dimensional box is the plane attractor.

diagonalized in a new coordinate system which is defined by the eigenvectors of connection matrix. By defining a new norm in the spaces of $R^n$, we established some interesting sufficient conditions to represent the continuous attractors of CNNs with asymmetric connection weights explicitly. The method in this paper appears firstly in the continuous attractors study.

# References

1. Zhang, R., Zeng, D., Zhong, S.: Event-triggered sampling control for stability and stabilization of memristive neural networks with communication delays. Appl. Math. Comput. **310**, 57–74 (2017)
2. Li, D., Leng, J., Huang, T., Sun, G.: On sum and stability of G-frames in Hilbert spaces. Linear Multilinear Algebra **66**(8), 1578–1592 (2018)
3. Hopfield, J.J.: Neurons with graded response have collective computational properties like those of two-state neurons. Proc. Natl. Acad. Sci. USA **81**(10), 3088–3092 (1984)
4. Seung, H.S.: Continouous attractors and oculomotor control. Neural Netw. **11**, 1253–1258 (1998)

5. Xu, F., Yi, Z.: Continuous attractors of a class of neural networks with a large number of neurons. Comput. Math. Appl. **62**(10), 3785–3795 (2011)
6. Wu, S., Wong, K.Y., Fung, C.C., Mi, Y., Zhang, W.: Continuous attractor neural networks candidate of a canonical model for neural information representation. F1000Research **5**, 1–9 (2016)
7. Yoon, K.J., Buice, M.A., Barry, C., Hayman, R., Burgess, N., Fiete, I.R.: Specific evidence of low-dimensional continuous attractor dynamics in grid cells. Nature Neurosci. **16**(8), 1077–1084 (2013)
8. Yi, Z.: Foundations of implementing the competitive layer model by Lotka-Volterra recurrent neural networks. IEEE Trans. Neural Networks **21**, 494–507 (2010)
9. Shuang, W., Lan, S.: Maximum principle for partially-observed optimal control problems of stochastic delay systems. J. Syst. Sci. Complexity **30**, 316–328 (2017)
10. Zhang, K.C.: Representation of spatial orientation by the intrinsic dynamics of the head-direction cell ensemble: a theory. J. Neurosci. **16**, 2112–2126 (1996)
11. Seung, H.S., Lee, D.D.: The manifold ways of perception. Science **290**, 2268–2269 (2000)
12. Stringer, S.M., Rolls, E.T., Trappenberg, T.P., Araujo, I.E.T.: Self-organizing continuous attractor networks and motor function. Neural Netw. **16**, 161–182 (2003)
13. Robinson, D.A.: Integrating with neurons. Annu. Rev. Neurosci. **12**, 33–45 (1989)
14. Koulakov, A., Raghavachari, S., Kepecs, A., Lisman, J.E.: Model for a robust neural integrator. Nature Neurosci. **5**(8), 775–782 (2002)
15. Stringer, S.M., Trappenberg, T.P., Rolls, E.T., Araujo, I.E.T.: Self-organizing continuous attractor networks and path integration: one-dimensional models of head direction cells. Network Comput. Neural Syst. **13**, 217–242 (2002)
16. Samsonovich, A., McNaughton, B.L.: Path integration and cognitive mapping in a continuous attractor neural network model. J. Neurosci. **17**, 5900–5920 (1997)
17. Pouget, A., Dayan, P., Zemel, R.: Information processing with population codes. Nat. Rev. Neurosci. **1**, 125–132 (2000)
18. Wu, S., Hamaguchi, K., Amari, S.: Dynamics and computation of continuous attractors. Neural Comput. **20**(4), 994–1025 (2007)
19. Miller, P.: Analysis of spike statistics in neuronal systems with continuous attractors or multiple, discrete attractor states. Neural Comput. **18**, 1268–1317 (2006)
20. Oliver, S., Lukas, S., Nolan, M.: Continuous attractor network models of grid cell firing based on excitatory-inhibitory Interactions. J. Physiol. **594**(22), 6547–6557 (2016)
21. Dehyadegary, L., Seyyedsalehi, S., Nejadgholi, I.: Nonlinear enhancement of noisy speech, using continuous attractor dynamics formed in recurrent neural networks. Neurocomputing **74**(17), 2716–2724 (2011)
22. Yu, J., Yi, Z., Zhang, L.: Representations of continuous attractors of recurrent neural networks. IEEE Trans. Neural Networks. **20**, 368–372 (2009)
23. Machens, C.K., Brody, C.D.: Design of continuous attractor networks with monotonic tuning using a symmetry principle. Neural Comput. **20**, 452–485 (2008)
24. Li, D., Leng, J., Huang, T.: Some Properties of G-frames for hilbert space operators. Oper. Matrices **11**(4), 1075–1085 (2017)

# Optimization

# Two Matrix-Type Projection Neural Networks for Solving Matrix-Valued Optimization Problems

Lingmei Huang[1], Youshen Xia[1(✉)], and Songchuan Zhang[2]

[1] College of Mathematics and Computer Science, Fuzhou University, Fuzhou, China
ysxia@fzu.edu.cn
[2] College of Mathematics and Computer Science, Wuyi University, Nanping, China
zsc_1977@126.com

**Abstract.** In recent years, matrix-valued optimization algorithms have been studied to enhance the computational performance of vector-valued optimization algorithms. This paper presents two matrix-type projection neural networks, continuous-time and discrete-time models, for solving matrix-valued optimization problems. The proposed continuous-time neural network may be viewed as a significant extension to the vector-type double projection neural network. More importantly, the proposed discrete-time projection neural network can be parallelly implemented in terms of matrix state space. Under pseudo-monotonicity condition and Lipschitz continuous condition, it is guaranteed that the two proposed matrix-type projection neural networks are globally convergent to the optimal solution. Finally, computed examples show that the two proposed matrix-type projection neural networks are much superior to the vector-type projection neural network in computation speed.

**Keywords:** Matrix-type neural network
Matrix-valued optimization · Global convergence · Computation time

## 1 Introduction

Most of engineering application problems, such as signal and image processing and pattern recognition [1], may be reformulated as optimization problems. Many solution methods were developed in past decades. They may be classified as numerical optimization algorithms [2–4] and recurrent neural networks [5–20]. The numerical optimization algorithms mainly include the gradient projection algorithm, the disciplined convex programming algorithm, and the interior point algorithm. By contrast, the recurrent neural network (RNN)-based algorithm has a potential capacity in solving constrained optimization problems under a weak condition for global convergence [14,15].

This work is supported by the National Natural Science Foundation of China under Grant No. 61473330.

© Springer Nature Switzerland AG 2018
L. Cheng et al. (Eds.): ICONIP 2018, LNCS 11302, pp. 405–416, 2018.
https://doi.org/10.1007/978-3-030-04179-3_36

Many of recurrent neural networks were proposed for solving different constrained optimization problems ranging from linear programming to nonlinear optimization, convex optimization to non-convex optimization, smooth optimization to non-smooth optimization and references therein. For example, Kennedy and Chua [10] proposed a neural network for nonlinear convex programming. To avoid using penalty parameters, Rodriguez-Vazquez et al. [11] proposed a switched-capacitor neural network for solving a class of nonlinear optimization problems by determining the feasible solution set. Zhang and Constantinides [12] developed a Lagrangian neural network for solving nonlinear convex optimization problems with linear equality constraints only. Xia et al. [13] developed one projection neural network for constrained optimization problems. Xia and Feng [14] proposed another projection neural network for the constrained optimization problems. Xia [15] and Xia and Wang [16,17] proposed several extended projection neural networks for nonlinearly constrained optimization problems. Furthermore, Xia [18] proposed a cooperative projection neural network for solving monotone variational inequality problems and a class of nonmonotone variational inequality problems with linear and nonlinear constraints. In summary, the state dimension of existing projection neural networks is in general larger than the dimension of the original optimization problem to be solved. As a result, the computation time is very dependent on the solution dimension of the optimization problem. When the solution dimension is large, they will have slow convergence. There are two ways to overcome this difficulty. One way is to develop a dimension-reduced neural network. Another way is to develop a matrix-type recurrent neural network. This paper is going to focus on the matrix-type recurrent neural network. In recent years, matrix-valued optimization algorithms were studied to enhance the computational performance of vector-valued optimization algorithms in speed [21,23–27]. At present, there are several matrix-type recurrent neural networks which are used to solve a class of matrix-valued optimization problems, such as the minimization problems of matrix Sylvester equations and matrix inverse problems [21–23].

This paper presents two matrix-type recurrent neural networks, continuous-time and discrete-time models, for solving matrix-valued nonlinear optimization problems. The proposed continuous-time projection neural network may be viewed as a significant extension to existing vector-valued projection neural network. The proposed discrete-time neural network can be parallely implemented in terms of matrix state space. Under pseudo-monotonicity condition and Lipschitz continuous condition, the proposed two matrix-type recurrent neural networks are able to globally converge to an optimal solution. Finally, numerical examples show the effectiveness of the two proposed matrix-type projection neural networks.

## 2   Matrix-Valued Optimization Problem

In this paper, we are concerned with the following optimization problem with matrix variables as follows:

$$\min \quad f(X) \qquad s.t. \quad X \in \Omega \tag{1}$$

where $f : \Re^{n \times n} \to \Re$ is a continuous function, $X \in \Re^{n \times n}$ is the matrix variable, and $\Omega$ is a closed and convex set. Since (1) includes the matrix variable, we call it as the matrix-valued optimization problem. According to the KKT optimality condition [22] we can see that $X^*$ is an optimal solution of (1) if and only if $X^*$ is a solution to the following variational inequality problem denoted as $VI(\nabla f, \Omega)$:

$$tr\big(\nabla f(X^*)^T(Y - X)\big) \geq 0, \forall Y \in \Omega \tag{2}$$

where $\nabla f(X)$ is the gradient of $f(X)$. Furthermore, by the projection property [22] we know that $X^*$ is also the solution of the following nonlinear projection equation:

$$P_\Omega[X - \alpha \nabla f(X)] = X \tag{3}$$

where $\alpha > 0$ and $P_\Omega(\cdot)$ is a projection operator defined by

$$P_\Omega(X) = \arg \min_{Y \in \Omega} \|X - Y\| \tag{4}$$

In this paper, we consider three sets: $\Omega_1 = \{X \in \Re^{n \times n} : X \geq 0\}$, $\Omega_2 = \{X \in \Re^{n \times n} : L \leq X \leq U\}$, and $\Omega_3 = \{X \in \Re^{n \times n} : \|X\| \leq \varepsilon\}$ where $X \geq 0$ means its each entry $x_{ij} \geq 0$, $L \leq X \leq U$ means $l_{ij} \leq x_{ij} \leq u_{ij}$, $\varepsilon > 0$ is a given scalar, $\|\cdot\|$ is the Frobenius norm, and $\|X\|^2 = tr(X^T X) = <X, X>$. It is seen that three projection operators [7] can be given by $P_{\Omega_1}(X) = \max(0, X)$ and

$$P_{\Omega_2}(x_{ij}) = \begin{cases} l_{ij}, & x_{ij} < l_{ij}, \\ x_{ij}, & l_{ij} \leq x_{ij} \leq u_{ij}, \\ u_{ij}, & x_{ij} > u_{ij}. \end{cases} \quad P_{\Omega_3}(X) = \begin{cases} X, & \|X\| \leq \varepsilon, \\ \dfrac{\varepsilon}{\|X\|} \cdot X, & \|X\| > \varepsilon, \end{cases}$$

respectively.

## 3   Matrix-Type Projection Neural Network

Based on Eq. (3) and the design method developed in [8], we propose a continuous-time matrix-type projection neural network as follows:

$$\frac{dX(t)}{dt} = P_\Omega(X(t) - \lambda g(X(t), \beta)) - X(t) \tag{5}$$

where $X(t) \in \Re^{n \times n}$ is a matrix state variable, $\lambda > 0$ and $\beta > 0$ are design parameters, and

$$g(X, \beta) = e(X, \beta) - \beta[\nabla f(X) - \nabla f(X - e(X, \beta))] \tag{6}$$

$$e(X, \beta) = X - P_\Omega[X - \beta \nabla f(X)] \qquad (7)$$

Furthermore, following (5), a discrete-time matrix-type projection neural network is proposed as follows:

$$X^{k+1} = (1 - h)X^k + hP_\Omega(X^k - \lambda g(X^k, \beta)) \qquad (8)$$

where $h > 0$ is a fixed step length. It is seen that the proposed two projection neural networks can be parallelly implemented in terms of matrix state space.

To illustrate the performance of the proposed matrix-type neural network, we first analyze its computational complexity in terms of the state space and the total number of multiplications/divisions per iteration. It is seen that the proposed matrix-type neural networks in (5) and (8) have the state space being $n \times n$, and the computational complexity being $O(n^3)$ and $O(n^3)$, respectively. By contrast, we consider the existing vector-type projection neural network for solving (1) with bounded set $\Omega = \Omega_2$. To do that, let us define a function $\varphi(x) = f(X)$ where $x = vec(X) = [x_1, x_2, \ldots, x_n]^T \in R^{n^2}$, $X = [x_1, \ldots, x_n]$, and $vec(\cdot)$ denotes an operator from a two-dimensional array into a one-dimensional vector [7]. Then (1) can be rewritten as the following vector-valued optimization problem:

$$\min \quad \varphi(x) \qquad s.t. \quad x_i \in \Omega_i \qquad (9)$$

where $\Omega_i = \{x_i \mid l_i \leq x_i \leq h_i\}$, $L = [l_1, \ldots, l_n]$, and $H = [h_1, \ldots, h_n]$ for $i = 1, \ldots, n$. To solve (9), Xia and Wang [8,13] proposed two projection neural networks. Recently, Eshaghnezhad et al. [20] also presented a double projection neural network as below:

$$\frac{dx(t)}{dt} = P_\Omega(x(t) - \lambda g(x(t), \beta)) - x(t) \qquad (10)$$

where $x(t) \in \Re^{n^2}$ is a vector type variable, $\Omega = \Omega_1 \times \Omega_2 \times \ldots \times \Omega_n$, $0 < \lambda \leq 1$ and $\beta > 0$ are design parameters,

$$\frac{dx(t)}{dt} = [\frac{dx_1(t)}{dt}, \frac{dx_2(t)}{dt}, \ldots, \frac{dx_n(t)}{dt}]^T, \quad \nabla\varphi(x) = [\frac{\partial\varphi(x)}{\partial x_1}, \frac{\partial\varphi(x)}{\partial x_2}, \ldots, \frac{\partial\varphi(x)}{\partial x_n}]^T,$$

$g(x, \beta) = e(x, \beta) - \beta[\nabla\varphi(x) - \nabla\varphi(x - e(x, \beta))]$, and $e(x, \beta) = x - P_\Omega[x - \beta\nabla\varphi(x)]$. It is seen that the double projection neural network has the state space being $n^2 \times 1$, and its computational complexity is $O(2n^4 + n^3)$. Table 1 lists the comparison of the algorithm complexity of (5), (8), and (10). From Table 1 we see that the proposed matrix-type projection neural networks have the lower complexity than the vector-type double projection neural network in (10). Also, the proposed continuous-time matrix-type projection neural network includes the double projection neural network as its spacial case.

Next, we illustrate that solving the matrix-valued optimization problem (1) is superior to solving the vector-valued optimization problem (9). Without

**Table 1.** Complexity of two models.

| Algorithm | Model of (10) | Model of (5) | Model of (8) |
|---|---|---|---|
| State space | $n^2 \times 1$ | $n \times n$ | $n \times n$ |
| Complexity | $O(2n^4 + n^3)$ | $O(n^3)$ | $O(n^3)$ |

**Table 2.** Computational complexity of solving (11) and (13).

| Objective function | $\|Hx - g\|_2$ | $\|H_1 X H_2 - G\|$ |
|---|---|---|
| State space | $N^2 \times N^2$ | $N \times N$ |
| Complexity | $O(N^4)$ | $O(N^3)$ |

loss of generality, let us consider the following minimization problem of image restoration:

$$\min_{v^{(1)} \leq x \leq v^{(2)}} \|Hx - g\|_2 \tag{11}$$

where $\| \cdot \|_2$ denotes $l_2$ norm, $x \in \Re^{N^2}$ is the image to be estimated, $H \in \Re^{N^2 \times N^2}$ is a blurring matrix, and $g \in \Re^{N^2}$ is an observed image [27]. In general, the blurred kernel is separable. Thus blurred matrix $H$ can be decomposed as Kronecker product of two low order matrices, that is, $H = H_1 \otimes H_2$ where $H_1, H_2 \in \Re^{N \times N}$. The problem (11) is rewritten as

$$\min_{L \leq X \leq U} \|(H_2 \otimes H_1)vec(X) - vec(G)\| \tag{12}$$

where $vec(X) = x, vec(G) = g$. Using the properties of the Kronecker product, we can equally convert the problem (12) into

$$\min \quad \|H_1 X H_2 - G\| \qquad s.t. \quad V_1 \leq X \leq V_2 \tag{13}$$

where $V_1 = vec^{-1}(v^{(1)})$ and $V_2 = vec^{-1}(v^{(2)})$. Table 2 lists the algorithm complexity for solving both (11) and (13), respectively. It is seen that solving (11) requires much higher complexity than solving (13).

## 4    Stability Results

First, we introduce the monotone definition of the matrix-valued function.

**Definition 1.**(1) A mapping $\nabla f : \Omega \to \Re^{n \times n}$ is Lipschitz continuous if there is a constant $L > 0$, such that

$$\|\nabla f(X) - \nabla f(Y)\| \leq L\|X - Y\|, \forall X, Y \in \Omega.$$

(2) A mapping $\nabla f$ is pseudo-monotone with respect to a matrix $Y \in \Omega$, if

$$tr\big((X - Y)^T \nabla f(Y)\big) \geq 0 \Rightarrow tr\big((X - Y)^T \nabla f(X)\big) \geq 0, \forall X \in \Omega.$$

(3) A mapping $\nabla f$ is strongly pseudo-monotone with respect to a matrix $Y \in \Omega$, if there is a constant $\gamma > 0$ such that

$$tr\big((X - Y)^T \nabla f(Y)\big) \geq 0 \Rightarrow tr\big((X - Y)^T \nabla f(X)\big) \geq \gamma \|X - Y\|^2, \forall X \in \Omega.$$

By extending the analysis of both [8,20], we have the following results:

**Theorem 1.** Let $\nabla f(\cdot)$ be a pseudo-monotone and Lipschitz continuous function with the constant $L$. Then the proposed two neural networks in (5) and (8) with $0 < h < 1$ are stable in the sense of Lyapunov and globally convergent to the solution of (1) if $0 < \beta \leq \frac{1}{5L}$ and $0 < \lambda \leq 1$.

**Theorem 2.** Let $\nabla f(\cdot)$ be strongly pseudo-monotone with the constant $\gamma$ and Lipschitz continuous function with constant $L$. Then the proposed two neural networks in (5) and (8) with $0 < h < 1$ are globally exponentially stable if $0 < \beta \leq \frac{1}{5L}$, $0 < \lambda \leq 1$, and $\gamma \geq 2L$.

We omit the proof of Theorems 1 and 2 for limitation of paper length.

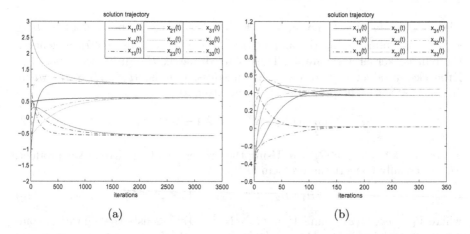

(a)                                                (b)

**Fig. 1.** (a) shows the trajectory behavior of model (8) in Example 1 with bounded constraint. And (b) displays the trajectory behavior of model (8) in Example 1 with bounded constraint.

## 5    Computed Results

In this section, we give several numerical examples to show the effectiveness of the proposed matrix-type neural networks. The experimental results were conducted in Maltab of Windows desktop with Intel core i5 processor 3.20 GHz and 4 GB RAM.

**Example 1.** Consider the following quadratic optimization problem:

$$min \quad \|AX - B\|^2 \quad s.t. \quad X \in \Omega$$

where

$$A = \begin{bmatrix} 1 & 6 & 3 \\ 3 & 7 & 6 \\ 4 & 3 & 0 \end{bmatrix}, B = \begin{bmatrix} 3 & 3 & 3 \\ 4 & 4 & 4 \\ 6 & 6 & 6 \end{bmatrix}.$$

For our testing, we consider two cases of set $\Omega$. First, we study the case of $\Omega = \{X \in \Re^{3 \times 3} : -10 \leq x_{ij} \leq 10\}, i, j = 1, 2, 3$. Without loss of generality, we perform the proposed matrix-type neural network in (8) where parameters are taken as $\lambda = 1, \beta = 0.0005$ and $h = 0.8$, respectively. All simulation results show that the matrix-type neural network in (8) is always globally convergent. For example, Fig. 1(a) displays the trajectory behavior of the matrix-type neural network in (8) with an initial point, which gives the unique solution:

$$X_1^* = \begin{bmatrix} 1.0435 & 1.0435 & 1.0435 \\ 0.6087 & 0.6087 & 0.6087 \\ -0.5652 & -0.5652 & -0.5652 \end{bmatrix}.$$

Next, we study the case of $\Omega = \{X \in \Re^{3 \times 3} : \|X\| \leq 1\}$. Let us take parameters $\lambda = 1, \beta = 0.0005$, and $h = 1$. Figure 1(b) shows the trajectory behavior of the matrix-type neural network in (8) with an initial point. It gives the solution:

$$X_2^* = \begin{bmatrix} 0.4407 & 0.4407 & 0.4407 \\ 0.3725 & 0.3725 & 0.3725 \\ 0.0190 & 0.0190 & 0.0190 \end{bmatrix}.$$

Finally, for a comparison, we also perform the continuous-time recurrent neural network in (10). Figure 2(a) displays the convergence behavior of $\|X(t) - X_1^*\|$ based on (8) and (10). From Fig. 2(a) we see that the proposed matrix-type neural network in (8) has a much faster speed than the vector-type neural network in (10).

**Example 2.** Consider a large-sized quadratic optimization problem:

$$min \quad \|AX - B\|^2 \quad s.t. \quad X \in \Omega$$

where $\Omega = \{X \in \Re^{n \times n} : 0 \leq x_{ij} \leq 1\}$, matrix $A$ is an $n \times m$ random matrix and the ideal solution $X^*$ is a $n \times n$ random matrix. The observed matrix $B$ is produced by adding noise random matrix $V$ in $AX^*$. All simulation results show that the matrix-type neural network in (8) is always globally convergent. For our comparison, we also perform the vector-type projection neural network in (10). Let $n = 10, \lambda = 0.8, \beta = 0.0005$, and $h = 1$. Figure 2(b) shows that the convergence behavior of $\|X(t) - X^*\|$ based on (8) and (10), respectively. From Fig. 2(b) we see that the proposed matrix-type neural network (8) has a much faster speed than the vector-type neural network (10).

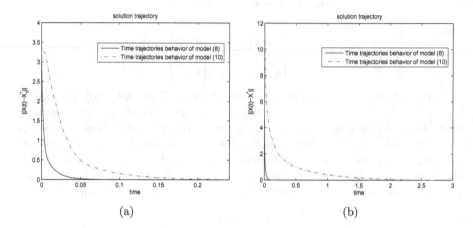

(a)                                                (b)

**Fig. 2.** (a) shows the convergence behavior of $\|X(t) - X_1^*\|$ based on (8) and (10) in Example 1 with bounded constraint. And (b) displays the convergence behavior of $\|X(t) - X^*\|$ based on (8) and (10) in Example 2 with bounded constraint.

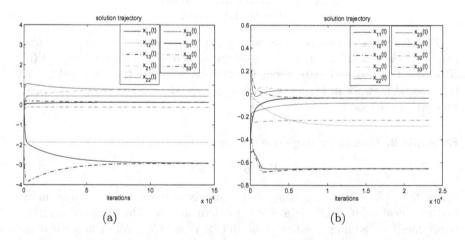

(a)                                                (b)

**Fig. 3.** (a) shows the trajectory behavior of model (8) in Example 3 with bounded constraint.(b) displays the trajectory behavior of model (8) in Example 3 with sphere constraint.

**Example 3.** Consider the optimization problem

$$min \quad \|AXB - C\|^2 \qquad s.t. \quad X \in \Omega$$

where

$$A = \begin{bmatrix} 0.2 & -1 & 0 \\ -1 & 0.2 & -1 \\ 0 & -1 & 0.2 \end{bmatrix}, B = \begin{bmatrix} 1 & 2 & 2 \\ 2 & 5 & 6 \\ 2 & 6 & 0.9 \end{bmatrix}, C = \begin{bmatrix} 1 & 2 & 1 \\ 2 & 2 & 2 \\ 1 & 2 & 1 \end{bmatrix}.$$

For our testing, we consider two cases of set $\Omega$. First, we study the case of $\Omega = \{X \in \Re^{3 \times 3} : -5 \le x_{ij} \le 5\}, i, j = 1, 2, 3$. Without loss of generality, we perform the proposed matrix-type neural network in (8) where parameters are taken as $\lambda = 1, \beta = 0.0005$ and $h = 0.8$, respectively. All simulation results show that the matrix-type neural network in (8) is always globally convergent. For example, Fig. 3(a) displays the trajectory behavior of the matrix-type neural network in (8) with an initial point, which gives the solution:

$$X_3^* = \begin{bmatrix} -2.9080 & 0.7628 & 0.1296 \\ -1.8626 & 0.4340 & -0.1150 \\ -2.9012 & 0.7606 & 0.1291 \end{bmatrix}.$$

Next, we study the case of $\Omega = \{X \in \Re^{3 \times 3} : \|X\| \le 1\}$. Let us take parameters $\lambda = 1, \beta = 0.0005$, and $h = 1$. Figure 3(b) displays the trajectory behavior of the matrix-type neural network (8) with an initial point. It gives the solution:

$$X_4^* = \begin{bmatrix} -0.6545 & 0.0313 & -0.0363 \\ -0.2814 & -0.0798 & -0.2302 \\ -0.6545 & 0.0313 & -0.0363 \end{bmatrix}.$$

Finally, for a comparison, we also perform the vector-type projection neural network in (10). Figure 4(a) displays the convergence behavior of $\|X(t) - X_3^*\|$ based on (8) and (10). From Fig. 4(a) we see that the proposed matrix-type neural network in (8) has a much faster speed than the vector-type neural network in (10).

**Example 4.** Consider a large-sized quadratic optimization problem:

$$min \quad \|AXB - C\|^2 \qquad s.t. \quad X \in \Omega$$

where $\Omega = \{X \in \Re^{n \times n} : 0 \le x_{ij} \le 1\}$ and $A$ and $B$ are two $n \times n$ random matrices and the ideal solution $X^*$ is a $n \times n$ random matrix. The observed matrix $C$ is produced by adding noise random matrix $V$ in $AX^*B$. All simulation results show that the matrix-type neural network in (8) is always globally convergent. For our comparison, we also perform the vector-type projection neural network in (10). Let $n = 10, \lambda = 0.8, \beta = 0.0005$, and $h = 1$. Figure 4(b) displays that the convergence behavior of $\|X(t) - X^*\|$ based on (8) and (10), respectively. From Fig. 4(b) we see that the proposed matrix-type neural network in (8) has a much faster speed than the vector-type projection neural network in (10).

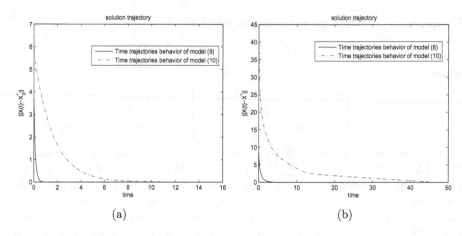

**Fig. 4.** (a) shows the convergence behavior of $\|X(t) - X_3^*\|$ based on (8) and (10) in Example 3 with bounded constraint.(b) displays the convergence behavior of $\|X(t) - X^*\|$ based on (8) and (10) in Example 4 with bounded constraint.

## 6    Conclusion

This paper has presented two matrix-type projection neural networks, continuous-time and discrete-time models, for solving matrix-valued optimization problems. The proposed continuous-time neural network model can significantly extend the vector-type double projection neural network model. The proposed discrete-time projection neural network can be fast implemented in terms of matrix state space. Under pseudo-monotonicity condition and Lipschitz continuous condition, it is shown that the two proposed matrix-type projection neural networks are globally convergent to the optimal solution. Finally, numerical examples are provided to show the effectiveness of the two proposed matrix-type projection neural networks.

## References

1. Kalouptsidis, N.: Signal Processing Systems: Theory and Design. Wiley-Interscience, New York (1997)
2. Mohammed, J.L., Hummel, R.A., Zucker, S.W.: A gradient projection algorithm for relaxation methods. IEEE Trans. Pattern Anal. Mach. Intell. **5**, 330–332 (1983)
3. Grant, M., Boyd, S., Ye, Y.: Disciplined Convex Programming. Springer, Boston (2006). https://doi.org/10.1007/0-387-30528-9_7
4. Vanderbei, R.J., Shanno, D.F.: An interior-point algorithm for nonconvex nonlinear programming. Comput. Optim. Appl. **13**, 231–252 (1999)
5. Xia, Y.: A new neural network for solving linear programming problems and its application. IEEE Trans. Neural Netw. **7**, 525–529 (1996)

6. Xia, Y., Wang, J.: A recurrent neural network for nonlinear convex optimization subject to nonlinear inequality constraints. IEEE Trans. Circ. Syst. I Regul. Pap. **51**, 1385–1394 (2004)

7. Xia, Y.: A compact cooperative recurrent neural network for computing general constrained $L_1$ norm estimators. IEEE Trans. Sig. Process. **57**(9), 3693–3697 (2009)

8. Xia, Y., Wang, J.: A general methodology for designing globally convergent optimization neural networks. IEEE Trans. Neural Netw. **9**, 1331–1343 (1998)

9. Liu, Q., Wang, J.: A one-layer recurrent neural network for non-smooth convex optimization subject to linear equality constraints. In: Köppen, M., Kasabov, N., Coghill, G. (eds.) ICONIP 2008. LNCS, vol. 5507, pp. 1003–1010. Springer, Heidelberg (2009). https://doi.org/10.1007/978-3-642-03040-6_122

10. Kennedy, M.P., Chua, L.O.: Neural networks for nonlinear programming. IEEE Trans. Circ. Syst. **35**, 554–562 (1988)

11. Rodriguez-Vazquez, A., Dominguez-Castro, R., Rueda, A., Huertas, J.L.: Nonlinear switched capacitor neural networks for optimization problems. IEEE Trans. Circ. Syst. **37**, 384–398 (1990)

12. Zhang, S., Constantinides, A.G.: Lagrange programming neural networks. IEEE Trans. Circ. Syst. II Analog Digit. Sig. Process. **39**, 441–452 (1992)

13. Xia, Y., Leung, H., Wang, J.: A projection neural network and its application to constrained optimization problems. IEEE Trans. Circ. Syst. I Fundam. Theory Appl. **49**, 447–458 (2002)

14. Xia, Y., Feng, G., Wang, J.: A novel recurrent neural network for solving nonlinear optimization problems with inequality constraints. IEEE Trans. Neural Netw. **19**, 1340–53 (2008)

15. Xia, Y.: An extended projection neural network for constrained optimization. Neural Comput. **16**, 863–883 (2004)

16. Xia, Y., Wang, J.: Solving variational inequality problems with linear constraints based on a novel recurrent neural network. In: Liu, D., Fei, S., Hou, Z., Zhang, H., Sun, C. (eds.) ISNN 2007. LNCS, vol. 4493, pp. 95–104. Springer, Heidelberg (2007). https://doi.org/10.1007/978-3-540-72395-0_13

17. Xia, Y., Wang, J.: A bi-projection neural network for solving constrained quadratic optimization problems. IEEE Trans. Neural Netw. Learn. Syst. **27**, 214–224 (2016)

18. Xia, Y.S.: New cooperative projection neural network for nonlinearly constrained variational inequality. Sci. China **52**, 1766–1777 (2009)

19. Cheng, L., Hou, Z.G., Lin, Y., et al.: Recurrent neural network for non-smooth convex optimization problems with application to the identification of genetic regulatory networks. IEEE Trans. Neural Netw. **22**, 714–726 (2011)

20. Eshaghnezhad, M., Effati, S., Mansoori, A.: A neurodynamic model to solve nonlinear pseudo-monotone projection equation and its applications. IEEE Trans. Cybern. **47**, 3050–3062 (2016)

21. Xia, Y., Chen, T., Shan, J.: A novel iterative method for computing generalized inverse. Neural Comput. **26**(2), 449–465 (2014)

22. Bertsekas, D.P., Tsitsiklis, J.N.: Parallel and Distributed Computation: Numerical Methods. Prentice Hall, Upper Saddle River (1989)

23. Li, Z., Cheng, H., Guo, H.: General recurrent neural network for solving generalized linear matrix equation. Complexity **3**, 1–7 (2017)

24. Bouhamidi, A., Jbilou, K., Raydan, M.: Convex constrained optimization for large-scale generalized Sylvester equations. Comput. Optim. Appl. **48**(2), 233–253 (2011)
25. Shi, Q.B., Xia, Y.S.: Fast multi-channel image reconstruction using a novel two-dimensional algorithm. Multimedia Tools Appl. **71**, 2015–2028 (2014)
26. Li, J.F., Li, W., Huang, R.: An efficient method for solving a matrix least squares problem over a matrix inequality constraint. Comput. Optim. Appl. **63**(2), 393–423 (2016)
27. Bouhamidi, A.: A Kronecker approximation with a convex constrained optimization method for blind image restoration. Optim. Lett. **6**, 1251–1264 (2012)

# An Adaptive Ant Colony System for Public Bicycle Scheduling Problem

Di Liang[1], Zhi-Hui Zhan[1(✉)], and Jun Zhang[2]

[1] School of Computer Science and Engineering,
South China University of Technology,
Guangzhou 510006, People's Republic of China
zhanapollo@163.com

[2] Guangdong Provincial Key Laboratory of Computational Intelligence
and Cyberspace Information, South China University of Technology,
Guangzhou 510006, People's Republic of China

**Abstract.** Public bicycle scheduling problem (PBSP) is a kind of problem that how to design a reasonable transportation route in order to reduce cost or improve user satisfaction under certain constraints. PBSP can be regarded as a specific combinatorial optimization problem that ant colony system (ACS) can solve. However, the performance of conventional ACS is sensitive to its parameters. If the parameters are not properly set, ACS may have the disadvantage of being easy to fall into local optimum, resulting in poor accuracy and poor robustness. This paper proposes an adaptive ACS (AACS) to efficiently solve PBSP. Instead of fixed parameters in ACS, each ant is configured with own different parameters automatically to construct solutions in AACS. In each generation, AACS regards the parameters in the well-performed ants as good parameters and spreads these parameters among the ant colony via selection, crossover, and mutation operators like in genetic algorithm (GA). This way, the key parameters of ACS can be evolved into a more suitable set to solve PBSP. We applied AACS to solve PBSP and compared AACS with conventional ACS and greedy algorithm. The results show that AACS will improve the accuracy of the solution and achieve better robustness.

**Keywords:** Public Bicycle Scheduling Problem · Ant Colony System
Genetic Algorithm

## 1 Introduction

Inspired by the sharing economy, public bicycle sharing systems (PBSS) are bringing a revolutionary new environmental protection method to smog-ridden city traffic [1]. Despite their obvious success as a new form of public transportation, problems with borrowing bicycles and high bicycle scheduling cost still arise in the actual operation. The emergence of these problems is largely due to the irrationality of bicycle scheduling, which makes the public bicycle scheduling problem (PBSP) an important subject worth studying. As a variant of vehicle routing problem (VRP) [2], the PBSP belongs to combinatorial optimization problems (COPs), which can be solved by meta-heuristic methods [3].

© Springer Nature Switzerland AG 2018
L. Cheng et al. (Eds.): ICONIP 2018, LNCS 11302, pp. 417–429, 2018.
https://doi.org/10.1007/978-3-030-04179-3_37

As a kind of intelligent bionic optimization algorithm, ant colony optimization (ACO), especially its ant colony system (ACS) [4] variant has been successfully applied to many COPs [5], such as aircraft arrival sequencing and scheduling problem [6], virtual machine placement in cloud computing [7], and multiobjective cloud workflow scheduling [8]. Furthermore, ACO performs well on vehicle routing issues [2, 9]. A distributed ACS algorithm was proposed to solve school bus scheduling problem in [9]. Liu et al. [10] used ACO to optimize the scheduling path in PBSP, but did not consider the various demands of users in different periods of the day. Zhang et al. [11] considered the time factor, but simply used ACO to solve the problem, which was easy to converge to local optimal solution.

Given that the parameters such as $\beta$, $\varepsilon$, $\rho$, which are the heuristic factor and two parameters of pheromone updating rules in ACS respectively, have significant effect on the performance of ACS [12], many studies have been conducted to analyze parametric relation and find the best parameter values of ant-based algorithms [12–15]. Yu et al. [12] proposed to adjust the vales of $\rho$ and $\varepsilon$ dynamically in different states of the optimization process. In this paper, we expand parameters to be optimized (i.e., $\alpha$, $\beta$, $\rho$, $\varepsilon$, and $q_0$) and adopt the evolutionary mechanisms in GA to adaptively control these parameters. Two models are established to minimize transportation cost and maximize user satisfaction, and an adaptive ACS (AACS) is proposed to solve these models in this research. The results show that the AACS method can effectively improve the quality of solutions.

The rest of this paper is organized as follows. Section 2 describes PBSP. Section 3 introduces AACS. In Sect. 4 we compare AACS algorithm with other general algorithms: conventional ACS and greedy algorithm. Section 5 summarizes the study and points out the research direction in the future.

## 2    Description of PBSP

In PBSP, the location of each bicycle rental station is fixed and known. The scheduling vehicle departs from one of these stations, collects or delivers a certain number of bicycles according to the needs of each station passing by, and finally returns to the starting station to complete a dispatch task.

The problem can be modeled under the environments of flat-peak period or peak period [11]. In the flat-peak period, users need fewer bicycles and each rental station can basically meet the needs of users. The vehicle mainly considers how to plan the route to minimize scheduling cost. It is assumed that the cost is positively correlated with the distance of the scheduling path, that is, a longer path means a higher cost. Therefore, the objective can be transformed to minimize the path length. While in the peak period, users have greater demand, and their demand time is more concentrated. Therefore, time factor should be taken into account in the scheduling process. The primary consideration is to maximize user satisfaction. According to two different situations above, PBSP can be modeled as ‘*distance model*’ in flat-peak period and as ‘*time window model*’ in peak period, respectively. We will describe these two models in the following of this section.

## 2.1  Distance Model

Undirected graph $G = (V, E)$ denotes the scheduling domain, where $V = \{1, 2, \ldots, n\}$ denotes the set of rental stations, $E = \{(i,j)|i,j \in V, i \neq j\}$ denotes the edges between these stations. $d_i$ denotes the demand for bicycles in station $i(i \in V)$. Herein, $d_i > 0$ means that $i$ has extra bicycles that need to be taken away, $d_i > 0$ means $i$ needs to be placed bicycles, and $d_i = 0$ means that $i$ does not need to be served. Here the vehicle departs from the starting station with no bicycle. A complete scheduling route $R = \{r_1, r_2, \ldots, r_{n+1}\}, r_i \in V$ is formed when the vehicle returns to the starting station after visiting all the stations. Our goal is to minimize the length of $R$, thereby reducing the scheduling cost.

Three constraints need to be met when choosing the next station to serve. (i) The station has not been accessed before. (ii) If the station's demand is positive, the number of bicycles that need to be taken away should be less than or equal to the left capacity of the vehicle. (iii) If the station's demand is negative, the number of bicycles required should be less than or equal to the available number of bicycles on the vehicle this moment.

Above, distance model can be expressed as follows.

Firstly, the optimization objective is as

$$\min L = \sum_{r_x, r_{x+1} \in R} l(r_x, r_{x+1}) \quad x = 1, \ldots, n \tag{1}$$

Then, the constraints are as

$$0 \leq q_w = d_i + q_{w-1} \leq Q \quad w = 2, \ldots, n, i \in V \tag{2}$$

$$r_1 = r_{n+1} \tag{3}$$

$$r_x \neq r_y \quad x, y = 1, \ldots, n, x \neq y \tag{4}$$

$$\sum_{i \in V} d_i \geq 0 \tag{5}$$

where $l(r_x, r_{x+1})$ is the distance between stations $r_x$ and $r_{x+1}, q_w$ is the number of bicycles on the vehicle at time $w$, and $Q$ is the capacity of the vehicle. Equation (2) shows the condition that the demand of next station should be satisfied. Equation (3) shows that the vehicle finally returns to the starting station after finishing a scheduling task. Equation (4) indicates that each station has only one chance to be served in a scheduling process. Equation (5) indicates that the algebraic sum of all stations' demand should be greater than or equal to zero. Otherwise, this problem can not be solved.

## 2.2  Time Window Model

Time window model is based on distance model. In this model, the vehicle also completes a scheduling route $R$, and during the scheduling process, a corresponding number of bicycles are collected or put into service according to the needs of each station. Unlike distance model, the objective is no longer to minimize the path length, but to maximize user satisfaction.

In time window model, each station maintains a time window that represents the period when users expect to be served at that station. The vehicle should try to reach the station within the specified time window to reduce the possibility that the user is rejected when borrowing or returning a bicycle. The calculation of user satisfaction at each station $i$ follows the function of visiting time $t_i$ shown in Fig. 1.

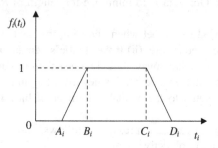

**Fig. 1.** User satisfaction function

In $(B_i, C_i)$, user satisfaction is 1, which is called the best visiting period. In $(A_i, B_i)$ and $(C_i, D_i)$, user satisfaction changes linearly with $t_i$ within $(0,1)$. In $(A_i, D_i)$, user satisfaction is positive, which is called acceptable visiting time. Outside $(A_i, D_i)$, user satisfaction is 0, which is unacceptable visiting period. The user satisfaction function at station $i$ can be expressed as:

$$f_i(t_i) = \begin{cases} 1, & \text{if } t_i \in (B_i, C_i) \\ \frac{t_i - A_i}{B_i - A_i}, & \text{if } t_i \in (A_i, B_i) \\ \frac{D_i - t_i}{D_i - C_i}, & \text{if } t_i \in (C_i, D_i) \\ 0, & \text{otherwise} \end{cases} \tag{6}$$

Above, time window model is expressed as follows, with the objective as

$$\max S = \sum_{i \in V} f_i(t_i) \tag{7}$$

where $S$ is the sum of user satisfaction at all stations. The constraints are set the same as those in distance model.

## 3   AACS to Solve PBSP

In general, the parameters in ACS are set based experience or through trial and error, which may result in low precision and low efficiency respectively. Different from conventional ACS that each ant shares the same parameters to build solutions, AACS assigns different parameters for each ant and combines the evolutionary mechanisms in GA (i.e., selection, crossover, and mutation) to adjust these parameters online. In AACS, each ant constructs solutions with independent parameters. According to this method, different ants may have different performance on this problem, mainly due to their different parameters. In order to deliver the good parameters in the well-performed ants to the poorly-performed ants, evolutionary mechanisms, such as selection, crossover, and mutation, are used to improve these parameters in ant colony. This way, the parameters in the ant are treated as a chromosome in GA.

When AACS is applied to solve PBSP, an ant can be considered as a vehicle with a constant capacity. First of all, a set of gene sequence values $\{\alpha_i, \beta_i, \rho_i, \varepsilon_i, q_{0i}\}$ is initialized randomly in certain ranges for each ant $i$. The ants construct solutions according to ACS rule under the control of their own parameters. After a certain number of iterations $I$, calculate the fitness of ants and update their parameters (or genes) by selecting some good parameters in the well-performed ants. Cross and mutate these parameters to generate better parameters for the colony.

## 3.1 Key Components in Designing AACS

**Pheromone Initialization.** The pheromone on the edge between each station is initialized to

$$\tau_0 = (N \cdot L_{nn})^{-1} \tag{8}$$

where $N$ is the number of stations, $L_{nn}$ is the tour length produced by the nearest neighbor heuristic search.

**Constraints Processing.** For constraints (2) and (4), a set of candidate stations is set up when choosing the next station, and only stations that satisfy both of these conditions can join the set.

**Solution Construction.** In one iteration, each ant randomly chooses a departure station and determines the next station with the guidance of a state transition rule. The transition probability that ant $k$ moves from station $i$ to station $j$ is calculated according to

$$p_k(i,j) = \begin{cases} \dfrac{[\tau(i,j)]^\alpha \cdot [\eta(i,j)]^\beta}{\sum\limits_{u \in allowed_k} [\tau(i,u)]^\alpha \cdot [\eta(i,u)]^\beta}, & \text{if } j \in allowed_k \\ 0, & \text{otherwise} \end{cases} \tag{9}$$

where $allowed_k$ is the set of candidate stations that $k$ is allowed to access next, $\tau(i,j)$ is the amount of pheromone, and $\eta(i,j)$ is the heuristic factor. Herein, $\eta(i,j) = 1/l(i,j)$ in distance model and $\eta(i,j) = C \cdot f_j(t_i + T_0 + l(i,j)/v_0)$ in time window model, where $t_i$ is the time when the ant reaches $i$, $T_0$ is the time it takes to load or unload bicycles, $v_0$ is the speed of movement, and $C$ is a parameter.

In AACS, the state transition rule is as follows: ant $k$ positioned on station $i$ chooses the next station $j$ from the candidate stations $allowed_k$ by applying the rule given by

$$j = \begin{cases} \arg\max_{u \in allowed_k}\{[\tau(i,u)^{\alpha}] \cdot [\eta(i,u)^{\beta}]\}, & \text{if } q \leq q_0 \\ J, & \text{otherwise} \end{cases} \quad (10)$$

where $q$ is a random number uniformly distributed in [0,1], $J$ is a random variable selected from $allowed_k$ according to the probability distribution given in (9), and $q_0$ $(0 \leq q_0 \leq 1)$ is a parameter which is used to balance exploration of new edges and exploitation of accumulated knowledge about the problem.

**Pheromone Updating Rule.** Pheromone local updating and global updating are performed in the optimization process. The significance of local updating is to explore more potential solutions, while the significance of global updating lies in exploitation of the optimal solutions. When an ant has built a path, local updating rule (11) is applied on each edge of the path to decay the pheromone, which leads to a greater probability of exploring other edges. In contrast, only the globally best path so far is allowed to deposit pheromone according to global updating rule (12) after all the ants have completed their tours

$$\tau(i,j) = (1 - \rho) \cdot \tau(i,j) + \rho \cdot \tau_0 \quad (11)$$

$$\tau(i,j) = (1 - \varepsilon) \cdot \tau(i,j) + \varepsilon \cdot \Delta\tau(i,j), \forall(i,j) \in T^{gb} \quad (12)$$

where $\rho(0 < \rho < 1)$ and $\varepsilon(0 < \varepsilon < 1)$ are the decay parameters of local pheromone and global pheromone respectively. Herein, $\Delta\tau(i,j) = (L_{gb})^{-1}$ in distance model, where $L_{gb}$ is the length of the globally best Tour $T_{gb}$; $\Delta(i,j) = C_1 \cdot S_{gb}$ in time window model, where $S_{gb}$ is the maximum user satisfaction value, and $C_1$ is a parameter.

**Fitness Function.** The fitness function is designed to evaluate the adaptability of ants. Since that ACS is a stochastic search method and each ant chooses the starting station randomly, which affects the evaluation of solution, especially in time window model, we allow an ant to construct solutions for number of iterations $I$ and take the value of the best solution produced by the ant as its fitness. The fitness function in distance model and time window model is implemented as (13) and (14) respectively.

$$F = (L_b)^{-1} \quad (13)$$

$$F = S_b \quad (14)$$

where $L_b$ is the length of the optimal scheduling route that the ant has passed and $S_b$ is the maximal user satisfaction value in $I$ iterations.

## 3.2   Complete AACS Algorithm

The complete AACS algorithm contains the following eight steps.

Step 1: Initialize parameters such as population size $M$, iteration times $I$, maximum generation $G$, crossover probability $p_c$ and mutation probability $p_m$, and initialize the amount of pheromone $\tau_0$. Randomly Generate $\{\alpha_i, \beta_i, \rho_i, \varepsilon_i, q_{0i}\}$ for each ant $i$ and set its initial fitness value to 0.

Step 2: Set the iteration $t = 1$.

Step 3: Each ant randomly chooses the starting station and serves it. Determine the next station according to state transition rule. The ant goes on like this until it finishes serving all the stations and then goes back to the starting station. Perform local pheromone updating rule.

Step 4: For each ant, calculate $L^{-1}$ (or $S$) and compare it with the original fitness value. The better will be the new fitness value.

Step 5: Until all the ants have constructed their tours, perform global pheromone updating rule. Set $t = t+1$.

Step 6: Judge whether $t$ is equal to $I$. If so, turn to Step 7, otherwise turn to Step 3.

Step 7: Update parameters of these ants by selection, crossover, and mutation operations.

Step 8: Judge whether the number of evolutionary generations is equal to $G$. If so, output the optimal solution, otherwise turn to Step 2.

The flowchart of AACS is shown in Fig. 2.

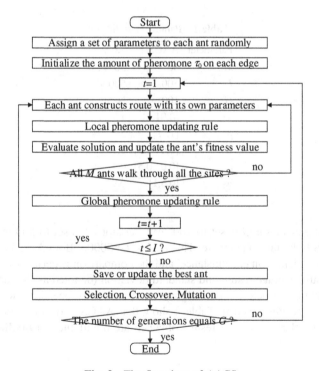

**Fig. 2.** The flowchart of AACS

## 4 Experiments and Discussions

In this section, we compare AACS with other general algorithms. Considering the existence of multiple constraints in PBSP, it is easy to generate an illegal solution in calculating process if using GA to solve. Therefore, we take conventional ACS and greedy algorithm as competitors in the experiments. We report three sets of experiments. Firstly, AACS is compared with the conventional ACS and GA on some symmetric TSP instances, include fri26, eil51, eil76, and eil101 in TSPLIB (https://wwwproxy.iwr.uni-heidelberg.de/groups/comopt/software/TSPLIB95/) to test the performance of each algorithm without bicycle restrictions. The second set and the third set compare their performance in distance model and time window model. The AACS performs well in all these cases. In addition, the time consumption of AACS is basically the same as that of ACS, which is verified through experiments.

### 4.1 Comparison Without Bicycle Constraints

The parameter settings are listed in Table 1. The parameters of ACS are set as originally suggested in [4, 16].

**Table 1.** Parameter settings

| Parameter | AACS | ACS |
|---|---|---|
| $M$ | 50 | 50 |
| $I$ | 500 | 50000 |
| $G$ | 100 | N/A |
| $\alpha$ | [0.01,4] | 1 |
| $\beta$ | [1, 7] | 5 |
| $\rho$ | [0.01,0.5] | 0.1 |
| $\varepsilon$ | [0.01,0.5] | 0.1 |
| $q_0$ | [0.5,1] | 0.9 |

Maximum generation $G$ is set to 100. Ant iterations $I$ is set to 500 in AACS and 50000 in ACS to be fair to compare. That is, $I \times G$=50000. Both AACS and ACS run 20 times independently on each instance, and the comparison is performed with respect to the best result, average result, and standard deviation (denoted as 'Std.dev.'). As the greedy algorithm is a deterministic algorithm, it runs only one time, with the result comparing. The results given in Table 2 show that AACS performs better than ACS and greedy. For clarity, the best results obtained are highlighted in **boldface**.

Table 2. Experimental results without bicycle constraints

| Instance | AACS | | | ACS | | | Greedy | Optimum |
|---|---|---|---|---|---|---|---|---|
| | Best | Average | Std.dev. | Best | Average | Std.dev. | | |
| fri26 | **937** | **937.00** | 0.00 | **937** | **937.00** | 0.00 | 965 | 937 |
| eil51 | **426** | **427.15** | 0.49 | **426** | 428.10 | 2.90 | 482 | 426 |
| eil76 | **538** | **543.70** | 4.38 | **538** | 546.45 | 5.16 | 608 | 538 |
| eil101 | **629** | **639.65** | 6.48 | 630 | 640.10 | 5.39 | 746 | 629 |

## 4.2 Experimental Results in Distance Model

In this experiment, we conduct two PBSP instances based on eil51 and eil76. The locations of the stations are the same as the corresponding ones in TSPLIB. The difference is that the PBSP instance contains more information that each station has a number of bicycle demand. Tables 3 and 4 give the demand for bicycles at each station in eil51 and eil76 respectively, where $i$ denotes the index of the station and $d_i$ is the number of bicycles needed in station $i$. The parameters of ACS and AACS are also set as Table 1, and the results are obtained by 20 independent runs on each instance. The results given in Table 5 show that AACS has general better performance than both ACS and greedy algorithm.

Table 3. Demand of each station in eil51

| $i$ | 1 | 2 | 3 | 4 | 5 | 6 | 7 | 8 | 9 | 10 | 11 | 12 | 13 |
|---|---|---|---|---|---|---|---|---|---|---|---|---|---|
| $d_i$ | −2 | 2 | −10 | −9 | −2 | −15 | −13 | −15 | −5 | 17 | 5 | 20 | 4 |
| $i$ | 14 | 15 | 16 | 17 | 18 | 19 | 20 | 21 | 22 | 23 | 24 | 25 | 26 |
| $d_i$ | 20 | 10 | 3 | −8 | 11 | 7 | −13 | 8 | −10 | −11 | −1 | 1 | −9 |
| $i$ | 27 | 28 | 29 | 30 | 31 | 32 | 33 | 34 | 35 | 36 | 37 | 38 | 39 |
| $d_i$ | 15 | 5 | −4 | −3 | 6 | 15 | −9 | −7 | −12 | 14 | 18 | 6 | −7 |
| $i$ | 40 | 41 | 42 | 43 | 44 | 45 | 46 | 47 | 48 | 49 | 50 | 51 | |
| $d_i$ | 5 | −2 | −18 | −20 | 9 | 13 | 16 | −12 | −8 | 19 | −12 | −6 | |

## 4.3 Experimental Results in Time Window Model

To simulate real-road conditions more accurately, we adopt floating-point to record the distance between various stations and set $T_0 = 2$ min and $v_0 = 20$ km/h to correspondingly expand the distance to 20 times the original. The best visiting period and the acceptable visiting period of each station are set 5 min and 9 min respectively (i.e., $[B_i, C_i] = 5$ min, $[A_i, D_i] = 9$ min). For the choice of next station in time window model has a significant relationship with current time, the accumulated knowledge is not particularly valuable for ants. Therefore, the range of $\beta$ is adjusted to [3, 7] to increase the relative importance of desirability heuristic information. Tables 6 and 7 give the start time of the best visiting period (i.e., $B_i$) in eil51 and eil76, which use the bicycle constraints data in Tables 3 and 4 respectively. We set $I = 1000$ and $G = 100$ because of the complexity of the problem. Correspondingly, $I = 100000$ in ACS to be fair to compare. The results are based on 20 independent runs and are given in Table 8.

**Table 4.** Demand of each station in eil76

| $i$ | 1 | 2 | 3 | 4 | 5 | 6 | 7 | 8 | 9 | 10 | 11 | 12 | 13 |
|---|---|---|---|---|---|---|---|---|---|---|---|---|---|
| $d_i$ | 4 | –5 | –10 | –9 | 20 | –15 | –13 | 8 | –7 | 17 | –2 | 20 | 4 |
| $i$ | 14 | 15 | 16 | 17 | 18 | 19 | 20 | 21 | 22 | 23 | 24 | 25 | 26 |
| $d_i$ | 3 | 10 | –12 | –8 | 22 | 7 | –11 | 9 | –10 | –20 | 1 | 25 | –1 |
| $i$ | 27 | 28 | 29 | 30 | 31 | 32 | 33 | 34 | 35 | 36 | 37 | 38 | 39 |
| $d_i$ | 3 | 15 | 5 | –22 | 15 | 5 | –14 | –7 | 12 | 14 | 6 | 18 | –16 |
| $i$ | 40 | 41 | 42 | 43 | 44 | 45 | 46 | 47 | 48 | 49 | 50 | 51 | 52 |
| $d_i$ | 5 | –2 | –18 | 16 | 9 | –25 | 13 | –20 | –5 | –8 | 19 | –17 | –4 |
| $i$ | 53 | 54 | 55 | 56 | 57 | 58 | 59 | 60 | 61 | 62 | 63 | 64 | 65 |
| $d_i$ | –20 | 6 | 19 | –21 | –10 | 11 | 8 | 5 | –24 | –5 | 15 | 13 | 6 |
| $i$ | 66 | 67 | 68 | 69 | 70 | 71 | 72 | 73 | 74 | 75 | 76 | | |
| $d_i$ | –8 | 2 | –25 | –3 | 16 | 6 | 9 | 25 | –15 | –3 | 10 | | |

**Table 5.** Experimental results in distance model

| Instance | AACS | | | ACS | | | Greedy |
|---|---|---|---|---|---|---|---|
| | Best | Average | Std.dev. | Best | Average | Std.dev. | |
| eil51 | **426** | **431.45** | 2.52 | **426** | 431.6 | 3.25 | 499 |
| eil76 | 558 | **566.20** | 4.00 | **555** | 567.20 | 7.51 | 648 |

**Table 6.** Start time of the best visiting period in eil51

| $i$ | Time | $i$ | Time | $i$ | Time | $i$ | Time | $i$ | Time | $i$ | Time |
|---|---|---|---|---|---|---|---|---|---|---|---|
| 1 | 07:04 | 10 | 07:29 | 19 | 07:35 | 28 | 08:25 | 37 | 08:35 | 46 | 08:08 |
| 2 | 07:08 | 11 | 07:17 | 20 | 07:36 | 29 | 08:22 | 38 | 08:42 | 47 | 08:48 |
| 3 | 07:40 | 12 | 07:27 | 21 | 07:39 | 30 | 08:15 | 39 | 08:50 | 48 | 08:53 |
| 4 | 08:20 | 13 | 07:30 | 22 | 07:43 | 31 | 08:28 | 40 | 08:45 | 49 | 08:02 |
| 5 | 07:13 | 14 | 07:55 | 23 | 07:45 | 32 | 08:06 | 41 | 07:58 | 50 | 09:00 |
| 6 | 07:05 | 15 | 08:30 | 24 | 07:48 | 33 | 08:05 | 42 | 08:00 | 51 | 08:33 |
| 7 | 07:20 | 16 | 07:00 | 25 | 07:52 | 34 | 08:10 | 43 | 08:18 | | |
| 8 | 07:10 | 17 | 07:30 | 26 | 08:30 | 35 | 08:27 | 44 | 07:50 | | |
| 9 | 07:23 | 18 | 07:32 | 27 | 08:37 | 36 | 08:23 | 45 | 07:56 | | |

It can be seen from the above results that AACS outperforms conventional ACS generally. Table 2 shows that AACS has a stronger ability to find better solutions than according to 'Best' column and 'Average' column. Greedy algorithm performs poorly compared to AACS and ACS especially in large-scale problem. AACS generates smaller average tour lengths in Table 5. In Table 8 we can see that AACS performs better on satisfying the demand of users in time window model, in terms of both solution quality and stability. This is because the conventional parameter settings suggested for TSP are not suitable enough to solve PBSP due to its complexity,

**Table 7.** Start time of the best visiting period in eil76

| $i$ | Time | $i$ | Time | $i$ | Time | $i$ | Time | $i$ | Time | $i$ | Time |
|---|---|---|---|---|---|---|---|---|---|---|---|
| 1 | 07:04 | 14 | 07:30 | 27 | 08:30 | 40 | 08:42 | 53 | 08:27 | 66 | 08:47 |
| 2 | 07:08 | 15 | 08:30 | 28 | 08:37 | 41 | 08:50 | 54 | 07:28 | 67 | 08:08 |
| 3 | 08:20 | 16 | 07:00 | 29 | 08:06 | 42 | 07:58 | 55 | 08:16 | 68 | 08:58 |
| 4 | 07:40 | 17 | 07:36 | 30 | 08:22 | 43 | 08:00 | 56 | 07:35 | 69 | 08:50 |
| 5 | 07:05 | 18 | 07:32 | 31 | 08:25 | 44 | 07:50 | 57 | 07:45 | 70 | 08:52 |
| 6 | 07:56 | 19 | 07:35 | 32 | 08:15 | 45 | 08:08 | 58 | 08:30 | 71 | 08:55 |
| 7 | 07:20 | 20 | 07:55 | 33 | 08:28 | 46 | 08:48 | 59 | 08:34 | 72 | 08:21 |
| 8 | 07:10 | 21 | 07:45 | 34 | 08:05 | 47 | 08:18 | 60 | 08:26 | 73 | 08:30 |
| 9 | 07:13 | 22 | 07:39 | 35 | 08:10 | 48 | 08:02 | 61 | 08:25 | 74 | 08:45 |
| 10 | 07:23 | 23 | 07:30 | 36 | 08:23 | 49 | 08:33 | 62 | 08:36 | 75 | 07:05 |
| 11 | 07:29 | 24 | 07:43 | 37 | 08:45 | 50 | 09:00 | 63 | 08:40 | 76 | 08:09 |
| 12 | 07:17 | 25 | 07:48 | 38 | 08:27 | 51 | 08:53 | 64 | 08:43 | | |
| 13 | 07:27 | 26 | 07:52 | 39 | 08:35 | 52 | 08:00 | 65 | 08:10 | | |

**Table 8.** Experimental results in time window model

| Instance | AACS | | | ACS | | | Greedy |
|---|---|---|---|---|---|---|---|
| | Best | Average | Std.dev. | Best | Average | Std.dev. | |
| eil51 | **35.19** | **34.46** | 0.19 | 34.56 | 34.39 | 0.24 | 30.11 |
| eil76 | **39** | **38.68** | 0.36 | **39** | 38.35 | 0.64 | 32.14 |

especially in the time window model. In contrary, the parameters in AACS can be dynamically adjusted with the problem in the optimization process, which means that the best parameter values are problem-independent. Therefore, the proposed AACS has better adaptability to different COPs than ACS.

## 5  Conclusion

This paper presents an adaptive ACS to solve PBSP, which is based on evolutionary mechanisms to optimize the key parameters of ACS instead of the direct assignment based on past experience. The adaptive convergence of parameters enables AACS better to find out the global optimal solution. The experimental results show that the proposed method can effectively overcome the drawbacks of ACS in solution quality and stability. Furthermore, for the parameters can converge adaptively with the problem to be solved, this method has good generality which can also be used to solve other COPs like PBSP.

In the future, we will combine parallel computing to reduce runtime, especially when dealing with problems with large scale. Real-time updating of bicycle demand to realize dynamic scheduling is also an aspect to research.

**Acknowledgments.** This work was partially supported by the Outstanding Youth Science Foundation with No. 61822602, the National Natural Science Foundations of China (NSFC) with No. 61772207 and 61332002, the Natural Science Foundations of Guangdong Province for Distinguished Young Scholars with No. 2014A030306038, the Project for Pearl River New Star in Science and Technology with No. 201506010047, the GDUPS (2016), and the Fundamental Research Funds for the Central Universities.

# References

1. Labadi, K., Benarbia, T., Barbot, J.P., Hamaci, S., Omari, A.: Stochastic petri net modeling, simulation and analysis of public bicycle sharing systems. IEEE Trans. Autom. Sci. Eng. **12**(4), 1380–1395 (2015)
2. Wang, X., Choi, T.M., Liu, H., Yue, X.: Novel ant colony optimization methods for simplifying solution construction in vehicle routing problems. IEEE Trans. Intell. Transp. Syst. **17**(11), 3132–3141 (2016)
3. Rabbani, M., Tahaei, Z., Farrokhi-Asl, H., Saravi, N.A.: Using meta-heuristic algorithms and hybrid of them to solve multi compartment vehicle routing problem. In: 2017 IEEE International Conference on Industrial Engineering and Engineering Management (IEEM), pp. 1022–1026. IEEE, Singapore (2017)
4. Dorigo, M., Gambardella, L.M.: Ant colony system: a cooperative learning approach to the traveling salesman problem. IEEE Trans. Evol. Comput. **1**(1), 53–66 (1997)
5. Zhang, Z., Zou, K.: Simple ant colony algorithm for combinatorial optimization problems. In: 36th Chinese Control Conference (CCC), pp. 9835–9840. IEEE, Dalian (2017)
6. Zhan, Z.H., Zhang, J., Li, Y., et al.: An efficient ant colony system based on receding horizon control for the aircraft arrival sequencing and scheduling problem. IEEE Trans. Intell. Transp. Syst. **11**(2), 399–412 (2010)
7. Liu, X.F., Zhan, Z.H., Deng, J.D., Li, Y., Gu, T.L., Zhang, J.: An energy efficient ant colony system for virtual machine placement in cloud computing. IEEE Trans. Evol. Comput. **22**(1), 113–128 (2018)
8. Chen, Z.G., Zhan, Z.H., Gong, Y.J., et al.: Multiobjective cloud workflow scheduling: a multiple populations ant colony system approach. IEEE Trans. Cybern. (2018). https://doi.org/10.1109/tcyb.2018.2832640
9. Mouhcine, E., Khalifa, M., Mohamed, Y.: Route optimization for school bus scheduling problem based on a distributed ant colony system algorithm. In: Intelligent Systems and Computer Vision (ISCV), pp. 1–8. IEEE, Fez (2017)
10. Liu, Z.P., Li, K.P., Zhu, X.H.: Optimal dispatch between stations for public bicycle based on ant colony algorithm. J. Transp. Inf. Saf. **30**(4), 71–74 (2012)
11. Zhang, J.G., Wu, T., Jiang, Y.S.: Study on scheduling algorithm for public bicycle system based on ant colony algorithm. J. Xihua Univ. Nat. Sci. **33**(3), 70–76 (2014)
12. Yu, W.J., Hu, X.M., Zhang, J., Huang, R.Z.: Self-adaptive ant colony system for the traveling salesman problem. In: 2009 IEEE International Conference on Systems, Man and Cybernetics, pp. 1399–1404. IEEE, San Antonio (2009)
13. Zecchin, A.C., Simpson, A.R., Maier, H.R., Nixon, J.B.: Parametric study for an ant algorithm applied to water distribution system optimization. IEEE Trans. Evol. Comput. **9**(2), 175–191 (2005)

14. Jangra, R., Kait, R.: Analysis and comparison among ant system; ant colony system and max-min ant system with different parameters setting. In: 2017 3rd International Conference on Computational Intelligence & Communication Technology (CICT), pp. 1–4. IEEE, Ghaziabad (2017)
15. Gong, Y.J., Xu, R.T., Zhang, J., Liu, O.: A clustering-based adaptive parameter control method for continuous ant colony optimization. In: 2009 IEEE International Conference on Systems, Man and Cybernetics, pp. 1827–1832. IEEE, San Antonio (2009)
16. Dorigo, M., Maniezzo, V., Colorni, A.: Ant system optimization by a colony of cooperating agents. IEEE Trans. Syst. Man Cybern. Part B (Cybern.) 26(1), 29–41 (1996)

# An Artificial Neural Network for Distributed Constrained Optimization

Na Liu[1], Wenwen Jia[1], Sitian Qin[1(✉)], and Guocheng Li[2]

[1] Department of Mathematics, Harbin Institute of Technology, Weihai, China
qinsitian@163.com
[2] Department of Mathematics, Beijing Information Science and Technology
University, Beijing, China

**Abstract.** This paper studies the distributed convex optimization problems, where the objective function can be expressed as the sum of nonsmooth local convex objective functions. By the virtue of KKT conditions, an artificial neural network is presented to solve the distributed convex optimization problems with inequality and equality constraints. And it is shown that the state solution of the artificial neural network converges to the optimal solution to the original optimization problem. Compared with the existing continuous time algorithms, the provided algorithm has the advantages of lower model complexity and easy implementation. Finally, a numerical example displays the practicality of the algorithm.

**Keywords:** Distributed optimization · Artificial neural network
Consensus · Lyapunov function · Global convergence

## 1 Introduction

Recently, the multi-agent network has attracted the attention of many researchers due to its wide application in artificial intelligence [16], machine learning [2] and social engineering system [20]. In the practical applications of the multi-agent systems, the complexity of solving the original optimization problems can be reduced by constructing the corresponding agent for each subsystem.

Distributed optimization is an efficient framework for solving the distributed optimization problems. Many efforts on distributed optimization have been made in recent years. The most existing methods are discrete-time algorithms [9,19]. For example, the discrete multi-agent consensus algorithm was proposed by Nedic and Ozdaglar [9]. In [8], a discrete subgradient algorithm over a time-varying network was proposed for unconstrained optimization problems. As a class of parallel computational model for solving optimization problems, artificial neural networks have the advantages of fast calculation and real-time application. Hence, more and more researchers have shifted their focus to distributed optimization problems with continuous time algorithms ([11–14]). For instance, for the distributed convex optimization, Lu et al. [7] presented a continuous time

© Springer Nature Switzerland AG 2018
L. Cheng et al. (Eds.): ICONIP 2018, LNCS 11302, pp. 430–441, 2018.
https://doi.org/10.1007/978-3-030-04179-3_38

zero-gradient-sum algorithm. The literature [17] and its extension [15] investigated a continuous-time version of the work in literature [9]. Based on the work in [18], two kinds of distributed optimization problems over and balanced graphs were studied in [3,10]. Based on the weight balanced and strongly connected topology, a continuous-time algorithm to solve an unconstrained non-differentiable convex distributed optimization was presented in [1]. An adaptive convex distributed optimization problem over directed graph by continuous-time algorithm was considered in [4].

Inspired by the aforementioned researches, a continuous time optimization algorithm for distributed optimization problems with inequality and equality constraints is proposed in this paper. Compared with the existing continuous time algorithm, the algorithm provided in this paper has lower model complexity and displays a great potential for real-time applications. Moreover, the stability and convergence of the dynamic system are analyzed by Lyapunov method and it is shown that the state solution of the dynamic system converges to the optimal solution of the corresponding optimization problem.

## 2   Problem Statement and Optimization Algorithm

Consider a network of $m$ agents on a graph $\mathcal{G}$, whose objective is to solve the following constrained nonsmooth optimization problem in a distributed way:

$$
\begin{aligned}
\min \quad & f(x) = \sum_{i=1}^{m} f_i(x) \\
\text{s.t.} \quad & A_i x = b_i \\
& g_i(x) \leq 0, \quad i \in \{1, 2, \ldots, m\}
\end{aligned}
\tag{1}
$$

where $x \in \mathbb{R}^n$, $f_i : \mathbb{R}^n \to \mathbb{R}$ is the $i$th local objective function, $A_i \in \mathbb{R}^{p_i \times n}$ is row full rank, $b_i \in \mathbb{R}^{p_i}$ ($0 \leq p_i < n$), $g_i : \mathbb{R}^n \to \mathbb{R}^{r_i}$ is the inequality constraints of agent $i$. Moreover, the following two assumptions are needed in this paper.

**Assumption 1.** 1. $f_i$ and $g_i$ in (1) are convex in $\mathbb{R}^n$.
2. There is $\hat{x} \in \mathbb{R}^n$ satisfying $A_i \hat{x} = b_i$ and $g_i(\hat{x}) < 0, i = 1, 2, \ldots, m$.

**Assumption 2.** Graph $\mathcal{G}$ is undirected and connected.

**Lemma 1** [5]. *Based on Assumption 2 described above, problem (1) can be reformulated as the following optimization problem:*

$$
\begin{aligned}
\min \quad & \boldsymbol{f}(\boldsymbol{x}) = \sum_{i=1}^{m} f_i(x_i) \\
\text{s.t.} \quad & \boldsymbol{g}(\boldsymbol{x}) \leq 0 \\
& \boldsymbol{A}\boldsymbol{x} = \boldsymbol{b}, \quad \boldsymbol{L}\boldsymbol{x} = 0
\end{aligned}
\tag{2}
$$

*where* $\boldsymbol{x} = (x_1^\mathsf{T}, x_2^\mathsf{T}, \cdots, x_m^\mathsf{T})^\mathsf{T}$, $\boldsymbol{A} = \mathrm{blkdiag}\{A_1, A_2, \cdots, A_m\}$, $\boldsymbol{b} = (b_1^\mathsf{T}, b_2^\mathsf{T}, \cdots, b_m^\mathsf{T})^\mathsf{T}$, *and* $\boldsymbol{g}(\boldsymbol{x}) = (g_1(x_1)^\mathsf{T}, g_2(x_2)^\mathsf{T}, \cdots, g_m(x_m)^\mathsf{T})^\mathsf{T}$, $\boldsymbol{L} = L_m \otimes I_n \in \mathbb{R}^{mn \times mn}$, $\otimes$ *denotes the Kronecker product, and* $L_m$ *is the Laplacian matrix of graph* $\mathcal{G}$.

Let $p = p_1 + p_2 + \cdots + p_m$, $r = r_1 + r_2 + \cdots + r_m$. Then, in (2), one has $\boldsymbol{A} \in \mathbb{R}^{p \times mn}$, $\boldsymbol{b} \in \mathbb{R}^p$, and $\boldsymbol{g} : \mathbb{R}^{mn} \to \mathbb{R}^r$. Projection operator has a wide range of applications in many optimization problems, especially for the optimization problems with set constraints. Throughout this paper, let $\boldsymbol{P} = \boldsymbol{A}^\mathsf{T}(\boldsymbol{A}\boldsymbol{A}^\mathsf{T})^{-1}\boldsymbol{A}$ and $\boldsymbol{q} = \boldsymbol{A}^\mathsf{T}(\boldsymbol{A}\boldsymbol{A}^\mathsf{T})^{-1}\boldsymbol{b}$. Then, the matrix $I - \boldsymbol{P}$ is the projection matrix of $\boldsymbol{x} \in \mathbb{R}^{mn}$ onto $\{\boldsymbol{x} : \boldsymbol{A}\boldsymbol{x} = 0\}$. Since $\boldsymbol{A} = \text{blkdiag}\{A_1, A_2, \cdots, A_m\}$, we can write $\boldsymbol{P}$ and $\boldsymbol{q}$ as $\boldsymbol{P} = \text{blkdiag}\{P_1, P_2, \cdots, P_m\}$, and $\boldsymbol{q} = \text{blkdiag}\{q_1^\mathsf{T}, q_2^\mathsf{T}, \cdots, q_m^\mathsf{T}\}$ with $P_i = A_i^\mathsf{T}(A_i A_i^\mathsf{T})^{-1}A_i$ and $q_i = A_i^\mathsf{T}(A_i A_i^\mathsf{T})^{-1}b_i$.

**Lemma 2** [5]. *Suppose that the matrix $\boldsymbol{A} \in \mathbb{R}^{p \times mn}$ is row full rank, then for all $\boldsymbol{x} \in \mathbb{R}^{mn}$, $\boldsymbol{P}\boldsymbol{x} = \boldsymbol{q}$ if only if $\boldsymbol{A}\boldsymbol{x} = \boldsymbol{b}$.*

By the Karush-Kuhn-Tucker (KKT) conditions, the following theorem can be given.

**Theorem 1** [5]. *Based on Assumption 1 described above, $\boldsymbol{x}^* \in \mathbb{R}^{mn}$ is an optimal solution to problem (2) if and only if there are $\boldsymbol{y}^* \in \mathbb{R}^r$, $\boldsymbol{z}^* \in \mathbb{R}^{mn}$ satisfying*

$$\begin{cases} 0 \in -\boldsymbol{P}\boldsymbol{x}^* + \boldsymbol{q} - (I - \boldsymbol{P})(\partial \boldsymbol{f}(\boldsymbol{x}^*) + (\partial \boldsymbol{g}(\boldsymbol{x}^*))^\mathsf{T}\boldsymbol{y}^* + \boldsymbol{L}\boldsymbol{z}^*) \\ \boldsymbol{y}^* = (\boldsymbol{y}^* + \boldsymbol{g}(\boldsymbol{y}^*))^+ \\ \boldsymbol{L}\boldsymbol{x}^* = 0, \end{cases} \tag{3}$$

*where $\partial \boldsymbol{f}(\boldsymbol{x}^*)$ is the subdifferential of $\boldsymbol{f}$ at $\boldsymbol{x}^*$ and is given by*

$$\partial \boldsymbol{f}(\boldsymbol{x}^*) = \{\xi \in \mathbb{R}^n : \boldsymbol{f}(\boldsymbol{x}) - \boldsymbol{f}(\boldsymbol{x}^*) \le \langle \xi, \boldsymbol{x} - \boldsymbol{x}^* \rangle, \forall \boldsymbol{x} \in \mathbb{R}^n\}.$$

From the above analysis, we present the following neural network to solve problem (1):

$$\begin{cases} \dfrac{d\boldsymbol{x}(t)}{dt} \in -2\big[(I - \boldsymbol{P})(\partial \boldsymbol{f}(\boldsymbol{x}(t)) + (\partial \boldsymbol{g}(\boldsymbol{x}(t)))^\mathsf{T}(\boldsymbol{y}(t) + \boldsymbol{g}(\boldsymbol{x}(t)))^+ + \boldsymbol{L}\big(\boldsymbol{x}(t) \\ \qquad\qquad + \displaystyle\int_0^t \boldsymbol{x}(s)ds)\big)\big] - \partial \|\boldsymbol{A}\boldsymbol{x}(t) - \boldsymbol{b}\|_1 \\ \dfrac{d\boldsymbol{y}(t)}{dt} = -\boldsymbol{y}(t) + (\boldsymbol{y}(t) + \boldsymbol{g}(\boldsymbol{x}(t)))^+. \end{cases} \tag{4}$$

Moreover, the dynamic equation of the $i$th RNN can be given as:

$$\begin{cases} \dfrac{dx_i(t)}{dt} \in -2\big[(I - P_i)(\partial f_i(x_i(t)) + \partial g_i(x_i(t))^\mathsf{T}(y_i(t) + g_i(x_i(t)))^+ \\ \qquad\qquad + \displaystyle\sum_{j=1, j \ne i}^{m} a_{ij}\big(x_i(t) - x_j(t) + \int_0^t (x_i(s) - x_j(s))ds\big)\big] \\ \qquad\qquad - \partial \|A_i x_i(t) - b_i\|_1 \\ \dfrac{dy_i(t)}{dt} = (y_i(t) + g_i(x_i(t)))^+ - y_i(t), \end{cases} \tag{5}$$

where $a_{ij}$ is the connection weight between the $i$th RNN and $j$th RNN, $y_i$ is the hidden state vector of the $i$th RNN.

*Remark 1.* Recently, distributed optimization problems have attracted increasing attention (see [6,7,21]). While in references [7,21], the conclusions are based on the hypothesis that the objective function is strictly convex and twice continuously differentiable. Though the objective functions in [6] just need to be convex, the presented neural networks in [6] only can solve the distributed optimization problems with no constraints or bound constraints. Therefore, the proposed neural network (4) has wider applications than the other existing neural networks.

## 3   Consensus Analysis

Similar to the proof of Theorem 3.1 in [12], we give the following Theorem.

**Theorem 2.** *For any initial point $\boldsymbol{x}(0) = \boldsymbol{x}_0$, the state solution $\boldsymbol{x}(t)$ of network (4) reaches the region $\{\boldsymbol{x} : \boldsymbol{A}\boldsymbol{x} = \boldsymbol{b}\}$ in finite time and remain there thereafter.*

*Remark 2.* From Theorem 2, there is $T_{\mathcal{E}} \geq 0$ satisfying $\boldsymbol{x}(t) \in \mathcal{E}$, $\forall t \geq T_{\mathcal{E}}$. Hence, in the following analysis, we can always suppose that

$$\boldsymbol{x}(t) \in \mathcal{E} = \{\boldsymbol{x} : \boldsymbol{A}\boldsymbol{x} = \boldsymbol{b}\}, \quad \forall t \geq 0.$$

**Theorem 3.** *Based on Assumptions 1 and 2, the state $x_i(t)(i = 1, 2, \cdots, m)$ of neural network (5) with any initial point will reach consensus at an optimal solution to optimization problem (1).*

*Proof.* Let $K(s) = (\tilde{k}(s_1), \tilde{k}(s_2), ..., \tilde{k}(s_m))^{\mathsf{T}}$ and its component is defined as

$$\tilde{k}(s_i) = \begin{cases} 1, & \text{if } s_i > 0 \\ [-1, 1], & \text{if } s_i = 0 \\ -1, & \text{if } s_i < 0. \end{cases} \tag{6}$$

Then, there exists $\vartheta(t) \in K\big(\boldsymbol{A}\boldsymbol{x}(t) - \boldsymbol{b}\big)$ satisfying $\gamma(t) = \boldsymbol{A}^{\mathsf{T}}\vartheta(t)$, $\forall t \geq 0$. By Remark 2, $\boldsymbol{A}(I - \boldsymbol{P}) = 0$, it follows that $0 = \dfrac{\mathrm{d}}{\mathrm{d}t}\boldsymbol{A}\boldsymbol{x}(t) = -\boldsymbol{A}\boldsymbol{A}^{\mathsf{T}}\vartheta(t)$, for a.e. $t \geq 0$. Since $\boldsymbol{A}$ is row full rank, $\boldsymbol{A}\boldsymbol{A}^{\mathsf{T}}$ is invertible, we have $\vartheta(t) \equiv 0$. Thus, (4) can be reduced to

$$\begin{cases} \dfrac{\mathrm{d}\boldsymbol{x}(t)}{\mathrm{d}t} \in -2\big[(I - \boldsymbol{P})\big(\partial \boldsymbol{f}(\boldsymbol{x}(t)) + (\partial \boldsymbol{g}(\boldsymbol{x}(t)))^{\mathsf{T}}(\boldsymbol{y}(t) + \boldsymbol{g}(\boldsymbol{x}(t)))^+ \\ \qquad\qquad + \boldsymbol{L}\big(\boldsymbol{x}(t) + \displaystyle\int_0^t \boldsymbol{x}(s)ds\big)\big)\big] \\ \dfrac{\mathrm{d}\boldsymbol{y}(t)}{\mathrm{d}t} = (\boldsymbol{y}(t) + \boldsymbol{g}(\boldsymbol{x}(t)))^+ - \boldsymbol{y}(t), \end{cases} \tag{7}$$

for a.e. $t \geq 0$. We let $\boldsymbol{x}^*$ be the optimal solution of problem (2). From Theorem 1, there are $\boldsymbol{y}^* \in \mathbb{R}^r$, $\boldsymbol{z}^* \in \mathbb{R}^{mn}$ such that (3) holds. Let

$$\phi(t) = \boldsymbol{f}(\boldsymbol{x}(t)) + \frac{1}{2}\|(\boldsymbol{y}(t) + \boldsymbol{g}(\boldsymbol{x}(t)))^+\|^2$$
$$+ \frac{1}{2}\Big(\int_0^t \boldsymbol{x}(s)ds + \boldsymbol{x}(t)\Big)^{\mathsf{T}}\boldsymbol{L}\Big(\int_0^t \boldsymbol{x}(s)ds + \boldsymbol{x}(t)\Big). \tag{8}$$

By chain rule, for any $\xi(t) \in \partial \boldsymbol{f}(\boldsymbol{x}(t))$, $G(t) \in \partial g(\boldsymbol{x}(t))$, we get

$$
\begin{aligned}
\frac{\mathrm{d}}{\mathrm{d}t}\phi(t) = {}&\xi(t)^{\mathsf{T}}\dot{x}(t) + (\dot{x}(t) + \boldsymbol{x}(t))^{\mathsf{T}}\boldsymbol{L}\Big(\int_0^t \boldsymbol{x}(s)\mathrm{d}s + \boldsymbol{x}(t)\Big) + ((\boldsymbol{y}(t) + \boldsymbol{g}(\boldsymbol{x}(t)))^+)^{\mathsf{T}} \\
&\times (G(t)\dot{x}(t) + \dot{y}(t)),
\end{aligned} \tag{9}
$$

for a.e. $t \geq 0$. Denote $\eta^* = (\boldsymbol{y}^* + \boldsymbol{g}(\boldsymbol{x}^*))^+$. We introduce a Lyapunov function as:

$$
\begin{aligned}
V(t) ={}& \phi(t) - \boldsymbol{f}(\boldsymbol{x}^*) + \frac{1}{2}(\boldsymbol{x}^* + \boldsymbol{z}^*)^{\mathsf{T}}\boldsymbol{L}(\boldsymbol{x}^* + \boldsymbol{z}^*) + \frac{1}{2}\|(\boldsymbol{y}^* + \boldsymbol{g}(\boldsymbol{x}^*))^+\|^2 - (\boldsymbol{x}(t) \\
&- \boldsymbol{x}^*)^{\mathsf{T}}(\xi^* + (G^*)^{\mathsf{T}}\eta^* + \boldsymbol{L}\boldsymbol{z}^*) - (\boldsymbol{y}(t) - \boldsymbol{y}^*)^{\mathsf{T}}\eta^* - \Big(\int_0^t \boldsymbol{x}(s)\mathrm{d}s - \boldsymbol{z}^*\Big)^{\mathsf{T}} \\
&\times \boldsymbol{L}\boldsymbol{z}^* + \frac{1}{2}(\|\boldsymbol{x}^* - \boldsymbol{x}(t)\|^2 + \|\boldsymbol{z}^* - \boldsymbol{z}(t)\|^2) + \frac{1}{2}\Big(\int_0^t \boldsymbol{x}(s)\mathrm{d}s - \boldsymbol{z}^*\Big)^{\mathsf{T}} \\
&\times \boldsymbol{L}\Big(\int_0^t \boldsymbol{x}(s)\mathrm{d}s - \boldsymbol{z}^*\Big).
\end{aligned} \tag{10}
$$

Let $\eta(t) = (\boldsymbol{g}(\boldsymbol{x}(t)) + \boldsymbol{y}(t))^+$ and

$$
\varsigma(t) = (I - P)\Big(-\xi(t) - G(t)^{\mathsf{T}}\eta(t) + \boldsymbol{x}(t) - \boldsymbol{L}\Big(\int_0^t \boldsymbol{x}(s)\mathrm{d}s + \boldsymbol{x}(t)\Big)\Big) + \boldsymbol{q} \tag{11}
$$

and denote $\Psi = (\partial \boldsymbol{f}(\boldsymbol{x}(t)), \partial g(\boldsymbol{x}(t)))$. Then differentiating $V$ along the solution of (4), combining with (7), Remark 2 and Lemma 2, one has

$$
\begin{aligned}
\frac{\mathrm{d}}{\mathrm{d}t}V(t) \leq{}& 2\sup_{(\xi(t),G(t))\in\Psi}\Big\{\Big[\xi(t) + \eta(t)^{\mathsf{T}}G(t) + \boldsymbol{L}\Big(\int_0^t \boldsymbol{x}(s)\mathrm{d}s + \boldsymbol{x}(t)\Big) - \xi^* - (G^*)^{\mathsf{T}}\eta^* \\
&- \boldsymbol{L}\boldsymbol{z}^* - \boldsymbol{x}^* + \boldsymbol{x}(t)\Big]^{\mathsf{T}}\big[\boldsymbol{q} - P\boldsymbol{x}(t) - (I - P)(\xi(t) + G(t)^{\mathsf{T}}\eta(t) \\
&+ \boldsymbol{L}\Big(\int_0^t \boldsymbol{x}(s)\mathrm{d}s + \boldsymbol{x}(t)\Big)\big)\Big]\Big\} + (\eta(t) - \eta^* + \boldsymbol{y}(t) - \boldsymbol{y}^*)^{\mathsf{T}}(\eta(t) - \boldsymbol{y}(t)) \\
&+ \big(\boldsymbol{L}\boldsymbol{x}(t) + 2\boldsymbol{L}\Big(-\boldsymbol{z}^* + \int_0^t \boldsymbol{x}(s)\mathrm{d}s\Big)\big)^{\mathsf{T}}\boldsymbol{x}(t) \\
={}& 2\sup_{(\xi(t),G(t))\in\Psi}\Big\{\Big[\xi(t) + G(t)^{\mathsf{T}}\eta(t) + \boldsymbol{L}\Big(\int_0^t \boldsymbol{x}(s)\mathrm{d}s + \boldsymbol{x}(t)\Big) - \xi^* - (G^*)^{\mathsf{T}}\eta^* \\
&- \boldsymbol{L}\boldsymbol{z}^* - \boldsymbol{x}^* + \boldsymbol{x}(t)\Big]^{\mathsf{T}}(-\boldsymbol{x}(t) + \varsigma(t))\Big\} + (\eta(t) - \eta^* + \boldsymbol{y}(t) - \boldsymbol{y}^*)^{\mathsf{T}} \\
&\times (\eta(t) - \boldsymbol{y}(t)) + \big(\boldsymbol{L}\boldsymbol{x}(t) + 2\boldsymbol{L}\Big(\int_0^t \boldsymbol{x}(s)\mathrm{d}s - \boldsymbol{z}^*\Big)\big)^{\mathsf{T}}\boldsymbol{x}(t).
\end{aligned} \tag{12}
$$

Let

$$
\begin{aligned}
W_1 &= 2\big[G(t)^\mathsf{T}\eta(t) + \xi(t) + L\Big(\int_0^t x(s)\mathrm{d}s + x(t)\Big) - \xi^* - (G^*)^\mathsf{T}\eta^* - Lz^* + x(t) \\
&\quad - x^*\big]^\mathsf{T}\big(-x^* + \varsigma(t)\big) \\
W_2 &= 2\big[\xi(t) + L\Big(\int_0^t x(s)\mathrm{d}s + x(t)\Big) - \xi^* - Lz^* + x(t) - x^*\big]^\mathsf{T}\big(x^* - x(t)\big) \\
&\quad + 2\big(L\big(\int_0^t x(s)\mathrm{d}s - z^*\big)\big)^\mathsf{T}x(t) + x(t)^\mathsf{T}Lx(t) \\
W_3 &= \big(\eta(t) - \eta^* + y(t) - y^*\big)^\mathsf{T}\big(\eta(t) - y(t)\big) + 2\big(G(t)^\mathsf{T}\eta(t) - (G^*)^\mathsf{T}\eta^*\big)^\mathsf{T} \\
&\quad \times \big(x^* - x(t)\big).
\end{aligned}
$$

It follows that $\frac{\mathrm{d}}{\mathrm{d}t}V(t) \le \sup\limits_{(\xi(t),G(t))\in\Psi}(W_1 + W_2 + W_3)$. For $W_1$, we derive that

$$
\begin{aligned}
W_1 &= -2\big[x(t) - \xi(t) - \eta(t)^\mathsf{T}G(t) - L\Big(\int_0^t x(s)\mathrm{d}s + x(t)\Big) - \varsigma(t)\big]^\mathsf{T}\big(\varsigma(t) - x^*\big) \\
&\quad - 2(\xi^* + (G^*)^\mathsf{T}\eta^* + Lz^*)^\mathsf{T}(\varsigma(t) - x^*) - 2\|\varsigma(t) - x(t)\|^2 + 2\|x^* - x(t)\|^2
\end{aligned}
$$

From (11), one has

$$
\begin{aligned}
&- \xi(t) - \eta(t)^\mathsf{T}G(t) - L\Big(\int_0^t x(s)\mathrm{d}s + x(t)\Big) - \varsigma(t) + x(t) \\
=&P\big(-\xi(t) - G(t)^\mathsf{T}\eta(t) - L\Big(x(t) + \int_0^t x(s)\mathrm{d}s\Big) + x(t)\big) - q.
\end{aligned}
\tag{13}
$$

On the other hand, considering $x^*$ is an optimal solution and by (3), there are $y^* \in \mathbb{R}^r$, $z^* \in \mathbb{R}^{mn}$, $G^* \in \partial g(x^*)$ and $\xi^* \in \partial f(x^*)$ satisfying

$$
x^* = (I - P)(-\xi^* - (G^*)^\mathsf{T}\eta^* - Lz^* + x^*).
\tag{14}
$$

Then, we have

$$
\begin{aligned}
\varsigma(t) - x^* &= (I - P)\big(-\xi(t) - G(t)^\mathsf{T}\eta(t) - L\big(x(t) + \int_0^t x(s)\mathrm{d}s\big) + x(t)\big) - x^* \\
&\quad + \xi^* + (G^*)^\mathsf{T}\eta^* + Lz^*\big).
\end{aligned}
\tag{15}
$$

Since $P(I - P)^2 = 0$ and $(I - P)q = 0$, one has

$$
\big[-\xi(t) - G(t)^\mathsf{T}\eta(t) - L\Big(x(t) + \int_0^t x(s)\mathrm{d}s\Big) - \varsigma(t) + x(t)\big]^\mathsf{T}(\varsigma(t) - x^*) = 0.
\tag{16}
$$

By $(I - P)^2 = I - P$, $\varsigma(t) - x^* = (I - P)(\varsigma(t) - x^*)$. Combining with (14) and the fact that $Px^* = q$, we get

$$
(\xi^* + (G^*)^\mathsf{T}\eta^* + Lz^*)^\mathsf{T}(-x^* + \varsigma(t)) = 0.
\tag{17}
$$

Thus, $W_1 = -2\|\boldsymbol{x}(t) - \varsigma(t)\|^2 + 2\|\boldsymbol{x}^* - \boldsymbol{x}(t)\|^2$.

For $W_2$, by $\boldsymbol{L}\boldsymbol{x}^* = 0$, one has

$$W_2 = 2(\xi(t) - \xi^*)^\mathsf{T}(\boldsymbol{x}^* - \boldsymbol{x}(t)) - 2\|\boldsymbol{x}(t) - \boldsymbol{x}^*\|^2 - \boldsymbol{x}(t)^\mathsf{T}\boldsymbol{L}\boldsymbol{x}(t). \tag{18}$$

Since $f$ is convex, we get

$$(\xi(t) - \xi^*)^\mathsf{T}(\boldsymbol{x}^* - \boldsymbol{x}(t)) \le 0, \quad \forall \xi(t) \in \partial f(\boldsymbol{x}(t)), \ \xi^* \in \partial f(\boldsymbol{x}^*). \tag{19}$$

Therefore, $W_2 \le -2\|\boldsymbol{x}(t) - \boldsymbol{x}^*\|^2 - \boldsymbol{x}(t)^\mathsf{T}\boldsymbol{L}\boldsymbol{x}(t)$.

For $W_3$, by (3), we get $\eta^* = \boldsymbol{y}^*$, then

$$\begin{aligned}
W_3 &= 2\big(G(t)^\mathsf{T}\eta(t) - (G^*)^\mathsf{T}\eta^*\big)^\mathsf{T}(\boldsymbol{x}^* - \boldsymbol{x}(t)) + (\eta(t) + \boldsymbol{y}(t) - 2\eta^*)^\mathsf{T}\big(-\boldsymbol{y}(t) + \eta(t)\big) \\
&= 2(\eta(t) - \eta^*)^\mathsf{T}(-\boldsymbol{y}(t) + \eta(t)) - (\boldsymbol{y}(t) - \eta(t))^\mathsf{T}(\boldsymbol{y}(t) - \eta(t)) + 2\big(G(t)^\mathsf{T}\eta(t) \\
&\quad - (G^*)^\mathsf{T}\eta^*\big)^\mathsf{T}(\boldsymbol{x}^* - \boldsymbol{x}(t)) \\
&= 2(\eta(t))^\mathsf{T}\big(-\boldsymbol{y}(t) + \eta(t) - G(t)(\boldsymbol{x}(t) - \boldsymbol{x}^*)\big) - 2(\eta^*)^\mathsf{T}(-\boldsymbol{y}(t) \\
&\quad + \eta(t) - G^*(\boldsymbol{x}(t) - \boldsymbol{x}^*)) - \|\boldsymbol{y}(t) - \eta(t)\|^2.
\end{aligned} \tag{20}$$

Noting that $\boldsymbol{g}(\boldsymbol{x}(t)) + \boldsymbol{y}(t) = \big(\boldsymbol{g}(\boldsymbol{x}(t)) + \boldsymbol{y}(t)\big)^+ - (-\boldsymbol{g}(\boldsymbol{x}(t)) - \boldsymbol{y}(t))^+$, it follows that $-\boldsymbol{y}(t) + \eta(t) = \boldsymbol{g}(\boldsymbol{x}(t)) + (-\boldsymbol{g}(\boldsymbol{x}(t)) - \boldsymbol{y}(t))^+$. Then, we can get that

$$\begin{aligned}
W_3 &= 2(\eta(t))^\mathsf{T}\big(\boldsymbol{g}(\boldsymbol{x}(t)) - \boldsymbol{g}(\boldsymbol{x}^*) - G(t)(\boldsymbol{x}(t) - \boldsymbol{x}^*)\big) - \|\boldsymbol{y}(t) - \eta(t)\|^2 \\
&\quad + 2(\eta(t))^\mathsf{T}(\boldsymbol{g}(\boldsymbol{x}^*) + (\boldsymbol{y}(t) - \boldsymbol{g}(\boldsymbol{x}(t)))^+ - 2(\eta^*)^\mathsf{T}(\boldsymbol{g}(\boldsymbol{x}(t)) - \boldsymbol{g}(\boldsymbol{x}^*) \\
&\quad - G^*(\boldsymbol{x}(t) - \boldsymbol{x}^*)) - 2(\eta^*)^\mathsf{T}(\boldsymbol{g}(\boldsymbol{x}^*) + (-\boldsymbol{y}(t) - \boldsymbol{g}(\boldsymbol{x}(t)))^+.
\end{aligned} \tag{21}$$

Since $g$ is convex, we get

$$\boldsymbol{g}(\boldsymbol{x}^*) - \boldsymbol{g}(\boldsymbol{x}(t)) \le (G^*)^\mathsf{T}(\boldsymbol{x}^* - \boldsymbol{x}(t)), \quad \forall G^* \in \partial g(\boldsymbol{x}^*) \tag{22}$$

and

$$\boldsymbol{g}(\boldsymbol{x}^*) - \boldsymbol{g}(\boldsymbol{x}(t) \ge (G(t))^\mathsf{T}(\boldsymbol{x}^* - \boldsymbol{x}(t)), \quad \forall G(t) \in \partial g(\boldsymbol{x}(t)). \tag{23}$$

Combining with $\eta(t) = (\boldsymbol{y}(t) + \boldsymbol{g}(\boldsymbol{x}(t)))^+$, one has

$$(\eta(t))^\mathsf{T}(\boldsymbol{g}(\boldsymbol{x}(t)) - \boldsymbol{g}(\boldsymbol{x}^*)) - (G^*)^\mathsf{T}(\boldsymbol{x}(t) - \boldsymbol{x}^*)) \le 0, \tag{24}$$

and

$$-(\eta^*)^\mathsf{T}(\boldsymbol{g}(\boldsymbol{x}(t)) - \boldsymbol{g}(\boldsymbol{x}^*) - (G^*)^\mathsf{T}(\boldsymbol{x}(t) - \boldsymbol{x}^*)) \le 0. \tag{25}$$

By the definition of $\eta(t)$ and $\boldsymbol{g}(\boldsymbol{x}^*) \le 0$, we have $(\eta(t))^\mathsf{T}\boldsymbol{g}(\boldsymbol{x}^*) \le 0$. Moreover, one has $(\eta(t))^\mathsf{T}(-\boldsymbol{g}(\boldsymbol{x}(t)) - \boldsymbol{y}(t)) = ((\boldsymbol{g}(\boldsymbol{x}(t)) + \boldsymbol{y}(t))^+)^\mathsf{T}(-\boldsymbol{g}(\boldsymbol{x}(t)) - \boldsymbol{y}(t)) = 0$. From (3), we can obtain that $\boldsymbol{g}(\boldsymbol{x}^*)^\mathsf{T}\eta^* = \boldsymbol{g}(\boldsymbol{x}^*)^\mathsf{T}\boldsymbol{y}^* = 0$ and $(\eta^*)^\mathsf{T}(-\boldsymbol{g}(\boldsymbol{x}(t)) - \boldsymbol{y}(t))$. Thus, one has $W_3 \le -\|\eta(t) - \boldsymbol{y}(t)\|^2$.

Consequently, one has

$$W_1 + W_2 + W_3 \le -\boldsymbol{x}(t)^\mathsf{T}\boldsymbol{L}\boldsymbol{x}(t) - \|\boldsymbol{y}(t) - \eta(t)\|^2 - 2\|\boldsymbol{x}(t) - \varsigma(t)\|^2. \tag{26}$$

Then, substituting (11) into the above inequality, we have

$$\frac{\mathrm{d}}{\mathrm{d}t}V(t) \leq - \inf_{(\xi(t),G(t))\in\Psi} \{2\|(I-P)(\xi(t)+G(t)^\mathsf{T}(g(x(t))+y(t))^+ + L\left(\int_0^t x(s)ds \right. \tag{27}$$
$$\left. +x(t)))\|^2\} - \|(y(t)+g(x(t)))^+ - y(t)\|^2 - x(t)^\mathsf{T} Lx(t) \leq 0.$$

And by the definition of $V$ and the convexity of $\phi$, we have

$$\frac{1}{2}\left(\|z(t)-z^*\|^2 + \|x(t)-x^*\|^2\right) + \frac{1}{2}\left(\int_0^t x(s)\mathrm{d}s - z^*\right)^\mathsf{T} L\left(\int_0^t x(s)\mathrm{d}s - z^*\right) \tag{28}$$
$$\leq V(t) \leq V(0).$$

Therefore, $\{x(t), y(t), L\int_0^t x(s)\mathrm{d}s\}$ is bounded. Then, there exists an increasing sequence $\{t_k\}$ satisfying that $\lim_{k\to\infty} x(t_k) = \bar{x}$, $\lim_{k\to\infty} y(t_k) = \bar{y}$ and $\lim_{k\to\infty} \int_0^{t_k} Lx(s)\mathrm{d}s = L\bar{z}$. Let

$$Q(\bar{x},\bar{y},\bar{z}) = \inf\{2\|(I-P)(\bar{\xi}+(\bar{G})^\mathsf{T}(\bar{y}+g(\bar{x}))^+ + L(\bar{x}+\bar{z})\|^2 + \bar{x}^\mathsf{T} L\bar{x} + \|\bar{y}-(\bar{y} \tag{29}$$
$$+g(\bar{x}))^+\|^2 : \bar{G}\in\partial g(x), \bar{\xi}\in\partial f(\bar{x})\}.$$

It is clear that $Q(\bar{x},\bar{y},\bar{z}) = 0$ equals that $(\bar{x},\bar{y},\bar{z})$ satisfies (3), which implies that $Q(\bar{x},\bar{y},\bar{z}) = 0$ equals that $\bar{x}$ is an optimal solution.

We claim that $Q(\bar{x},\bar{y},\bar{z}) = 0$. If not, then $Q(\bar{x},\bar{y},\bar{z}) > 0$. Since $Q$ is lower semicontinuous at $(\bar{x},\bar{y},\bar{z})$. Hence, there are $\varepsilon > 0$ and $\delta > 0$ satisfying

$$Q(x,y,z) > \varepsilon, \ \forall(x,y,z)\in B(\bar{x},\bar{y},\bar{z};\delta). \tag{30}$$

Since $\lim_{k\to\infty} x(t_k) = \bar{x}$, $\lim_{k\to\infty} y(t_k) = \bar{y}$ and $\lim_{k\to\infty} \int_0^{t_k} Lx(s)\mathrm{d}s = L\bar{z}$, there is a integer $N > 0$ satisfying

$$\|x(t_k)-\bar{x}\| + \|y(t_k)-\bar{y}\| + \left\|\int_0^{t_k} x(s)\mathrm{d}s - \bar{z}\right\| \leq \frac{\delta}{2}, \tag{31}$$

for all $k \geq N$. From (7) and $\{x(t), y(t), L\int_0^t x(s)\mathrm{d}s\}$ is bounded, there is $M > 0$ satisfying $\|\dot{x}(t)\| + \|\dot{y}(t)\| + \|x(t)\| \leq M$. Therefore, for any $t \in \left[t_k - \frac{\delta}{4M}, t_k + \frac{\delta}{4M}\right]$ and $k \geq N$, one has

$$\|\bar{z}-z(t)\| + \|\bar{x}-x(t)\| + \left\|\bar{z}-\int_0^t x(s)\mathrm{d}s\right\|$$
$$\leq \|z(t_k)-z(t)\| + \|\bar{z}-z(t_k)\| + \|x(t_k)-x(t)\| + \|\bar{x}-x(t_k)\|$$
$$+ \left\|\int_0^{t_k} x(s)\mathrm{d}s - \int_0^t x(s)\mathrm{d}s\right\| + \left\|\bar{z}-\int_0^{t_k} x(s)\mathrm{d}s\right\|$$
$$\leq M|t-t_k| + \frac{\delta}{2} \leq \delta. \tag{32}$$

By (30), we get $Q\big(\boldsymbol{x}(t), \boldsymbol{y}(t), \int_0^t \boldsymbol{x}(s)\mathrm{d}s\big) > \varepsilon$, $\forall t \in \left[t_k - \frac{\delta}{4M}, t_k + \frac{\delta}{4M}\right]$ and $k \geq N$. Since the Lebesgue measure of the set $\bigcup_{k \geq N}\left[t_k - \frac{\delta}{4M}, t_k + \frac{\delta}{4M}\right]$ is infinite, hence

$$\int_0^\infty Q\Big(\boldsymbol{x}(t), \boldsymbol{y}(t), \int_0^t \boldsymbol{x}(s)\mathrm{d}s\Big)\mathrm{d}t \geq \int_{\bigcup_{k \geq N}\left[t_k - \frac{\delta}{4M}, t_k + \frac{\delta}{4M}\right]} \varepsilon\,\mathrm{d}t = \sum_{k \geq N} \frac{\delta}{2M}\varepsilon = \infty. \tag{33}$$

Since $V(t)$ is non-increasing and $V(t) \geq 0$, then there is $\bar{V}$ satisfying that $\lim_{t \to \infty} V(t) = \bar{V}$. Hence, from (27), one has

$$\int_0^\infty Q\Big(\boldsymbol{x}(t), \boldsymbol{y}(t), \int_0^t \boldsymbol{x}(s)\mathrm{d}s\Big)\mathrm{d}t = \lim_{s \to \infty} \int_0^s Q\Big(\boldsymbol{x}(t), \boldsymbol{y}(t), \int_0^t \boldsymbol{x}(s)\mathrm{d}s\Big)\mathrm{d}t$$
$$\leq -\lim_{s \to \infty} \int_0^s \dot{V}(t)\mathrm{d}t = -\bar{V} + V(0) < +\infty, \tag{34}$$

which leads to a contradiction. Thus, one has $Q(\bar{\boldsymbol{x}}, \bar{\boldsymbol{y}}, \bar{\boldsymbol{z}}) = 0$ and $\bar{\boldsymbol{x}}$ is an optimal solution.

Next, we claim $\lim_{t \to +\infty} \boldsymbol{x}(t) = \bar{\boldsymbol{x}}$. Constructing another function as follows:

$$\bar{V}(t) = \phi(t) - f(\bar{\boldsymbol{x}}) + \frac{1}{2}(\bar{\boldsymbol{x}} + \bar{\boldsymbol{z}})^\mathsf{T} L(\bar{\boldsymbol{x}} + \bar{\boldsymbol{z}}) + \frac{1}{2}\|(g(\bar{\boldsymbol{x}}) + \bar{\boldsymbol{y}})^+\|^2 - (\boldsymbol{x}(t)$$
$$- \bar{\boldsymbol{x}})^\mathsf{T}(\bar{\xi} + (\bar{G})^\mathsf{T}\bar{\eta} + L\bar{\boldsymbol{z}}) - (\boldsymbol{y}(t) - \bar{\boldsymbol{y}})^\mathsf{T}\bar{\eta} - \Big(\int_0^t \boldsymbol{x}(s)\mathrm{d}s - \bar{\boldsymbol{z}}\Big)^\mathsf{T}$$
$$\times L\bar{\boldsymbol{z}} + \frac{1}{2}\Big(\int_0^t \boldsymbol{x}(s)\mathrm{d}s - \bar{\boldsymbol{z}}\Big)^\mathsf{T} \times L\Big(\int_0^t \boldsymbol{x}(s)\mathrm{d}s - \bar{\boldsymbol{z}}\Big)$$
$$+ \frac{1}{2}\big(\|\boldsymbol{z}(t) - \bar{\boldsymbol{z}}\|^2 + \|\boldsymbol{x}(t) - \bar{\boldsymbol{x}}\|^2\big). \tag{35}$$

By similar analysis, we have $\frac{\mathrm{d}}{\mathrm{d}t}\bar{V}(t) \leq 0$. From the continuity of $\bar{V}$, for any $\varepsilon > 0$, there exists $\bar{\delta} > 0$ such that $\bar{V}(t) < \varepsilon$, when $\big\|\big(\boldsymbol{x}(t), \boldsymbol{y}(t), \int_0^t \boldsymbol{x}(s)\mathrm{d}s\big) - (\bar{\boldsymbol{x}}, \bar{\boldsymbol{y}}, \bar{\boldsymbol{z}})\big\| \leq \bar{\delta}$. By $\big(\boldsymbol{x}(t_k), \boldsymbol{y}(t_k), \int_0^{t_k} \boldsymbol{x}(s)\mathrm{d}s\big) \to (\bar{\boldsymbol{x}}, \bar{\boldsymbol{y}}, \bar{\boldsymbol{z}})$, there exists $t_N$ such that

$$\|\bar{\boldsymbol{x}} - \boldsymbol{x}(t_N)\| + \|\bar{\boldsymbol{y}} - \boldsymbol{y}(t_N)\| + \Big\|\int_0^{t_N} \boldsymbol{x}(s)\mathrm{d}s - \bar{\boldsymbol{z}}\Big\| \leq \bar{\delta}. \tag{36}$$

Thus, by (35), for any $\varepsilon > 0$, we can see that

$$\frac{1}{2}\|\bar{\boldsymbol{x}} - \boldsymbol{x}(t)\|^2 \leq \bar{V}(t) \leq \bar{V}(t_N) \leq \varepsilon, \ \forall t > t_N. \tag{37}$$

Thus, $\lim_{t \to +\infty} \boldsymbol{x}(t) = \bar{\boldsymbol{x}}$.

Finally, by Lemma 1 and under Assumption 2, the state solution $x_i(t)$ of (5) with any initial point achieves consensus at an optimal solution to (1).

## 4  Numerical Example

*Example 1.* Consider a network of 5 agents interacting on a undirected and connected graph to cooperatively minimize the following problem:

$$
\begin{aligned}
\min \quad & \|Cx - d\|_1 \\
\text{s.t.} \quad & A_j x = b_j, \; j \in \{1, 2, 3\} \\
& -1 \leq x_k \leq 1, \; k \in \{1, 2, \cdots, 5\},
\end{aligned}
$$

where $C = \begin{pmatrix} 0.1 & 0.2 & 0.3 & 0.4 & 0.5 \\ 0.6 & 0.7 & 0.8 & 0.9 & 1.0 \\ 1.5 & 1.2 & 1.3 & 1.4 & 1.5 \\ 0.2 & 0.2 & 0.32 & 0.75 & 0.1 \\ 0.32 & 0.1 & 0.3 & 1.2 & 0.5 \end{pmatrix}$, $d = \begin{pmatrix} -0.1 \\ 0.1 \\ -0.1 \\ 0.1 \\ -0.1 \end{pmatrix}$, $A_j$ and $b_j$ are the $j$th row of the following matrix and vector, respectively:

$$
A = \begin{pmatrix} 0.4 & 0.2 & 0.3 & 0.4 & 0.5 \\ 0.6 & 0.5 & 0.2 & 0.9 & 0.6 \\ 0.4 & 1.2 & 0.5 & 1.4 & 0.9 \end{pmatrix}, \; b = \begin{pmatrix} 1 \\ 2 \\ 3 \end{pmatrix}.
$$

Noting that $\|Cx - d\|_1$ is the sum of $|C_i x - d_i|$ $(i \in \{1, 2, \cdots, 5\})$. Five RNNs, representing the five nodes, are applied to solve the above problem. And the connection weight between two RNNs is 1 if they are connected and 0 otherwise. One undirected graph with five multi-agent networks is shown in Fig. 1. And the Fig. 2 shows that the state solution of neural network (4) converges to the optimal solution $\mathbf{x}^* = 0$.

**Fig. 1.** One undirected graph with five multi-agent networks.

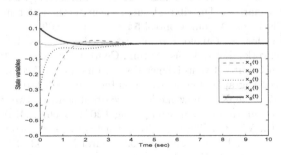

**Fig. 2.** Transient behaviors of the state $x_i(t)(i = 1, 2, \cdots, 5)$ based on network (4).

## 5    Conclusion

This paper proposes a continuous-time neurodynamic approach for constrained distributed optimization problems. It is shown that each agent converges to the optimal solution to the considered optimization problem. Moreover, the presented algorithm has the advantages of lower model complexity and real-time application. Numerical result displays the utility and efficacy of the presented neural network.

**Acknowledgments.** This research is supported by the National Natural Science Foundation of China (61773136, 11471088) and the NSF project of Shandong province in China with granted No. ZR2014FM023.

## References

1. Gharesifard, B., Cortés, J.: Distributed continuous-time convex optimization on weight-balanced digraphs. IEEE Trans. Autom. Control **59**(3), 781–786 (2014)
2. Kazakov, D., Kudenko, D.: Machine learning and inductive logic programming for multi-agent systems. In: Luck, M., Mařík, V., Štěpánková, O., Trappl, R. (eds.) ACAI 2001. LNCS (LNAI), vol. 2086, pp. 246–270. Springer, Heidelberg (2001). https://doi.org/10.1007/3-540-47745-4_11
3. Kia, S.S., Cort, J., Martnez, S.: Distributed convex optimization via continuous-time coordination algorithms with discrete-time communication. Automatica **55**, 254–264 (2015)
4. Li, Z., Ding, Z., Sun, J., Li, Z.: Distributed adaptive convex optimization on directed graphs via continuous-time algorithms. IEEE Trans. Autom. Control **63**(5), 1434–1441 (2018)
5. Liu, Q., Yang, S., Wang, J.: A collective neurodynamic approach to distributed constrained optimization. IEEE Trans. Neural Netw. Learn. Syst. **28**(8), 1747–1758 (2017)
6. Liu, Q., Wang, J.: A second-order multi-agent network for bound-constrained distributed optimization. IEEE Trans. Autom. Control **60**(12), 3310–3315 (2015)
7. Lu, J., Tang, C.: Zero-gradient-sum algorithms for distributed convex optimization: the continuous-time case. IEEE Trans. Autom. Control **57**(9), 2348–2354 (2011)
8. Nedic, A., Ozdaglar, A.: Distributed subgradient methods for multi-agent optimization. IEEE Trans. Autom. Control **54**(1), 48–61 (2009)
9. Nedic, A., Ozdaglar, A., Parrilo, P.A.: Constrained consensus and optimization in multi-agent networks. IEEE Trans. Autom. Control **55**(4), 922–938 (2010)
10. Nowzari, C.: Distributed Event-Triggered Coordination for Average Consensus on Weight-Balanced Digraphs. Pergamon Press Inc., Oxford (2016)
11. Qin, S., Bian, W., Xue, X.: A new one-layer recurrent neural network for nonsmooth pseudoconvex optimization. Neurocomputing **120**, 655–662 (2013)
12. Qin, S., Fan, D., Wu, G., Zhao, L.: Neural network for constrained nonsmooth optimization using Tikhonov regularization. Neural Netw. **63**, 272–281 (2015)
13. Qin, S., Feng, J., Song, J., Wen, X., Xu, C.: A one-layer recurrent neural network for constrained complex-variable convex optimization. IEEE Trans. Neural Netw. Learn. Syst. **99**, 1–11 (2016)

14. Qin, S., Yang, X., Xue, X., Song, J.: A one-layer recurrent neural network for pseu-
    doconvex optimization problems with equality and inequality constraints. IEEE
    Trans. Cybern. **47**(10), 3063–3074 (2017)
15. Qiu, Z., Liu, S., Xie, L.: Distributed constrained optimal consensus of multi-agent
    systems. Automatica **68**, 209–215 (2016)
16. Rich, E.: Artificial Intelligence. E. Horwood (1985)
17. Shi, G., Johansson, K.H., Hong, Y.: Reaching an optimal consensus: dynamical
    systems that compute intersections of convex sets. IEEE Trans. Autom. Control
    **58**(3), 610–622 (2013)
18. Wang, J., Elia, N.: A control perspective for centralized and distributed convex
    optimization. In: Decision and Control and European Control Conference, pp.
    3800–3805 (2011)
19. Wei, E., Ozdaglar, A., Jadbabaie, A.: A distributed newton method for network
    utility maximization. In: 2010 49th IEEE Conference on Decision and Control
    (CDC), pp. 1816–1821 (2010)
20. White, S.M.: Social engineering. In: IEEE International Conference and Workshop
    on the Engineering of Computer-Based Systems, 2003 Proceedings, pp. 261–267
    (2003)
21. Yang, S., Liu, Q., Wang, J.: Distributed optimization based on a multiagent system
    in the presence of communication delays. IEEE Trans. Syst. Man Cybern. Syst.
    **47**(5), 717–728 (2017)

# An Estimation of Distribution Algorithm for Large-Scale Optimization with Cooperative Co-evolution and Local Search

Jia-Ying Lin, Wei-Neng Chen[(✉)], and Jun Zhang

South China University of Technology, Guangzhou, China
cwnraul634@aliyun.com

**Abstract.** Cooperative co-evolution (CC) is an effective framework for evolutionary algorithms (EAs) to solve large-scale optimization problems. By combining a divide-and-conquer strategy and the classic evolutionary algorithms (EA) like genetic algorithm (GA), CC has shown promising performance in many fields. As a family of EAs, the estimation of distribution algorithm (EDA) is good at search diversity maintenance, but its capability in solving large-scale problems has not been fully explored. In this paper, we aim to propose a new estimation of distribution algorithm with the cooperative co-evolution framework (EDACC). The proposed EDACC has the following features. (1) The differential grouping (DG) strategy is applied for variable decomposition. (2) A combination of the Gaussian and Cauchy distributions are adopted to generate offspring. (3) A local search method is performed in promising domains to accelerate the search. To verify the performance of EDACC, experiments are conducted on 20 single-objective functions in the CEC 2010 benchmarks. The experimental results show that EDACC can still achieve competitive performance in spite of the weakness of the original EDAs like the low accuracy in global optima searching compared with classical EAs.

**Keywords:** Estimation of distribution algorithm (EDA)
Large-scale optimization · Cooperative Co-evolution (CC)

## 1 Introduction

Many complex and difficult real-world problems such as traffic control [21] and data mining [19] in different areas can be reduced to some large-scale optimization problems. Classical evolutionary algorithms (EAs) like PSO [10], DE [6] and CSO [5], cannot solve these high-dimensional problems perfectly in terms

This work was supported in part by the National Natural Science Foundation of China under Grant 61622206, 61332002, and the Natural Science Foundation of Guangdong under Grant 2015A030306024.

© Springer Nature Switzerland AG 2018
L. Cheng et al. (Eds.): ICONIP 2018, LNCS 11302, pp. 442–452, 2018.
https://doi.org/10.1007/978-3-030-04179-3_39

of effectiveness and feasibility since they lack some specific framework for handling them. Therefore, it is desirable to find some general solution for large-scale optimization problems.

Cooperative Co-evolution [1,7,14] is a framework for improving the efficacy of various evolutionary algorithms in large-scale optimization problems. It employs a divide-and-conquer strategy to divide a large-scale optimization problem into some small-scale problems then uses the methods supposed to solve them adaptively and independently. Hence, as an effective and powerful framework for large-scale optimization problems, cooperative co-evolutionary is widely used in some variations of EAs including Cooperative Co-evolutionary Genetic Algorithm (CCGA) [14] and CPSO [3], taking the advantage of EAs in solving function optimization problems. Recently, some new methods that using new decomposition method called differential grouping have been proposed and performed well in benchmark problems like DECC-DG [13] by efficient decomposition strategy. Many satisfying experimental results have been achieved by embedding evolutionary algorithm into Cooperative Co-evolution framework. However, how to maintain the search diversity under CC framework is still a challenging problem, since the number of local optima will rapidly increase as the dimension of solution space grows higher in a large-scale optimization problem.

To maintain good search diversity, a family of EAs called estimation of distribution algorithms (EDAs) [9,12] has attracted much attention as it can maintain good search diversity at the population level working with probabilistic models. Comparing with EDAs, conventional EAs use crossover and mutation operators to generate new population and trial solutions which may easily get closer to the parents but far away from global solutions. The population level diversity of EDAs allows itself to be a critical role in solving high-dimensional problems among all variants of EAs.

However, keeping diversity at the population level makes EDA hard to directly control similarities among offspring and parent at the population level. To deal with this weakness of EDAs, some efforts have been made by using some local search methods to refine the offspring solution in promising areas of the optimal locations when some solutions are found [23].

In this paper, we aim to propose a new estimation of distribution algorithm with the cooperative co-evolution framework (EDACC). The differential grouping (DG) strategy applied in EDACC performs decomposition of variables. Thus, a large-scale optimization problem can be divided into several small-scale problems and solved efficiently under CC framework. With regard to possibility distribution used in EDA, a combination of the Gaussian and Cauchy distributions are adopted in EDACC to generate offspring with high diversity. In order to improve solution accuracy and accelerate the search, a local search method is performed in promising domains found by EDACC. The experimental results show that EDACC can still achieve competitive performance in spite of the weakness of the original EDAs like the low accuracy in global optima searching compared with classical EAs.

The rest of this paper is organized as follows. Section 2 introduces some related works and the background of this research. Section 3 introduces our estimated distribution algorithm with CC framework and Sect. 4 shows our experiment result.

## 2    Relate Work

### 2.1    Estimation of Distribution Algorithm

EDA is a family of evolutionary algorithm firstly proposed in [12], combining classic EAs with probabilistic models to increase diversity in EA offspring generating at the population level. Unlike traditional EAs, EDAs do not use crossover or mutation operators. Instead, the probability model sampled in population level is a role to control the generation of offspring and this model of promising solutions can help keep and improve the diversity in offspring generation. A general framework of EDA is built as Algorithm 1.

---

**Algorithm 1. EDA**

---

**Input:** population size $M$, the number of selected individuals from parent $N$
**Output:** the best solution and the corresponding fitness
 1: **while** termination criterion is not met  **do**
 2:    Select N individuals from population
 3:    Estimate the new probability distribution of selected N individuals
 4:    Generate offspring by partly replacing parent with individuals sampled by the
       new probability distribution
 5: **end while**

---

EDA has shown its power in both discrete and continuous domains, single and multiple optimization problems [2,11,22]. Recently, some research fields like multi-policy insurance investment [18] planning and protein folding problem [16] start to adapt EDA model to enhance their solution performance due to the feasibility and effectiveness of EDAs. However, few attempts have been made to enhance EDA performance with cooperative co-evolution framework.

### 2.2    Cooperative Co-evolution

Cooperative Co-evolution is a framework to solve large-scale optimization problems by divide-and-conquer strategy, decomposing the complex, high dimensional problem into simpler, lower dimensional subproblems. The performance of CC is directly influenced by its grouping strategy, which is the principle of generating suitable subproblems. On the basis of the CC framework, different evolutionary algorithms can be used to solve the large-scale optimization problem effectively. Algorithm 2 presents the framework of CC.

---

**Algorithm 2.** Cooperative Co-evolution

---
**Input:** Original population $P$
**Output:** the best solution and the corresponding fitness
1: Decompose original problem into $K$ subproblems, which $K < P$
2: **while** termination criterion is not met **do**
3:   **for** $i = 1 : K$ **do**
4:     Solve the subproblem by optimizer
5:     Update current best solution
6:   **end for**
7: **end while**

---

## 2.3   Differential Grouping Strategy

Recently, some grouping strategies are proposed to find the dependence among variables in specific optimization problems under CC framework. These grouping strategies aim to generate a reasonable subset of variables for problem optimizer. Differential Grouping (DG) is a dynamic grouping method dividing the variables of original problems according to the interaction between any two different variables. DG keeps the correlation among subproblems and helps the problem optimizer perform later work.

In DG, the dependence between two variable $x_i$ and variable $x_j$ are detected by following inequation:

$$f(x_i + \delta_i, x_j) - f(x_i, x_j) \neq f(x_i + \delta_i, x_j + \delta_j) - f(x_i, x_j + \delta_j),$$

which $\delta_i, \delta_i \neq 0$. The satisfaction of above inequation demonstrates that two variable $x_i$ and $x_j$ interact.

# 3   Estimation of Distribution Algorithm with CC Framework

## 3.1   Distribution Estimation

In EDACC, to preserve the diversity at the population level and make sure promising solution converge, the probabilistic distribution models we use are not only traditional probabilistic model such as Gaussian distribution but also Cauchy distribution model, which can expand the offspring's range because of its long fat tail.

As mentioned in Algorithm 1 line 3, we adopt both Gaussian distribution and Cauchy distribution to EDACC and choose one of them for offspring generation decided by the current promising solution. In EDACC, these two distributions are assigned according to the current decision variable variance of all individuals. If current variance of variables is too small, Cauchy distribution should be assigned since it can offer high diversity for later population and avoid dropping in local optima. In order to accelerate the convergence speed in EDACC, the

---

**Algorithm 3.** Offspring generation on EDA

---

**Input:** Population $P$, Smallest variance $\sigma_s$ that applied Gaussian distribution
**Output:** Offspring generated by Gaussian distribution or Cauchy distribution
 1: Calculate the mean $\mu_{ij}$ and the variance $\sigma_{ij}$ of $j^{th}$ variable
 2: **if** $\sigma_{ij} < \sigma_s$ **then**
 3:     Generate corresponding variable offspring by Cauchy($\mu_{ij}, scale$), where scale is a parameter of Cauchy distribution.
 4: **else**
 5:     Generate corresponding variable offspring by Gaussian distribution($\mu_{ij}, \sigma_{ij}$)
 6: **end if**
 7: Update current population

---

threshold of variance assigning these two distributions should be set to a smaller value.

$\mu_i$ stands for the mean value of $x_i$ decision variable among all $j$ individuals and $\sigma_i$ stands for the variance of $x_i$ decision variable among all $j$ individuals. It should be noted that because of the characteristic of Cauchy distribution, we select $\mu_i$ as its median and $\sigma_i$ as its scale parameter.

## 3.2   Cooperative Co-evolution

One important factor of EDA offspring generation is the way we select individuals and decision variables for probabilistic distribution model simulation. The problem of selecting individuals can be solved by original EDA like PBILc, while the selection of decision variables is still required developed in EDA algorithm. CC framework gives us a new insight for selecting subproblems in the original problem and DG offers a great grouping strategy for CC framework.

Taking all these factors into consideration, we propose a new generating offspring method in EDA under integrating CC framework and DG grouping strategy. In population updating procedure, EDACC doesn't take all decision variables in each selected individuals for the offspring generation. According to the decision variable division result using DG grouping method, we generate new offspring for each decision variable in one group together, while different decision variable groups are updated separately.

We denote that $x_{ij}$ stands for the $j^{th}$ decision variables in $i^{th}$ individual. In offspring generation, for $K$ selected individuals like $(x_1, x_2, \ldots, x_K)$, each individual contains $M$ decision variables and the decision variables only required for probabilistic model parameters updating are those variables in the same subgroup generated by DG strategy. Since DG strategy can reallocate interacting decision variables into the same subgroup, it's necessary to generate offspring one group by one group for EDA offspring generating integrating CC framework. The interaction among population and the characteristic of EDA is suitable for CC framework and DG strategy in solving a large-scale optimization problem by a divide-and-conquer method.

### 3.3   Local Search

Some previous researches [18,22] in EDA have shown that original EDA may be poor at finding a high accuracy solution due to its offspring generating strategy. In EDACC, since the promising solution area can be found approximately, an efficient local search method can be used in the final procedure of generating the best solution when the current solution can be guaranteed converged. Powell method [8] is a classic optimization method without derivatives. It uses bi-directional search along each search vector and ensures efficiency during its searching. This characteristic is suitable for some method that solving large-scale optimization problems with high diversity but low accuracy like EDA. The total EDACC method is presented as Algorithm 4.

---

**Algorithm 4.** EDACC method

---

**Input:** Population $P$, Smallest variance $\sigma_s$ that applied Gaussian distribution
**Output:** the best solution and the corresponding fitness
 1: Decompose original problem into $K$ subproblems by DG strategy, which $K < P$
 2: **while** termination criterion is not met **do**
 3:   **for** $i = 1 : K$ **do**
 4:     **for** $j = 1 : M$, where $M$ is the problem size of $i$(the number of variables in specific subproblem $i$) **do**
 5:       Calculate the mean $\mu_{ij}$ and the variance $\sigma_{ij}$ of $j^{th}$ variable
 6:       **if** $\sigma_{ij} < \sigma_s$ **then**
 7:         Generate corresponding variable offspring by Cauchy$(\mu_{ij}, scale)$, where scale is a parameter of Cauchy distribution.
 8:       **else**
 9:         Generate corresponding variable offspring by Gaussian distribution$(\mu_{ij}, \sigma_{ij})$
10:       **end if**
11:       Update current best solution
12:     **end for**
13:     Use local search method
14:     Update current best solution
15:   **end for**
16: **end while**

---

## 4   Experiment Studies

In this section, a series of experiments are conducted on a large-scale optimization benchmark called CEC2010 [4] to verify the performance of EDACC. Then we compare the result of EDACC with other classic methods in different aspects of optimizer and grouping strategy. Some experimental analysis of EDACC will be demonstrated. All the algorithms are implemented in C and executed in Hasee K660 with i5-4210M CPU @ 2.60 GHz, 8.00 GB RAM, and Windows 10.

### 4.1    Parameter Setting

In EDACC, population size we set is 150 and the maximum number of function evaluation is 3.0E+6. The optimizer we use in EDACC is PBILc [17], which has an offspring update formula:

$$p_{l+1}(x) = (1 - \alpha)p_l(x) + \frac{\alpha}{N} \sum_{k=1}^{K} x_k,$$

denoted that $l$ is $l^{th}$ iteration and $x_k$ is the $k^{th}$ best individuals. The reason why we use PBILc as optimizer is that PBILc is a simple and classical method among variants of EDA. In EDACC, $\alpha$ is set to 0.5 and $K$ is set to 10. Besides, taking diversity into consideration, we add a truncated ratio $r_t$ equals to 0.3 on the population level updated by EDA. The threshold of Gaussian distribution and Cauchy distribution on the variable variance is set to 1.0E−9, which is a small value that makes sure to convergence of the solution.

In local search part, we use a modified Powell method to make CEC2010 enhance solution accuracy and help EDACC converge quickly in a promising solution area. Comparing with traditional Powell method proposed in [15], the parameter in Powell method controlling accuracy is adjusted to a smaller value. For instance, the parameter TOL (a distance tol from a point already evaluated) are set to 2.0E−10.

### 4.2    Experiment Result

Table 1 presents the average results over 25 independent runs of EDACC, DECC-DG and DECC-XDG on CEC2010 benchmark. What should be highlighted is that the reason we compare EDACC with DECC-DG is to study how different evolutionary algorithms such as EDA and SaNSDE under CC framework. We use a traditional optimizer of EDA in EDACC, PBILc, to decrease the effect of the optimizer in EDA. Table 1 shows that EDACC significantly performs well in non-separable function like $f_1$, $f_2$ and $f_{20}$, especially in $f_{20}$ which DG performs a bad grouping result. To study how EDA can overcome the weakness in grouping of DG, some functions that are hard to do grouping, are especially studied as follows. Table 1 also demonstrates that EDACC has a significantly better performance than DECC-DG especially on those benchmark functions which has bad grouping result by DG such as $f_{13}$, $f_{18}$ and $f_{20}$. These benchmark functions are hard to adjust $\epsilon$ to get a good grouping result, according to the study of DG. Due to the low sensitivity of differential grouping result, EDACC can obtain a better result in spite of low efficacy of its optimizer. Besides, as argued in XDG [20], XDG can achieve 100% grouping accuracy on all of CEC2010 20 benchmark functions, so the comparison on the experiment result between EDACC and DECC-DG is more believable on the study of different grouping result. Although EDACC perform a little worse than DECC-XDG, it still can achieve a better result on $f_1$, $f_2$, $f_4$, $f_9$, $f_{14}$, $f_{17}$ and $f_{20}$, where also outperforms DECC-DG. For fully separable functions, EDACC also outperforms significantly

**Table 1.** Comparison Result of EDACC, DECC-DG and DECC-XDG on CEC10'

| Functions | EDACC | | DECC-DG | | DECC-XDG | |
|---|---|---|---|---|---|---|
| | Mean | Std | Mean | Std | Mean | Std |
| $f_1$ | **2.84E−26** | 5.98E−26 | 5.47E+03 | 2.02E+04 | 2.23E+04 | 8.01E+04 |
| $f_2$ | **2.66E+03** | 2.17E+02 | 4.39E+03 | 1.97E+02 | 4.44E+03 | 1.64E+02 |
| $f_3$ | 1.94E+01 | 4.95E−02 | **1.66E+01** | 3.34E−01 | 1.68E+01 | 4.12E−01 |
| $f_4$ | **4.89E+11** | 2.30E+11 | 4.79E+12 | 1.44E+12 | 7.84E+11 | 1.67E+11 |
| $f_5$ | 3.69E+08 | 7.77E+07 | **1.55E+08** | 2.17E+07 | 1.68E+08 | 1.74E+07 |
| $f_6$ | 1.97E+07 | 8.80E+04 | **1.64E+01** | 2.71E−01 | **1.63E+01** | 3.28E−01 |
| $f_7$ | 1.98E+09 | 2.65E+09 | 1.16E+04 | 7.41E+03 | **1.39E+03** | 2.61E+03 |
| $f_8$ | 1.18E+08 | 1.63E+08 | 3.04E+07 | 2.11E+07 | **4.78E+05** | 1.32E+06 |
| $f_9$ | **6.92E+06** | 1.79E+06 | 5.96E+07 | 8.18E+06 | 1.12E+08 | 1.13E+07 |
| $f_{10}$ | 8.52E+03 | 2.41E+02 | **4.52E+03** | 1.41E+02 | 5.31E+03 | 1.55E+02 |
| $f_{11}$ | 2.18E+02 | 2.36E−01 | **1.03E+01** | 1.01E+00 | **1.04E+01** | 1.15E+00 |
| $f_{12}$ | 4.94E+04 | 2.71E+04 | **2.52E+03** | 4.86E+02 | 1.24E+04 | 2.32E+03 |
| $f_{13}$ | 3.00E+05 | 2.92E+05 | 4.54E+06 | 2.13E+06 | **1.12E+03** | 2.25E+02 |
| $f_{14}$ | **1.45E+07** | 2.98E+06 | 3.41E+08 | 2.41E+07 | 5.83E+08 | 4.11E+07 |
| $f_{15}$ | 1.45E+04 | 4.29E+02 | **5.88E+03** | 1.03E+02 | **5.91E+03** | 7.56E+01 |
| $f_{16}$ | 3.97E+02 | 3.73E−01 | **7.39E−13** | 5.70E−14 | 1.81E−08 | 1.57E−09 |
| $f_{17}$ | **1.24E+04** | 2.65E+04 | 4.01E+04 | 2.85E+03 | 1.26E+05 | 7.47E+03 |
| $f_{18}$ | 5.78E+05 | 3.33E+05 | 1.11E+10 | 2.04E+09 | **1.41E+03** | 1.88E+02 |
| $f_{19}$ | 6.53E+06 | 2.16E+06 | 1.74E+06 | 9.54E+04 | **1.59E+06** | 4.96E+04 |
| $f_{20}$ | **1.04E+02** | 1.10E+02 | 4.87E+07 | 2.27E+07 | 5.55E+05 | 1.75E+06 |

in $f_1$ and $f_2$ although putting all of the separable decision variables into the same subgroup is not a good choice as XDG studies. The reason may be that EDACC can give a reliable promising solution area and the way generating offspring can maintain high diversity in population level. Besides, as argued in DG, on instances of rotated elliptic function such as $f_4$, $f_9$ and $f_{14}$, an observation is that EDACC shows a better performance though DG and XDG also can get an optimal grouping.

Figure 1 shows the converging curves of EDACC on $f_{20}$. From Fig. 1, as can be seen, the fitness value evaluated by EDACC outperforms other methods significantly. EDACC can have achieved lower fitness in the 5.0e+05 evaluation comparing with the final fitness evaluated by DECC-DG and DECC-XDG. This result can show EDACC can also obtain good performance in some complex fully-nonseparable function in large-scale optimization problems, which may be associated with its high diversity in population level and CC framework's divide-and-conquer strategy used in it.

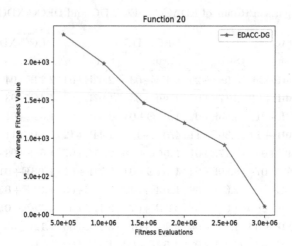

**Fig. 1.** The converging curves on $f_{20}$ for EDACC

## 5    Conclusion

In this paper, we have proposed an estimation of distribution algorithm for large-scale optimization with cooperative co-evolution and local search (EDACC). Taking advantages of EDA, EDACC, which firstly combines CC framework with EDA can outperform in specific problems with a competitive final result comparing with DECC-DG and DECC-XDG. In EDACC, a simple optimizer used in its EDA and grouping strategy used in its CC framework, with low accuracy in some functions, can also keep outperforming in some large-scale optimization problems.

This work has shown its given competitive result in large-scale optimization problems. However, taking consideration of EDA we used in EDACC, the optimizer and the grouping strategy are still required to be investigated how to achieve higher performance incorporating CC framework with EDA.

## References

1. Antonio, L.M., Coello, C.A.C.: Use of cooperative coevolution for solving large scale multiobjective optimization problems. In: 2013 IEEE Congress on Evolutionary Computation (CEC), pp. 2758–2765. IEEE (2013)
2. Bengoetxea, E., Larrañaga, P., Bloch, I., Perchant, A.: Estimation of distribution algorithms: a new evolutionary computation approach for graph matching problems. In: Figueiredo, M., Zerubia, J., Jain, A.K. (eds.) EMMCVPR 2001. LNCS, vol. 2134, pp. 454–469. Springer, Heidelberg (2001). https://doi.org/10.1007/3-540-44745-8_30
3. Van den Bergh, F., Engelbrecht, A.P.: A cooperative approach to particle swarm optimization. IEEE Trans. Evol. Comput. 8(3), 225–239 (2004)

4. Chen, W., Weise, T., Yang, Z., Tang, K.: Large-scale global optimization using cooperative coevolution with variable interaction learning. In: Schaefer, R., Cotta, C., Kołodziej, J., Rudolph, G. (eds.) PPSN 2010. LNCS, vol. 6239, pp. 300–309. Springer, Heidelberg (2010). https://doi.org/10.1007/978-3-642-15871-1_31

5. Cheng, R., Jin, Y.: A competitive swarm optimizer for large scale optimization. IEEE Trans. Cybern. **45**(2), 191–204 (2015)

6. Das, S., Suganthan, P.N.: Differential evolution: a survey of the state-of-the-art. IEEE Trans. Evol. Comput. **15**(1), 4–31 (2011)

7. Fan, J., Wang, J., Han, M.: Cooperative coevolution for large-scale optimization based on kernel fuzzy clustering and variable trust region methods. IEEE Trans. Fuzzy Syst. **22**(4), 829–839 (2014)

8. Fletcher, R.: Practical Methods of Optimization. Wiley, Hoboken (2013)

9. Hauschild, M., Pelikan, M.: An introduction and survey of estimation of distribution algorithms. Swarm Evol. Comput. **1**(3), 111–128 (2011)

10. Kennedy, J.: Particle swarm optimization. In: Sammut, C., Webb, G.I. (eds.) Encyclopedia of Machine Learning, pp. 760–766. Springer, Boston (2011)

11. Luo, N., Qian, F.: Estimation of distribution algorithm sampling under Gaussian and Cauchy distribution in continuous domain. In: 2010 8th IEEE International Conference on Control and Automation (ICCA), pp. 1716–1720. IEEE (2010)

12. Mühlenbein, H., Paaß, G.: From recombination of genes to the estimation of distributions I. Binary parameters. In: Voigt, H.-M., Ebeling, W., Rechenberg, I., Schwefel, H.-P. (eds.) PPSN 1996. LNCS, vol. 1141, pp. 178–187. Springer, Heidelberg (1996). https://doi.org/10.1007/3-540-61723-X_982

13. Omidvar, M.N., Li, X., Mei, Y., Yao, X.: Cooperative co-evolution with differential grouping for large scale optimization. IEEE Trans. Evol. Comput. **18**(3), 378–393 (2014)

14. Potter, M.A., De Jong, K.A.: A cooperative coevolutionary approach to function optimization. In: Davidor, Y., Schwefel, H.-P., Männer, R. (eds.) PPSN 1994. LNCS, vol. 866, pp. 249–257. Springer, Heidelberg (1994). https://doi.org/10.1007/3-540-58484-6_269

15. Press, W.H., Teukolsky, S.A., Vetterling, W.T., Flannery, B.P.: Numerical Recipes 3rd Edition: The Art of Scientific Computing. Cambridge University Press, New York (2007)

16. Santana, R., Larrañaga, P., Lozano, J.A.: Protein folding in simplified models with estimation of distribution algorithms. IEEE Trans. Evol. Comput. **12**(4), 418–438 (2008)

17. Sebag, M., Ducoulombier, A.: Extending population-based incremental learning to continuous search spaces. In: Eiben, A.E., Bäck, T., Schoenauer, M., Schwefel, H.-P. (eds.) PPSN 1998. LNCS, vol. 1498, pp. 418–427. Springer, Heidelberg (1998). https://doi.org/10.1007/BFb0056884

18. Shi, W., Chen, W.N., Lin, Y., Gu, T., Kwong, S., Zhang, J.: An adaptive estimation of distribution algorithm for multi-policy insurance investment planning. IEEE Trans. Evol. Comput. (2017)

19. Srinivasa, K., Venugopal, K., Patnaik, L.M.: A self-adaptive migration model genetic algorithm for data mining applications. Inf. Sci. **177**(20), 4295–4313 (2007)

20. Sun, Y., Kirley, M., Halgamuge, S.K.: Extended differential grouping for large scale global optimization with direct and indirect variable interactions. In: Proceedings of the 2015 Annual Conference on Genetic and Evolutionary Computation, pp. 313–320. ACM (2015)

21. Teklu, F., Sumalee, A., Watling, D.: A genetic algorithm approach for optimizing traffic control signals considering routing. Comput. Aided Civ. Infrastruct. Eng. **22**(1), 31–43 (2007)
22. Yang, Q., Chen, W.N., Li, Y., Chen, C.P., Xu, X.M., Zhang, J.: Multimodal estimation of distribution algorithms. IEEE Trans. Cybern. **47**(3), 636–650 (2017)
23. Zhou, A., Sun, J., Zhang, Q.: An estimation of distribution algorithm with cheap and expensive local search methods. IEEE Trans. Evol. Comput. **19**(6), 807–822 (2015)

# A Collaborative Neurodynamic Approach to Symmetric Nonnegative Matrix Factorization

Hangjun Che[1,2(✉)] and Jun Wang[1,2(✉)]

[1] Department of Computer Science, City University of Hong Kong,
Kowloon Tong, Hong Kong
hjche2-c@my.cityu.edu.hk, jwang.cs@cityu.edu.hk
[2] Shenzhen Research Institute, City University of Hong Kong, Shenzhen, China

**Abstract.** This paper presents a collaborative neurodynamic approach to symmetric nonnegative matrix factorization (SNMF). First, a formulated nonconvex optimization problem of SNMF is described. To solve this problem, a neurodynamic model based on an augmented Lagrangian function is proposed and proven to be convergent to a strict local optimal solution under the second-order sufficiency condition. Next, a group of neurodynamic models are employed to search for an optimal factorized matrix by using particle swarm algorithm to update the initial neuronal states. The efficacy of the proposed approach is substantiated on two datasets.

**Keywords:** Symmetric nonnegative matrix factorization
Collaborative neurodynamic approach
Augmented Lagrangian function

## 1 Introduction

Nonnegative matrix factorization (NMF) aims to decompose a nonnegative matrix $V \in R_+^{m \times n}$ into two low-rank matrices such that $V \approx WH^T$, where $W \in R_+^{m \times r}$, $H \in R_+^{n \times r}$ and $0 < r < \min(m, n)$ [1]. NMF has found numerous applications, such as data dimension reduction [2], clustering [3,4] and information retrieval [5], just to name a few. As an extension of NMF, symmetric nonnegative matrix factorization (SNMF) aims to find a nonnegative matrix $H \in R_+^{n \times r}$ such that $V \approx HH^T$. It has been proven that SNMF is equivalent to kernel K-means clustering if $H$ satisfies $H^T H = I$ [6]. Unfortunately, the formulated problem of SNMF is nonconvex. Moreover, the orthogonality constraint is nonlinear equality which makes difficult to optimize the problem.

This work was supported in part by the Research Grants Council of the Hong Kong Special Administrative Region of China, under Grants 14207614 and 11208517, and in part by the National Natural Science Foundation of China under grant 61673330.

© Springer Nature Switzerland AG 2018
L. Cheng et al. (Eds.): ICONIP 2018, LNCS 11302, pp. 453–462, 2018.
https://doi.org/10.1007/978-3-030-04179-3_40

As a parallel computing optimization approach, neurodynamic approach has been successfully applied in many fields, such as pattern classification [7], k-winners-take-all [8], nonlinear model predictive control [9], image restoration [10], adaptive beamforming [11]. Many neurodynamic models are developed for solving constrained optimization problems in the past decades [12–21].

In recent years, collaborative neurodynamic optimization (CNO) approach shows effectiveness for solving some nonconvex optimization problems [25, 26]. In CNO, a group of neurodynamic models are employed to search the global optimum. Each neurodynamic model is convergent to an optimal solution in local search and exchanges searching information through meta-heuristic algorithm.

In this paper, as the formulated problem of SNMF is nonconvex, a neurodynamic model based on an augmented Lagrangian function is proposed. Next, a group of neurodynamic models are employed to search for an optimal factorized matrix $H$. In Sect. 2, problem formulation and preliminaries are reviewed. In Sect. 3, a neurodynamic model is described. In Sect. 4, experimental results are discussed. The conclusion is made in Sect. 5.

## 2 Preliminaries

### 2.1 Problem Formulation

The symmetric nonnegative matrix factorization is formulated as the following constrained nonconvex problem [6]:

$$\min \quad \|V - HH^T\|_F^2$$
$$\text{s.t.} \quad H^T H = I \quad H \in \Re_+^{n \times r}. \tag{1}$$

where $V \in \Re_+^{n \times n}$, $I \in \Re^{r \times r}$ is the identity matrix, $\| \cdot \|_F$ is the Frobenius norm, and $r < n$. For clustering, $V = X^T X$, where $X \in \Re^{m \times n}$ is the data matrix.

**Definition 1:** $x^*$ is called a strict local minimum if $f(x^*) < f(x)$, $\forall x \in \mathcal{N}(x^*, \epsilon) \cap \mathcal{S}$, where $\mathcal{N}(x^*, \epsilon)$ is a neighborhood of $x^*$ with the radius $\epsilon > 0$ and $\mathcal{S}$ is the feasible region of the problem.

**Definition 2:** $\tilde{x}$ is called a regular point if $\tilde{x}$ is a feasible solution and the gradients of the active inequality constraints and the gradients of the equality constraints are linearly independent at $\tilde{x}$.

For a nonnegative matrix $H$, the Lagrangian function associated with problem (1) is defined as follows:

$$L(h, \lambda) = f(h) + \lambda^T g(h), \tag{2}$$

where $h$ is the vector of the vectorized matrix $H$, $f(h)$ is the objective function of problem (1), $g(h)$ is the vector value function of $(H^T H - I) = 0$, $\lambda$ is the Lagrangian multiplier.

The Karush-Kuhn-Tucker (KKT) condition for problem (1) can be written as [28]:

$$\nabla f(h) + \nabla g(h)\lambda = 0,$$
$$g(h) = 0, h \geq 0, \tag{3}$$

where $\nabla f(h)$ is the vector of the vectorized matrix $\nabla f(H)$, $\nabla f(H) = -2(V^T H + VH) + 3(HH^T H)$, $\nabla g(h)$ is the vector of the vectorized matrix $\nabla g(H)$, $\nabla g(H) = (I_{n \times n} + \Phi)(I_n \otimes H^T)$, where $\otimes$ is the kronecker product, $\Phi \in \Re^{n \times n}$ is the vectorized transpose matrix defined as follows:

$$\Phi_{ij} = \begin{cases} 1 & j = 1 + n(i-1) - (n^2 - 1)\lfloor (i-1)/n \rfloor, \\ 0 & \text{otherwise.} \end{cases}$$

## 2.2 Particle Swarm Optimization

Particle swarm optimization (PSO) is a class of meta-heuristic optimization algorithms. It searches the solution via exchanging information among multiple state vectors (particles). Let $x_i \in \Re^n$ denote the state vector (position) of the $i$th state (particle). $x_i^p$ denotes the best previous position yielding the minimum value of $f(x_i)$. $x_g$ denotes the best position of the swarm. $v_i \in \Re^n$ denotes the velocity of the $i$th particle. $w$ denotes the inertia weight, determining how much of the previous velocity of the particle is preserved. $\eta_1$ denotes the weight of difference between $x_i$ and $x_i^p$. $\eta_2$ denotes the weight of difference between $x_i$ and $x_g$. The updating rule of PSO is as follows [24]:

$$v_i(k+1) = wv_i(k) + \eta_1 r_1(x_i^p - x_i(k)) + \eta_2 r_2(x_g(k) - x_i(k)),$$
$$x_i(k+1) = x_i(k) + v_i(k+1), \tag{4}$$

where $k$ is the iterative index, $r_1$ and $r_2$ are two random variables in $[0, 1]$.

## 2.3 Collaborative Neurodynamic Optimization

Collaborative neurodynamic optimization (initially called collective neurodynamic optimization) approach is a paradigm of hybrid intelligence based on integrating the merits of neurodynamic optimization and swarm intelligence.

Due to the existence of local minima in nonconvex problem, single neurodynamic model is often stuck in local minima around their initial solutions. It is necessary to have some mechanisms to employ multiple neurodynamic models with updating their diverse initial states to jump out of the local minimum. In view of the need, the proposed collaborative neurodynamic approach consists of two interactive levels in a hierarchy. In the lower level, each neurodynamic model is used to perform the local search to get a local minimum. In the upper level, neurodynamic models exchange information between each other through updating rules (4) of PSO to update the individual initial state.

Collaborative neurodynamic optimization approaches with the particle swarm optimization algorithm at the upper level for updating the initial neuronal states of multiple recurrent neural networks work well for global optimization [25,26]. Figure 1 shows a sketch of a collaborative neurodynamic optimization approach. In theory, if the neuronal states are diversified and the updated objective function value is monotonically nonincreasing, then a collaborative neurodynamic approach is globally convergent to a global optimal solution with probability one [26].

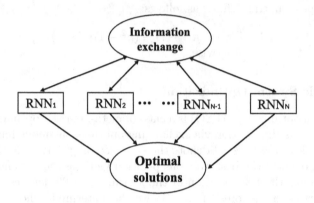

**Fig. 1.** Sketch of the collaborative neurodynamic optimization approach.

## 3 Main Results

Figure 1 shows that collaborative neurodynamic optimization (CNO) consists of a group of neurodynamic models. The stability of an individual neurodynamic model is crucial in CNO. In this section, the stability of the proposed neurodynamic model is discussed.

### 3.1 Augmented Lagrangian Method

As (1) is a global optimization problem with nonlinear equality constraint, similar to [27], an augmented Lagrangian function is introduced for problem (1):

$$\Psi(h, \lambda) = L(h, \lambda) + \frac{\alpha}{2} \sum_{j=1}^{m} (\lambda_j g_j(h))^2, \tag{5}$$

where $\alpha$ is a nonnegative parameter. The gradient of $\Psi(h, \lambda)$ is

$$\nabla_h \Psi(h, \lambda) = \nabla_h L(h, \lambda) + \alpha \sum_{i=1}^{m} \lambda_i^2 g_i(h) \nabla g_i(h). \tag{6}$$

As $g(h^*) = 0$, $\nabla_h \Psi(h^*, \lambda^*) = \nabla_h L(h^*, \lambda^*)$. (3) is equivalent to

$$\nabla_h L(h, \lambda) + \alpha \sum_{i=1}^{m} \lambda_i^2 g_i(h) \nabla g_i(h) = 0$$
$$g(h) = 0, \quad h \geq 0. \tag{7}$$

## 3.2  Model Analysis

**Lemma 1** [28]: Suppose that $x^*$ is a feasible and regular solution to problem (1). $x^*$ is a strict local minimum of problem (1) if there exist $\lambda^* \in \Re^m$ and $\nu^* \in \Re^q$, such that $(x^*, \lambda^*, \nu^*)$ is a KKT point and $\nabla_{xx} L(x^*, \lambda^*, \nu^*)$ is positive definite on an open cone:

$$\mathcal{C} = \{d \in R^n | \nabla g_j(x^*)^T d = 0, \forall j \in \mathcal{I}^1,$$
$$\nabla g_j(x^*)^T d \leq 0, \forall j \in \mathcal{I}^2, \tag{8}$$
$$\nabla h_i(x^*)^T d = 0, \forall i = 1, ..., q\}.$$

where $d \neq 0$, $\mathcal{I} = \{j | g_j(x) \leq 0\}$, $\mathcal{I}^1 = \{j \in \mathcal{I} | g_j(x) = 0\}$ and $\mathcal{I}^2 = \{j \in \mathcal{I} | g_j(x) < 0\}$.

The conditions in Lemma 1 are called second-order sufficiency conditions (SOSC) [28].

**Lemma 2** [29]: Let $P \in \Re^{n \times n}$ be symmetric and $Q \in \Re^{n \times n}$ be symmetric and positive semidefinite. If $\forall x \neq 0$, $x^T Q x = 0$, but $x^T P x > 0$, then there exists a scalar $c$ such that $P + cQ$ is positive definite.

Based on (7), following neurodynamic model is proposed for (1):

$$\begin{cases} \epsilon \frac{dh}{dt} = -h + P_\Omega(h - (\nabla f(h) + \nabla g(h)\lambda + \alpha \nabla g(h)\Gamma(\lambda^2)g(h))), \\ \epsilon \frac{d\lambda}{dt} = -g(h), \end{cases} \tag{9}$$

where $\epsilon$ is the time constant, and $P_\Omega$ is a piecewise activation function defined as:

$$P_\Omega(\zeta_i) = \begin{cases} l_i, & \zeta_i < l_i \\ \zeta_i, & l_i \leq \zeta_i \leq u_i \\ u_i, & \zeta_i > u_i \end{cases}.$$

In particular for SNMF, $u_i$ is $\infty$ and $l_i$ is 0. Therefore, $P_\Omega(\cdot)$ is modified as follows:

$$P_\Omega(\zeta_i) = \begin{cases} 0, & \zeta_i < 0 \\ \zeta_i, & \zeta_i \geq 0. \end{cases}$$

$\Gamma(\lambda^2)$ is a diagonal matrix defined as follows:

$$\Gamma(\lambda^2) = \begin{bmatrix} \lambda_1^2 & 0 & \cdots & 0 \\ 0 & \lambda_2^2 & \cdots & 0 \\ \vdots & \vdots & \ddots & \vdots \\ 0 & 0 & \cdots & \lambda_m^2 \end{bmatrix} \tag{10}$$

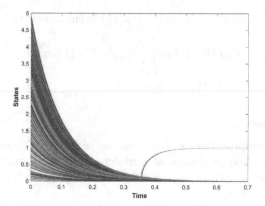

**Fig. 2.** A snapshot of the transient states of a neurodynamic model with the IRIS dataset where $\epsilon = 10^{-2}$.

In contrast with the neurodynamic model in [27], the neurodynamic model (9) is able to solve the nonconvex problem with equality constraints.

**Theorem 1:** *Let $u^* = ((h^*)^T, (\lambda^*)^T)^T$ be a KKT point of problem (1) satisfying the second-order sufficiency conditions in Lemma 1. There exists $c > 0$ such that the neurodynamic model (9) is asymptotically stable at $u^*$, where $h^*$ is a strict local minimum of the problem (1).*

*Proof:* Similar to the proof of Theorem 2 in [27].

## 4    Experimental Results

This section is to show the efficacy of the collaborative neurodynamic approach. Two datasets are used to construct the original matrix $V$. Next, four SNMF algorithms are used to decompose $V$ to compare the convergent orthogonality errors. Three neurodynamic models are used in the proposed collaborative method, $\epsilon = 10^{-2}$.

Figures 2 and 3 show a snapshot of the transient states of a neurodynamic model in the collaborative neurodynamic approach with two datasets. Figures 4 and 5 show the convergence of the orthogonality error of a neurodynamic model in the collaborative approach. Figures 6 and 7 show the convergence of the collaborative approach with two datasets. The objective function value deceases if the proposed approach jumps out from a local minimum at that iteration. Table 1 records the convergent orthogonality errors of four algorithms to symmetric nonnegative matrix factorization. According to Table 1, the proposed method achieves the minimum orthogonality error.

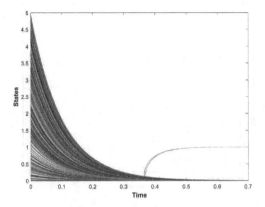

**Fig. 3.** A snapshot of the transient states of a neurodynamic model with the WINE dataset where $\epsilon = 10^{-2}$.

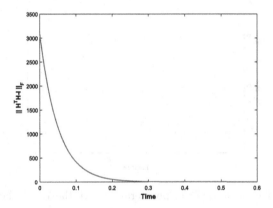

**Fig. 4.** Orthogonality error over time with the IRIS dataset.

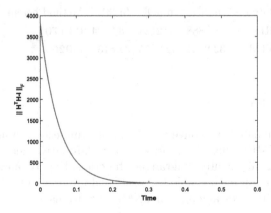

**Fig. 5.** Orthogonality error over time with the WINE dataset.

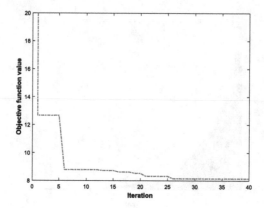

**Fig. 6.** Factorization error per iteration with the IRIS dataset.

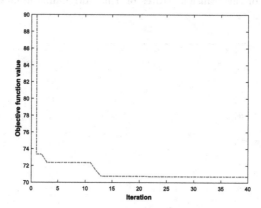

**Fig. 7.** Factorization error per iteration with the WINE dataset.

**Table 1.** Convergent orthogonality errors of four algorithms

| Datesets | In [6] | In [30] | In [31] | Method herein |
|----------|--------|---------|---------|---------------|
| IRIS     | 8.1888 | 8.2124  | 8.2124  | **0.0170**    |
| WINE     | 32.8131| 32.8155 | 32.8155 | **0.0266**    |

## 5   Conclusions

To solve the formulated nonconvex problem of symmetric nonnegative matrix factorization (SNMF), this paper presents a collaborative neurodynamic approach based on an augmented Lagrangian function. Experimental results show that the proposed approach is able to avoid local minima and achieves the minimum orthogonality error among four SNMF algorithms.

# References

1. Lee, D.D., Seung, H.S.: Learning the parts of objects by non-negative matrix factorization. Nature **401**(6755), 788 (1999)
2. Che, H., Wang, J.: A nonnegative matrix factorization algorithm based on a discrete-time projection neural network. Neural Netw. **103**, 63–71 (2018)
3. Kuang, D., Ding, C., Park, H.: Symmetric nonnegative matrix factorization for graph clustering. In: 12th SIAM International Conference on Data Mining, pp. 106–117. SIAM Press (2012)
4. Fan, J., Wang, J.: A collective neurodynamic optimization approach to nonnegative matrix factorization. IEEE Trans. Neural Netw. Learn. Syst. **28**(10), 2344–2356 (2017)
5. Shahnaz, F., Berry, M.W., Pauca, V.P., Plemmons, R.J.: Document clustering using nonnegative matrix factorization. Inf. Process. Manag. **42**(2), 373–386 (2006)
6. Ding, C., He, X., Simon, H. D.: On the equivalence of nonnegative matrix factorization and spectral clustering. In: 5th SIAM International Conference on Data Mining, pp. 606–610. SIAM Press (2005)
7. Xia, Y., Wang, J.: A one-layer recurrent neural network for support vector machine learning. IEEE Trans. Syst. Man Cybern. Part B (Cybern.) **34**(2), 1261–1269 (2004)
8. Liu, S., Wang, J.: A simplified dual neural network for quadratic programming with its KWTA application. IEEE Trans. Neural Netw. **17**(6), 1500–1510 (2006)
9. Yan, Z., Wang, J.: Model predictive control of nonlinear systems with unmodeled dynamics based on feedforward and recurrent neural networks. IEEE Trans. Ind. Inform. **8**(4), 1500–1510 (2012)
10. Xia, Y., Sun, C., Zheng, W.X.: Discrete-time neural network for fast solving large linear $L_1$ estimation problems and its application to image restoration. IEEE Trans. Neural Netw. Learn. Syst. **23**(5), 812–820 (2012)
11. Che, H., Li, C., He, X., Huang, T.: A recurrent neural network for adaptive beamforming and array correction. Neural Netw. **80**, 110–117 (2016)
12. Tank, D., Hopfield, J.J.: Simple'neural'optimization networks: an A/D converter, signal decision circuit, and a linear programming circuit. IEEE Trans. Circ. Syst. **33**(5), 533–541 (1986)
13. Kennedy, M.P., Chua, L.O.: Neural networks for nonlinear programming. IEEE Trans. Circ. Syst. **35**(5), 554–562 (1988)
14. Zhang, S., Constantinides, A.G.: Lagrange programming neural networks. IEEE Trans. Circ. Syst. II Analog. Digit. Signal Process. **39**(7), 441–452 (1992)
15. Zhang, Y., Wang, J.: A dual neural network for convex quadratic programming subject to linear equality and inequality constraints. Phys. Lett. A **298**(4), 271–278 (2002)
16. Xia, Y., Leung, H., Wang, J.: A projection neural network and its application to constrained optimization problems. IEEE Trans. Circ. Syst. I Fundam. Theory Appl. **49**(4), 447–458 (2002)
17. Hu, X., Wang, J.: Solving pseudomonotone variational inequalities and pseudoconvex optimization problems using the projection neural network. IEEE Trans. Neural Netw. **17**(6), 1487–1499 (2006)
18. Le, X., Wang, J.: A two-time-scale neurodynamic approach to constrained minimax optimization. IEEE Trans. Neural Netw. Learn. Syst. **28**(3), 620–629 (2017)
19. Qin, S., Le, X., Wang, J.: A neurodynamic optimization approach to bilevel quadratic programming. IEEE Trans. Neural Netw. Learn. Syst. **28**(11), 2580–2591 (2017)

20. Yang, S., Liu, Q., Wang, J.: A collaborative neurodynamic approach to multiple-objective distributed optimization. IEEE Trans. Neural Netw. Learn. Syst. **29**(4), 981–992 (2018)

21. Leung, M.F., Wang, J.: A collaborative neurodynamic approach to multiobjective optimization. IEEE Trans. Neural Netw. Learn. Syst. **29**(11), 5738–5748 (2018). https://doi.org/10.1109/TNNLS.2018.2806481

22. Kinderlehrer, D., Stampacchia, G.: An Introduction to Variational Inequalities and Their Applications, vol. 31. SIAM, Philadelphia (1980)

23. Xia, Y.: An extended projection neural network for constrained optimization. Neural Comput. **16**(4), 863–883 (2004)

24. Clerc, M., Kennedy, J.: The particle swarm-explosion, stability, and convergence in a multidimensional complex space. IEEE Trans. Evol. Comput. **6**(1), 58–73 (2002)

25. Yan, Z., Wang, J., Li, G.: A collective neurodynamic optimization approach to bound-constrained nonconvex optimization. Neural Netw. **55**, 20–29 (2014)

26. Yan, Z., Fan, J., Wang, J.: A collective neurodynamic approach to constrained global optimization. IEEE Trans. Neural Netw. Learn. Syst. **28**(5), 1206–1215 (2017)

27. Hu, X., Wang, J.: Convergence of a recurrent neural network for nonconvex optimization based on an augmented lagrangian function. In: Liu, D., Fei, S., Hou, Z., Zhang, H., Sun, C. (eds.) ISNN 2007. LNCS, vol. 4493, pp. 194–203. Springer, Heidelberg (2007). https://doi.org/10.1007/978-3-540-72395-0_25

28. Bazaraa, M.S., Sherali, H.D., Shetty, C.M.: Nonlinear Programming: Theory and Algorithms. Wiley, Hoboken (2013)

29. Bertsekas, D.P.: Constrained Optimization and Lagrange Multiplier Methods. Academic Press, New York (2014)

30. Long, B., Zhang, Z.M., Wu, X., Yu, P.S.: Relational clustering by symmetric convex coding. In Proceedings of the 24th International Conference on Machine Learning, pp. 569–576 (2007)

31. Long, B., Zhang, Z.M., Yu, P.S.: Co-clustering by block value decomposition. In: 11th ACM SIGKDD International Conference on Knowledge Discovery in Data Mining, pp. 635–640. ACM (2005)

# Modularity Maximization for Community Detection Using Genetic Algorithm

Hu Lu$^{(\boxtimes)}$ (iD) and Qi Yao

School of Computer Science and Communication Engineering,
Jiangsu University, Zhenjiang 212003, China
luhu@ujs.edu.cn

**Abstract.** Modularity function is a widely-used criterion to evaluate the strength of community structure in community detection. In this paper, we propose a modularity maximization method for detecting communities, based on genetic algorithm and random walk model, and propose a new community structure encoding method for networks. First, the random walk model was applied to calculate the similarity between nodes, resulting in a weighted matrix as derived from the original adjacency matrix. According to the nearest neighbor-based similarity representation provisional, a weighted network connection structure was then coded into a chromosome. The genetic algorithm modified the structure of a predefined number of chromosomes and computed the corresponding modularity, ultimately yielding the maximum value of modularity as it corresponds to community structure and number of communities. We tested this method on a series of real social networks. Compared with several state-of-the-art methods, the novel method obtained both greater modularity value. Thus, results by the proposed method are more practical, since this method does not require specified number of communities at the outset of community partition. Here, the optimal number of communities and community structures are automatically determined.

**Keywords:** Community detection · Genetic algorithm · Random walk

## 1 Introduction

The real-world relationships of interacting entities or individuals can be represented by the structures of a network connection diagram, which networks have an important feature in common, complexity, and which include social collaboration networks, biological networks, World Wide Web, and other networks. Complex networks can be found everywhere. Community structure detection forms critical content in the analysis of complex networks, which divides a network into several smaller modules according to internal connections. Community detection has attracted the interest of multi-disciplinary researchers.

With increasing need to analyze the social networks, many different community structure detection approaches have been proposed. Girvan and Newman first proposed a community detection method as based on calculation of edge betweenness [1]. Later, an evaluation function for modularity was proposed by Newman to measure the

© Springer Nature Switzerland AG 2018
L. Cheng et al. (Eds.): ICONIP 2018, LNCS 11302, pp. 463–472, 2018.
https://doi.org/10.1007/978-3-030-04179-3_41

strength of divided communities [2, 14]. When a network is divided into a specified number of communities, modularity value for this network can be calculated. The greater the modularity value is, the better the partition is. Generally, values between 0.3 and 0.7 indicate the existence of strong network modularity. Modularity quickly became the partition standard in community detection, receiving wide use [3]. In recent years, Langone et al., proposed a new community detection method that also uses modularity Q value as the evaluation function for division, based on the kernel spectral clustering method, called SKSC [4, 5]. The best point for network partition is taken as that which maximizes the modularity value. New approaches to obtaining maximal modularity value have become hot in community detection research.

Chan and Yeung proposed a convex formulation for modularity maximization in community detection [6]. In fact, there does not appear to be an existing best method for solving this optimization problem. Brandes proved maximization of modularity is an NP-hard problem [7]. Therefore, the network partition problem is rendered one of optimization. Various external optimization methods are used to calculate modularity values [8]. Genetic algorithm, a simple and superior-performance evolutionary algorithm, subsequently became the preferred community detection method [9]. Pizzuti proposed a community detection method as based on genetic algorithm, which algorithm does not require a pre-specified number of communities, using instead locus-based adjacency representation to construct chromosomes [10, 11]. Furthermore, reducing the number of graph edges by applying the nearest neighbor of each node has been proposed by Amelio and Pizzuti in different papers [12, 13]. Community structure detection currently focuses on new methods to obtain maximum modularity value for a given network.

In this paper, we propose a new community detection method, which combines the random walk model and genetic algorithm to maximize the modularity Q to realize the community partitioning, called GAcut. First, we use the random walk model to calculate the transition probabilities of nodes. The adjacency matrix of a network is then converted to a weighted matrix, so the nearest neighbor node of each node is unique. We propose a new weighted network community structures encoding method. A nearest neighbor-based weighted similarity representation was proposed to realize the chromosome encoding. We use the genetic algorithm to change the structure of chromosomes. By maximizing the value of modularity, we obtain the corresponding division of networks into groups. Unlike many existing methods, the algorithm does not require a prior setting of the number of communities. Compared with other traditional community detection methods, or modularity optimization methods, experiments on real networks with the true ground truth show the GAcut partitioning method gets a larger value of modularity than other state-of-the-art methods. In addition, we tested the method on other real large networks which have community structures that are unknown in advance. Results indicate that the GAcut method can automatically detect the community structure.

# 2   Our Proposed Method

## 2.1   Problem Definition

Community detection originated from graph partitioning techniques. A network structure can be represented by a graph model, $G = (V, E)$. $V$ is a set of nodes and $E$ is the set of edges which connect the two nodes of $V$. A set of edges is usually represented by a symmetric adjacency matrix, $A = [n \times n]$, where $n$ is the number of nodes, $n = |V|$. At present, research of community detection mainly focuses on binary networks. Namely, if there is a connection between nodes $v_i$ and $v_j$, then $A_{ij} = A_{ji} = 1$; otherwise they equal zero. Community detection generally divides the $n$ nodes into k communities, $C = \{C_1, C_2, \ldots, C_k\}$. The aim of partitioning is to make the number of connections of inter-communities greater than the connections between communities. The problem of detecting communities in a network can then be transformed to that of optimization of a quality evaluation function $f(C)$. When the value of $f(C)$ achieves the optimal, it represents the best community division obtained. The most popularly used evaluation function is the modularity Q. Modularity was firstly proposed by Newman as a performance measure for the quality of the community structure [14]. Here, we also use the $Q$, $f(C) = Q$. Modularity $Q$ is defined as

$$Q = \frac{1}{2m} \sum_{ij} (A_{ij} - \frac{k_i k_j}{2m}) \delta(C_i, C_j),$$ (1)

where A is the adjacency matrix, $k_i$ and $k_j$ are the degree of nodes i and j, $C_i$ is the community node to which i belongs. The variable $m$ is the number of edges. $\delta(C_i, C_j)$ is a kronecker delta function; the higher the $Q$, the better result. In this way, community detection becomes a procedure of finding the maximum value of $Q$. Therefore, several community detection methods mainly centered on the maximization of modularity $Q$ have been proposed. In this paper, we propose a modularity maximization method based on a combination of the random walk distance and the genetic algorithm. We get a better performance.

## 2.2   Random Walk Distance

To realize the coding of chromosomes and divide the network into communities, we use the random walk model to convert the adjacency matrix of network into a weighted matrix. Given an adjacency matrix A, a walker will start from node i to node j with a probability of $P_{ij} = \{X_{t+1} = j | X_t = i\}$. One can also write $P_{ij} = A_{ij}/D$. Let $P_{ij}$ be the transition probability from node i to node j, where $A_{ij}$ denotes the adjacency matrix and D is the degree matrix; $D = diag(d_1, d_2, \ldots, d_n)$. It can also be expressed as $P = D^{-1}A$, where P is the 1-step transition matrix. The t-step transition matrix for the random walk is $P^t = P^{t-1} \times P$. The random walk distance of node i and node j is defined as

$$d(x, y) = \sum_{k=1}^{t} p_{xy}^k$$ (2)

The metric used to quantify the structural similarity between nodes i and j is shown

$$s_{ij} = \frac{\sqrt{\sum_{k \neq i,j}^{n} [d(i,k) - d(j,k)]^2}}{n-2} \tag{3}$$

The value of $s$ is between 0 and 1. The smaller the value $s$, the more similarity exists between the two nodes. S is the matrix having elements $s_{ij}$. S is a symmetric weighted matrix. Random walk distance has been used to detect the community because of its good performances [15, 16]. In this paper, based on the matrix $S$, we propose a new chromosome representation method and encode community structure into a chromosome.

## 2.3  Community Structure Representation

Traditional graph partitioning and community detection algorithms, such as the Ncut algorithm and the GN algorithm, need to specify the number of groups before the network partitioning [17]. Choosing the number of groups always uses trial and error, and cannot achieve the truly unsupervised analysis. In this paper, we proposed a chromosome encoding based on the nearest neighbor-based similarity. A possible connecting structure of the network can be encoded into a chromosome. Figure 1A is the original network having 10 nodes. It is an undirected binary network. Given the adjacency matrix A, we calculated the node similarity between nodes according to the random walk model, resulting in a weighted matrix S. We found out the nearest neighbor node of each of the nodes based on matrix S and each gene can assume allele values j in the range $\{1, \ldots, N\}$, shown in Fig. 1B. A value j assigned to the i$th$ gene is interpreted as a link between the nodes i and j. The 1$th$ position of gene is 3, representing that the nearest neighbor of node 1 is node 3, so this is constructing a chromosome of N nodes. In the decoding step, we constructed a network linking graph accordance with a single chromosome. If the 1th gene is 3, this means that in the clustering solution there is a link between nodes 1 and 3, forming a network structure as shown in Fig. 1C. Compared to Fig. 1A, the number of edges in the Fig. 1C are greatly reduced and only the N edges remain. Shown in Fig. 1C, the original network of 10 nodes can be clearly divided into two communities. A main advantage of this representation is that the number of clusters is automatically determined by the coding and decoding of the chromosome. This method does not require specification of the number of groups at the outset.

## 2.4  Genetic Algorithm

We used genetic algorithm to change the structure of chromosomes in the partitioning process and to obtain the maximum value of modularity. Genetic algorithm as a classic evolutionary algorithm is widely believed to be effective on NP-complete global optimization problems [18]. In the genetic algorithm, a possible solution of communities is represented by a chromosome. Genetic algorithm uses the crossover and mutation to produce a series of candidate solutions.

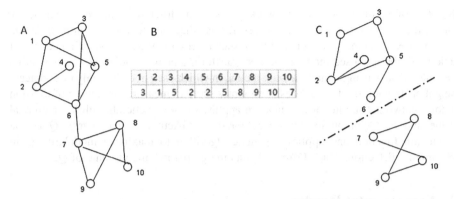

**Fig. 1.** Chromosome encoding and decoding process. (A) The original network connection diagram. (B) The nearest neighbor-based similarity representation of a chromosome with 10 nodes, the first line is position and the second line is genotype. (C) The chromosome can be encoded and split into two modules.

In the initialization process, we generate $m$ chromosomes as the initial solutions. Through the decoding step, the $m$ chromosomes can produce $m$ partition results. We calculated the value of modularity corresponding to each solution. We then chose two chromosomes with maximum modularity as the parents to achieve uniform crossover of the genetic algorithm, and applied the mutation operator of chromosomes to produce the new populations.

The crossover operation is to produce a new generation of chromosomes, which can generate a new network community structure based on the new chromosome. In this paper, we always take as the parents the two optimal chromosomes corresponding with the biggest value of modularity. Given two chromosomes, we randomly select the different positions of two chromosomes and exchange the values of the corresponding position of them, thus forming two new chromosomes.

In the genetic algorithm, the mutation operation prevents the early-maturing of chromosomes and stops searching early. Here, given a chromosome, we randomly select a position i and search for the nearest neighbor j of node i. The position i allele value j, because similarity $s_{ij}$ is the smallest, the nodes i and j may be in the same community.

In the partitioning step, when a chromosome is decoded, a network is divided into k modules $C = \{C_1, C_2, \ldots, C_k\}$. We need an objective function to evaluate the quality of partitioning. The goal of this paper is to find the maximum value of modularity, so we used the modularity proposed by Newman as the objective function. The objective function is thus $f(C) = Q$.

## 2.5   Comparison of Proposed Method and Ga-Net

First proposed by Pizzuti, Ga-net is a community detection method, based on genetic algorithm. In this case, the algorithm did not need to specify the number of communities, and used locus-based adjacency representation to represent the chromosome.

Number of communities and network connection structures can be automatically encoded in a given chromosome. However, this adjacency representation method can only be used in a binary network, while in such a network, the neighbor of a node is not unique. In the proposed method, we first transform a network adjacency matrix into a weighted matrix using a random walk model, in which each node has only one adjacent neighbor. In addition, Ga-net uses community score to measure the quality of division among network communities, which are applied to the genetic algorithm for optimal value and best community division. Moreover, GAcut uses modularity $Q$ as the evaluation function and applies a genetic algorithm to maximize modularity $Q$ in community detection. Thus, GAcut obtains the global maximum value of $Q$.

## 3    Experimental Results

In this paper, the GAcut algorithm was implemented by using Matlab. In a genetic algorithm, crossover rate and mutation rate all are fixed at 0.5. The population size was 50 and the maximum number of generations was 20. In order to show the effectiveness of the proposed algorithm, we presented experimental results of proposed algorithm and compare them with some related methods in the community detection literature. First of all, we used the four standard network datasets. (downloaded from http://www-personal.umich.edu/~mejn/netdata/). These datasets are the social network of friendships between 34 members of a karate club at a US university in the 1970s (karate), an undirected social network of frequent associations between 62 dolphins in a community living off Doubtful Sound (dolphin), a network of books about US politics published around the time of the 2004 presidential election and sold by the online bookseller Amazon.com (kreb), and a network of American football games between Division I colleges during regular season Fall 2000 (football). These four datasets gave the initial partitions in advance. The number of nodes and edges of each data set are shown in Table 1.

**Table 1.** Description of four benchmark networks

|      | Karate (N1) | Dolphin (N2) | Kreb (N3) | Football (N4) |
|------|-------------|--------------|-----------|---------------|
| Size | 34          | 62           | 105       | 115           |
| Edge | 78          | 159          | 441       | 613           |

The good performance of the proposed algorithm with respect to the some modularity maximization community detection algorithms and other famous community partitioning algorithms is shown in Table 2. Each algorithm performed 10 independent runs.

In Table 2, $Q_{max}$ is the maximum modularity value obtained by the different algorithms and $Q_{avg}$ is the mean result of the ten times. NC is the corresponding number of communities of maximum modularity value.

GAcut is our proposed algorithm in this paper. GN algorithm was proposed by Newman and Girvan, which is based on edge betweenness measurements [1]. Later,

**Table 2.** Best modularity results and corresponding number of communities of different algorithms

|     |            | GAcut  | Fast_mo | Fast_newman | NMF    | Qcut   | HQcut  | Ga-net | Cdp    | SKSC   |
|-----|------------|--------|---------|-------------|--------|--------|--------|--------|--------|--------|
| N1  | $Q_{max}$  | **0.4198** | 0.4198 | 0.3974      | 0.3990 | 0.4198 | 0.4198 | 0.4198 | 0.4174 | 0.3715 |
|     | NC         | 4      | 4       | 4           | 2      | 4      | 4      | 4      | 4      | 2      |
|     | $Q_{avg}$  | **0.4189** | 0.4158 | 0.3974      | 0.3280 | 0.4198 | 0.4198 | 0.4109 | 0.4174 | 0.3715 |
| N2  | $Q_{max}$  | **0.5277** | 0.5273 | 0.5149      | 0.5083 | 0.5040 | 0.5175 | 0.4992 | 0.5267 | 0.4754 |
|     | NC         | 5      | 5       | 5           | 4      | 5      | 5      | 7      | 4      | 3      |
|     | $Q_{avg}$  | **0.5269** | 0.5193 | 0.5149      | 0.4819 | 0.5040 | 0.5082 | 0.4680 | 0.5267 | 0.4754 |
| N3  | $Q_{max}$  | 0.5270 | **0.5272** | 0.4992      | 0.5240 | 0.5028 | 0.5045 | 0.5035 | **0.5272** | 0.4987 |
|     | NC         | 5      | 5       | 3           | 4      | 4      | 6      | 7      | 5      | 3      |
|     | $Q_{avg}$  | 0.5247 | 0.5266  | 0.4992      | 0.5207 | 0.5028 | 0.5045 | 0.4757 | **0.5272** | 0.4987 |
| N4  | $Q_{max}$  | **0.6046** | **0.6046** | 0.5782      | 0.6032 | 0.6000 | 0.6032 | 0.5638 | 0.5372 | 0.5830 |
|     | NC         | 10     | 10      | 8           | 8      | 9      | 11     | 11     | 4      | 8      |
|     | $Q_{avg}$  | 0.6033 | **0.6038** | 0.5782      | 0.5966 | 0.6000 | 0.6032 | 0.5215 | 0.5372 | 0.5754 |

Newman proposed a fast modular division method (Fast_newman) [14]. To measure the quality of the network division found by an algorithm, they introduce the concept of modularity. Fast_mo algorithm is a partition algorithm based on the optimization of modularity [19]. NMF is a partitioning algorithm based on non negative matrix factorization, using the modularity to evaluate partitioning [20]. Qcut and HQcut algorithms are local modularity maximization algorithms proposed by Ruan based on the spectral graph partitioning [21]. Ga-net is an evolution of fitness function community partitioning algorithm based on genetic algorithm [10]. The Cdp algorithm is a k-partition partitioning method based on the convex relaxation, performing modularity maximization for tackling the general k partition problem (Convex Formulation) [6]. The last algorithm is SKSC [4, 5].

We can see that GAcut outperforms the majority of algorithms for all datasets. For dolphin network, GAcut obtain a modularity of 0.5277 and is the biggest over all other algorithms.

For some networks, partitions obtained by the maximization of modularity may obtain a more rational partition. We use the karate network to illustrate it. We gave the several different partitioning results corresponding to maximum modularity obtained by different community detection algorithms.

We can see from Fig. 2 that the community structure in Fig. 2A is the original partition, represented by different shapes and colors. The value of modularity corresponding to this partition is 0.371. The value of modularity is less than for other algorithms. Although the non-negative matrix factorization method obtained the results in good agreement with the original partitioning results, this algorithm needs to specify the number of communities in advance. The GN algorithm obtained the maximum modularity of 0.385, when the network was divided into 6 communities (see Fig. 2B). Our proposed algorithm obtained the biggest value of $Q$. $Q$ is 0.4198. The result is listed in Table 2. This is the biggest value that these algorithms can obtain. The network was divided into 4 communities. Shown in Fig. 2C, this partition is more reasonable.

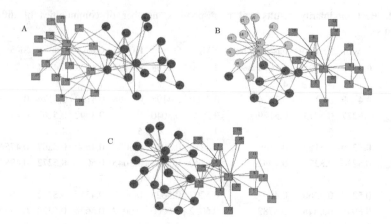

**Fig. 2.** Karate network is divided into several different network structures. (A) The network is divided into 2 groups. (B) The network is divided into 6 groups. (C) The network is divided into 4 groups.

In addition, for some networks, such as large-size-of-nodes networks, even if we have found the maximum modularity, it cannot guarantee that the best community structure can be found. Fortunato has shown that maximum modularity has a resolution limit [23]. It is not possible to detect communities with sizes smaller than a certain threshold. Two small modules may be merged and thus cannot be detected. It may be practical to propose new methods which can obtain the optimal value of modularity and detect more hidden small modules. Meanwhile, GAcut obtain a larger $Q$ value, which shows that the algorithm is more reasonable.

We further tested the algorithm on other networks. We used three large networks. The three networks were the Erdös collaboration network (Erdös), a network of email contacts at a university (email) [24], and a yeast network (yeast) [25]. The number of nodes and experimental results are shown in Table 3.

**Table 3.** Experimental results of three algorithms on three large networks

|        | Size | GAcut | | Ga-net | | SKSC | |
|--------|------|-------|----|--------|----|-------|----|
|        |      | Q | NC | Q | NC | Q | NC |
| Erdös  | 433  | **0.4541** | 31 | 0.3365 | 72 | 0.308 | 58 |
| Email  | 1133 | **0.4541** | 41 | 0.2811 | 95 | 0.402 | 12 |
| Yeast  | 2284 | **0.4725** | 132 | 0.3781 | 247 | 0.4792 | 17 |

Since the networks concerned here contain the numerous nodes absent advance knowledge of community number, the NMF algorithm, the Convex Formulation algorithm (Cdp), or related community detection algorithms cannot be used directly in these large networks. Such algorithms typically require specification of group number at the outset.

We compared GAcut with algorithms that automatically determine the number of communities: Ga-net and SKSC. Shown in Table 3, in these large networks, the number of communities is also great. There may be a resolution limit problem where maximization of modularity fails to further define smaller communities. Compared with Ga-net and SKSC algorithms, our method obtains greater modularity $Q$ value and number of communities. Although Ga-net detected more communities, the modularity is less than that of GAcut, thus indicating that the modularity of the network is weak. Some indistinct modules are also being detected. In these large networks of unknown structure, the kinds of partitions that are significant needs further study.

## 4  Conclusion

Here, we propose a partitioning algorithm, based on genetic evolution of modularity maximization, which applies new community structure encoding and decoding methods that automatically determine logical division of community structure. Combined with the random walk model, results of community detection are more accurate. Results demonstrate that this method does not require specification of community number for the partitioning process, and is thus a completely unsupervised community partitioning method. Moreover, the novel approach automatically determines best modularity value and corresponding optimal number of communities. Therefore, the novel algorithm is very practical for large actual networks without *a priori* defined structures.

**Acknowledgments.** This study was supported by the National Natural Science Foundation of China (Project No. 61375122 and Project No. 61572239). Scientific Research Foundation for Advanced Talents of Jiangsu University (Project No. 14JDG040). Postgraduate Research & Practice Innovation Program of Jiangsu Province (Project No. SJCX18_0741).

## References

1. Girvan, M., Newman, M.E.J.: Community structure in social and biological networks. Proc. Natl. Acad. Sci. U.S.A. **99**, 7821–7826 (2002)
2. Newman, M.E.J., Girvan, M.: Finding and evaluating community structure in networks. Phys. Rev. E **69**(2), 026113 (2004)
3. Humphries, M.D.: Spike-train communities: finding groups of similar spike trains. J. Neurosci. **31**(6), 2321–2336 (2011)
4. Langone, R., Mall, R., Suykens, J.A.K.: Soft Kernel spectral clustering. In: Proceedings of the IJCNN, Dallas, Texas, pp. 1028–1035 (2013)
5. Langone, R., Mall, R., Vandewalle, J., Suykens, J.A.K.: Discovering cluster dynamics using kernel spectral methods. In: Lü, J., Yu, X., Chen, G., Yu, W. (eds.) Complex Systems and Networks. UCS, pp. 1–24. Springer, Heidelberg (2016). https://doi.org/10.1007/978-3-662-47824-0_1
6. Chan, E.Y.K., Yeung, D.Y.: A convex formulation of modularity maximization for community detection. In: International Joint Conference on Artificial Intelligence, pp. 2218–2225. AAAI Press (2011)

7. Brandes, U., et al.: Maximizing modularity is hard. arXiv preprint physics/0608255 (2006)
8. Duch, J., Arenas, A.: Community detection in complex networks using extremal optimization. Phys. Rev. E **72**(2), 027104 (2005)
9. Tasgin, M., Herdagdelen, A., Bingol, H.: Community detection in complex networks using genetic algorithms. arXiv preprint arXiv:0711.0491 (2007)
10. Pizzuti, C.: GA-Net: a genetic algorithm for community detection in social networks. In: Rudolph, G., Jansen, T., Beume, N., Lucas, S., Poloni, C. (eds.) PPSN 2008. LNCS, vol. 5199, pp. 1081–1090. Springer, Heidelberg (2008). https://doi.org/10.1007/978-3-540-87700-4_107
11. Park, Y.J., Song, M.S.: A genetic algorithm for clustering problems. In: 3rd Annual Conference, pp. 568–575. Morgan, Kaufmann (1989)
12. Amelio, A., Pizzuti, C.: A genetic algorithm for color image segmentation. In: Esparcia-Alcázar, A.I. (ed.) EvoApplications 2013. LNCS, vol. 7835, pp. 314–323. Springer, Heidelberg (2013). https://doi.org/10.1007/978-3-642-37192-9_32
13. Amelio, A., Pizzuti, C.: A new evolutionary-based clustering framework for image databases. In: Elmoataz, A., Lezoray, O., Nouboud, F., Mammass, D. (eds.) ICISP 2014. LNCS, vol. 8509, pp. 322–331. Springer, Cham (2014). https://doi.org/10.1007/978-3-319-07998-1_37
14. Newman, M.E.J.: Fast algorithm for detecting community structure in networks. Phys. Rev. E **69**, 066133 (2004)
15. Pons, P., Latapy, M.: Computing communities in large networks using random walks. In: Yolum, P., Güngör, T., Gürgen, F., Özturan, C. (eds.) ISCIS 2005. LNCS, vol. 3733, pp. 284–293. Springer, Heidelberg (2005). https://doi.org/10.1007/11569596_31
16. Valencia, M., Pastor, M.A., Artieda, J., Martinerie, J., Chavez, M.: Complex modular structure of large-scale brain networks. Chaos Interdiscip. J. Nonlinear Sci. **19**, 023119 (2009)
17. Shi, J., Malik, J.: Normalized cuts and image segmentation. IEEE Trans. Pattern Anal. Mach. Intell. **22**, 888–905 (2000)
18. Hruschka, E.R.: A genetic algorithm for cluster analysis. Intell. Data Anal. **7**, 15–25 (2003)
19. Martelot, E.L., Hankin, C.: Multi-scale community detection using stability as optimisation criterion in a greedy algorithm. KDIR, pp. 216–225 (2011)
20. Psorakis, I., Roberts, S., Ebden, M., Sheldon, B.: Overlapping community detection using Bayesian non-negative matrix factorization. Phys. Rev. E **83**(2), 066114 (2011)
21. Ruan, J., Zhang, W.: Identifying network communities with a high resolution. Phys. Rev. E **77**, 016104 (2008)
22. Good, B.H., Montjoye, Y.A.D., Clauset, A.: Performance of modularity maximization in practical contexts. Phys. Rev. E Stat. Nonlin. Soft Matter. Phys. **81**(2), 046106 (2009)
23. Fortunato, S., Barthélemy, M.: Resolution limit in community detection. Proc. Natl. Acad. Sci. **104**, 36–41 (2007)
24. Guimerà, R., Danon, L., Díazguilera, A., Giralt, F., Arenas, A.: Self-similar community structure in a network of human interactions. Phys. Rev. E Stat. Nonlin. Soft Matter. Phys. **68**, 065103 (2003)
25. Bu, D., et al.: Topological structure analysis of the protein-protein interaction network in budding yeast. Nucleic Acids Res. **31**(9), 2443–2450 (2003)

# Continuous Trade-off Optimization Between Fast and Accurate Deep Face Detectors

Petru Soviany[1] and Radu Tudor Ionescu[1,2(✉)]

[1] University of Bucharest, 14 Academiei, Bucharest, Romania
petru.soviany@yahoo.com, raducu.ionescu@gmail.com
[2] SecurifAI, 24 Mircea Vodă, Bucharest, Romania

**Abstract.** Although deep neural networks offer better face detection results than shallow or handcrafted models, their complex architectures come with higher computational requirements and slower inference speeds than shallow neural networks. In this context, we study five straightforward approaches to achieve an optimal trade-off between accuracy and speed in face detection. All the approaches are based on separating the test images in two batches, an *easy* batch that is fed to a faster face detector and a *difficult* batch that is fed to a more accurate yet slower detector. We conduct experiments on the AFW and the FDDB data sets, using MobileNet-SSD as the fast face detector and S³FD (Single Shot Scale-invariant Face Detector) as the accurate face detector, both models being pre-trained on the WIDER FACE data set. Our experiments show that the proposed difficulty metrics compare favorably to a random split of the images.

**Keywords:** Face detection · Deep neural networks · MobileNet-SSD S³FD

## 1 Introduction

Face detection, the task of predicting where faces are located in an image, is one of the most well-studied problems in computer vision, since it represents a prerequisite for many other tasks such as face recognition [22], facial expression recognition [9,16], age estimation, gender classification and so on. Inspired by the recent advances in deep object detection [12,24], researchers have proposed very deep neural networks [4,18,23,28,30] as a solution to the face detection task, providing significant accuracy improvements. Although deep models [11,12] generally offer better results than shallow [13,19] or handcrafted models [8,27], their complex architectures come with more computational requirements and slower inference speeds. As an alternative for environments with limited resources, e.g. mobile devices, researchers have proposed shallower neural networks [13] that provide fast but less accurate results. In this context, we believe it is relevant to propose and evaluate an approach that allows to set the trade-off between

© Springer Nature Switzerland AG 2018
L. Cheng et al. (Eds.): ICONIP 2018, LNCS 11302, pp. 473–485, 2018.
https://doi.org/10.1007/978-3-030-04179-3_42

accuracy and speed in face detection on a continuous scale. Based on the same principles described in [26], we hypothesize that using more complex and accurate face detectors for *difficult* images and less complex and fast face detectors for *easy* images will provide an optimal trade-off between accuracy and speed, without ever having to change anything about the face detectors. The only problem that prevents us from testing our hypothesis in practice is finding an approach to classify the images into easy or hard. In order to be useful in practice, the approach also has to work fast enough, e.g. at least as fast as the faster face detector. To this end, we propose and evaluate five simple and straightforward approaches to achieve an optimal trade-off between accuracy and speed in face detection. All the approaches are based on separating the test images in two batches, an *easy* batch that is fed to the faster face detector and a *hard* (or *difficult*) batch that is fed to the more accurate face detector. The difference between the five approaches is the criterion used for splitting the images in two batches. The first approach assigns a test image to the easy or the hard batch based on the class-agnostic image difficulty score, which is estimated using a recent approach for image difficulty prediction introduced by Ionescu et al. [15]. The image difficulty predictor is obtained by training a deep neural network to regress on the difficulty scores produced by human annotators. The second approach is based on a person-aware image difficulty predictor, which is trained only on images containing the class *person*. The other three approaches used for splitting the test images (into easy or hard) employ a faster single-shot face detector, namely MobileNet-SSD [13], in order estimate the number of faces and each of their sizes. The third and the fourth approaches independently consider the number of detected faces (images with less faces go into the easy batch) and the average size of the faces (images with bigger faces go into the easy batch), while the fifth approach is based on the number of faces divided by their average size (images with less and bigger faces go into the easy batch). If one of the latter three approaches classifies an image as easy, there is nothing left to do (we can directly return the detections provided by MobileNet-SSD). Our experiments on the AFW [31] and the FDDB [17] data sets show that using the class-agnostic or person-aware image difficulty as a primary cue for splitting the test images compares favorably to a random split of the images. However, the other three approaches, which are frustratingly easy to implement, can also produce good results. Among the five proposed methods, the best results are obtained by the class-agnostic image difficulty predictor. This approach shortens the processing time nearly by half, while reducing the Average Precision of the Single Shot Scale-invariant Face Detector ($S^3FD$) [30] from 0.9967 to no less than 0.9818 on AFW. Moreover, all our methods are simple and have the advantage that they allow us to choose the desired trade-off on a continuous scale.

The rest of this paper is organized as follows. Recent related works on face detection are presented in Sect. 2. Our methodology is described in Sect. 3. The face detection experiments are presented in Sect. 4. Finally, we draw our conclusions in Sect. 5.

## 2  Related Work

To our knowledge, there are no previous works that study the trade-off between accuracy and speed for deep face detection. However, there are works [14,26] that study the trade-off between accuracy and speed for the more general task of object detection. Huang et al. [14] have tested different configurations of deep object detection frameworks by changing various components and parameters in order to find optimal configurations for specific scenarios, e.g. deployment on mobile devices. Different from their approach, Soviany et al. [26] treat the various object detection frameworks as black boxes. Instead of looking for certain configurations, they propose a framework that allows to set the trade-off between accuracy and speed on a continuous scale, by specifying the point of splitting the test images into easy versus hard, as desired. We build our work on top of the work of Soviany et al. [26], by considering various splitting strategies for a slightly different task: face detection.

In the rest of this section, we provide a brief description of some of the most recent deep face detectors, in chronological order. CascadeCNN [20] is one of the first models to successfully use convolutional neural networks (CNN) for face detection. Its cascading architecture is made of three CNNs for face versus non-face classification and another three CNNs for bounding box calibration. At each step in the cascade setup, a number of detections are dropped, while the others are passed to the next CNN. By changing the thresholds required in this process, a certain trade-off between accuracy and speed can be obtained. Jian et al. [18] use Faster R-CNN to detect faces. Faster R-CNN [24] is a very accurate region-based deep detection model which improves Fast R-CNN [10] by introducing the Region Proposal Networks (RPN). It uses a fully convolutional network that can predict object bounds at every location in order to solve the challenge of selecting the right regions. In the second stage, the regions proposed by the RPN are used as an input for the Fast R-CNN model, which will provide the final object detection results. $S^3FD$ [30] is a highly accurate real-time face detector, based on the anchor model used initially for object detection [21,24]. In order to solve the limitations on small objects (faces) of these methods, $S^3FD$ introduces a scale compensation anchor matching strategy to improve recall, and a max-out background label to reduce the false positive detections. In order to handle different scales of faces, it uses a scale-equitable face detection framework, tiling anchors on a wide range of layers, while also designing many different anchor scales. MobileNets [13] are a set of lightweight models that can be used for classification, detection and segmentation tasks. They are built on depth-wise separable convolutions with a total of 28 layers and can be further parameterized, making them very suitable for mobile devices. The fast speeds and the low computational requirements of MobileNets make up for the fact that they do not achieve the accuracy of the very-deep models. The experimental results show they can also be successfully used for face detection.

# 3   Methodology

Humans learn much better when the examples are not randomly presented, but organized in a meaningful order which gradually illustrates more complex concepts. Bengio et al. [1] have explored easy-to-hard strategies to train machine learning models, showing that machines can also benefit from learning by gradually adding more difficult examples. They introduced a general formulation of the easy-to-hard training strategies known as *curriculum learning*. However, we can hypothesize that an *easy-versus-hard* strategy can also be applied at test time in order to obtain an optimal trade-off between accuracy and processing speed. For example, if we have two types of machines (one that is simple and fast but less accurate, and one that is complex and slow but more accurate), we can devise a strategy in which the fast machine is fed with the easy test samples and the complex machine is fed with the difficult test samples. This kind of strategy will work as desired especially when the fast machine can reach an accuracy level that is close to the accuracy level of the complex machine for the easy test samples. Thus, the complex and slow machine will be used only when it really matters, i.e. when the examples are too difficult for the fast machine. The only question that remains is how to determine if an example is easy or hard in the first place. If we focus our interest on image data, the answer to this question is provided by the recent work of Ionescu et al. [15], which shows that the difficulty level of an image (with respect to a visual search task) can be automatically predicted. With an image difficulty predictor at our disposal, we have a first way to test our hypothesis in the context of face detection from images. However, if we further focus our interest on the specific task of face detection in images, we can devise additional criteria for splitting the images into easy or hard. One criterion is to consider an image difficulty predictor that is specifically trained on images with people, i.e. a person-aware image difficulty predictor. Other criteria can be developed by considering the output of a very fast single-shot face detector, e.g. MobileNet-SSD [13]. These criteria are the number of detected faces in the image, the average size of the detected faces, and the number of detected faces divided by their average size.

To obtain an optimal trade-off between accuracy and speed in face detection, we propose to employ a more complex face detector, e.g. $S^3FD$ [30], for difficult test images and a less complex face detector, e.g. MobileNet-SSD [13], for easy test images. Our simple easy-versus-hard strategy is formally described in Algorithm 1. Since we apply this strategy at test time, the face detectors as well as the image difficulty predictors can be independently trained beforehand. This allows us to directly apply state-of-the-art pre-trained face detectors [13,30], essentially as black boxes. It is important to note that we use one of the following five options as the criterion function $C$ in Algorithm 1:

1. a class-agnostic image difficulty predictor that estimates the difficulty of the input image;
2. a person-aware image difficulty predictor that estimates the difficulty of the input image;

---

**Algorithm 1.** Easy-versus-Hard Face Detection

---

1 **Input:**
2 $I$ – an input test image;
3 $D_{fast}$ – a fast but less accurate face detector;
4 $D_{slow}$ – a slow but more accurate face detector;
5 $C$ – a criterion function used for dividing the images;
6 $t$ – a threshold for dividing images into easy or hard;

7 **Computation:**
8 **if** $C(I) \leq t$ **then**
9 $\quad \lfloor \; B \leftarrow D_{fast}(I);$

10 **else**
11 $\quad \lfloor \; B \leftarrow D_{slow}(I);$

12 **Output:**
13 $B$ – the set of predicted bounding boxes.

---

3. a fast face detector that returns the number of faces detected in the input image (*less faces* is easier);
4. a fast face detector that returns the average size of the faces detected in the input image (*bigger faces* is easier);
5. a fast face detector that returns the number of detected faces divided by their average size (*less and bigger faces* is easier).

We note that if either one of the last three criteria are employed in Algorithm 1, and if the fast face detector used in the criterion function $C$ is the same as $D_{fast}$, we can slightly optimize Algorithm 1 by applying the fast face detector only once, when the input image $I$ turns out to be easy. Another important note is that, for the last three criteria, we consider an image to be *difficult* if the fast detector does not detect any face. Our algorithm has only one parameter, namely the threshold $t$ used for dividing images into easy or hard. This parameter depends on the criterion function and it needs to be tuned on a validation set in order to achieve a desired trade-off between accuracy and time. While the last three splitting criteria are frustratingly easy to implement when a fast pre-trained face detector is available, we have to train our own image difficulty predictors as described below.

**Image Difficulty Predictors.** We build our image difficulty prediction models based on CNN features and linear regression with $\nu$-Support Vector Regression ($\nu$-SVR) [2]. For a faster processing time, we consider a rather shallow pre-trained CNN architecture, namely VGG-f [3]. The CNN model is trained on the ILSVRC benchmark [25]. We remove the last layer of the CNN model and use it to extract deep features from the fully-connected layer known as $fc7$. The 4096 CNN features extracted from each image are normalized using the $L_2$-norm. The normalized feature vectors are then used to train a $\nu$-SVR model to regress to the ground-truth difficulty scores provided by Ionescu et al. [15] for the PASCAL VOC 2012 data set [5]. We use the learned model as a continuous measure to automatically predict image difficulty. We note that Ionescu et al. [15] showed that the resulted image difficulty predictor is class-agnostic. Since our focus is on face detection, it is perhaps more useful to consider a class-specific image difficulty predictor. For this reason we train a different image difficulty predictor by selecting only the PASCAL VOC 2012 images that contain the class *person*. As these images are likely to contain faces, the person-aware difficulty predictor could be more appropriate for the task at hand. Both image difficulty predictors are based on the VGG-f architecture, which is faster than the considered face detectors, including MobileNet-SSD [13], and it reduces the computational overhead at test time.

## 4    Experiments

### 4.1    Data Sets

We perform face detection experiments on the AFW [31] and the FDDB [17] data sets. The AFW data set consists of 205 images with 473 labeled faces, while the FDDB data set consists of 2845 images that contain 5171 face instances.

### 4.2    Evaluation Details

**Evaluation Measures.** On the FDDB data set, the performance of the face detectors is commonly evaluated using the area under the discrete ROC curve (DiscROC) or the area under the continuous ROC curve (ContROC), as defined by Jain et al. [17]. On the other hand, the performance of the face detectors on the AFW data set is typically evaluated using the Average Precision (AP) metric, which is based on the ranking of detection scores [7]. The Average Precision is given by the area under the precision-recall (PR) curve for the detected faces. The PR curve is constructed by mapping each detected bounding box to the most-overlapping ground-truth bounding box, according to the Intersection over Union (IoU) measure, but only if the IoU is higher than 50% [6].

**Models and Baselines.** We use $S^3FD$ [30] as our accurate model for predicting bounding boxes, and experiment with the pre-trained version available at https://github.com/sfzhang15/SFD. As our fast detector, we choose the pre-trained version of MobileNet-SSD from https://github.com/yeephycho/tensorflow-face-detection, slightly modified. For both models, which are pre-trained on the WIDER FACE data set [29], we set the confidence threshold to 0.5.

The main goal of the experiments is to compare our five different strategies for splitting the images between the fast detector (MobileNet-SSD) and the accurate detector ($S^3FD$) with a baseline strategy that splits the images randomly. To reduce the accuracy variation introduced by the random selection of the baseline strategy, we repeat the experiments for 5 times and average the resulted scores. We note that all standard deviations are lower than 0.5%. We consider various splitting points starting with a 100%−0% split (equivalent with applying the fast MobileNet-SSD only), going through three intermediate splits (75%−25%, 50%−50%, 25%−75%) and ending with a 0%−100% split (equivalent with applying the accurate $S^3$FD only).

**Table 1.** Average Precision (AP) and time comparison between MobileNet-SSD [13], $S^3$FD [30] and various combinations of the two face detectors on AFW. The test data is partitioned based on a random split (baseline) or five easy-versus-hard splits given by: the class-agnostic image difficulty score, the person-aware image difficulty score, the number of faces ($n$), the average size of the faces ($avg$), and the number of faces divided by their average size ($n/avg$). For the random split, we report the AP over 5 runs to reduce bias. The reported times are measured on a computer with Intel Core i7 2.5 GHz CPU and 16 GB of RAM.

| | MobileNet-SSD (left) to $S^3$FD (right) | | | | |
| | 100%−0% | 75%−25% | 50%−50% | 25%−75% | 0%−100% |
|---|---|---|---|---|---|
| *Splitting criterion* | *Average Precision (AP)* | | | | |
| (1) Random (baseline) | 0.8910 | 0.9116 | 0.9355 | 0.9640 | 0.9967 |
| (2) Class-agnostic difficulty | 0.8910 | 0.9327 | 0.9818 | 0.9923 | 0.9967 |
| (3) Person-aware difficulty | 0.8910 | 0.9268 | 0.9804 | 0.9900 | 0.9967 |
| (4) Number of faces ($n$) | 0.8910 | 0.9591 | 0.9741 | 0.9912 | 0.9967 |
| (5) Average face size ($avg$) | 0.8910 | 0.9250 | 0.9565 | 0.9747 | 0.9967 |
| (6) $n/avg$ | 0.8910 | 0.9571 | 0.9776 | 0.9873 | 0.9967 |
| *Component* | *Time* (seconds) | | | | |
| (2, 3) Image difficulty | - | 0.05 | 0.05 | 0.05 | - |
| (4, 5, 6) Estimation of $n$, $avg$ | - | 0.28 | 0.28 | 0.28 | - |
| Face detection | 0.28 | 0.68 | 1.08 | 1.49 | 1.89 |
| Face detection + (2, 3) | 0.28 | 0.73 | 1.13 | 1.54 | 1.89 |
| Face detection + (4, 5, 6) | 0.28 | 0.75 | 1.22 | 1.70 | 1.89 |

## 4.3  Results and Discussion

Table 1 presents the AP scores and the processing times of MobileNet-SSD [13], $S^3$FD [30] and several combinations of the two face detectors, on the AFW data set. Different model combinations are obtained by varying the percentage of images processed by each detector. The table includes results starting with a 100%−0% split (equivalent with MobileNet-SSD [13] only), going through three intermediate splits (75%−25%, 50%−50%, 25%−75%) and ending with a 0%−100% split (equivalent with $S^3$FD [30] only). In the same manner, Table 2 shows the results for the same combinations of face detectors on the FDDB data set. While the results of various model combinations are listed on different columns in Tables 1 and 2, the results of various splitting strategies are listed on separate rows.

We first analyze the detection accuracy and the processing time of the two individual face detectors, namely MobileNet-SSD [13] and $S^3FD$ [30]. On AFW, $S^3FD$ reaches an AP score of 0.9967 in about 1.89 s per image, while on FDDB, it reaches a DistROC score of 0.9750 in about 1.17 s per image. MobileNet-SDD is more than four times faster, attaining an AP score of 0.8910 on AFW and a DistROC score of 0.8487 on FDDB, in just 0.28 s per image. We next analyze the average face detection times per image of the various model combinations on AFW. As expected, the time improves by about 19% when running MobileNet-SSD on 25% of the test set and $S^3FD$ on the rest of 75%. On the 50%−50% split, the processing time is nearly 40% shorter than the time required for processing the entire test set with $S^3FD$ only (0%−100% split). On the 75%−25% split, the processing time further improves by 63%. As the average time per image of $S^3FD$ is shorter on FDDB, the time improvements are close, but not as high. The improvements in terms of time are 15% for the 25%−75% split, 34% for the 50%−50% split, and 55% for the 75%−25% split. We note that unlike the random splitting strategy, the easy-versus-hard splitting strategies require additional processing time, either for computing the difficulty scores or for estimating the number of faces and their average size. The image difficulty predictors run in about 0.05 s per image, while the MobileNet-SSD detector (used for estimating the number of faces and their average size) runs in about 0.28 s per image. Hence, the extra time required by the two splitting strategies based on image difficulty is almost insignificant with respect to the total time required by the various combinations of face detectors. For instance, in the 50%−50% split with MobileNet-SSD and $S^3FD$, the difficulty predictors account for roughly 4% of the total processing time (0.05 out of 1.13 s per image) for an image taken from AFW.

Regarding our five easy-versus-hard strategies for combining face detectors, the empirical results indicate that the proposed splitting strategies give better performance than the random splitting strategy, on both data sets. Although using the number of faces or the average size of the faces as splitting criteria is better than using the random splitting strategy, it seems that combining the two measures into a single strategy $(n/avg)$ gives better and more stable results. On the AFW data set, the $n/avg$ strategy gives the best results for the 75%−25% split (0.9571), while the class-agnostic image difficulty provides the best results for the 50%−50% split (0.9818) and the 25%−75% split (0.9923). The highest improvements over the random strategy can be observed for the 50%−50% split. Indeed, the results for the 50%−50% split shown in Table 1 indicate that our strategy based on the class-agnostic image difficulty gives a performance boost of 4.63% (from 0.9355 to 0.9818) over the random splitting strategy. Remarkably, the AP of the MobileNet-SSD and $S^3FD$ 50%−50% combination is just 1.49% under the AP of the standalone $S^3FD$, while the processing time is reduced by almost half. On the FDDB data set, the $n/avg$ strategy gives the best DiscROC score for the 75%−25% split (0.9214), while the class-agnostic image difficulty provides the best DiscROC score for the 25%−75% split (0.9673). The $n/avg$ strategy and the class-agnostic image difficulty provide equally good DiscROC

a) Easy images according to our class-agnostic image difficulty predictor.

b) Bounding boxes of MobileNet-SSD for easy images.

c) Bounding boxes of S³FD for easy images.

d) Hard images according to our class-agnostic image difficulty predictor.

e) Bounding boxes of MobileNet-SSD for hard images.

f) Bounding boxes of S³FD for hard images.

**Fig. 1.** Examples of easy (top three rows) and hard images (bottom three rows) from AFW according to the class-agnostic image difficulty. For each set of images, the bounding boxes predicted by the MobileNet-SSD [13] and the S³FD [30] detectors are also presented. The correctly predicted bounding boxes are shown in green, while the wrongly predicted bounding boxes are shown in red. Best viewed in color. (Color figure online)

**Table 2.** Area under the discrete and the continuous ROC curves and time comparison between MobileNet-SSD [13], $S^3FD$ [30] and various combinations of the two face detectors on FDDB. The test data is partitioned based on a random split (baseline) or five easy-versus-hard splits given by: the class-agnostic image difficulty score, the person-aware image difficulty score, the number of faces ($n$), the average size of the faces ($avg$), and the number of faces divided by their average size ($n/avg$). For the random split, we report the scores over 5 runs to reduce bias. The reported times are measured on a computer with Intel Core i7 2.5 GHz CPU and 16 GB of RAM.

| | MobileNet-SSD (left) to $S^3FD$ (right) | | | | |
| | 100%−0% | 75%−25% | 50%−50% | 25%−75% | 0%−100% |
|---|---|---|---|---|---|
| *Splitting criterion* | *Area under the discrete ROC curve (DiscROC)* | | | | |
| (1) Random (baseline) | 0.8487 | 0.8775 | 0.9110 | 0.9432 | 0.9750 |
| (2) Class-agnostic difficulty | 0.8487 | 0.9131 | 0.9493 | 0.9673 | 0.9750 |
| (3) Person-aware difficulty | 0.8487 | 0.9139 | 0.9441 | 0.9659 | 0.9750 |
| (4) Number of faces ($n$) | 0.8487 | 0.9102 | 0.9379 | 0.9551 | 0.9750 |
| (5) Average face size ($avg$) | 0.8487 | 0.9187 | 0.9475 | 0.9638 | 0.9750 |
| (6) $n/avg$ | 0.8487 | 0.9214 | 0.9493 | 0.9626 | 0.9750 |
| | *Area under the continuous ROC curve (ContROC)* | | | | |
| (1) Random (baseline) | 0.7092 | 0.7425 | 0.7756 | 0.8096 | 0.8432 |
| (2) Class-agnostic difficulty | 0.7092 | 0.7729 | 0.8096 | 0.8302 | 0.8432 |
| (3) Person-aware difficulty | 0.7092 | 0.7737 | 0.8052 | 0.8291 | 0.8432 |
| (4) Number of faces ($n$) | 0.7092 | 0.7741 | 0.8032 | 0.8221 | 0.8432 |
| (5) Average face size ($avg$) | 0.7092 | 0.7776 | 0.8082 | 0.8281 | 0.8432 |
| (6) $n/avg$ | 0.7092 | 0.7824 | 0.8111 | 0.8275 | 0.8432 |
| *Component* | *Time* (seconds) | | | | |
| (2, 3) Image difficulty | - | 0.05 | 0.05 | 0.05 | - |
| (4, 5, 6) Estimation of $n$, $avg$ | - | 0.27 | 0.27 | 0.27 | - |
| Face detection | 0.27 | 0.51 | 0.73 | 0.94 | 1.17 |
| Face detection + (2, 3) | 0.27 | 0.56 | 0.78 | 0.99 | 1.17 |
| Face detection + (4, 5, 6) | 0.27 | 0.58 | 0.86 | 1.14 | 1.17 |

scores on the 50%−50% split (0.9493). As indicated in Table 2, the best DiscROC score for the 50%−50% split (0.9493) on FDDB is 3.83% above the DiscROC score (0.9110) of the baseline strategy. The ContROC scores are generally lower for all models, but the same patterns occur in the results presented in Table 2, i.e. the best results are provided either by the $n/avg$ strategy when MobileNet-SSD has a higher contribution in the combination or by the class-agnostic image difficulty when $S^3FD$ has to process more images.

To understand why our splitting strategy based on the class-agnostic image difficulty scores gives better results than the random splitting strategy, we randomly select a few easy examples (with less and bigger faces) and a few difficult examples from the AFW data set, and we display them in Fig. 1 along with the bounding boxes predicted by the MobileNet-SSD and the $S^3FD$ models. On the easy images, both detectors are able to detect the faces without any false positive detections, and the bounding boxes of the two detectors are almost identical. Nevertheless, we can perceive a lot more differences between MobileNet-SSD and $S^3FD$ on the hard images. In the left-most hard image, MobileNet-SSD is not able to detect the profile face of the man in the near right side of the image. In the second image, MobileNet-SSD wrongly detects the dog's face as a human face and it fails to detect the face of the boy sitting in the right. In the third image, MobileNet-SSD wrongly detects the small snowman's face sitting in the background and it fails to detect the face of the baby. In the right-most hard image, MobileNet-SSD fails to detect the face of the person looking down, which is difficult to detect because of the head pose. In the same image, MobileNet-SSD also fails to detect the profile face of the man in the far right side of the image. Remarkably, the $S^3FD$ detector is able to correctly detect all faces in the hard images illustrated in Fig. 1, without any false positive detections. We thus conclude that the difference between MobileNet-SSD and $S^3FD$ is only noticeable on the hard images. This could explain why our splitting strategy based on the class-agnostic image difficulty scores is effective in choosing an optimal trade-off between accuracy and speed.

## 5   Conclusion

In this paper, we have presented five easy-versus-hard strategies to obtain an optimal trade-off between accuracy and speed in face detection from images. Our strategies are based on dispatching each test image either to a fast and less accurate face detector or to a slow and more accurate face detector, according to the class-agnostic image difficulty score, the person-aware image difficulty score, the number of faces contained in the image, the average size of the faces, or the number of faces divided by their average size. We have conducted experiments using state-of-the-art face detectors such as $S^3FD$ [30] or MobileNet-SSD [13] on the AFW [31] and the FDDB [17] data sets. The empirical results indicate that using either one of the image difficulty predictors for splitting the test images compares favorably to a random split of the images. However, our other easy-versus-hard strategies also outperform the random split baseline. Since all the proposed splitting strategies are simple and easy to implement, they can be immediately adopted by anyone that needs a continuous accuracy versus speed trade-off optimization strategy in face detection.

**Acknowledgments.** The work of Petru Soviany was supported through project grant PN-III-P2-2.1-PED-2016-1842. The work of Radu Tudor Ionescu was supported through project grant PN-III-P1-1.1-PD-2016-0787.

# References

1. Bengio, Y., Louradour, J., Collobert, R., Weston, J.: Curriculum learning. In: Proceedings of ICML, pp. 41–48 (2009)
2. Chang, C.C., Lin, C.J.: Training $\nu$-support vector regression: theory and algorithms. Neural Comput. **14**, 1959–1977 (2002)
3. Chatfield, K., Simonyan, K., Vedaldi, A., Zisserman, A.: Return of the devil in the details: delving deep into convolutional nets. In: Proceedings of BMVC (2014)
4. Chen, D., Hua, G., Wen, F., Sun, J.: Supervised transformer network for efficient face detection. In: Leibe, B., Matas, J., Sebe, N., Welling, M. (eds.) ECCV 2016. LNCS, vol. 9909, pp. 122–138. Springer, Cham (2016). https://doi.org/10.1007/978-3-319-46454-1_8
5. Everingham, M., Van Gool, L., Williams, C., Winn, J., Zisserman, A.: The PASCAL Visual Object Classes Challenge 2012 Results (2012)
6. Everingham, M., Eslami, S.M., Gool, L., Williams, C.K., Winn, J., Zisserman, A.: The Pascal visual object classes challenge: a retrospective. Int. J. Comput. Vis. **111**(1), 98–136 (2015)
7. Everingham, M., van Gool, L., Williams, C.K., Winn, J., Zisserman, A.: The Pascal visual object classes (VOC) challenge. Int. J. Comput. Vis. **88**(2), 303–338 (2010)
8. Felzenszwalb, P.F., Girshick, R.B., McAllester, D., Ramanan, D.: Object detection with discriminatively trained part-based models. IEEE Trans. Pattern Anal. Mach. Intell. **32**(9), 1627–1645 (2010)
9. Georgescu, M., Ionescu, R.T., Popescu, M.: Local learning with deep and handcrafted features for facial expression recognition. arXiv preprint arXiv:1804.10892 (2018)
10. Girshick, R.: Fast R-CNN. In: Proceedings of ICCV, pp. 1440–1448 (2015)
11. He, K., Zhang, X., Ren, S., Sun, J.: Deep residual learning for image recognition. In: Proceedings of CVPR, pp. 770–778 (2016)
12. He, K., Gkioxari, G., Dollár, P., Girshick, R.: Mask R-CNN. In: Proceedings of ICCV, pp. 2961–2969 (2017)
13. Howard, A.G., et al.: MobileNets: efficient convolutional neural networks for mobile vision applications. arXiv preprint arXiv:1704.04861 (2017)
14. Huang, J., et al.: Speed/accuracy trade-offs for modern convolutional object detectors. In: Proceedings of CVPR, pp. 7310–7319 (2017)
15. Ionescu, R., Alexe, B., Leordeanu, M., Popescu, M., Papadopoulos, D.P., Ferrari, V.: How hard can it be? estimating the difficulty of visual search in an image. In: Proceedings of CVPR, pp. 2157–2166 (2016)
16. Ionescu, R.T., Popescu, M., Grozea, C.: Local learning to improve bag of visual words model for facial expression recognition. In: Workshop on Challenges in Representation Learning, ICML (2013)
17. Jain, V., Learned-Miller, E.: FDDB: a benchmark for face detection in unconstrained settings. Technical report UM-CS-2010-009, University of Massachusetts, Amherst (2010)
18. Jiang, H., Learned-Miller, E.: Face detection with the faster R-CNN. In: Proceedings of FG, pp. 650–657 (2017)
19. Krizhevsky, A., Sutskever, I., Hinton, G.E.: ImageNet classification with deep convolutional neural networks. In: Proceedings of NIPS, pp. 1106–1114 (2012)
20. Li, H., Lin, Z., Shen, X., Brandt, J., Hua, G.: A convolutional neural network cascade for face detection. In: Proceedings of CVPR, pp. 5325–5334 (2015)

21. Liu, W., et al.: SSD: single shot multibox detector. In: Leibe, B., Matas, J., Sebe, N., Welling, M. (eds.) ECCV 2016. LNCS, vol. 9905, pp. 21–37. Springer, Cham (2016). https://doi.org/10.1007/978-3-319-46448-0_2

22. Parkhi, O.M., Vedaldi, A., Zisserman, A., et al.: Deep face recognition. In: Proceedings of BMVC, pp. 6–17 (2015)

23. Qin, H., Yan, J., Li, X., Hu, X.: Joint training of cascaded CNN for face detection. In: Proceedings of CVPR, pp. 3456–3465 (2016)

24. Ren, S., He, K., Girshick, R., Sun, J.: Faster R-CNN: towards real-time object detection with region proposal networks. In: Proceedings of NIPS, pp. 91–99 (2015)

25. Russakovsky, O., et al.: ImageNet large scale visual recognition challenge. Int. J. Comput. Vis. **115**, 211–252 (2015)

26. Soviany, P., Ionescu, R.T.: Optimizing the trade-off between single-stage and two-stage deep object detectors using image difficulty prediction. In: Proceedings of SYNASC (2018)

27. Viola, P., Jones, M.J.: Robust real-time face detection. Int. J. Comput. Vis. **57**(2), 137–154 (2004)

28. Yang, S., Luo, P., Loy, C.C., Tang, X.: From facial parts responses to face detection: a deep learning approach. In: Proceedings of ICCV, pp. 3676–3684 (2015)

29. Yang, S., Luo, P., Loy, C.C., Tang, X.: WIDER FACE: a face detection benchmark. In: Proceedings of CVPR, pp. 5525–5533 (2016)

30. Zhang, S., Zhu, X., Lei, Z., Shi, H., Wang, X., Li, S.Z.: $S^3$FD: single shot scale-invariant face detector. In: Proceedings of ICCV, pp. 192–201 (2017)

31. Zhu, X., Ramanan, D.: Face detection, pose estimation, and landmark localization in the wild. In: Proceedings of CVPR, pp. 2879–2886 (2012)

# Multi-Dimensional Optical Flow Embedded Genetic Programming for Anomaly Detection in Crowded Scenes

Zeyu Mi[1(✉)], Lin Shang[1(✉)], and Bing Xue[2(✉)]

[1] State Key Laboratory for Novel Software Technology, Department of Computer Science and Technology, Nanjing University, Nanjing, China
mizeyu@smail.nju.edu.cn, shanglin@nju.edu.cn
[2] Victoria University of Wellington, Wellington, New Zealand
Bing.Xue@ecs.vuw.ac.nz

**Abstract.** Anomaly detection is an important issue in the fields of video behavior analysis and computer vision. Genetic Programming (GP) performs well in those applications. Histogram of Oriented Optical Flow (HOOF), which is based on Optical Flow (OF), is a significant method to extract features of frames in videos and has been widely used in computer vision. However, OF may produce a large number of features that will lead to high computational cost and poor performance. Moreover, HOOF accumulates optical flow values for moving objects in a region, so motion information of these objects may not be well represented. Especially in crowding scenes, common anomaly detection methods usually have a poor performance. Aiming to address the above issues, we propose a new feature called Multi-Dimensional Optical Flow (MDOF) and a new method GP-MDOF which embeds HOOF features and optimizes the structure of GP, and makes better use of the change information between consecutive frames. In this paper, we apply GP-MDOF to classify frames into abnormality or not. Experimental evaluations are conducted on the public dataset UMN, UCSD Ped1 and Ped2. Our experimental results indicate that the proposed feature extraction method MDOF and the new method GP-MDOF can outperform the popular techniques such as OF and Social Force Model in anomaly detection in crowded scenes.

**Keywords:** Anomaly detection · Optical Flow · Genetic Programming

## 1 Introduction

Problems in densely crowded environments are increasing because of the large population and high diversity of human activities. Extreme crowded scenes have an excessive number of individuals and their activities, which bring challenge for human behaviors analysis. Crowded scene analysis [8,12,21] is an important branch in computer vision and has been attracting increasing attention in recent years. It focuses on numbers of moving objects in the crowded scenes. More and

© Springer Nature Switzerland AG 2018
L. Cheng et al. (Eds.): ICONIP 2018, LNCS 11302, pp. 486–497, 2018.
https://doi.org/10.1007/978-3-030-04179-3_43

more researchers constructed models and analyzed crowded behaviors depending on data from surveillance cameras equipped in public areas [4, 9, 14, 15, 24, 25].

Various approaches have been proposed for anomaly detection. They can be divided into two kinds according to the type of scene representation. One is based on trajectory, i.e. tracking each object in the scene and constructing models for the detected objects. Dee et al. [6] proposed a method for determining whether a moving person was normal or abnormal according to the fact that people usually moved along regular paths. In [17], Robertson et al. added additional information to the tracking-based information for tennis analysis in videos. This kind of approaches often depend on the accuracy of tracking, while it is infeasible to track objects in densely crowded scenes.

Many researchers have turned to other alternative approaches based on motion representation, of which the most popular one is Optical Flow (OF) [11], more robust than trajectory-based approaches. Some modified approaches based on OF have been proposed, such as Histogram of gradient (HOG) [5], Histogram of Optical Flow (HOOF) [2], and Multi-Histogram of Optical Flow (MHOF) [25].

Genetic Programming (GP) [10] is an evolutionary computation method and has many applications. GP aims at automatically building programs to solve problems without using any domain knowledge. Zhang et al. [3] used GP for various problems in object detection. GP was employed by Howard et al. [7] to detect vehicles in different indoor and outdoor scenes such as factory, urban and rural areas. A GP-based motion detection method was introduced by Song et al. in [19] and have been widely utilized in various real-world scenarios [16].

In this paper we propose a new method (named GP-MDOF) considering how to extract representative features, making better use of the change information between consecutive frames and how to construct an effective model. Firstly, we propose MDOF, which can extract frame-level features, and contains spatial-temporal information in videos and comparatively fewer dimensions of a vector. Then we use MDOF embedded GP to learn a model to generate an anomaly detector in which a training sample is labeled as negative or positive. Finally, on the public dataset UMN, UCSD Ped1 and Ped2, we validate our approach. Experimental results indicate that the proposed approach (GP-MDOF) performs better than traditional methods in anomaly detection in crowded scenes.

## 2   Related Work

### 2.1   Genetic Programming

In anomaly detection, a program tree in Genetic Programming (GP) [10] is regarded as a classifier [20], or a detector, in which features of small sub-regions from instances are the input, and the class it belongs to is generated by the output. Figure 1 shows a simple GP program tree representing the detector for anomaly detection.

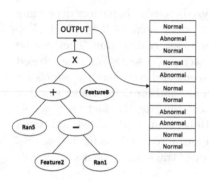

**Fig. 1.** An example of program tree in GP.

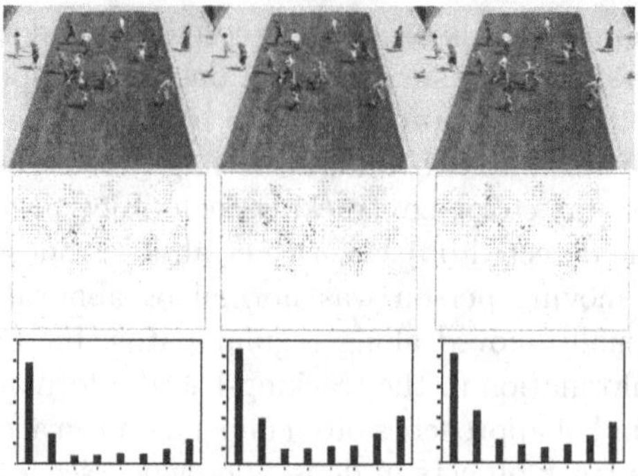

**Fig. 2.** The process of extraction for the Histogram of Oriented Optical Flow.

## 2.2    Histogram of Oriented Optical Flow

Histogram of Oriented Optical Flow (HOOF) [2], based on OF, is a non-parameter estimation method to perform statistics on optical flow values according to the corresponding directions. Figure 2 shows the HOOF extraction process. To obtain optical flow feature in a region, optical flow directions are divided into 8 sectors.

## 3    Our Approach for Anomaly Detection in Crowed Scenes

In this section, we will present the details of our proposed approach based on MDOF embedded GP. The overview of our proposed method is illustrated in Fig. 3.

There are two main steps in our method:

- **Evolution Phase.** Given the dataset, we divide it into a training set and a testing set with manual labels and then features are extracted from videos using our method. After that, there are two stages: Evolution and Evaluation. In the process of evolution, GP operations will be employed to create a new generation. Then the labels of training samples are used as the calculations of fitness to evaluate the generated detector and one of the best-performing program is then selected as the final detector.
- **Application Phase.** The new-coming video streams are input into the anomaly detector trained from GP evolution above, and we can evaluate and achieve their real labels.

### 3.1    MDOF and Feature Extraction

Traditional HOOF accumulates optical flow values for moving objects in a region and motion information of these objects may not be represented well. So we propose a new feature extraction approach based on MDOF in this paper.

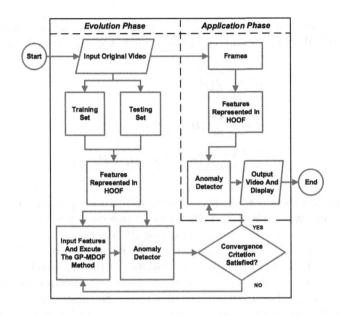

**Fig. 3.** Overview of our methods.

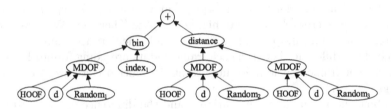

**Fig. 4.** Structure of the GP-MDOF method.

The structure of the GP-MDOF method is showed in Fig. 4. The steps are as follows:

1. Perform traditional HOOF for each frame. The details have been introduced in Sect. 2 and the HOOF is obtained the form of:

$$H = (G_1, G_2, ..., G_b) \in \mathbb{R}^{1 \times b}. \tag{1}$$

2. Compute the mean value, variance and the difference for a frame, after achieving the HOOF of this frame and its previously consecutive $n$ frames.

$$\mu = \frac{1}{n} \sum_{i=1}^{n} H_i,$$

$$s = \frac{1}{n} \sum_{i=1}^{n} (H_i - \mu)^2, \tag{2}$$

$$v = \frac{\sum_{i=1}^{n-1} (n-i) \times (|H_i - H_{i+1}|)}{n-1},$$

where $H_i$ indicates the $i_{th}$ histogram of HOOF. To be specific, it indicates the current frame when $n$ is set to 1, the former frame when $n$ is set to 2, and so on. After that, we have $\mu = (\mu_1, \mu_2, ..., \mu_b) \in \mathbb{R}^{1 \times b}$, $s = (s_1, s_2, ..., s_b) \in \mathbb{R}^{1 \times b}$, $v = (v_1, v_2, ..., v_b) \in \mathbb{R}^{1 \times b}$, where $\mu_k$ indicates the mean value of the $k_{th}$ direction interval of OF value, $s_k$ indicates the variance of the $k_{th}$ direction interval of OF value, and $v_k$ indicates the difference of the $k_{th}$ direction interval of OF value. The larger the variance of one direction is, the more evidently the speed of OF of this direction varies, in other words, rather larger the accelerated speed is. The accelerated speed of the $k_{th}$ direction $a_k \propto s_k, k = 1, 2, 3, ..., b$. Short-term motion clues are important but insufficient to represent crowed events because crowed events are highly complex. Therefore, We consider the current frame and its previous n frames, and calculate the sum of absolute value of the difference between each frame and its preceding frame. The closer the current frame is, the more important the motion clues are, so we add a weight: $n - i$.

3. Combine the mean value, the variance and the difference of each histogram of HOOF we can convert the above with original HOOF into a new frame-level feature called MDOF. Multi-Dimensional $(M)$ of Optical Flow is in the form of:

$$M = (H, \mu, s, v). \tag{3}$$

MDOF can describe motion information in videos in a better way of integrating the statistical features of optical flow distribution in traditional HOOF with the features of accelerated speed representing an optical flow value in the corresponding direction. It is better for MDOF embedded GP to represent motion features. Figure 5 illustrates the process of MDOF feature extraction.

### 3.2   MDOF Embedded GP Evolution

**Function Set.** Parameter settings are shown in Table 1. The inputs represent the values and their types. The outputs represent the returned value, which will be achieved after the operators deal with inputs.

In our method, we design arithmetic operators, *disatance*, *bin* and *MDOF* nodes. Arithmetic operators contain addition $(+)$, subtraction $(-)$, multiplication $(\times)$ and division $(\div)$.

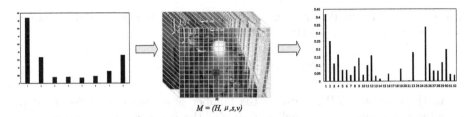

$M = (H, \mu, s, v)$

**Fig. 5.** Representation of MDOF. The process of transforming from HOOF to MDOF.

**Table 1.** Function set

| Name | Output | Input | Details |
|------|--------|-------|---------|
| $+, -, \times, \div$ | Double | Double, double | Arithmetic operators |
| $MDOF$ | Histogram | HOOF, d | Perform the MDOF algorithm and return a histogram |
| $distance$ | Double | Histogram, histogram | Return the distance between two histograms |
| $bin$ | Double | Histogram, index | Return the value of the specified index |

$MDOF$ is the most important new function in our method. $MDOF$ uses the $HOOF$, d (including $\mu$, $s$, $v$) value from the terminal nodes and uses random weights to evolve $MDOF$. And $distance$ is the Euclidean distance between two histograms.

**Terminal Set.** Table 2 shows the terminal set. Apart from the terminals that represent the $HOOF$ feature of a video frame, the others are values in $Rand$, which are random numbers between 0 and 1 and serve as coefficients in normal functions and the weight affecting change between survival frames. In Table 2, d represents the histogram of $(\mu, s, v)$, and $index$ is a random integer value in range of $[0, 31]$, which indicates the bin index of a histogram.

**Table 2.** Terminal set

| Terminal | Type | Value |
|----------|------|-------|
| $Rand$ | Double | Random double in the half-closed interval $[0, 1)$ |
| $HOOF$ | Histogram | The features being represented as HOOF |
| $d$ | Histogram | $\mu$, $s$, $v$ represented as histogram of optical flow |
| $index$ | Integer | Random integer value in $[0, 31]$, which represents the bin index of a histogram |

**Fitness.** In our experiments, the classification accuracy for training data is regarded as the fitness, as calculated in Eq. (4).

$$Fitness = \frac{TP + TN}{NUM} \times 100. \tag{4}$$

In the equation, $TP$ represents the number of correctly classified positive samples which means abnormality class. On the contrary, $TN$ represents the number of correctly classified negative samples which means non-abnormality class. $NUM$ is the total number of samples in the training set.

## 4    Experiments

In this section, We verify the effectiveness of our approach on three real world datasets.

**Fig. 6.** Kinds of different crowded scenes in the UMN dataset: Normal (left) and abnormal (right).

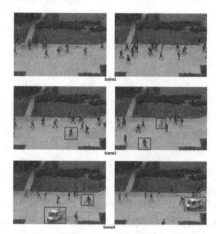

**Fig. 7.** Normal and Abnormal events (in red boxes) in UCSD dataset. (Color figure online)

### 4.1    Dataset

The **UMN**[1] dataset includes 11 videos which have 7740 frames, and their resolutions are all $320 \times 240$. As is shown in Fig. 6, there are three kinds of different crowded scenes in these videos consisting lawn, hall and square, represented by Scene1, Scene2, Scene3. And the left scenes are all normal scenes, while the right scenes are all abnormal scenes. To be specific, they are composed of 1450, 4415

---

[1] http://mha.cs.umn.edu/Movies/Crowd-Activity-All.avi.

and 2145 frames respectively. In this dataset, walking normally is regarded as normal behavior, while the sudden scatter or panic running away when people encounter emergency is regarded as abnormal behavior.

The **UCSD Pedestrian**[2] dataset includes 2 subsets: Ped1 and Ped2. There are 34 training and 36 testing image clips in Ped1 dataset, respectively. Each clip has 200 frames, and their resolutions are all 238 × 158. The UCSD Ped2 dataset includes 16 training and 12 testing image clips with about 120 to 180 frames each folder. Ped1 and Ped2 have the same definition of anomaly. And we evaluate our method on these two datasets separately. Both normal and abnormal events of this video are shown in Fig. 7.

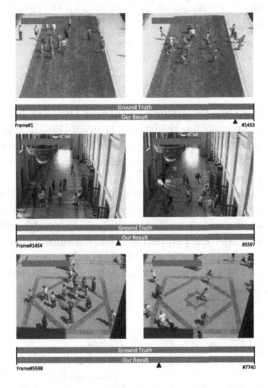

**Fig. 8.** The contrast between our results and the ground truth in three videos of UMN dataset. (Color figure online)

## 4.2 Results on the UMN Data Set

**Global Abnormal Event Detection.** The results based on our method in all three videos are shown in Fig. 8. Each row represents the results for a video in the dataset. For each video, the ground truth bar represents the true label of each frame and the detection bar represents the detection labels of each frame,

---

[2] http://www.svcl.ucsd.edu/projects/anomaly/dataset.htm.

respectively. Normal frames are represented by green while abnormal frames corresponds to red. The black triangle indicates the time of the displayed abnormal frames. We can see that we have accurately predicted most of the anomaly behaviors. It is worth mentioning that the training set of this classifier uses only one of the three scenes. And it still have a good performance in other two different scenes.

**Table 3.** UMN dataset: comparison (AUC) of our proposed method with the state-of-the-art methods

| Scene | AUC | | | | | | |
|---|---|---|---|---|---|---|---|
| | Ours | HOFO [23] | Sparse [4] | STCOG [18] | NN [4] | Optical Flow [1] | SF [15] |
| Lawn | **99.71%** | 98.45% | 99.5% | 93.62% | 93% | 84% | 96% |
| Indoor | **98.72%** | 90.37% | 97.5% | 77.59% | | | |
| Plaza | 97.22% | 98.15% | 96.4% | 96.61% | | | |

**Comparisons with the Traditional State-of-the-Art Methods.** We also compare our method with other traditional state-of-the-art approaches: Optical Flow [1], Social Force Model [15], NN [4], STCOG [18], Sparse [4], and HOFO [23] in all the three scenes on the same dataset. The experimental results and comparison are represented in Table 3.

As shown in Table 3, the performance of our method in Scene3 is close to that of the HOFO and is better than that of the others. However, HOFO has a poor performance in Scene2 while the proposed method perform well in all the three scenes. Our method can extract the motion information between the consecutive frames well, Then it is more stable in different scenes. Furthermore, the proposed method is far superior to all other methods including the Sparse in the anomaly detection task.

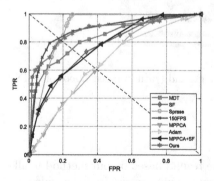

**Fig. 9.** Frame-level ROC curve

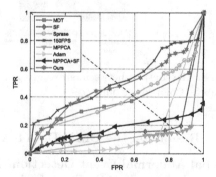

**Fig. 10.** Pixel-level ROC curve

### 4.3  Results on the UCSD Data Set

On the UCSD dataset, we compare our approach with several state-of-the-art methods. Pixel-level evaluation can measure the accuracy of abnormal localization. Usually, if at least 40% of the ground truth abnormal pixels are predicted correctly, the frame is counted as a true positive. In order to achieve more comprehensive results, we carry out both a frame-level and pixel-level evaluation.

**Table 4.** UCSD dataset: comparison (AUC and EER) of our proposed method with the state-of-the-art methods

| Method | Ped1 (frame) | | Ped2 (pixel) | | Ped2 (frame) | |
|---|---|---|---|---|---|---|
| | EER | AUC | EER | AUC | EER | AUC |
| MPPCA [9] | 40% | 59.0% | 81% | 20.5% | 30% | 69.3% |
| Social force [15] | 31% | 67.5% | 79% | 19.7% | 42% | 55.6% |
| Social force+MPPCA [14] | 32% | 66.8% | 71% | 21.3% | 36% | 61.3% |
| Sparse reconstruction [4] | 19% | 86.0% | 54% | 45.3 | - | - |
| Unmasking [22] | - | 68.4% | - | 52.4% | - | 82.2% |
| Detection at 150 FPS [13] | 15% | 91.8 | 43% | 63.8 | - | - |
| **Ours** | **18%** | **87.7%** | **44%** | **61.5%** | **17%** | **90%** |

**Comparisons with the Traditional State-of-the-Art Methods.** A quantitative comparison of different methods using both EER (Equal Error Rate) and AUC (Area Under Curve) is shown in Table 4. A frame-level ROC curve is reported in Fig. 9 and a pixel-level ROC curve is reported in Fig. 10 by altering the detection threshold. The results show that proposed approach overcomes most previous methods from both the frame-level and the pixel-level evaluation and have more stable performances on both Ped1 and Ped2.

## 5  Conclusions

In this paper, we proposed a new Multi-Dimensional Optical Flow based anomaly detection method in crowded scenes embedded genetic programming named GP-MDOF. We carried out several experimental evaluations, considering three publicly video anomaly detection datasets (UMN, USCD Ped1 and USCD Ped2) which are widely used. Firstly, we develop a new feature MDOF, which has better extraction and use of motion and change information between consecutive frames. The GP-MDOF method is then proposed based on GP and MDOF. GP-MDOF has achieved better and more stable performance in different scenes. Furthermore, we optimize the structure of GP, making it have more space for improvement in anomaly detection. Finally, we evaluated the anomaly detector using the testing data and the best-performing program can still have a good performance in unseen videos.

**Acknowledgement.** This work is supported by the National Natural Science Foundation of China (No. 61672276) and Natural Science Foundation of Jiangsu, China (BK20161406).

# References

1. Bertini, M., Del Bimbo, A., Seidenari, L.: Scene and crowd behaviour analysis with local space-time descriptors. In: 2012 5th International Symposium on Communications Control and Signal Processing (ISCCSP), pp. 1–6. IEEE (2012)
2. Chaudhry, R., Ravichandran, A., Hager, G., Vidal, R.: Histograms of oriented optical flow and Binet-Cauchy kernels on nonlinear dynamical systems for the recognition of human actions. In: IEEE Conference on Computer Vision and Pattern Recognition, CVPR 2009, pp. 1932–1939. IEEE (2009)
3. Chin, B., Zhang, M.: Object detection using neural networks and genetic programming. In: Giacobini, M., et al. (eds.) EvoWorkshops 2008. LNCS, vol. 4974, pp. 335–340. Springer, Heidelberg (2008). https://doi.org/10.1007/978-3-540-78761-7_34
4. Cong, Y., Yuan, J., Liu, J.: Sparse reconstruction cost for abnormal event detection. In: 2011 IEEE Conference on Computer Vision and Pattern Recognition (CVPR), pp. 3449–3456. IEEE (2011)
5. Dalal, N., Triggs, B.: Histograms of oriented gradients for human detection. In: IEEE Computer Society Conference on Computer Vision and Pattern Recognition, CVPR 2005, vol. 1, pp. 886–893. IEEE (2005)
6. Dee, H.M., Hogg, D.C.: Detecting inexplicable behaviour. In: BMVC, pp. 1–10 (2004)
7. Howard, D., Roberts, S.C., Ryan, C.: Pragmatic genetic programming strategy for the problem of vehicle detection in airborne reconnaissance. Pattern Recognit. Lett. **27**(11), 1275–1288 (2006)
8. Kim, B., Lee, G.-G., Yoon, J.-Y., Kim, J.-J., Kim, W.-Y.: A method of counting pedestrians in crowded scenes. In: Huang, D.-S., Wunsch, D.C., Levine, D.S., Jo, K.-H. (eds.) ICIC 2008. LNCS (LNAI), vol. 5227, pp. 1117–1126. Springer, Heidelberg (2008). https://doi.org/10.1007/978-3-540-85984-0_134
9. Kim, J., Grauman, K.: Observe locally, infer globally: a space-time MRF for detecting abnormal activities with incremental updates. In: IEEE Conference on Computer Vision and Pattern Recognition, CVPR 2009, pp. 2921–2928. IEEE (2009)
10. Koza, J.R.: Genetic programming as a means for programming computers by natural selection. Stat. Comput. **4**(2), 87–112 (1994)
11. Kratz, L., Nishino, K.: Anomaly detection in extremely crowded scenes using spatio-temporal motion pattern models. In: IEEE Conference on Computer Vision and Pattern Recognition, CVPR 2009, pp. 1446–1453. IEEE (2009)
12. Li, W., Mahadevan, V., Vasconcelos, N.: Anomaly detection and localization in crowded scenes. IEEE Trans. Pattern Anal. Mach. Intell. **36**(1), 18–32 (2014)
13. Lu, C., Shi, J., Jia, J.: Abnormal event detection at 150 fps in MATLAB. In: 2013 IEEE International Conference on Computer Vision (ICCV), pp. 2720–2727. IEEE (2013)
14. Mahadevan, V., Li, W., Bhalodia, V., Vasconcelos, N.: Anomaly detection in crowded scenes. In: 2010 IEEE Conference on Computer Vision and Pattern Recognition (CVPR), pp. 1975–1981. IEEE (2010)

15. Mehran, R., Oyama, A., Shah, M.: Abnormal crowd behavior detection using social force model. In: IEEE Conference on Computer Vision and Pattern Recognition, CVPR 2009, pp. 935–942. IEEE (2009)

16. Pinto, B., Song, A.: Detecting motion from noisy scenes using genetic programming. In: 24th International Conference on Image and Vision Computing New Zealand, IVCNZ 2009, pp. 322–327. IEEE (2009)

17. Robertson, N., Reid, I.: Behaviour understanding in video: a combined method. In: Tenth IEEE International Conference on Computer Vision, ICCV 2005, vol. 1, pp. 808–815. IEEE (2005)

18. Shi, Y., Gao, Y., Wang, R.: Real-time abnormal event detection in complicated scenes. In: 2010 20th International Conference on Pattern Recognition (ICPR), pp. 3653–3656. IEEE (2010)

19. Song, A., Fang, D.: Robust method of detecting moving objects in videos evolved by genetic programming. In: Proceedings of the 10th Annual Conference on Genetic and Evolutionary Computation, pp. 1649–1656. ACM (2008)

20. Song, A., Zhang, M.: Genetic programming for detecting target motions. Connect. Sci. 24(2–3), 117–141 (2012)

21. Sun, X., Yao, H., Ji, R., Liu, X., Xu, P.: Unsupervised fast anomaly detection in crowds. In: Proceedings of the 19th ACM International Conference on Multimedia, pp. 1469–1472. ACM (2011)

22. Tudor Ionescu, R., Smeureanu, S., Alexe, B., Popescu, M.: Unmasking the abnormal events in video. In: Proceedings of the IEEE Conference on Computer Vision and Pattern Recognition, pp. 2895–2903 (2017)

23. Wang, T., Snoussi, H.: Detection of abnormal visual events via global optical flow orientation histogram. IEEE Trans. Inf. Forensics Secur. 9(6), 988–998 (2014)

24. Xu, D., Yan, Y., Ricci, E., Sebe, N.: Detecting anomalous events in videos by learning deep representations of appearance and motion. Comput. Vis. Image Underst. 156, 117–127 (2017)

25. Yang, C., Yuan, J., Liu, J.: Abnormal event detection in crowded scenes using sparse representation. Pattern Recogn. 46(7), 1851–1864 (2013)

# Robust Regression with Nonconvex Schatten $p$-Norm Minimization

Deyu Zeng, Ming Yin$^{(\boxtimes)}$, Shengli Xie, and Zongze Wu

School of Automation, Guangdong University of Technology,
Guangzhou 510006, China
deyuzeng@hotmail.com, {yiming,shlxie,zzwu}@gdut.edu.cn

**Abstract.** Linear regression classification known as a classical supervised learning algorithm has been widely used in face recognition, image alignment, pose estimation and so on. Unfortunately, this algorithm often suffers from significant degradation in prediction accuracy when the presence of outlier or gross errors in the training data. To handle this issue, a novel robust regression is proposed by exploiting a nonconvex schatten $p$-norm minimization in this paper. Concretely, the $\ell_p$-norm and nonconvex schatten $p$-norm are adopted to better approximate the $\ell_0$-norm and rank minimization problem, respectively. Experimental results on several datasets demonstrate the superiority of our proposed model on classification, against to the state-of-the-art methods.

**Keywords:** Robust regression
Robust Principal Component Analysis · Schatten $p$-norm
Face recognition

## 1 Introduction

Supervised learning methods, such as Linear Regression (LR) and Support Vector Machine (SVM), have been widely used in face recognition [13], image alignment [5], pose estimation [10] and so on. Suppose there are $n$ $d$-dimensional data belonging to $k$ classes, Linear Regression aims to learn a function that predicts the unknown by the linear combination of features, i.e.,

$$f(\mathbf{x_i}) = b_0 + b_1 x_{i1} + b_2 x_{i2} + ... + b_d,$$

where $\mathbf{x}_i = \{x_{i1}; x_{i2}; ...; x_{id}\}$ is the $i$-th data of input data $X$ and $x_{ij}$ is the $j$-th feature of $x_i$. Rewrite into matrix form as follows,

$$f(X) = B[X; \mathbf{1}^T],$$

where $X \in \Re^{d \times n}$, $B \in \Re^{k \times (d+1)}$ and $\mathbf{1}$ is a all 1 vector with suitable dimension. Mathematically, linear regression is formulated by the following optimization problem.

$$\min_B \|Y - B[X; \mathbf{1}^T]\|_F^2, \tag{1}$$

© Springer Nature Switzerland AG 2018
L. Cheng et al. (Eds.): ICONIP 2018, LNCS 11302, pp. 498–508, 2018.
https://doi.org/10.1007/978-3-030-04179-3_44

where $Y \in \Re^{k \times n}$ is ground truth of $X$ with $k$ possible labels and $\|\cdot\|_F$ is Frobenius norm of a matrix defined by $\|A\|_F^2 = \sum_{ij} a_{ij}^2$. Generally, Eq. 1 can seek the optimal solution when the error of $Y - BX$ is normally distributed. Moreover, the input training data $X$ is usually assumed to be noise free in standard linear regression. However, in fact, data are often contaminated by the varying noise and interference. This scenario is very common for image datasets, due to the light and other reasons. As such, this will greatly degrade the prediction accuracy.

To handle this issue, plenty of the algorithms have been developed recently. In order to deal with unbound covariate corruptions robustly, Chen *et al.* [11] proposed a novel linear regression algorithm that replaces the standard vector inner product with a trimmed one. In particular, they utilized a pre-screening procedure to remove possible outliers based on their magnitude and then the standard linear regression is applied. In paper [7], the authors consider a novel robust linear regression with $\ell_0$-norm regularization for robust outlier support identification (AROSI). By combining manifold regularization, Wu and Souvenir [6] proposed a new robust regression method to handle the ordered (e.g., real-valued, ordinal) label denoising problem. Different from the above pre-screening strategy, Huang *et al.* [2] presented a robust regression with an effective convex formulation that exploited the recent advances on rank minimization [9] in which the noisy input data are directly used in training model. Specifically, the authors adopted the trace norm (or nuclear norm) as the convex surrogate of the rank and $\ell_1$-norm to relax $\ell_0$-norm respectively. However, these relaxations may result in the sub-optimal solution of the model. Concretely, this relaxation may deviate the solution from the original solution. Thus, Nie *et al.* [3] proposed to use schatten $p$-norm and $\ell_p$-norm to better approximate the rank constraint and $\ell_0$-norm respectively, leading to the robustness of the approach to outliers.

In this paper, we focus on the errors in input data instead of the labelling errors. Inspired by the above work, we propose a novel robust regression by solving a nonconvex schatten $p$-norm minimization problem. In summary, the main contributions of this paper are two-fold.

- We formulate a novel objective function of robust regression by introducing both schatten $p$-norm and $p$-norm.
- The alternating direction method of multipliers (ADMM) method is utilized to optimize the proposed problem, where the convergence of the algorithm is guaranteed.

The remains of the paper is organized as follows. In Sect. 2, we briefly review the robust regression model and introduce schatten $p$-norm and $\ell_p$-norm. Section 3 presents a novel robust regression method using nonconvex constraints and its optimization by ADMM. Section 4 gives experimental results on several datasets to show the efficiency. Finally, the conclusion is drawn in Sect. 5.

## 2    Related Work

### 2.1    Robust Regression

In order to deal with the noisy input data used in training model, the robust regression [2] was recently proposed based on the advance of Robust Principal Component Analysis (RPCA). RPCA aims at recovering a clean part $D$ from corrupted data $X = D + E$ by the low-rank property, where $E$ is assumed to be sparsely supported [8]. Figure 1 shows corrupted $X$ can be recovered to a clean $D$ and error $E$ via low-rank and sparsity constraints respectively. In combination with RPCA, the author presented the following model, namely RR [2],

$$\min_{B,D,E} \frac{\eta}{2}\|Y - B\hat{D}\|_F^2 + \text{rank}(D) + \lambda\|E\|_0, \text{ s.t. } X = D + E, \hat{D} = [D; \mathbf{1}^T], \quad (2)$$

where $B \in \Re^{k \times (d_x+1)}$ is the mapping matrix (the extra dimension is for the regression bias term) and $\|\cdot\|_0$ is $\ell_0$ norm[1] of the matrix. $\eta$ and $\lambda$ are the penalty parameters for balancing the effects of all terms.

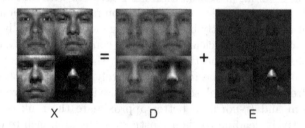

**Fig. 1.** $X$ shows the data that contains some gross noise and it can be divided into clean $D$ and interference $E$ via low-rank and sparse constraints.

However, solving the above problem is not easy as the rank minimization problem is known as NP-hard. Thus the author relaxes these rank and cardinality operators respectively to their convex surrogates, i.e., the trace norm[2] (or nuclear norm) and the $\ell_1$-norm[3]. Then the problem 2 is equivalent to the followings.

$$\min_{B,D,E} \frac{\eta}{2}\|Y - B\hat{D}\|_F^2 + \|D\|_* + \lambda\|E\|_1, \text{ s.t. } X = D + E, \hat{D} = [D; \mathbf{1}^T]. \quad (3)$$

It is noteworthy that RR is not equivalent to training linear regression using clean data that is firstly processed by RPCA, since RR recovered the clean part from the input data $X = D + E$ in a *supervised* manner. Thus, the $D$ of RR will preserve the subspace of $X$ that is maximally correlated with $Y$.

---

[1] $\ell_0$-norm of a matrix is the number of non-zero elements of a matrix.

[2] Trace norm or nuclear norm of a matrix $A \in \Re^{m \times n}$ denotes the sum of the singular values of the matrix. $\|A\|_* = \sum_{i=1}^{\min\{m,n\}} \sigma_i$, where $\sigma_i$ is the singular value of A.

[3] $\ell_1$ norm of a matrix is defined by the sum of absolute value of all elements of the matrix. $\|A\|_1 = \sum_{i=1}^m \sum_{j=1}^n |a_{ij}|$.

## 2.2 Schatten $p$-Norm and $\ell_p$-Norm

Although the trace norm and the $\ell_1$-norm are widely used for relaxation of low-rank and $\ell_0$-norm minimization, this relaxation may make the solution deviate from the original solution. To overcome this deficiency, Nie *et al.* [3] used joint schatten $p$-norm and $\ell_p$-norm to better approximate the rank constraint and $\ell_0$-norm respectively. This enhanced the robustness of the recovery performance to outliers.

$\ell_p$-norm of a matrix is defined by $\|A\|_p = (\sum_{ij} a_{ij}^p)^{\frac{1}{p}}$ $(0 < p < \infty)$. When $p = 2$, the $\ell_p$ norm becomes the standard Frobenius norm. When $p = 1$, the $\ell_p$ norm becomes $\ell_1$ norm. When $p = 0$, and if we define $0^0 = 0$, $\|\cdot\|_p^p$ norm is equal to $\ell_0$-norm. Then, $p$-norm is extended by computing $p$-th power of $\ell_p$-norm. Mathematically, it is formulated by $\|A\|_p^p = \sum_{ij} a_{ij}^p$ $(0 < p < \infty)$.

On the other hand, the Schatten $p$-norm of a matrix $A \in \Re^{m \times n}$ denotes the $\frac{1}{p}$-th power of the sum of $p$-th singular value of $A$. Its definition is given by,

$$\|A\|_{S_p} = \left( \sum_{i=1}^{min\{m,n\}} \sigma_i^p \right)^{\frac{1}{p}} \quad (0 < p < \infty),$$

where $\sigma_i$ is the $i$-th singular value of $A$. Similarly, the definition of the Schatten $p$ norm is the $p$-th power of the extended Schatten $p$ norm, and its definition is given as follows,

$$\|A\|_{S_p}^p = \sum_{i=1}^{min\{m,n\}} \sigma_i^p.$$

Obviously, when $p = 1$, the Schatten $p$ norm is equal to the trace norm or nuclear norm, namely $\|\cdot\|_*$. And it becomes the rank of matrix for $p = 0$. The Schatten $p$ norm at $p < 1$ provides the desirable faithful low-rank solution, while the nuclear norm using p $= 1$ also provides a low-rank solution, but the low-rank part is more suppressed [1].

## 3 Robust Regression Combined with Joint Schatten $p$-Norm and $\ell_p$-Norm

### 3.1 The Proposed Model

Based on the aforementioned observations, we propose a more general robust regression by solving the following problem.

$$\min_{B,D,E} \frac{\eta}{2}\|Y - B\hat{D}\|_F^2 + \|D\|_{S_p}^p + \lambda\|E\|_p^p + \frac{\gamma}{2}\|B\|_F^2,$$

$$\text{s.t. } X = D + E, \ \hat{D} = [D; 1^T]. \tag{4}$$

When $p \to 0$, $\|D\|_{S_p}^p$ will approach the rank of the matrix $D$ [4]. For $\|D\|_{S_p}^p$ term, the value of $p$ should be selected from $(0, 1]$. Especially when $p = 1$, $\|D\|_{S_p}^p$ is the same as $\|D\|_*$ in Eq. 4. Thus, if the value of $p$ is smaller than 1, the solution of Eq. 4 will better approximate that of the original problem 2, than that of Eq. 3. In addition, in order to improve the robustness of the algorithm against sparse noise, the value of $p$ of $\|E\|_p^p$ term should be smaller than 1 as well. Note that, in a sense, the RR is in fact a special case of our proposed method.

## 3.2 Optimization

Optimizing problem Eq. 4 with joint variables $B, D, E$ is difficult, Alternating Direction Method of Multipliers (ADMM)[14] is often applied to solve it. ADMM solve the optimization problem by optimizing a subproblem of a variables while the other variable are fixed. Accordingly, the augmented Lagrangian function of the proposed problem is given as follows.

$$\min_{B,D,E} \frac{\eta}{2}\|Y - B\hat{D}\|_F^2 + \|D\|_{S_p}^p + \lambda\|E\|_p^p + \frac{\gamma}{2}\|B\|_F^2 + \langle \Gamma_1, X - D - E\rangle$$

$$+ \langle \Gamma_2, \hat{D} - [D; \mathbf{1}^T]\rangle + \frac{\mu}{2}(\|X - D - E\|_F^2 + \|\hat{D} - [D; \mathbf{1}^T]\|_F^2), \quad (5)$$

where $\Gamma_1$ and $\Gamma_2$ are the Lagrange multiplier matrices with compatible dimension, and $\mu$ is the penalty parameter.

For each of the three variables $\{B, D, E\}$ to be solved in Eq. 5, we can efficiently solve one while keeping others fixed.

The optimization algorithm is elaborated as follows.

– Update $B$
  The subproblem is thus given by,

$$\arg\min_B J(B) = \arg\min_B \frac{\eta}{2}\|Y - B\hat{D}\|_F^2 + \frac{\gamma}{2}\|B\|_F^2. \quad (6)$$

The subproblem has the closed-form solution as it is a standard linear regression of $\hat{D}$. That is,

$$\frac{\partial J(B)}{\partial B} = -\eta Y\hat{D} + B\hat{D}\hat{D}^T + \gamma B = 0,$$

$$B^* = \eta Y\hat{D}^T(\hat{D}\hat{D}^T + \gamma\mathbf{I}_d)^{-1}. \quad (7)$$

– Update $\hat{D}$
  Similarly, $\hat{D}$ can be updated by solving the following problem.

$$\arg\min_{\hat{D}} J(\hat{D}) = \arg\min_{\hat{D}} \frac{\eta}{2}\|Y - B\hat{D}\|_F^2 + \frac{\mu}{2}\|\hat{D} - [D; (1)^T]\|_F^2 + \frac{\Gamma_2}{\mu}\|_F^2 \quad (8)$$

Its closed-form solution is achieved by conducting derivation w.r.t. $\hat{D}$ and letting be zeros. Specifically,

$$\hat{D} = (\eta B^T B + \mu)^{-1}(\eta B^T Y + \mu[D; \mathbf{1}^T] + \Gamma_2) \quad (9)$$

– Update $D$

The corresponding subproblem of $D$ is written by,

$$\arg\min_{D} \|D\|_{S_p}^p + \mu(\|D - A\|_F^2),\tag{10}$$

where $P = X - E + \frac{1}{\mu}\Gamma_1$, $Q = \hat{D}_{(1:d_x,:)} + \frac{1}{\mu}\Gamma_2$, $A = \frac{P+Q}{2}$.

Thus, by dividing the set of the following sub-problems, the problem (10) can be solved efficiently.

$$\arg\min_{\delta_i \geq 0} \frac{1}{2}(\delta_i - \sigma_i)^2 + \frac{1}{2\mu}|\delta_i|^p,\tag{11}$$

where $\sigma_i$ and $\delta_i$ are the $i$-th singular value of $A$ and $D$ respectively. Then $D$ can be computed by [3],

$$D = Q_A \triangle R_A^T, \ (\triangle = diag(\delta_1, \delta_2, ..., \delta_n)),\tag{12}$$

where $Q_A$ and $R_A$ contain the left and right singular vectors of $A$, respectively.

– Update $E$

The sub-problem of $E$ is updated as follows,

$$\arg\min_{E} \lambda\|E\|_p^p + \frac{\mu}{2}\|E - W\|_F^2, \ W = X - D + \frac{1}{\mu^{(k)}}\Gamma_1^{(k)}.\tag{13}$$

For each element of $E$, we only need to solve the following problem.

$$\min_{e} \frac{1}{2}(e - b)^2 + \hat{\lambda}|e|^p,\tag{14}$$

where $\hat{\lambda} = \frac{1}{\mu}\lambda$. The solution to the above problem can be computed by [3],

$$e^* = \begin{cases} 0, & -v_1 \leq b \leq v_1 \\ \arg\min_{e \in \{0, \hat{e}_1\}} \frac{1}{2}(e - b)^2 + \hat{\lambda}|e|^p, & b > v_1 \\ \arg\min_{e \in \{0, \hat{e}_2\}} \frac{1}{2}(e - b)^2 + \hat{\lambda}|e|^p, & b < -v_1 \end{cases}\tag{15}$$

where $v_1 = v + p|v|^{p-1}, v = (p(1-p))^{\frac{1}{2-p}}$ and $\hat{e}_1, \hat{e}_2$ is the root of $h'(e) = e - b + p|x|^{p-1}\text{sgn}(e) = 0$ at $v < e < b$ and $b < e < -v$, respectively. The roots can be easily obtained with Newton method initialized at $b$.

– Update $\Gamma_1, \Gamma_2$ and $\mu$

$$\Gamma_1 = \Gamma_1 + \mu(X - D - E), \ \Gamma_2 = \Gamma_2 + \mu(\hat{D} - [D; 1^T]), \ \mu = \min(\rho\mu, \mu_{\max}),\tag{16}$$

where $\rho > 1$ is a constant and $\mu_{\max}$ is the upper bound of $\mu$.

The overall algorithm is summarized in the Algorithm 1.

---

**Algorithm 1.** Algorithm for solving problem 4.

---

**Input** $\{\mathbf{x}_i, \mathbf{y}_i\}_{i=1}^N$, parameter $\eta$ (a positive scalar weights term $\|Y - B\hat{D}\|_F^2$) and $\lambda$ (a positive scalar weights term $\|E\|_p^p$), $\gamma$ (a positive scalar for regularizing the solution to $B$).

**Initialization:** $D_0 = X$, $E_0 = X - D_0$, $B_0 = 0$, $\mu_0 = 1.2$, $\rho = 1.01$; $\varepsilon_1 = 10^{-4}$, $\varepsilon_2 = 10^{-5}$.

**While** not converged ($t = 0, 1, ...$) **do**

    1. Update $B$ according to Eq. 7 ;
    2. Update $\hat{D}$ according to Eq. 9;
    3. Update $D$ according to Eq. 12;
    4. Update $E$ according to Eq. 15;
    5. Updating $\Gamma_1, \Gamma_2$ and $\mu$ according to Eq. 16;
    6. Check convergence

**End while**
**Output** $B^*, \hat{D}^*, D^*, E^*$

---

### 3.3 Convergence Analysis

The Algorithm 1 will converge globally for any sufficiently large $\mu$. That is, starting from any $B^0, D^0, E^0$, it generates a sequence that is bounded, having at least one limit point, and that each limit point $(B^*, D^*, E^*)$ is stationary point of $J_\mu$, namely, $0 \in \partial J_\mu(B^*, D^*, E^*)$ [12].

## 4 Experiment

In this section, two experiments are performed to test the effectiveness of the proposed approach. Seven state-of-the-art approaches of classification are selected to compare with ours. For the sake of clarity, we denote the $p$ of $\|D\|_{S_p}^p$ by $p_1$ and the $p$ of $\|E\|_p^p$ by $p_2$, respectively. **RPCA+LDA** aims to first perform RPCA [8] on the input data, and then learn from the clean data using standard linear discriminant analysis. Similarly, **RPCA+LSR** learns the mapping by the pre-cleaned data using RPCA. We name Robust Regression [2] as RR, while ours as $RR_{Sp}$. The values of $p_1$ and $p_2$ are empirically chosen as 0.1 in the experiments. Meanwhile the parameters of other compared methods are given in the tables too. Note that the parameters for $RR_{Sp}$ are $\eta$ and $\lambda$ respectively. The same case is to RR.

### 4.1 Face Recognition on YaleB Dataset

We firstly test our approach on the Extended YaleB dataset[4], which contains 2414 different face images, that belonging to 38 classes. Each class of the dataset is randomly divided into two parts, half for training and the rest for testing.

---

[4] http://www.cad.zju.edu.cn/home/dengcai/.

**Fig. 2.** Samples of YaleB dataset with 5%–40% artificial interference of random pixel corruption.

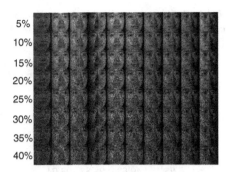

**Fig. 3.** Samples of PIE dataset with 5%–40% artificial interference of random pixel corruption.

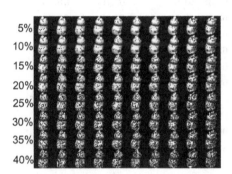

**Fig. 4.** Samples of COIL20 dataset with 5%–40% artificial interference of random pixel corruption.

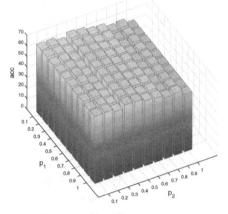

**Fig. 5.** Performance of different value of $p$ on YaleB dataset with 30% random pixel corruption.

The classification results are shown in Table 1. As can be seen, the classification performance deteriorates naturally, along with the increasing of noise. Compared to others, our method achieved the best score, showing the best robustness. While for RR, it achieved the second best gains as it has the capability of recovering the clean data in a *supervised* manner. This is further confirmed by ours in that the capability of better approximation to the low-rank property (Fig. 2).

### 4.2 Face Recognition on PIE Dataset

Next, the PIE dataset is adopted to test our approach in the second experiment. Similar to that of the experiment on YaleB dataset, in particular, artificial interference of random pixel corruption is appended to dataset showing in Fig. 3.

**Table 1.** The recognition rate on YaleB dataset and PIE dataset with different pixel corruptions. The percentages in the backets denotes the portion of pixel corruption. Bigger value (%) indicates better performance. Best results are in bold.

|  |  | K-NN | LDA | $RPCA + LDA$ | LSR | RPCA+LSR | RR | $RR_{Sp}$ |
|---|---|---|---|---|---|---|---|---|
| YaleB | Para. | (1.0) | - | (0.03) | - | (0.03) | (1e7, 100) | (1e8, 1e5) |
|  | 5 | 75.48 | 64.95 | 80.03 | 89.31 | 93.21 | 94.20 | **94.28** |
|  | 10 | 70.34 | 52.11 | 64.79 | 76.14 | 80.28 | 89.64 | **90.39** |
|  | 15 | 62.47 | 43.58 | 43.83 | 59.65 | 60.98 | 81.86 | **84.01** |
|  | 20 | 57.75 | 33.97 | 34.22 | 48.88 | 51.37 | 75.48 | **80.79** |
|  | 25 | 49.63 | 28.83 | 28.94 | 41.59 | 42.92 | 69.01 | **71.25** |
|  | 30 | 42.59 | 24.03 | 23.94 | 32.73 | 31.40 | 59.40 | **63.96** |
|  | 35 | 35.88 | 19.80 | 19.88 | 23.53 | 22.45 | 49.21 | **57.83** |
|  | 40 | 29.08 | 16.90 | 16.98 | 22.29 | 20.80 | 39.76 | **51.95** |
| PIE | Para. | (6.0) | - | (0.03) | - | (0.03) | (1e5, 100) | (1e8, 1e10) |
|  | 5 | 64.54 | 60.40 | 61.78 | 63.17 | 63.17 | 64.16 | **68.71** |
|  | 10 | 61.98 | 61.78 | 61.58 | 63.76 | 63.96 | 64.75 | **65.94** |
|  | 15 | 62.77 | 58.42 | 59.01 | 62.38 | 61.98 | 63.76 | **65.74** |
|  | 20 | 64.95 | 59.01 | 59.80 | 63.96 | 63.56 | 65.35 | **66.34** |
|  | 25 | 62.38 | 58.22 | 58.61 | 63.56 | 62.57 | 64.95 | **66.73** |
|  | 30 | 62.57 | 55.45 | 54.65 | 63.96 | 63.76 | 65.35 | **66.53** |
|  | 35 | 59.60 | 50.69 | 50.50 | 60.40 | 60.59 | 62.57 | **63.56** |
|  | 40 | 60.99 | 48.52 | 49.11 | 60.99 | 59.41 | 61.39 | **62.77** |
| COIL20 | Para. | (1.0) | - | (0.03) | - | (0.03) | (1e3, 100) | (1e8, 1e6) |
|  | 5 | 91.94 | 35.28 | 92.08 | 92.36 | 93.75 | 93.89 | **95.14** |
|  | 10 | 91.81 | 37.78 | 84.44 | 90.56 | 89.31 | 93.06 | **93.47** |
|  | 15 | 89.17 | 34.58 | 75.69 | 82.92 | 83.75 | 89.31 | **90.24** |
|  | 20 | 87.22 | 26.39 | 68.61 | 81.39 | 76.81 | 87.64 | **89.44** |
|  | 25 | 86.39 | 35.42 | 56.81 | 74.03 | 65.42 | 86.81 | **87.64** |
|  | 30 | 85.56 | 26.53 | 49.03 | 69.86 | 53.89 | 86.26 | **86.81** |
|  | 35 | 85.42 | 29.72 | 41.11 | 63.61 | 43.47 | 85.69 | **86.25** |
|  | 40 | 84.58 | 28.06 | 42.36 | 60.69 | 41.53 | 85.42 | **85.56** |

The results of the classification accuracy of each algorithm are shown in Table 1. Obviously, our method gets the best performance over the other methods.

## 4.3   Object Recognition on COIL20 Dataset

Then the COIL20 dataset is chosen in our third experiment. Since the COIL20 dataset is relatively clean, we have to add noise to the dataset to compare the effectiveness of the algorithm. Same as the experiment on two dataset above, we add artificial interference of random pixel corruption to dataset and show in Fig. 4. Half of each class of the dataset is for training, the rest for testing. The results of recognition accuracy are shown in Table 1. The result shows that as the noise increases, the superiority of our algorithm is higher.

### 4.4 Performance of Varying $p$

In this section, we investigate the effect of $p$ to the performance of our algorithm. The experiments are conducted on YaleB dataset with 30% random pixel corruption. Undoubtedly, our method will degrade to the robust regression model [2] when $p = 1$. While for $p < 1$, the performance of our method is usually better than robust regression model, illustrated by Fig. 5. Thus it confirms that the solution of robust regression model with convex surrogates may seriously diverge from the original solution of rank minimization problem. In contrast, our approach with nonconvex regularizers can archive better approximation to the rank, such that the better performance can be attained.

## 5 Conclusion

In this paper, we aim to address the deviation of robust regression using convex surrogate in the presence of noise in data. Specifically, we adopt joint schatten $p$ norm and $p$ norm instead of the rank constraint and $\ell_0$-norm respectively. Experimental results show that our method outperforms other state-of-the-art methods, including robust regression [2]. This is attributed to the fact that the combination of schatten $p$ norm and $\ell_p$ norm is a better way to approach the original solution, than using the nuclear norm and the $\ell_1$-norm, respectively.

**Acknowledgement.** This work was supported by the National Natural Science Foundation of China through Grant 61703114, Grant 61673126, Grant 61876042, and Grant 61773130, the Science and Technology Plan Project of Guangdong (2015B010131014) and the Educational Commission of Guangdong Province, China (2017KTSCX059).

## References

1. Kong, D., Zhang, M., Ding, C.: Minimal shrinkage for noisy data recovery using Schatten-$p$ norm objective. In: Blockeel, H., Kersting, K., Nijssen, S., Železný, F. (eds.) ECML PKDD 2013. LNCS (LNAI), vol. 8189, pp. 177–193. Springer, Heidelberg (2013). https://doi.org/10.1007/978-3-642-40991-2_12
2. Huang, D., Cabral, R., De la Torre, F.: Robust regression. IEEE Trans. Pattern Anal. Mach. Intell. **38**(2), 363–375 (2016)
3. Nie, F., Wang, H., Cai, X.: Robust matrix completion via joint Schatten p-norm and lp-norm minimization. In: 12th International Conference on Data Mining (ICDM), pp. 566–574. IEEE (2012)
4. Nie, F., Huang, H., Ding, C.H.Q.: Low-rank matrix recovery via efficient Schatten p-norm minimization. In: AAAI, pp. 655–661 (2012)
5. Huang, D., Storer, M., De la Torre, F.: Supervised local subspace learning for continuous head pose estimation. In: IEEE Conference on Computer Vision and Pattern Recognition (CVPR), pp. 2921–2928. IEEE (2011)
6. Wu, H., Souvenir, R.: Robust regression on image manifolds for ordered label denoising. In: IEEE Conference on Computer Vision and Pattern Recognition, pp. 305–313. IEEE (2015)

7. Liu, J., Cosman, P.C., Rao, B.D.: Robust linear regression via $\ell_0$ regularization. IEEE Trans. Signal Process. **66**(3), 698–713 (2018)
8. Wright, J., Ganesh, A., Rao, S.: Robust principal component analysis: Exact recovery of corrupted low-rank matrices via convex optimization. In: Advances in Neural Information Processing Systems, pp. 2080–2088 (2009)
9. Yin, M., Gao, J., Lin, Z.: Laplacian regularized low-rank representation and its applications. IEEE Trans. Pattern Anal. Mach. Intell. **38**(3), 504–517 (2016)
10. Liu, X.: Discriminative face alignment. IEEE Trans. Pattern Anal. Mach. Intell. **31**(11), 1941 (2009)
11. Chen, Y., Caramanis, C., Mannor, S.: Robust sparse regression under adversarial corruption. In: International Conference on Machine Learning, pp. 774–782 (2013)
12. Wang, Y., Yin, W., Zeng, J.: Global convergence of ADMM in nonconvex nonsmooth optimization. arXiv:1605.02408 (2018)
13. Yin, M., Cai, S., Gao, J.: Robust face recognition via double low-rank matrix recovery for feature extraction. In: 20th IEEE International Conference on Image Processing (ICIP), pp. 3770–3774. IEEE (2013)
14. Lin, Z., Chen, M., Ma, Y.: The augmented Lagrange multiplier method for exact recovery of corrupted low-rank matrices. arXiv preprint arXiv:1009.5055 (2010)

# Adaptive Crossover Memetic Differential Harmony Search for Optimizing Document Clustering

Ibraheem Al-Jadir[1,2(✉)], Kok Wai Wong[1(✉)], Chun Che Fung[1(✉)], and Hong Xie[1(✉)]

[1] School of Engineering and Information Technology, Murdoch University, Perth 6163, Australia
{I.Al-Jadir, K.Wong, L.Fung, H.Xie}@murdoch.edu.au
[2] College of Science, Baghdad University, Baghdad, Iraq

**Abstract.** An Adaptive Crossover Memetic Differential Harmony Search (ACMDHS) method was developed for optimizing document clustering in this paper. Due to the complexity of the documents available today, the allocation of the centroid of the document clusters and finding the optimum clusters in the search space are more complex to deal with. One of the possible enhancements on the document clustering is the use of Harmony Search (HS) algorithm to optimize the search. As HS is highly dependent on its control parameters, a differential version of HS was introduced. In the modified version of HS, the Band Width parameter (BW) has been replaced by another pitch adjustment technique due to the sensitivity of the BW parameter. Thus, the Differential Evolution (DE) mutation was used instead. In this paper the DE crossover was also used with the Differential HS for further search space exploitation, the produced global search is named Crossover DHS (CDHS). Moreover, DE crossover (Cr) and mutation (F) probabilities are dynamically tuned through generations. The Memetic optimization was used to enhance the local search capability of CDHS. The proposed ACMDHS was compared to other document clustering techniques using HS, DHS, and K-means methods. It was also compared to its other two variants which are the Memetic DHS (MDHS) and the Crossover Memetic Differential Harmony Search (CMDHS). Moreover, two state-of-the-art clustering methods were also considered in comparisons, the Chaotic Gradient Artificial Bee Colony (CGABC) and the Differential Evolution Memetic Clustering (DEMC). From the experimental results, it was shown that CMDHS variant (the non-adaptive version of ACMDHS) and ACMDHS were highly competitive while both CMDHS and ACMDHS were superior to all other methods.

**Keywords:** Clustering · Memetic · Optimization · Harmony Search

© Springer Nature Switzerland AG 2018
L. Cheng et al. (Eds.): ICONIP 2018, LNCS 11302, pp. 509–518, 2018.
https://doi.org/10.1007/978-3-030-04179-3_45

# 1 Introduction

Document clustering is a necessary process for efficient document management such as retrieval, archival, topic extraction, and summarization [1]. Due to the drawbacks of the traditional methods of clustering, optimization methods have been used [2]. Evolutionary Algorithms (EA) were introduced to optimize the selection of the centroids in document clustering [3]. In the last few years, a large number of optimization-based methods have been proposed [4]. As an improvement to the evolutionary algorithms, Memetic algorithm (MA) [5] have been proposed to combine the advantages of the evolutionary algorithms with problem-specific optimization methods. The powerful global search ability is combined with some local improver methods in the memetic algorithms. Such a combination have been successfully applied to global optimization of numerical functions [6] and have been utilized to solve many real-world optimization problems [7]. MA has been introduced in many other applications but has not been investigated in detail for optimizing document clustering.

For making MA more powerful in terms of the exploration of the search space, modifications to MA's global search could also be productive. For instance, combining two or three global search methods might be more efficient than the old methods. For example, the use of Differential Evolution (DE) Mutation with Harmony Search instead of the pitch adjustment step can produce better optimizing performance. Furthermore, the use of the adaptive parameter settings can also enhance the performance in comparison to the performance of the static parameter setting [8, 9]. These improvements can all be combined to produce an efficient method for document clustering.

In this paper, an intelligent document clustering method is proposed that uses the Differential HS as an optimization method to better search centroids of the document clusters. DHS is used as a modification to the original HS that uses bandwidth technique for the harmonies' pitch adjustment. In this paper, DHS is used to eliminate the need for manually setting the Bandwidth (BW) parameter as it might affect document clustering performance negatively. Also in this paper, DHS was further enhanced by incorporating the DE crossover along with the mutation; the modified version was named Crossover DHS (CDHS). The mutation scaling factor (F) and the crossover probability rate (Cr) are adaptively set. In that case, whenever an optimal solution is generated, the best control parameters contributed to the best solution are admitted to the next generations. Finally, the memetic optimization is used with CDHS to reduce the chances of entrapment in local optima. The resulted method is an Adaptive Crossover Memetic Differential Harmony Search (ACMDHS). Moreover, the other two variant of the ACMDHS resulted in this paper are the Memetic DHS (MDHS) and the non-adaptive version of the ACMDHS named CMDHS, which both have slight differences in the way they navigate through the search space using different set of parameters.

## 2  Related Work

Genetic Algorithm (GA) was first used as an efficient optimization method in different optimization problems. However, due to the drawbacks of GA, many other optimization methods have been devised in the last two decades. For example, the swarm intelligence methods were deployed to enhance the traditional document clustering. Two swarm methods have been adopted as document clustering algorithms which are the Artificial Bee Colony (ABC) [10] and the Bee Colony Optimization (BCO) [11]. The ABC is recently combined with two local search methods used for document clustering. The two local search methods are the Gradient and the Chaotic methods, which are used to enhance the performance of local searches [12]. The BCO is another swarm-based method which is modified in [13] by introducing two new heuristic operators which are *fairness* and *cloning*. This method is also combined with a local search method. According to their proposal, the BCO is combined with k-means, and the k-means is used to perform the local search instead of clustering. In both ABC and BCO methods, it can be noted that using the global search is insufficient in the exploitative aspect to find the best solution for document clustering.

Besides the above mentioned examples, recently the Differential Evolution (DE) has been successfully combined with the HS. The resulted method is named Differential HS (DHS) which is a combination of HS and DE. DHS has been empirically experimented and it outperformed both DE and HS [14]. In the DHS, the DE mutation operator was used instead of the original pitch adjustment process of the HS [15].

Before DHS, HS was used as an efficient document clustering along with the k-means in [16]. The k-means was combined in three different positions, namely, interleaved, sequential, and one-step. The position of using the k-means differs from one to another. The tests revealed that the hybrid approaches outperformed the HS, GA, and k-means as the search algorithms. Furthermore, a Global-Best HS combined with the k-means was introduced in [17] to cluster Web documents. The hybrid approach outperformed both methods separately. The same combination described previously can also be seen in [18], but in this case the GA was integrated with k-means. The solutions produced from the GA-k-means were used to construct the HS memory (population). The authors claimed that with the combination of HS and GA performed better than using just the GA or the HS with the k-means separately.

Although HS has been successful, it suffers from slow convergence and stagnation problems [15]. DE can be considered another variant of the HS [19]. In contrast to the HS, DE has not been utilized extensively in document clustering. Owing to its promising results seen in many other combinatorial problems, DE was used in [20] to enhance the k-means clustering and to prevent the production of weak solutions. Later, the same combination was improved so that the K cluster number could be set automatically [21].

One of the attempts to integrate the prospects of both HS and DE was proposed in [15] where the tests showed that using the DE operators instead of the pitch adjustment step can outperform both the HS and the DE. However, the proposed method still needs an accurate tuning of the DE parameters as the DE is sensitive to the setting of its control parameters [8, 9] in the same way as many other global search methods. Therefore in [22], an adaptive method named Self-adaptive Differential Evolution algorithm with Improved Mutation Mode (IMMSADE) was proposed. In IMMSADE, the control parameters were dynamically modified on the basis of the diversity of the population. As a result, it can be concluded that the combination of the adaptive way of parameter tuning with the DHS can lead to further improvement to the DHS.

The majority of the above-mentioned optimization methods have a drawback in the local search. Therefore, the Memetic optimization was proposed. Moreover, due to the sensitivity of the Bandwidth (BW) parameter in the pitch adjustment of the Harmony Memory (HM) in the traditional HS, the DHS is used instead with the document clustering problem. In this way, the need to manually set the BW parameters can be eliminated. In the DHS, the use of the DE operators replaces the need to use the traditional Pitch Adjustment technique that uses the bandwidth operator. Moreover, the static settings of both DE crossover and mutation operators represented by the $Cr$ and $F$ parameters were also adaptively set. Due to the gaps noted earlier, the aim of this work was intended to overcome the limitations of both HS and DE using the DHS first as global search in MA. Later, the DHS was enhanced to modify its parameters adaptively.

## 3   Document Clustering Using ACMDHS Optimization

The proposed method is based on the combination of the global search using DHS once and CDHS once again with the k-means local search. The main steps of the proposed methods are listed below:

1. The first step begins by transforming the text documents into a numeric format using the same techniques explained in [23]. These techniques involve the text tokenization, stop words removal, stemming and weighting keywords.
2. The harmony memory initialization, the initial population of the harmony memory contains a random assignment of documents to cluster centroids. Each row of the harmony memory is solution whereas the length of each solution is fixed to the number of documents. In this step the evaluation of each random solution will be calculated using the fitness function explained in step 5.
3. The harmony memory update using mutation, in the differential harmony search DHS method, the DE mutation used instead of the traditional pitch adjustment step in HM as illustrated in Eq. (1).

$$v_{j,i,G+1} = x_{j,r1,G} + F.\left(x_{j,r2,G} - x_{j,r3,G}\right), \qquad i = 1, 2, .., N; j = 1, 2, .., D \qquad (1)$$

where, $v_{j,i,G+1}$ is the newly generated trial vector, $x_{j,r1,G+1}$, $x_{j,r2,G+1}$, and $x_{j,r3,G+1}$ are three randomly selected vectors, $j$ is the solution inedex and $i$ is the bit index with the jth solution. $G$ is the generation number, and $F$ is a scaling factor [0,1] ($F$ is adaptively tuned. If the new solution has a higher fitness, the current $F$ value will be retained. Otherwise; it will be discarded and substituted by a newly generated value).

4. The harmony memory update using crossover, this step is only applied with the CDHS. Unlike DHS, that only makes use of the mutation for the pitch adjustment, a modified DHS version used in this paper that uses DE crossover after applying the mutation. The new solution is created based on Eq. (2).

$$U_i^g = \begin{cases} v_i^g & rand(0, 1) < Cr \\ x_i & otherwise \end{cases} \tag{2}$$

where $U_i^g$ is the newly generated solution and $x_i$ is the target solution from the HMS and $v_i^g$ is the mutant vector obtained from Eq. (1) and $Cr$ is the crossover probability.

5. The fitness function used to evaluate the viability of solutions is the Average Distance of Document to Centroid (ADDC). Equation (3) shows how the ADDC is calculated.

$$ADDC = \left[ \sum_{i=1}^{k} \frac{1}{n_i} \sum_{j=1}^{m_i} D(c_i, d_j) \right] \Big/ k \tag{3}$$

where $k$ clusters and $m$ represent the total number of documents, and the $D$ is the cosine similarity measure calculated in Eq. (4), $d$ is a particular document and $c$ is a centroid calculated in Eq. 10.

$$s(d_1, d_2) = \frac{d_1 \cdot d_2}{\|d_1\| \cdot \|d_2\|} = \frac{\sum_{i=0}^{n-1} d_{1i} d_{2i}}{\sqrt{\sum_{i=0}^{n-1} (d_{1i})^2} \times \sqrt{\sum_{i=0}^{n-1} (d_{2i})^2}} \tag{4}$$

where $d_1$ and $d_2$ are a pair of documents and $n$ is the entire number of documents in the corpus.

6. After improvising the best solutions in steps 4 and 5, the local search using k-means is applied. The k-means can adjust the current best solution by modifying the current centroids positions. A comparison between the solution resulted from the local search and its original version is then conducted. The solution that achieves the best ADDC value will replace the inferior.
7. Substitution: after every fitness evaluation process, the solutions resulted in the current generation are compared with those of the older generation. If their fitness is better, they will replace them.

In Table 1, the datasets used in this research are explained in terms of the documents number and the number of classes in each of the datasets mentioned above is reported.

## 4 Initial Parameters Setting and Test Results

In this section, an explanation of the parameters used and their setting is give. In addition, the test results of the proposed method and other methods are also provided.

**Table 1.** Datasets

| Dataset | D# | #Classes | Instances | Features |
|---|---|---|---|---|
| 6 event crimes | D1 | 6 | 223 | 3864 |
| 10 types crime | D2 | 10 | 2422 | 15601 |
| Reuters | D3 | 10 | 2277 | 13310 |
| Pair 20news Groups | D4 | 2 | 1071 | 9497 |
| 20 news Groups | D5 | 20 | 1489 | 6738 |

## 4.1   Parameter Settings

Although an adaptive mechanism of parameter tuning is being followed in this research, still an initial tuning of the DE parameters $F$ and $Cr$ should be performed in prior. The tuning is based on three values for each single parameter. For the $Cr$, the three selected values were 0.2, 0.5 and 0.9 whereas for the $F$ the values tested were 0.8, 0.1 and 0.5. These numbers are selected randomly. These two parameter values should be selected between zero and one. Therefore, the tested value for each parameter can show at which value the performance increases with all used datasets. Even though the performance might not be similar in all cases, a studied selection of these parameters will be preferable than a random one. For instance, if the parameter generated the highest performance with the majority of datasets, it will be used for the later tests. The parameter tests were conducted using these values with all datasets. After the selection of the best value of the $Cr$ parameter, the $F$ is then tested. As shown in Table 2, the value that helped to obtain the highest F-measure and the lowest ADDC was 0.5. Therefore, the other two values were discarded. In Table 2, it becomes evident that the use of $Cr = 0.5$ has four out of five highest F-measure scores. When it comes to ADDC measure, the values of all the runs are almost consistent with only minimal differences. Thus, using the F-measure values to determine the best parameter value seems to be more appropriate in that case.

When it comes to the HS parameters, the Harmony Memory Size (HMS) is set to twice the number of centroids. According to [16], it was discovered that this is the best value of this parameter. The PAR and the HMCR were set based on [15], and their values were PAR = 0.9 while HMCR = 0.99. It is important to note that other candidate values of the Cr and F can be used, but at this stage using three experimental values seem to be sufficient.

## 4.2   Test Results

In this section, the comparison results of the proposed method with other methods are given. The methods used for the comparison are the Harmony Search (HS) method [16], Differential Harmony Search (DHS) [15], the Memetic HS [24] and the k-means. Two state-of-the-art methods were also used which are CGABC [25] and DEMC [26]. Besides two variants of the ACMDHS were also tested which are the MDHS and the CMDHS.

**Table 2.** Cr and F parameter Tuning Table

| Dataset | Cr = 0.2 | | Cr = 0.5 | | Cr = 0.9 | |
|---------|-----------|------|-----------|------|-----------|------|
| | F-Measure | ADDC | F-Measure | ADDC | F-Measure | ADDC |
| D1 | 83.90945 | 0.718938 | **85.81185** | **0.714983** | 79.866677 | 0.714957 |
| D2 | 94.540028 | 0.748749 | **98.429393** | **0.74538** | 96.452157 | 0.74342 |
| D3 | 88.05235 | 0.711795 | 89.205819 | 0.711147 | **92.794232** | **0.707603** |
| D4 | 94.953637 | 0.842767 | **96.090954** | **0.843154** | 94.857401 | 0.839821 |
| D5 | 66.480562 | 0.847179 | **99.547975** | **0.846073** | 66.666667 | 0.848331 |
| Dataset | F = 0.1 | | F = 0.5 | | F = 0.8 | |
| | F-Measure | ADDC | F-Measure | ADDC | F-Measure | ADDC |
| D1 | **89.732655** | **0.721584** | 77.772432 | 0.719301 | 85.81185 | 0.714983 |
| D2 | 97.834008 | 0.749086 | 96.460589 | 0.747813 | **98.429393** | **0.74538** |
| D3 | **92.757559** | **0.711581** | 90.263957 | 0.708986 | 89.205819 | 0.711147 |
| D4 | **98.230616** | **0.842119** | 90.055755 | 0.823152 | 94.953637 | 0.842767 |
| D5 | **99.646309** | **0.846778** | 99.176776 | 0.846605 | 99.547975 | 0.846073 |

### 4.2.1 The Internal and External Evaluation

In order to evaluate the performance of each one of the competent algorithms, the internal and external evaluation measures are used. The ADDC internal measure serves as a fitness function with the aim of minimizing the distance among documents within one cluster. As mentioned before, it is used as a clustering compactness measure. The smaller the ADDC, the more compact the clusters are. The F-measure is utilized to express an expert view of the resulted clusters. In other words, this measure uses a truth table of the original representation of documents and it compares it with the newly resulted clusters. Thus, the F-measure needs the original class labels to be used in evaluations. The ADDC values needs to be minimized while the F-measure values need to be maximized. Both measures have their values listed in Tables 3 and 4. Table 3 depicts the external measure values using F-measure while Table 4 shows the internal measure values using ADDC.

**Table 3.** F-measure values

| Runs | HS | DHS | MHS | MDHS | KM | CGABC | DEMC | CMDHS | ACMDHS |
|------|-----|------|------|-------|------|--------|-------|--------|---------|
| D1 | 76.98 | 50.06 | 78.62 | 63.84 | 63.62 | 82.65 | 86.00 | **88.50** | 85.97 |
| D2 | 90.50 | 80.85 | 90.75 | 80.31 | 35.52 | 94.28 | 94.00 | **98.69** | 96.03 |
| D3 | 91.94 | 83.84 | 88.01 | 88.75 | 16.37 | 91.95 | 80.30 | 96.94 | **97.56** |
| D4 | 96.72 | 90.76 | 95.17 | 89.53 | 67.72 | 95.48 | 88.75 | 97.56 | **98.84** |
| D5 | 98.16 | 93.86 | 97.23 | 95.69 | 0.60 | 98.96 | 53.88 | **99.91** | 98.92 |

It is apparent from Table 3 that the performance of HS method was relatively better than the performance of DHS with all datasets while in Table 4 their ADDC values were almost the same. In that case the ADDC values will be ignored and depending

**Table 4.** ADDC values

| Runs | HS | DHS | MHS | MDHS | CGABC | DEMC | CMDHS | ACMDHS |
|------|------|------|------|------|-------|------|-------|--------|
| D1 | 0.72 | 0.71 | 0.71 | 0.72 | 0.72 | 0.72 | 0.72 | 0.72 |
| D2 | 0.71 | 0.71 | 0.68 | 0.71 | 0.86 | 0.86 | 0.74 | 0.71 |
| D3 | 0.84 | 0.82 | 0.82 | 0.82 | 0.72 | 0.67 | 0.73 | 0.71 |
| D4 | 0.74 | 0.74 | 0.74 | 0.74 | 0.83 | 0.82 | 0.82 | 0.75 |
| D5 | 0.73 | 0.73 | 0.73 | 0.73 | 0.84 | 0.83 | 0.72 | 0.73 |

only on F-measure values. The performance of the modified version of harmony search was supposed to be better than the native version due to the elimination of the need to use the Bandwidth parameter [15]. This comparison indicates that the modification of DHS was insufficient to empower it to outperform its ancestor HS. Therefore, two Memetic versions were based on the HS and the differential HS were tested; (i) Memetic HS (MHS) and (ii) Memetic Differential HS (MDHS). For the MHS, the performance is still similar to native HS whereas an improvement can be observed after using the MDHS in comparison to the DHS. The single most striking observation to emerge from the comparison of both DHS and MDHS was the local optima problem could have been solved using the local search.

However, the MDHS method performance has a different attitude when it is compared to other two state-of-the-art-methods which are the chaotic gradient-based artificial bee colony method CGABC and the differential evolution memetic clustering DEMC. The results reveal that MDHS underperformed the CGABC method with all. When it comes to the DEMC, another scenario can be observed. The performance of MDHS was better than DEMC with almost all datasets. Based on that comparison, the use of DHS as a global search is better than the differential evolution in DEMC. On the other hand, it is highly suspected that the use of more than one local search in CGABC has a positive effect on the performance. However, using more than one local search will add more complexity to the DHS. Thus, modifying the global search further could be worthy for another test. Thus, the Crossover-added version of DHS was tested in two modes one in an adaptively-based parameter tuned mode and another is based on statically-based parameter tuned mode after using the best of them in Table 2.

As Table 3 shows, the F-measure values of both the crossover memetic MDHS (CMDHS) and the Adaptive CMDHS (ACMDHS) outperformed all of the two states-of-the-art methods and other versions of the harmony search. What is interesting in this comparison is that the enhancement of the global search has a remarkable effect on the performance even when it is compared with another Memetic scheme that incorporates two local search methods as is seen with CGABC. Now, by restricting the comparison between CMDHS and ACMDHS, it becomes clear that the statically-based parameter tuned version (CMDHS) outperformed the dynamically-based one (ACMDHS). The single most striking observation to emerge from that comparison is the tuned parameters i.e. the $F$ for the mutation and the Cr for the crossover have only minor effect on the performance of the centroids allocation.

Finally, in Table 4 the ADDC values of the K-means were not listed because the K-means is not an optimization-based method. However, in the ACMDHS the k-means is used as a local search. Through Table 4, it can be noted that the general trends in all results are compatible. The stability of the ADDC in comparison to the F-measure did not mean all methods performing equally. That is because the F-measure values were changing when ADDC values were almost steady.

## 5  Conclusions

In this paper a Memetic-based clustering method named Adaptive Crossover Memetic Differential Harmony Search (ACMDHS) is proposed. Other two variants of ACMDHS resulted in this paper are the Memetic DHS (MDHS) and the non-adaptive version of the ACMDHS named CMDHS, which both have slight differences in the way they navigate through the search space using different set of parameters. The experimental results showed that the proposed ACMDHS provided the best F-measure results in comparison to other document clustering methods: Harmony Search (HS) method, Differential Harmony Search (DHS) [15], the Memetic HS, Memetic DHS and the k-means. Moreover, it was also compared to its other two variants: MDHS and CMDHS. The test results showed that the CMDHS performed the same or slightly better than the Adaptive CMDHS indicating to the minor effect of the differential evolution parameters on the performance. Finally among the other two state-of-the-art methods which are CGABC and DEMC, the ACMDHS has achieved the highest F-measure.

**Acknowledgements.** Ibraheem would like to express his gratitude to the Higher Committee of Education Development in Iraq (HECD) for the scholarship he has received to fund his PhD study.

## References

1. Saiyad, N.Y., Prajapati, H.B., Dabhi, V.K: A survey of document clustering using semantic approach. In: International Conference on Electrical, Electronics, and Optimization Techniques (ICEEOT). IEEE (2016)
2. Feng, A.: Document clustering: an optimization problem. In: Proceedings of the 30th Annual International ACM SIGIR Conference on Research and Development in Information Retrieval. ACM (2007)
3. Hruschka, E.R., et al.: A survey of evolutionary algorithms for clustering. IEEE Trans. Syst. Man Cybern. Part C Appl. Rev. 39(2), 133–155 (2009)
4. Yang, Y., Kamel, M., Jin, F.: A model of document clustering using ant colony algorithm and validity index. In: 2005 Proceedings of the IEEE International Joint Conference on Neural Networks, IJCNN 2005. IEEE (2005)
5. Smith, J.E.: Coevolving memetic algorithms: a review and progress report. IEEE Trans. Syst. Man Cybern. Part B Cybern. 37(1), 6–17 (2007)
6. Nguyen, Q.H., Ong, Y.-S., Lim, M.H.: A probabilistic memetic framework. IEEE Trans. Evol. Comput. 13(3), 604–623 (2009)
7. Neri, F., Mininno, E.: Memetic compact differential evolution for cartesian robot control. IEEE Comput. Intell. Mag. 5(2), 54–65 (2010)

8. Reynoso-Meza, G., et al.: Hybrid DE algorithm with adaptive crossover operator for solving real-world numerical optimization problems. In: 2011 IEEE Congress on Evolutionary Computation (CEC) (2011)

9. Zhang, J., Sanderson, A.C.: JADE: adaptive differential evolution with optional external archive. IEEE Trans. Evol. Comput. **13**(5), 945–958 (2009)

10. Karaboga, D., Ozturk, C.: A novel clustering approach: Artificial Bee Colony (ABC) algorithm. Appl. Soft Comput. **11**(1), 652–657 (2011)

11. Lučić, P., Teodorović, D.: Computing with bees: attacking complex transportation engineering problems. Int. J. Artif. Intell. Tools **12**(3), 375–394 (2003)

12. Bharti, K.K., Singh, P.K.: Chaotic gradient artificial bee colony for text clustering. Soft. Comput. **20**(3), 1113–1126 (2016)

13. Forsati, R., Keikha, A., Shamsfard, M.: An improved bee colony optimization algorithm with an application to document clustering. Neurocomputing **159**, 9–26 (2015)

14. Chakraborty, P., et al.: An improved harmony search algorithm with differential mutation operator. Fundamenta Informaticae **95**(4), 401–426 (2009)

15. Abedinpourshotorban, H., et al.: A differential-based harmony search algorithm for the optimization of continuous problems. Expert Syst. Appl. **62**, 317–332 (2016)

16. Forsati, R., et al.: Efficient stochastic algorithms for document clustering. Inf. Sci. **220**, 269–291 (2013)

17. Cobos, C., et al.: Web document clustering based on global-best harmony search, K-means, frequent term sets and Bayesian information criterion. In: 2010 IEEE Congress on IEEE Evolutionary Computation (CEC) (2010)

18. Rafi, M., et al.: Towards a soft computing approach to document clustering. In: Proceedings of the 2017 International Conference on Machine Learning and Soft Computing, pp. 74–81. ACM, Ho Chi Minh City (2017)

19. Vakil-Baghmisheh, M.-T., Ahandani, M.A.: A differential memetic algorithm. Artif. Intell. Rev. **41**(1), 129–146 (2014)

20. Abraham, A., Das, S., Konar, A.: Document clustering using differential evolution. In: IEEE Congress on Evolutionary Computation, CEC (2006)

21. Das, S., Abraham, A., Konar, A.: Automatic clustering using an improved differential evolution algorithm. IEEE Trans. Syst. Man Cybern. Part A Syst. Hum. **38**(1), 218–237 (2008)

22. Wang, S., Li, Y., Yang, H.: Self-adaptive differential evolution algorithm with improved mutation mode. Appl. Intell. 1–15 (2017)

23. Uysal, A.K., Gunal, S.: The impact of preprocessing on text classification. Inf. Process. Manage. **50**(1), 104–112 (2014)

24. Gao, X.Z., Wang, X., Zenger, K.: A memetic-inspired harmony search method in optimal wind generator design. Int. J. Mach. Learn. Cybernet. **6**(1), 43–58 (2015)

25. Bharti, K.K., Singh, P.K.: Opposition chaotic fitness mutation based adaptive inertia weight BPSO for feature selection in text clustering. Appl. Soft Comput. **43**, 20–34 (2016)

26. Al-Jadir, I., et al.: Differential evolution memetic document clustering using chaotic logistic local search. In: Liu, D., Xie, S., Li, Y., Zhao, D., El-Alfy, E.S. (eds.) ICONIP 2017, Part I. LNCS, vol. 10634, pp. 213–221. Springer, Cham (2017). https://doi.org/10.1007/978-3-319-70087-8_23

# Neurodynamics-Based Nonnegative Matrix Factorization for Classification

Nian Zhang$^{(\boxtimes)}$ and Keenan Leatham

Department of Electrical and Computer Engineering,
University of the District of Columbia, Washington, D.C. 20008, USA
{nzhang,keenan.lamtham}@udc.edu

**Abstract.** This paper contributes to study the influence of various NMF algorithms on the classification accuracy of each classifier as well as to compare the classifiers among themselves. We focus on a fast nonnegative matrix factorization (NMF) algorithm based on a discrete-time projection neural network (DTPNN). The NMF algorithm is combined with three classifiers in order to find out the influence of dimensionality reduction performed by the NMF algorithm on the accuracy rate of the classifiers. The convergent objective function values in terms of two popular objective functions, Frobenius norm and Kullback-Leibler (K-L) divergence for different NMF based algorithms on a wide range of data sets are demonstrated. The CPU running time in terms of these objective functions on different combination of NMF algorithms and data sets are also shown. Moreover, the convergent behaviors of different NMF methods are illustrated. In order to test its effectiveness on classification accuracy, a performance study of three well-known classifiers is carried out and the influence of the NMF algorithm on the accuracy is evaluated.

**Keywords:** Nonnegative matrix factorization
Discrete-time projection neural network · Dimensional reduction
Feature selection · Classification

## 1 Introduction

Rapid growing modern technologies, such as remote explosive detection, 4D CT imaging, and DNA microarrays have generated an explosion of massive data. It is estimated that by 2020, over 40 trillion gigabytes of data will be generated, reproduced, and consumed [1]. The rapid growth of complex and heterogeneous data has posed great challenges to data processing and management. Established data processing technologies are becoming inadequate given the growth of data in volumes the digital world is currently generating. Advanced machine learning technologies are urgently needed to overcome big data challenges. They can help to ascertain valued insights for enhanced decision-making process in critical fields such as healthcare, economy, smart energy systems, and natural catastrophes prediction, etc.

One of the biggest challenges that traditional classification methods face is that when the dimensionality of data is high but with very few data, a large number of class prototypes existing in a dynamically growing dataset will lead to inaccurate classification

© Springer Nature Switzerland AG 2018
L. Cheng et al. (Eds.): ICONIP 2018, LNCS 11302, pp. 519–529, 2018.
https://doi.org/10.1007/978-3-030-04179-3_46

results. Therefore, selection of effective dimensionality reduction techniques is of great importance. Feature selection is one of the powerful dimensionality reduction techniques that choose the most distinct features according to the cost function values to acquire high prediction rate but without losing the best predictive accuracy. Although numerous combinations of feature selection algorithms and classification algorithm have been demonstrated, we explore an emerging and increasingly popular technique in analyzing multivariate data - non-negative matrix factorization (NMF) technique, and combine it with three state-of-the-art classifier, namely Gaussian process regression, Support Vector Machine, and Enhanced K-Nearest Neighbor (ENN), in order to study the influence of NMF on the classification accuracy.

NMF is a popular low-rank approximation technique that is ideal for dimensionality reduction. However, unlike other dimensionality reduction techniques, NMF incorporates non-negative constraints and thus, obtains part-based representation [2] which has remarkable semantic interpretability [3]. Nevertheless, since it was first introduced, NMF and its varied forms were primarily studied in image retrieval and classification [4–9]. However, the effectiveness of NMF for classifying numerical features other than images was still under investigation. In this paper, we therefore explore this aspect to find out if NMF can significantly improve the classification accuracy. Moreover, there is lack of study on the performance of a combined NMF with classifiers to our best knowledge. Thus, we extend research concerning integrate NMF with different classifiers with the goal to determine appropriate ones. A discrete-time projection neural network will be used develop the NMF algorithm due to the power of global convergence and fast convergence rate [10].

The rest of the paper is organized as follows. In Sect. 2, related work about non-negative matrix factorization (NMF) and different classifiers are discussed. In Sect. 3, continuous-time projection neural network and discrete-time projection neural network are introduced. In Sect. 4, the NMF algorithm based on the discrete-time projection neural network (DTPNN) are described. In Sect. 5, the comparison of convergent objective function values and CPU running time on different NMF based algorithms in terms of the two objective functions are presented. The comparison of different classifiers is also demonstrated. Finally, the paper is concluded in Sect. 5.

## 2  Related Works

### 2.1  Non-negative Matrix Factorization

Non-negative matrix factorization (NMF), also known as non-negative matrix approximation is an emerging technique where a matrix $V$ is factorized into two matrices, $W$ and $H$, with all three matrices containing no negative elements in them. Part of the reason is because the non-negativity will make the new matrices easier to investigate [11]. Assume matrix $V$ be the product of the matrices $W$ and $H$,

$$V = W \times H$$

Each column of $V$ can be computed as a product of the column vectors in $W$ and the coefficients supplied by columns of $H$, as follows:

$$v_i = W \times h_i$$

where $v_i$ is the $i$-th column vector of the product matrix $V$ and $h_i$ is the $i$-th column vector of the matrix $H$.

An appealing advantage by adopting NMF is dimensional reduction. Because the dimensions of those factor matrices will be significantly lower than the original matrix. For example, if $V$ is an $m \times n$ matrix, $W$ is an $m \times p$ matrix, and $H$ is a $p \times n$ matrix, then $p$ can be significantly smaller than both $m$ and $n$ (Fig. 1).

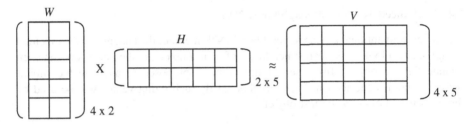

**Fig. 1.** Representation of non-negative matrix factorization. The matrix $V$ is factorized into two reduced matrices, $W$ and $H$. When multiplied, they approximately reconstruct $V$.

## 2.2 Gaussian Process Regression (GPR)

Gaussian process regression (GPR) are nonparametric kernel-based probabilistic techniques with infinite-dimensional generalization of multivariate normal distributions. Gaussian processes are utilized in statistical modeling, regression to multiple target values, and analyzing mapping in higher dimensions. There are four varied models with different kernels. The rational quadratic GPR kernel allows us to model data varying at multiple scales. Square exponential GPR is a function space expression of a radial basis function regression model with infinitely many basis functions. A fascinating feature is it replaces inner products of basis functions with kernels. The advantage to this feature is handling large data sets in higher dimensions will unlikely produce huge errors. Also, it handles discontinuities well. The matern 5/2 kernel takes spectral densities of the stationary kernel and create Fourier transforms of RBF kernel. Exponential GPR is identical to the Squared Exponential GPR except that the Euclidean distance is not squared. Exponential GPR replaces inner products of basis functions with kernels slower than the Squared Exponential GPR. It handles smooth functions well with minimal errors, but with discontinuities it does not handle well. A comprehensive comparison of classification performance among them is shown in terms of various model statistics. The classification error rates of these four models are also compared to the extended nearest neighbor (ENN), classic k-nearest Neighbor (KNN), naive Bayes, linear discriminant analysis (LDA), and the classic multilayer perceptron (MLP) neural network [12].

## 2.3  Support Vector Machine (SVM)

Support vector machine (SVM) is currently a popular supervised machine learning technique for classification and regression, first identified by Vapnik et al. in 1992. SVM regression is well-known for its nonparametric capability and has various kernel models. Linear kernel linearly scales the size of the training data. If data are not linearly separable, Quadratic kernel is adopted to find out the maximum margin between two classes. It is implemented by mapping the original feature space to a higher dimensional feature space where the training data is separable that allows a linear classifier. Gaussian kernel only depends on the Euclidean distance to decide if two points belong to the same class or not. The comparisons of their performance on the photo-thermal infrared imaging spectroscopy classification are demonstrated in [13].

## 2.4  Enhanced K-Nearest Neighbor (ENN)

Unlike the conventional k-nearest neighbor (KNN) method, the enhanced KNN method a. find out also look for k nearest neighbors of an test object, b. the neighbors can even include the test object [14]. The validity rating is proposed to measure the similar between samples and their nearest neighbors [15]. Such rating values will be used by the classifier to classify the test object.

# 3  Background

## 3.1  Continuous-Time Projection Neural Network

An optimization problem can be generally formulated as

$$\text{Min} f(u) \rightarrow \text{s.t } l \leq u \leq h \tag{1}$$

This problem can be solved by the following one-layer continuous-time projection neural network solution [16]

$$\epsilon \frac{du}{dt} = -u + g(u - \nabla f(u)) \tag{2}$$

Where $\epsilon > 0$ is a time constant, $\nabla f(u)$ denotes the gradient of $f$, and $g(\cdot)$ represents a piecewise linear activation function.

$$g(\xi_i) = \begin{cases} l_i, & \xi_i < l_i \\ \xi_i, & l_i \leq \xi_i \leq h_i \\ h_i, & \xi_i > h_i \end{cases}$$

To customize to the NMF algorithms, $l_i$ will be 0 and $h_i$ will be $\infty$. Accordingly, $g(\cdot)$ has become a rectified linear unit (ReLU) activation function.

$$g(\xi_i) = \begin{cases} 0, & \xi_i < 0 \\ \xi_i, & \xi_i \geq 0 \end{cases} \tag{3}$$

## 3.2 Discrete-Time Projection Neural Network

A discrete-time projection neural network will be used develop the NMF algorithm due to the power of global convergence and fast convergence rate. By applying Euler discretization to the continuous-time projection neural network in (2), it will be transformed into a discrete-time projection neural network (DTPNN).

$$x_{k+1} = x_k + \lambda_k[-x_k + g(x_k - \nabla f(x_k))] \tag{4}$$

where $\lambda_k$ is a step size.

# 4 Neurodynamics-Based Non-negative Matrix Factorization Method

## 4.1 Neurodynamics Equation of Discrete-Time Projection Neural Network (DTPNN)

The dynamic equations of DTPNN for the two factorization matrices are formulated based on (4):

$$\begin{aligned} w_{k+1} &= w_k + \lambda_k[-w_k + g(w_k - \nabla f(w_k))] \\ h_{k+1} &= h_k + \lambda_k[-h_k + g(h_k - \nabla f(h_k))] \end{aligned} \tag{5}$$

where $\lambda_k$ is a step size.

The selection of step size $\lambda_k$ is extremely important. The stability of the DTPNN will be unstable if $\lambda_k$ equals or exceeds a certain bound [17]. The procedure of the selection of step size $\lambda_k$ can be found in Sect. 4.2.

## 4.2 Backtracking Line Search

In order to minimize $f(x_k + \lambda_k p_k)$ in (5), we use the following procedure to find the step size $\lambda_k$.

---

**Algorithm 1.** Backtracking Line Search Algorithm

---

**Given** $\lambda_{init} > 0$, *i.e.* $\lambda_{init} = 1$, $\alpha \in \left(0, \frac{1}{2}\right)$, $\beta \in (0,1)$, *i.e.* $\beta = 1/2$

   Set $\lambda_0 = \lambda_{init}$

   **Repeat** $\lambda_{k+1} = \beta\lambda_k$ **Until**

   $f(x_k + \lambda_k p_k) \leq f(x_k) + \alpha\lambda_k \nabla f(x_k)^T p_k \tag{6}$

---

## 4.3 Neurodynamics-Based Non-negative Matrix Factorization Algorithm

A non-negative matrix factorization algorithm named PN$^3$MF based on biconvex optimization formulation is developed in [17].

---

**Algorithm 2.** The PN$^3$MF algorithm

---

**Initialization**

Set $k = 0$, $\alpha, \beta, w_0, h_0, \lambda_k^w, \lambda_k^h$, error tolerance $\epsilon$ and maximum iteration $K$.

**while** $k < K$ **and** $|f(w_{k+1}, h_{k+1}) - f(w_k, h_k)| > \epsilon$ **do**

   **while** (6) is not satisfied **do**

   $\lambda_k^w = \lambda_k^w \cdot \beta$

   $\lambda_{k+1}^w = \lambda_k^w$

   $w_{k+1} = w_k + \lambda_{k+1}^w[-w_k + g_w(w_k - \nabla_w f(w_k, h_k))]$ $\qquad$ (7)

   **end while**

   **while** (6) is not satisfied **do**

   $\lambda_k^h = \lambda_k^h \cdot \beta$

   $\lambda_{k+1}^h = \lambda_k^h$

   $h_{k+1} = h_k + \lambda_{k+1}^h[-h_k + g_w(h_k - \nabla_h f(w_{k+1}, h_k))]$ $\qquad$ (8)

   **end while**

   $k = k + 1$

**end while**

**return** $w_k, h_k$

---

## 4.4 Combined NMF and Classification Algorithm

In this paper, we combine the NMF algorithm with different classifier to explore the efficiency of the PN$^3$MF algorithm.

---

**Algorithm 3.** Combined PN$^3$MF-Classification Algorithm

---

**Input**: $V$: training set

$\qquad$ $r$: cluster numbers

$\qquad$ $S$: $p$ unknown samples without labels

**Output**: $c$: predicted class labels of the $p$ unknown samples

**Training Procedure:**

1.  Normalize the training set
2.  Solve the NMF optimization problem:

$$[W, H] = PN^3MF(V, r)$$

**Test Procedure:**

1.  Normalize the test set
2.  Solve the NMF optimization problem:

$$\min f(W, H) = \frac{1}{2} \|V - WH\|_F^2$$

3.  Predict the class label, $c_i$
4.  Return $c$

---

# 5  Experimental Results

In this section, we intend to study the influence of various NMF algorithms on the classification accuracy of each classifier as well as to compare the classifiers among themselves. NMF algorithms are used to decompose original data set $V$ according to the cluster number $r$. MUR [18], ALS [19], PG [20], AS [21], BBP [22], NeNMF [23], and PN$^3$MF [17] algorithms are compared. Three classifiers are applied to both the original and reduced dimensionality. Nine commonly used real-world datasets from UCI Machine Learning Repository are chosen to conduct the experiments [24].

## 5.1  Initialization

The error tolerance and the maximum number of iterations are initialized to $10^{-7}$ and 5,000, respectively. Let $\alpha \in \left(0, \frac{1}{2}\right)$ and $\beta \in (0, 1)$. The initial value of $\lambda_{init}$ for $f_1$ (Frobenius-norm) and $f_2$ (Kullback-Leibler divergence) is set to 2.0 and 1.0, respectively.

## 5.2  Convergent Objective Function Values

Two objective functions, Frobenius-norm and Kullback-Leibler (K-L) divergence are adopted to evaluate the optimization performance of factorization. Table 1 shows the convergent values when the Frobenius-norm function is adopted. Compared to six popular NMF algorithms, most of the time PN$^3$MF reaches the lowest objective function value. Similarly, Table 3 records convergent values in terms of the Kullback-Leibler (K-L) divergence function. PN$^3$MF gets the best results on most data sets.

## 5.3  CPU Running Time

Table 2 presents the CPU time of these algorithms when Frobenius-norm function is used. Although MUR and ALS algorithms consume less time on the breast tissue data set, they fails to achieve the minimum objective function value.

Table 4 provides the CPU time of those algorithms when Kullback-Leibler (K-L) divergence function is used. It shows that PN$^3$MF always consumes less CPU time than other NMF algorithms.

**Table 1.** Convergent objective function values in terms of Frobenius-norm function

| | Breast cancer | Haberman | Breast tissue | Movement libras | ILPD | Ionosphere | Vowel | Segmentation | Pen digits |
|---|---|---|---|---|---|---|---|---|---|
| MUR | **15.84** | 1.59 | 1.58 | **0.98** | **15.91** | 2.00 | 2.00 | **1.40** | **0.99** |
| ALS | **15.84** | **1.33** | 1.58 | **0.98** | **15.91** | 1.58 | 1.11 | **1.40** | **0.99** |
| PG | **15.84** | 1.59 | 1.00 | 0.99 | **15.91** | 1.11 | 1.11 | **1.40** | **0.99** |
| AS | **15.84** | 1.44 | **0.99** | 1.00 | **15.91** | 1.11 | 1.11 | **1.40** | **0.99** |
| BBP | **15.84** | **1.33** | **0.99** | 1.28 | **15.91** | **1.00** | 1.11 | **1.40** | **0.99** |
| NeNMF | **15.84** | **1.33** | **0.99** | 1.11 | **15.91** | **1.00** | **1.00** | **1.40** | **0.99** |
| PN$^3$MF | **15.84** | **1.33** | **0.99** | 1.11 | 16.00 | **1.00** | **1.00** | **1.40** | **0.99** |

**Table 2.** CPU running time in seconds when Frobenius-norm function is used

|        | Breast cancer | Haberman | Breast tissue | Movement libras | ILPD | Ionosphere | Vowel | Segmentation | Pen digits |
|--------|--------|--------|--------|--------|--------|--------|--------|--------|--------|
| MUR    | 0.0434 | 0.0250 | **0.0100** | **0.099** | 0.12 | **0.112** | **0.111** | 0.113 | **0.099** |
| ALS    | 0.0400 | **0.0200** | **0.0100** | **0.099** | 0.12 | **0.112** | **0.111** | 0.113 | **0.099** |
| PG     | 0.0233 | 0.0240 | 0.0150 | **0.099** | **0.11** | **0.112** | **0.111** | 0.113 | **0.099** |
| AS     | 0.0233 | **0.0200** | **0.0100** | **0.099** | **0.11** | 0.113 | **0.111** | **0.112** | **0.099** |
| BBP    | **0.0200** | 0.0300 | 0.0340 | **0.099** | **0.11** | **0.112** | **0.111** | **0.112** | **0.099** |
| NeNMF  | **0.0200** | **0.0200** | 0.0240 | **0.099** | **0.11** | **0.112** | **0.111** | **0.112** | **0.099** |
| PN$^3$MF | **0.0200** | **0.0200** | **0.0100** | **0.099** | **0.11** | **0.112** | **0.111** | **0.112** | **0.099** |

**Table 3.** Convergent objective function values in terms of Kullback-Leibler (K-L) divergence

|        | Breast cancer | Haberman | Breast tissue | Movement libras | ILPD | Ionosphere | Vowel | Segmentation | Pen digits |
|--------|--------|--------|--------|--------|--------|--------|--------|--------|--------|
| MUR    | 1.11 | 1.40 | 1.20 | **0.50** | **0.32** | 1.00 | 1.00 | 1.12 | **0.90** |
| ALS    | 1.11 | 1.11 | 1.20 | **0.50** | **0.32** | 1.00 | **0.50** | **0.60** | **0.90** |
| PG     | **0.99** | 1.11 | 1.00 | 1.00 | 0.50 | **0.40** | **0.50** | **0.60** | **0.90** |
| AS     | **0.99** | 1.33 | 1.00 | 1.00 | 0.50 | **0.40** | **0.50** | **0.60** | **0.90** |
| BBP    | **0.99** | **0.20** | 1.00 | 1.00 | 1.00 | 1.11 | 1.00 | **0.60** | **0.90** |
| NeNMF  | **0.99** | **0.20** | **0.20** | **0.50** | **0.32** | 1.11 | 1.00 | 1.00 | **0.90** |
| PN$^3$MF | **0.99** | **0.20** | **0.20** | **0.50** | **0.32** | 1.11 | 1.00 | 1.00 | **0.90** |

**Table 4.** CPU running time in seconds when Kullback-Leibler (K-L) divergence is used

|        | Breast cancer | Haberman | Breast tissue | Movement libras | ILPD | Ionosphere | Vowel | Segmentation | Pen digits |
|--------|--------|--------|--------|--------|--------|--------|--------|--------|--------|
| MUR    | 0.0144 | **0.04** | **0.030** | **0.0009** | 0.02 | **0.012** | **0.01** | 0.03 | **0.009** |
| ALS    | 0.0144 | **0.04** | **0.030** | **0.0009** | 0.02 | **0.012** | **0.01** | 0.03 | **0.009** |
| PG     | 0.0155 | **0.04** | **0.030** | **0.0009** | **0.01** | **0.012** | **0.01** | 0.03 | **0.009** |
| AS     | 0.0155 | **0.04** | 0.033 | **0.0009** | **0.01** | 0.013 | **0.01** | **0.02** | **0.009** |
| BBP    | 0.0155 | **0.04** | 0.033 | **0.0009** | **0.01** | **0.012** | **0.01** | **0.02** | **0.009** |
| NeNMF  | **0.0100** | **0.04** | **0.030** | **0.0009** | **0.01** | **0.012** | **0.01** | **0.02** | **0.009** |
| PN$^3$MF | **0.0100** | **0.04** | **0.030** | **0.0009** | **0.01** | **0.012** | **0.01** | **0.02** | **0.009** |

## 5.4   Convergent Objective Function Values vs. Iterations

We compare the convergent behaviors values on the wine data set among five NMF algorithms in terms of Frobenius-norm function. Figure 2 shows that PN$^3$MF algorithm takes the minimum number of iterations to converge.

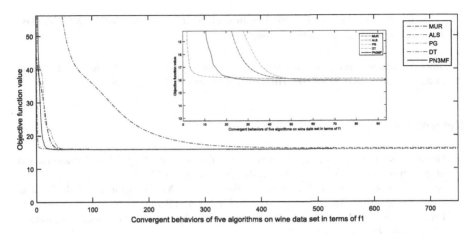

**Fig. 2.** Convergent behaviors of five algorithms on wine data using Frobenius-norm function.

## 5.5 Classification Results

We further investigate the influence of various NMF algorithms on the classification accuracy of each classifier as well as to compare the classifiers among themselves, as shown in Table 5. The experimental results demonstrate that PN³MF can improve the classification accuracy on most data sets. In addition, the combination of PN³MF+SVM performs better than other combinations.

**Table 5.** Classification accuracy comparison (percentage)

|  | Breast cancer | Haberman survival | Breast tissue | Movement libras | Vowel | Pen digits |
|---|---|---|---|---|---|---|
| GPR | 96.35 | 97.45 | 95.63 | 98.64 | 97.45 | 100 |
| SVM | 96.45 | 97.45 | 95.45 | 98.45 | 97.35 | 100 |
| ENN | 96.35 | 97.45 | 95.35 | 98.45 | 97.45 | 97.84 |
| PN³MF +GPR | 98.75 | **100** | **100** | 98.56 | 98.65 | **100** |
| PN³MF +SVM | **100** | 98.75 | 98.75 | **100** | **100** | **100** |
| PN³MF +ENN | 98.75 | 98.75 | 98.75 | 98.75 | 98.65 | **100** |

## 6 Conclusions

In this paper, the NMF algorithm is combined with three classifiers in order to find out the influence of dimensionality reduction performed by the NMF algorithm on the accuracy rate of the classifiers, as well as to compare the classifiers among themselves. The results show that the classification accuracy has been improved after applying the

NMF algorithm. In addition, the combination of NMF algorithm with the SVM classifier performs better than other combinations.

**Acknowledgements.** This work was supported in part by the National Science Foundation (NSF) under Grants HRD #1505509, HRD #1533479, and DUE #1654474.

# References

1. Gantz, J., Reinsel, D.: The digital universe in 2020: big data, bigger digital shadows, and biggest growth in the far east. IDC – EMC Corporation (2012)
2. Xiao, Y., Zhu, Z., Zhao, Y., Wei, Y., Wei, S., Li, X.: Topographic NMF for data representation. IEEE Trans. Cybern. **44**(10), 1762–1771 (2014)
3. Lee, D.D., Seung, H.S.: Learning the parts of objects by nonnegative matrix factorization. Nature **401**(6755), 788–791 (1999)
4. Liu, X., Zhong, G., Dong, J.: Natural image illuminant estimation via deep non-negative matrix factorization. IET Image Process. **12**(1), 121–125 (2018)
5. Li, X., Cui, G., Dong, Y.: Graph regularized non-negative low-rank matrix factorization for image clustering. IEEE Trans. Cybern. **47**(11), 3840–3853 (2017)
6. Wang, S., Deng, C., Lin, W., Huang, G.B., Zhao, B.: NMF-based image quality assessment using extreme learning machine. IEEE Trans. Cybern. **47**(1), 232–243 (2017)
7. He, W., Zhang, H., Zhang, L.: Sparsity-regularized robust non-negative matrix factorization for hyperspectral unmixing. IEEE J. Sel. Top. Appl. Earth Obs. Remote Sens. **9**(9), 4267–4279 (2016)
8. Babaee, M., Yu, X., Rigoll, G., Datcu, M.: Immersive interactive SAR image representation using non-negative matrix factorization. IEEE J. Sel. Top. Appl. Earth Obs. Remote Sens. **9**(7), 2844–2853 (2016)
9. Xu, R., Li, Y., Xing, M.: Fusion of multi-aspect radar images via sparse non-negative matrix factorization. Electron. Lett. **49**(25), 1635–1637 (2013)
10. Xu, B., Liu, Q., Huang, T.: A discrete-time projection neural network for sparse signal reconstruction with application to face recognition. IEEE Trans. Neural Netw. Learn. Syst. **99**, 1–12 (2018)
11. Gong, M., Jiang, X., Li, H., Tan, K.C.: Multiobjective sparse non-negative matrix factorization. IEEE Trans. Cybern. **99**, 1–14 (2018)
12. Zhang, N., Xiong, J., Zhong, J., Leatham, K.: Gaussian process regression method for classification for high-dimensional data with limited samples. In: The 8th International Conference on Information Science and Technology (ICIST 2018), Cordoba, Granada and Seville, Spain (2018)
13. Zhang, N., Leatham, K.: Feature selection based on SVM in photo-thermal infrared (IR) imaging spectroscopy classification with limited training samples. WSEAS Trans. Sig. Process. **13**(33), 285–292 (2017)
14. Tang, B., He, H.: ENN: extended nearest neighbor method for pattern recognition. IEEE Comput. Intell. Mag. **10**(3), 52–60 (2015)
15. Zhang, N., Karimoune, W., Thompson, L., Dang, H.: A between-class overlapping coherence-based algorithm in KNN classification. In: The 2017 IEEE International Conference on Systems, Man, and Cybernetics (SMC2017), Banff, Canada (2017)
16. Xia, Y., Wang, J.: On the stability of globally projected dynamical systems. J. Optim. Theory Appl. **106**(1), 129–150 (2000)

17. Che, H., Wang, J.: A nonnegative matrix factorization algorithm based on a discrete-time projection neural network. Neural Netw. **103**, 63–71 (2018)
18. Lee, D.D., Seung, H.S.: Algorithms for non-negative matrix factorization. In: Advances in Neural Information Processing Systems, vol. 13, pp. 556–562. MIT Press, Cambridge (2001)
19. Berry, M.W., Browne, M., Langville, A.N., Pauca, V.P., Plemmons, R.J.: Algorithms and applications for approximate nonnegative matrix factorization. Comput. Stat. Data Anal. **52** (1), 155–173 (2007)
20. Lin, C.J.: Projected gradient methods for nonnegative matrix factorization. Neural Comput. **19**(10), 2756–2779 (2007)
21. Kim, H., Park, H.: Nonnegative matrix factorization based on alternating nonnegativity constrained least squares and active set method. SIAM J. Matrix Anal. Appl. **30**(2), 713–730 (2008)
22. Kim, H., Park, H.: Toward faster nonnegative matrix factorization. A new algorithm and comparisons. In: Proceedings of the Eighth IEEE International Conference on Data Mining, pp. 353–362 (2008)
23. Guan, N.Y., Tao, D.C., Luo, Z.G., Yuan, B.: NeNMF: an optimal gradient method for nonnegative matrix factorization. IEEE Trans. Signal Process. **60**(6), 2882–2898 (2012)
24. Lichman, M.: UCI Machine Learning Repository. School of Information and Computer Science, University of California, Irvine, CA (2013). http://archive.ics.uci.edu/ml/

# A Multi-kernel Semi-supervised Metric Learning Using Multi-objective Optimization Approach

Rakesh Kumar Sanodiya[✉], Sriparna Saha, and Jimson Mathew

Indian Institute of Technology Patna, Patna, India
rakesh.pcs16@iitp.ac.in

**Abstract.** A kernel-matrix based distance measure is utilized for computing the similarities between the data points. The available few labeled data is used as constraints to project on initial kernel-matrix using Bregman projection. Since the projection of constraints onto the matrix is not orthogonal, we need to identify an appropriate subset of constraints subject to objective functions, measuring the quality of partitioning of the data. As the kernel-space is large in size, we have divided the original kernel space into multiple kernel sub-spaces so that each kernel can be processed independently and parallelly in advance GPU and kernel semi-supervised metric learning using multi-objective approach is applied on individual kernels parallelly. The multi-objective framework is used to select the best subset of constraints to optimize multiple objective functions for grouping the available data. Our approach outperforms the state of the art algorithms on the various datasets with respect to different validity indices.

**Keywords:** Semi-supervised · Multi-objective optimization
Classification · Clustering · Graphics processing unit (GPU)

## 1 Introduction

With the rapid development of wirelessly connected devices such as sensors, Raspberry Pi, ZigBee, etc. placed everywhere in our daily life like in smart home, wearables devices, connected cars, etc. a massive amount of unlabeled data are being generated. Challenges lie in converting such huge data to useful information to provide a more comfortable environment to the users. For this, machine learning algorithms are required to be deployed. Such algorithms provide a way to mine the useful information hidden in this data, which can later be used for enhancing the performance of user's environment. Supervised learning approaches require huge amount of labeled data for generating a predictive model and unsupervised approaches such as k-means algorithm require appropriate distance functions for clustering the unlabeled data. As discussed, the generated data by such devices is in general unlabeled in nature. Annotating

© Springer Nature Switzerland AG 2018
L. Cheng et al. (Eds.): ICONIP 2018, LNCS 11302, pp. 530–541, 2018.
https://doi.org/10.1007/978-3-030-04179-3_47

this data manually is a time-consuming process. Thus semi-supervised machine learning approaches, which utilize a large collection of unlabeled data and few labeled data, could be of use for handling this available huge unlabeled data. While studying semi-supervised learning approaches, we have found that metric based semi-supervised learning approaches such as Information-Theoretic Metric Learning [1], a kernel-learning approach to semi-supervised clustering with relative distance comparisons [2] can play very important roles for labeling this data when available labeled data is very few.

Many of the machine learning approaches (examples include k-Nearest neighbor classifier and K-Means clustering) need appropriate distance/similarity matrices for their computations. For such algorithms, kernel matrix based distance measures play important roles. The main goal of the kernel matrix learning is to learn appropriate distance function with the help of side information which is given in the form of constraints such as Must-link/Cannot-link and Relative distance.

As the unlabeled data is huge, kernel matrix computation requires huge computation as well as time for labeling the data. One property of this kernel matrix is independent distance function calculation between the data. Therefore, we can divide the huge data into multiple parts and assume each part as a separate kernel matrix. For each part we can choose appropriate constraints with the help of MOO approach. Each kernel matrix can be parallelly executed on different GPUs. Thus, the performance and time can be improved.

There have been some exciting recent works on semi-supervised learning approaches to label the data, but such methods don't consider multiple cluster quality measures during their computations. However, the use of different cluster quality measures or objective functions during clustering of the data can identify more precisely different shaped clusters well. Optimizing all objective functions at the same time can help in the better evaluation of the quality of labeled data. This motivates us to propose a method that incorporates some internal and external cluster validity index based objective functions within the framework of multi-objective optimization (MOO). There have been many MOO approaches, some of them based on Pareto optimality. A popular MOO technique, namely Non-dominated Sorting Genetic Algorithm (NSGA) is proposed by Deb et al. Deb et al. again proposed NSGA-II which is the upgraded version of NSGA. We have used NSGA-III [3] as a multi-objective framework to optimize multiple objective functions over NSGA-II [4] because it uses concepts of reference points instead of crowding distance. Therefore, NSGA-III is better for optimizing more than two objective functions [5].

Contributions of the current work are as follows:

- The proposed approach divides the data set into multiple parts, and each part is treated as a separate independent kernel matrix. Each kernel can be executed on different grids of the GPU. Thus, the overall performance of computing and time can be improved.
- The proposed approach utilizes few labeled data for generating some equality and inequality constraints; the best subset of constraints are selected using

the search capability of MOO for each kernel matrix. The selected subset of constraints for each kernel matrix are utilized for adjusting the corresponding sub-kernel matrix for distance computation. Finally, K-means algorithm is applied to the adjusted sub-kernel matrix to label the unannotated data.

- The existing semi-supervised approaches, available for labeling a huge collection of unlabeled data, do not ensure the quality of resultant clusters. However, our proposed approach ensures the quality of labeled data generated during the process by optimizing different cluster quality measures simultaneously. Several internal and external cluster validity indices capturing the quality of obtained partitioning are optimized during the search process.

## 2  Background

This section introduces basics of kernel learning with relative distance technique.

### 2.1  Kernel Learning with Relative Distance

**Notation Description.** The dataset D, containing n data points, consists of a set of labeled data points $\{(x_j, y_j)\}_{j=1}^{l}$ and a set of unlabeled data points $\{(x_j)\}_{j=l+1}^{n}$ here $j^{th}$ point is $x_j \in R^d$ and its output $y_j \in R$. Typically, we have $n >> l$. The distance comparisons between data points in D are unknown distance functions. This function is given in terms of $\Delta : D * D \rightarrow R$. The function $\Delta$ reveals some knowledge in terms of Euclidean distance or Mahalanobis distance, which is not sufficient to quantify precisely, and cannot be calculated using only the features available in the dataset $D$. Thus, there is a requirement of some external information about the data points in addition to the features in the dataset $D$. This external information is given in the form of relative distance comparisons in the set $C$ by human evaluators or from other sources. Our main objective is to find a kernel matrix K that has more accurate and precise distance measures between all the data points in the dataset D. This kernel can be used for several purposes for accurate labeling or clustering.

**Pairwise Relative Constraints.** With the help of labeled data, relative distance constraint set, C, is generated. The constraints present in C infer some similarities/distances between data points in the data set D. For the current work, exact distances between data points in D are not required. Instead constraints satisfying some relative distance measurements are generated, for example $\Delta(a, b) \leq \Delta(b, c)$ for some $a, b, c \in D$ and a distance function $\Delta$. There are two types of constraint sets in C. First is equality constraint set, $C_{eq}$, and second is inequality constraint set, $C_{neq}$. The set $C_{eq}$ contains constraints where all three points belong to the same class, i.e., pair-wise distances between all three points are assumed to be same. The set $C_{neq}$ contains constraints where two points belong to the same class and other point belongs to the different class, i.e., the distance of the other class point to the two points belonging to the same class is always larger than the distance between two points of the same class. In general $C_{neq}$ is a set of rows of the form $(a, b|c)$, where every row is interpreted as "Point c belongs to other class and points a and b belong to the same class". Every row of set $C_{neq}$ implies the following two equations.

$$(a \leftarrow b|c) : \Upsilon\Delta(a,b) \leq \Delta(a,c) \ \ and \tag{1}$$

$$(b \leftarrow a|c) : \Upsilon\Delta(b,a) \leq \Delta(b,c) \tag{2}$$

where $\Upsilon$ is a constant parameter.

Similarly, $C_{eq}$ is the set of rows of the form (a,b,c), where each row is represented as "all points a, b, and c belong to the same class". Each row of the set $C_{eq}$ implies the following equation.

$$(a \leftrightarrow b; c) : \Upsilon(a,b) = \Upsilon(a,c) \tag{3}$$

$$(b \leftrightarrow a; c) : \Upsilon(b,a) = \Upsilon(b,c) \tag{4}$$

$$(c \leftrightarrow a; b) : \Upsilon(c,a) = \Upsilon(c,b) \tag{5}$$

## 3 Problem Statement

All the constraints generated after application of Eqs. 1 to 5 on the available few labeled dataset may not be helpful in obtaining the final partitioning of the entire dataset. All the constraints (one by one) are satisfied by using Bregman projection, but previously satisfied constraints may get unsatisfied because of projection of current constraints. Therefore, an appropriate subset of constraints is required to be identified automatically which will be suitable for partitioning the unlabeled data as shown in Fig. 1(b)(c). Again because of high volume of unlabeled dataset, concept of "divide and conquer" is applied to impose parallelism. The available data is divided into multiple parts and each part can be executed independently in different grids of GPU. Each part is independent of each other and requires a subset of constraints to obtain the best partitioning. Thus, for each part, only that subset of constraints should be selected, which can help in identifying the optimal partitioning. The optimality of the partitioning can be checked with respect to different cluster validity indices. The selected constraint subset of each part is utilized to transform the corresponding initial kernel matrix into an optimal kernel matrix. Let, the dataset D be divided into n parts, i.e., $T_1, T_2, T_3, \ldots, T_n$.

Thus the problem statement of the current work is as follows:

$$\forall_i^n \underset{K_i \preceq 0}{\text{minimize}} \quad D_{ld}(K_i, K_{0i})$$

$$\text{subject to} \quad \bar{C}_{i=1}^n \subseteq C \tag{6}$$

$$\forall_j^m obj_j \in O.$$

where $K_{0i}$ is initial matrix , $\bar{C}_i^n$ is optimal subset of constraints corresponding to $i^{th}$ part, and $O = \{obj_1, obj_2, obj_3, ..., obj_m\}$ is a set of some internal and external cluster validity indices. These objective functions are simultaneously optimized using the search capability of MOO-based technique.

# 4  Multi-kernel Based Semi-supervised Metric Learning Using Multi-objective Optimization

We propose a Multi-Kernel based semi-supervised metric learning using multi-objective algorithm. The flow chart of the proposed approach is shown in Fig. 4(a). As the proposed approach aims to divide the dataset into multiple parts and optimizes a set of cluster quality measures simultaneously while selecting the appropriate subset of constraints for different parts, a popular MOO technique called NSGA-III (non-dominated sorting genetic algorithm) [6] is utilized as the underlying optimization strategy. Initially, in order to reduce the time-complexity of calculating distances between all pairs of points present in the whole dataset, we divide the whole dataset into different smaller subparts and each subpart is treated as sub-kernel. This division is possible because distance calculation is an independent task. With each sub-kernel, labeled data is added in order to generate the relative constraints. Now each sub-kernel is treated as independent kernel and we apply our proposed Algorithm 1 on it. To understand the algorithm in detail, let us consider $i^{th}$ generation in which population set P is having N solutions (each solution is composed of subset of constraints and corresponding objective functions). Next, two solutions are randomly selected from the set $P$ and crossover operator is applied on them as discussed in Sect. 4.2 to generate two new chromosomes. After that each chromosome also undergoes through some mutation operation. Application of crossover and mutation operators helps in generating a new population Q of size $N$. Next the old population, $P$ and the new population, $Q$ are combined to generate a merged population (R) of size $2 \times N$. Non-Dominated sorting algorithm is applied to sort $R$ in different non-domination levels $(F_1, F_2, \ldots, F_m)$. Next, each population member is normalized and then each normalized population member is associated with reference points generated. Finally niche operation discussed in is applied. All of the above expressions are discussed in Sect. 4.4 Then, one by one each non-dominated level is selected to construct the new population $T$, starting from $F_1$, until size of $|T| \geq N$. If the last included level is L, then all the solutions from level (L+1) onward are rejected from R. If $|T| = N$ then no further operation is needed and set $T$ is kept for next generation. If $|T| \leq N$, then members from 1 to L-1 are selected and the remaining solutions are chosen from the last front L. At the end we choose final solution from the top most Pareto front as discussed in Sect. 4.4.

## 4.1  Chromosome Representation

All the constraints induced from the Eqs. 1, 2, 3, 4 and 5 are encoded into a constraint matrix called a chromosome. All the induced constraints are categorized as two types, equality constraint $(C_{eq})$ and inequality constraint $(C_{neq})$. To define these constrains, let us consider a dataset that have 3 points i, j and k. In inequality constraint $(C_{neq})$, points i and j belong to the same class and remaining point k belongs to other class. But, in equality constraint $(C_{eq})$, all

**Algorithm 1.** Kernel-NSGA-III

1: *Initialize all the parameters such as $i=1$, $Division=p$, $Maxit=N$, $Npop=m$,*
   *$pCrossover=0.5$, $pMutation=0.5$*
2: *Compute the number of reference points based on the number of divisions*
3: *Generate Npop initial population members with constraints and objective cost functions*
4: *Apply Non-Dominated sorting*
5: **while** $i \leq Maxit$ **do**
6:    *Choose two random population members $s_1$ and $s_2$ from the population members*
7:    *Apply crossover operator between selected members with crossover probability*
      *pCrossover*
8:    *Apply Mutation operator between selected members with mutation probability*
      *pmutation*
9:    *Apply Non-Dominated sorting*
10:   *Normalize all the population members*
11:   *Associate population members to the reference points*
12:   *Apply niche preservation operation*
13:   *Store the niche obtained population members for the next generation*

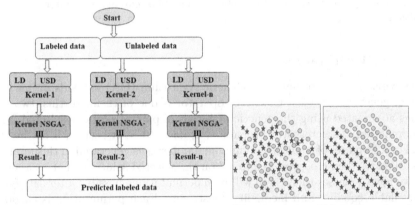

(a) Flow chart of proposed architecture  (b) 2 dimensional  (c) 2 dimensional
                                         representation      representation
                                         before projection  after projection

**Fig. 1.** Showing flow chart of our proposed architecture and 2 dimensional representation of data before or after Bregman projection

the points i, j and k belong to the same class. If we have three classes, namely a, b, and c, then different permutations possible are shown in Fig. 2(a).

For each class of different permutations, n-random data points are considered depending upon the user requirement. Let us take an example: for a permutation (a, a and b) n random data points are chosen from class a, class a and from class b, respectively, thus, there will be total (n* number of total permutation) rows possible corresponding to this permutation.

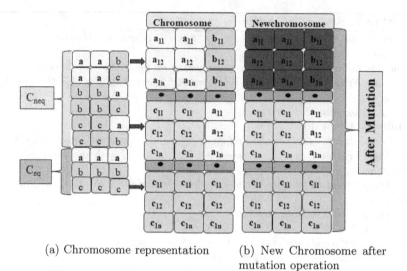

(a) Chromosome representation

(b) New Chromosome after mutation operation

**Fig. 2.** Showing chromosome representation and new chromosome after mutation operation

## 4.2   Genetic Operators

**Crossover.** The performance of Genetic Algorithm (GA) depends on various types of genetic operators. Crossover operator is one of them. In this, two off-springs are created using the genetic materials of parent-solutions selected from the population. The crossover probability is denoted by $\alpha_c$, which is set by the user. For a given set of solutions, a random number $c_0$ is generated; if this number is less than $\alpha_c$ then crossover is applied on the selected solutions. Let us assume that we have 3 classes then there will be $p$ permutations possible after satisfying the available equality and inequality constraints. Each permutation is encoded with n-combinations, which may vary. Two chromosomes, chromosome-1 and chromosome-2 shown in Fig. 3(a) are selected to perform crossover operation. If the generated random value $c_0$ is less than crossover probability $\alpha_c$ then another random number $c_1$ between 1 to p is generated. If the value of $c_1$ is p then all n combinations of last permutations of chromosome-1 and chromosome-2 are swapped with each other. The generated new chromosome-1 and chromosome-2 are shown in Fig. 3(b).

**Mutation.** In a simple way, mutation is defined as a small random change in the chromosome to generate the new solution. It is used to maintain the genetic diversity from one generation of a population of the genetic algorithm to the next. In this, the mutation probability is represented by $\mu_0$. Here, a random numbers $m_0$ is generated. If $m_0$ is less than the mutation probability $\mu_0$, then another random number $m_1$ generated between 1 and total number of permutations. Let us take an example: if $p_1$ is 1 then all combinations corresponding to first

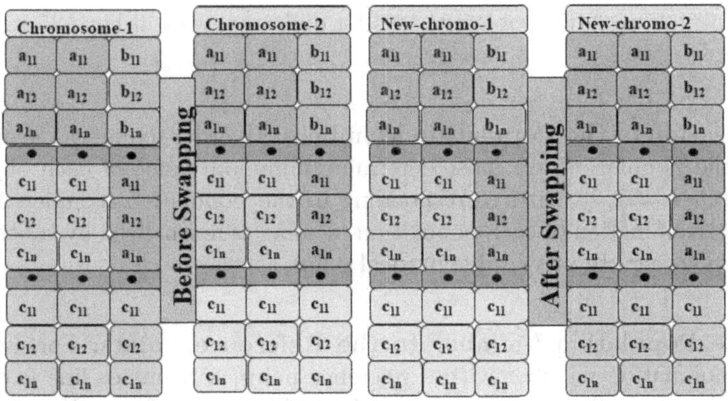

(a) Chromosomes representation before crossover

(b) Chromosome representation after crossover

**Fig. 3.** Showing crossover operation between the chromosomes

permutation of the chromosome shown in Fig. 2(a) are selected and replaced by random instances of the corresponding class, and the results are shown in Fig. 2(b).

### 4.3 Objective Functions

In order to obtain good partitioning of data, we have used both types of cluster validity indices, internal cluster validity measures and external cluster validity measures. Internal validity indices are based on information intrinsic to the data itself where as external validity indices are based on previous knowledge about data. In this paper, we have used total 4 validity indices namely, Accuracy (Acc), Adjusted Rand Index (ARI) [7], Silhouette Index (SIL) [8], and Dunn Index (DI) [8]. Two (Accuracy (Acc) and Adjusted Rand Index (ARI)) of them belong to external validation indices and remaining two (Silhouette Index (SI) and Dunn Index (DI)) belong to internal validation indices.

### 4.4 NSGA-III

We have used NSGA-III framework because we have more than two objective functions and NSGA-III performs better while optimizing more than two objective functions. Following are the steps of the NSGA-III algorithm.

**Determination of Reference Points on Hyper-Plan.** we choose a set of reference points in order to maintain the diversity of the obtained solutions and place them on normalized hyper-planes. If we have m objective functions, m dimensional hyper-plane is possible and each dimension is divided into k divisions. Then, reference points are calculated using the formula: $^{m+k-1}C_k$.

For example, as we have 4 objectives and 4 divisions, we will have 35 reference points.

**Normalization of the Population Members.** In order to determine an ideal point of the population, we must determine minimum value of each objective function $(O_i, i = 1, 2, ..., m)$ corresponding to the population members. Then, each objective function of each population member is normalized by subtracting the corresponding objective value of ideal point.

**Associate Population Member to the Reference Points.** For associating each population member with a reference point, a reference line is defined corresponding to each reference to the origin of the hyperplane. We calculated perpendicular distance of each population member to the reference line. Then a population member is assigned to that reference point whose reference line is the nearest.

**Niche Preservation Operation.** A reference point may be associated with one or more population members or may not be associated with anyone. we count that how many population members are associated with each reference point. Let $\sigma_i$ denote niche count for $i^{th}$ reference point. With the help of niche preservation, first we select a set of reference point which has minimum niche count. If we have more than one reference point in the set, then one of them is randomly selected.

### 4.5  Selection of a Single Solution for Reporting

After application of NSGA-III algorithm, at the end of generation we get a set of solutions on final Pareto front. We select one of the solutions which is having some optimal values with respect to all the objective functions. Then, we only consider subset of constraints of that best solution for projection.

## 5  Experimental Results

In order to demonstrate the effectiveness of our proposed method, we have compared our work with two existing semi-supervised based techniques, namely, a kernel semi-supervised clustering with relative distance comparisons (SKLR)[1] [2] and Information-Theoretic Metric Learning (ITML)[2] [1].

---

[1] https://github.com/eamid/sklr.
[2] http://www.cs.utexas.edu/~pjain/itml.

## 5.1    Data Sets Used

UCI Human Activity Recognition using Smart phones Dataset along with three popular datasets of UCI library like Iris, Thyroid and Wine are utilized for validating our method. Detailed descriptions of all these data sets in terms of total number of points present, dimension of the data set and the number of clusters are presented in Table 1.

**Table 1.** Shows description of the datasets

| Data set | Points | Dimension (d) | Actual number of clusters (K) |
|---|---|---|---|
| Activity activity | 10299 | 561 | 6 |
| Iris | 150 | 4 | 3 |
| Thyroid | 215 | 5 | 3 |
| Wine | 178 | 13 | 3 |

## 5.2    Evaluation Measures

To check the performance of our system in comparison to other existing systems (SKLR and ITML), we have used accuracy as an evolution measure. It is important to note that accuracy is also taken as one of the objective function in our proposed framework.

## 5.3    Discussion of Results

The activity dataset is divided into training (7532 samples) and testing (2947 samples) parts. We used 25% dataset of training for generating pairwise constrains and divided testing data into 3 equal parts. Then, we applied our approach on each part individually and got a set of Pareto optimal solutions. From each set of Pareto optimal solutions, we chose solution that have higher accuracy. Accuracies obtained for each part are 98.97%, 96.58%, and 95.53% respectively. We achieved on an average 97% Acc and 93% ARI, which are higher compared to corresponding objective functions of SKLR and ITML. But in terms of other objective functions like SS and DI we achieved less values because all the objective are conflicting in nature and it is difficult to optimize all the objective function simultaneously.

Further, we also divided the iris dataset into 3 equal parts (kernel) and use 25% labeled data for generating constraints. Here also we have done the same job as done for activity dataset. We obtained accuracies are 98.86%, 98.86%, and 94.31%, respectively. On an average we obtained 98% Acc, 96% ARI, and 47% SS which are higher compared to corresponding objective functions of SKLR and ITML but our approach achieved low DI.

For wine dataset, we use the same experimental setting as we used for Iris. After application of our proposed approach with individual kernels, our proposed method obtained 100%, 99.04%, and 99.029% accuracy, respectively. On an average our approach achieved 0.99% accuracy, 97% ARI, 20% SS, and 69% DI while SKLR and ITML achieved lower values.

For thyroid dataset also, same experimental setting is used as used for Iris. The obtained individual accuracies are 96.03%, 97.61%, and 97.60%, respectively, with different kernels by our proposed approach. Our system achieved on an average 0.9708% Acc, and 90% ARI but with respect to SS and DI our results are poor compared to those by SKLR and ITML.

Comparison results shown in Fig. 4 clearly illustrate that our approach is better than SKLR and ITML for activity, iris, wine and thyroid datasets in terms of external validation indices like Accuracy and Adjusted Rand Index. However, results are slightly less in terms of internal validation indices like silhouetted Index and Dunn Index because our approach optimizes all the objective functions while clustering. For reporting we have selected only a single solution from the final Pareto front. This solution may correspond to optimization of some other objective functions apart from silhouetted Index and Dunn Index

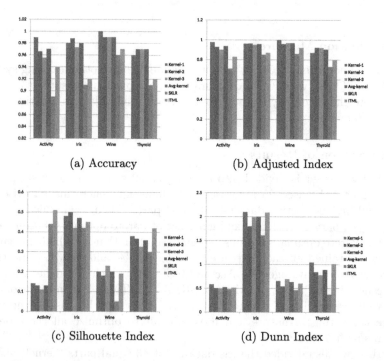

(a) Accuracy          (b) Adjusted Index

(c) Silhouette Index          (d) Dunn Index

**Fig. 4.** Showing clustering results obtained after application of proposed method, SKLR, ITML on human activity, iris, wine, and thyroid datasets.

## 6   Conclusion

In this work, a new semi-supervised multi-kernel based metric learning algorithm is proposed for clustering huge amount of unlabeled data with the help of minimal amount of labeled data. In order to deploy the concept of parallelism as well as to reduce the time complexity, the original dataset is divided into multiple kernels and each kernel is executed independently. For each kernel, a set of objective functions, comprising of internal and external cluster validity indices, are optimized simultaneously for generating optimal partitioning. As these functions are multiple conflicting objective functions, the MOO-based framework is utilized to simultaneously optimize them. A reasonable assumption that only a few labeled data (25%) is available while proposing the current approach. A set of constraints are generated from the available labeled data. For each kernel, a suitable subset of constraints is selected from the generated constraint set using the search capability of MOO-based approach utilizing the set of objective functions. The selected constraints for each kernel are finally utilized to develop an adjusted kernel matrix which is given as an input to the K-means clustering technique. The partitioning obtained by $K$-means can be viewed as a labeling of the whole dataset. In future, we will target to solve multiple tasks in multi-task learning framework with respect to various objective functions by using multi-objective optimization framework.

# References

1. Davis, J.V., Kulis, B., Jain, P., Sra, S., Dhillon, I.S.: Information-theoretic metric learning. In: Proceedings of the 24th International Conference on Machine Learning, pp. 209–216. ACM (2007)
2. Amid, E., Gionis, A., Ukkonen, A.: A kernel-learning approach to semi-supervised clustering with relative distance comparisons. In: Appice, A., Rodrigues, P.P., Santos Costa, V., Soares, C., Gama, J., Jorge, A. (eds.) ECML PKDD 2015. LNCS (LNAI), vol. 9284, pp. 219–234. Springer, Cham (2015). https://doi.org/10.1007/978-3-319-23528-8_14
3. Deb, K., Jain, H.: An evolutionary many-objective optimization algorithm using reference-point-based nondominated sorting approach, part I: solving problems with box constraints. IEEE Trans. Evol. Comput. 18(4), 577–601 (2014)
4. Deb, K., Pratap, A., Agarwal, S., Meyarivan, T.A.M.T.: A fast and elitist multi-objective genetic algorithm: NSGA-II. IEEE Trans. Evol. Comput. 6(2), 182–197 (2002)
5. Ciro, G.C., Dugardin, F., Yalaoui, F., Kelly, R.: A NSGA-II and NSGA-III comparison for solving an open shop scheduling problem with resource constraints. IFAC-PapersOnLine 49(12), 1272–1277 (2016)
6. Li, H., Zhang, Q.: Multiobjective optimization problems with complicated Pareto sets, MOEA/D and NSGA-II. IEEE Trans. Evol. Comput. 13(2), 284–302 (2009)
7. Robert, V. Vasseur, Y.: Comparing high dimensional partitions, with the Coclustering Adjusted Rand Index. ArXiv Preprint ArXiv:1705.06760 (2017)
8. Rendón, E., Abundez, I., Arizmendi, A., Quiroz, E.M.: Internal versus external cluster validation indexes. Int. J. Comput. Commun. 5(1), 27–34 (2011)

# Software-as-a-Service Composition in Cloud Computing Using Genetic Algorithm

Samuel Yu Toh and Maolin Tang(✉)

School of Electrical Engineering and Computer Science,
Queensland University of Technology, 2 George Street,
Brisbane, QLD 4001, Australia
yu.toh@connect.qut.edu.au, m.tang@qut.edu.au

**Abstract.** Cloud computing is a new IT paradigm. Over the last few years, there has been a trend of increasing adoption of a new software delivery model called Software-as-a-Service (SaaS) in the new IT paradigm. While the availability of SaaS in cloud computing has yet created a challenge for the service-oriented computing community, we believe it is only a matter of time that SaaS will grow exponentially to a stage where manual SaaS composition becomes impossible. In order to better prepare ourselves for this challenge, this paper proposes a multi-tenant enabled SaaS composition framework for cloud computing. While there have already been studies involved in tackling service composition in cloud computing, most of them ignore a key feature that is specific to cloud computing, that is multi-tenancy. This paper proposes a SaaS composition framework that can be used to automatically build SaaS in cloud computing.

**Keywords:** SaaS · Cloud computing · Service composition
Genetic algorithm

## 1 Introduction

Cloud computing is a new business model driven by IT utility service for IT industry. It offers great flexibility that allows users to scale and shrink IT resources at minimal effort, risk and cost. Generally, for cloud computing it is distinguished into three primary types of services, Infrastructure-as-a-Service (IaaS), Platform-as-a-Service (PaaS) and Software-as-a-Service (SaaS). IaaS is an on-demand service for providing third party with virtualized computing infrastructures like disk storages, computer processing powers and network connections. PaaS is a facility for software developers to build and deploy software applications over the cloud, e.g. Google AppEngine. SaaS is an alternative software delivery model for traditional on-premise software service where its ownership and management are outsourced to a cloud service provider.

© Springer Nature Switzerland AG 2018
L. Cheng et al. (Eds.): ICONIP 2018, LNCS 11302, pp. 542–551, 2018.
https://doi.org/10.1007/978-3-030-04179-3_48

These services together created a novel IT paradigm for which enterprises can utilise for better cost efficiency and scalability on their IT solutions, without having to worry about the underlying hardware problems. This in return created a trend on accelerated cloud computing adoption which was observed by many over the recent years [1–4]. This emergence has also attracted many researchers to study the possibility on shifting conventional software development methods into the cloud. To this end, many recent studies have just begun to address SaaS composition in cloud computing [5, 6, 13–16]. However, some important features of SaaS composition in could computing are not considered in these existing SaaS composition methods.

Motivated by this, we will propose a new framework for SaaS composition in cloud computing. In this work we attempt to create a new SaaS by considering the properties of cloud computing. Firstly, we incorporate cloud software design principals, such as multi-tenancy, into the SaaS composition problem. Secondly, we also consider the deployment of the SaaS in the SaaS composition problem.

The remaining paper is organized as follows. Firstly, a literature review on the state of the art about web service composition research, in cloud and in non-cloud contexts, is presented in Sect. 2. Then, in Sect. 3 the SaaS composition problem is discussed. Our solution to the SaaS composition problem is proposed in Sect. 4 and evaluated in Sect. 5. Finally we conclude the research on the SaaS composition problem in Sect. 6.

## 2   Related Work

As of today, web service remains to be the dominating technology for building composite services, and to the best of our knowledge using web service components to create a new value added web service as an SaaS, which is called *SaaS composition*, is a new topic to the cloud computing domain. While SaaS composition is not the same as web service composition, the concept for aggregating web service components into a new web service is similar to SaaS composition. Thus, we start this literature review on web service composition.

Since the beginning of web service composition, numerous theoretical models have been proposed. Zeng et al. [7] was the first group who proposed the idea of quality-driven selection for web service composition, to which the problem was formalized as an integer programming problem. While linear programming was proven to be an optimal method in terms of solution quality, its scalability remains the greatest deficiency due to the exponential time required for solving problems with large search space. In [8], the web service composition problem was converted into a knapsack problem and the authors proposed two methods for solving the knapsack problem: a combinatorial optimization problem method and a graph theory based method. Considering that the web service composition problem is NP-hard, evolutionary computation methods were then proposed by [9–11]. Although these methods consider various QoS metrics and constraints, such as the inter-service constraints and dependencies between the web service components, they cannot be directly used for the SaaS composition problem as they do not consider the features of the software development and deployment of software in the cloud.

In cloud computing, a multi-tenant software model is usually adopted. Instead of serving a single end-user, a SaaS in cloud computing provides the services to multiple users at the same time, but with different QoS values. Thus, the SaaS composition needs to consider different QoS requirements from different stakeholders. In addition, the SaaS composition needs to optimize the overall performance and to minimize the composition and running costs of the SaaS. Thus, existing web service composition methods cannot be immediately used for the SaaS composition problem in cloud computing.

In [12], a multi-tenant SaaS composition problem was proposed and solved using an integer programming method. Instead of deigning a service plan for a sole customer, their work involved creating an optimized solution for multiple users known as tenants. The proposed integer programming method can maximize the overall SaaS utility and minimize the overall cost of resource usage simultaneously. The advantage with the integer programming method is that it always produce an optimal solution. The deficiency with the method is that it is not scalable. When the size of the SaaS composition problem increases, the computation time will increases exponentially.

In addition to the integer programming method, many other attempts have been made to the SaaS composition problem [5,6,12–16]. These methods addressed some issues specific to cloud computing. However, they did not consider the multi-tenant issue. In [5], for example, the researchers highlighted the importance of having a network awareness model for a cloud based service composition by incorporating network latency and data transfer rate attributes into their work.

## 3    The SaaS Composition Problem

To build a new SaaS, it begins with a lucrative business idea. Then a preliminary study is made to understand the functional requirements for the new SaaS. Next, a decision is made to decide how the service will be built. Typically in the software development practise there are two options that could go about getting the SaaS done. It is by either building from scratch or by SaaS composition. If the decision is to go by the latter, then a business analyst works out the workflow of the new SaaS. At the same time, a study of the non-functional requirements for the new SaaS is also carried out to map out the QoS requirements. Then, the SaaS developer discovers the potential SaaS components that can fill into each of the defined tasks in the workflow. For each task in the workflow there could be multiple SaaS components, all of which can fulfil the functionality of the task, but may have different QoS values. Thus, a decision is made by the SaaS developer to pick a SaaS component among all the candidate SaaS components for each of the tasks in the workflow such that the overall QoS of the new SaaS is optimal for all stakeholders. This is so called *SaaS composition*.

There are many QoS metrics that must be considered in SaaS composition. Apart from *response time, reputation, reliability, price* and *availability*, which have been considered in web service composition, there is a new QoS metric that must be considered in SaaS composition, that is, *scalability*, which will be discussed next.

Multi-tenancy is the key design principal for cloud computing software development. This concept refers to having a single instance of software serving multiple users known as tenants in virtualized instances that are tailored for different requirements. For a SaaS to be considered as multi-tenant, a population of stakeholder's QoS requirements must be taken into consideration. This is different from traditional service composition where the service plan was designed explicitly to suit a sole stakeholder's needs. In this paper, we categorise tenants into a set of tenant types, namely *VIP Users, Elite Users, Standard Users*, and *Minor Users*. Note that this set of categories is extensible and can be customized for different SaaS.

Each tenant's type and its QoS requirements are modelled into a tenant profile, which is then used in our multi-tenant SaaS composition modelling. The purpose of tenant types is to differentiate the importance of different tenants. VIP users have higher priority than standard users, standard users have higher priority than elite users, and elite users have higher priority than minor users. The importance for each of the user categories may be different for different SaaS. The following is an instance of importance values for different tenant types.

**Table 1.** Importance for different tenant types

| Tenant type | Importance value |
| --- | --- |
| Minor users | 0.1 |
| Standard users | 0.2 |
| Elite users | 0.3 |
| VIP users | 0.4 |

Given a workflow for a new SaaS, a collection of selective SaaS components for each of the tasks in the workflow, a set of constraints between the selective SaaS components, and a set of tenant profiles, the SaaS composition problem is to select a SaaS component for each of the tasks in the workflow for each type of tenants such that the response time and price of the new SaaS is minimised, and the reputation, reliability and availability of the new SaaS is maximised while guaranteeing the scalability of the new SaaS.

This SaaS composition problem is different from the SaaS composition problem discussed in [12] in the following aspects. The first and most important is that the SaaS composition problem addressed in [12] is to build a SaaS for a tenant on the fly when receiving a request for the SaaS. In contrast, this SaaS composition problem is to build a SaaS before receiving any request for such a SaaS. In addition, the QoS requirements and constraints considered are different in the two SaaS composition problems.

# 4  Our Genetic Algorithm

Our genetic algorithm (GA) is a penalty-based GA. Thus, in our GA, infeasible chromosomes are allowed. However, we give penalties to them such that they have a less chance to be selected for reproduction. There are two advantages in our GA. Firstly, the genes in the population are more diverse. In addition, the design of our GA is simple as we do not have to explicitly handle the infeasible solution issue.

## 4.1  Genetic Encoding

A chromosome in our GA is encoded by a sequence integer, which is broken into two segments. The first segment is a sequence of SaaS component IDs, representing the SaaS component selection for each of the tasks in the workflow and the second segment is a sequence of physical server IDs, representing the deployment of the corresponding SaaS component.

## 4.2  Genetic Operators

For each iteration, our GA attempts to mimic the genetic operations on a pool of chromosomes to improve the overall quality of these chromosomes. The genetic operators adopted in our GA are a classical two-point crossover and a random mutation.

For the crossover, the first segment of SaaS genes are selected from the first chromosome and this is also repeated on the other chromosome, then the selected genes from both chromosomes are swapped across from one another to create two new offspring chromosomes.

Hence the crossover operator creates two new chromosomes. Our GA evaluates the new chromosomes using a fitness function and selects two best chromosomes among the two parent chromosomes and two child chromosomes.

To maintain the genetic diversity from one generation to the next. The mutation is randomly applied onto those chromosomes in the population at a very low probability rate to yield some new chromosomes. When mutation occurs on a chromosome, one or more randomly selected genes in the first segment and only one gene in the last segment is randomly replaced with an alternative value to create new chromosomes with new genes.

## 4.3  Fitness Function

To effectively deal with potential infeasible chromosomes generated by the genetic operators, we introduce a penalty strategy into the fitness function in our GA. The strategy allows infeasible solutions in the population of our GA, but gives penalties to them. The basis ideas behind the penalty strategy are: firstly any infeasible solution must have a fitness value that is always less than that of any feasible solution; secondly, the more constraint violations in an infeasible solution, the harsher punishment will be given to its fitness value. With these two considerations in mind, we define the fitness function as below:

$$Fitness(x) = \begin{cases} 0.5 + 0.5 \times F_{obj}(x), & \text{if no violation;} \\ 0.5 \times F_{obj}(x) - \frac{violate(x)}{violateMax}, & \text{otherwise.} \end{cases} \tag{1}$$

where $F_{obj}(x)$ is the value of the weighted objective function of the SaaS composition solution $x$. The above fitness function guarantees that feasible solutions always have a fitness value greater 0.5 and infeasible chromosomes always have a fitness value less than 0.5.

## 5   Evaluation

Our genetic algorithm was evaluated by experiments, which were conducted on a 64-bit Windows 7 computer equipped with 8 GB of RAM and a multi-threading enabled CPU that has 8 cores (4 logical cores and 4 physical cores) running at 2.8 GHz. Our genetic algorithm was implemented in C# and was executed on .NET environment version 4.5.

For proving the effectiveness of our genetic algorithm, we designed a hill climbing algorithm as a benchmark algorithm. The experiments were conducted through a systematic manner by first generating a set of business workflows with randomized control structures, and then generating potential SaaS services for each of these tasks in a workflow, with their non-functional QoS property values and the location. After that, the constraints and dependencies between these candidate SaaS services were randomly generated across multiple and different tasks. After that the QoS requirements and the types of $n$ tenants were generated. To ensure that our tests are fair both the algorithms were given the same test data to run, these data were bundled into a set of text files which can be used by our test engine for testing. Prior to these examinations, a preliminary performance tuning were made on our genetic algorithm to ensure it to perform at optimal level. Table 2 shows the parameters we used for running the genetic algorithm.

Table 2. SaaS composition GA parameters

| Parameter | Value |
| --- | --- |
| Population size ($PopSize$) | 90 |
| Crossover rate ($p_c$) | 0.70 |
| Mutation rate ($p_m$) | 0.18 |

Even though these actions can ensure a certain degree of fairness for an experiment set, the result set produced by our genetic algorithm may still vary from each run. This is mainly contributed by the stochastic nature of our genetic algorithm. To overcome this, all tests were run 20 times and a snapshot of the median results are used for consideration. Our experimental results reveal the followings.

In the following experiments, the test parameters used were set at, when applicable, and we assumed that the composite SaaS service was designed for 50 tenants in a cloud environment with 200 physical servers with each network switch capable of holding 10 servers.

## 5.1 Performance and Scalability with Number of Tasks

In this experiment we incrementally scaled the number of tasks in a SaaS composition problem by 10 per new test. Our goal was to test the performance of our genetic algorithm to determine its scalability in handling problem of various sizes. Table 3 shows the experimental results.

**Table 3.** Experimental results for performance and scalability with number of tasks

| Number of tasks | GA | | HA | |
|---|---|---|---|---|
| | Comp. time (sec) | Fitness value | Comp. time (sec) | Fitness value |
| 10 | 0.903 | 0.844 | 0.003 | 0.606 |
| 20 | 1.338 | 0.787 | 0.014 | 0.589 |
| 30 | 1.878 | 0.799 | 0.030 | 0.604 |
| 40 | 2.341 | 0.769 | 0.057 | 0.591 |
| 50 | 2.866 | 0.746 | 0.096 | 0.584 |
| 60 | 3.209 | 0.749 | 0.129 | 0.586 |
| 70 | 3.548 | 0.735 | 0.164 | 0.582 |
| 80 | 3.931 | 0.741 | 0.207 | 0.582 |
| 90 | 4.286 | 0.734 | 0.243 | 0.585 |
| 100 | 5.106 | 0.729 | 0.335 | 0.586 |

In the experiments, the number of tasks varied from 10 to 100, the number of selective SaaS for each of the tasks was fixed to 30, and the number of constraints was fixed to 20.

There are 10 rows and 5 columns in Table 3. Each row shows the statistics for the 20 repeated runs for a test problem. The first column shows the number of tasks in the new SaaS, the second column and the third column are the average computation time and the average quality of the solutions of the 20 runs for our GA, respectively, and the fourth column and the fifth column are the computation time of the HA and the quality of the solution generated by the HA, respectively. The quality of a solution is measured by the fitness value of the solution, which is defined in Eq. 1. The fitness value is between 0 and 1. The bigger its fitness value, the better the solution.

It can be seen from Table 1 that our GA constantly generated a significantly better solution than the HA for all the test problems. In addition, the computation time increased linearly when the number of tasks in the workflow of the SaaS increased.

## 5.2 Performance and Scalability with Number of Selectable SaaS Components per Task

In this section we aimed to test the ability of our genetic algorithm in coping small to large scales of selectable SaaS components for the defined tasks in a SaaS business workflow. Table 4 shows the experimental results.

**Table 4.** Experimental results for performance and scalability with number of selective SaaS components per task

| Number of componenets | GA | | HA | |
|---|---|---|---|---|
| | Comp. time (sec) | Fitness value | Comp. time (sec) | Fitness value |
| 10 | 3.107 | 0.784 | 0.019 | 0.611 |
| 20 | 2.623 | 0.799 | 0.029 | 0.603 |
| 30 | 2.739 | 0.791 | 0.039 | 0.600 |
| 40 | 2.603 | 0.812 | 0.071 | 0.601 |
| 50 | 2.602 | 0.801 | 0.074 | 0.607 |
| 60 | 2.746 | 0.814 | 0.105 | 0.607 |
| 70 | 2.737 | 0.806 | 0.128 | 0.600 |
| 80 | 2.703 | 0.807 | 0.113 | 0.606 |
| 90 | 2.541 | 0.804 | 0.134 | 0.608 |
| 100 | 3.165 | 0.800 | 0.180 | 0.604 |

In the experiments, the number of selective SaaS for each task varied from 10 to 100, the number of tasks was fixed to 30, and the number of constraints was fixed to 20.

There are 10 rows and 5 columns in Table 4. Each row shows the statistics for the 20 repeated runs for a test problem. The first column shows the number of selective SaaS componenets per task in the new SaaS, the second column and the third column are the average computation time and the average quality of the solutions of the 20 runs for our GA, respectively, and the fourth column and the fifth column are the computation time of the HA and the quality of the solution generated by the HA, respectively. The quality of a solution is measured by the fitness value of the solution, which is defined in Eq. 1. The fitness value is between 0 and 1. The bigger its fitness value, the better the solution.

It can be seen from Table 2 that our GA constantly generated a significantly better solution than the HA for all the test problems. In addition, the computation time increased linearly when the number of selective SaaS componenets per task increased.

### 5.3  Summary of the Experimental Results

With our experimentation results, we concluded our SaaS Composition approach has the following merits:

- It is efficient and effective. For all tested problems, the genetic algorithm produced a better solution than the benchmark algorithm;
- It is scalable. The experimental results have shown that computation time incremented linearly when the number of tasks, or the number of selective SaaS components per task increased.

## 6  Conclusion

In this paper we have discussed the SaaS composition problem in cloud computing. We have also discussed the similarities and differences between the web service composition problem and the SaaS composition problem in this paper. The SaaS composition problem has been formulated as a multi-objective and multi-constraint combinatorial optimisation problem and a genetic algorithm has been developed for solving the combinatorial optimisation problem. We have evaluated the genetic algorithm by experiments and the experimental results have shown that the genetic algorithm has better performance than a heuristic algorithm, which was used as a benchmark algorithm. The experimental results have also demonstrated that the genetic algorithm is scalable.

## References

1. Wu, Q., Zhou, M., Zhu, Q., Xia, Y.: VCG auction-based dynamic pricing for multi-granularity service composition. IEEE Trans. Autom. Sci. Eng. **15**(2), 796–805 (2017)
2. Kritikos, K., Plexousakis, D.: Multi-cloud application design through cloud Service composition. In: Proceeding of IEEE 8th International Conference on Cloud Computing, New York, USA, pp. 686–693 (2015)
3. Buyya, R., Yeo, C.S., Venugopal, S., Broberg, J., Brandic, I.: Cloud computing and emerging IT platforms: vision, hype, and reality for delivering computing as the 5th utility. Futur. Gener. Comput. Syst. **25**(6), 599–616 (2009)
4. Banerjee, P., et al.: Everything as a service: powering the new information economy. IEEE Comput. **44**(3), 36–43 (2011)
5. Klein, A., Ishikwa, F., Honiden, S.: Towards network-aware service composition in the cloud. In: Proceedings of the 21st International Conference on World Wide Web, New York, USA, pp. 959–968 (2012)

6. Ai, L., Tang, M., Fidge, C.: QoS-oriented resource allocation and scheduling of multiple composite web services in a hybrid cloud using random key genetic algorithm. Aust. Joournal Intell. Inf. Process. Syst. **12**(1), 29–34 (2010)

7. Zeng, L., Benatallah, B., Dumas, M., Kalagnamam, J., Chang, H.: QoS-aware middleware for Web services composition. IEEE Trans. Softw. Eng. **30**(5), 311–327 (2004)

8. Yu, T., Zhang, V., Lin, K.: Efficient algorithms for Web services selection with end-to-end QoS constraints. ACM Trans. Web **1**(1), Article 6 (2007)

9. Canfora, G., Penta, M.D., Esposito, R., Villani, M.L.: An approach for QoS-aware service composition based on genetic algorithms. In: Proceedings of the Genetic and Evolutionary Computation Conference, Seattle, USA, pp. 1069–1075 (2005)

10. Gao, Y., Zhang, B., Na, J., Yang, L., Dai, Y., Gong, Q.: Optimal selection of Web services with end-to-end constraints. In: Proceedings of the 1st International Multi-Symposiums on Computer and Computational Sciences, Hanzhou, China, pp. 460–467 (2006)

11. Tang, M., Ai, L.: A hybrid genetic algorithm for the optimal constrained Web service selection problem in Web service composition. In: Proceeding of the 2010 IEEE Congress on Evolutinary Computation, Barcenlona, Spain, pp. 268–275 (2010)

12. He, Q., Han, J., Yang, Y., Grundy, J., Jin, H.: QoS-driven service selection for multi-tenant SaaS. In: Proceedings of the IEEE 5th International Conference on Cloud Computing, Hawaii, USA, pp. 566–573 (2012)

13. Ye, Z., Mistry, S., Bouguettaya, A., Dong, H.: Long-term QoS-aware cloud service composition using multivariate time series analysis. IEEE Trans. Serv. Comput. **9**(3), 382–393 (2016)

14. Vakili, A., Navimipour, N.J.: Comprehensive and systematic review of the service composition mechanisms in the cloud environments. J. Netw. Comput. Appl. **81**, 24–36 (2017)

15. Jula, A., Sundararajan, E., Othman, Z.: Cloud computing service composition: a systematic literature review. Expert. Syst. Appl. **41**(8), 3809–3824 (2014)

16. Dastjerdi, V., Buyya, R.: Compatibility-aware cloud service composition under fuzzy preferences of users. IEEE Trans. Cloud Comput. **2**(1), 1–13 (2014)

# Evolving Computationally Efficient Hashing for Similarity Search

David Iclanzan[1(✉)], Sándor Miklós Szilágyi[2], and László Szilágyi[1,3]

[1] Faculty of Technical and Human Sciences, Sapientia University, Târgu Mureş,
Romania
david.iclanzan@gmail.com
[2] Department of Informatics, Petru Maior University, Târgu Mureş, Romania
[3] John von Neumann Faculty of Informatics, Óbuda University, Budapest, Hungary

**Abstract.** Finding nearest neighbors in high-dimensional spaces is a
very expensive task. Locality-sensitive hashing is a general dimension
reduction technique that maps similar elements closely in the hash space,
streamlining near neighbor lookup.

In this paper we propose a variable genome length biased random key
genetic algorithm whose encoding facilitates the exploration of locality-
sensitive hash functions that only use sparsely applied addition operations instead of the usual costly dense multiplications.

Experimental results show that the proposed method obtains highly
efficient functions with a much higher mean average precision than standard methods using random projections, while also being much faster to
compute.

**Keywords:** Locality-sensitive hashing · Optimal design
Genetic algorithms · Variable length representation

## 1 Introduction

The objective of nearest neighbor search or similarity search is to find the item
that is closest to what has been queried, from a search (reference) database.
Closeness or proximity is evaluated by some distance measure and the item
found is called nearest neighbor. If the reference database is very large or the
distance computation between the query and database item is costly, finding the
exact nearest neighbor is often not feasible computationally. This has fuelled
great research efforts towards an alternative approach, the approximate nearest
neighbor search, which proves to be not only more efficient, but also sufficient
for many practical problems.

Hashing is a popular, widely-studied solution for approximate nearest neighbor search [11], locality sensitive hashing [3,8] and learning to hash are the two
main classes of hashing algorithms. Locality sensitive hashing (LSH) is data-
independent. It has been adopted in many applications, e.g., fast object detection
[6], image matching [4]. Learning to hash is a data-dependent hashing approach.

© Springer Nature Switzerland AG 2018
L. Cheng et al. (Eds.): ICONIP 2018, LNCS 11302, pp. 552–563, 2018.
https://doi.org/10.1007/978-3-030-04179-3_49

It learns hash functions for specific types of data or datasets, providing a short encoding of the original data with the property that the nearest neighbor search result in the hash coding space are also close to the query in the original space [10].

In a recent work [5] the authors showed that a computationally efficient hashing method based on sparse random binary projections has strong distance preserving properties. While the sparse tags can be computed efficiently, their big size prevents a fast nearest neighbor search in the hash space.

In this paper we develop hash function instances based on similarly efficient sparse binary projections. However, instead of a long sparse binary tags the method returns short tags that contain a combination of indices. For nearest neighbor search in the hash space the Jaccard distance is used. Instead of relying on random projections we propose a method for learning the best performing projections using a new random key Genetic Algorithm (GA) that can handle variable length genomes. The new method is used to efficiently explore the sparse binary projection hash functions space, searching for solutions that are both short, therefore very efficient to compute, and also provide high precision for nearest neighbor search.

## 2    Overview

### 2.1    Nearest Neighbor Search and Hashing

Nearest neighbor search (NNS) is the optimization problem of locating in a particular space the point that is closest to a query point. To measure closeness or proximity, most commonly a dissimilarity function is used which maps the less similar pair of points to larger values.

The term "locality-sensitive hashing" (LSH) originates from 1998, denoting a "randomized hashing framework for efficient approximate nearest neighbor (ANN) search in high dimensional space. It is based on the definition of LSH family H, a family of hash functions mapping similar input items to the same hash code with higher probability than dissimilar items" [11]. Multiple communities from science and industry alike have been studying LSH, with focus on different aspects and goals.

The theoretical computer science community developed different LSH families for various distances, sign-random-projection (or sim-hash) for angle-based distance [3], min-hash for Jaccard coefficient [2] explored the theoretical boundary of the LSH framework and improved the search scheme.

Learning to hash approaches revolve around the following concepts: the hash function, the similarity in the coding space, the similarity measure in the input space, the loss function for the optimization objective, and the optimization technique. Linear projection, kernels, spherical function, (deep) neural networks, non-parametric functions etc. can all be used as hash functions.

In this paper for the hash function we use sparse binary projections, the similarity measure in the input space is the Euclidean distance while the one in the coding space is the Jaccard distance. The optimisation technique used is an extension of BRKGA, for which we provide an overview in the following.

## 2.2   Random Key Genetic Algorithms

A class of random key genetic algorithms (RKGA) was first introduced by Bean [1] for solving combinatorial optimization problems. In a RKGA, chromosomes are encoded as strings, vectors, of random real numbers in the interval $[0, 1]$. Chromosomes, on the other hand, represent solutions to the combinatorial optimization problem for which an objective value or fitness can be computed. The translation from chromosomes to solutions and fitness is done by a problem dependent, deterministic algorithm called decoder.

The initial population is created randomly by generating $p$ vectors, where $p$ is the population size. Each gene of each individual is independently, randomly sampled from the uniform distribution over $[0, 1]$. In the next step the fitness of each individual is computed (by the decoder) and the population is divided into two groups: a smaller group of $p_e$ elite individuals (those with the best fitness values), and the rest, $p - p_e$ non-elite individuals.

The population evolves into new generations based on a number of principles. First, all the elite individuals are copied unchanged from generation $t$ to generation $t + 1$, resulting in a monotonically improving heuristic.

Mutation in Evolutionary Algorithms is usually used for exploration, enabling the method to escape from entrapment in local minima. In RKGA alleles are not mutated, however, at each generation a number $p_m$ of mutants are introduced into the population the same way that an element of the initial population are generated (i.e. randomly from the uniform distribution). Having $p_e$ elite individuals and $p_m$ mutants in population $t + 1$, $p - p_e - p_m$ additional individuals need to be produced to maintain the same population size $p$ in generation $t + 1$. This is done by the process of mating, where two parents are selected from the entire population and a new offspring is created by crossover. The crossover is repeatedly applied until the size of population reaches $p$.

A biased random key genetic algorithm, or BRKGA [7], differs from a RKGA in the way parents are selected for crossover. Each offspring is generated by mating an elite individual with a non-elite one, both taken at random. The bias can be further reinforced by the parametrized uniform crossover [9] and the fact that one parentis always selected from the elite group.

BRKGA searches the continuous n-dimensional unit hypercube (representing the populations), using the decoder to translate search results in the hypercube to solutions in the solution space of the combinatorial optimization problem.

The method and the unbiased variant differ in how they select parents for mating and how crossover favours elite individuals. Offsprings produced by the BRKGA have a higher chance inheriting traits of elite solutions. This seemingly small difference most of the times has a large impact on the performance of these variants. BRKGAs have a stronger exploitation power and tend to find better solutions faster than RKGAs.

# 3   Materials and Methods

## 3.1   Distance Preserving Sparse Projections

In [5] it is shown that uniformly distributed, random sparse binary projections preserve neighborhood structure if the number of projections $m$ is sufficiently large.

For the data-dependent hashing, we wish to learn the minimal number of projections $m$ needed and the most discriminatory sparse projections.

Let $x \in \mathbb{R}^d$ denote an input that needs to be hashed. $y = (y_1, \ldots, y_m) \in \mathbb{R}^m$ projections are computed as

$$y_i = \sum_{j \in P_i} x_j \tag{1}$$

where $P_i$ is a projection vector defining which components from $x$ are summed for the $i$-th projection.

The hash is formed by retaining the indexes of the $k$ highest scoring projections: $Z = (z_1, \ldots, z_k) \in \mathbb{Z}^k$ where for all $z_i$ it holds that $y_{z_i}$ is one of the largest entries in $y$.

We use the Jaccard distance to measure how close two hashes $Z_1$ and $Z_2$ are to each-other:

$$d_J(Z_1, Z_2) = 1 - J(Z_1, Z_2) = \frac{|Z_1 \cup Z_2| - |Z_1 \cap Z_2|}{|Z_1 \cup Z_2|} \tag{2}$$

For computing a discrete bin index, where one can store similar items for fast reference, we can exploit the fact the computed tags represent a combination of $k$ elements taken from $\{1, \ldots, d\}$. The combinatorial number system of degree $k$ defines a correspondence between natural numbers and k-combinations. The number $N$ corresponding to a combination $Z = (z_k, \ldots, z_2, z_1)$ is:

$$N = \binom{z_k}{k} + \cdots + \binom{z_2}{2} + \binom{z_1}{1} \tag{3}$$

## 3.2   Biased Variable Length Random Key Genetic Algorithm

BRKGA is restricted to searching a continuous $n$-dimensional unit hypercube, therefore it is not suitable for exploring open-ended design spaces.

For our problem, the optimal number of projections $m$ and the size of the projection vectors $P_i$ must be determined by the learning algorithm. The method has to explore solutions of various complexities, that cannot be efficiently encoded in a unit hypercube of fixed size.

To enable an efficient exploration of different solution complexities, we extend the BRKGA framework with the following modifications:

- Each chromosome begins with a fixed number of so called interpreter or signalling alleles that can carry information on the complexity of the encoded solution, enabling the decoder to interpret the random keys differently for separate solution classes.

- Genomes can have various lengths, and they can be as long or as short as needed, to encode highly complex solutions or contrary, simple ones, like the empty set.
- A new 2-parents 2-offsprings crossover operator that can change genome lengths.
- A re-encode operator that can "translate" alleles between different genomes by analysing their interpreter alleles.

The new method, the Biased Variable Length Random Key Genetic Algorithm (BVLRKGA), is detailed in the following subsections.

### 3.3    Interpreter Alleles and Encoding

A solution for the proposed hash model must encode the number of projections $m$ and the projection vectors $P_i$. Assuming an upper bound $\theta$ for the number of projections, and knowing that the elements of the projection vectors are integer numbers between 1 and $d$ the following encoded can be used.

The first allele on each chromosome is reserved for encoding $m$. The unit interval is partitioned in $\theta$ subintervals of size $1/\theta$, the upper bound of each subinterval being $b_i = i/\theta, i \in \{1, \ldots, \theta\}$. For $r \in [0,1]$ the first $b_i \geq r$ defines the number of projections: $m(r) = \lceil r \times \theta \rceil$.

The rest of the genes encode elements of the projection vectors. By dividing each subinterval $[b_{i-1}, b_i)$ assigned to a projection in $d$ subintervals, we reserve a smaller subinterval for each index from $\{1, \ldots, d\}$. To decode a random key $r \in [0,1]$ first we compute in which projection it encodes an index:

$$i(r) = 1 + \lfloor m(r) \times r \rfloor \tag{4}$$

then we compute the index itself:

$$j(r) = 1 + \lfloor (r - (i(r) - 1) \times m(r)) \times m(r) \times d \rfloor \tag{5}$$

For this problem, the order of the genomes is irrelevant, each random key is decoded the same way, independently of its position.

There is no upper bound on the genome length as each index in each projection vector can appear any number of times. In such cases those $x_{j(r)}$ elements are summed multiple times, hence the number of times they appear in a solution can be regarded as the weight assigned to that dimension.

This encoding can also represent the empty solution when there are no genes besides the interpreter ones. The decoder must take this into account and assign a proper fitness value for empty solutions (usually $-\inf$, 0 or $+\inf$).

### 3.4    Biased Variable Length Crossover

BRKGA uses a parametrised uniform crossover [9] where alleles from an elite parent are inherited with a higher probability. This operator produces a fixed length offspring, therefore is not suited for variable length encodings.

We propose a new biased crossover operator where 2 offsprings of potentially various lengths are produced. Each allele from each parent can be inherited by both offsprings with different probabilities. The key is in setting up the probabilities in such a way that (i) offsprings are more likely to inherit alleles from the elite parent; (ii) the sum of offsprings genome lengths is a random variable whose expected value equals the sum of parents genome lengths. By (ii) the crossover is not inherently biased toward in producing shorter or longer offsprings, it produces both with certain variance. The mean average length of offsprings and parents is the same after crossover. It is the role of the selection operator to favour shorter or longer solutions on average.

The crossover operation's steps are detailed in Algorithm 1.

---

**Algorithm 1.** Biased variable length crossover

---

**Data:** elite and other parents: *elite, indiv*
**Result:** two offsprings: $o_1, o_2$
1  eliteBias = 0.2 ;    // set up the bias that favours the inheritance of genomes from the elite parent
   /* compute probabilities of inheritance                           */
2  eliteP1 = eliteBias + rand*(1-eliteBias); eliteP2 = (1-eliteBias) -eliteP1;
3  indivP1 = rand*(1-eliteBias); indivP2 = (1-eliteBias) -indivP1;
   /* inherit the interpreter alleles from one of the parents        */
4  $o_1 \leftarrow$ elite.interpreterAlleles(); $o_2 \leftarrow$ indiv.interpreterAlleles();
   /* inherit the genes according to the 4 probabilities             */
5  **foreach** allele $\in$ elite.randomKeyAlleles() **do**
6  | **if** rand < eliteP1 **then**
7  | | $o_1$.addAllele(allele);
8  | **if** rand < eliteP2 **then**
9  | | $o_2$.addAllele(reenc(allele,indiv.interpreterAlleles()));

10 **foreach** allele $\in$ indiv.randomKeyAlleles() **do**
11 | **if** rand < indivP1 **then**
12 | | $o_1$.addAllele(reenc(allele,elite.interpreterAlleles()));
13 | **if** rand < indivP2 **then**
14 | | $o_2$.addAllele(allele);

---

The elite bias variable is an element of $[0, 1]$ and controls how much the genes of the elite parents are favoured. A value of 0 means that there is no bias while a value of 1 restricts the inheritance only to the genes of the elite individual. The value 0.2 from the listing will result in the offsprings inheriting on average 60% of their alleles from the elite parent.

The first offspring inherits the interpreter genes on the elite parent while the second inherits from the other individual. As they use different interpretations of random keys, we must make sure then when a allele is inherited from a parent that uses a different solution design, the meaning is preserved. Therefore, a re-encoding happens, where the allele is decoded and if possible remapped in the solution space of the offspring. In cases when this is not possible, for example the to be inherited allele encodes an index in the 1000-th projection, while the offspring works with less than 1000 projections, the allele is discarded.

## 3.5   BVLRKGA

The proposed method uses an elitist approach but there are some notable differences compared to other random key GA. As discussed in the previous section, crossover produce two offsprings of varying lengths. Also, in BVLRKGA we don't restrict crossover to be between elite and non-elite individuals. While the first parent is always an elite the second one is randomly chosen. This allows mating between elite individuals, that enables the rapid adaptation of ideal genome length in the elite subpopulation. If selection favours short genomes, mating two elite short genomes will produce even shorter solutions with higher probability than mating a short and a long genome. The converse is also true, if long genomes perform better.

In this work, for the hash learning task, random solutions are built by first randomly choosing a projection number $m$ between the hash length $k$ and an upper bound $\theta = 2000$. In a second step a spareness factor $\gamma \in [\frac{3}{d}, \frac{9}{d}]$ is randomly chosen, then we decide for each projection if each input dimension index is included in the vector, by flipping a $\gamma$-loaded coin. After encoding, $m$ is implicitly encoded in the interpretation allele, while the sparseness is indirectly reflected in the genome length.

# 4   Experiments

For the experiments, we used a population size of 100 individuals and the evolution was performed over 100 generations. The proportion of elite individuals in the next population was set to $p_e = 0.2$. The fraction of the population to replaced by mutants was set to $p_m = 0.4$. Therefore, the fraction of the new population obtained by crossover is $1 - p_e - p_m = 0.4$. For crossover the parameter $eliteBias = 0.2$. Each experiment was run 10 times.

## 4.1   Genome Length Adaptation

In a first experiment we checked how fast BVLRKGA can grow or shrink genomes to converge to an optimal target length under ideal selection pressure. The absolute difference between genome length and target length was used as fitness, in a minimisation problem context. While most real-world problems will not have a noise free, unimodal feedback regarding the lengths of the solutions, these experiments are very useful in helping assess the effectiveness of the proposed operators in providing solutions of various lengths.

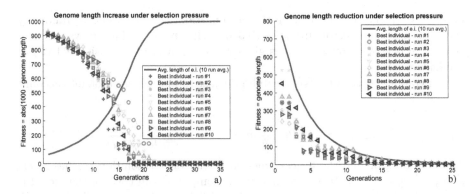

**Fig. 1.** Under selection pressure BVLRKGA is able to quickly reduce (a) and increase (b) the genome length of the best solutions.

In a first test we set the random individual seeder to build solutions of length 1000, while the best solution is the empty one. Therefore by exploiting the crossover and selection the method will eliminate alleles gradually. Figure 1(a) depicts the results of the 10 runs. We can see that the average chromosome length of the elite individuals quickly drops in only 15 generations. Every run reached the empty solution in 24.7 generations on average, while the longest run required 32 generations.

In a second setup we experimented with the converse problem, increasing the genome length beyond what the random individual seeder provides. Here the seeder returned solutions of length 50, while the target solution length was set to 1000. A twentyfold increase in genome length in needed to reach the ideal solution length. The results of 10 runs are shown in Fig. 1(b). Again, every run found a best solution, on average in 40.2 generations, the longest run requiring 75 generations to find a solution with a length of exactly 1000. This problem was harder for the BVLRKGA because the optimisation could overshoot, generating solutions with a length beyond 1000, while in the first setup undershooting was not possible.

## 4.2  Learning to Hash

For testing the BVLRKGA ability to learn hash function we used the MNIST[1] dataset that contains handwritten digits encoded as $28 \times 28$ pixel images ($d = 28 \times 28 = 784$). We followed the same procedure from [5], using from the dataset a subset of 10.000 vectors, that we pre-processed to have zero mean. During 4 experiments we evolved models that produced hash tags of lengths $k \in \{4, 8, 16, 32\}$.

For a candidate hash model we first calculated the average precisions (AP: average precision at different recall levels) by predicting the nearest 200 vectors

---

[1] http://yann.lecun.com/exdb/mnist/.

for a set of query vectors. During the optimisation, this was done by randomly choosing 200 query vectors from the 10.000, each time an evaluation was needed. Averaging over the 200 AP values gives the mean average precision (MAP) of the hash model.

For computing the fitness of a candidate hash model we also took into account the length of the genome:

$$fitness(c) = MAP(c) + min(0, 1000 \times \frac{MAP(c) - bt_k}{length(t)}) \qquad (6)$$

The reasoning behind the fitness function is that once a solution reaches a certain precision, compactness (implicitly low computational cost) of the model is rewarded. The bonus thresholds $bt_k = \{0.1, 0.2, 0.3, 0.45\}$ were set high enough to reward compactness, shorter solutions once the MAP of the solution is higher as the one of dense projection based LSH functions.

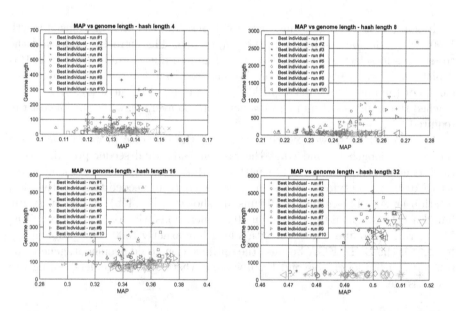

**Fig. 2.** Development of the two optimisation criteria, MAP and solution length.

The evolution over 100 generations of the best individual is depicted in Fig. 2 for each hash length $k$ and each of the 10 runs. The size and color intensity of the markers are proportional to the generation numbers. Small and dark markers correspond to early generations, while individuals depicted from later generations are displayed with a linearly increasing marker size and lighter colors. The scatter plots depict the two relevant metrics in the computation of the fitness, MAP and genome lengths. For hash lengths $k \in \{4, 8, 16\}$ we can observe the same trend: in early generations the genome lengths are large but with time both the compactness and precision is improved. The genome length is drastically reduced, the biggest markers corresponding to the last generations from each run are very close to $x$-axis.

For $k = 32$ we can observe two clusters for the final generation markers. In some runs the method is not able to reduce the length of the best solution while also keeping the precision high. These final solutions have genome lengths around 3000. In other runs BVLRKGA is able to find very compact hashes, encoded in genomes with lengths between 200–300, while also keeping MAP above 0.5.

Because all final solution are well above the bonus threshold $b_{32} = 0.45$ we suspect that in some runs the genome lengths are not reduced because the bonus reward is to small and does not compensate the drop in MAP for a shorter solution. This can be mitigated by emphasising, rewarding more the compact solutions.

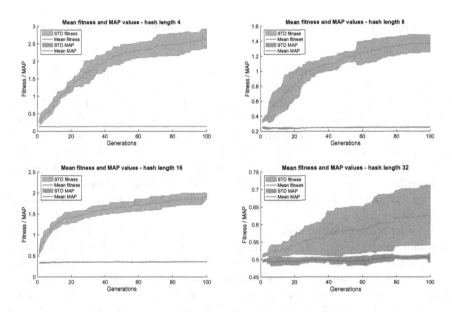

**Fig. 3.** Fitness development.

Figure 3 presents the average fitness and MAP values of the best individuals, over the 100 generations. The fitness increases with a much bigger slope than the MAP curve, meaning that the fitness improvements mostly came from the BVLRKGA discovering shorter solutions that do not compromise precision. For $k = 32$ we can again observe that the standard deviation of the fitness is very large. This comes from the fact that for some runs the fitness curve stayed very close of the MAP, as the genome lengths were high, the compactness bonus was small. In other runs by finding short high-quality solutions, the fitness curve was much higher than the MAP curve.

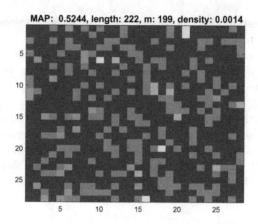

**Fig. 4.** The pixel locations used by the best model obtained for $k = 32$.

The details of the best solution found for $k = 32$ is shown in Fig. 4. As during optimisation the MAP is computed for 200 random queries, we recomputed the MAP by querying and averaging al 10.000 images. The solution has a $MAP$ of 0.5244, it uses 222 out the $d = 768$ input pixels to compute $m = 199$ projections.

The pixels that are part of projection sets with the same size were depicted with the same color in the figure. There is one projection set containing 5 pixel indices (brightest pixels in the figure) and two sets of size 4. Only 31 projection set sizes have more than 1 element. It is interesting to note that the projection sets are mutually exclusive, there is no pixel input that is used in more than one projection.

In order to check if the high MAP value is a results of this very sparse model, with short projections or the location of the inputs discovered by the BVLRKGA has also a big influence, we performed the following experiment: 10.000 random models were generated, using the same parameters, $m = 199$ projections and a density of 0.001423 to obtain on average 222 indices. The average MAP values of these models was 0.4481 and none of them reached a MAP value of 0.5. The experiments show that even if the model parameters are fixed it is very hard to replicate the performance of the proposed method with random sampling.

Therefore, we conclude that BVLRKGA not only explores the model design space efficiently, obtaining very compact solutions, but it also combines meaningfully solutions to reach models with high MAP.

## 5   Conclusions

We proposed a method for obtaining computationally efficient locality sensitive hash functions based on sparse binary projections. The tags are not binary, instead they encode a short combination of indices. For these tags, nearest neighbor search in the hash space can is performed by using the Jaccard distance.

The paper introduces the Biased Variable Length Random Key Genetic Algorithm (BVLRKGA) that is able to explore a solution design space in order to find the appropriate solution complexity that provide both high precision and good computational efficiency - short solutions. By using a crossover operator that can produce offsprings of different lengths and enabling mating between elite individuals, under the appropriate fitness pressure the method can quickly converge to an ideal genome length.

The method was used to find computationally very efficient locality-sensitive hash functions for the MNIST dataset. Hashing with the evolved solutions is order of magnitudes faster than standard LSH functions that use many multiplications. The evolved functions only require a few addition operations to compute.

**Acknowledgments.** This paper was partially supported by the Sapientia Institute for Research Programs (KPI). The work of Sándor Miklós Szilágyi and László Szilágyi was additionally supported by the Hungarian Academy of Sciences through the János Bolyai Fellowship program.

# References

1. Bean, J.C.: Genetic algorithms and random keys for sequencing and optimization. ORSA J. Comput. **6**(2), 154–160 (1994)
2. Broder, A.Z.: On the resemblance and containment of documents. In: Proceedings of Compression and Complexity of Sequences, pp. 21–29. IEEE Computer Society (1997)
3. Charikar, M.S.: Similarity estimation techniques from rounding algorithms. In: Proceedings of the 34th Annual ACM Symposium on Theory of Computing, pp. 380–388. ACM, New York (2002)
4. Chum, O., Matas, J.: Large-scale discovery of spatially related images. IEEE Trans. Pattern Anal. Mach. Intell. **32**(2), 371–377 (2010)
5. Dasgupta, S., Stevens, C.F., Navlakha, S.: A neural algorithm for a fundamental computing problem. Science **358**(6364), 793–796 (2017)
6. Dean, T., Ruzon, M.A., Segal, M., Shlens, J., Vijayanarasimhan, S., Yagnik, J.: Fast, accurate detection of 100,000 object classes on a single machine. In: IEEE Conference on Computer Vision and Pattern Recognition (CVPR), pp. 1814–1821. IEEE (2013)
7. Gonçalves, J.F., Resende, M.G.C.: Biased random-key genetic algorithms for combinatorial optimization. J. Heuristics **17**(5), 487–525 (2011)
8. Indyk, P., Motwani, R.: Approximate nearest neighbors: towards removing the curse of dimensionality. In: Proceedings of the 30th Annual ACM Symposium on Theory of Computing, pp. 604–613. ACM (1998)
9. Spears, W., De Jong, K.: On the virtues of parameterized uniform crossover. In: Proceedings of the Fourth International Conference on Genetic Algorithms, pp. 230–236 (1991)
10. Wang, J., Zhang, T., Song, J., Sebe, N., Shen, H.T.: A survey on learning to hash. IEEE Trans. Pattern Anal. Mach. Intell. **40**(4), 769–790 (2018)
11. Wang, J., Shen, H.T., Song, J., Ji, J.: Hashing for similarity search: a survey. CoRR abs/1408.2927 (2014)

# Combining Two-Phase Local Search with Multi-objective Ant Colony Optimization

Chun-Wa Leung[(✉)], Sin-Chun Ng, and Andrew K. Lui

School of Science and Technology, The Open University of Hong Kong,
Kowloon, Hong Kong
derek38352@gmail.com, {scng,alui}@ouhk.edu.hk

**Abstract.** Multi-objective Ant Colony Optimization (MOACO) is a popular algorithm in solving the multi-objective combinational optimization problem. Many variants were introduced to solve different types of multi-objective optimization problem. However, MOACO does not guarantee to generate a good approximation of solution in a predefined termination time. In this paper, Two-Phase Local Search (TPLS) was introduced to cooperate with the MOACO, a two-phase strategy that creates a very good approximation of Pareto Front at the beginning of the algorithm and then further explores the Pareto Front iteratively. We also propose Iterated Local Search – Variable Neighborhood Search (ILS-VNS) as the first phase in TPLS, an iterative improvement process that allows finding improving solutions from adaptively sized neighborhood space. A series of experiments were performed to investigate the performance improvement on the solutions. At the same time, we studied the effect of Weighted Local Search (WLS) and Pareto Local Search (PLS) in the proposed algorithm. The results showed that the newly proposed algorithm obtains a larger area on hypervolume space and exhibits a significantly larger accuracy rate in obtaining the true Pareto-optimal solutions. Additionally, a statistical testing was also performed to verify the significance of the result.

**Keywords:** Multi-objective ant colony optimization · Two-phase local search
Iterated local search – variable neighborhood search
Bi-objective traveling salesman problem · Hybridization

## 1 Introduction

Multi-objective Optimization Problem (MOOP) is an important field in operational research. It relates to various real-world applications such as routing [1] and cost minimization. Traveling Salesman Problem is one of the crucial benchmarks in reflecting the performance of the algorithm design.

Ant Colony Optimization (ACO) [2] is a kind of popular swarm intelligence in solving the combinational optimization problem (COPs), which was proposed to solve the combinational optimization problem. Local Search can be implemented in aid of ACO. This is regarded as a daemon action to improve solution quality from the ants' solution construction process.

© Springer Nature Switzerland AG 2018
L. Cheng et al. (Eds.): ICONIP 2018, LNCS 11302, pp. 564–576, 2018.
https://doi.org/10.1007/978-3-030-04179-3_50

Multi-Objective Ant Colony Optimization (MOACO), an ACO variant solving multi-objective problem [1]. Generally, MOACO makes use of the information in multi-pheromone and heuristic matrices to satisfy multiple objective functions. It constructs solution from a random solution set. Such that it requires a longer time to reach local optimal solution. Thus, poor quality solutions returned when the algorithm terminates before it converges. With respect to the problem, Two-Phase Local Search (TPLS) [3] is suggested. TPLS is a two-stage local search strategy that creates a very good quality approximation of Pareto Front and further improves the solutions in the second phase iteratively.

Three objectives were carried out to come up with the research. First, in order to compensate for the ability to find good initial solutions in MOACO, we proposed a hybridization of TPLS with MOACO. This algorithm generates a good quality solution in the first phase and passes the solution to ants' solution construction and second phase local search such that the solution will be further improved. Essentially, the ants employ *pseudo-random-proportional rule* [4], a probabilistic strategy exploring solution with pheromone and heuristic and random neighborhoods. Second, we proposed Iterated Local Search – Variable Neighborhood Search (ILS-VNS) [5], an iterated procedure that invokes a variable-sized search on neighborhood space and finds other improving solutions. We have put this as a component in the first phase of the local search. The algorithms help to improve the searching ability to find good initial approximations. As to improve the quality of supported solution obtained in the first phase local search, restarting strategy was implemented. The method allows the first phase local search to start a new search space to find a good solution with a higher probability. Finally, an investigation on the performance of the proposed algorithm with different settings of local search was performed. To specify, the comparison of TPLS using WLS and PLS were also considered. Hypervolume indicators were used to examine the performance of the result. An experimental study was carried out to investigate the performance of the result. The algorithm instances were tested with Bi-objective Traveling Salesman Problem (bTSP), they were tested with Euclidean bTSP instances. A statistical testing was carried out to evaluate the significance of the result.

In this paper, a background on related topics and the proposed algorithm for bTSP, TPLS algorithm will be described in the next section. Then, a description of the implementation of TPLS into MOACO algorithms will be given in Sect. 3. Moreover, an empirical testing which compares the choice of local search in TPLS with MOACO is presented in Sect. 3.4. The test results will be evaluated with 2-way ANOVA. Finally, Sect. 4 will conclude the relationship between local search algorithm and MOACO in different sizes of bTSP instances and carry out the best and the worst component choice of the proposed algorithm.

## 2  Background

In this section, the descriptive information reviewing on existing algorithms were presented. Several reviews on the algorithms were also included.

## 2.1    Two-Objective Traveling Salesman Problem (bTSP)

To elaborate the topic, beginning with single-objective Traveling Salesman Problem. Given that G (V, E) is a complete graph with V as a set of nodes {1…n} representing cities and E as a set of edges fully connecting cities.

$$\min f(x) = p_{(x_n,x_1)} + \sum_{i=1}^{n-1} p_{(x_i,x_{i+1})} \tag{1}$$

The objective of TSP is to construct a Hamiltonian tour $x = (x_1…x_n)$ that minimizes the total cost for all objective component. Bi-Objective Traveling Salesman Problem (bTSP) extends the objective optimizing the route from one to two. The cost to two objectives were given $p_{x_i,x_j}^1$ and $p_{x_i,x_j}^2$ respectively, where $i \neq j$.

$$\min F = (f_1, f_2) \tag{2}$$

Assume there is no prior knowledge given by decision makers. Let F consist of two TSP. To minimize F, in other words, to find a set of Hamiltonian tours that have a good trade-off between two objective functions.

## 2.2    Iterated Local Search – Variable Neighborhood Search (ILS-VNS)

Iterated Local Search – Variable Neighborhood Search (ILS-VNS) [5] features from both ILS and VNS and further improve the original ILS. The motivation in ILS-VNS is to perform a good exploration of solution space and to have a higher possibility to obtain global optima.

The main idea of the algorithm is to make large step perturbation with ILS and then perform an adaptive neighborhood searching process with VNS.

A random solution is initialized by the first improvement of local search. Then the solution is further optimized in local search phrase perform by VNS. If the new solution passes the acceptance criteria, which is usually defined to accept a better solution than the current best solution (or adopt the acceptance criteria from simulated annealing in long run [6]), then the new solution replaces the best solution and reset the iteration counter to be zero. Otherwise, the current best solution will be perturbed as a newly copied solution and perform VNS again until the max iteration reached.

## 2.3    Two-Phase Local Search (TPLS)

Two-Phase Local Search (TPLS) is created by [7] solving bi-objective traveling problem. The author compared the design of combinations in the first phase and second local search components. As he mentioned, the first phase of local search algorithm applied to create a powerful initial solution, the second phase local search then further improve the quality of the solution. Another version of TPLS from [8], he has precisely compared the algorithmic component of TPLS and PLS.

The first phase local search involves more intensive local search algorithms, such as Iterated Local Search (3-opt), Lin-Kernighan [9] and Concorde. The second phase local search is an iterative process to improve the algorithm. It improves the solution with lightweight local search algorithms, greedy algorithm, Tabu search are the possible choice.

In a latter research [10], first phase local search also assigned with a set of uniformly distributed weight vector and generate a solution with the weighted scalarization. This approach manages a better trade-off on search space.

## 3 Proposed Algorithm

Our proposed algorithm combined TPLS and MOACO together to archive two aims. First, improve the searching ability. Second, return a good enough quality Pareto Front in any termination condition. The iterative MOACO algorithm to further improve the solutions. The MOACO contains three main components: solution construction, second phase local search and pheromone update. On every iteration of MOACO, ants' tour (solutions) were based on the output of first phase local search. They construct solutions based on the pheromone and heuristic information. The solutions were further improved by second phase local search, either Pareto Local Search (PLS) or Weighted Local Search (WLS). After all ants have constructed the solutions, global pheromone update decides the amount of pheromone to deposit and evaporate to the trails with respect to the ant's tour.

### 3.1 Related Works

Lust [8] concerned time complexity on TPLS solving large size instances. A hybrid algorithm of TPLS and Variable Neighborhood Search was introduced. The proposed algorithm was implemented to solve Multi-Objective Set Covering, (MOSCP) and Multi-objective multi-dimension knapsack problem (MOMKP). An experimental study indicated the new proposed algorithm improves the searching time for non-supported solutions.

Jaszkiewicz [10] noticed advanced the TPLS + VNS algorithm introduced previously. The author improved the algorithm in three aspects: too much solution is generated, unclear stop condition on first phase local search and sensitivity of the solution obtained. To restrict the number of the solutions, a set of weight vector is proposed to the scalarization to restrict Pareto archive to expand largely.

López [11] introduced a hybrid strategy for population-based optimization algorithm in solving the bi-objective Quadratic Assignment Problem (bQAP). In specific, the author introduced MOACO and SPEA2 [12] algorithm cooperating with Robust Tabu Search (RoTS) algorithm [13]. A performance assessment compared the iterations performed on RoTS with two population-based algorithms were tested.

In our proposed algorithm, the proposed algorithm is the combination of the above hybridization approaches. The algorithm organized in both sequential and iterative hybridization. To put it another way, the algorithm invokes the first phase local search in TPLS and MOACO sequentially, and the second phase local search improves the solution iteratively inside MOACO.

## 3.2  Hybrid MOACO + TPLS

Algorithm 1 is the pseudo code for the MOACO + PLS framework. To begin with, the ants and other information were initialized (Line 1), then the algorithm performs first phase local search (Line 2). The output generated were passed as the current tour of ants in MOACO. Under the iterative procedure, an empty set of iteration-best Pareto archive is initialized (Line 4). Ants construct solution with current tour makes use of pheromone and heuristic information (Line 7). The tour of ants will be further improved by the second phase local search (Line 8). A local pheromone update is performed immediately when every single ant has completed their next node construction. A Global pheromone update is performed to update the best ant's tour in the iteration (Line 12).

---

**Algorithm 1 TPLS with MOACO**

1: Initialize Pareto, Pheromone and Heuristic Information
2: FirstPhraseLocalSearch()
3: **while** stop criterion **do**
4:     $A_{iter} := \emptyset$
5:     **for** each colony $c \in \{1, \cdots, N^{col}\}$ **do**
6:         **for** each ant $k \in \{1, \cdots, N^a\}$ **do**
7:             ConstructSolution(a)
8:             LocalSearch()
9:             PheromoneUpdate()
10:        **end for**
11:    **end for**
12:    ParetoUpdate($A_{best}, A_{iter}$)
13: **end while**
14: **Output:**$A_{best}$

---

During the solution construction process, ants apply uniformly distributed weight setting, they have evenly distributed a weight to search the neighborhood space.

A uniformly distributed weight vectors are configured for each ant on solution construction. The weighting acts in two functions: construct solution with the decision rule, perform scalarization on local search. First, a solution is randomly generated on each ant in the colony, then improved by the first phase local search based on scalarization of objective functions, for instance, $F = \lambda f_1 + (1 - \lambda)f_2$. Next, each ant constructs pheromone paths and perform local pheromone update. The solution is further improved with the second phase local search. Following the solution construction process, global pheromone update is performed. All solutions are updated into iteration-best Pareto archive $A_{iter}$, all dominated solutions were removed inside the archive. The archive union with the global best archive $A_{best}$ when the iteration ends. When the stop condition has reached, the algorithm returns the global best Pareto archive $A_{best}$. The detailed description of MOACO was on previous section.

## 3.3  TPLS + WLS / PLS

After the first phase, local search is performed. A less intensive local search algorithm is invoked for further improvement as in the second phase. These local search algorithms improve the previously constructed tour from ants. Essentially, this algorithm cooperates with MOACO to improve the solution.

In the first phase local search, ILS-VNS algorithm is chosen with the 3-opt local search algorithm. At the beginning of the algorithm, ILS-VNS improves the solution iteratively with random restart, this help obtains more high-quality solutions.

For the second phase local search, there is two options of local search, Weighted Local Search (WLS) and Pareto Local Search (PLS).

PLS improves solutions by using the Pareto dominance relation, non-dominated solutions are updated to Pareto archive and replace the ant's solution immediately. In the proposed design, the PLS version from Angel [14] is adopted in the algorithm.

WLS scalarizes the objective functions using predefined weight settings mentioned before. It is assigned every single ant with a weight vector $\lambda_a, a \in \{1 \ldots N^a\}$ from a set of weighted vectors for $\Lambda > 0$, $N^a$ denoted the number of ants in the colony. Then, the objective functions are combined into a single-objective optimization problem based on given weight vector. In case of two-objective optimization, written as:

$$F = \lambda_a \psi_1(x) + (1 - \lambda_a)\psi_2(x): \ a \in \{1 \ldots N^a\} \tag{3}$$

WLS uses 2-opt local search operator to improve the solutions. If the evaluated value is smaller than the current solution, the current solution would be replaced. The process ends when all ants of each colony have performed the local search.

## 3.4   ILS-VNS

To improve the searching intensity of first phase local search in TPLS, we introduced ILS-VNS. It is an intensive local search algorithm on searching neighborhood solutions. To begin with, the main process of ILS-VNS is an iterative improvement method invoking perturbation and VNS. Double bridge move was chosen for the perturbation of the solution. Every time ILS-VNS receive ant's solution, it is first improved by the local search algorithm, afterward, the solution is improved by the VNS operator. If the new solution is non-dominated to the current solution, the old one comes up with replacement by the new solution, and the iteration counter resets. Secondly, perturbation, the best solution is perturbed with a random double bridge move for what suggested in [10]. The tour is decomposed into four random segments (a, a + 1), (b, b + 1), (c, c + 1), (d, d + 1) and then reconnected to a single tour (a, c + 1), (d, b + 1), (c, a + 1), (b, d + 1). Acceptance Criteria, best solution always accepts a solution if it has a better value. The improvement ends when there is no improvement after specified subsequent iterations.

---

**Algorithm 2** First Phase Local Search

---
1:  ... Initization()
2:  **for** each colony $c \in \{1, \cdots, N^{col}\}$ **do**
3:      **for** each ant $a \in \{1, \cdots, N^a\}$ **do**
4:          AssignWeight($a, \lambda_a \in \Lambda$)
5:          value := $\lambda y_1(x) + (1 - \lambda)y_2(x)$
6:          $a' :=$ VNS(a)
7:          new_value := $\lambda y_1(x') + (1 - \lambda)y_2(x')$
8:          **if** new_value $<$ value **then**
9:              $a := a'$
10:             UpdateSolution($A_{best}$, a)
11:         **end if**
12:     **end for**
13: **end for**
14: .. MOACO Phrase()
15: **Output:**$a_{improved}$

---

Summarized in Algorithm 2, in the very beginning, every ant is initialized with a random solution and assigned a set of weights $\lambda$ (Line 2–4). The objective functions are combined into a single objective function (Line 5). The objective function is used improved by VNS algorithm (Line 6). If an improving solution is found, it replaces the original ants' tour and update it to the global best archive $A_{best}$ (Line 7–10). When the first phase local search has completed, it returns all ants that having optimized tour. The ants' further improve the solution in MOACO and second phase local search (Line 15).

For the VNS process, it first creates a list of neighborhood structure. While the neighborhood list is not empty, the algorithm tends to search on next neighbors, if there is an improvement solution, it replaces the current one and continues until all neighbors are explored. It stops when all the neighbors are explored.

**Experimental Analysis.** In this section, we compared the use of different choice of local search operator in both the first phase and second phase local search under MOACO framework. We chose Pareto Ant Colony Optimization, one of the MOACO variant, to perform the test. The parameters of PACO were precisely configured.

For the first phase in TPLS, 3-opt were used as the local search operator in ILS and ILS-VNS. And 2-opt was used as the local search operator in WLS. For those algorithms, Don't look bits [15] and Nearest Neighborhood list [16] are equipped to improve the efficiency instead of using the random exchange.

**Parameter.** On the first phase local search, it ran 10 times random start. For the ILS-VNS algorithm, define $iter_{max}$ to 5. In WLS algorithm, 2-opt exchange neighborhood is used to improve the tour. Moreover, the size of the candidate list was set to 20, also, employ don't look bit, fixed radius strategies to improve the searching results.

The ACO algorithm used in PACO was Ant Colony System (ACS) [4]. Under the pseudo-random proportional rule, $q_0$ said the probability for the best choice in tour construction, which is $q_0 = 0.1$. $prob_{best}$ is the probability of constructing the best solution, was set to $prob_{best} = 0.98$. The population size of ants was 100, with $\alpha = 3$,

$\beta = 11$, corresponding to the degree of weight on pheromone and heuristic information used in probabilistic transition rule. $\rho$ represents the pheromone evaporation rate, $\rho = 0.3$. $\tau_{init}$ refers to the pheromone deposit on all pheromone on algorithm start. For other parameter of PACO, default settings were used described in [17].

Furthermore, the neighborhood size of PLS was configured to 20, with neighborhood exploration first-improvement strategy. The algorithm accepts any non-dominating solutions.

**Test Setting.** All experiments were run on OS Ubuntu 16.04.1 (64bits) with FX-6300 @ 4.0 GHz, 8 GB Memory environment. We instantiate the MOACO algorithm with the framework described in [18], available online http://lopez-ibanez.eu/moaco. The program was written in C and compiled with GCC version 5.4.0.

Basically, the testing includes Euclidean TSP with 3 different sizes of bTSP. And there are totally 6 instances to be tested. The size was: 100, 300 and 500. The source of the bTSP instances is from https://eden.dei.uc.pt/~paquete/tsp/.

All testing scenarios were tested in 20 trials for the stability of results. Here is the equation compute the time running on the testing instance $\left(\frac{n}{100}\right)^2 \times 100$ s, n represents the problem size. Hypervolume measure [19], a unary measure as the mean of result evaluation was applied to examine the result. It is the volume of the area of the non-dominated solution to given solution. The higher value of indicates the more cover area to the true solution space.

*Performance Indicator.* The result can be reflected by performance indicator of how the distribution of Pareto front covered the true Pareto-optimal solution. *Hypervolume Ratio Performance Indicator (HVR)* [20] calculates the ratio between the approximated hypervolume value and the reference hypervolume volume value, is represented by:

$$HVR = \frac{HV(Q)}{HV(P)} \qquad (4)$$

The reference sets were obtained from the best solution set from https://eden.dei.uc.pt/~paquete/tsp/. The reference set of euclidAB100 and euclidCD100 are "161934 167169" and "177713 165188" respectively. For euclidAB300 and euclidCD300, they are "500163 484988" and "485909 502006". And "858929 822318" and "808104 848343" for euclidAB500 and euclidCD500.

**Results and Discussion.** The goal of the test is to compare the local search in both the first phase and second phase. In particular, the candidate options of TPLS were, ILS with 3-opt and ILS-VNS 3-opt, came with a different choice of the second phase local search: WLS and PLS. Furthermore, the simple MOACO algorithm, without first phase local search, was employed to participate in the test as the baseline on the experiment.

*Hypervolume Measure.* For ease of representation, the algorithm instances were abbreviated. **AA** referred to the one only implemented with WLS. **AB** referred to the algorithm with PLS only. **BA** represented the algorithm implemented with ILS in the first phase and WLS. **BB** was same as BA in the first phase, which implemented with

PLS. **CA** involved ILS-VNS in the first phase and WLS in the second phase. **CB** used ILS-VNS for the first phase TPLS and PLS for the second phase TPLS. For the results in detail, illustrated in Tables 1, 2 and 3. The following tables reveal the result of the 6 algorithms instances on the bTSP benchmark. On problem size n = 100, euclidAB100 and euclidCD100, illustrated in Table 1. Two instances {ILS-VNS + PLS} and {ILS-VNS + WLS} perform the highest contribution on HVR to both euclidAB100 and euclidCD100. In the same time, the algorithm instance PLS is the worst choice in contributing the average hypervolume value.

**Table 1.** Euclidean bTSP Test 1, N = 100

| | | Iterations | 1-LS time (sec) | LS time (sec) | Hypervolume | HVR |
|---|---|---|---|---|---|---|
| euclidAB100 | WLS | 704.70 | – | 2.41 | 1.67E+10 | 98.9786% |
| | PLS | 743.50 | – | 0.87 | 1.27E+10 | 75.2539% |
| | ILS, WLS | 709.85 | 0.18 | 2.43 | 1.68E+10 | 99.1643% |
| | ILS, PLS | 767.00 | 0.17 | 0.83 | 1.67E+10 | 99.0064% |
| | ILS-VNS, WLS | 665.80 | 3.20 | 2.15 | 1.69E+10 | 99.8501% |
| | ILS-VNS, PLS | 697.40 | 3.20 | 0.76 | 1.69E+10 | 99.8462% |
| euclidCD100 | WLS | 704.90 | – | 2.41 | 1.82E+10 | 99.1173% |
| | PLS | 743.50 | – | 0.86 | 1.40E+10 | 75.9086% |
| | ILS, WLS | 710.50 | 0.17 | 2.40 | 1.83E+10 | 99.2506% |
| | ILS, PLS | 768.35 | 0.16 | 0.85 | 1.82E+10 | 99.0558% |
| | ILS-VNS, WLS | 663.60 | 3.32 | 2.17 | 1.84E+10 | 99.8412% |
| | 1LS-VNS, PLS | 694.10 | 3.32 | 0.76 | 1.84E+10 | 99.8363% |

In sense of time consumption, the best performing ILS-VNS algorithm instances required much more time on first phase local search. Notably, the time required for second phase local search also decreased on both WLS and PLS instances. On the other hand, the algorithm with PLS as the second phase local search requires less time than WLS.

**Table 2.** Euclidean bTSP Test 2, N = 300

|  |  | Iterations | 1-LS time (sec) | LS time (sec) | Hypervolume | HVR |
|---|---|---|---|---|---|---|
| euclidAB300 | WLS | 580.55 | – | 14.21 | 1.81E+11 | 98.6257% |
|  | PLS | 685.75 | – | 3.03 | 1.83E+11 | 66.5822% |
|  | ILS, WLS | 613.90 | 0.74 | 13.45 | 1.82E+11 | 98.8113% |
|  | ILS, PLS | 684.15 | 0.74 | 3.05 | 1.82E+11 | 98.7607% |
|  | ILS-VNS, WLS | 525.90 | 15.53 | 11.39 | 1.84E+11 | 99.7597% |
|  | 1LS-VNS, PLS | 590.40 | 15.42 | 2.59 | 1.84E+11 | 99.7618% |
| euclidCD300 | WLS | 576.85 | – | 14.09 | 1.83E+11 | 98.5868% |
|  | PLS | 682.25 | – | 3.02 | 1.25E+11 | 67.3138% |
|  | ILS, WLS | 611.15 | 0.75 | 13.49 | 1.83E+11 | 98.7680% |
|  | ILS, PLS | 679.40 | 0.74 | 3.03 | 1.83E+11 | 98.7092% |
|  | ILS-VNS, WLS | 516.70 | 16.58 | 11.27 | 1.85E+11 | 99.7445% |
|  | 1LS-VNS, PLS | 577.30 | 16.69 | 2.55 | 1.85E+11 | 99.7460% |

On the problem size n = 300, euclidAB300 and euclidCD300, details were described in Table 2. {ILS-VNS + WLS} and {ILS-VNS + PLS} implementation perform the best average HVR. The worst instance is PLS, which has the lowest HVR. On the other hand, the algorithm instance equipped with ILS-VNS required the first phase local search time, likewise, the ILS-VNS implementations reduced the time required in second phase local search.

On problem size n = 500, euclidAB500 and euclidCD500, summarized in Table 3. {ILS-VNS + WLS} and {ILS-VNS + PLS} implementations performed the best average HVR. The worst design is {PLS}. Besides, the time required by the ILS-VNS required the most time between the first phase local search options. That said, to compare with smaller problem size n = 100, 300, the time required by ILS-VNS scales linearly varies the problem size. On the contrary, both WLS and PLS components required much more time to compare with previous problem size, even more than the first phase ILS-VNS. WLS contributed around one-fourth of the total running time required in second phase local search reduced in ILS-VNS to compare with ILS and simple MOACO implementations.

**Table 3.** Euclidean bTSP Test 3, N = 500

|  |  | Iterations | 1-LS time (sec) | LS time (sec) | Hypervolume | HVR |
|---|---|---|---|---|---|---|
| euclidAB500 | WLS | 1987.80 | – | 181.73 | 5.66E+11 | 98.8318% |
|  | PLS | 2315.30 | – | 34.81 | 3.76E+11 | 65.6091% |
|  | ILS, WLS | 1995.35 | 2.42 | 180.70 | 5.67E+11 | 98.9243% |
|  | ILS, PLS | 2294.40 | 2.55 | 35.10 | 5.66E+11 | 98.8708% |
|  | ILS-VNS, WLS | 1870.70 | 39.52 | 179.84 | 5.71E+11 | 99.7355% |
|  | ILS-VNS, PLS | 2209.70 | 41.48 | 32.10 | 5.71E+11 | 99.7368% |
| euclidAB500 | WLS | 1700.20 | – | 192.49 | 5.34E+11 | 97.0424% |
|  | PLS | 1928.20 | – | 34.05 | 3.68E+11 | 66.7475% |
|  | 1LS, WLS | 1681.10 | 2.68 | 194.44 | 5.44E+11 | 98.8295% |
|  | [LS, PLS | 2289.30 | 2.57 | 35.24 | 5.44E+11 | 98.8272% |
|  | 1LS-VNS, WLS | 1873.90 | 39.70 | 179.75 | 5.49E+11 | 99.7144% |
|  | ILS-VNS, PLS | 2183.05 | 41.77 | 32.37 | 5.49E+11 | 99.7137% |

Summarized from three problem size above, that said ILS-VNS time to consume to optimize solution, it improved the solution finding efficiently. Also, reminded that the ILS-VNS with restarting strategy in fixed iteration maintain a good scaling capability on larger problem size, that is, that time required were scaled up linearly. Besides, PLS algorithm required much less time than WLS algorithm and perform a not difference performance when the first phase local search was involved.

*Two-Way ANOVA.* In this evaluation, statistical testing was performed to verify the significance of the result on the choice of local search designs. 2-way ANOVA was chosen and evaluate the result in terms of hypervolume value, with the level of significance $\alpha = 0.05$. The aim of the test is to find the best design of MOACO + TPLS. For that end, algorithms that did not use first phase local search were not included in the test. The statistical test resulted in a p-value of 6.2364E−192 on first phase local search. The p-value for the comparison of WLS and PLS was 0.001830102, in which concluded that the WLS implementation was better than PLS in sense of performance. The p-value for the intersection between the instances was 0.002307393. This refers to the algorithm instances were significantly different from each other. The rank from the worst to the best performing algorithm instances: {ILS + PLS}, {ILS + WLS}, {ILS-VNS + PLS}, {ILS-VNS + WLS}.

Above all, the best performing instance was {ILS-VNS + WLS}, and the {PLS} is the worst performing instance among the six variants. This is what mentioned in previous research that PLS perform a bad anytime behavior [21] when PLS starts at an initial solution with less supported efficient solutions, an early termination upon completion may return a very poor quality of approximation. Although ILS-VNS requires a quite amount of time to complete to the whole algorithm, we can see ILS-VNS implementations were still performed the best with the same time limit. This concludes the performance of MOACO could be improved by sharing a proportion of time with TPLS.

## 4 Conclusion

In this paper, TPLS were implemented with MOACO algorithm. A different choice of second phase local search, WLS or PLS, were chosen to improve the solutions. The result showed that the algorithm significantly improved the solution in terms of hypervolume and accuracy. And we discovered that the choice on either WLS or PLS did not affect the performance of the Pareto Front significantly. Besides, WLS consumed more time to finish an iteration. In addition, ILS-VNS algorithm was implemented as the first phase of TPLS to obtain a better supported efficient solution. And a restarting strategy was adapted in the first phase of TPLS to produce a larger number of initial solutions. The result showed the implementation of ILS-VNS with TPLS generated a better quality of Pareto Front than the originally proposed TPLS-MOACO algorithm.

Since our work has only focused on the bTSP problem, we would like to expand the experiment to solve other different kinds of problem with three or more objectives. In this way, we can see how the algorithm is applicable to solving real-world problems. In addition, instead of 3-opt, a more advanced local search algorithm can be implemented with the ILS-VNS algorithms, such as EAX and Lin–Kernighan [9]. This could help ILS-VNS algorithm to obtain a higher quality approximation of initial solutions. On the other hand, it is desirable to investigate the time effect of the hypervolume contribution of the proposed algorithm. This would provide a more detailed view to analyze different algorithm components.

## References

1. Gambardella, L.M., Taillard, É., Agazzi, G.: MACS-VRPTW: a multiple colony system for vehicle routing problems with time windows. In: New Ideas in Optimization, pp. 63–76. McGraw-Hill (1999)
2. Dorigo, M., Stützle, T.: Ant Colony Optimization, pp. 33–38. Bradford Company, Scituate (2004)

3. Paquete, L., Stützle, T.: A two-phase local search for the biobjective traveling salesman problem. In: Fonseca, C.M., Fleming, P.J., Zitzler, E., Thiele, L., Deb, K. (eds.) Evolutionary Multi-Criterion Optimization, pp. 479–493. Springer, Heidelberg (2003). https://doi.org/10.1007/3-540-36970-8_34
4. Dorigo, M., Gambardella, L.M.: Ant colony system: a cooperative learning approach to the traveling salesman problem. IEEE Trans. Evol. Comput. 1, 53–66 (1997)
5. Ren, L., Duhamel, C., Quilliot, A.: A hybrid ILS/VND Heuristic for the one-commodity pickup-and-delivery traveling salesman problem. Presented at the International Workshop on Green Supply Chain–GSC, Arras, France (2012)
6. Hoos, H.H., Stützle, T.: Stochastic Local Search: Foundations and Applications. pp. 298–358. Elsevier, Amsterdam (2004)
7. Dubois-Lacoste, J., López-Ibáñez, M., Stützle, T.: Combining two search paradigms for multi-objective optimization: two-phase and pareto local search. In: Talbi, E.-G. (ed.) Hybrid Metaheuristics, pp. 97–117. Springer, Heidelberg (2013). https://doi.org/10.1007/978-3-642-30671-6_3
8. Lust, T., Teghem, J., Tuyttens, D.: Very large-scale neighborhood search for solving multiobjective combinatorial optimization problems. In: Takahashi, R.H.C., Deb, K., Wanner, E.F., Greco, S. (eds.) Evolutionary Multi-Criterion Optimization, pp. 254–268. LNCS, vol. 6576, Springer, Heidelberg (2011). https://doi.org/10.1007/978-3-642-19893-9_18
9. Lin, S., Kernighan, B.W.: An effective heuristic algorithm for the traveling-salesman problem. Oper. Res. 21, 498–516 (1973)
10. Jaszkiewicz, A., Lust, T.: Proper balance between search towards and along pareto front: biobjective TSP case study. Ann. Oper. Res. 254(1–2), 111–130 (2017)
11. López-Ibáñez, M., Paquete, L., Stützle, T.: Hybrid population-based algorithms for the bi-objective quadratic assignment problem. J. Math. Model. Algorithms 5, 111–137 (2006)
12. Zitzler, E., Laumanns, M., Thiele, L.: SPEA2: improving the strength pareto evolutionary algorithm. TIK-report. 103, Computer Engineering and Networks Laboratory (TIK), Swiss Federal Institute of Technology (ETH), Zurich, Switzerland (2001)
13. Taillard, E.: Robust taboo search for the quadratic assignment problem. Parallel Comput. 17, 443–455 (1991)
14. Angel, E., Bampis, E., Gourvès, L.: Approximating the pareto curve with local search for the bicriteria TSP(1, 2) problem. Theoret. Comput. Sci. 310, 135–146 (2004)
15. Bentley, J.J.: Fast algorithms for geometric traveling salesman problems. ORSA J. Comput. 4, 387–411 (1992)
16. Johnson, D., Mcgeoch, L.: The traveling salesman problem: a case study in local optimization. In: Local Search in Combinatorial Optimization (1997)
17. Doerner, K., Gutjahr, W.J., Hartl, R.F., Strauss, C., Stummer, C.: Pareto ant colony optimization: a metaheuristic approach to multiobjective portfolio selection. Ann. Oper. Res. 131, 79–99 (2004)
18. Lopez-Ibanez, M., Stutzle, T.: The automatic design of multiobjective ant colony optimization algorithms. IEEE Trans. Evol. Comput. 16, 861–875 (2012)
19. Zitzler, E.: Evolutionary algorithms for multiobjective optimization: methods and applications. Ph.D. thesis, Swiss Federal Institute of Technology (ETH), Zurich, Switzerland (1999)
20. Ariyasingha, I., Fernando, T.: Performance analysis of the multi-objective ant colony optimization algorithms for the traveling salesman problem. Swarm Evol. Comput. 23, 11–26 (2015)
21. Dubois-Lacoste, J., López-Ibáñez, M., Stützle, T.: Anytime pareto local search. Eur. J. Oper. Res.Eur. J. Oper. Res. 243, 369–385 (2015)

# A Neural Network Based Global Optimal Algorithm for Unconstrained Binary Quadratic Programming Problem

Shenshen Gu$^{(\boxtimes)}$ and Xinyi Chen

School of Mechatronic Engineering and Automation, Shanghai University,
Shanghai, China
gushenshen@shu.edu.cn

**Abstract.** Unconstrained binary quadratic programming problem is a classical integer optimization problem and is well known to be NP-hard. In order to improve the performance of global optimal algorithms for unconstrained binary quadratic programming problem, in this paper, we proposed a new exact solution method. By investigating the geometric properties of the original problem, the quality of the algorithms for calculating the upper bound and lower bound are improved. And then, for the new derived upper bound algorithm and lower bound algorithm, their recurrent neural network models are proposed and applied respectively in order to speed up the computation. The numerical results shows that the proposed algorithm of a branch-and-bound type is quite effective and efficient.

**Keywords:** Binary quadratic problem · Branch-and-bound
Recurrent neural network

## 1 Introduction

The unconstrained binary quadratic programming problem is defined as follows:

$$(P) \quad \min_{x \in \{0,1\}^n} f(x) = \frac{1}{2}x^T Q x + c^T x. \tag{1}$$

where $Q = (q_{ij})_{n \times n}$ is a symmetric matrix with zero elements in the main diagonal and $c \in \mathbb{R}^n$. It is a very famous integer optimization problem which is NP-hard [1].

The unconstrained binary quadratic programming has many applications in the real world [2,3]. For its exact solution methods, most algorithms are based on the branch-and-bound framework (see, e.g., [4–8] and the references therein).

In this paper, We focus on a global optimal algorithm of a branch-and-bound type implemented by recurrent neural network model for solving (P). For the branch-and-bound type algorithm, the efficiency mainly depends on the quality of the upper and lower bound estimation and the corresponding computational

© Springer Nature Switzerland AG 2018
L. Cheng et al. (Eds.): ICONIP 2018, LNCS 11302, pp. 577–587, 2018.
https://doi.org/10.1007/978-3-030-04179-3_51

complexity. As a particularly important case of integer programming, unconstrained binary quadratic programming problem processes rich geometric properties. However, such important features hidden behind are only explored by a few papers, e.g., [9,10] until recently.

By introducing a positive parameter $\mu$, we can get the following perturbed quadratic objective function,

$$f_\mu(x) = \frac{1}{2}x^T(Q - \mu I)x + (c + \frac{\mu e}{2})^T x, \tag{2}$$

where $I$ is an $n \times n$ identity matrix, and $e$ is an $n$-dimensional vector with all elements equal to 1.

Since $x_i^2 = x_i$ for any $x_i \in \{0,1\}$, it can be easily seen that $f(x) = f_\mu(x)$ on $\{0,1\}^n$. Thus, the following perturbed problem is equivalent to $(P)$:

$$(P_\mu) \quad \min_{x \in \{0,1\}^n} f_\mu(x).$$

When setting $\mu$ large enough, e.g., $(\lambda_n - \mu) < 0$, where $\lambda_n$ is the biggest eigenvalue of $Q$, or

$$\mu - q_{ii} \geq \sum_{j=1, j \neq i}^{n} |q_{ij}|, \quad i = 1, \ldots, n,$$

matrix $(Q - \mu I)$ will be negative definite. Therefore in the following sections, we always assume $f(x)$ is concave.

Note that $\{x \in R^n | f(x) \geq v\}$ forms an ellipsoid for any $v < -\frac{1}{2}c^T Q^{-1}c$ and the center of the ellipse $f(x) = v$ is $x^0$. It is clear that any point in $\{0,1\}^n$ outside the ellipse $f(x) = v$ possesses an objective value less than $v$. We can prove that the point $x \in \{0, 1\}^n$ is the solution of (P) if only there is no point in the box $\{0, 1\}$ outside of the ellipsoid $E(f(x))$.

By taking advantage of geometric properties, this paper proposed the upper bound algorithm and the lower bound algorithm to improve the quality of the upper bound and the lower bound calculation. Then we applied recurrent neural network models to implement the upper bound algorithm and lower bound algorithm to speed up the computation.

The rest of the paper is organized as follows: First, a new method based on neural network to find the upper bound of the optimal value of (P) is addressed in Sect. 2. To compute a good lower bound, in Sect. 3, a new lower bound algorithm by taking advantage of the $k$th shortest path algorithm is discussed. And in Sect. 4, the neural network model to implement this algorithm is addressed. Then, the whole algorithm is summarized in Sect. 5 and the effectiveness and efficiency are illustrated. Finally, we conclude the paper in Sect. 6.

## 2   The Upper Bound Algorithm and Its Recurrent Neural Network Model

We introduced our new method to find an upper bound here. Suppose that the incumbent, the best solution found so far, is $\tilde{x}$ and the current best upper bound is $\tilde{v} = f(\tilde{x})$. If the incumbent is not the optimal solution to the original problem, there must exist an intersection between the ellipse contour $f(x) = \tilde{v}$ and the box $[0,1]^n$. Note that any outer normal vector on the ellipse $f(x) = \tilde{v}$ represents a descent direction. Therefore, if we can find a point $\bar{x}$ located on the ellipse contour $f(x) = \tilde{v}$ and within the box $[0,1]^n$, the normal vector of this point directs to a better integer point.

If any point on the objective contour $f(x) = \tilde{v}$ is inside the box $[0,1]^n$, this point provides a descent direction. We thus propose to solve the following problem to identify the point on $f(x) = \tilde{v}$ that is closest to the center of the box $[0,1]^n$.

$$
\begin{aligned}
\min \quad & \sum (x_i - 0.5)^2 \\
s.t. \quad & f(x) = \tilde{v}, \\
& x_i \in [0,1].
\end{aligned}
\tag{3}
$$

After finding the point $\bar{x}$, which is on the ellipse $f(x) = \tilde{v}$ and within the box $[0,1]^n$, the gradient of $f(x) = \tilde{v}$ at $\bar{x}$ can be calculated as $Q\bar{x} + c$. The next incumbent, $\hat{x}$, can be identified as follows,

$$
\hat{x}_j = \begin{cases}
= 1, & \text{if } (Q\bar{x} + c)_j < 0, \\
= 0 \text{ or } 1, & \text{if } (Q\bar{x} + c)_j = 0, \ j = 1, \ldots, n. \\
= 0, & \text{if } (Q\bar{x} + c)_j > 0,
\end{cases}
\tag{4}
$$

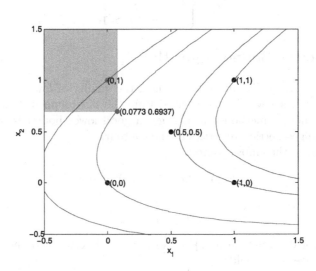

**Fig. 1.** Normal vector on the ellipse and better solution

Figure 1 shows the normal vectors with an example. It is clear that the optimal solution is $(0,1)^T$. Suppose that the incumbent is $(0,0)^T$. The normal vector at the point $\bar{x} = (0.0773, 0.6937)^T$, $(74.5923, -47.0012)^T$, defines the shadow region. Note that any integer point inside the shadow region, in our case $(0,1)^T$, is a better solution.

Some conventional methods such as iterative method can be used to find out the point $\bar{x}$. However, when the problem is very large, find out $\bar{x}$ with the traditional methods will be very time consuming. Neural network model is a promising method to solve this problem due to its parallel property. Based on the formulation of Eq. 3, some recurrent neural network models are proposed by us [11,12].

In this paper, we apply the result of [12] to find out $\bar{x}$ and then calculate the upper bound. The dynamic equation of the neural network model is described as follows (see [11,12] and the references therein for the detail).

$$
\begin{aligned}
\dot{x}(t) = &- \mu x(t) + \mu P_\Omega \{ x(t) - \nabla f(x(t)) + \nabla h(x(t)) \\
&[-\mu \nabla h(x(t))^T \nabla f(x(t)) + \rho h(x(t))] \\
&[\mu \nabla h(x(t))^T \nabla h(x(t))]^{-1} \}
\end{aligned}
\tag{5}
$$

where $x$ is the state vector of the neural network, $\mu$ is positive weighted coefficient. Both the objective function $f(x)$ and the constraint function $h(x)$ are assumed to be continuously differentiable. $\nabla f$ is the gradient of $f$, $\nabla h$ is the gradient vector of $h$ and $P_\Omega$ is a projection operator defined as

$$
P_\Omega =
\begin{cases}
0, & U_i < 0 \\
U_i, & 0 < U_i < 1 \\
1, & U_i > 1
\end{cases}
\tag{6}
$$

## 3   The Lower Bound Algorithm

In addition to the quality of upper bounds, the quality of lower bounds is also important. Here, a lower bound derived from the maximum distance sphere is introduced. Then, a method that can get a good lower bound is proposed by exploiting the geometric properties of the contour.

The center of the ellipse contour

$$
E(\tilde{v}) = \{ x \in R^n | f(x) = \tilde{v} \}
$$

is

$$
x^0 = -Q^{-1} c.
$$

Ranking the 0-1 points in $\{0,1\}^n$ according to a descending order with respect to their distance to $x^0$ yields

$$
x^1, x^2, x^3, \ldots
$$

with corresponding distance

$$
r^1 \geq r^2 \geq r^3 \geq \ldots
$$

where

$$
r^i = \| x^i - x^0 \|.
$$

Let $\underline{f}^i$ be the contour level, then $E(\underline{f}^i) = \{x \in R^n | f(x) = \underline{f}^i\}$ is the minimum circumscribed contour containing the ball $S_i = \{x \in R^n | \|x - x^0\|^2 = r_i^2\}$. It can be proved that $\underline{f}^i = \frac{\lambda_1}{2} r_i^2 - \frac{1}{2} c^T Q^{-1} c$, for $i = 1, 2, \ldots$. In addition, $\underline{f}^1$ is a lower bound.

The range of the lower bound derived from the above algorithm based on maximum distance sphere is too large. Therefore, we discuss a new algorithm providing a lower bound to problem $(P)$ which is better than $\underline{f}^1$. The algorithm based on the $k$th farthest point to center point of the ellipse $(k \geq 2)$ can be eventually transformed to the shortest path algorithm based on neural network. The algorithm is given as follows.

**Algorithm 1:**
    **Step 1:** Calculate $x^1$, $r^1$ and $f^1$. Let $x^* = x^1$ and $\bar{f} = f(x^*)$. Let $k = 2$.

    **Step 2:** Calculate $x^k$, $r^k$ and $f^k$. If $\bar{f} \leq f^k$, stop and $x^*$ is the optimal solution.

    **Step 3:** Let

$$x^* = \begin{cases} x^*, & \text{if } f(x^*) \leq f(x^k) \\ x^k, & \text{if } f(x^k) < f(x^*) \end{cases}$$

$$\bar{f} = f(x^*)$$

    **Step 4:** If $k = 2^n$, stop and $x^*$ is the optimal solution. Let $k = k+1$ and go back to Step 2.

In real implementation, we do not implement Algorithm 1 as the exact algorithm because it costs a lot of time for large scale problem but use this algorithm to provide us with a lower bound. For the latter case, we implement the algorithm for a few steps. When we stop the algorithm at the $k$th iteration, the lower bound is the corresponding $f(x^*)$.

## 4 Neural Network Based Algorithm to Sort $x^k$

The key point of the lower bound algorithm we proposed in previous section is that we need to sort the zero-one point with respect to their distance to the center point in a descending order, which is $(x^1, x^2, \ldots, x^k)$. Here we introduce an algorithm to find the $k$th farthest zero-one point. Suppose that the nearest zero-one points to $x^0$ is $\tilde{x}$, then the distance of any other zero-one points to $x^0$ can be written as:

$$\begin{aligned}
\|x - x^0\|^2 &= \sum_{i=1}^n (x_i - x_i^0)^2 \\
&= \sum_{i=1}^n (x_i - x_i^1 + x_i^1 - x_i^0)^2 \\
&= \sum_{i \notin I} (x_i - x_i^1 + x_i^1 - x_i^0)^2 + \sum_{i \in I} (x_i - x_i^1 + x_i^1 - x_i^0)^2 \\
&= \sum_{i \notin I} (x_i^1 - x_i^0)^2 + \sum_{i \in I} (1 - x_i^1 - x_i^0)^2 \\
&= \sum_{i=1}^n (x_i^1 - x_i^0)^2 - \sum_{i \in I} (x_i^1 - x_i^0)^2 + \sum_{i \in I} (1 - x_i^1 - x_i^0)^2 \\
&= \|x^1 - x^0\|^2 + \sum_{i \in I} (1 - 2x_i^0)(1 - 2x_i^1)
\end{aligned} \tag{7}$$

where $I = \{i | x_i \neq x_i^1, \ i = 1, \ 2, \ ..., \ n\}$ is an index set. If $x_i^0 < 0.5$, then $x_i^1 = 1$ and $(1 - 2x_i^0)(1 - 2x_i^1) = -(1 - 2x_i^0) < 0$; otherwise, $x_i^1 = 0$ and $(1 - 2x_i^0)(1 - 2x_i^1) = (1 - 2x_i^0) < 0$. Therefore, $(1 - 2x_i^0)(1 - 2x_i^1) = -|1 - 2x_i^0|$. Then from (7), we get

$$d^2(x, x^0) = \|x^1 - x^0\|^2 - \sum_{i \in I}(1 - 2x_i^0)$$

**Definition 1:** *Let* $S = \{\alpha_i | \alpha_i = |1 - 2x_i^0|, \ i = 1, \ 2, \ ..., \ n\}$, $S_K$ *is the K-subset of S, where, the K-subset of S means one subset of S which contains K elements, and* $S_0 = \{\emptyset\}$.

Obviously, the set $S$ containing $n$ elements has $C_n^K$ kinds of $K$-subset: $S_K$ and $2^n$ kinds of subsets (including the empty subset $\{\emptyset\}$). Assumed that the sum of a subset is the sum of the elements of the subset (the sum of the empty subset $\{\emptyset\}$ is $-\infty$), we get the following definition:

**Definition 2:** *Problem* $(P_k)$ *is the problem of solving the first k subsets of S when ranking all the subsets of S in an ascending order.*

For example, if $x^0 = (0.4, \ 0.3, \ 0.2, \ 0.1)^T$, then $\alpha_1 = 0.2$, $\alpha_2 = 0.4$, $\alpha_3 = 0.6$, $\alpha_4 = 0.9$, $S = 0.2, 0.4, 0.6, 0.9$, solve $(P_{12})$.

All the subsets of $S$ is: $\{\emptyset\}$, $\{0.2\}$, $\{0.4\}$, $\{0.6\}$, $\{0.9\}$, $\{0.2, 0.4\}$, $\{0.2, 0.6\}$, $\{0.2, 0.9\}$, $\{0.4, 0.6\}$, $\{0.4, 0.9\}$, $\{0.6, 0.9\}$, $\{0.2, 0.4, 0.6\}$, $\{0.2, 0.4, 0.9\}$, $\{0.2, 0.6, 0.9\}$, $\{0.4, 0.6, 0.9\}$, $\{0.2, 0.4, 0.6, 0.9\}$.

The solution of $(P_{12})$ is: $\{\emptyset\}$, $\{0.2\}$, $\{0.4\}$, $\{0.6\}$, $\{0.9\}$, $\{0.2, 0.4\}$, $\{0.2, 0.6\}$, $\{0.2, 0.9\}$, $\{0.4, 0.6\}$, $\{0.4, 0.9\}$, $\{0.6, 0.9\}$, $\{0.2, 0.4, 0.6\}$.

Let

$$x_i = \begin{cases} 0, & \text{if } \alpha_i \notin S_k \\ 1, & \text{if } \alpha_i \in S_k \end{cases}$$

then, the solution of $(P_{12})$ is equivalent to: $\{0,0,0,0\}$, $\{1,0,0,0\}$, $\{0,1,0,0\}$, $\{0,0,1,0\}$, $\{1,1,0,0\}$, $\{1,0,1,0\}$, $\{0,0,0,1\}$, $\{0,1,1,0\}$, $\{1,0,0,1\}$, $\{1,1,1,0\}$, $\{0,1,0,1\}$, $\{0,0,1,1\}$.

Obviously: $x_i = 1$ implies $x_i \neq x_i^1$, $x_i = 0$ implies $x_i = x_i^1$, and $x^1 = (1,1,1,1)^T$. Therefore, $x^1 = (1,1,1,1)^T$, $x^2 = (0,1,1,1)^T$, $x^3 = (1,0,1,1)^T$, ..., $x^{16} = (0,0,0,0)^T$.

From the above example it is clearly that the problem of finding $(x^1, x^2, ..., x^k)$ is equivalent to solving the problem $(P_k)$. However, when $n$ is very large, performing this sorting task is very difficult and time consuming. In addition, in most cases we just need to sort a small portion of these zero-one points. In [15], we proposed a novel algorithm to deal with this problem. First, a specific network structure is created to represent all $k$-subsets of set $S$. As shown in Fig. 2, each network has one source node $n_{start}$ (i.e., start point), one sink node $n_{end}$ (i.e., ending point) and several intermediate nodes such as $n_{1.1}$ and $n_{2.1}$. It is clear that finding the $K$ smallest $k$-subsets sum is equivalent to searching $K$ shortest paths in a network.

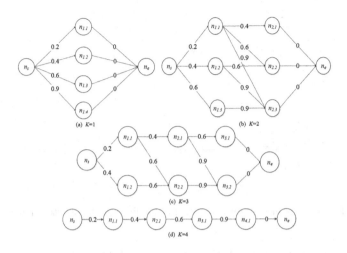

**Fig. 2.** Networks representing 1-subsets, 2-subsets and 3-subsets of $\{0.2, 0.4, 0.6, 0.9\}$.

Then, the algorithm proposed by Yen [13] is used to find $K$ shortest paths this network. Yen's algorithm begins with the original graph but update the graph structure in each following steps. For each step, finding the shortest path is performed on the current graph. After $K$ steps, the first $K$ shortest paths can be found out. As for finding the shortest path, There are many famous algorithms such as dynamic programming [14]. However, for applications which are real-time or large-scale, these series algorithms may not be efficient. Therefore, parallel computational models are more desirable.

For this reason, in [15], by taking advantage of the efficiency of the recurrent neural network methods in finding the shortest path, we integrated the finite time convergent recurrent neural network with the $K$ shortest path algorithm to solve the $K$ smallest $k$-subsets problem.

Here we apply the algorithm proposed in [15] (see [13,15] and the references therein for the detail) to the computation of the lower bounds of problem $(P)$. The block diagram of the specific neural network is shown in Fig. 3. The circuit implementation of the neural network is illustrated in Fig. 4.

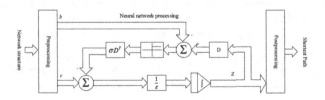

**Fig. 3.** The block diagram of neural network for finding the shortest path

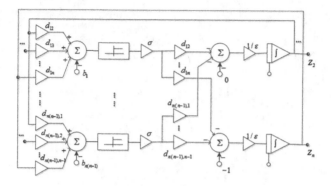

**Fig. 4.** The circuit implementation for finding the shortest path

## 5    The Algorithm and Experimental Results

Integrating the upper bounding method and the lower bounding method, we present a global optimal algorithm combining the branch-and-bound solution method and neural network models. And the following is the main algorithm.

**Algorithm 2 (Branch-and-bound method for $(P)$):**
   **Step 0:** *(Initialization).*
   *(i) Choose $x^0 = -Q^{-1}c$ and the farthest point $\tilde{x}$ of $\{0, 1\}^n$ to $x^0$. Let $f_{opt} = f(\tilde{x})$ and $x_{opt} = \tilde{x}$.*
   *(ii) If $I_{free} = \phi$, set $k = 0$. Otherwise, set $L = \{(I_{fix}^0, I_{fix}^1, I_{free}, l, u)\}$ and $k = 1$.*
   **Step 1:** *If $k = 0$, stop and $x_{opt}$ is the optimal solution. Otherwise select a node from $L$ and construct the subproblem at the selected node by setting $\tilde{Q} = (Q)_{I_{free} \times I_{free}}$ and $\tilde{c}_i = c_i + \sum_{j \in I_{fix}^1} q_{ij}$ for $i \in I_{free}$. Let*

$$v_{fix} = \frac{1}{2} \sum_{i \in I_{fix}^1} \sum_{j \in I_{fix}^1} q_{ij} + \sum_{i \in I_{fix}^1} c_i$$

*Set $k := k - 1$*
   **Step 2:** *(Upper bounding, lower bounding and fathoming) Choose a suitable $\mu \geq 0$.*
   *(i) Compute the upper bound $u(\mu)$ defined in Eq. (4) for the subproblem at this node. If $u(\mu) + v_{fix} < f_{opt}$, update $f_{opt} = u(\mu) + v_{fix}$.*
   *(ii) If $l(\mu) \geq f_{opt} - v_{fix}$, fathom the current node and goto Step 1.*

   **Step 3:** *(Brunching). Choose an $i \in I_{free}$ and set $I_{free} := I_{free} \setminus \{i\}$. Set $I_{fix}^0 := I_{fix}^0 \cup \{i\}$ and $I_{fix}^1 := I_{fix}^1 \cup \{i\}$, respectively, to generate two new nodes. Add these two nodes to $L$ and set $k := k + 2$. Goto Step 1.*

In order to test the efficiency of the new upper bounding and lower bounding methods discussed in the previous sections, we perform computational experiments and present numerical results for the proposed algorithm. The upper bounding is realized through nonfeasible gradient projection neural network

**Table 1.** Comparison of computation time (sec.) of traditional algorithm and our proposed algorithm

| n | 10 | | 15 | | 20 | | 25 | | 30 | |
|---|---|---|---|---|---|---|---|---|---|---|
| | $T_A$ | $T_N$ | $T_A$ | $T_N$ | $T_A$ | $T_N$ | $T_A$ | $T_N$ | $T_A$ | $T_N$ |
| *problem* 1 | 0.23 | 0.23 | 6.24 | 5.99 | 83.28 | 34.23 | 1151.58 | 235.31 | 1050.02 | 177.88 |
| *problem* 2 | 0.36 | 0.36 | 5.94 | 5.16 | 60.15 | 11.02 | 1175.45 | 4.00 | 2151.26 | 83.16 |
| *problem* 3 | 0.29 | 0.28 | 7.73 | 7.71 | 66.80 | 41.15 | 816.78 | 83.35 | 2083.84 | 541.02 |
| *problem* 4 | 0.37 | 0.37 | 7.87 | 4.98 | 79.34 | 25.41 | 1487.02 | 556.78 | 1075.47 | 2.66 |
| *problem* 5 | 0.49 | 0.48 | 5.40 | 4.65 | 72.38 | 59.63 | 595.48 | 1.99 | 3273.50 | 1269.94 |
| *problem* 6 | 0.23 | 0.22 | 8.86 | 8.07 | 66.05 | 16.41 | 464.85 | 1.60 | 2469.82 | 159.14 |
| *problem* 7 | 0.38 | 0.32 | 7.97 | 6.01 | 40.30 | 11.02 | 1433.32 | 111.11 | 3409.36 | 313.85 |
| *problem* 8 | 0.28 | 0.27 | 3.07 | 1.33 | 107.66 | 23.82 | 628.52 | 463.80 | 2052.78 | 137.28 |
| *problem* 9 | 0.34 | 0.33 | 5.54 | 5.52 | 78.51 | 47.19 | 1317.03 | 518.23 | 726.03 | 135.13 |
| *problem* 10 | 0.55 | 0.54 | 8.35 | 5.18 | 89.95 | 54.77 | 772.47 | 13.47 | 3980.94 | 110.83 |
| *Ave time* | 0.35 | 0.34 | 6.70 | 5.46 | 74.44 | 32.46 | 984.25 | 198.96 | 2227.30 | 293.09 |

based algorithm and the lower bounding is applied a neural network based $k$th shortest path algorithm. The algorithm was coded by VS 2013 and run on a Intel(R) Core(TM)i7-7700K CPU@4.20 GHz PC. The detailed simulation results are shown in Table 1.

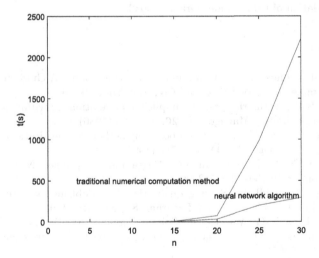

**Fig. 5.** The computational time of two algorithm

Table 1 shows the ideal computation time of our algorithm compared with the traditional numerical computation algorithm. All the simulated examples are generated randomly, with $q_{ij} \in [-50, 50]$ for $i \neq j, q_{ii} = 0$, and $c_i \in [-50, 50]$.

$T_N$ is the ideal computation time of our neural network based algorithm and $T_A$ is the computation time based on traditional numerical computation method. And the *Ave time* represents the average time of the algorithms.

Since both the upper bound algorithm and the lower bound algorithm are based on neural networks, the computation time is very when neural networks are implemented through parallel circuit. For calculating $T_N$, we set the computation time of upper bound algorithm and lower bound algorithm for each sub-problem to $3 \times 10^{-4}$ s and $10 \times 10^{-4}$ s respectively. The visual comparison of $T_N$ and $T_A$ is given in Fig. 5, which further illustrate the effectiveness and efficiency of our proposed algorithm.

## 6   Conclusion

In this paper, we proposed a new upper and lower bounding method for solving unconstrained binary quadratic programming problem by investigating its geometric features. Integrating these new bounding schemes into a proposed solution algorithm of a branch-and-bound type, we propose a global optimal algorithm for solving unconstrained binary quadratic programming problem with promising preliminary computational results. Neural network models are proposed and applied to these new bounding schemes to speed up the algorithm. The numerical results shows that our algorithm is efficient and accurate.

**Acknowledgments.** The work described in the paper was supported by the National Science Foundation of China under Grant 61503233.

## References

1. Garey, M.R., Johnson, D.S.: Computers and Intractability: A Guide to the Theory of NP-Completeness. WH Freeman. Co., New York (1979)
2. Mcbride, R.D., Yormark, J.S.: An implicit enumeration algorithm for quadratic integer programming. Manage. Sci. **26**, 282–296 (1980)
3. Chardaire, P., Sutter, A.: A decomposition method for quadratic zero-one programming. J. Manage. Sci. **41**, 704–712 (1995)
4. Li, D., Sun, X.L.: Nonlinear Integer Programming. Springer, New York (2006). https://doi.org/10.1007/0-387-32995-1
5. Helmberg, C., Rendl, F.: Solving quadratic (0,1)-problems by semidefinite programs and cutting planes. Math. Program. **82**, 291–315 (1998)
6. Rendl, F., Rinaldi, G., Wiegele, A.: Solving max-cut to optimality by intersecting semidefinite and polyhedral relaxation. Lecture Notes Computer Science, vol. 4513, pp. 295–309 (2007)
7. Pardalos, P.M., Rodgers, G.P.: Computational aspects of a branch-and-bound algorithm for quadratic zero-one programming. Computing **45**, 131–144 (1990)
8. Barahona, F., Jünger, M., Reinelt, G.: Experiments in quadratic 0–1 programming. Math. Program. **44**, 127–137 (1989)
9. Boros, E., Hammer, P.L., Tavares, G.: Local search heuristics for unconstrained quadratic binary optimization. Technical report, RUTCOR, Rutgers University, Rut-cor Research Report (2005)

10. Li, D., Sun, X.L., Liu, C.L.: An exact solution method for quadratic 0–1 programming: a geometric approach. Technical report, Chinese University of Hong Kong, Department of Systems Engineering and Engineering Management (2006)
11. Gu, S., Peng, J.: A neural network based algorithm to compute the distance between a point and an ellipsoid. In: 2015 Seventh International Conference on Advanced Computational Intelligence (ICACI), pp. 294–299. IEEE (2015)
12. Gu, S., Peng, J., Zhang, J.: A projection based recurrent neural network approach to compute the distance between a point and an ellipsoid with box constraint. In: Youth Academic Annual Conference of Chinese Association of Automation (YAC), pp. 459–462. IEEE (2016)
13. Yen, Y.: Finding the K shortest loopless paths in a network. Manag. Sci. **17**(11), 712–716 (1971)
14. Bellman, R.: On a routing problem. Quart. Appl. Math. **16**, 87–90 (1958)
15. Gu, S., Cui, R.: An efficient algorithm for the subset sum problem based on finite-time convergent recurrent neural network. Neurocomputing **149**, 13 (2014)

# Supervised Learning

# Stochasticity-Assisted Training in Artificial Neural Network

Adedamola Wuraola[(✉)] and Nitish Patel

Department of Electrical and Computer Engineering, The University of Auckland,
Auckland, New Zealand
awur978@aucklanduni.ac.nz, nd.patel@auckland.ac.nz

**Abstract.** Strategically injected noise can speed up convergence during
Neural Network training using backpropagation algorithm. Noise injection during Neural Network training have been proven empirically to
improve convergence and generalizability. In this work, a new methodology proven to be efficient for speeding up learning convergence using
weight noise in Single Layer Feed-forward Network (SLFN) architecture is presented. We present efficient and effective methods in which
local minimum entrapment can be avoided. Our proposed controlled
introduction of noise is based on 4 proven analytical and experimental
methods. We show that criteria-based mini-batch noise injection to the
weights during training often outperforms the noiseless weights as well
as fixed noise introduction as seen in literature both in network generalization and convergence speed. The effectiveness of this methodology has
been empirically shown as well as it achieving on an average 15%–25%
improvement in convergence speed when compared to fixed and noiseless networks. The proposed method is evaluated on the MNIST dataset
and other datasets from UCI repository. The comparative analysis confirms that the proposed method achieves superior performance regarding
convergence speed.

**Keywords:** Activation function · Artificial Neural Networks
Stochasticity · Weight noise

## 1 Introduction

State-of-the-art software simulation, hardware and high availability of data have
led to the fast and continuous growth of Artificial Neural Networks (ANN). ANN
is finding applications in almost all facets of life ranging from medicine to mun-
dane day-to-day activities. They continuously recorded equal to and most time
better than human level intelligence. Convergence and generalizability are two of
the essential characteristics required during training and usage of ANN models
respectively. A network model able to reach convergence quickly without negatively impacting the accuracy is highly desirable. However, due to the nature of
the commonly used stochastic gradient descent and its associated issues, achieving quick convergence seems to require external factors. On the other hand,

© Springer Nature Switzerland AG 2018
L. Cheng et al. (Eds.): ICONIP 2018, LNCS 11302, pp. 591–602, 2018.
https://doi.org/10.1007/978-3-030-04179-3_52

accuracy during training is essential, but the main hurdle and capability of the ANN are tested on unseen data. Slow convergence and unacceptable generalization ability of ANN is based on but not limited to the following; learning algorithm, activation function, and network architecture. The trial and error method associated with network architecture definition has contributed to the identified issues, and work has been done in literature to give the best way of determining architecture size and structure [1,2]. In this paper, the work is based on the learning algorithm induced slow convergence and low generalization. Therefore, the importance of these two characteristics cannot be overemphasized.

The effects of injecting random noise into network inputs and parameters (weights, activations and so on) of a multilayer neural network and deep neural network architectures when training using backpropagation algorithm has been widely discussed in the literature. Apart from having a positive effect on convergence, noise injection has been proven analytically and experimentally to improve generalizability and network's robustness. As described in [3], backpropagation algorithm has been identified as a special case of the generalized expectation-maximization (EM) algorithm hence, the reason for training speed up when noise is injected. Due to this observation, researchers have injected noise into the training dataset. Data noise constitute the most common type of noise injection in literature, and the noise can be added to inputs or outputs of the ANN [3]. The major impact data noise has on neural network training is robustness capability which helps the trained network model to generalize very well on unseen data [4] but in some cases data noise have been shown to increase network convergence [4]. Furthermore, weight noise [5,6] has been greatly explored as well as weight change noise [7]. [5,6] introduced analog noise to the weights by expanding the cost function to include noise-mediated terms. Recently, noise has been added to the activation function to improve network performance as well as the network's computed activations.

Apart from various parts of a neural network that noise can be injected, over the years, researchers have proposed several methodologies of injecting noise. The most common method being the fixed noise injection into the desired aspect of the network model. An addictive and multiplicative noise was injected into a training set and weight respectively in [8–10] and results in better generalization. [11] injected fixed Gaussian noise type with a mean of zero and finite variance into the input dataset as well as network parameters. Furthermore, comprehensive noise introduction was discussed in [12–14]. Finally, authors in [15] improve learning and reduce local minimum entrapment by using additive noise in the desired/target output. Using annealing schedule to control the beginning and the later end of training. In other words at the beginning of training, random perturbation of noise was imposed but towards the end the variance of the perturbation is made to decrease to zero.

To the best of our knowledge, the additive noise and multiplicative noise introduced to the network are either fixed or based on the loss function and also universally added. However, as described in [10], any unsuitable noise injection may result into slow convergence and even oscillation. Particularly, in deep neural network, noise introduction is seen as the import of uncertainty into the networks and a large uncertainty will result in negative effect in the overall network performance. Hence the motivation behind our methodology of mini-batch criteria based noise injection protocols. During gradient descent optimization, the optimization process (weight update) is updated using a cost function. The process is further characterized by the global minimum cost, learning rate, and the gradient. So the decision on how much the weight is increased or decreased is based on those characteristics. In this work, we are proposing the decision to be based on one extra parameter called noise. We propose to exploit the injection of criteria-based noise that will help alleviate the above-mentioned gradient descent training issues. Our approach is criteria-based because the training of any ANN is heavily dependent on data. Some data are "good" and training will proceed as necessary without getting stuck in local minimum whereas the reverse is the case in some others. Criteria-based noise injection helps to eliminate the problem of how much noise is needed to improve network performance and not degrade performance. Just like the way in which stochastic gradient descent training algorithms starts from random network parameters (weights), the noise also starts from a set of small random numbers. This will be tunable during the cause of network training alongside the network learning. It has been proven that stochastic gradient descent (SGD) which is one of the variant of the gradient descent algorithm due to it randomness introduces noise to the weights by default [7]. In this work, we show that by strategically introducing criteria-based weight noise, the convergence speed of SGD can improve further. This work aims to eliminate network re-training by controlled introduction of noise based on four proven analytical and experimental methods. We proposed criteria-based noise injection into the network weights instead of universal introduction of noise in which the same noise is added all through training. The contribution of our current work is listed as follows:

- Introduction of batch by batch noise injection during training.
- Four methods in which learnable noise is introduced to the network weights was proposed. The injected noise act like an hyperparameter which is controlled using two criteria at the end of the predefined training batches.
- Introduced one more decision to increase or decrease this learnable noise after each batch. This is based on either the loss function, the rate of change of error or combination of both.
- We show its practical use in SLFNs. Particularly, we use different numerical experiments for supervised classification. In comparison with noiseless training and non-learnable noise injection, our approach is shown to improve convergence speed by about 15%–25% for the MNIST dataset and the other selected benchmark datasets.

The next section presents the concept of our criteria-based weight noise. Section 3 presents the algorithm weight noise injection during MLP training. The experimental results and comparison was presented in Sect. 4 while we concluded in Sect. 5.

## 2     Concepts

The training of an MLP neural network model using gradient descent is based on minimizing the training error. In network learning, the network is presented with a training pattern, and corresponding desired targets which are passed through the layers to compute the output values. The error between the computed output and the desired targets is some scalar function of the network weights. These network weights are adjusted at each training pass to reduce the error value. The training error on a pattern can be described in the form:

$$J(w) = \frac{1}{2} \sum_{k=1}^{c} (O_i - T_i)^2 \tag{1}$$

where $O_i$ and $T_i$ are the network output and target vectors of length $c$ and $w$ is the network weight. Whenever training starts with a "horrible" weights given a much worse calculated output, the ANN learning is much slower and needs quite a bit of time to start learning properly. It has been reported that if a bit of stochasticity is introduced during that slow learning phase, the model can avoid the slow path [3]. This process is pictorially illustrated in Fig. 1a where the navigation to global minimum C is the ultimate goal. During training, the minimization of the cost function aims to find the smallest error possible - in our case shallow point C. Points A and B are not desirable, but most of the time, the training may not be able to find its way out of this points thereby leading to a bad performance as well as a need for re-training. At the moment, the training algorithm does not know the implication of these two points and may never know. Adding the same amount of noise to points A and B is not "reasonable". This is because, for the training to get out of point A, a "small" noise relative to noise in point B is required due to their different depths. Figure 1b depicts the pictorial navigation during training to get out of shallow point A. There is different trajectory in which the blue circle can navigate to get out of this shallow end. As shown initially to get it to the red circle point, it can be anywhere on the red cloud during training trying to get out of the shallow point. The right amount of stochasticity can either push it further away from the purple circle or closer to it. The change in error between red and green circle will not be enough to decide on the amount of noise required to get out of this shallow point. Hence, the motivation for noise injection based on not only absolute $J(w)$ but also the rate of change of $J(w)$. This will assist in knowing how much of noise is required for each local minimum points during training. $J(w)$ will be referred to as error in subsequent sections.

We propose the injection of noise to the network training using adaptive methods instead of fixed techniques found in the literature. Adaptive noise introduction to any part of the NN model during training allows for criteria-based noise addition instead of blind addition. During the cause of analyzing the training of ANN model, we discovered that two parameters/criteria would influence the addition of noise namely the absolute error and rate of change of this error.

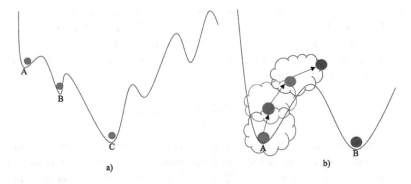

**Fig. 1.** Concept showing noise injection assisting learning

The overview of the noise injection technique employed in this work is using multiplicative noise. The noise is normally distributed pseudorandom in nature. This noise is not injected directly into the network weights but is controlled based on some well-defined techniques.

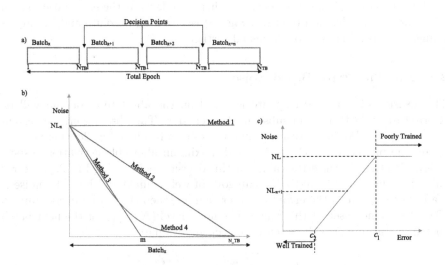

**Fig. 2.** Noise introduction methods

We defined several heuristics rules to control the injection of noise during training. The epoch is divided into batches as shown Fig. 2a. In each batch, noise is injected into the network weights as learning progresses based on some control mechanism. We defined four methods in which noise injection can be introduced during training in a particular mini-batch. Figure 2b shows as a diagram these noise injection methods. The way noise is injected in each batch is dependent on the noise method selected. Given any noise level in a particular batch, the increase or decrease in this noise level is dependent on four methods. In method 1, in each batch, constant noise is injected. The noise level to be injected into the weights for Method 2 is based on (2). Where $NL_{old}$ and $NL_{new}$ are the old/initial and updated noise level respectively, $N_{TB}$ is the epoch batch size and $C$ is the current epoch value.

$$NL_{new} = \frac{NL_{old}}{N_{TB}}C + NL_{old} \tag{2}$$

Method 3 is based on (3). Where $m$ is a constant between the start and end of a particular batch. $m$ is user defined and experiment have shown that the values of $m$ greater than or equal to $N_{TB}/2$, gives better convergence performance for method 3.

$$NL_{new} = \frac{NL_{old}}{m}C + NL_{old} \tag{3}$$

Finally Method 4 is using an exponential function approach to control the noise injection level as described in (4).

$$NL_{new} = NL_{old}e^{-C/\tau} \tag{4}$$

The decision to increase or decrease this noise level at the end of each batch (as shown at the decision points in Fig. 2a) is based on absolute and/or rate of change of error. This will be discussed further.

## 2.1   Absolute Error Based Noise

The decision of noise level update is based on the absolute error and will be referred to as AEB noise in subsequent discussion. The absolute error is calculated using (5). We heuristically define values as a result of Fig. 2c where $c_1$ and $c_2$ are the predefined error criteria and maximum allowable activation respectively, $E$ is the current error value. At this decision point using Fig. 2c, an error value less than $c_2$ falls under the category of well trained and here the noise is made zero whereas if the error is greater than $c_1$, then these are poorly trained. The noise is increased at this range. The noise level ($NL_{x+1}$) for the next batch (batch$_{x+1}$) at any error between $c_2$ and $c_1$ are evaluated using (5).

$$AEB = \frac{NL_{new}}{c_2 - c_1}E - \frac{NL_{new}}{c_2 - c_1}c_1 \tag{5}$$

## 2.2   Rate of Change of Error

In calculating the rate of change of error, we begin by making the backpropagation algorithm make a prediction on the training data, the error on this prediction is known as the target profile ($TR$) as shown in Fig. 3a. We defined the Current profile $CR$, as the ratio of the difference between previous error and current error to the batch epoch as described in (6). As shown in Fig. 3a, if the rate of change of error is zero and the absolute error is large then the noise introduced is large. Otherwise, with a large rate of change of error and medium absolute error the injected noise level is small. The relationship between current profile and target profile of Fig. 3a is computed using (7). The rate of change of error is evaluated using (8) which is as a result of Fig. 3c.

$$CR = \frac{E_2 - E_1}{N_{TB}} \qquad (6)$$

$$CRTR = \frac{CR}{TR} \qquad (7)$$

$$RCEB = -NL_{new} * CRTR + NL_{new} \qquad (8)$$

Finally, using the illustrated error types, we defined two protocols namely protocol 1 and 2. Protocol 1 has the noise decision point of Fig. 2a based strictly on absolute error. Protocol 2, on the other hand, is based on the combination of absolute error and the rate of change of error.

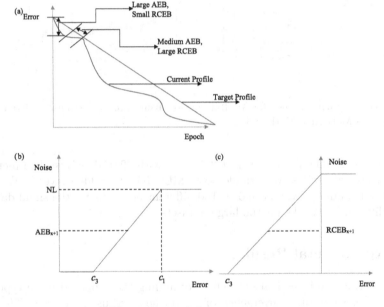

**Fig. 3.** Rate of change operation

## 3   Implementation Details

We used five distinct datasets: small to large. We used the mini-batch stochastic gradient descent to train our MLP classifier for the large datasets and batch gradient descent for small datasets. We ran the simulation in Matlab Environment. Our system is a 64-bit windows server 2012 R2 standard with 8 processors

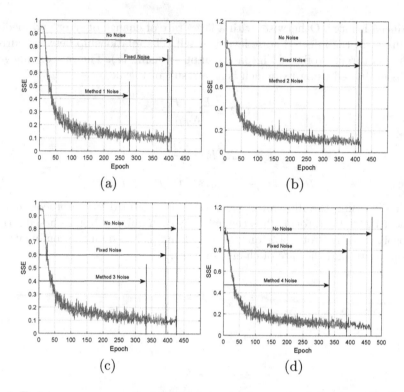

**Fig. 4.** Training convergence for MNIST dataset using protocol 1 noise vs fixed noise vs no noise - Method 1–Method 4

and the Intel Xeon CPU running at 2.90 GHz with 32.0 GB of installed memory (RAM). Our classifiers were single layer MLP (single hidden layer). We compared the training convergence over 100 different networks for the small dataset and 5 different networks for the large datasets.

## 4    Experimental Results

In this section, we investigate the effect of adding the proposed noise types on the generalization and convergence of neural network using the Iris, Thyroid, Letters, APS [16] and MNIST dataset [17]. The Iris dataset is made up of 150 samples of three classes of flowers. We used 10 hidden units with hyperbolic tangent sigmoid as the activation function. The experiment was repeated a total of 100 times, averaged and the results are presented in Table 1. The medium sized dataset (Thyroid) on the other hand comprises of 7200 samples of 3 classes. We used the same network set up as the Iris dataset. The APS dataset has a total of 60000 instances and 171 attributes of data. This is a binary output class problem set. The APS dataset used an architecture of 10 hidden nodes and 2 output nodes. The Letters dataset consisting of 26 letters made up of

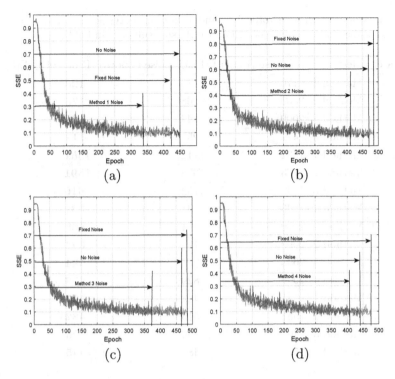

**Fig. 5.** Training convergence for MNIST dataset using protocol 2 noise vs fixed noise vs no noise - Method 1–Method 4

black-and-white rectangular pixel displays of the English alphabet was selected. The dataset is made up of 26 attributes and 20000 instances. A network of hidden layer of 25 neurons of hyperbolic tangent sigmoid and 26 output nodes of hyperbolic tangent sigmoid was used. Finally, the MNIST dataset which is a collection of 60,000 images of handwritten numbers ranging from 0 to 9 was used. The dataset also came with a validation set of 10,000 images. We converted the $28 \times 28$ grey scale images of handwritten numbers to vectors resulting in the dimension $784 \times 60,000$. The output was encoded using one hot encoding and the dimension is $10 \times 60,000$. A single hidden layer MLP network of 100 hidden neurons gave us an accuracy of 94.50%, using a learning rate of 0.3 and batch size of 100. This network was chosen for our experiment to achieve uniformity. The result presented is based on the following parameter values: $NL = 0.01$, $m = 55$, $LL = 0.01$, $N_{TB} = 100$, $c_1 = 2 * Error$, $c_2 = 2 * Maximum \ Activation$, $c_3 = -0.001$ and $E = 0.06$.

### 4.1 Protocol 1 Noise Decision Update

Here, we used (5) for our weight noise decision inference. Table 1 shows the comparison between the convergence speed up for Protocol 1 noise, fixed noise and noiseless network for the datasets. On an average, injection of noise speeds up

**Table 1.** Comparison of Protocol 1 and other noise types on selected datasets. Reported is the convergence speed in epoch. Best results are in bold

| Dataset | Method | No noise (Epoch) | Fixed noise (Epoch) | Protocol 1 (Epoch) |
|---------|--------|------------------|---------------------|--------------------|
| APS | Method 1 | 176 | 184 | **128** |
| | Method 2 | 2000 | 268 | **178** |
| | Method 3 | 2000 | 144 | **139** |
| | Method 4 | 412 | 257 | **111** |
| Letters | Method 1 | 1632 | 1712 | **1561** |
| | Method 2 | 1703 | 1673 | **1591** |
| | Method 3 | 1662 | 1752 | **1610** |
| | Method 4 | 1633 | 1721 | **1531** |
| Thyroid | Method 1 | 2000 | 741 | **668** |
| | Method 2 | 2000 | 797 | **664** |
| | Method 3 | 2000 | 860 | **594** |
| | Method 4 | 2000 | 829 | **658** |
| Iris | Method 1 | 533 | 485 | **397** |
| | Method 2 | 551 | 513 | **482** |
| | Method 3 | 463 | 456 | **426** |
| | Method 4 | 522 | 481 | **445** |

**Table 2.** Comparison of Protocol 2 and other noise types on selected datasets. Reported is the convergence speed in epoch. Best results are in bold

| Dataset | Method | No noise (Epoch) | Fixed noise (Epoch) | Protocol 2 (Epoch) |
|---------|--------|------------------|---------------------|--------------------|
| APS | Method 1 | 2000 | 131 | **131** |
| | Method 2 | 200 | **131** | 144 |
| | Method 3 | 2000 | 198 | **120** |
| | Method 4 | 2000 | 256 | **146** |
| Letters | Method 1 | 1600 | 1697 | **1537** |
| | Method 2 | 1554 | 1728 | **1438** |
| | Method 3 | 1721 | 1766 | **1608** |
| | Method 4 | 1645 | 1608 | **1519** |
| Thyroid | Method 1 | 2000 | 820 | **723** |
| | Method 2 | 2000 | 817 | **558** |
| | Method 3 | 2000 | 714 | **669** |
| | Method 4 | 2000 | 768 | **746** |
| Iris | Method 1 | 519 | 485 | **446** |
| | Method 2 | 575 | 486 | **441** |
| | Method 3 | 458 | 454 | **442** |
| | Method 4 | 564 | 561 | **479** |

the convergence in all datasets consistently for all our defined methods. Likewise, for the Mnist dataset, the four methods of noise defined consistently outperforms fixed noise and the no noise methods as shown in Fig. 4(a)–(d). This proves that injecting noise based on absolute error have a great impact on how fast convergence is reached without any negative effect on the generalization ability which on an average is recorded at 94.39% classification accuracy for Mnist dataset.

### 4.2  Protocol 2 Noise Decision Update

The amount of noise is calculated based on equation proposed in Sect. 2. The result presented here is based on the rate of change of error and the absolute error as the noise level decision inference. Firstly for all the four methods, we saw a more significant improvement in convergence speed when compared with the no noise as well as when compared with protocol 1 noise type for the Thyroid and MNIST dataset. For Method 4, protocol 2 Noise is approximately twice as fast as using only protocol 1 Noise as shown in Fig. 5(d). The Iris dataset, however, did not show a large significant improvement as recorded for MNIST, APS, Letters and Thyroid for the no noise network. We aim to perform further analysis to determine the underlying reasons behind this (Table 2).

## 5  Conclusion

We described a new method of introducing noise by using mini-batches. Noise updates within a batch is based on 4 new methods. These methods have been proven experimentally to improve convergence speed and avoid local minimum entrapment. Absolute Error Based (AEB) Noise and Rate of Change of Error Based (RCEB) Noise are the criteria in which the noise methods are updated after each predefined epoch batch. We tested the algorithm on small to large datasets in a repeated number of experiments. The results show that protocol 1 and protocol 2 Noise for the four methods outperformed Noiseless and fixed noise found in literature training regarding convergence speed. We see an increase in protocol 2 noise as compared to protocol 1 noise for all the four methods. On an average, we observed that using protocol 2 is 2 times faster than protocol 1 noise. We aim to further explore in the future introduction of noise into other parameters of NN using our defined methods and further improve our heuristics to be more robust.

## References

1. Srivastava, N., Hinton, G.E., Krizhevsky, A., Sutskever, I., Salakhutdinov, R.: Dropout: a simple way to prevent neural networks from overfitting. J. Mach. Learn. Res. **15**, 1929–1958 (2014)
2. Russell, R.: Pruning algorithms-a survey. IEEE Trans. Neural Netw. **4**, 740–747 (1993)

3. Kartik, A., Osoba, O., Kosko, B.: Noise-enhanced convolutional neural networks. Neural Net. **78**, 15–23 (2016)
4. Reed, R., Marks, R.J., Oh, S.: Similarities of error regularization, sigmoid gain scaling, target smoothing, and training with jitter. IEEE Trans. Neural Netw. **6**(3), 529–538 (1995)
5. Murray, A.F., Edwards, P.J.: Enhanced MLP performance and fault tolerance resulting from synaptic weight noise during training. IEEE Trans. Neural Netw. **5**(5), 792–802 (1994)
6. Murray, A.F., Edwards, P.J.: Synaptic weight noise during multilayer perceptron training: fault tolerance and training improvements. IEEE Trans. Neural Netw. **4**(4), 722–725 (1993)
7. Jim, K.C., Giles, C.L., Horne, B.G.: Synaptic noise in dynamically-driven recurrent neural networks: convergence and generalization (1998)
8. Holmstrom, L., Koistinen, P.: Using additive noise in back-propagation training. IEEE Trans. Neural Netw. **3**(1), 24–38 (1992)
9. Motaz, S., Kurita, T.: Effect of additive noise for multi-layered perceptron with autoencoders. IEICE Trans. Inf. Syst. **100**(7), 1494–1504 (2017)
10. Di, X., Yu, P.: Multiplicative noise channel in generative adversarial networks. In: Proceedings of the IEEE Conference on Computer Vision and Pattern Recognition, pp. 1165–1172 (2017)
11. Chandra, P., Singh, Y.: Regularization and feedforward artificial neural network training with noise. In: IEEE International Joint Conference on Neural Networks (IJCNN), vol. 3, pp. 2366–2371 (2003)
12. An, G.: The effects of adding noise during backpropagation training on a generalization performance. Neural Comput. **8**(3), 643–674 (1996)
13. Jim, K.C., Giles, C.L., Horne, B.G.: An analysis of noise in recurrent neural networks: convergence and generalization. IEEE Trans. Neural Netw. **7**(6), 1424–1438 (1996)
14. Max, W., Teh, Y.W.: Bayesian learning via stochastic gradient Langevin dynamics. In: Proceedings of the 28th International Conference on Machine Learning (ICML), pp. 681–688 (2011)
15. Wang, C., Principe, J.C.: Training neural networks with additive noise in the desired signal. IEEE Trans. Neural Networks. **10**(6), 1511–1517 (1999)
16. University of California Machine Learning Repository. https://archive.ics.uci.edu/ml/
17. MNIST handwritten digit database. http://yann.lecun.com/exdb/mnist/

# An Independent Approach to Training Classifiers on Physiological Data: An Example Using Smiles

Md Zakir Hossain$^{(\boxtimes)}$ and Tom D. Gedeon

The Australian National University (ANU), Canberra 2601, Australia
zakir.hossain@anu.edu.au, tom@cs.anu.edu.au

**Abstract.** Training neural network and other classifiers on physiological signals has challenges beyond more traditional datasets, as the training data includes data points which are not independent. Most obviously, more than one sample can come from a particular human subject. Standard cross-validation as implemented in many AI tools gives artificially high results as the common human subject is not considered. This is handled by some papers in the literature, by using leave-one-subject-out cross-validation. We argue that this is not sufficient, and introduce our independent approach, which is leave-one-subject-and-one-stimulus-out cross-validation. We demonstrate our approach using KNN, SVM and NN classifiers and their ensemble, using an extended example of physiological recordings from subjects observing genuine versus posed smiles, which are the two kinds of the nicest smiles and hard for people to differentiate reliably. We use three physiological signals, 20 video stimuli and 24 observers/participants, achieving 96.1% correct results, in a truly robust fashion.

**Keywords:** Observers · Smilers · Physiological features
Independent approach · Affective computing

## 1 Introduction

Classification is an important task to classify new instances, and the use of 10 fold cross-validation is common. Particularly in the context of data from human measurements, leave-one-out cross validation is a useful technique for evaluating the performance of classification algorithms, which estimates out of sample predictive accuracy using within-sample fits. This technique is generally task dependent otherwise it may provide noisy outcomes [1] and/or inconsistent accuracies [2]. In this paper, first we introduce our example domain with background and then our *Independent Approach* in Sect. 3.

In this example domain, we choose observers' physiological responses while watching real and posed smile videos. As in practical situation, life provides many reasons to smile that generally indicate pleasure, appreciation, happiness, or satisfaction. A smile face evokes positive feelings and conveys different messages to others. Because people can smile in different situations either positive or negative, such as a polite smile, false smile, acted smile, or smile to hide something [3]. We mean the later

© Springer Nature Switzerland AG 2018
L. Cheng et al. (Eds.): ICONIP 2018, LNCS 11302, pp. 603–613, 2018.
https://doi.org/10.1007/978-3-030-04179-3_53

types of smiles as posed smiles and smile that signifies happiness as real/genuine smiles. The correct reading of smilers' faces seems to guarantee the understanding of smiler's affective states appropriately. In this connection, the potential application of this system are not limited to the finding the relationship between smilers' facial changes with the observers' physiology, but also this system can be applied to design sensing technologies from care-givers' peripheral physiology to understand patients' mental states, detecting avatars' emotional realism, and so on. Even the performance of face recognition systems can be improved when trained by smiling faces [4].

In the past, researchers focused on analyzing smilers' faces directly to differentiate between real and posed smiles. Dibeklioğlu et al. [6] implemented their own computer vision technique on the smilers' facial features. Hoque et al. [7] used similar features to discriminate between delighted and frustrated smiles. Gan et al. [8] applied two-layer deep Boltzmann machine on smiler's images. Gunadi et al. [9] used linear support vector machine to detect fake smiles from smiler's faces. Cohn et al. [10] measured the timing of face motion during smiles on images. Frank et al. [5] and Hoque et al. [7] considered observers' verbal responses to recognize real smiles. It is usually very hard to discriminate real and posed smiles by relying on observers' verbal responses [5], even though people generally emotionally react to others' emotion during face to face interaction [11]. This is because the smile is one of the easiest facial expressions that can be faked voluntarily [6]. But observer's peripheral physiology is associated with emotional states [11] and can be used in classifying smiler's affective state. In general, this concept can be applicable in security systems. For example, suspect can smile during verifying their truthfulness in a hearing, interrogation, or customs. Then we may apply this concept to identify the genuineness of a smile from lawyers, police, or custom officers' peripheral physiology.

Understanding a smiler's affective conditions can be achieved from observers' physiological states. This is because smilers show their smile using facial expressions, and observers will have certain impressions or feelings caused by the smiler facial actions, which makes a change to the observers' peripheral physiology. The physiological signals are not voluntarily controllable and have the ability to vary differently in different circumstances [12]. Thus, physiological signals [13–15] are considered in several studies to be useful to understand facial expressions. Here, we considered three of these signals to discriminate between real and posed smiles, namely blood volume pulse (BVP), galvanic skin response (GSR), and pupillary response (PR).

The pupillary responses can change for many reasons, including memory load, stress, pain, watching videos, face to face interactions etc., and would offer a good method for classifying real and posed real smiles, because it does not require to attach any sensors either to the observer or to the smiler [13]. GSR is an automatic reaction that causes continuous electrical changes in sweat gland activity of human skin and is considered one of the strongest signals in emotion detection [14]. BVP is another physiological signal that uses infrared light to measure blood volume changes in arteries of the human body and the shape of BVP reflects the emotional changes of the human [15]. Due to the involuntary nature of the physiological signals, and observers lower verbal response rate for classifying real and posed smiles, we considered observers' physiology here. It is also worth mentioning that smiles are chosen due to the universal role in presenting emotion of happiness for the smiler and in being understood as signs that observers note as sociality or closeness. In the end, we extract six temporal features from each of the above physiological signals and measure classification performances by leave-one-out-cross validation techniques.

In this connection, the motivation of this research comes from the usefulness of the Assistive Context Aware Toolkit (ACAT), which was designed for Hawking to enable him to control his computer and communicate with others [22]. We are hoping to design similar technologies as Professor Hawking's from the care-givers' peripheral physiology in future, and we choose smiles at the first instance. In previous research, people considered a leave-one-out approach because they were using either smilers' facial expression [6, 16] or observers' verbal responses [7]. But none of them analyzed observers' (such as care givers) physiological responses to recognize smilers' facial expressions. In this context, leave-one-observer-out or leave-one-smiler-out approach is not fully noise free or robust, because when leave-one-observer-out approach is used then the training data is contaminated by the smilers' information, and vice versa. Thus, our work initiates a new approach called an Independent Approach that highlights the main idea of noise free test data by removing potentially contaminating data from the training data.

## 2 Smiles and Observers' Physiology

This section addresses the example data (physiological signal) collection and the processing procedures. This is the initialization section before using our Independent Approach, as well as input data preparation section for our classifiers.

### 2.1 Smilers

Twenty smilers' videos were randomly collected from four databases (five from each), namely UvA-NEMO [16], MAHNOB [17], MMI [18], and CK+ [19]. MAHNOB and NEMO databases were chosen to collect real smiles, because participants' smiles were elicited in these databases by showing a number of pleasant or funny video clips. CK+ and MMI databases were considered to collect posed smiles, because participants were asked instructed or requested to show a smile in these databases. MATLAB platform was considered to process the collected smile videos. The height, width, format, color, and duration of each smile video were processed into 336 pixels, 448 pixels, mp4, grey, and 5 s respectively. Due to smile time duration being 0.5 s to 4 s in general [3], we choose smiler' video length up to 5 s long that was adjusted by controlling the frame rate of videos. To avoid the effect of light/dark backgrounds, smilers' faces were masked before showing them to the observers. Finally, the MATLAB SHINE toolbox [20] were used to adjust the luminance (128 ALU (Arbitrary Linear Unit)) and contrast (32 ALU) of each smile videos.

### 2.2 Observers

Twenty-six (11 female, 15 male) right-handed participants participated voluntarily in this experiment (age: 30.7 ± 5.96 (mean ± Std.)). They signed written consent forms before starting the experiment. The experiment was approved by the Australian National University's Human Research Ethics Committee.

## 2.3 Data Recording

Smile videos were presented to the observers using an ASUS laptop, and observers were seated in front of the laptop comfortably, in a static chair before starting the experiment. Each smile video was followed by a question: How did this smile look to you? Happy (Real/Spontaneous/Genuine) or Fake (Posed/Acted). Observers used a computer mouse to input their answers. Three physiological signals, being pupillary responses, BVP, and GSR were recorded at a sampling rate of 60 Hz, 64 Hz, and 4 Hz respectively, according to the device specifications. The pupillary responses were recorded using The Eye Tribe (theeyetribe.com) remote eye-tracker system, and BVP and GSR were recorded using the Empatica E4 (www.empatica.com). Before starting the experiment, the eye tracker was calibrated and observers were requested to limit their body movements to minimize the noise in the signals. The smile videos were presented to the observers in an order balanced way. It is worth noting that the results are reported in this paper by analyzing twenty-four observers' data due to very noisy physiological responses from the other two observers.

## 2.4 Signal Processing

To reduce the undesired noise from the peripheral physiological signals, eye blink points (which show up as zero in pupillary responses) were reconstructed using interpolation technique (cubic spline), and pupil data were smoothed using moving average filtering (Hanning window) [11]. Butterworth filter (low-pass, order = 6) was considered to filter out the noises from BVP and GSR signals [12]. To reduce between-observer differences, the maximum value normalization technique was applied to keep the signals and their extracted features in the range between 0 and 1. In this sense, each value of a particular observer's specific physiological signal is divided by the maximum value of that observer's specified physiological signal [19]. Before normalizing, signals were kept over the positive axis by changing their dc label.

## 2.5 Signal Extraction

Six temporal features are calculated from each physiological signal relevant to each smile video for an observer. These features convey the information of observers' physiological behaviours as well as their thinking to the smilers' affective states as typical range, variation and gradient like characteristics [21]. In this specific case, we extracted 120 features (20 smile videos × 6 features) from an observer (50% from real smiles and the rest from posed smiles) and 2880 features in total (120 features × 24 observers) for all observers, considering each signal (BVP or GSR or PR). Let $y(n)$ represents the value of the $n^{th}$ sample of the processed physiological signals $n = 1, \ldots \ldots \ldots N$.

1. Means

$$\mu_y = 1/N \sum_{n=1}^{N} y(n) \tag{1}$$

2. Maximum

$$M_y = \max(y(n)) \tag{2}$$

3. Minimum

$$m_y = \min(y(n)) \tag{3}$$

4. Standard Deviations

$$\tilde{\sigma}_y = \sqrt{\frac{1}{N-1} \sum_{n=1}^{N} (y(n) - \mu_y)^2} \tag{4}$$

5. Means of the absolute values of the first differences

$$\tilde{\delta}_y = \frac{1}{N-1} \sum_{n=1}^{N-1} |\tilde{y}(n+1) - y(n)| \tag{5}$$

6. Means of the absolute values of the second differences

$$\tilde{\gamma}_y = \frac{1}{N-2} \sum_{n=1}^{N-2} |\tilde{y}(n+2) - y(n)| \tag{6}$$

## 3  An Independent Approach

We propose a novel approach we call an *independent approach*. This is independent in the sense that the test data is fully free from training data: for each test where observer '*Om*' watches the video of smiler '*Sn*', the classifier is not contaminated by those observers' physiological features and it is not contaminated by other observers' physiological features while watching that smiler. So, beyond the normal leave-one-observer-out cross-validation, we are also performing leave-one-smiler-observer-out at the same time. We consider this fully independent approach to be necessary to validly conclude than a classifier is not contaminated during training. This level of rigor is not matched in the literature. For example, suppose observer 1 (*O1*) (when watching the n[th] smiler (*Sn*)) is considered as test data, then any other data related to *O1* is not used to either train or test the classifier as illustrated in Fig. 1.

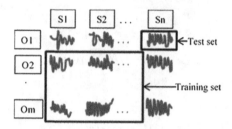

**Fig. 1.** An independent approach to compute classification accuracies, S = Smiler, O = Observer, n = 20, and m = 24. One of 437 (i.e. 19 × 23) training + test sets shown.

In total, there are n × m sets of physiological data, for each of n Smilers, being watched by m Observers, i.e. 480 in this case. A smiler independent (leave-one-video-out) process would train using n − 1 Smiler videos from m Observers, and test using the $n^{th}$ video, repeatedly, so this 19 × 24 size of training data would be repeated 20 times. An observer independent (leave-one-observer-out) process would train using n Smiler videos from m − 1 Observers, and test using the $m^{th}$ observer, repeatedly, so the 20 × 23 size of training data would be repeated 24 times. In our fully Independent Approach, we train using n − 1 Smiler videos from m − 1 Observers, and test using the $n^{th}$ video from the $m^{th}$ observer, repeatedly, so the 19 × 23 size of training data is repeated 480 times. Finally, average classification accuracies are reported from all these executions.

An advantage of this approach is that it is quite computationally intensive; it is as robust as possible. It also ensures the quality of training data by removing redundant and irrelevant data, which is not possible in the other two approaches, because they are not fully independent as discussed in previous paragraphs. In the case of leave-one-observer-out approach, trained data is contaminated by smilers, and vice versa in case of leave-one-smiler-out approach.

## 4   Results and Discussion

The observed smiles are classified into real smile and posed smile. The classification accuracies are computed using k-nearest neighbour (KNN), support vector machine (SVM), neural network (NN), and ensemble over the decision of these three classifiers. We considered default parameter settings in MATLAB as Euclidean distance matric and 5 nearest neighbours for KNN, sequential minimal optimization method and Gaussian radial basis kernel function with scaling factor of 1 for SVM, Levenberg-Marquardt training function with 10 hidden nodes for NN classifiers respectively. The mean square error performance function is considered to compute classification accuracies from each classifier.

The features are divided according to the test smiler identifications, such as S1, S2 all the way to S20. When test smiler is S1 and other smilers' (S2 to S20) data is used to train the classifiers, we call it S1 and so on. In a similar fashion, test observers are identified by O1, O2 all the way to O24. According to the independent approach, the final outcome of O1 is the average value over 20 executions (S1 to S20) for each set of physiological features. The outcomes from observer's PR features are explored in Table 1.

**Table 1.** Individual classification accuracies (%) of 24 observers

|       | KNN  | SVM  | NN   | Ensemble |       | KNN  | SVM  | NN   | Ensemble |
|-------|------|------|------|----------|-------|------|------|------|----------|
| O1    | 69.6 | 72.1 | 77.6 | 83.2     | O13   | 68.0 | 73.3 | 78.1 | 84.8     |
| O2    | 67.6 | 72.4 | 69.4 | 83.2     | O14   | 69.2 | 72.6 | 70.5 | 84.3     |
| O3    | 69.4 | 71.2 | 75.1 | 84.1     | O15   | 69.0 | 71.5 | 82.9 | 84.9     |
| O4    | 69.0 | 73.1 | 76.3 | 85.1     | O16   | 69.2 | 72.1 | 82.0 | 84.7     |
| O5    | 68.7 | 71.2 | 74.7 | 82.9     | O17   | 69.0 | 74.0 | 82.9 | 85.3     |
| O6    | 69.2 | 71.7 | 73.7 | 83.5     | O18   | 69.9 | 70.5 | 75.6 | 84.1     |
| O7    | 68.5 | 71.5 | 76.3 | 83.6     | O19   | 68.7 | 71.9 | 68.3 | 83.7     |
| O8    | 69.2 | 71.7 | 79.0 | 83.8     | O20   | 69.4 | 71.9 | 73.3 | 83.7     |
| O9    | 69.0 | 71.9 | 72.8 | 84.0     | O21   | 69.9 | 71.5 | 80.1 | 85.0     |
| O10   | 69.0 | 71.7 | 76.5 | **82.5** | O22   | 68.3 | 71.0 | 77.9 | 84.4     |
| O11   | 70.3 | 70.3 | 71.9 | 83.3     | O23   | 69.6 | 71.5 | 79.0 | 85.0     |
| O12   | 70.3 | 71.0 | 76.3 | 84.4     | O24   | 70.3 | 72.4 | 79.7 | **85.6** |

It is obvious from Table 1 that the ensemble classifier shows higher classification accuracies compared to the other classifiers. The classification accuracies are quite similar for each observer at discriminating between posed and real smiles, with highest value of 85.6% for O24 and lowest of 82.5% for O10. Table 1 depicts the results from PR features. The variation of the outcomes changes in a similar fashion, when training and testing with GSR or BVP features. The comparative results of GSR, BVP and PR features from the ensemble classifier using our independent approach are shown in Table 2.

**Table 2.** Ensemble classification results for 24 observers (individual)

|       | GSR  | BVP  | PR   |       | GSR  | BVP  | PR   |
|-------|------|------|------|-------|------|------|------|
| O1    | 84.5 | 83.5 | 83.2 | O13   | 84.9 | 84.1 | 84.8 |
| O2    | 83.0 | 84.6 | 83.2 | O14   | 84.1 | 83.7 | 84.3 |
| O3    | 85.8 | 85.1 | 84.1 | O15   | 83.5 | 84.5 | 84.9 |
| O4    | 84.5 | 84.9 | 85.1 | O16   | 83.3 | 84.1 | 84.7 |
| O5    | 84.3 | 83.2 | 82.9 | O17   | 84.5 | 85.2 | 85.3 |
| O6    | 84.2 | 84.0 | 83.5 | O18   | 85.0 | 82.7 | 84.1 |
| O7    | 83.2 | 85.3 | 83.6 | O19   | 83.4 | 84.1 | 83.7 |
| O8    | 85.2 | **85.9** | 83.8 | O20   | 82.7 | 83.3 | 83.7 |
| O9    | 82.7 | 85.1 | 84.0 | O21   | **85.2** | 84.4 | 85.0 |
| O10   | 83.8 | 85.0 | **82.5** | O22   | 84.0 | 83.4 | 84.4 |
| O11   | 84.3 | 85.4 | 83.3 | O23   | **81.1** | 84.1 | 85.0 |
| O12   | 83.1 | **82.5** | 84.4 | O24   | 85.0 | 84.2 | **85.6** |

There are no large differences among observers' peripheral physiological features to discriminate between posed and real smiles as shown in Table 2. In comparison, GSR shows the highest classification accuracy of 85.2% for O21, BVP shows 85.9% for O8 and PR shows 85.6% for O24 respectively, where the lowest accuracies of 81.1% for O23, 82.5% for O12, and 82.5% for O10 are found in the case of GSR, BVP, and PR features respectively. The average accuracies over all observers are shown in Fig. 2. Standard deviations are represented by error bars.

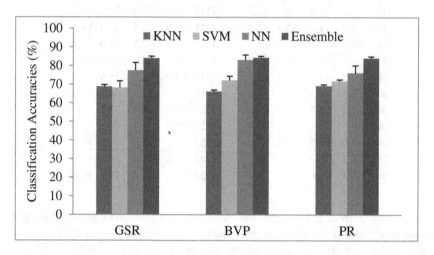

**Fig. 2.** Average classification accuracies using independent approach.

It can be seen from Fig. 2 that higher accuracies are reported for the ensemble classifier, and then for NN, SVM, and KNN classifiers respectively. We also test another two possible approaches, namely leave-one-smiler-out (means that the classifiers have seen no physiological features from any observers on that smiler, i.e. results are smiler independent) and leave-one-observer-out (that means the classifiers have not seen any physiological features from any smilers of that observer, i.e. results are observer independent). The results of three approaches using the ensemble classifier are explored in Fig. 3 where standard deviations are denoted by error bars.

It can be seen from Fig. 3 that higher accuracies of 97.1% is found from PR features using the observer independent approach. Although we leave out the data of the test observer from training, there was information of similar smilers' videos that were observed by the other observers. In the case of the smiler independent approach, higher accuracy of 92.8% is found from PR features where the information of similar observers were seen to the training data, although data of smilers' videos were not considered to test the classifiers. On the other hand, our independent approach shows a bit lower accuracy of 84.1% compared to the other two approaches. It is expected that other two approaches show higher accuracies compared to ours, because they use test data relevant information during training, but it does not occur in our case. Thus our Independent Approach is robustly applicable in various situations, such as verifying trustworthiness from judge's physiological signal, making decision of patients from care-givers' physiology, and so on; specially who wish to make decision from observers' physiology.

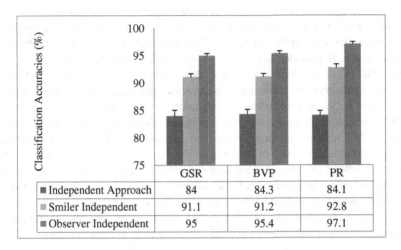

Fig. 3. Average classification accuracies from ensemble classifier.

To improve the classification accuracies, using our Independent Approach, in discriminating between posed and real smiles from observers' physiological features, a feature level fusion (concatenating all features from PR, GSR, and BVP) technique is employed. It does improve the average classification accuracy to 96.1% (±0.25) with the ensemble classifier. The same information fusion approach has much less benefit on the less robust observer-independent (97.2% ± 0.49) and smiler-independent (93.7% ± 0.82) approaches. The lower smiler-independent result implies there is more information in the smilers than in the observers, further suggesting that the observer-independent approach is contaminated with this extra information. Our independent approach achieves 96.1% without this contamination.

On the other hand, observers were averagely 59.0% (±11.13) correct according in their verbal responses. A final accuracy of 96.1% from our Independent Approach demonstrated that observers' automatic physiological responses are strong indicators to discriminate between posed and real smiles with a significant degree of accuracy. In comparison, Dibeklioğlu et al. [6], Hoque et al. [7], Gan et al. [8], and Cohn et al. [10] used leave-one-subject-out approach and reported 89.84%, 92.30%, 91.73%, and 93.0% correctness respectively. It is also worth noting that they used the approach on the features that were extracted from smilers' facial expressions, but we extracted features from observers' physiology while watching the smilers' video. Thus our system is more robust and effective in this specific or similar type of cases.

# 5 Conclusion

We have overcome the effect of biasing on testing set from training physiological features using an independent approach and showed that high accuracy results are achievable using a highly robust cross-validation approach. We considered four classifiers to discriminate between posed and real smiles from observers' physiological features using this independent approach. The ensemble classifier performs better than other classifiers. It provides accuracies of about 84% from individual physiological features (PR, BVP, or GSR), where two other approaches, called smiler independent and observer independent, show higher accuracies compared to independent approach. Feature level fusion improves the classification accuracy of 96.1% using a simple ensemble technique. In this context, we perform the analysis on observers' physiological features without hassling smilers and can perform this analysis on historical data. The final accuracy figure obtained from observers' fused physiological features to distinguish smilers' affective states into real or posed, shows that this system could be applicable in many situations, such as patients' mental state monitoring, verifying trustworthiness during questioning, relationship management, and so on. This is in agreement with the physiological features of observers in affective computing area indicating that smilers leak their intentions through their facial behaviors.

# References

1. Vehtari, A., Gelman, A., Gabry, J.: Practical Bayesian model evaluation using leave-one-out cross-validation and WAIC. J. Stat. Comput. **27**(5), 1413–1453 (2017)
2. Wong, T.: Performance evaluation of classification algorithms by k-fold and leave-one-out cross validation. Pattern Recognit. **48**(9), 2839–2846 (2015)
3. Ekman, P., Friesen, W.V.: Felt, false, and miserable smiles. J. Nonverbal Behav. **6**(4), 238–252 (1982)
4. Yacoob, Y., Davis, L.: Smiling faces are better for face recognition. In: 5th International Proceedings on Proceedings, International Conference on Automatic Face and Gesture Recognition, pp. 52–57. IEEE, Washington, DC, USA (2002)
5. Frank, M.G., Ekman, P., Friesen, W.V.: Behavioral markers and recognizability of the smile of enjoyment. J. Pers. Soc. Psychol. **64**(1), 83–93 (1993)
6. Dibeklioğlu, H., Salah, A.A., Gevers, T.: Recognition of genuine smiles. IEEE Trans. Multimed. **17**(3), 279–294 (2015)
7. Hoque, M.E., McDuff, D.J., Picard, R.W.: Exploring temporal patterns in classifying frustrated and delighted smiles. IEEE Trans. Affect. Comput. **3**(3), 323–334 (2012)
8. Gan, Q., Wu, C., Wang, S., Ji, Q.: Posed and spontaneous facial expression differentiation using deep Boltzmann machines. In: 6th International Proceedings on Proceedings, International Conference on Affective Computing and Intelligent Interaction, pp. 643–648. IEEE, Xi'an, China (2015)
9. Gunadi, I.G.A., Harjoko, A., Wardoyo, R., Ramdhani, N.: Fake smile detection using linear support vector machine. In: International Proceedings on Proceedings, International Conference on Data and Software Engineering, pp. 103–107. IEEE, Yogyakarta, Indonesia (2015)

10. Cohn, J.F., Schmidt, K.L.: The timing of facial motion in posed and spontaneous smiles. Int. J. Wavelets Multi Resolut. Inf. Process. **2**(2), 1–12 (2004)

11. Kim, J., Andre, E.: Emotion recognition based on physiological changes in music listening. IEEE Trans. Pattern Anal. Mach. Intell. **30**(12), 2067–2083 (2008)

12. Gong, P., Ma, H. T., Wang, Y.: Emotion recognition based on the multiple physiological signals. In: International Proceedings on Proceedings, International Conference on Real-time Computing and Robotics, pp. 140–143. IEEE, Angkor Wat, Cambodia (2016)

13. Hossain, M.Z., Gedeon, T., Sankaranarayana, R., Apthorp, D., Dawel, A.: Pupillary responses of Asian observers in discriminating real from fake smiles: a preliminary study. In: 10th International Proceedings on Proceedings, International Conference on Methods and Techniques in Behavioral Research, pp. 170–176, Dublin, Ireland (2016). Measuring Behavior

14. Hossain, M.Z., Gedeon, T., Sankaranarayana, R.: Observer's galvanic skin response for discriminating real from fake smiles. In: 27th International Proceedings on Proceedings, Australian Conference on Information Systems, pp. 1–8, Wollongong, Australia (2016)

15. Hossain, M.Z., Gedeon, T.: Classifying posed and real smiles from observers' peripheral physiology. In: 11th International Proceedings on Proceedings, EAI International Conference on Pervasive Computing Technologies for Healthcare, pp. 460–463. PervasiveHealth, Barcelona, Spain (2017)

16. Dibeklioglu, H., Salah, A.A., Gevers, T.: Are you really smiling at me? Spontaneous versus posed enjoyment smiles. In: Fitzgibbon, A., Lazebnik, S., Perona, P., Sato, Y., Schmid, C. (eds.) Computer Vision – ECCV 2012. LNCS, vol. 7574, pp. 525–538. Springer, Berlin, Heidelberg (2012)

17. Soleymani, M., Lichtenauer, J., Pun, T., Pantic, M.: A multimodal database for affect recognition and implicit tagging. IEEE Trans. Affect. Comput. **3**(1), 42–55 (2012)

18. Pantic, M., Valstar, M., Rademaker, R., Maat, L.: Web-based database for facial expression analysis. In: International Proceedings on Proceedings, International Conference on Multimedia and Expo, pp. 317–321. IEEE, Amsterdam, Netherlands (2005)

19. Lucey, P., Cohn, J. F., Kanade, T., Saragih, J., Ambadar, Z., Matthews, I.: The extended Cohn-Kanade dataset (CK+): a complete expression dataset for action unit and emotion-specified expression. In: International Proceedings on Proceedings, IEEE Computer Society Conference on Computer Vision and Pattern Recognition - Workshops, pp. 94–101. IEEE, San Francisco, CA (2010)

20. Willenbockel, V., Sadr, J., Fiset, D., Horne, G.O., Gosselin, F., Tanaka, T.W.: Controlling low-level image properties: the SHINE toolbox. Behav. Res. Methods **42**(3), 671–684 (2010)

21. Hossain, M.Z., Gedeon, T.: Effect of parameter tuning at distinguishing between real and posed smiles from observers' physiological features. In: Liu, D., Xie, S., Li, Y., Zhao, D., El-Alfy, E.S. (eds.) Neural Information Processing, ICONIP 2017. LNCS, vol. 10637, pp. 839–850. Springer, Cham (2017)

22. Denman, P., Nachman, L., Prasad, S.: Designing for "a" user: Stephen Hawking's UI. In: 14th International Proceedings on Proceedings, Participatory Design Conference, pp. 94–95. ACM, New York, USA (2016)

# Mixed Precision Weight Networks: Training Neural Networks with Varied Precision Weights

Ninnart Fuengfusin[✉] and Hakaru Tamukoh

Graduate School of Life Science and Systems Engineering,
Kyushu Institute of Technology,
2-4 Hibikino, Wakamatsu-ku, Kitakyushu, Fukuoka 808-0196, Japan
fuengfusin.ninnart553@mail.kyutech.jp, tamukoh@brain.kyutech.ac.jp

**Abstract.** We propose Mixed Precision Weight Networks (MPWNs), neural networks that jointly utilize weights with varied precision in the layers. MPWNs constrain the weight layers to either 1-bit binary $\{-1, 1\}$, 2-bit ternary $\{-1, 0, 1\}$, or the original 32-bit full precision weights. Each weight space contains unique properties for instance, high classification accuracy, small number of bit, and high sparsity. Hence, the properties of MPWNs can be adjusted by varying the combinations and orders of the weight layers. In this study, we identify three heuristic rules for effectively setting each of the weight layers. Therefore, MPWNs successfully utilize the robust properties from each weight space while avoiding their disadvantages. We evaluated MPWNs with MNIST, CIFAR-10, and CIFAR-100 training datasets. Our evaluation revealed that MPWNs models trained on CIFAR-10 and CIFAR-100 achieved the best overall properties comparing to conventional methods.

**Keywords:** Deep learning · Neural networks · Model compression

## 1 Introduction

Neural networks (NNs) have been applied in a wide-range of applications, including image classification [13], objection detection [16] and object segmentation [9]. NNs tend to go deeper in the number of layers, causing the number of weights within NNs to rapidly increase. In 1998, handwritten digit recognition networks called LeNet-5 [14], were proposed with five weight layers with around 60K trainable weight parameters. In 2015, ResNet-152 [10], the winner of the ImageNet [5] Large Scale Visual Recognition Challenge 2015, greatly increased the number of weight layers to 152 layers with 60M trainable weight parameters. Although the increasing the number of weight layers leads to better classification accuracy, it significantly increases the power consumption from higher memory usage. To deploy NNs in mobile devices, which have limited memory storage and finite battery capacities, NNs must be modified.

© Springer Nature Switzerland AG 2018
L. Cheng et al. (Eds.): ICONIP 2018, LNCS 11302, pp. 614–623, 2018.
https://doi.org/10.1007/978-3-030-04179-3_54

Recently, two research fields have been focusing on solving this problem. The first research field is sparse neural networks (SNNs) [7,8]. To reduce the model size, SNNs first prune weights that are lower than the threshold to zero. Next, SNNs retrain the unpruned weights to regain classification accuracy. After the pruning, SNNs have a large proportion between the number of zero-valued weights and all weights. This property is called sparsity. As proved by a previous study [6], higher sparsity leads to lower power consumption by the model. The other research field is quantized neural networks (QNNs), which are NNs that constrain 32-bit floating point weight into fewer amount of bits. For this reason, the weights can be stored using less memory. In an extreme case, BinaryConnect (BC) [4] quantizes the weights into the binary space $\{1, -1\}$. More recently, ternary weight networks (TWNs) [15] were used to quantize the weights into the ternary space with the scaling factor $\{-S, 0, S\}$, where $S$ is a positive real number. We have been motivated to create NNs that balance the properties of different QNNs. Here, we propose mixed precision weight networks (MPWNs), which are NNs that utilize several types of weight spaces together. In general, MPWNs constrain weight layers into either the binary space $\{-1, 1\}$ or, ternary space $\{-1, 0, 1\}$, or the layers remain as full precision weights.

The main contributions of this paper are as following:

- We introduce MPWNs, the networks that utilize the binary, ternary and full precision weights during both the training and inference phase.
- We show how to explore the model attributes and present the three heuristic rules for effectively setting the weight space for each layer in MPWNs.
- We show that MPWNs have the best overall properties compared to those of conventional methods.

## 2  Related Works

### 2.1  BinaryConnect

BC discretizes the 32-bit floating point weights $W$ into binary weights $W_b$, as shown in Eq. 1. Because $W_b$ can be either $-1$ or $1$, two possibilities, a weight can be stored with a bit fixed point.

$$W_b = sign(W) = \begin{cases} 1, & W \geq 0 \\ -1, & W < 0 \end{cases} \tag{1}$$

Because the derivation of the sign function, or signum function, is zero except for the discrete-time impulse function at $W = 0$, the derivation of the sign function cannot be used during back propagation. The estimator of the gradient for the sign function is assigned instead as $\frac{\partial C}{\partial W^{t-1}} = \frac{\partial C}{\partial W_b^{t-1}}$. The weight $W$ can be updated via using $W^t = clip(W^{t-1} - \eta \frac{\partial C}{\partial W_b^{t-1}})$, where $\eta$ is the learning rate, $C$ is the cost function, $W^{t-1}$ is the weight before an iteration of training, $W^t$ is the weight after an iteration of training and $clip$ is the function that limits the input to the range of $[-1, 1]$. BC achieves the compression rate 32 times in exchange for a loss in classification accuracy.

## 2.2 Ternary Weight Networks

TWNs constrain the full precision weights into ternary weights with the scaling factor $\{-S, 0, S\}$. With the extra zero state in the weight space as compared to BC, TWNs need a two-bit fixed point to store three possibilities for a weight value. This requirement causes the compression rate to 16 times. However, TWNs can use sparsity to reduce the model size in the same manner that SNNs can. TWNs find the scaling factor $S$, as shown in Eq. 2, where $E$ is the function to calculate the mean and $S$ is used to minimize the quantization loss.

$$S = E_{i \in \{i \mid |W_i| > t\}}(|W_i|) \tag{2}$$

The threshold $t$ can be determined through Eq. 3, where $\alpha$ is a hyper-parameter in the range $(0, 1)$. $\alpha$ has a negative correlation with the sparsity. Hence, $\alpha$ can be used to set the sparsity to the desired amount.

$$t = \alpha \times E(|W_i|) \tag{3}$$

TWNs then quantize the weights by inputting the threshold $t$ and scaling factor $S$ into Eq. 4.

$$W_t = \begin{cases} S, & W > t \\ 0, & W \le t \\ -S, & W < -t \end{cases} \tag{4}$$

For the back propagation, TWNs assign $\frac{\partial C}{\partial W^{t-1}} = \frac{\partial C}{\partial W_t^{t-1}}$. The real valued weight $W$ can be updated via $W^t = W^{t-1} - \eta \frac{\partial C}{\partial W_t^{t-1}}$. One of the differences between TWNs and BC is that TWNs do not use the *clip* function during updates.

# 3   Mixed Precision Weight Networks

MPWNs are the combination of the weight layers from the binary weight space $\{-1, 1\}$, ternary weight space $\{-1, 0, 1\}$ and full precision or 32-bit floating point weights. MPWNs receive the properties from each combined weight. The properties of each weight space are summarized in Table 1. MPWNs constrain the input weights to binary weights using Eq. 1, which is from BC. We quantize the ternary weights using Eqs. 3 and 4, which are from TWNs. To make the methods from BC comparable with TWNs, we do not apply the scaling factor $S$ to the ternary weight space and the *clip* function during back propagation. An overview of MPWNs is shown in Fig. 1.

For notation of MPWNs, the bold capitalized text represents the structure of MPWNs. For example, **FBT** indicates MPWNs with three weight layers. The first layer is full precision weights (**F**), the second layer is binary (**B**) weights, and last layer is ternary weights (**T**). The combination of MPWNs can be accomplished through two approaches: the exploring, and heuristic rules.

**Table 1.** Properties of each weight space.

| | Accuracy | Sparsity | Bits per weight (bit) |
|---|---|---|---|
| Full precision weights (**F**) | High | None | 32 |
| Ternary weights (**T**) | Low | High | 2 |
| Binary weight (**B**) | Mid | None | 1 |

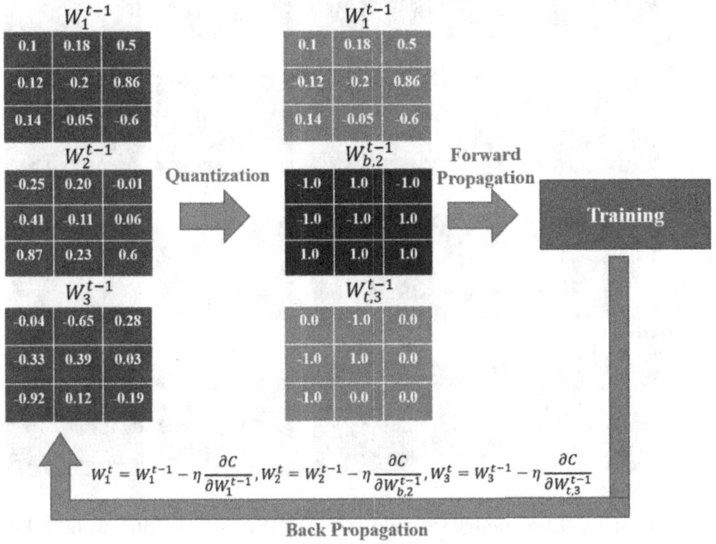

$$W_1^t = W_1^{t-1} - \eta \frac{\partial C}{\partial W_1^{t-1}}, W_2^t = W_2^{t-1} - \eta \frac{\partial C}{\partial W_{b,2}^{t-1}}, W_3^t = W_3^{t-1} - \eta \frac{\partial C}{\partial W_{t,3}^{t-1}}$$

**Back Propagation**

**Fig. 1.** Overview of MPWN processes.

**Exploring:** We explored the model attributes by using the following concepts. If we want the model to have higher sparsity, we include a higher amount of **T** layers in the model. If we desire enhanced classification accuracy, we apply the greater amount of **F** layers in the model. If we want fewer bits of weight value, we include more **B** layers into the model. We evaluated the outcomes of the model and exploring with different combinations until the model contained all attributes with the desired values.

**Heuristic Rules:** After exploring different combinations of MPWNs, we discovered, that the weight layers of NNs are not equally important because of differences in the order and number of parameters in each weight layer. From that concept, we found the heuristic rules for optimally setting the combination of MPWNs. The heuristic rules can be summarized into three rules. The rule is to apply **T** in the layers with a large number of weights. This rule causes the model to obtain high sparsity. The second rule is to use a small number of weight layers as **B** because **B** increases in the classification accuracy compared to **T**, in which accuracy suffers because of the high sparsity. The last rule is to assign high precision weights or **F** to the first and last layers of the NNs. This rule comes from the assumption that the first and last layers are more significant than the other layers. The three rules are represented in Fig. 2.

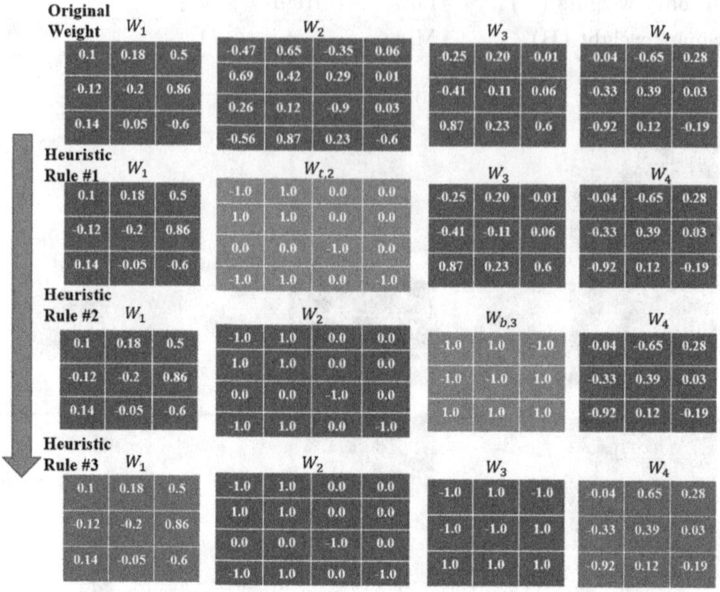

**Fig. 2.** Illustration of three heuristic rules on MPWNs with four weight layers. The highlighted weights represent the weights that were quantized according to each heuristic rule. In this example, the combining of the of three rules together results in **FTBF**.

## 4    Experimental Results

Using MNIST, CIFAR-10 and CIFAR-100 datasets, we analyzed the performance of our proposed MPWNs method by comparing it with the conventional methods of BC, TWNs, and full precision NNs. In this section, to make methods on the same scale as MPWNs, we applied TWNs with only to Eqs. 3 and 4, without using the scaling factor $S$. We did not apply *clip* function to the gradient in BC. Our models were implemented with the TensorFlow [3] framework. In our analysis, we evaluated on three important properties after training: the test accuracy, which indicates the performance of the models in the classification task; sparsity, which shows the ability of the model to be compressed; and bit of weights which shows the number of bits that were used for storing all weight variables.

### 4.1    MNIST

MNIST [2] is a hand-written numerical digit dataset with the digits ranging from 0 to 9. The dataset can be separated into 60,000 training images and 10,000 test images. Each MNIST image has the size of $28 \times 28 \times 1$ pixels. For MNIST, we used the multi-layer perceptrons (MLPs) with the structure $200FC - 100FC - 60FC - 30FC - 10Softmax$, where FC is a fully connected layer and Softmax is applied to the output layer. We trained the models for 150 epochs with an initial learning rate of $10^{-3}$ that was reduced to $10^{-4}$ at 50 epochs. We used rectified linear unit (ReLU) as the activation function and set the training batch size to 128. Batch Normalization [11] was applied to the model which was trained with Adam optimization [12]. We tended to select a big value of $\alpha$ to maximize the sparsity. However, the sparsity needed to be in which the model is trainable. The experimental results are shown in Table 2.

**Table 2.** Experimental results of MLPs on MNIST

|  | Full precision | BC | TWNs | **TBBBB** | **FBBBF** | **FTBBF** | **TFBBF** |
|---|---|---|---|---|---|---|---|
| Accuracy | 0.986 | 0.982 | 0.977 | 0.980 | 0.985 | 0.984 | 0.9815 |
| Sparsity | None | None | 0.8548 | 0.6917 | None | 0.0887 | 0.674 |
| Bit of weights (bit) | 5.92M | 0.18M | 0.37M | 0.342M | 5.06M | 5.08M | 0.971M |
| $\alpha$ | None | None | 0.7 | 0.7 | None | 0.7 | 0.7 |

To evaluate the effects of the heuristic rules, we applied the heuristic rules to MLPs one at a time. For the first rule, the results for **TBBBB** show that although, only the first layer is **T**, the sparsity reaches 0.6917, which closes to TWNs. For the second rule, **B** can regain test accuracy as compared between TWNs and **TBBBB**. For the last rule, **FBBBF** causes an increase in accuracy that is close to full precision NNs even though the three middle layers have the binary weights. The three heuristic rules have a shortcoming when it comes to this structure of MLPs because of rule 1 and 3 conflict with each other. According to the first and third rule, the first weight layer of MLPs must be converted into **T** and **F**. Subsequently, we attempted to create the closest combinations with three heuristic rules. **FTBBF** was designed by prioritizing rule 3 over rule 1. As a consequence, **FTBBF** has a classification accuracy that is close to that of full precision NNs. However, the sparsity of **FTBBF** is close to zero. **TFBBF** is the opposite of **FTBBF**. **TFBBF** prioritizes rule 1 over rule 3. Therefore, the sparsity of **FTBBF** is close to that of TWNs. Nonetheless, **TFBBF** suffers from a loss in classification accuracy.

## 4.2   CIFAR-10

CIFAR-10 [1] is an image dataset that consists of 60,000 images belonging to 10 different classes: airplanes, automobiles, birds, cats, deer, dogs, frogs, horses, ships and trucks. The data are separated into 50,000 training images and 10,000 test images. The size of each image is $32 \times 32 \times 3$ pixels. For prepossessing, we normalized the images with the mean and standard deviation of the training dataset for each red, green, blue (RGB) layer of images. In our evaluation with CIFAR-10, we used LeNet-5 [14] with the structure as $64C5 - MP2 - 64C5 - MP2 - 384FC - 192FC - 10Softmax$, where C5 is a $5 \times 5$ convolution layer, MP2 is a $2 \times 2$ max pooling layer, FC is a fully connected layer and Softmax is an output layer. We applied the model with an L2 weight decay of $10^{-5}$ along with a Dropout [18] of $p = 0.7$ on the fully connected layers. For the other settings, we used the same as those in the MNIST experiment. The experimental results are reported in Table 3 and, Figs. 3 and 4.

**Table 3.** Experimental results of LeNet-5 on CIFAR-10

|  | Full precision | BC | TWNs | **BBTBB** | **BFFFB** | **FBBBF** | **FBTBF** |
|---|---|---|---|---|---|---|---|
| Accuracy | 0.778 | 0.7366 | 0.7202 | 0.7307 | 0.7485 | 0.7707 | 0.7727 |
| Sparsity | None | None | 0.956 | 0.684 | None | None | 0.713 |
| Bit of weights (bit) | 25.5M | 0.8M | 1.59M | 1.41M | 25.3M | 1M | 1.62M |
| $\alpha$ | None | None | 0.5 | 0.5 | None | None | 0.5 |

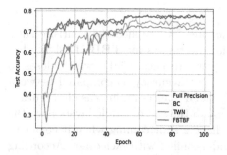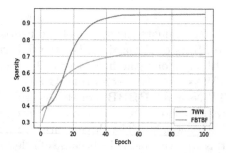

**Fig. 3.** A comparison between the conventional methods and MPWNs with three rules with LeNet-5 on the CIFAR-10 dataset. Left: the test accuracy across the training epochs. Right: sparsity indexed by epochs.

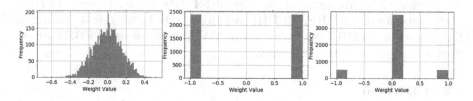

**Fig. 4.** Distribution of weight values in the first convolution layer, the left figure is **F**, the middle figure is **B** and the right one is **T**.

In the CIFAR-10 experiment, the third rule positively affects LeNet-5. Despite the fact that the number of bits of weights in **BFFFB** is 25 times that of **FBBBF**, the test accuracy of **FBBBF** is higher approximately 0.0222. We successfully applied all of the heuristic rules into LeNet-5. With the three rules together, **FBTBF** in Table 3 achieves the best overall performance. **FBTBF** has an accuracy close to that of full precision NNs, a sparsity close to that of TWNs and the number of bits close to that of BC.

## 4.3 CIFAR-100

CIFAR-100 [1] is an image benchmark dataset similar to CIFAR-10. CIFAR-100 consists of 100 categories for example, the beavers, seas, and willows. Each category has 500 training images and 100 testing images therefore, CIFAR-100 contains a total of 60K images. In this experiment, we applied the model that was inspired from by VGG-16 [17]. The structure of the model is $(2 \times 64C3) - MP2 - (2 \times 128C3) - MP2 - (3 \times 256C3) - MP2 - (3 \times 512C3) - MP2 - (3 \times 512C3) - MP2 - 100Softmax$. The model was regularized with the dropout $p = (0.7, 1.0, 0.6, 1.0, 0.6, 0.6, 1.0, 0.6, 0.6, 1.0, 0.6, 0.6, 1.0, 0.5)$ for each individual layer of the model. We further added the l2 weight decay of $10^{-5}$ and Batch Normalization to the model. The model was trained using 300 epochs, with a training batch size of 128. We applied Adam optimization with an initial learning rate of $10^{-3}$ and reduce it by a factor of 0.5 every 40 epochs. The exponential learning rate decay was used to decay $10^{-4}$ of the remaining learning rate in every epoch. We set the minimum learning rate to $10^{-6}$. The experimental results are shown in Table 4.

**Table 4.** Experimental results of the modified VGG-16 on CIFAR-100 dataset

|  | Full precision | BC | TWNs | **F(7×B)(5×T)F** |
|---|---|---|---|---|
| Accuracy | 0.6515 | 0.5711 | 0.5388 | 0.6319 |
| Sparsity | None | None | 0.7819 | 0.64 |
| Bit of weights (bit) | 472.4M | 14.76M | 29.52M | 24.25M |
| $\alpha$ | None | None | $3 \times 10^{-3}$ | $3 \times 10^{-3}$ |

VGG-16 require longer training time compared to the models in the previous experiments. Hence, an exploration to find the optimal combination of MPWNs is not suitable because it takes considerable times to train the combinations. The three heuristic rules can be directly used for an optimized model. With the three heuristic rules, **F(7×B)(5×T)F** or **FBBBBBBBBTTTTTF** are the optimized combination for this model. Compared with the conventional models, comparisons, **F(7×B)(5×T)F** supports that MPWNs with three rules have the best overall properties as those in the LeNet-5 experiment.

# 5    Conclusion and Future Works

We have proposed MPWNs, which are NNs that potentially consist of binary, ternary and full precision weight layers. The properties of MPWNs can be changed depending on the combinations of weight layers. Three heuristic rules were proposed to optimize the settings and properties of the MPWNs. The MPWNs successfully used all of the heuristic rules on LeNet-5 in the CIFAR-10 and CIFAR-100 benchmarks, this leading to the best overall properties. However, the three rules cannot be fully applied to MLPs in the MNIST experiment.

Future work will extend our work to apply the various quantization methods to the outputs of the activation function or the activation. Our goal for future work is to replace the multiplication between the activation and weights with the operations that consume the less hardware resources, for example, shift operations and logic gate operations.

**Acknowledgement.** This research was supported by JSPS KAKENHI Grant Numbers 17H01798 and 17K20010.

# References

1. Cifar-10 and cifar-100 datasets. https://www.cs.toronto.edu/kriz/cifar.html. Last Accessed 21 Sep 2018
2. Mnist handwritten digit database. http://yann.lecun.com/exdb/mnist. Last Accessed 21 Sep 2018
3. Abadi, M., Barham, P., Chen, J., Chen, Z., Davis, A., et al.: Tensorflow: a system for large-scale machine learning. arXiv preprint arXiv:1603.04467 (2016)
4. Courbariaux, M., Bengio, Y., David, J.P.: Binaryconnect: training deep neural networks with binary weights during propagations. In: Advances in Neural Information Processing Systems, pp. 3123–3131 (2015)
5. Deng, J., Dong, W., Socher, R., Li, L.J., Li, K., Fei-Fei, L.: Imagenet: a large-scale hierarchical image database. In: Proceedings of the IEEE Conference on Computer Vision and Pattern Recognition, pp. 248–255 (2009)
6. Han, S., et al.: EIE: efficient inference engine on compressed deep neural network. In: Proceedings of the 43rd Annual International Symposium on Computer Architecture, pp. 243–254 (2016)
7. Han, S., Mao, H., Dally, W.: Deep compression: compressing deep neural networks with pruning, trained quantization and Huffman coding. arXiv preprint arXiv:1510.00149 (2015)
8. Han, S., Pool, J., Tran, J., Dally, W.: Learning both weights and connections for efficient neural network. In: Advances in Neural Information Processing Systems, pp. 1135–1143 (2015)
9. He, K., Gkioxari, G., Dollár, P., Girshick, R.: Mask R-CNN. In: Proceedings of the IEEE Conference on Computer Vision, pp. 2980–2988 (2017)
10. He, K., Zhang, X., Ren, S., Sun, J.: Deep residual learning for image recognition. In: Proceedings of the IEEE Conference on Computer Vision and Pattern Recognition, pp. 770–778 (2016)
11. Ioffe, S., Szegedy, C.: Batch normalization: accelerating deep network training by reducing internal covariate shift. arXiv preprint arXiv:1502.03167 (2015)

12. Kingma, D.P., Ba, J.: Adam: a method for stochastic optimization. arXiv preprint arXiv:1412.6980 (2014)
13. Krizhevsky, A., Sutskever, I., Hinton, G.: Imagenet classification with deep convolutional neural networks. In: Advances in Neural Information Processing Systems, pp. 1097–1105 (2012)
14. LeCun, Y., Bottou, L., Bengio, Y., Haffner, P.: Gradient-based learning applied to document recognition. Proc. IEEE **86**(11), 2278–2324 (1998)
15. Li, F., Zhang, B., Liu, B.: Ternary weight networks. arXiv preprint arXiv:1605.04711 (2016)
16. Redmon, J., Farhadi, A.: Yolo9000: better, faster, stronger. arXiv preprint arXiv:1612.08242 (2016)
17. Simonyan, K., Zisserman, A.: Very deep convolutional networks for large-scale image recognition. arXiv preprint arXiv:1409.1556 (2014)
18. Srivastava, N., Hinton, G., Krizhevsky, A., Sutskever, I., Salakhutdinov, R.: Dropout: a simple way to prevent neural networks from overfitting. J. Mach. Learn. Res. **15**(1), 1929–1958 (2014)

# Policy Space Noise in Deep Deterministic Policy Gradient

Yan Yan[1] and Quan Liu[1,2,3(✉)]

[1] School of Computer Science and Technology, Soochow University, Suzhou
215006, Jiangsu, China
20165227005@stu.suda.edu.cn, quanliu@suda.edu.cn
[2] Collaborative Innovation Center of Novel Software Technology and
Industrialization, Nanjing 210000, China
[3] Key Laboratory of Symbolic Computation and Knowledge Engineering of
Ministry of Education, Jilin University, Changchun 130012, China

**Abstract.** Deep deterministic policy gradient (DDPG) algorithm is an attractive reinforcement learning method, which directly optimizes the policy and has good performance in many continuous control tasks. In DDPG, the agent explores the environment by using the Gaussian white noise in the action space. In this paper, to further improve the efficiency of exploration, we inject the factorized Gaussian noise straightly to the policy space and propose a novel dithering perturbation way that can affect subsequent states in the future, resulting in more abundant trajectories. The noise is sampled from a Gaussian noise distribution and learned together with the network weight. In order to guarantee the effectiveness of the perturbation, the same perturbation ratio is used on all layers. Our method does not require augment the environment's reward signal with additional intrinsic motivation term and the agent can learn directly from the environment. The proposed DDPG with exploratory noise in policy space was tested on the continuous control tasks. The experimental results demonstrate that it achieved better performance than the methods with no noise or action space noise.

**Keywords:** Deep reinforcement learning · Exploratory noise
Actor-critic algorithm · Deterministic policy gradient

## 1 Introduction

Recently, the deep deterministic policy gradient (DDPG) algorithm, which combined with neural network, has achieved great success in many continuous control tasks [1]

However, DDPG still faces serious challenges in the exploration. Exploration ensures the policy that learned by the agent when interacting with the environment will not quickly converge to the local optimum. Although a lot of methods are suggested to solve the problem of exploration in continuous-action Markov decision process (MDP), they usually rely on complex external structures such as well-designed reward functions or learned dynamic models [2–4]. On the one hand, these methods are limited by the linear approximation. On the other hand, the well-designed reward mechanism breaks the original rules of the environment and do not have a good generalization. The

© Springer Nature Switzerland AG 2018
L. Cheng et al. (Eds.): ICONIP 2018, LNCS 11302, pp. 624–634, 2018.
https://doi.org/10.1007/978-3-030-04179-3_55

dynamic model requires many duplicates of sections of the network, which greatly increases the complexity of the network and need for high time costs.

A reasonable way to improve exploration is adding extra Gaussian noise [5]. The noise does not break the original environmental rules and have lower calculation cost. In the DDPG algorithm, Ornstein-Uhlenbeck (OU) process is used to deal with the noise in action space to generate temporally correlated exploration that ensures agent's behavior does not prematurely converge to a local optimum [1]. The OU process gives the agent power to explore but its efficiency still needs further improvement.

In DDPG, the policy space is represented by the policy network. We add exploratory noise in the policy space, which can increase the potential for exploration. The main motivation is that small changes to weights in the neural network will lead to huge output differences. This difference will drive the agent to explore different behaviors. At the same time, in an episode, changes in one time step can have profound effects on multiple time steps in the future. The noise is derived from a noise distribution and it will be resampled in different episodes. Policy space noise will lead to the perturbation and more trajectories will be explored. However, it is necessary to control the noise scale to maintain the effectiveness of perturbed exploration. In this paper, we use the layer normalization to make sure that the agent can learn stable optimal policy. In addition, gradient descent method is adopted to train the scale of the perturbation through the loss function. The agent controls the degree of exploration by adjusting the noise scale adaptively.

We summarize our main contributions as below: (1) A way to inject the noise in the policy space. The agent can directly learn effective behaviors when interacting with the environment. (2) A method of using factorized Gaussian noise to reduce noise variables. (3) We tested our ideas on continuous control tasks and the results indicate that noise in policy space learns more efficiently than no noise or action space noise.

## 2 Background

### 2.1 Markov Decision Process and Reinforcement Learning

In a standard reinforcement learning task, an agent interacts with the environment in order to obtain information [6]. During the interaction, the agent follows a Markov decision process (MDP), which is described by a four-element tuple $(S, A, R, P)$, where $S$ is state space, $A$ is action space, $R : S \times A \to \mathbb{R}$ is a reward function and $P : S \times A \times S \to [0, 1]$ is the state transition probability. The goal of the agent is to find the maximum cumulative reward $G_t = \sum_{t'=t}^{T} \gamma^{t'-t} R(s_{t'}, a_{t'})$ through the policy $\pi : S \to A$, where $T$ is the termination time step, and $\gamma \in (0, 1]$ is a discount factor [7]. The state-action value function is the cumulative reward that the agent takes the action $a_t$ under the current state $s_t$ and then has always been followed by the policy $\pi$:

$$Q^{\pi}(s, a) = E[G_t | s_t = s, a_t = a, \pi] \tag{1}$$

The state-action value function can be solved by using the recursive form of the Bellman equation:

$$Q^\pi(s_t, a_t) = E_{r_t \sim R, s_{t+1} \sim S}[R(s_t, a_t) + \gamma E_{a_{t+1} \sim \pi}[Q^\pi(s_{t+1}, a_{t+1})]] \tag{2}$$

However, the above method is limited by the fact that the action space is discrete. The actor-critic method provides a good way to work around this problem. In the actor-critic method, two separate structures are used to explicitly represent the value function and the policy. The portion of the policy used to select the action is called the actor, and the part of the evaluation function is called the critic. The actor-critic method learns through policy gradient algorithm [6].

## 2.2  Deep Deterministic Policy Gradient

Different from the stochastic policy gradient, which generates the probability of each action, the actor only output a deterministic action $a_t$ under the state $s_t$ in the DDPG. Compared with the stochastic policy gradient method, deterministic policy gradient has smaller variance and higher efficiency [8, 9].

In DDPG, deep neural networks are used as approximations to evaluate the actor $a = \pi(s|\theta^\pi)$ and the critic $Q(s, a|\theta^Q)$, where $\theta^\pi$ and $\theta^Q$ are parameters in networks. There is no strict proof of convergence if the nonlinear neural network is used as an approximation, and learning instability occurs when combined with reinforcement learning [10]. The target network and experience replay mechanism have been put forward to solve this problem [11]. We use $\hat{\theta}^\pi$ and $\hat{\theta}^Q$ represent the parameters of target policy network and the target value function network respectively. For the value function network, loss function is used to measure the difference between predicted value and optimized objectives:

$$L(\theta^Q) = E_{s_t, a_t, r_t, s_{t+1} \sim D}[(y_t - Q(s_t, a_t|\theta^Q))^2] \tag{3}$$

In general, using $y_t$ as an approximate optimal goal of the value function:

$$y_t = R(s_t, a_t) + \gamma Q(s_{t+1}, \pi(s_{t+1}|\hat{\theta}^\pi)|\hat{\theta}^Q) \tag{4}$$

The parameters of the actor network are updated by the policy gradient method:

$$\begin{aligned}
&\nabla_{\theta^\pi} Q(s, a|\theta^Q)|_{s=s_t, a=\pi(s_t|\theta^\pi)} \\
&= \nabla_a Q(s, a|\theta^Q)|_{s=s_t, a=\pi(s_t|\theta^\pi)} \nabla_{\theta^\pi} \pi(s|\theta^\pi)|_{s=s_t}
\end{aligned} \tag{5}$$

The target network uses "*soft*" update method. The weight of the target network is slowly updated by tracking the weight of the online network: $\hat{\theta} \leftarrow \tau\theta + (1 - \tau)\hat{\theta}$.

Another way to make learning more stable is to use the experience replay mechanism. The agent stores the samples in the sample pool, which obtained by interacting with the environment at each time step. During the experience replay, a *mini-batch* of samples were taken from the pool each time and were updated using stochastic gradient descent.

Trajectories generated by DDPG is fixed because of the determined policy. The problem of exploration has become the primary challenge in DDPG

## 3   Exploratory Noise in DDPG

In this section, we will describe the action space noise in DDPG and then explain the problems of noise in action space. To solve these problems, we propose a kind of exploratory noise in policy space, which can combine with DDPG.

### 3.1   DDPG with Action Space Noise

DDPG is an algorithm based on the actor-critic method. The actor is represented by the policy network and the critic is represented by the value network. We use three layers of fully connected network to represent the policy network and the value function network respectively. The forward propagation of the network is shown in Fig. 1, where the number in the rectangle represents the number of hidden units, we use the *relu* function as the activation function. In the forward calculation, the actor takes an action $a$ according to the current state $s$, and the critic evaluates $Q$ values according to the state-action pair $(s, a)$. The backward propagation starts with the critic by calculating error and then spreads the error to the actor.

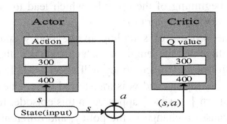

**Fig. 1.** Forward propagation in the Actor-Critic method.

**Action Space Noise.** The decision made by the agent is deterministic in DDPG. In order to make the agent has the ability to explore, random Gaussian white noise with mean $\mu = 0$ is added to the action space. When using random Gaussian white noise, the decision made by the agent at the time $t$ is $a_t = \pi(s_t) + N(0, \sigma^2 I)$. However, this method does not consider the time correlation in the episode. We use time-dependent Ornstein-Uhlenbeck (OU) process, in which the Wiener Process is simulated by time-dependent Gaussian white noise instead of random Gaussian white noise, to establish time correlation exploration. The action at the time $t$ is concatenated the OU process with the original actor $\pi(s_t)$, as shown in Eq. 6:

$$a_t = \pi(s_t) + OU(\pi(s_t)) \tag{6}$$

## 3.2  DDPG with Policy Space Noise

There is a crucial difference between action space noise and policy space noise. The noise in the action space has the following disadvantages. (1) The optimal action made by the current policy may not be selected, due to the OU process. (2) If the state $s_t$ repeats in one episode, the action cannot maintain the consistency when exploring. In some continuous control tasks, the consistency of the action will have better performance. (3) Making decisions in the action space will not cause changes in the policy, whereas these changes in the policy can enable the agent to explore more samples.

**Policy Space Noise.** The noise in the action space can only explore part of the environment, so we add noise to the policy space to increase the effectiveness of exploration. At the time $t$, the action made by the policy space noise can be represented as $a_t = \tilde{\pi}(s_t)$. Unlike doing additional processing in the action space, exploring directly in the policy space ensures that the optimal action $a_t$ can be selected. At the same time, the relationship between time and action has also been established. In DDPG, because the action made by the policy is deterministic, the consistency of action is guaranteed in an episode. The noise is sampled in a Gaussian noise distribution and the agent has the ability to explore by resampling noise in different episodes. Injecting noise into the policy network at the beginning of the episode, which lead to changes in the policy, increasing the potential of the exploration.

Figure 2 shows the process of sampling in the policy space noise, where the number of episodes is three and the length of each episode is $n$, the circle represents the state. As can be seen from Fig. 2, each policy will affect a whole trajectory. In the first episode, the state $s_1$, $s_2$, $s_3$ repeats and $s_3$ is transferred to $s_1$. Due to the consistency of the action, the same pattern $\{s_1, s_2, s_3\}$ appeared in this episode. In the second episode, resampling the noise causes a change in the policy, and different action will be taken from the state $s_1$, leading to the new state $s_4$. In the third episode, owing to the change of policy, the state $s_6$ is transferred to the state $s_8$ instead of the $s_7$.

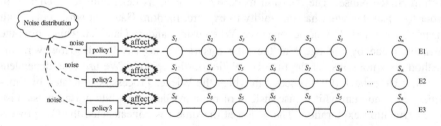

**Fig. 2.** The process of sampling in the policy space noise.

Adding the exploratory Gaussian noise with mean $\mu = 0$ in the action space results in little difference in the action, so the exploration is inefficient. If Gaussian exploratory noise with mean $\mu \neq 0$, it is difficult to select the optimal action under the current policy $\pi$, which will lead the policy converging to the local optimum. In addition, using the same policy under different episodes also affects the efficiency of exploration. Exploratory noise in the policy space will greatly deepen the exploration. Because of the noise resampling, the policy under different episodes has changed, which will affect subsequent states in an episode, and more trajectories will be explored.

**Factorized Gaussian Noise.** We use the Factorized Gaussian noise to reduce the number of noise variables. For the input vector $x$ and the output vector $y$, we abstract the policy network as $y = f_\theta(x)$, where $\theta$ indicates the noised parameters in the network: $\theta = \mu + \varepsilon \odot \psi$. It specifies that $\zeta = (\mu, \varepsilon)$ is a series of learnable parameters, where $\mu$ is the weight of the network, $\psi$ is a white noise vector with a mean of 0 and $\varepsilon$ is responsible for controlling the scale of the noise, $\odot$ represents the multiplication between matrix elements. Training the network is to optimize the parameters $\zeta$.

The full connection layer with $p$ input units and $q$ output units can be abstracted as:

$$y = w \cdot x + b \tag{7}$$

It is necessary for the perturbed network to assign a noise variable to each weight, then $p \times q + q$ noise variables are needed. To reduce the cost of the network, we use the factorized Gaussian noise. Specifically, we decompose the noise matrix as follows: for the noise matrix $\psi^w$ and $\psi^b$ between two layers, generate the noise vector $\psi_i$ with $p$ elements and the noise vector $\psi_j$ with $q$ elements, then introduce the decomposition function to generate noise variables $\psi_{i,j}^w$ and $\psi_j^b$:

$$\psi_{i,j}^w = f(\psi_i)f(\psi_j) \tag{8}$$

$$\psi_j^b = f(\psi_j) \tag{9}$$

where $f = \mathrm{sgn}(x)\sqrt{|x|}$. This means that only $p + q$ noise variables are needed. The structure of exploratory noise in the layer is shown in Fig. 3.

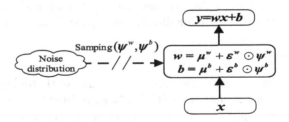

**Fig. 3.** Network with exploratory noise.

As the time increases, network layers will have different sensitivity to different noise scale. In this paper, the layer normalization is used in the perturbed network as shown in Eq. 10. Even if different layers may have different sensitivity to the noise, the same perturbation ratio is used on all layers to ensure that the agent can learn stable optimal policy. In Eq. 10, $\lambda$ and $\beta$ are parameters, $E[x]$ represents the expectation of the $x$, and var$[x]$ represents the variance of the $x$.

$$y = \lambda \frac{x - E[x]}{\sqrt{\text{Var}[x]}} + \beta \qquad (10)$$

The parameter $\varepsilon$, which controls the noise scale, trained with the weight of the network. The agent adaptively adjusts the scale of noise according to the degree of perturbation to ensure the effectiveness of exploration.

## 4   Experiment

### 4.1   Experimental Settings

In this paper, experiments are based on the Mujoco environment in OpenAI Gym platform. Mujoco provides AI researchers with challenging continuous control tasks [12].

For the continuous task without termination conditions, we set the maximum time step $T_{\max} = 1000$. If the time steps exceed 1,000, the task will be terminated and restarted.

We initialize online network weights by uniformly selecting values between 0 and 1. The replay buffer size is set to 1,000,000 and the value of *mini-batch* is fixed to 32. The network will not be trained until the samples in the experience pool reach 50,000, to improve the effect of training. In the layer regularization method, $\lambda$ and $\beta$ are set to 1 and 0.01 respectively. A momentum-based algorithm called Adam [13] is adopted to train the networks, with a learning rate of $10^{-4}$ for actor network and $10^{-3}$ for the critic network. In addition, we use the target network to reduce the fluctuations during the learning period, the initial weight in the target network is consistent with the online network. For the "*soft*" update, the parameter $\tau$ is set to 0.01.

The parameters of OU process are the same as [1]. For the Wiener process, the time variation was chosen to obey a Gaussian distribution with a mean of 0 and a variance of 0.1 that are consistent with the noise distribution in policy space noise.

### 4.2   Main Evaluation

We choose three kinds of exploration methods to evaluate: (1) No exploratory noise. Exploratory noise is not applied in DDPG (No Noise in DDPG, NN-DDPG). Without the noise, the agent explores shallowly by updating the policy network parameters through the back propagation algorithm. (2) Action space noise. Picked up the OU process in the action space to obtain time correlation exploration (Action Space Noise in DDPG, ASN-DDPG). (3) Policy space noise. We add factorized Gaussian exploratory noise directly to the policy space and use the gradient descent method to update the network weight (Policy Space Noise in DDPG, PSN-DDPG).

We evaluate the performance on eight continuous control tasks. Figure 4 depicts the result of the three kinds of exploration method. Each agent is trained for 100 epochs, where 1 epoch consist of 10,000 timesteps. The simulation results indicate the score status of the agent.

We found that PSN-DDPG achieved higher returns in most tasks than NN-DDPG and ASN-DDPG. Two reasons are responsible for this result. On the one hand, the perturbation in the policy space has deeper exploration, and on the other hand, a reasonable dithering method makes the agent learning more effective.

*Halfcheetah* is a task that require to maintain the consistency of the action. In this task, the result of the PSN-DDPG is significantly higher than the ASN-DDPG. The ASN-DDPG quickly converges to the local optimum, while the PSN-DDPG broke out the local optimum and achieved higher performance. It is worth noting that NN-DDPG has better performance than ASN-DDPG in the *InvertedPendulum* and *InvertedDoublePendulum* task, indicating that these environments don't need much exploration. ASN-DDPG adds time-correlated noise to the action space, which cannot be trained together with other parameters of the network to adapt the changes of the environment. PSN-DDPG also got higher score because the goal of reducing exploration was learned by decreasing the noise scale.

Notice, however, that PSN-DDPG did not show its advantages in *Hopper* and *Reacher*. In the *Hopper* environment, the task is not very difficult. The small changes in the weight have led to great differences in policy. Even though PSN-DDPG can adjust the noise scale adaptively, the injection of noise still increases the complexity of the network due to the relatively simple task, leading to very noisy result. In *Reacher*, the environment provides enough reward information, the well-shape reward function is sufficient to meet the need of exploration. The noise in the policy space disturbs the decision, so the score of PSN-DDPG is not very satisfactory. Therefore, we have make a conclusion that policy space noise does not necessarily in a simple environment with sufficient information.

### 4.3 Model Test

The optimal policy and the optimal value function reflect the ultimate learning effect of the agent. When training is completed, the agent gets a reusable model. Under the guidance of the optimal policy, the agent can obtain the optimal action in the environment. In this paper, we evaluate the quality of the model by comparing the average reward in episode.

In the test, 100 episodes were used to evaluate. We selected *HalfCheeah*, *HumanoidStanup*, *Reacher*, *Swimmer*, *InvertedDoublePendulum* to test. The result is given in Table 1.

As can be seen from Table 1, the average score and the max score of PSN-DDPG in the continuous control task were improved compared with NN-DDPG and ASN-DDPG except for the Reacher task when the training is finished. However, our method does not show its superiority in the standard deviation, because the noise in the policy space has deepened the complexity of the model to some extent.

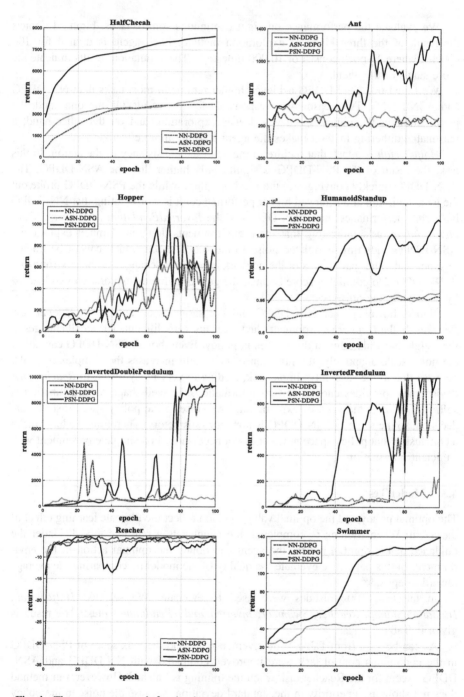

**Fig. 4.** The return per epoch for the three agents on eight continuous control environments.

**Table 1.** The test results for the three agents on five continuous control environments

| Name | Algorithm | Average | Max | Standard deviation |
|---|---|---|---|---|
| *HalfCheetah* | NN-DDPG | 3645.54 | 3663.54 | 77.65 |
| | ASN-DDPG | 4041.05 | 4852.01 | 586.56 |
| | PSN-DDPG | 8212.11 | 8408.76 | 739.35 |
| *HumanoidStanup* | NN-DDPG | 94360.30 | 144203.78 | 176002.99 |
| | ASN-DDPG | 86994.66 | 139004.32 | 246024.61 |
| | PSN-DDPG | 187629.76 | 217925.87 | 266520.05 |
| *Reacher* | NN-DDPG | −5.71 | −1.09 | 18.44 |
| | ASN-DDPG | −6.04 | −5.98 | 1.26 |
| | PSN-DDPG | −6.18 | −6.10 | 0.78 |
| *Swimmer* | NN-DDPG | 12.92 | 30.63 | 84.36 |
| | ASN-DDPG | 80.95 | 81.28 | 104.63 |
| | PSN-DDPG | 144.26 | 149.29 | 38.17 |
| *InvertedDoublePendulum* | NN-DDPG | 9043.65 | 9340.97 | 210.66 |
| | ASN-DDPG | 1008.54 | 1270.54 | 180.63 |
| | PSN-DDPG | 9343.78 | 9344.64 | 83.49 |

## 5 Conclusion

The DDPG method provides a good idea for solving the continuous control tasks. However, exploration has always been one of the challenges in DDPG. In this paper, we propose a method of exploration in the policy space and design a reasonable dithering method to drive the agent makes the optimal decision. Our results show that in the complex task, noise in the policy space can achieve better results than no noise and action space noise. These preliminary results may provide empirical clues for further research, especially the problem of reducing the standard deviation during the learning process.

**Acknowledgement.** This work was funded by National Natural Science Foundation (61272005, 61303108, 61373094, 61502323, 61272005, 61303108, 61373094, 61472262). We would also like to thank the reviewers for their helpful comments. Natural Science Foundation of Jiangsu (BK2012616), High School Natural Foundation of Jiangsu (13KJB520020), Key Laboratory of Symbolic Computation and Knowledge Engineering of Ministry of Education, Jilin University (93K172014K04), Suzhou Industrial application of basic research program part (SYG201422).

## References

1. Lillicrap, T.P., Hunt, J.J., Pritzel, A., et al.: Continuous control with deep reinforcement learning. In: Proceedings of the International Conference on Learning Representations, San Juan, Puerto Rico (2016)
2. Singh, S.P., Barto, A.G., et al.: Variational information maximisation for intrinsically motivated reinforcement learning. In: NIPS, pp. 2125–2133 (2015)

3. Osband, I., Blundell, C., Pritzel, A., et al.: Deep exploration via bootstrapped DQN. In: NIPS, pp. 4026–4034 (2016)
4. Fix, J., Geist, M.: Monte-Carlo swarm policy search. In: ICAISC, pp. 75–83 (2012)
5. Plappert, M., Houthooft, R., Dhariwal, P., et al.: Parameter space noise for exploration. In: Proceedings of the International Conference on Learning Representations, Vancouver, Canada (2018)
6. Sutton, R.S., Barto, A.G.: Reinforcement Learning: An Introduction. MIT Press, Cambridge (1998)
7. Putta, S.R., Tulabandhula, T.: Pure exploration in episodic fixed-horizon markov decision processes. In: Proceedings of the International Joint Conference on Autonomous Agents and Multi-Agent Systems, pp. 1703–1704 (2017)
8. Chou, P.-W., Maturana, D., Scherer, S.: Improving stochastic policy gradients in continuous control with deep reinforcement learning using the beta distribution. In: ICML, pp. 834–843 (2017)
9. Silver, D., Lever, G., Heess, N., et al.: Deterministic policy gradient algorithms. In: Proceedings of the 31th International Conference on Machine Learning, pp. 387–395 (2014)
10. Busoniu, L., Babuska, R., De Schutter, B., et al.: Reinforcement Learning and Dynamic Programming Using Function Approximators. CRC press, Boca Raton (2010)
11. Mnih, V., Kavukcuoglu, K., Silver, D., et al.: Human-level control through deep reinforcement learning. Nature **518**(7540), 529–533 (2015)
12. Duan, Y., Chen, X., Houthooft, R., et al.: Benchmarking deep reinforcement learning for continuous control. In: Proceedings of the 33rd International Conference on Machine Learning, pp. 1329–1338 (2016)
13. Kingma, D., Ba, J.: Adam: a method for stochastic optimization. In: Proceedings of the 3th International Conference for Learning Representations, San Diego, USA (2015)

# Mixup of Feature Maps in a Hidden Layer for Training of Convolutional Neural Network

Hideki Oki[✉] and Takio Kurita

Department of Information Engineering, Graduate School of Engineering,
Hiroshima University, Higashihiroshima, Japan
{m181021,tkurita}@hiroshima-u.ac.jp

**Abstract.** The deep Convolutional Neural Network (CNN) became very popular as a fundamental technique for image classification and objects recognition. To improve the recognition accuracy for the more complex tasks, deeper networks have being introduced. However, the recognition accuracy of the trained deep CNN drastically decreases for the samples which are obtained from the outside regions of the training samples. To improve the generalization ability for such samples, Krizhevsky et al. proposed to generate additional samples through transformations from the existing samples and to make the training samples richer. This method is known as data augmentation. Hongyi Zhang et al. introduced data augmentation method called mixup which achieves state-of-the-art performance in various datasets. Mixup generates new samples by mixing two different training samples. Mixing of the two images is implemented with simple image morphing. In this paper, we propose to apply mixup to the feature maps in a hidden layer. To implement the mixup in the hidden layer we use the Siamese network or the triplet network architecture to mix feature maps. From the experimental comparison, it is observed that the mixup of the feature maps obtained from the first convolution layer is more effective than the original image mixup.

## 1 Introduction

After the deep Convolutional Neural Network (CNN) proposed by Krizhevsky et al. [1] won the ILSVRC 2012 with higher score than the conventional methods, it became very popular for image classification and object recognition. Usually the parameters of the deep CNN are estimated by minimizing the empirical loss defined on the training samples. Since the number of parameters in the deep CNN is very large, the use of the regularization techniques is usually necessary. Also, the prediction accuracy of the trained deep CNN drastically decreases when the samples obtained from the outside regions of the training samples [2].

To improve the generalization ability for such samples, Krizhevsky et al. [1] proposed to generate the additional samples through transformations from the existing samples and to make the training samples richer. This method is known

© Springer Nature Switzerland AG 2018
L. Cheng et al. (Eds.): ICONIP 2018, LNCS 11302, pp. 635–644, 2018.
https://doi.org/10.1007/978-3-030-04179-3_56

as data augmentation. For example small shifts in location, small rotations or shears, changes in intensity, changes in stroke thickness, changes in size etc. are used to generate the additional samples for handwritten character recognition because the labels should be invariant to such perturbations.

It is also possible to incorporate the invariance directly into a classification function. Simard et al. [3] proposed a modification of the error back-propagation algorithm to train the transformation invariant classification function. The algorithm is called tangent propagation in which the invariance are learned by gradient descent.

We can virtually generate the perturbation by introducing additive noises in the hidden layers in the neural networks [4,5]. Inayoshi et al. proposed to combine the neural network classifier with auto-encoder [6]. The sum of the squared reconstruction errors of the auto-encoder is minimized in addition to the supervised objective function for classification while injecting noise in the hidden layer of the auto-encoder. By introducing the reconstruction error and the injected noise, we can virtually generate the perturbation along the principal directions of the variations in the training samples.

Recently Hongyi Zhang et al. [7] introduced an data augmentation method called mixup in which new samples are generated by mixing pairs of the training samples. Mixing of the two images is implemented with simple image morphing. This simple mixup achieves the state-of-the art performance in various datasets.

In this paper, we propose to apply mixup to the feature maps in a hidden layer instead of the input images. To apply mixup to the feature maps in a hidden layer, we have to extract the features in the hidden layer for a pair of the training samples. In the proposed method, we use the Siamese Network [10–12] or the triplet network [13] architecture to extract the feature maps in the hidden layer.

The effectiveness of the mixup of the feature maps in a hidden layer is confirmed by the experiments on the image classification using CIFAR-10 dataset.

## 2    Related Works

### 2.1    Deep Convolutional Neural Network

The deep convolutional neural network (CNN) is effective for image classification tasks. The computation within the convolution layers is regarded as a filtering process of the input image as

$$f_{p,q}^{(l)} = h\Big( \sum_{s=0}^{convy-1} \sum_{t=0}^{convx-1} w_{s,t}^{(l)} f_{p+s,q+t}^{(l-1)} + b^{(l)} \Big), \tag{1}$$

where $w_{s,t}^{(l)}$ is the weight of the neuron indexed as $(s,t)$ in the $l$-th convolution layer and $b^{(l)}$ is the bias of the $l$-th convolution layer. The size of the convolution filter is given $convx \times convy$. The activation function of each neuron is denoted as $h$.

Usually, pooling layers are added after the convolution layers. The pooling layer performs downsampling for reducing computational costs and enhancing against micro position changes.

Fully-connected layers like multi layer Perceptron is connected to the convolution layers which is used to construct the classifier.

## 2.2   Mixup

Mixup was introduced by Hongyi Zhang et al. [7] as an data augmentation method (Fig. 1). The samples are generated by mixing two different training samples by simple weighted average. It is reported that this simple method achieves the state-of-the art performance in various datasets [7]. And, the similar method was introduced by Yuji Tokozume et al. [8,9]. They conducted a detailed analysis of this method.

Let $x_1$ and $x_2$ be the images randomly extracted from the training samples and $t_1$ and $t_2$ are the corresponding teacher vectors. Then the new image $x$ and the corresponding teacher vector $t$ is generated by

$$x = \lambda x_1 + (1 - \lambda)x_2 \tag{2}$$
$$t = \lambda t_1 + (1 - \lambda)t_2 \tag{3}$$

where $\lambda$ is random number generated from the beta distribution $\beta(\alpha, \alpha)$ ($0 \leq \lambda \leq 1$), and $\alpha$ is a hyper parameter ($\alpha > 0$). The random number $\lambda$ is generated for each training pair.

It is notice that the new image $x$ is generated as a linear interpolation of the two images $x_1$ and $x_2$. It is expected that the linear interpolation gives incentives for smooth network operation and can successfully interpolate pair of the samples. The teacher vectors are similarly mixed by the linear interpolation.

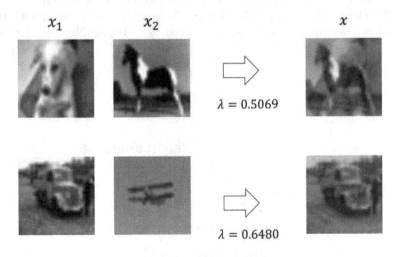

**Fig. 1.** Mixup applied for training samples

Therefore, intermediate classes can be represented by a mixed teacher vector when the classes of two samples are different.

We can consider mixup for three or more samples by a weighted interpolation with random numbers generated from the Dirichlet distribution. New sample from three or more samples can be generated as

$$x = u_1 x_1 + u_2 x_2 + u_3 x_3 + \cdots \tag{4}$$
$$t = u_1 t_1 + u_2 t_2 + u_3 t_3 + \cdots \tag{5}$$

where $u_1, u_2, u_3, \ldots$ are random numbers generated from the Dirichlet distribution $Dir(\alpha, \alpha, \alpha, \cdots)$ $(u_1 + u_2 + u_3 + \cdots = 1)$, and $\alpha$ is a hyper parameter $(\alpha > 0)$. It is reported that mixing of three or more samples works well as the mixing of two samples but the computational cost increases [7].

In this paper, we apply the weighted linear interpolation for the feature maps in a hidden layer instead of the original input images. To do this, we have to extract the feature maps in the hidden layer. This is possible by using Siamese Network.

### 2.3   Siamese Network and Triplet Network

The Siamese Network [10–12] consists of two identical sub-networks joined at their outputs as shown in Fig. 2. The two sub-networks extract feature vectors from two different samples. The objective function of the optimization for training the parameters of the networks is defined by using these extracted feature vectors. This network architecture is usually used for metric learning and the contrastive loss is often used for error function. The parameters of the Siamese Network are trained to distinguish between similar and dissimilar pairs of the training samples. Usually a contrastive loss over the metric defined on the trained representation is used as the objective function for the optimization.

Hoffer et al. [13] extended the Siamese Network to the network with three identical sub-networks. The network is called as a Triplet Network. By using

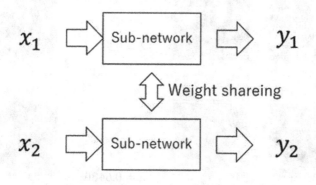

**Fig. 2.** Siamese network

three sub-networks, we can treat three input images simultaneously and extract better feature vector than the Siamese Network.

In this paper, we use these architectures to extract feature maps in a hidden layer.

## 3    Mixup of Feature Maps in a Hidden Layer

The mixup proposed by H. Zhang et al. generates intermediate images by mixing the pairs of the original training images as shown in Fig. 3. However, the original image often contains irrelevant information that is not necessary for image classification such as brightness changes or difference in subtle color. Mixing such irrelevant information may cause some bad influence for classification. Since the convolution layers are trained to extract class-specific feature maps, the important information for classification are extracted in the feature map. This means that irrelevant information such as the brightness of the image and the difference in subtle color are reduced in the feature maps in the hidden layers. Therefore, we propose to apply a mixup to the feature map extracted from the hidden layers.

In order to extract feature maps of a pair of images in the training phase, we use Siamese network architecture which has two sub-networks with sharing weights as shown in Fig. 4. For the pair of the input images $x_1$ and $x_2$, each of the sub-networks computes the feature maps $f_1$ and $f_2$ respectively. Then these feature maps are mixed with the mixing weight $\lambda$ as

$$f = \lambda f_1 + (1 - \lambda) f_2. \tag{6}$$

Then the mixed feature map $f$ is used as the input of the third CNN as shown in Fig. 3. The mixing parameter $\lambda$ is generated as a random number from the $\beta$ distribution defined by

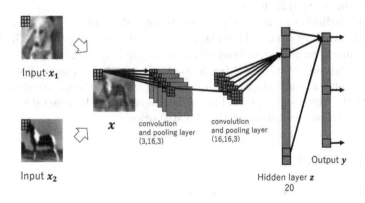

**Fig. 3.** Conventional mixup

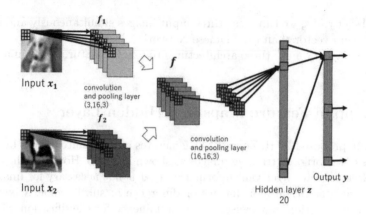

**Fig. 4.** Apply mixup to 1st convolution layer

$$\beta(x; \alpha, \alpha) = \frac{x^{\alpha-1}(1 - x)^{\alpha-1}}{\int_0^1 x^{\alpha-1}(1 - x)^{\alpha-1}dx}. \tag{7}$$

The corresponding teacher vectors $t_1$ and $t_2$ are also mixed with the same mixing weight $\lambda$ as

$$t = \lambda t_1 + (1 - \lambda)t_2. \tag{8}$$

The weights of the sub-networks and the third CNN of the proposed network architecture are trained by minimizing the cross-entropy loss defined by using the mixed feature maps $f$ and the mixed teacher vectors $t$. By changing the number of layers of the Siamese network and the third CNN, we can introduce the mixup in the middle layers.

In the prediction phase, one of the sub-networks and the CNN are directly connected as one network and the class of the input image is estimated by feeding the input image to the input layer of the connected network.

We can extend the mixing of two feature maps to three or more. For example, to extract three feature maps, we can used the triplet network architecture [13] instead of the Siamese network. The mixing weights can be generated using the Dirichlet distribution instead of the $\beta$ distribution.

## 4  Experiment

### 4.1  Dataset and Network Architectures

To confirm the effectiveness of the proposed mixup of feature maps in the hidden layer, we have performed experiments to compare the classification accuracies and the obtained feature maps in the hidden layer by using CIFAR10 dataset. CIFAR10 dataset includes 60,000 labeled small images for image classification. The size of each image is $32 \times 32$ pixels. The number of classes is 10. Usually they are divided into the training samples with 50,000 images (5,000 images per each class) and the test samples with 10,000 images (1,000 images per each class). In the following experiments, the number of training samples was reduced to 5,000 (500 samples per each class) to make the improvement of the generalization ability of the mixup approaches clear. The 10,000 test samples were used for evaluation of the classification accuracy.

In the proposed method, the feature maps from the first and the second convolution layers were extracted. To extract the feature maps from the first convolution layer, the Siamese network and the triplet network with only one convolution layer were used. The networks with two convolution layers were used to extract the feature maps from the second convolution layer.

For the comparisons, we also evaluated the classification accuracies of the network trained without mixup and the network trained with the original mixup.

To train the parameters of the networks, the standard stochastic gradient descent (SGD) learning algorithm were used for all the network architectures. The learning rate of SGD was started at 0.01 and was gradually reduced by multiplying 0.1 after 100 epochs. In addition, ReLU function is used as activation function for output of each convolution layer and output of hidden layer of Fully-connected layer. To prevent the over fitting to the training samples, we used the weight decay. The weight decay parameter was changed depending on the over fitting. For the original CNN without mixup the weight decay parameter was set to 0.04 and it was set to 0.02 for the mixup of two images or two feature maps. In addition, it was set to 0.02 for the mixup of three feature maps of the second convolution layer. The weight decay parameter 0.01 was used for the mixup of three images or three feature maps of the first convolution layer. The mixing parameter $\alpha$ in the mixup was changed from 0.2 to 1.0.

## 4.2  Comparison of the Classification Accuracy

The classification accuracy of the model is shown in Table 5. In this table, "cnn (original)" and "cnn (mixup)" denote the network that was trained without mixup and the network that was trained with the mixup of the two images, namely the original mixup. The mixup with the three input images is also denoted as "cnn (mixup3)". The proposed mixup of the two feature maps in the hidden layers are shown as "cnn (conv1-mixup)" and "cnn (conv2-mixup)" where "conv1" and "conv2" are used for the feature maps extracted from the first convolution layer and the second convolution layer respectively. So "cnn (conv1-mixup)" means that the network is trained by using the mixup of two feature maps extracted the first convolution layer. Similarly the proposed mixup of the three feature maps are denoted as "cnn (conv1-mixup3)" or "cnn (conv2-mixup3)".

From the comparison experiments, we observed that the effect of mixup increase for this dataset as the value of $\alpha$ becomes larger. Namely the accu-

| model | accuracy | $\alpha = 0.7$ | $\alpha = 1.0$ |
|---|---|---|---|
| cnn (original) | 50.01% | × | × |
| cnn (mixup) | × | 52.00% | 52.22% |
| cnn (mixup3) | × | 51.39% | 52.22% |
| cnn (conv1-mixup) | × | **52.39%** | **52.93%** |
| cnn (conv2-mixup) | × | 51.69% | **52.26%** |
| cnn (conv1-mixup3) | × | **54.67%** | **55.31%** |
| cnn (conv2-mixup3) | × | **52.78%** | 51.50% |

**Fig. 5.** Accuracy of classification

racies for the cases where the mixing parameter $\alpha$ was less than 0.7 were less than the cases with $\alpha = 0.7$ or $\alpha = 1.0$. In the this Table 5, we show only the accuracies for the cases with $\alpha = 0.7$ and $\alpha = 1.0$.

From this Table, it is noticed that the classification accuracies of "cnn (conv1-mixup3)" are highest for both cases with $\alpha = 0.7$ and $\alpha = 1.0$. Also, the classification accuracies of "cnn (conv1-mixup2)" are better than the original mixup of two images "cnn (mixup)". These results shows that the mixup of the feature maps is more effective than the mixup of the input images.

For any $\alpha$, the accuracy of the model "cnn (conv2-mixup)" in which mixup is applied to the feature maps extracted from the second convolution layer is almost same as the "cnn (mixup)" or "cnn (mixup3)". In addition, the accuracy of the model "cnn (conv2-mixup3)" is lower than the "cnn (mixup)" or "cnn (mixup3)".

It is expected that the first convolution layer is working to suppress the general image variations and the second convolution layer is extracting more class specific information. Thus we think that the mixup of the feature maps in the first convolution layer can generate the reasonable intermediate feature maps but the mixup of the feature maps in the second convolution layer maybe destroy the class specific information.

## 4.3   Comparison of the Feature Maps in the Hidden Layer

Furthermore, we compare the feature maps extracted by the first convolution layer of the model "cnn(mixup3)" and "cnn(conv1-mixup3)". Figure 6 shows the visualization of the extracted feature maps for the same input images.

From this figure, we can find that the features map obtained by the "cnn (mixup3)" model are influenced by the brightness of the original image. On the other hand, the shape edges are extracted in the feature maps obtained by the "cnn (conv1-mixup 3)" model regardless of the brightness of the original image. This result shows the some improvements of the feature map in the first convolution layer by the mixup of the feature maps.

cnn(mixup3)          cnn(conv1-mixup3)

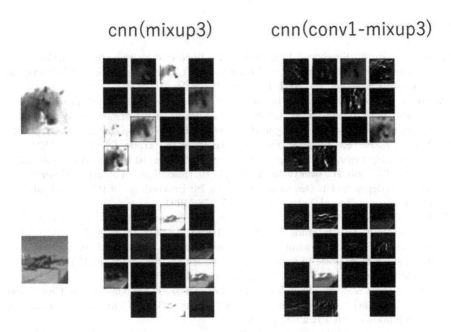

**Fig. 6.** Feature map extracted by the first convolution layer

## 5  Conclusion and Future Works

In this paper, we proposed an data augmented learning algorithm in which the
feature maps in the hidden layer are mixed. To extract the feature maps during
the training, the Siamese network or the triplet network architecture is used.
Experimental results show that some improvement of the classification accuracy
is achieved by applying the mixup to the feature maps extracted from the first
convolution layer. On the other hand, applying mixup to the second convolution
layer does not produce significant improvement. From the experiment results,
it is noticed that the classification accuracy depends on the mixing parameter
$\alpha$ but the tendency is the almost same for all the mixup models for CIFAR10
dataset.

In this paper we did not consider the effect of the distances between the mixing images or feature maps but we think that distances probably are important
factor to generate good intermediate image or feature maps. So we would like
to introduce some mechanism to control the probability of the mixup depending
on the distances.

**Acknowledgment.** This work was partly supported by JSPS KAKENHI Grant Number 16K00239.

# References

1. Krizhevsky, A., Sutskever, I., Hinton, G.E.: ImageNet classification with deep convolutional neural networks. In: Proceedings of Conference on Neural Information Processing Systems, pp. 1097–1105 (2012)
2. Szegedy, C., et al.: Intriguing properties of neural networks. arXiv:1312.6199 (2014)
3. Simard, P.Y., LeCun, Y.A., Denker, J.S., Victorri, B.: Transformation invariance in pattern recognition — tangent distance and tangent propagation. In: Orr, G.B., Müller, K.-R. (eds.) Neural Networks: Tricks of the Trade. LNCS, vol. 1524, pp. 239–274. Springer, Heidelberg (1998). https://doi.org/10.1007/3-540-49430-8_13
4. Kurita, T., Asoh, H., Umeyama, S., Akaho, S., Hosomi, A.: A structural learning by adding independent noises to hidden units. In: Proceedings of IEEE International Conference on Neural Networks, pp. 275–278 (1994)
5. Sabri, M., Kurita, T.: Effect of additive noise for multi-layered perceptron with autoencoders. IEICE Trans. Inf. Syst. **E100D**(7), 1494–1504 (2017)
6. Inayohsi, H., Kurita, T.: Improved generalization by adding both auto-association and hidden-layer noise to neural-network-based-classifiers. In: IEEE Workshop on Machine Learning for Signal Processing, pp. 141–146 (2005)
7. Zhang, H., Cisse, M., Dauphin, Y.N., Lopez-Paz, D.: mixup: Beyond empirical risk minimization. In: Proceedings of 2018 International Conference on Learning Representations (ICLR 2018) (2018)
8. Tokozume, Y., Ushiku, Y., Harada, T.: Learning from between-class examples for deep sound recognition. In: Proceedings of 2018 International Conference on Learning Representations (ICLR 2018) (2018)
9. Tokozume, Y., Ushiku, Y., Harada, T.: Between-class learning for image classification. In: Proceedings of 2018 IEEE computer society conference on Computer Vision and Pattern Recognition (CVPR 2018) (2018)
10. Bromley, J., Guyon, I., LeCun, Y., Säckinger, E., Shah, R.: Signature verification using a Siamese time delay neural network. In: Advances in Neural Information Processing Systems, vol. 6 (1993)
11. Chopra, S., Hadsell, R., LeCun, Y.: Learning a similarity metric discriminatively, with application to face verification. In: Proceedings of the IEEE Computer Society Conference on Computer Vision and Pattern Recognition (CVPR 2005), vol. 1, pp. 539–546 (2005)
12. Hadsell, R., Chopra, S., LeCun, Y.: Dimensionality reduction by learning an invariant mapping. In: Proceedings of 2006 IEEE Computer Society Conference on Computer Vision and Pattern Recognition (CVPR 2006), vol. 2, pp. 1735–1742 (2006)
13. Hoffer, E., Ailon, N.: Deep metric learning using triplet network. In: Feragen, A., Pelillo, M., Loog, M. (eds.) SIMBAD 2015. LNCS, vol. 9370, pp. 84–92. Springer, Cham (2015). https://doi.org/10.1007/978-3-319-24261-3_7

# Deep Deterministic Policy Gradient
# with Clustered Prioritized Sampling

Wen Wu[1], Fei Zhu[1,2(✉)], YuChen Fu[3], and Quan Liu[1]

[1] School of Computer Science and Technology, Soochow University,
Suzhou 215006, China
20164227051@stu.suda.edu.cn,
{zhufei,quanliu}@suda.edu.cn
[2] Provincial Key Laboratory for Computer Information Processing Technology,
Soochow University, Suzhou 215006, China
[3] School of Computer Science and Engineering,
Changshu Institute of Technology, Changshu 215500, China
yuchenfu@cslg.edu.cn

**Abstract.** As a famous deep reinforcement learning approach, deep deterministic policy gradient (DDPG) is able to deal with the problems in the domain of continuous control. To remove temporal correlations among the observed transitions, DDPG uses a sampling mechanism called experience reply which replays transitions at random from the replay buffer. Experience reply removes correlations among different transitions. However, random sampling does not consider the importance of transitions in replay buffer which leads to the longer training time and poor performance. In this paper, we propose a novel efficient sampling mechanism which we call deep deterministic policy gradient with clustered prioritized sampling (CPS-DDPG). CPS-DDPG clusters the transitions by K-means in order to reduce the complexity of the algorithm. In addition, CPS-DDPG samples transitions from different categories according to priorities so as to train targeted transitions. The key idea of CPS-DDPG is to set high priorities to the valuable categories and increase the priorities of the categories that have not been selected for long time appropriately in order to increase the diversity of the transitions. The experimental results show that the proposed model achieves better performance than the traditional deep reinforcement learning model in the continuous domain.

**Keywords:** Reinforcement learning · Deep reinforcement learning
Deep deterministic policy gradient · K-means · Prioritized sampling

## 1 Introduction

Deep reinforcement learning (DRL) is recently a new research direction which combines reinforcement learning (RL) and deep learning (DL) [1, 2]. One of the most noticeable deep reinforcement learning methods called Deep Q-Networks (DQN) which combines deep convolutional neural network with a Q-learning algorithm in reinforcement learning has achieved great success in high-dimensional sensory input problem [3]. However, DQN can only handle discrete and low-dimensional

© Springer Nature Switzerland AG 2018
L. Cheng et al. (Eds.): ICONIP 2018, LNCS 11302, pp. 645–654, 2018.
https://doi.org/10.1007/978-3-030-04179-3_57

action space tasks. Many tasks have continuous and high dimensional action space [4, 5]. A method called deep deterministic policy gradient (DDPG) performs well in the field of deep reinforcement learning in continuous domain [6]. DDPG is an actor-critic, model-free algorithm which can learn a policy directly from raw pixel input [7–9].

DDPG uses an actor-critic approach and neural network function approximators to learn policy in large action and state space. In order to alleviate the instability of neural network as function approximator, DDPG utilizes a replay buffer to address these issues. The agent stores the transitions tuple $e_t = (s_t, a_t, r_t, s_{t+1})$ to a replay buffer $D_t = \{e_1, \ldots e_t\}$ at each step. These transitions will be uniformly sampled for network training. Experience replay removes the correlations among the transitions so as to ensure the algorithm more stable.

In addition, actor-critic approach is applied in DDPG. $\mu(s|\theta^\mu)$ and $Q(s, a|\theta^Q)$ represent the actor network and critic network respectively. Actor network and Critic network have target network $\mu'(s|\theta^{\mu'})$ and $Q'(s, a|\theta^{Q'})$ with weight $\theta^{\mu'}$ and $\theta^{Q'}$. Critic network is updated at iteration $t$ uses the following loss function:

$$L(\theta^Q) = E_{s_t, a_t, r_t, s_{t+1} \sim D}[(r_t + \gamma Q'(s_{t+1}, \mu'(s_{t+1}|\theta^{\mu'})|\theta^{Q'}) - Q(s_t, a_t|\theta^Q))^2] \quad (1)$$

in which $\gamma$ is the discount factor. At each time step, agent selects action according to the current policy. Actor network is updated by sampled policy gradient:

$$\nabla_{\theta^\mu} J \approx \frac{1}{N} \sum_i \nabla_a Q(s, a|\theta^Q)|_{s=s_i, a=\mu(s_i)} \nabla_{\theta^\mu} \mu(s|\theta^\mu)|_{s_i} \quad (2)$$

In reinforcement learning, researchers describe a reinforcement learning environment as a Markov decision process (MDP). MDP problems can be represented by a sequence of state transitions $(s_t, a_t, r_t, s_{t+1})$. The action-value function is used to evaluate strategy in reinforcement learning. It describes the expected reward under the policy $\pi$ after executing action $a_t$ at state $s_t$:

$$Q^\pi(s, a) = E[R_t|s_t = s, a_t = a, \pi] \quad (3)$$

Bellman optimality equation provides theoretical basis for updating action-value function iteratively:

$$Q_{t+1}(s_t, a_t) = E_{s_{t+1} \sim S, \ a_{t+1} \sim A}\left[r + \gamma \max_{a_{t+1}} Q_t(s_{t+1}, a_{t+1})\right] \quad (4)$$

The value iteration algorithms will finally converge to the optimal action-value function $Q^*$ while the number of iterations approach to infinity.

Q-learning is one of the most important algorithms in reinforcement learning. It is an off-policy Temporal Difference (TD) learning algorithm. The $Q$-values are updated through the Bellman optimality equation iteratively as below:

$$Q(s_t, a_t) = Q(s_t, a_t) + \alpha[r + \gamma \max_{a_{t+1}} Q(s_{t+1}, a_{t+1}) - Q(s_t, a_t)] \qquad (5)$$

However, the high complexity of action-state space makes it impractical to estimate the optimal $Q$ value. Generalization is necessary in this case. One of the most famous methods is to use a function approximator such as neural network [10] although sometimes the algorithm appears to be unstable.

In the traditional reinforcement learning algorithms, only one parameter is updated each step after a state transition is observed, which causes two problems: (1) The parameter updates is relevant in time. (2) The state of small number of occurrences is easily overlooked. On the other hand, most deep learning algorithms need to satisfy two conditions: (1) The transitions are low correlation. (2) The transitions can be reused multiple times during training. To meet these requirements, experience replay [11, 12] is utilized. Experience replay is a technique which collects the historical transitions into the replay buffer and samples uniformly from it to get the mini-batch transitions each step in order to update the network parameters. Experience replay not only makes sampling efficiency, but also removes correlations among transitions.

However, there are some limitations in uniformly sampling. The importance of transitions in the replay buffer is different. Uniformly sampling cannot distinguish the different value among the transitions and always selects the transitions of the most recent moment because of the finite memory size. Allocating priorities for transitions is a feasible way. However, the memory size is typically a large fixed constant. Assigning priority to each transition in replay buffer requires multiple iterations to visit every element. As a result, it needs more training time and will make algorithm more complex. So in this paper, we present a novel sampling mechanism. Specifically, we cluster the transitions by K-means [13] after each episode. Then, we set priorities to categories according to upper confidence bound (UCB) which helps to learn optimal policies faster. On one hand, we use an encouragement item to increase the priorities of the categories which have not been selected for many steps. On the other hand, we add a penalty term to the categories which have been selected to reduce the priorities. More importantly, we assign more priorities for the categories which get positive average reward during training because of the more valuable information the categories with positive reward have.

This paper presents a novel sampling mechanism called CPS-DDPG which combines K-means and upper confidence bound sampling. The experimental results show that CPS-DDPG performs better and more stable than DDPG and PS-DDPG without clustering.

## 2 Model and Algorithm

### 2.1 Model Structure

We propose a novel sampling mechanism called CPS-DDPG. Our main contribution is to improve the sampling mechanism and make it more efficient. This mechanism has two main advantages. Firstly, it sets reasonable priorities for transitions without sweeping all replay buffer. Secondly, CPS-DDPG not only increases the priorities of transitions which have higher reward but also keeps the diversity of transitions. The architecture used in CPS-DDPG is demonstrated in Fig. 1.

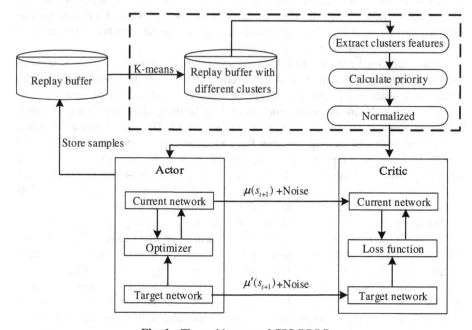

**Fig. 1.** The architecture of CPS-DDPG.

As we can see from Fig. 1, Actor network generates an action every step according to the current policy and exploratory noise. After executing the action, transition $(s_t, a_t, r_t, s_t)$ is stored in replay buffer. At each step the actor network and critic network are updated by sampling a mini-batch from the buffer. However, there are some limitations in sampling uniformly from the replay buffer. For example, low-value transitions may be repeatedly used while valuable transitions are missed. It makes training time longer. One effective way is to set the priorities for the transitions in replay buffer. However, set priorities for all transitions requires sweeping the replay buffer every step which leads to high algorithm complexity. CPS-DDPG can alleviate this problem to some extent. The part of the dotted line should be paid attention to in that it is the focus of CPS-DDPG. It contains two main parts as follows.

**Cluster Transitions in Replay Buffer.** To allocate the priorities more efficiently, we cluster the transitions by K-means after every episode. Transitions with similarities are grouped together. In addition, we classify categories based on the average reward of transitions in different categories. Then we set priorities for the categories in order to improve sampling efficiency.

**Upper Confidence Bound Sampling.** Proper exploration is necessary in reinforcement learning since the environment is unknown. Auer [14] proposed a method named upper confidence bound (UCB) firstly applied in multi armed bandit problem. This method effectively balances exploration and exploitation. Inspired by UCB and to alleviate the problem uniformly sampling has, we propose an upper confidence bound sampling method.

Generally, researchers usually consider the categories which have achieved great reward can accelerate the training process. Therefore, we set higher priorities for these categories. However, as the priorities of these categories increased, the priorities of the remaining categories will become lower, then the algorithm will easily fall into sub-optimal. In addition, the diversity of the transitions cannot be guaranteed. Some fraction of categories may never be sampled before they drop out of the replay buffer. Furthermore, the transitions that are frequently chosen will be low valuable in that more chance to approximate the target network. Accordingly, the step and the sampled times are considered in priority. Specifically, the formula for updating the priority for each category is:

$$p_i = p_i + c(\bar{r}_i + \sqrt{\ln t / N_i}) \tag{6}$$

in which $N_i$ is the number of time the $i$th category is selected. $c$ is the parameter that controls the degree of priority. If $c = 0$, the method is the same as uniformly sampling. $p_i$ is the priority of the $i$th category. Priorities are set equal in initial, that is $p_i = p_j(i, j \in [1, N])$, where $N$ is the number of the categories in replay buffer. $\bar{r}_i$ is the average reward that the $i$th category gets. $t$ is time step which records the number of steps from the beginning to the current. The average reward $\bar{r}_i$ in Eq. (6) greatly influences the priority. Categories with positive reward transitions may get higher priorities than others. On the other hand, the categories that are not sampled in the current step may have higher priorities at the next step. As we can see from the Eq. (6), categories with great reward which have not been sampled will get higher priorities at the next step. In addition, the priority monotonously decreases with the increase of sampled time.

Each category in the replay buffer is assigned a priority. After each action is performed, the priorities of the categories are updated. However, greedy sampling without exploration will lead to the lack of diversity in training data. To alleviate the problem, we normalize the priorities as follows:

$$p_j^r = p_j \bigg/ \sum_{k=1}^{N} p_k \tag{7}$$

Where $p_j^r$ is the probability of the $j$th category to be chosen. Every step when the category is chosen, we uniformly sampling from the category because of the similarity transitions in category.

## 2.2 Algorithm

To solve the problems, we propose an approach called deep deterministic policy gradient with clustered prioritized sampling (CPS-DDPG) which combines K-means and upper confidence bound sampling method. The whole algorithm is presented in Algorithm 1.

---

**Algorithm 1.** Deep Deterministic Policy Gradient with Clustered Prioritized Sampling

---

1: Randomly initialize actor network $\mu(s\,|\,\theta^\mu)$ and critic network $Q(s,a\,|\,\theta^Q)$ with weights $\theta^\mu$ and $\theta^Q$; target network $\mu'$ and $Q'$ with weights $\theta^{\mu'} \leftarrow \theta^\mu, \theta^{Q'} \leftarrow \theta^Q$; replay buffer $D$; mini-batch $M$; $p_i = 1$

2: **for** episode $1, N$ **do**

3:    Initialize random process $O$ for action exploration

4:    Initialize observation states $s_1$

5:    **for** $t = 1, T$ **do**

6:        Select action $a_t = \mu(s_t\,|\,\theta^\mu) + O_t$

7:        Execute action $a_t$ and observe reward $r_t$ and observe new state $s_{t+1}$

8:        Store transition $(s_t, a_t, r_t, s_{t+1})$ in $D$

9:        **if** done **then**:

10:            Cluster transitions and save them category in $D$

11:        **for** $j = 1, M$ **do**

12:         **if** $\sum_{k=1}^{n} P_k \le random() < \sum_{k=1}^{n+1} P_k$ **then**

13:            Sample a transition $(s_i, a_i, r_i, s_{i+1})$ from the $k$ th category

14:        **end for**

14:    Set $y_i = r_i + \gamma Q'(s_{i+1}, \mu'(s_{i+1}\,|\,\theta^{\mu'})\,|\,\theta^Q)$

15:    Update categories priority: $p_i = p_i + c(\overline{r_i} + (\ln t / N_i)^{1/2})$

16:    Update the probability of categories to be chosen: $p_i^r = p_i \Big/ \sum_{k=1}^{N} p_k$

17:    Minimize the loss: $L(\theta^Q) = E_{s_i,a_i,r_i,s_{i+1} \sim D}[y_i - Q(s_i, a_i\,|\,\theta^Q))^2]$

18:    Update actor policy: $\nabla_{\theta^\mu} J \approx \frac{1}{N} \sum_i \nabla_a Q(s,a\,|\,\theta^Q)\,|_{s=s_i, a=\mu(s_i)} \nabla_{\theta^\mu} \mu(s\,|\,\theta^\mu)\,|_{s_i}$

19:    Update the target networks: $\theta^{\mu'} \leftarrow \tau\theta^\mu + (1-\tau)\theta^{\mu'}, \theta^{Q'} \leftarrow \tau\theta^Q + (1-\tau)\theta^{Q'}$

20: **end for**

21: **end for**

---

The actor network in CPS-DDPG selects actions by adding noise which sampled from a noise process $O$. CPS-DDPG clusters the transitions by K-means after each episode just as in 10th step in Algorithm 1. Then we set priorities for categories and normalize the priorities. We use soft update rather than copying the parameters of the current networks to the target networks directly. The parameters of these target networks are slowly updated: $\theta^{\mu'} \leftarrow \tau\theta^\mu + (1-\tau)\theta^{\mu'}$ with $\tau \ll 1$. It improves the stability of training.

# 3  Experiments

## 3.1  Experiments Details

The proposed algorithm is evaluated on several challenging continuous control problems, all of which are simulated by using the gym which provided by OpenAI. Gym provides a series of continuous control tasks. For better algorithm comparison, the same parameters are used in the experiments.

To soft update the target networks, the rate of soft updating $\tau$ is set to 0.001. We use a discount factor of $\gamma = 0.99$ to discount the $Q$. In addition, we use a replay buffer size of $10^6$ and mini-batch size of 64. Max training step is set to $10^6$. In the following experiments, we use the Adam algorithm with learning rate $10^{-4}$ in actor network and $10^{-3}$ in critic network. To make the training more efficiency, the max step in one episode is set to 1000. In order to guarantee the diversity of transitions at the beginning, agent selects action randomly in the first 50 thousand steps. For the hyper-parameter $a$ that controls the degree of priorities, we evaluate on the task of HalfCheetah and find that $c = 0.6$ performs best in our algorithm. All the experiments are evaluated on the computer with i7-7820x CPU.

## 3.2  Experiments Results

We evaluate CPS-DDPG in four continuous control problems: HalfCheetah, Hopper, Humanoid and HumanoidStandup. DDPG algorithm is used as a baseline. In addition, to compare the efficiency of the model with K-means, we evaluate the algorithm PS-DDPG without clustering either. Table 1 displays the running time per episode in CPS-DDPG and PS-DDPG.

**Table 1.** The time required per episode of CPS-DDPG and PS-DDPG in four tasks. All tests are carried out after 50 thousand steps.

|              | CPS-DDPG | PS-DDPG |
| ------------ | -------- | ------- |
| BipedalWalker | **10.48 s** | 12.85 s |
| HalfCheetah  | **14.88 s** | 15.44 s |
| Hopper       | **4.03 s**  | 5.32 s  |
| Humanoid     | **15.5 s**  | 18.53 s |

We find that PS-DDPG takes more training time per episode than CPS-DDPG in that PS-DDPG sweeps all transitions in replay buffer while CPS-DDPG only sweeps the categories after clustering the transitions. In addition, the experiment on Hopper takes less training time owing to the short steps per episode. In summary, CPS-DDPG allocates priorities more efficiently, especially in the tasks which have more steps in one episode.

In order to compare the effect of three models, we show the average reward per episode during training in Fig. 2.

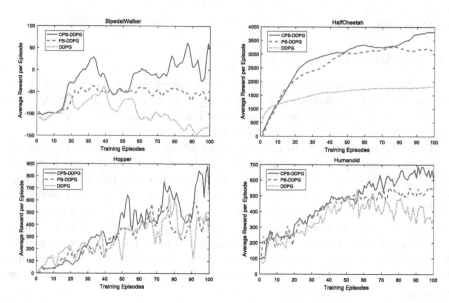

**Fig. 2.** Comparison of the average reward per episode on four continuous control problems. To reduce the fluctuation of data, we averaged reward every 5 data.

BipedalWalker is a task that the agent needs to learn how to walk for higher reward. The method with prioritized sampling achieves great reward per episode than DDPG. In addition, CPS-DDPG remains stable in the 21th episode while DDPG method gets much lower reward even in the 90th episode. The main reason is that CPS-DDPG can distinguish valuable transitions which make training more efficient. HalfCheetah is an environment in which the agent tries to make a cheetah robot run. As we can see from Fig. 2, adding prioritized sampling to DDPG give a great improvement in HalfCheetah in that CPS-DDPG can utilize the valuable transitions. In Hopper environment, the improvement is not very obvious. The main reason is that Hopper needs the agent hop to achieve high reward. It means that the agent must execute a series of actions repeatedly. However, CPS-DDPG contains exploration in sampling which leads to an unstable performance. Humanoid is a task to make a two-legged agent walk. The model with prioritized sampling performs better than DDPG while the average reward tends to be noisy. There are two main reasons. Firstly, the dimension of state in Humanoid is large (376) which makes the training slowly. Secondly, the random action can make the task unstable easily because small changes in training will disorder action sequence.

Nevertheless, CPS-DDPG has some limitations during training. The algorithm is unstable in some tasks as we can see in Fig. 2. It will lead to poor performance at some particular time. It is caused by overusing the high propriety samples which results in the inadequate training.

A good model performs well not only in training process but in testing process after training. We test the three algorithms in the tasks above after training. Every continuous problem is tested 10 times and there are 50 thousand steps in each test. The test results are shown in Table 2.

**Table 2.** The test results in four continuous tasks.

| Tasks | Models | Average reward | Standard deviation |
|---|---|---|---|
| BipedalWalker | CPS-DDPG | **17.46** | 25.76 |
| | PS-DDPG | −56.61 | 26.23 |
| | DDPG | −144.70 | 29.88 |
| HalfCheetah | CPS-DDPG | **3626.56** | 359.85 |
| | PS-DDPG | 3026.79 | 428.39 |
| | DDPG | 1802.73 | 192.73 |
| Hopper | CPS-DDPG | **780.16** | 323.99 |
| | PS-DDPG | 516.34 | 371.41 |
| | DDPG | 388.51 | 205.99 |
| Humanoid | CPS-DDPG | **680.19** | 242.82 |
| | PS-DDPG | 621.18 | 255.32 |
| | DDPG | 441.65 | 244.92 |

As we can see from Table 2, CPS-DDPG achieves the highest average reward in three models among four tasks. In addition, the smaller standard deviation in CPS-DDPG than PS-DDPG indicates that the CPS-DDPG is more stable.

# 4  Conclusion

In this work, a novel method CPS-DDPG which combines K-means and prioritized sampling according to upper confidence bound is proposed. CPS-DDPG clusters the transitions after each episode so that the transitions with similar features are grouped together. After that we set priorities to the categories and select transitions according to upper confidence bound sampling. Experiments show that CPS-DDPG performs well and effectively in four continuous control problems. A few limitations to our approach remain. The unstable performance in some tasks calls for new algorithms. One promising direction of future research would be to apply asynchronous methods [15] to continuous problems which may alleviate these limitations.

**Acknowledgements.** This work was supported by the National Natural Science Foundation of China (61303108, 61373094, 61772355); Jiangsu College Natural Science Research Key Program (17KJA520004); Program of the Provincial Key Laboratory for Computer Information Processing Technology (Soochow University) (KJS1524); China Scholarship Council Project (201606920013).

# References

1. Mnih, V., Kavukcuoglu, K., Silver, D., et al.: Playing Atari with deep reinforcement learning. In: Proceedings of Workshops at the 26th Neural Information Processing Systems, Lake Tahoe, USA (2013)
2. Mnih, V., Kavukcuoglu, K., Silver, D., et al.: Human-level control through deep reinforcement learning. Nature 518(7540), 529–533 (2015)
3. Silver, D., Huang, A., Maddison, C.J., et al.: Mastering the game of Go with deep neural networks and tree search. Nature 529(7587), 484–489 (2016)
4. Schulman, J., Moritz, P., Levine, S., et al.: High-dimensional continuous control using generalized advantage estimation. In: Proceedings of the International Conference on Learning Representations (2015)
5. Watter, M., Springenberg, J.T., Riedmiller, M., et al.: Embed to control: a locally linear latent dynamics model for control from raw images. In: Advances in Neural Information Processing System, pp. 2728–2736 (2015)
6. Lillicrap, T.P., Hunt, J.J., Pritzel, A., et al.: Continuous control with deep reinforcement learning. In: Proceedings of the International Conference on Learning Representations (2016)
7. Silver, D., Lever, G., Heess, N., et al.: Deterministic policy gradient algorithms. In: Proceedings of the International Conference on Machine Learning, pp. 387–395 (2014)
8. Heess, N., Wayne, G., Silver, D., et al.: Learning continuous control policies by stochastic value gradients. In: Advances in Neural Information Processing System, pp. 2926–2934 (2015)
9. Levine, S., Levine, S., Levine, S., et al.: Continuous deep Q-learning with model-based acceleration. In: Proceedings of the 33th International Conference on Machine Learning, pp. 2829–2838 (2016)
10. Levine, S., Abbeel, P.: Learning neural network policies with guided policy search under unknown dynamics. In: Advances in Neural Information Processing Systems, pp. 1071–1079 (2014)
11. O'Neill, J., Pleydell-Bouverie, B., Dupret, D., et al.: Play it again: reactivation of waking experience and memory. Trends Neurosci. 33(5), 220–229 (2010)
12. Wawrzyński, P., Tanwani, A.K.: Autonomous reinforcement learning with experience replay. Neural Netw. 41(5), 156 (2013)
13. Kang, S.H., Sandberg, B., Yip, A.M.: A regularized k-means and multiphase scale segmentation. Inverse Probl. Imaging 5(2), 407–429 (2017)
14. Auer, P., Cesa-Bianchi, N., Fischer, P.: Finite-time analysis of the multiarmed bandit problem. Mach. Learn. 47(2–3), 235–256 (2002)
15. Mnih, V., Badia, A.P., Mirza, M., et al.: Asynchronous methods for deep reinforcement learning. In: Proceedings of the International Conference on Machine Learning, New York, USA, pp. 1928–1937 (2016)

# Induced Exploration on Policy Gradients by Increasing Actor Entropy Using Advantage Target Regions

Alfonso B. Labao, Carlo R. Raquel, and Prospero C. Naval Jr.[✉]

Computer Vision and Machine Intelligence Group, Department of Computer Science,
University of the Philippines Diliman, Quezon City, Philippines
pcnaval@dcs.upd.edu.ph

**Abstract.** We propose a policy gradient actor-critic algorithm with a built-in exploration mechanism. Unlike existing policy gradient methods that use several actors asynchronously for exploration, our algorithm uses only a single actor that can robustly search for the optimal path. Our algorithm uses modified advantage targets that increase entropy in an actor's predicted advantage probability distribution. We do this using a two-step process, where the first step modifies advantage targets from points to regions, by sampling particles in neighborhoods along the direction of the critic value function. This step increases entropy in an actor's estimates and explicitly induces the actor to perform actions outside of past policies for exploration. The second step controls for variance increase due to sampling, where shortest-path dynamic programming selects particles from the regions with minimum inter-state movements. We present an analysis of our method and compare it with other exploration policy gradient algorithm, i.e. A3C, and report faster convergence in some VizDoom and Atari benchmarks given the same number of backpropagation steps on a deep network function approximator.

## 1 Introduction

Policy gradient methods [2,5,6,9] are a subset of reinforcement learning algorithms that approximate optimal policies through a latent function usually cast as soft-max. They are unlike Q-learning methods which directly estimates the value function of state-action pairs. Policy gradient methods are known to converge faster than Q-learning [9], since they prioritize searching for the optimal policy. However, as pointed out in [11], policy gradient methods have a propensity to incur high variances - as gradients reflect fluctuations in empirical rewards following changes in future state action pairs. In addition, policies may get stuck in local minima if there is insufficient exploration.

Many methods have been proposed to mitigate variances in policy gradients as well as encourage exploration in the context of deep network function approximation. Among the most popular methods are actor-critic algorithms [5,11] that use a critic value function to supervise actor training. Using a critic

© Springer Nature Switzerland AG 2018
L. Cheng et al. (Eds.): ICONIP 2018, LNCS 11302, pp. 655–667, 2018.
https://doi.org/10.1007/978-3-030-04179-3_58

value function instead of cumulative empirical rewards minimizes the variance of value function estimates. Another method is [9] called Generalized Advantage Estimation (GAE), which builds on the actor-critic algorithm by means of a dual parameter system to adjust bias and variance and proposes the use of advantage functions to mitigate variances while preserving unbiasedness. The popular A3C algorithm [6], focuses on exploration by using asynchronous training methods, where several actors simultaneously search for the optimal policy to encourage exploration and asynchronously update a global set of network parameters.

Our proposed algorithm falls under the family of actor-critic algorithms and incorporates a built-in exploration mechanism. However, compared to A3C [6], we only use a single actor (or worker in A3C terminology) to search for the optimal policy. Our strategy is to focus on increasing the entropy of an actor's estimated advantage probability distribution to encourage exploration. We do this by modifying advantage targets to advantage target regions by sampling several particles around a neighborhood along the direction of the critic value function. We then select via shortest path dynamic programming a configuration of particles with minimal inter-state movements to control gradient variances. The actor is trained from the modified advantage targets [9] derived from selected DP particles. By maintaining high entropy in the actor's predictions, the likelihood of constantly exploring for the optimal path is increased. Our method - which uses sampling in neighborhoods - may create more bias and variance, but experimental tests show that its underlying exploration gains lead to better performance. Some analysis regarding our method is presented.

We implement our method on several Atari games [8] and high-dimensional VizDoom AI benchmark scenarios [4, 10]. We compare our proposed algorithm with the exploration-intensive A3C algorithm [6] and a standard actor-critic algorithm [5] using deep network function approximation. From our experiments, our algorithm manages to converge faster than other methods. In difficult high-variance scenarios, our algorithm outperformed A3C and standard actor-critic. In summary, the following are the contributions of our experiment.

1. A deep-network policy gradient actor-critic algorithm with automatic exploration and an actor that estimates advantage probability distributions
2. A method to compute modified advantage targets for increasing entropy in actor's estimates, using advantage target regions and variance-control DP
3. Analysis on the benefits of our method
4. Experiments on VizDoom and Atari benchmarks.

## 2   Preliminaries

The components of reinforcement learning is an agent and an environment. The agent is situated in a current state $s_t$ provided by the environment for time $t = 0, 1, 2, ...T$. The agent performs actions $a_t$ in the environment based on a policy $\pi(s_t, a_t)$. Given $a_t$, the environment provides feedback to the agent through rewards $r_t$, along with the next state $s_{t+1}$.

Policy gradient methods directly approximate the optimal policy $\pi^*$ to arrive at a sequence of actions $a_0, a_1, ..., a_T$ given states $s_0, s_1, ..., s_T$ such that the cumulative reward is maximized under a discount parameter $\gamma$, or: $\sum_{t=0}^{\infty} \gamma^t r_t$. The optimal policy $\pi$ can be estimated using a deep network function approximator similar to deep Q-learning methods [1,8,12]. Policy gradient methods modify network parameters according to provided targets. The gradient $g_r$ could be directly estimated in a classical manner through the following equation, borrowing notation from [9], where policies $\pi(a_t|s_t)$ are cast in terms of probabilities.

$$g_r = \mathbb{E}[\Psi \nabla_\theta \log \pi(a_t|s_t)] \tag{1}$$

From [9], the $\Psi$ term in Eq. 1 serves as a weight to supervise the gradient, where larger weights are given to actions that provide higher rewards. $\Psi$ can be cumulative empirical rewards, but this is not desirable as cumulative empirical rewards provide high gradient variances. From [9], a preferable function for $\Psi$ are standard-form advantage targets, defined as:

$$A(s_t, a_t) = r_t + \mathop{\mathbb{E}}_{\pi, a_t} [\gamma v(s_{t+1})] - v(s_t) \tag{2}$$

where $v_t$ are the critic value functions that approximate cumulative empirical rewards. For each action $a_t$, advantages intuitively estimate the performance gain received by performing action $a_t$ over a baseline $V(s_t)$. Subtracting a baseline $V(s_t)$ provides desirable variance reduction properties while keeping unbiasedness. Our proposed algorithm in this paper uses advantage functions for $\Psi$, but we modify the form of the policy estimate $\pi(a_t|s_t)$. Instead of casting $\pi(a_t|s_t)$ as a probability taken from a distribution normalized over actions, we estimate a probability distribution $\rho(s_t, a_t)$ that is normalized over the range of possible advantage values for each action.

We also use the entropy function from information theory to describe our method of encouraging exploration. Entropy $H$ is defined as follows for a probability distribution $\rho$ over variables $\zeta$. The entropy function in Eq. 3 increases if probabilities are more dispersed across values.

$$H(\rho) = -\sum_i^n P(\zeta_i) \log P(\zeta_i) \tag{3}$$

## 3   Proposed Algorithm

Our proposed algorithm falls under the actor-critic family [5,11]. Actor-critic algorithms use a critic value function to reduce gradient variance. The novelty of our method lies in incorporating exploration mechanisms in the actor by means of modified advantage targets. The actor approximates the distribution of these advantage targets for policy computation. This is a change from the way policies are computed in standard actor-critic algorithms which estimate distributions over actions. We summarize the major components of our method as follows:

1. Actor: for each state-action pair, the actor estimates a probability distribution $\rho(s_t, a_t)$ over a range of possible advantage values
2. Training: Training of the actor uses modified advantage targets computed through the following steps
    (a) Step 1: Advantages for each state are transformed to advantage target regions by randomly sampling points (defined as particles) in a neighborhood along the direction of the unbiased critic value function
    (b) Step 2: Using shortest-path dynamic programming, particles for each state are selected on the basis of minimum inter-state distance
    (c) Modified advantage targets are computed from the selected particles.

We propose in Step 1 that using advantage target regions serves to increase the entropy in $\rho(s_t, a_t)$ and encourage exploration. The shortest-path dynamic programming step serves to reduce the variance brought about by using random particles in Step 1 above. Transforming targets to regions around it (Step 1) may create more bias and variance, but our method relies on underlying exploration gains to offset these. We will describe in more detail our algorithm as follows.

## 3.1    Policy Computations and Actor Training Mechanisms

**Policy Computation.** The actor in an actor-critic algorithm estimates a probability distribution (conditional on input state $s_t$) over a set of possible actions $[a_0, a_1, ...a_N]$, with $\sum_i^N p(a_{i,t}) = 1$. Actions that provide larger rewards have larger probabilities, and policies are formulated by choosing the action with the largest $p(a_t)$. In our algorithm however, the actor computes a probability distribution $\rho$ over a range of $N$ possible advantages values conditional on state $s_t$ and action $a_t$. We let $\zeta_i$ denote a certain advantage value, and $p(a_t, \zeta_i) \in \rho$ the probability that given action $a_t$ and state $s_t$, the advantage is $\zeta_i$. In the discrete case, $\rho$ is a probability vector estimated over a range of possible advantages under state-action pair $s_t, a_t$, i.e. $\rho(s_t, a_t) = [p(a_t, \zeta_0), p(a_t, \zeta_1), ..., p(a_t, \zeta_N)]$, where $\sum_i^N p(s_t, a_t) = 1$. To compute for the maximizing policy $\pi$ given a set of actions $[a_0, a_1, ...a_N]$, $\rho(s_t, a_t)$ is multiplied by an advantage value support vector $\mathbf{Z} = [\zeta_0, \zeta_1, ...\zeta_N]$:

$$\pi(s_t) = \underset{a_t}{\text{argmax}} \left[\rho(s_t, a_t) \times \mathbf{Z}\right] \tag{4}$$

**Actor Training.** With probability vector $\rho(s_t, a_t)$, an actor is trained using cross-entropy loss function l. Loss l is conditional on $s_t$, $a_t$, and $\mathbf{Z}$:

$$l = -\sum_t \sum_i^N \log p(s_t, a_t) I[\zeta_i, \hat{A}(s_t, a_t)] \tag{5}$$

In Eq. 5, we use $I[\zeta_i, \hat{A}(s_t, a_t)]$ as the $\Psi$ term in Eq. 1. It is an indicator function that assumes 1 if the computed target advantage function $\hat{A}(s_t, a_t)$ is near the $\zeta_i$ value. Much of our method relies on ensuring that $I[\zeta_i, \hat{A}(s_t, a_t)]$ is dispersed across discrete support $\mathbf{Z}$ to increase entropy. The next subsection describes the computation of the modified advantage targets for encouraging exploration.

## 3.2   Computation of Modified Advantage Targets

To encourage exploration, our method computes modified advantage targets $\hat{A}(s_t, a_t)$ for use in the indicator term $I$ in Eq. 5. The first step is to compute for advantage target regions by sampling points along the direction of the critic value function. We term these points as 'particles'. But since random sampling can increase variances, shortest-path DP is done afterwards for variance control.

**Modified Advantage Target Computation.** From Eq. 5, the loss function l is weighted by $I[\zeta_i, \hat{A}(s_t, a_t)]$. We define the modified advantage target parameter $\hat{A}(s_t, a_t)$ of $I$ as follows. Its standard non-modified form is described in [9]:

$$\hat{A}(s_t, a_t) = r_t + \gamma \tilde{\psi}^*(s_{t+1}) - v(s_t) \tag{6}$$

where $r_t$ is the empirical reward for state $s_t$, $\gamma$ is a discount factor, and $v_t$ is the critic value function from Eq. 2. This modified function differs from the standard-form advantages of [9] through $\tilde{\psi}^*(s_{t+1})$, which is computed using the following steps - comprising the novely of our algorithm from [6,9,11].

**Step 1:** Computation of Advantage Target Regions. Let $\psi_t$ denote on-policy cumulative empirical rewards. $\psi_t$ is computed as follows from $t = 0, 1...T$ (suppressing the underlying states $s_t$):

$$\psi_t = r_t + \gamma \psi_{t+1} \tag{7}$$

We can view $\psi_t$ as the unbiased (discounted) target approximated by the critic [9]. For our algorithm, we transform the single point $\psi_t$ to $K$ particles $\tilde{\psi}_{k,t}$, with $k \in K$. This is done by adding a random parameter $\epsilon_k \in N_\epsilon$ s.t. the $K$ particles are located in a neighborhood with maximum distance $|N_\epsilon|$. The formula is shown in Eq. 8, where $\alpha$ is a coefficient constant for scaling. In this experiment $\epsilon_k$ is drawn from Gaussian distribution $\mathcal{N}$, with $\mu$ as $\mathbb{E}[v_t]$. This creates a neighborhood along the direction of the critic value function $v_t$ and prevent inducing bias that go on an opposite direction from the critic.

$$\tilde{\psi}_{k,t} = \alpha \psi_t + \epsilon_k \tag{8}$$

**Step 2:** Variance Control via Dynamic Programming. The $K$ particles sampled from the Step 1 induce high variances as they are dispersed around the critic value function $v_t$. To control for variances, we implement a shortest-path dynamic programming approach in Eq. 9 that selects particles with shortest inter-state (i.e. $t$ to $t + 1$) distance:

$$D(i, t) = \min_j \left[ |\tilde{\psi}_{i,t} - \tilde{\psi}_{j,t+1}| + D(j, t+1) \right] \tag{9}$$

The cost function in Eq. 9 is the absolute distance $|\tilde{\psi}_{i,t} - \tilde{\psi}_{j,t+1}|$ between particles $i$ and $j$. The shortest path can be computed at each state $t$ by choosing the particle $i$ with minimum $D(i, t)$. We let $\tilde{\psi}_t^*$ denote the particle with the shortest

---

**Algorithm 1.** Proposed Algorithm

---

**Input 1:** State $s_t$
**Input 2:** Reward $r_t$
Intialize episode buffer $[s_t, a_t, r_t, s_{t+1}]$

1: **procedure** ALGORITHM 1
2:     Get current $s_t$ from environment
3:     Feed $s_t$ to actor, and compute score function $\rho(s_t, a_t)$
4:     Follow policy $\pi(s_t) = \text{argmax}_{a_t} \left[ \rho(s_t, a_t) \times \mathbf{Z} \right]$
5:     Retrieve reward $r_t$ and next state $s_{t+1}$
6:     Store sample batch $[s_t, a_t, r_t, s_{t+1}]$ in buffer
7:     **if** train actor == True **then**
8:         Randomly retrieve sample batch $B$ from buffer
9:         Compute critic baseline $v_t$ for each state $s_t$
10:        **for** $r_t$ in $B$ **do**
11:            compute $\psi_t = r_t + \gamma \psi_{t+1}$
12:            **for** no. of particles $i$ in neighborhood $N_\epsilon$ **do**
13:                $\tilde{\psi}_{k,t} = \alpha \psi_t + \epsilon_k$, where $\epsilon_k \in N_\epsilon$, with $N_\epsilon \sim \mathcal{N}(\mu, \sigma)$ and $\mu = v_t$
14:            Initialize dynamic programming (DP) to compute:
15:            $D(i,t) = \min_j \left[ |\tilde{\psi}_{i,t} - \tilde{\psi}_{j,t+1}| + D(j, t+1) \right]$
16:            Get sample $\tilde{\psi}_t^*$ that minimizes $D$
17:            Compute modified advantage targets $\hat{A}(s_t, a_t) = r_t + \gamma \tilde{\psi}_{t+1}^* - v_t$
18:            Train critic using $\psi_t$ and MSE loss for unbiasedness
19:            Train actor using $\hat{A}(s_t, a_t)$ using cross-entropy loss $l$ in Eq. 3.1

---

path at time $t$. Using the notation in Eq. 6, we can also write this as $\tilde{\psi}^*(s_t)$ to make the underlying state $s_t$ for a particle explicit. We can then compute the modified advantage targets in Eq. 6 using the derived $\tilde{\psi}^*(s_t)$. As a summary, we present the algorithm table Alg. 1.

### 3.3    Analysis

We present an analysis in this subsection to show how our algorithm (referred to as **Alg. 1**) encourages exploration. For comparison, we use a baseline algorithm (**Alg. 0**), which is a standard actor-critic that uses the distributional policy computation (as described in Sect. 3.1) but does not perform Steps 1 and 2 in Sect. 3.2. Figure 1 shows our analysis's intuition which is to increase entropy $H$ in the actor's advantage distribution $\rho$, and thereby have more probability to revise policies and deviate from sub-optimal paths. We start our analysis with Lemma 1 which states that Alg. 1 promotes larger entropy in $\rho$ than Alg. 0.

**Lemma 1.** *Suppose that the actor approximates the distribution of its advantage targets under Alg. 1. The entropy (H) of the actor's estimated advantage distribution $\rho$ under Alg. 1 is larger than that of Alg. 0.*

*Proof.* Given that the actor approximates its targets' distribution, if advantage targets in Alg. 1 are dispersed around a neighborhood region $N_\epsilon$ of $A_t$ (following

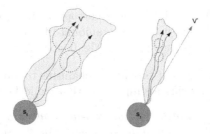

**Fig. 1.** A larger neighborhood (blue circles) around an advantage path (black lines) increases entropy of $\rho$ (gray region), and has higher probability to discover optimal path (red line). This figure will be experimentally verified in the results. (Color figure online)

Step 1 of Sect. 3.2), then its $\rho$ has higher entropy $H$ compared to another $\tilde{\rho}$ whose targets are fixed at a point $A_t$ (as in Alg. 0 which does not perform Step 1).

From Lemma 1, we state Proposition 1 which relates $\rho$'s entropy level to the ease of exploration in Alg. 1's actor. We then state a central Corollary 1 regarding the exploration benefits of Alg. 1. We use a simple case with 2 actions in Proposition 1's proof, but it could be generalized to more actions over a larger state space.

**Proposition 1.** *Given probabilities $p \in \rho$, with $\rho$ as the estimated advantage distribution, let $\delta$ refer to differences among values of $p$ as distribution $\rho$ varies during training. As $\rho$'s entropy increases, the amount of shifts $\delta$ in $p$ needed for an actor to perform actions outside past policy $\pi$ becomes smaller.*

*Proof.* Without loss of generality, let $\rho_1(a_1)$ and $\rho_2(a_1)$ be advantage distributions over 2 possible advantage values $\zeta_1$ and $\zeta_2$ conditional on state $s_t$ and action $a_1$, where $\zeta_1 > \zeta_2$. The environment allows another action $a_2$. Both $\rho_1(a_1)$ and $\rho_2(a_1)$ favor $\zeta_1$, where $p(\zeta_1, \rho_i) > p(\zeta_2, \rho_i)$ for $i = 1, 2$, and $p(\zeta_1, \rho_i) + p(\zeta_2, \rho_i) = 1$. Again, without loss of generality, let $\rho(a_2)$ for action $a_2$ have equal probabilities for both $\zeta_1$ and $\zeta_2$. Given this, the past policy $\pi_t$ always favors action $a_1$ since it has higher probability to pick $\zeta_1$ compared to $a_2$ (under both $\rho_1(a_1)$ and $\rho_2(a_2)$). Let $\rho_1(a_1)$'s distribution have lower entropy, i.e. $H(\rho_1) < H(\rho_2)$ s.t. $p(\zeta_1, \rho_2) = p(\zeta_1, \rho_1) - \lambda$ for some $\lambda > 0$ (the $a_1$ term is suppressed in $\rho_1$ and $\rho_2$). For an actor to explore, it has to deviate from $\pi_t$ at time $t + 1$ and choose action $a_2$. To pick $a_2$, the new probabilities $\tilde{p}$ under $\rho_1(a_1)$ and $\rho_2(a_1)$ have to favor $\zeta_2$, i.e. $\tilde{p}(\zeta_1, \rho_i) < \tilde{p}(\zeta_2, \rho_i)$. The shift needed in the old $p$ is $> \delta_i$, where $\delta_i = p(\zeta_1, \rho_i) - p(\zeta_2, \rho_i)$ for $i = 1, 2$, so that $(\rho_i(a_1) \times \mathbf{Z}) < (\rho(a_2) \times \mathbf{Z})$ given the new $\tilde{p}$. Let $\theta = p(\zeta_1, \rho_1)$, and w.l.o.g. set $\theta > 1 - \theta$. It follows that $\delta_1 = 2\theta - 1$, while $\delta_2 = 2\theta - 1 - 2\lambda$. Therefore $\delta_1 > \delta_2$ since $\lambda > 0$, and the shift $\delta_i$ needed for the actor to do action $a_2$ is less under the higher entropy $\rho_2(a_1)$ than in $\rho_1(a_1)$.

**Corollary 1.** *Alg. 1 encourages more exploration than Alg. 0.*

*Proof.* From Lemma 1, Alg. 1 has more entropy $H$ in its estimated advantage distribution $\rho$. Using Proposition 1, the amount of shifts $\delta$ in $p \in \rho$ needed for exploration under Alg. 1 is smaller than Alg. 0 since Alg. 1 has larger $H$.

Proposition 1 however does not provide mechanisms to induce exploration. This is done in Proposition 2 using a simple case that could be generalized.

**Proposition 2.** *Let $A_1$ and $A_2$ be standard-form advantage targets $\in \mathbb{R}$ for two actions $(a_1, a_2)$ conditional on state $s_t$. Under Alg. 1, suppose that given any sequence of advantage targets, the actor approximates the targets' probability distribution during training. Let $d = A_1 + \mu_1 - A_2 - \mu_2$, with $\mu$ as the center of $\epsilon$ (Eq. 8). If the modified advantage targets in Alg. 1 are used with $|N_\epsilon| > |d/2|$, $\exists$ a positive probability $\hat{p} > 0$ s.t. action $a_1$ is chosen by the actor despite $A_1 < A_2$.*

*Proof.* We recall from Eq. 2 that standard-form advantage target $A_i$ is computed as $A_i = r_t + \gamma v_{t+1} - v_t$, where $v_t$ is the critic value function baseline. W.l.o.g., given two sample advantages $i = 1, 2$, we can write down $A_1$ and $A_2$ as:

$$A_1 = r_t + \gamma v_{t+1}^1 - v_t \quad A_2 = r_t + \gamma v_{t+1}^2 - v_t \text{ with } (v_{t+1}^1 < v_{t+1}^2) \text{ since } A_1 < A_2 \tag{10}$$

Let $v_{t+1}^1$ be the expected value function at $t + 1$ if $a_1$ is performed at state $s_t$ and similarly, let $v_{t+1}^2$ be the expected value function at $t + 1$ after choosing $a_2$. Since the baseline $v_t$ is similar for both $A_1$ and $A_2$, their difference can be reduced to differences $v_{t+1}^2 - v_{t+1}^1$. Since $v_{t+1}^1 < v_{t+1}^2$, the standard-form advantage target of the past policy always dictates the actor to choose $a_2$. But with Alg. 1, advantage targets are modified and incorporate an added $\epsilon$ term brought about by Gaussian sampling of particles around neighborhood $N_\epsilon$ with center $v_t^i$ s.t. $\mathcal{N}(\mu_i) = v_t^i$, for $i = 1, 2$ and the maximum distance $|\hat{\epsilon}|$ from the center is $|N_\epsilon|$. Let $\hat{\epsilon}$ denote the distance of $\epsilon$ from its center $\mu$. It follows that $|\epsilon - \mu| = |\hat{\epsilon}| \leq |N_\epsilon|$. We can now write modified advantage targets $f(a_1)$ and $f(a_2)$ under Alg. 1 as:

$$f(a_1) = r_t + \gamma[v_{t+1}^1 + \epsilon_1] - v_t \quad f(a_2) = r_t + \gamma[v_{t+1}^2 + \epsilon_2] - v_t$$

Since particles are randomly sampled and $|N_\epsilon| > |d/2|$ for both $\epsilon_1$ and $\epsilon_2$, $\exists$ a positive probability $p > 0$ s.t. $\hat{\epsilon}_2 < d/2$ and $\hat{\epsilon}_1 > -d/2$. In this case:

$$\begin{aligned}
f(a_1) - f(a_2) &= v_{t+1}^1 - v_{t+1}^2 + \epsilon_1 - \epsilon_2 \\
&= A_1 - A_2 + \mu_1 - \mu_2 + (\epsilon_1 - \mu_1) - (\epsilon_2 - \mu_2) \\
&= d + (\epsilon_1 - \mu_1) - (\epsilon_2 - \mu_2) = (d/2 + \hat{\epsilon}_1) + (d/2 - \hat{\epsilon}_2) \\
&> 0 \qquad\qquad\qquad\qquad (\text{since } \hat{\epsilon}_2 < d/2 \text{ and } \hat{\epsilon}_1 > -d/2)
\end{aligned}$$

Hence at certain timesteps, $f_1 > f_2$ occurs with positive probability $p > 0$ despite $A_1 < A_2$. Using this fact, simulate a training schedule of $N$ steps, holding $v_{t+1}^1$, $v_{t+1}^2$, and $v_t$ fixed for state $s_t$ - without loss of generality. We then construct a sequence of advantage targets $F = [f^0(a_t), ... f^N(a_t)]$, conditional on state $s_t$. $F$ is constructed s.t. for each $f^i(a_t)$ for $i = 0, 1..N$, Alg. 1 provides a higher advantage target for on-policy action $a_1$ than $a_2$, i.e. $f(a_1) > f(a_2)$. Since $f(a_1) > f(a_2)$ occurs with probability $p > 0$ at each timestep, the probability of $F$ is $> 0$ as well. Let $\hat{p}$ denote the probability of $F$. Given that the actor is able to approximate the distribution of its targets with $\rho$, under $F$, we get a current policy $\pi$ with probability $\hat{p} > 0$ s.t. $\rho(s_t, a_1) \times \mathbf{Z} > \rho(s_t, a_2) \times \mathbf{Z}$. With this new policy, the maximizing action is $a_1$ despite $A_1 < A_2$ in the past policy.

Corollaries 2 and 3 follow from Proposition 2, where the neighborhood region $N_\epsilon$ under Alg. 1 has to be neither too large nor too small. If too large, the actor's actions behaves randomly. But if too small, the actor does not explore.

**Corollary 2.** *If $|N_\epsilon| >> |d|$, the actor's policy is close to random.*

*Proof.* Recall that for target $i$ under Alg. 1, $\hat{\epsilon}_i = \epsilon_i - \mu_i$, i.e. $\hat{\epsilon}_i$ is the distance of $\epsilon_i$ from its center $\mu_i$. It follows that $\mathbb{E}[\hat{\epsilon}_i] = 0$ and $|\hat{\epsilon}_i| < |N_\epsilon|$. W.l.o.g., let $A_1$ and $A_2$ denote standard advantage targets (Eq. 2) and let $d$ be the same as in Proposition 2. Recall from the previous example that Alg. 1's actor chooses $a_1$ if $f(a_1) > f(a_2)$ and o.w. chooses $a_2$. The discrepancy between $f(a_1)$ and $f(a_2)$ is: $f(a_1) - f(a_2) = d + \hat{\epsilon}_1 - \hat{\epsilon}_2$. Since $d$ is constant, and $\hat{\epsilon}_i$ are r.v. with $\mu_i = 0$, $f(a_1) - f(a_2) \sim \mathcal{N}(d, \sigma)$, with $\sigma = f(|N_\epsilon|)$, or that its standard deviation is a function of the radius of the neighborhood region. Now, let $P(a_1)$ denote the probability of choosing $a_1$ (given $d$ and $|N_\epsilon|$), and similar for $P(a_2)$. W.l.o.g, let $A_1 < A_2$ s.t. $d$ is negative. Under Alg. 1, we see that $P(a_1)$ is equal to the probability $P[f(a_1) - f(a_2)] > 0$, i.e. $P(a_1) = F([0, \infty])$, with $F$ being the CDF of $\mathcal{N}(d, \sigma)$. Conversely, $P(a_2) = F([-\infty, 0]) = F([-\infty, d]) + F([d, 0])$. We can apply the proof of Lemma 1 analogously here to see that as the neighborhood radius $|N_\epsilon|$ increases relative to $|d|$, the entropy of $\mathcal{N}(d, \sigma)$ likewise increases s.t. both $F([0, \infty])$ and $F([-\infty, d])$ approach 0.5 in the limit, and $F([d, 0])$ decreases -under a continuous $F$. In other words, as $|N_\epsilon|$ increases, the contribution of $d$ through $F([d, 0])$ to $P(a_1)$ and $P(a_2)$ diminishes, and $P(a_1) = P(a_2) = 0.5$. In this case, the actor chooses $a_1$ and $a_2$ both with close to probability 0.5.

**Corollary 3.** *If $|N_\epsilon| << |d|$, the actor's policy remains fixed. On the other hand, given sufficient $|N_\epsilon|$, if $|d|$ is small, the probability $\hat{p}$ for the actor under Alg. 1 to perform actions that deviate from past policies is larger.*

*Proof.* To prove the first statement, we use the example in Proposition 2, where $f(a_1) - f(a_2) = (d/2 + \hat{\epsilon}_1) + (d/2 - \hat{\epsilon}_2)$. If $|\hat{\epsilon}_1| < |d/2|$ and $|\hat{\epsilon}_2| < |d/2|$, then either $f(a_1) - f(a_2) > 0$ or $f(a_1) - f(a_2) \leq 0$ for all $t$, depending on whether $f_1 > f_2$ or $f_1 \leq f_2$ initially. For the second statement, using the same example in Proposition 2, with standard-form advantage targets $A_1 < A_2$, $a_1$ is chosen over $a_2$ iff $f(a_1) - f(a_2) > 0$ or $(d/2 + \hat{\epsilon}_1) + (d/2 - \hat{\epsilon}_2) > 0$. W.l.o.g, let $\hat{\epsilon}_2 = 0$ and $\epsilon_1$ be drawn from $\mathcal{N}(\mu_1, \sigma)$ with $\hat{\mu}_1 = 0$, and $\sigma = f(|N_\epsilon|)$ (where $|N_\epsilon|$ is sufficient s.t. the cases in Corollary 1 and 2 do not apply). In this case, $f(a_1) - f(a_2) > 0$ iff $\hat{\epsilon}_1 > -d/2$. Since $A_1 < A_2$, we have $d$ as negative, and $\hat{\epsilon}_1 > -d/2$ acquires the highest probability if $-d/2$ is a very small positive amount. The same applies if

$A_1 > A_2$, and $a_2$ is chosen. (In our experiments, we set $\mu = v_t \neq 0$ to prevent bias away from critic. But this proof can still be applied since it only takes into account $|d|$ in comparison with $|N_\epsilon|$.).

We now end our analysis with the following useful Corollary, which says that in 'uncertain' states where advantages have small discrepancies, i.e. $|d|$ is small, then Alg. 1 is bound to explore more than Alg. 0.

**Corollary 4.** *In states with small $|d|$ (i.e. advantages among actions are close), then Alg. 0's policy is fixed, while Alg. 1 explores more with sufficient $|N_\epsilon|$.*

*Proof.* Using the same example in Proposition 2, let standard-form advantage targets have the ordering $A_1 < A_2$, with small $|d|$. Here, Alg. 0's actor always follows the ordering $A_1 < A_2$ and performs $a_2$ in state $s_t$. But from Corollary 3, Alg. 1's probability $\hat{p}$ to explore and do $a_1$ is larger since $|d|$ is small relative to a sufficient $|N_\epsilon|$ s.t. cases in Corollaries 1 and 2 do not apply.

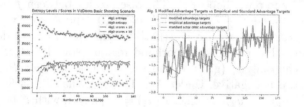

**Fig. 2.** (a) Entropy between Alg. 1 (blue crosses) and Alg. 0 (red crosses) in Basic Shooting relative to scores (blue and red solid line). (b) Alg. 1 modified advantage targets (blue line) are expanded around $N_\epsilon$ (red circles) of Alg. 0 targets (green line) (Color figure online)

## 4   Experiments

For our experiments, we use the VizDoom AI benchmarks [4] and Atari games [8]. Scenarios for VizDoom are (1) Easy Basic Shooting, (2) Medium Predict Position, and (3) Hard Health Gathering. Health Gathering has hard difficulty since there is no reward-shaping and the agent has to discover that it needs health kits to stay alive. We set (1) and (2) with 1.0 reward for kill, and $-0.01$ step penalty. For (3), we give reward of 1 per step and $-1.0$ for dying. For Atari, we use Pong and Asterix scenarios following its reward schemes. We first show experimental support for Corollary 1, by comparing entropy $H$ of Alg. 1 vs Alg. 0 in the VizDoom Basic Shooting Scenario. As mentioned in Sect. 3.1, Alg. 0 is equivalent to a standard actor-critic algorithm which estimates advantage probability distributions but without particle sampling + variance-control DP. From Fig. 2a, Alg. 1 shows higher entropy than Alg. 0. Alg. 1's $H$ stayed at the 40k level after $7M$ frames while Alg. 0's entropy lowered. From Fig. 2, it is also

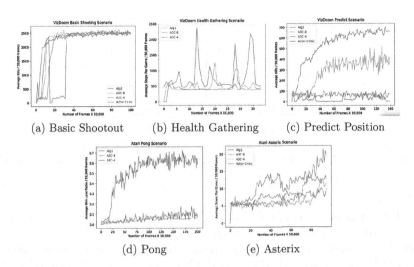

(a) Basic Shootout          (b) Health Gathering          (c) Predict Position

(d) Pong          (e) Asterix

**Fig. 3.** Average reward of models across frames given the same number of backprop-agation steps of Alg. 1 1 with 1 $A3C$ actor. This is to be fair with Alg. 1, since $A3C$ performs 8x or 4x more backpropagation given the same # of frames. (Color figure online)

apparent that continued exploration due to high $H$ results in higher performance (Alg. 1's reward (solid blue line) stayed above Alg. 0 (red line)).

For experiments on Atari and VizDoom, we set the $\alpha$ parameter as 0.8, and the max neighborhood distance $|N_\epsilon|$ as 0.2 for Basic Shooting, Health Gathering and Predict Position. For Pong and Asterix, we set $|N_\epsilon|$ as 0.5. We set $\mu$ of $N_\epsilon$ in Eq. 7 as the critic value function ($v_t$ in Eq. 6). For Atari games, we set episode buffer size as 1024, using e-greedy that tapers off from 1.0 to 0.01 after $1.2M$ steps for Atari and $2K$ steps for VizDoom. We compare our algorithm (referred to as Alg. 1) with $A3C$ models (4 and 8 actors) and with standard actor-critic. Our $A3C$ uses a GAE(1) base advantage from [9] which has nice bias/variance properties. For all models, we use deep network trunks in [3]. In Fig. 3, we compare Alg. 1 with $A3C - 8$ and $A3C - 4$ by taking results in Alg. 1 every 50k frames, and with $A3C - 8$ and $A3C - 4$ every $(50/8)$k and $(50/4)$k frames respectively. This is to be fair in comparison with Alg. 1 since $A3C - 8$ and $A3C - 4$ will have $8x$ or $4x$ more backpropagation given the same number of frames encountered by Alg. 1. We also had some experiments on Demon Attack, but variability of the results ($A3C$ has higher results on some steps than Alg. 1, but it also dips fast) makes it hard for comparison and we did not include it.

In our experiments, setting the value of $N_\epsilon$ is crucial since $N_\epsilon = 0.2$ does not provide as good a performance as $N_\epsilon = 0.5$. From the results in Fig. 3, the pro-posed algorithm Alg. 1 (blue line) converges faster than either $A3C - 8$, $A3C - 4$, or the standard actor-critic algorithm (given the same number of backpropaga-tion steps). In Fig. 3a, Alg. 1 reaches the 2000 reward mark first, but eventually other models converged as well. In the case of Asterix in Fig. 3e, $A3C - 4$ outper-

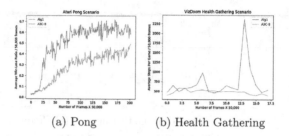

(a) Pong                        (b) Health Gathering

**Fig. 4.** Alg. 1 vs $A3C - 8$ with 8x more backpropagation for $A3C$ (all actors combined)

formed Alg. 1 after several steps, although Alg. 1 reached a higher reward level first. But Asterix is an easier scenario and converges quickly even with Q-learning [7]. For Fig. 3b and d, we do not show Actor-Critic since convergence is too slow. In more challenging scenarios such as Predict, Pong, or Health Gathering, the faster convergence of Alg. 1 is more noticeable. In the case of Atari Pong where there is high variability, the learning rate of $A3C$ algorithms is very slow relative to Alg. 1. In fact, as shown in Fig. 4, Alg. 1 maintains a higher performance (conditional on $N_\epsilon = 0.5$) than $A3C - 8$ even with much less backpropagation. The same can be seen for Health gathering where $A3C$ does not converge quickly.

## 5   Conclusion

We presented a policy gradient algorithm that performs automatic exploration. We use modified advantage functions formed from neighborhood particle sampling with a variance-controlled DP step to maintain high entropy in actor estimates of advantage probability distributions. From experimental tests, our algorithm manages to keep high entropy levels compared to other standard algorithms that do not use particle sampling. Implementation on VizDoom and Atari AI benchmarks showed that given the same number of backpropagation steps, our algorithm manages to outperform $A3C$ or standard actor-critic algorithms.

## References

1. Anschel, O., Baram, N., Shimkin, N.: Averaged-DQN: variance reduction and stabilization for deep reinforcement learning. In: International Conference on Machine Learning, pp. 176–185 (2017)
2. Gu, S., Lillicrap, T., Turner, R.E., Ghahramani, Z., Schölkopf, B., Levine, S.: Interpolated policy gradient: merging on-policy and off-policy gradient estimation for deep reinforcement learning. In: Advances in Neural Information Processing Systems 30
3. Hessel, M., et al.: Rainbow: combining improvements in deep reinforcement learning. arXiv preprint arXiv:1710.02298 (2017)

4. Kempka, M., Wydmuch, M., Runc, G., Toczek, J., Jaśkowski, W.: ViZDoom: a doom-based AI research platform for visual reinforcement learning. In: 2016 IEEE Conference on Computational Intelligence and Games (CIG), pp. 1–8. IEEE (2016)

5. Konda, V.R., Tsitsiklis, J.N.: Actor-critic algorithms. In: Advances in Neural Information Processing Systems, pp. 1008–1014 (2000)

6. Mnih, V., et al.: Asynchronous methods for deep reinforcement learning. In: International Conference on Machine Learning, pp. 1928–1937 (2016)

7. Mnih, V., et al.: Playing Atari with deep reinforcement learning. arXiv preprint arXiv:1312.5602 (2013)

8. Mnih, V., et al.: Human-level control through deep reinforcement learning. Nature 518(7540), 529–533 (2015)

9. Schulman, J., Moritz, P., Levine, S., Jordan, M., Abbeel, P.: High-dimensional continuous control using generalized advantage estimation. In: Proceedings of the International Conference on Learning Representations (ICLR) (2016)

10. Schulze, C., Schulze, M.: ViZDoom: DRQN with prioritized experience replay, double-Q learning, & snapshot ensembling. arXiv preprint arXiv:1801.01000 (2018)

11. Sutton, R.S., Barto, A.G.: Reinforcement Learning: An Introduction, vol. 1. MIT Press, Cambridge (1998)

12. Wang, Z., Schaul, T., Hessel, M., Hasselt, H., Lanctot, M., Freitas, N.: Dueling network architectures for deep reinforcement learning. In: Proceedings of The 33rd International Conference on Machine Learning, pp. 1995–2003 (2016)

# MCP Based Noise Resistant Algorithm for Training RBF Networks and Selecting Centers

Hao Wang[1], Andrew Chi Sing Leung[1(✉)], and John Sum[2]

[1] Department of Electronic Engineering, City University of Hong Kong, Kowloon Tong, Hong Kong
wanghaocityu@gmail.com, eeleungc@cityu.edu.hk
[2] Institute of Technology Management, National Chung Hsing University, Taichung, Taiwan
pfsum@dragon.nchu.edu.tw

**Abstract.** In the implementation of a neural network, some imperfect issues, such as precision error and thermal noise, always exist. They can be modeled as multiplicative noise. This paper studies the problem of training RBF network and selecting centers under multiplicative noise. We devise a noise resistant training algorithm based on the alternating direction method of multipliers (ADMM) framework and the minimax concave penalty (MCP) function. Our algorithm first uses all training samples to create the RBF nodes. Afterwards, we derive the training objective function that can tolerate to the existence of noise. Finally, we add a MCP term to the objective function. We then apply the ADMM framework to minimize the modified objective function. During training, the MCP term has an ability to make some unimportant RBF weights to zero. Hence training and RBF node selection can be done at the same time. The proposed algorithm is called the ADMM-MCP algorithm. Also, we present the convergent properties of the ADMM-MCP algorithm. From the simulation result, the ADMM-MCP algorithm is better than many other RBF training algorithms under weight/node noise situation.

**Keywords:** RBF · Center selection · ADMM · MCP
Multiplicative noise

## 1 Introduction

RBF node selection, or saying center selection, is an important issue in the training process of a radial basis function (RBF) neural network. Intuitively, we can directly use all training samples to create RBF nodes [1], or we can randomly select some training input vectors as the RBF centers [2]. These two schemes are simple, but the performance of them is not so good. Hence many other RBF center selection algorithms have been proposed, for instance, clustering algorithm and orthogonal least squares (OLS) approach [3]. However, the aforementioned

© Springer Nature Switzerland AG 2018
L. Cheng et al. (Eds.): ICONIP 2018, LNCS 11302, pp. 668–679, 2018.
https://doi.org/10.1007/978-3-030-04179-3_59

algorithms are only valid for the noise-free case. When we use hardware to realize a trained network, precision errors exist. Those precision errors can be described by the multiplicative noise model [4,5]. In the past two decades, several noise resistant RBF training algorithms have been proposed [6,7]. But, they usually need a separate process to determine the RBF centers first.

This paper proposes a noise resistant algorithm that is able to select and to train an RBF network at the same time. Since the minimax concave penalty (MCP) function [8], which is commonly used in sparse representation, has a good ability to control the number of non-zero elements, we add a MCP function into the training objective to remove the unnecessary RBF nodes. The alternating direction method of multipliers (ADMM) framework [9] and local linear approximation (LLA) method [10] are then used for minimizing the modified objective function. Since the behaviour of the MCP term is similar to an $\ell_0$-norm regularizer, the MCP term can effectively make some unnecessary RBF weights to zero. The proposed algorithm is called ADMM-MCP. In addition, we present the convergent property of the ADMM-MCP.

The rest of paper is organized as follows. Section 2 presents the background of RBF networks and ADMM. In Sect. 3, the proposed algorithm is developed, and its local convergence is proved. Simulation results are shown in Sect. 4. Finally, the conclusion is drawn in Sect. 5.

## 2   Background

### 2.1   RBF Networks Under Multiplicative Noise

This paper considers to use RBF networks for nonlinear regression. We are given a training set: $\mathcal{D} = \left\{ (\boldsymbol{x}_i, y_i) : \boldsymbol{x}_i \in \mathbb{R}^{K_1}, y_i \in \mathbb{R}, i = 1, \cdots, N \right\}$, where $N$ is the number of samples, and $\boldsymbol{x}_i$ and $y_i$ are the input and the desired output of the $i$-th sample. Similarly, there is a test set: $\mathcal{D}' = \left\{ (\boldsymbol{x}'_{i'}, y'_{i'}) : \boldsymbol{x}'_{i'} \in \mathbb{R}^{K_1}, y'_{i'} \in \mathbb{R}, i' = 1, \cdots, N' \right\}$, where $N'$ is the number of samples in the test set. The output of an RBF network with $M$ RBF nodes is given by

$$f(\mathbf{x}) = \sum_{j=1}^{M} w_j a_j(\boldsymbol{x}),  \tag{1}$$

where $a_j(\boldsymbol{x}) = \exp\left(-\|\boldsymbol{x} - \boldsymbol{c}_j\|_2^2 / s\right)$ is the output of the $j$-th RBF node, $w_j$ is the $j$-th RBF weight, $\boldsymbol{c}_j$ is the RBF center of the $j$-th RBF, and $s$ is the RBF width.

Center selection is an important issue in the construction of an RBF network. Usually, the RBF centers are selected form the input dataset $\{\boldsymbol{x}_1, \ldots, \boldsymbol{x}_N\}$. Suppose that we use all the training samples to construct an RBF network, i.e., $\boldsymbol{c}_j = \boldsymbol{x}_j, \forall j = 1, \cdots, N$. The network output is then given by $f(\boldsymbol{x}) = \sum_{j=1}^{N} w_j a_j(\boldsymbol{x})$. *Note that our proposed algorithm can automatically set some unnecessary weights to zero. Hence some unnecessary nodes are removed.*

For the noise-free case, the training set error is expressed as

$$\mathcal{E}_t = \frac{1}{N} \sum_{i=1}^{N} \left( y_i - \sum_{j=1}^{N} w_j a_j(\boldsymbol{x}_i) \right)^2 = \frac{1}{N} \|\boldsymbol{y} - \boldsymbol{A}\boldsymbol{w}\|_2^2, \tag{2}$$

where $\boldsymbol{w} = [w_1, \cdots, w_M]^{\mathrm{T}}$, $\boldsymbol{y} = [y_1, \cdots, y_N]^{\mathrm{T}}$, $\boldsymbol{A}$ is a $N \times N$ matrix, and the $(i,j)$ entry of $\boldsymbol{A}$ is given by $[\boldsymbol{A}]_{i,j} = a_j(\boldsymbol{x}_i) = \exp(-\|\boldsymbol{x}_i - \boldsymbol{c}_j\|^2 / s)$.

When we use the finite precision technology to implement RBF weights or RBF nodes, multiplicative noise are introduced [4,5]. Besides, when we use analog technology, the uncertainty of the value of an analog component is usually proportional to the nominal value of the analog component.

For multiplicative weight noise, a noisy weight is modeled as $\tilde{w}_j = w_j + b_j w_j$, where "$b_j w_j$" denotes the multiplicative noise of the $j$th weight, and $b_j$'s are independent and identically distributed (i.i.d.) random variables with mean equal to zero and variance equal to $\sigma_b^2$. For multiplicative node noise, the output of an RBF node is modeled as $\tilde{a}_j(\boldsymbol{x}) = a_j(\boldsymbol{x}) + b_j a_j(\boldsymbol{x})$, "$b_j a_j(\mathbf{x})$" denotes the multiplicative noise of the $j$-th node. From the properties of $b_j$'s (i.i.d. and zero mean), we deduce that $\langle b_j \rangle = 0$, $\langle b_j^2 \rangle = \sigma_b^2$, $\langle b_j b_{j'} \rangle = 0$, $\forall j \neq j'$, where $\langle \cdot \rangle$ denotes the expectation operator.

For a weight/node noise pattern, the training set error can be expressed as

$$\tilde{\mathcal{E}}_t = \frac{1}{N} \|\boldsymbol{y} - \boldsymbol{A}\tilde{\boldsymbol{w}}\|_2^2 \text{ (weight noise), or } \tilde{\mathcal{E}}_t = \frac{1}{N} \left\|\boldsymbol{y} - \tilde{\boldsymbol{A}}\boldsymbol{w}\right\|_2^2 \text{ (node noise),} \tag{3}$$

where $[\tilde{\boldsymbol{A}}]_{i,j} = \tilde{a}_j(\boldsymbol{x}_i) = a_j(\boldsymbol{x}_i) + b_j a_j(\boldsymbol{x}_i)$ is the element in the $i$th row and $j$th column of the matrix $\tilde{\boldsymbol{A}}$, and $\tilde{\boldsymbol{w}} = [\tilde{w}_1, \cdots, \tilde{w}_N]^{\mathrm{T}}$. Under weight noise or node noise, the training set error expressions have the identical form, given by

$$\tilde{\mathcal{E}}_t = \frac{1}{N} \left[ \sum_{i=1}^{N} y_i^2 - 2y_i \sum_{j=1}^{N} w_j a_j(\boldsymbol{x}_i)(1+b_j) + \sum_{j=1}^{N} \sum_{j'=j}^{N} w_j w_{j'}(1+b_j)(1+b_{j'}) a_j(\boldsymbol{x}_i) a_{j'}(\boldsymbol{x}_i) \right]. \tag{4}$$

From the statistics of $b_j$'s, we can further deduce the average training set error of noisy networks, given by

$$\psi(\boldsymbol{w}) := \overline{\tilde{\mathcal{E}}}_t = \frac{1}{N} \|\boldsymbol{y} - \boldsymbol{A}\boldsymbol{w}\|_2^2 + \boldsymbol{w}^{\mathrm{T}} \boldsymbol{R} \boldsymbol{w}, \text{ where } \boldsymbol{R} = \frac{1}{N} \sigma_b^2 \mathrm{diag}\left(\boldsymbol{A}^{\mathrm{T}} \boldsymbol{A}\right). \tag{5}$$

*It should be noticed that under weight noise or node noise we obtain the same the average training set error formula.*

## 2.2 ADMM

The ADMM framework is an iterative approach for solving optimization problems [9]. Consider the following constrained optimization problem:

$$\min_{\boldsymbol{w},\boldsymbol{u}} : \psi(\boldsymbol{w}) + g(\boldsymbol{u}) \quad s.t. \; \boldsymbol{C}\boldsymbol{w} + \boldsymbol{D}\boldsymbol{u} = \boldsymbol{v}, \tag{6}$$

where $\boldsymbol{w}$ and $\boldsymbol{u}$ are the decision variable vectors. First we construct its augmented Lagrangian function

$$L\left(\boldsymbol{w},\boldsymbol{u},\boldsymbol{\tau}\right) = \psi\left(\boldsymbol{w}\right) + g\left(\boldsymbol{u}\right) + \boldsymbol{\tau}^{\mathrm{T}}\left(\boldsymbol{Cw} + \boldsymbol{Du} - \boldsymbol{v}\right) + \frac{\rho}{2}\left\|\boldsymbol{Cw} + \boldsymbol{Du} - \boldsymbol{v}\right\|_2^2, \quad (7)$$

where $\rho > 0$ is a trade-off parameter, and $\boldsymbol{\tau}$ is a vector that contains all the Lagrange multipliers. With the augmented Lagrangian function, the updating equations for the decision variables and Lagrange multipliers are given by

$$\boldsymbol{u}^{k+1} = \underset{\boldsymbol{u}}{\operatorname{argmin}}\, L(\boldsymbol{w}^k, \boldsymbol{u}, \boldsymbol{\tau}^k), \quad (8)$$

$$\boldsymbol{w}^{k+1} = \underset{\boldsymbol{w}}{\operatorname{argmin}}\, L(\boldsymbol{w}, \boldsymbol{u}^{k+1}, \boldsymbol{\tau}^k), \quad (9)$$

$$\boldsymbol{\tau}^{k+1} = \boldsymbol{\tau}^k + \rho\left(\boldsymbol{Cw}^{k+1} + \boldsymbol{Du}^{k+1} - \boldsymbol{v}\right). \quad (10)$$

## 3    Development of ADMM-MCP

### 3.1    ADMM-MCP Algorithm

From the concept of sparse approximation [8,10], we introduce an additional MCP function into the objective function stated in (5). With this penalty term, we can remove some unnecessary centers during training. The new objective function is given by

$$Q(\boldsymbol{w}, \boldsymbol{\lambda}) = \frac{1}{N}\|\boldsymbol{y} - \boldsymbol{Aw}\|_2^2 + \boldsymbol{w}^{\mathrm{T}}\boldsymbol{Rw} + P_{\lambda,\gamma}(\boldsymbol{w}), \quad (11)$$

where $P_{\lambda,\gamma}(\boldsymbol{w}) = \sum_{i=1}^n P_{\lambda,\gamma}(w_i)$ $(\lambda > 0, \gamma > 0)$, and

$$P_{\lambda,\gamma}(w_i) = \begin{cases} \lambda|w_i| - \frac{w_i^2}{2\gamma}, & if \ |w_i| \le \gamma\lambda, \\ \frac{1}{2}\gamma\lambda^2, & if \ |w_i| > \gamma\lambda. \end{cases} \quad (12)$$

The shape of the MCP function with different parameter settings is given by Fig. 1(a). It can be seen that the MCP function is an approximation of the $l_0$-norm penalty term. The parameters $\lambda$ and $\gamma$ are used for tuning the shape and magnitude of the MCP function.

We can use the ADMM framework to solve the problem stated in (11). First we introduce a dummy variable $\boldsymbol{u} = [u_1, \ldots, u_N]^{\mathrm{T}}$ and transform the unconstrained problem into the standard ADMM form:

$$\underset{\boldsymbol{w},\boldsymbol{u}}{\min}\ \psi(\boldsymbol{w}) + P_{\lambda,\gamma}(\boldsymbol{u}), \quad s.t.\ \boldsymbol{u} = \boldsymbol{w}, \quad (13)$$

where $\psi(\boldsymbol{w})$ is given by (5). We then construct its augmented Lagrangian function:

$$L(\boldsymbol{w}, \boldsymbol{u}, \boldsymbol{\tau}) = \psi(\boldsymbol{w}) + P_{\lambda,\gamma}(\boldsymbol{u}) + \boldsymbol{\tau}^{\mathrm{T}}(\boldsymbol{u} - \boldsymbol{w}) + \frac{\rho}{2}\|\boldsymbol{w} - \boldsymbol{u}\|_2^2. \quad (14)$$

According to (8)–(10), the ADMM iteration is given by

$$\boldsymbol{u}^{k+1} = \operatorname*{argmin}_{\boldsymbol{u}} L(\boldsymbol{w}^k, \boldsymbol{u}, \boldsymbol{\tau}^k) = \operatorname*{argmin}_{\boldsymbol{u}} P_{\lambda,\gamma}(\boldsymbol{u}) + \frac{\rho}{2} \left\| \boldsymbol{w}^k - \boldsymbol{u} - \frac{1}{\rho}\boldsymbol{\tau}^k \right\|_2^2, \tag{15}$$

$$\boldsymbol{w}^{k+1} = \arg\min_{\boldsymbol{w}} L(\boldsymbol{w}, \boldsymbol{u}^{k+1}, \boldsymbol{\tau}^k) = \arg\min_{\boldsymbol{w}} \psi(\boldsymbol{w}) + \frac{\rho}{2} \left\| \boldsymbol{w} - \boldsymbol{u}^{k+1} - \frac{1}{\rho}\boldsymbol{\tau}^k \right\|_2^2$$

$$= \left[ \frac{2}{N}\boldsymbol{A}^{\mathrm{T}}\boldsymbol{A} + 2\boldsymbol{R} + \rho\boldsymbol{I} \right]^{-1} \left[ \frac{2}{N}\boldsymbol{A}^{\mathrm{T}}\boldsymbol{y} + \rho\boldsymbol{u}^{k+1} + \boldsymbol{\tau}^k \right], \tag{16}$$

$$\boldsymbol{\tau}^{k+1} = \boldsymbol{\tau}^k + \rho\left( \boldsymbol{u}^{k+1} - \boldsymbol{w}^{k+1} \right), \tag{17}$$

where $\boldsymbol{I}$ is an identity matrix. To solve (15), we utilize the LLA concept [10] and the update of $\boldsymbol{u}^{k+1} = [u_1^{k+1}, \ldots, u_N^{k+1}]^{\mathrm{T}}$ is given by

$$u_i^{k+1} = \begin{cases} \dfrac{S\left(w_i^k - v_i^k/\rho, \lambda\right)}{1 - 1/\gamma}, & if \ |w_i^k - v_i^k/\rho| \leq \gamma\lambda, \\[2mm] w_i^k - v_i^k/\rho, & if \ |w_i^k - v_i^k/\rho| > \gamma\lambda, \end{cases} \tag{18}$$

where $S$ is the soft-threshold operator [11], given by $S(z, \lambda) = sign(z)\max\{|z| - \lambda, 0\}$. For (18), when $\gamma \to \infty$, the $S$ function is a soft-threshold. When $\gamma \to 1$, the $S$ function is a hard-threshold function.

## 3.2   Convergence Analysis

This subsection analyzes the convergence of our proposed algorithm. We first prove that our proposed method satisfies the following three important properties.

**Property 1:** If $\rho$ is large enough, the proposed method satisfies the sufficient decrease condition: For each $k$, there is a constant $\phi_1 > 0$ such that

$$L(\boldsymbol{w}^{k+1}, \boldsymbol{u}^{k+1}, \boldsymbol{\tau}^{k+1}) - L(\boldsymbol{w}^k, \boldsymbol{u}^k, \boldsymbol{\tau}^k) \leq -\phi_1 \|\boldsymbol{w}^{k+1} - \boldsymbol{w}^k\|_2^2. \tag{19}$$

**Proof:** The Lagrangian function in (14) can be rewritten as

$$L(\boldsymbol{w}, \boldsymbol{u}, \boldsymbol{\tau}) = \psi(\boldsymbol{w}) + \frac{\rho}{2} \left\| \boldsymbol{w} - \boldsymbol{u} - \frac{1}{\rho}\boldsymbol{\tau} \right\|_2^2 + P_{\lambda,\gamma}(\boldsymbol{u}) - \frac{1}{2\rho}\|\boldsymbol{\tau}\|_2^2. \tag{20}$$

The second order gradient of $\psi(\boldsymbol{w})$ is

$$\frac{\partial^2\psi(\boldsymbol{w})}{\partial \boldsymbol{w}^2} = \frac{2}{N}\boldsymbol{A}^{\mathrm{T}}\boldsymbol{A} + 2\boldsymbol{R} = \frac{2}{N}[\boldsymbol{A}^{\mathrm{T}}\boldsymbol{A} + \sigma_b^2 diag(\boldsymbol{A}^{\mathrm{T}}\boldsymbol{A})]. \tag{21}$$

Obviously, $\frac{\partial^2\psi(\boldsymbol{w})}{\partial \boldsymbol{w}^2}$ is a positive definite matrix, hence $\psi(\boldsymbol{w})$ is strictly convex. Thus $L(\boldsymbol{w}, \boldsymbol{u}, \boldsymbol{\tau})$ in (20) is also strictly convex with respect to $\boldsymbol{w}$. Based on the definition of strictly convex function, we can deduce that there exists a constant $c > 0$, such that

$$L(\boldsymbol{w}^{k+1}, \boldsymbol{u}^{k+1}, \boldsymbol{\tau}^k) - L(\boldsymbol{w}^k, \boldsymbol{u}^{k+1}, \boldsymbol{\tau}^k) \leq -\frac{c}{2}\|\boldsymbol{w}^{k+1} - \boldsymbol{w}^k\|_2^2. \tag{22}$$

From (16), we have $\nabla\psi(\boldsymbol{w}^{k+1}) - \boldsymbol{\tau}^k + \rho(\boldsymbol{w}^{k+1} - \boldsymbol{u}^{k+1}) = \boldsymbol{0}$. Furthermore, from (17), we can deduce that $\nabla\psi(\boldsymbol{w}^{k+1}) = \boldsymbol{\tau}^{k+1}$ and $\boldsymbol{u}^{k+1} - \boldsymbol{w}^{k+1} = 1/\rho\left(\boldsymbol{\tau}^{k+1} - \boldsymbol{\tau}^k\right)$. Thus,

$$L(\boldsymbol{w}^{k+1}, \boldsymbol{u}^{k+1}, \boldsymbol{\tau}^{k+1}) - L(\boldsymbol{w}^{k+1}, \boldsymbol{u}^{k+1}, \boldsymbol{\tau}^k) = \left(\boldsymbol{\tau}^{k+1} - \boldsymbol{\tau}^k\right)^{\mathrm{T}}\left(\boldsymbol{u}^{k+1} - \boldsymbol{w}^{k+1}\right)$$

$$= \frac{1}{\rho}\|\boldsymbol{\tau}^{k+1} - \boldsymbol{\tau}^k\|_2^2 = \frac{1}{\rho}\|\nabla\psi(\boldsymbol{w}^{k+1}) - \nabla\psi(\boldsymbol{w}^k)\|_2^2 \leq \frac{l_\psi^2}{\rho}\|\boldsymbol{w}^{k+1} - \boldsymbol{w}^k\|_2^2, \qquad (23)$$

where $l_\psi$ is a Lipschitz constant of function $\psi(\boldsymbol{w})$. The last inequality in (23) is from the fact that $\psi(\boldsymbol{w})$ has Lipschitz continue gradient.

Finally, since $\boldsymbol{u}^{k+1}$ is an optimal solution of (15), we can deduce that

$$L(\boldsymbol{w}^k, \boldsymbol{u}^{k+1}, \boldsymbol{\tau}^k) - L(\boldsymbol{w}^k, \boldsymbol{u}^k, \boldsymbol{\tau}^k) \leq 0. \qquad (24)$$

Combining (22), (23) and (24), we have

$$L(\boldsymbol{w}^{k+1}, \boldsymbol{u}^{k+1}, \boldsymbol{\tau}^{k+1}) - L(\boldsymbol{w}^k, \boldsymbol{u}^k, \boldsymbol{\tau}^k) \leq \left(\frac{l_\psi^2}{\rho} - \frac{c}{2}\right)\|\boldsymbol{w}^{k+1} - \boldsymbol{w}^k\|_2^2. \qquad (25)$$

To ensure $l_\psi^2/\rho - c/2 < 0$, we need $\rho > 2l_\psi^2/c$. ∎

**Property 2:** If $\rho \geq l_\psi$, the Lagrangian function $L(\boldsymbol{w}^k, \boldsymbol{u}^k, \boldsymbol{\tau}^k)$ is bounded for all $k$, and $L(\boldsymbol{w}^k, \boldsymbol{u}^k, \boldsymbol{\tau}^k)$ converges when $k \to \infty$. Furthermore, the sequence $\{\boldsymbol{w}^k, \boldsymbol{u}^k, \boldsymbol{\tau}^k\}$ generated by the proposed method is also bounded.

**Proof:** Firstly, we prove that $L(\boldsymbol{w}^k, \boldsymbol{u}^k, \boldsymbol{\tau}^k)$ is lower bounded for all $k$.

$$L(\boldsymbol{w}^k, \boldsymbol{u}^k, \boldsymbol{\tau}^k) = \psi(\boldsymbol{w}^k) + P_{\lambda,\gamma}(\boldsymbol{u}^k) + \boldsymbol{\tau}^{k\mathrm{T}}(\boldsymbol{u}^k - \boldsymbol{w}^k) + \frac{\rho}{2}\|\boldsymbol{w}^k - \boldsymbol{u}^k\|_2^2,$$

$$= \psi(\boldsymbol{w}^k) + P_{\lambda,\gamma}(\boldsymbol{u}^k) + \nabla\psi(\boldsymbol{w}^k)^{\mathrm{T}}(\boldsymbol{u}^k - \boldsymbol{w}^k) + \frac{\rho}{2}\|\boldsymbol{w}^k - \boldsymbol{u}^k\|_2^2,$$

$$\geq \psi(\boldsymbol{u}^k) + \left(\frac{\rho}{2} - \frac{l_\psi}{2}\right)\|\boldsymbol{u}^k - \boldsymbol{w}^k\|_2^2 + P_{\lambda,\gamma}(\boldsymbol{u}^k). \qquad (26)$$

The inequality in (26) is from Lemma 3.1 in [12]. From Lemma 3.1 in [12], we can deduce that $\psi(\boldsymbol{w}^k) + \nabla\psi(\boldsymbol{w}^k)^{\mathrm{T}}(\boldsymbol{u}^k - \boldsymbol{w}^k) \geq \psi(\boldsymbol{u}^k) - \frac{l_\psi}{2}\|\boldsymbol{u}^k - \boldsymbol{w}^ik\|_2^2$. Hence, we obtain the inequality in (26).

Obviously, if $\rho \geq l_\psi$, then the left hand of the inequality (26) is greater than $-\infty$. Since $L(\boldsymbol{w}^k, \boldsymbol{u}^k, \boldsymbol{\tau}^k)$ is lower bounded. According to Property 1, $L(\boldsymbol{w}^k, \boldsymbol{u}^k, \boldsymbol{\tau}^k)$ is sufficient descent. Hence $L(\boldsymbol{w}^k, \boldsymbol{u}^k, \boldsymbol{\tau}^k)$ is upper bounded by $L(\boldsymbol{w}^0, \boldsymbol{u}^0, \boldsymbol{\tau}^0)$. Due to the sufficient descent property, $L(\boldsymbol{w}^k, \boldsymbol{u}^k, \boldsymbol{\tau}^k)$ converges, as $k \to \infty$. Next, we prove that the sequence $\{\boldsymbol{w}^k, \boldsymbol{u}^k, \boldsymbol{\tau}^k\}$ is bounded. From Property 1, we can easily obtain

$$\sum_{k=1}^{l}\|\boldsymbol{w}^{k+1} - \boldsymbol{w}^k\|_2^2 \leq \frac{1}{\phi_1}\left(L(\boldsymbol{w}^0, \boldsymbol{u}^0, \boldsymbol{\tau}^0) - L(\boldsymbol{w}^{l+1}, \boldsymbol{u}^{l+1}, \boldsymbol{\tau}^{l+1})\right) < \infty. \qquad (27)$$

Note that when $l \to \infty$, we still have $\sum_{k=1}^{\infty} \|\boldsymbol{w}^{k+1} - \boldsymbol{w}^k\|_2^2 < \infty$. Thus, $\boldsymbol{w}^k$ is bounded.

From (23), we have $\|\boldsymbol{\tau}^{k+1} - \boldsymbol{\tau}^k\|_2^2 \le l_\psi^2 \|\boldsymbol{w}^{k+1} - \boldsymbol{w}^k\|_2^2$. Hence we can also deduce that $\sum_{k=1}^{\infty} \|\boldsymbol{\tau}^{k+1} - \boldsymbol{\tau}^k\|_2^2 < \infty$. Thus, $\boldsymbol{\tau}^k$ is also bounded.

In addition, according to (17), we have

$$\|\boldsymbol{u}^{k+1} - \boldsymbol{u}^k\|_2^2 = \|\boldsymbol{w}^{k+1} - \boldsymbol{w}^k + \frac{1}{\rho}\left(\boldsymbol{\tau}^{k+1} - \boldsymbol{\tau}^k\right) + \frac{1}{\rho}\left(\boldsymbol{\tau}^{k-1} - \boldsymbol{\tau}^k\right)\|_2^2$$

$$\le 2\|\boldsymbol{w}^{k+1} - \boldsymbol{w}^k\|_2^2 + \frac{2}{\rho^2}\|\boldsymbol{\tau}^{k+1} - \boldsymbol{\tau}^k\|_2^2 + \frac{2}{\rho^2}\|\boldsymbol{\tau}^{k-1} - \boldsymbol{\tau}^k\|_2^2. \quad (28)$$

Thus $\sum_{k=1}^{\infty} \|\boldsymbol{u}^{k+1} - \boldsymbol{u}^k\|_2^2 < \infty$ and $\boldsymbol{\tau}^k$ is then bounded. To sum up, the sequence $\{\boldsymbol{w}^k, \boldsymbol{u}^k, \boldsymbol{\tau}^k\}$ is bounded. ∎

**Property 3:** There exists a vector $\boldsymbol{d}^{k+1} \in \partial L(\boldsymbol{w}^{k+1}, \boldsymbol{u}^{k+1}, \boldsymbol{\tau}^{k+1})$, and $\phi_2 > 0$ such that

$$\|\boldsymbol{d}^{k+1}\|_2^2 \le \phi_2 \|\boldsymbol{w}^{k+1} - \boldsymbol{w}^k\|_2^2. \quad (29)$$

**Proof:** Let $\boldsymbol{z}^{k+1} = (\boldsymbol{w}^{k+1}, \boldsymbol{u}^{k+1}, \boldsymbol{\tau}^{k+1})$,

$$\left.\frac{\partial L}{\partial \boldsymbol{w}}\right|_{\boldsymbol{z}^{k+1}} = \nabla \psi(\boldsymbol{w}^{k+1}) + \rho\left(\boldsymbol{w}^{k+1} - \boldsymbol{u}^{k+1}\right) - \boldsymbol{\tau}^{k+1} = \boldsymbol{\tau}^k - \boldsymbol{\tau}^{k+1}, \quad (30)$$

$$\left.\frac{\partial L}{\partial \boldsymbol{u}}\right|_{\boldsymbol{z}^{k+1}} = \partial P_{\lambda,\gamma}(\boldsymbol{u}^{k+1}) - \rho\left(\boldsymbol{w}^{k+1} - \boldsymbol{u}^{k+1}\right) + \boldsymbol{\tau}^{k+1} \ni \rho\left(\boldsymbol{w}^k - \boldsymbol{w}^{k+1}\right) + \boldsymbol{\tau}^{k+1} - \boldsymbol{\tau}^k, \quad (31)$$

$$\left.\frac{\partial L}{\partial \boldsymbol{\tau}}\right|_{\boldsymbol{z}^{k+1}} = \boldsymbol{u}^{k+1} - \boldsymbol{w}^{k+1} = \frac{1}{\rho}\left(\boldsymbol{\tau}^{k+1} - \boldsymbol{\tau}^k\right), \quad (32)$$

where the second equality in (31) is based on (15). Let

$$\boldsymbol{d}^{k+1} := \begin{bmatrix} \boldsymbol{\tau}^k - \boldsymbol{\tau}^{k+1} \\ \rho\left(\boldsymbol{w}^k - \boldsymbol{w}^{k+1}\right) + \boldsymbol{\tau}^{k+1} - \boldsymbol{\tau}^k \\ \frac{1}{\rho}\left(\boldsymbol{\tau}^{k+1} - \boldsymbol{\tau}^k\right) \end{bmatrix} \in \partial L\left(\boldsymbol{w}^{k+1}, \boldsymbol{u}^{k+1}, \boldsymbol{\tau}^{k+1}\right) \quad (33)$$

From (23) and (33), we obtain $\|\boldsymbol{d}^{k+1}\|_2^2 \le \phi_2 \|\boldsymbol{w}^{k+1} - \boldsymbol{w}^k\|_2^2$. ∎

Following the concept of the proofs in Proposition 2 in [13] and Theorem 2.9 in [12]. We have the following convergence result for our case, given by Theorem 1. *Due to page limit, we do not provide the details of the proof.*

**Theorem 1:** The proposed method satisfies the Property 1–3. Hence, the sequence $\{\boldsymbol{w}^k, \boldsymbol{u}^k, \boldsymbol{\tau}^k\}$ generated by the proposed method has at least one limit point $\{\boldsymbol{w}^*, \boldsymbol{u}^*, \boldsymbol{\tau}^*\}$ and any limit point $\{\boldsymbol{w}^*, \boldsymbol{u}^*, \boldsymbol{\tau}^*\}$ is a stationary point. In other words, the proposed method has local convergence.

## 4   Simulations

### 4.1   Settings and Properties of ADMM-MCP Algorithm

Three UCI datasets are used [14]. They are respectively Airfoil Self-Noise (ASN), Boston Housing (HOUSING), and Wine Quality White (WQW). For each dataset, its inputs and outputs are normalized to [0,1], and its RBF width is selected between 0.1 to 10. The basic settings of the three datasets are summarized in Table 1. The performance of the algorithms is evaluated by the average test set error of noisy networks. In our simulations, we consider three noise levels: $\{\sigma_b^2 = 0.005\}$, $\{\sigma_b^2 = 0.01\}$ and $\{\sigma_b^2 = 0.05\}$.

For the parameters of our proposed algorithm, we set $\gamma = 10$ and use $\lambda$ to control the number of nodes. Figure 2 shows the behaviour of the proposed algorithm under various values of $\lambda$. From Fig. 2, when the value of $\lambda$ is large, less RBF nodes will be used, but the performance of the algorithm will be poorer. Besides, we use the ASN dataset with noise level $\{\sigma_b^2 = 0.05\}$ as an example to show the convergence of our approach. The result is given by Fig. 1(b). It can be seen that our proposed algorithm converges within 50 to 100 iterations.

**Table 1.** Properties of the three datasets.

| Dataset | Number of features | Size of training set | Size of test set | RBF width |
|---------|--------------------|--------------------|-----------------|-----------|
| Airfoil Self-Noise (ASN) | 5 | 751 | 752 | 0.5 |
| Boston Housing (HOUSING) | 13 | 400 | 106 | 2 |
| Wine Quality White (WQW) | 12 | 2000 | 2898 | 1 |

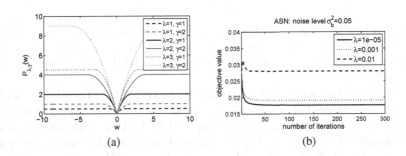

(a)    (b)

**Fig. 1.** (a) The shape of MCP penalty function under different parameter settings. (b) The convergence behaviour of the proposed algorithm.

## 4.2  Performance Comparison

We compare our proposed method with three other algorithms. They are the OLS regularization approach [7], the support vector regression (SVR) algorithm [15], and the Homotopy method (HOM) [16]. For each dataset, the simulation is carried out 10 trials. In each trial, the samples of dataset are randomly split for training set and testing set.

For the proposed ADMM-MCP algorithm, the OLS algorithm, and the HOM algorithm, we can adjust the number of the used nodes by tuning some parameters. Hence we can plot the MSE versus the number of nodes. Figure 3 shows

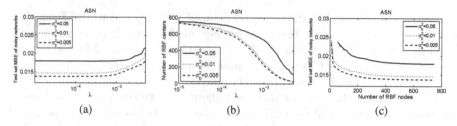

**Fig. 2.** (a) The plots of MSE versus λ. (b) The plots of number of nodes versus λ. (c) From (a) and (b), we obtain the plots of MSE versus the number of nodes.

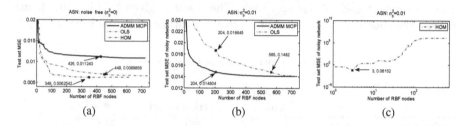

**Fig. 3.** Performance of different algorithms under multiplicative noise. (a) Noise-free. (b)–(c) The noise level is equal to 0.01. From (a), under noise-free condition. The HOM is the best. However, under noisy condition, from (b)–(c), the proposed ADMM-MCP is the best.

the typical results under noise-free and noisy conditions. For the noise-free situation, the HOM algorithm is the best, as shown Fig. 3(a). However, under noisy situation, our proposed method is the best and the HOM is very poor, as shown in Fig. 3(b)–(c).

The average test set performance is shown in Table 2. For the SVR algorithm and the HOM algorithm, their best case MSE values are shown. To make the comparison in a more fair way, for the proposed ADMM-MCP algorithm and the fault tolerant OLS algorithm, their numbers of the used RBF nodes are equal.

We use the ASN dataset, with noise level $\sigma_b^2 = 001$, to discuss the performance of the proposed ADMM-MCP algorithm. From the table, the HOM algorithms fails to tolerate multiplicative noise. Its MSE value under the noisy condition is very poor. Using the OLS algorithm and the SVR algorithm, we can lower the MSE values to 0.017149 and 0.016072, respectively. When the ADMM-MCP algorithm is used, we can further lower the average MSE to 0.014571. Clearly, our proposed algorithm is the best.

To further confirm that the proposed ADMM-MCP algorithm is better than the comparison algorithms, we perform the paired $t$-test. Since the performance of HOM is very poor, the paired $t$-test is not carried on it. The results are shown in Table 3. For 95% level of confidence with 10 trials, the critical $t$-value is equal to 1.833. Hence if the computing $t$ value is greater than that critical value, there is a strong evidence to conclude that the ADMM-MCP algorithm is better than the comparison algorithms. From Table 3, the computed $p$-values are less than

**Table 2.** Average test set MSE over 10 trails under multiplicative noise.

| Dataset | Noise level | ADMM MCP | | OLS | | SVR | | HOM | |
|---|---|---|---|---|---|---|---|---|---|
| | | AVG MSE | AVG no. of nodes | AVG MSE | AVG no. of nodes | AVG MSE | AVG no. of nodes | AVG MSE | AVG no. of nodes |
| ASN | 0.005 | 0.013495 | 200.3 | 0.015439 | 201.1 | 0.015074 | 426.3 | 0.062044 | 2.9 |
| | 0.01 | 0.014571 | 201.5 | 0.017149 | 200.1 | 0.016072 | 493.2 | 0.069529 | 2.9 |
| | 0.05 | 0.018631 | 195.6 | 0.02216 | 200.9 | 0.019846 | 466.6 | 0.087925 | 2.7 |
| HOUSING | 0.005 | 0.013825 | 61.1 | 0.01552 | 59 | 0.015408 | 91.1 | 0.048118 | 6.6 |
| | 0.01 | 0.015008 | 62.2 | 0.01681 | 60.5 | 0.017457 | 84 | 0.057920 | 4.8 |
| | 0.05 | 0.018772 | 60.2 | 0.021545 | 62.5 | 0.022312 | 106.1 | 0.075845 | 3.4 |
| WQW | 0.005 | 0.015961 | 253.4 | 0.016598 | 246.9 | 0.016658 | 1400.2 | 0.031266 | 5.9 |
| | 0.01 | 0.016217 | 249.5 | 0.016940 | 250.6 | 0.016952 | 1221.2 | 0.036170 | 5.8 |
| | 0.05 | 0.016955 | 252.9 | 0.017908 | 248.8 | 0.017609 | 693.6 | 0.059512 | 5 |

**Table 3.** Results of paired t-test between our proposed method, OLS and SVR.

Paired t-test between our proposed method and OLS

| Dataset | Noise level | AVG OLS MSE | AVG MCP MSE | AVG improvement | t-value | p-value | Confidence interval of AVG improvement |
|---|---|---|---|---|---|---|---|
| ASN | 0.005 | 0.015439 | 0.013495 | 0.001943 | 11.6 | $5.16 \times 10^{-7}$ | [0.001592,0.002294] |
| | 0.01 | 0.017148 | 0.014571 | 0.002578 | 12.1 | $3.57 \times 10^{-7}$ | [0.002132,0.003023] |
| | 0.05 | 0.02216 | 0.018631 | 0.003525 | 15.9 | $3.40 \times 10^{-8}$ | [0.003060,0.003989] |
| HOUS -ING | 0.005 | 0.015521 | 0.013825 | 0.001695 | 8.7 | $5.72 \times 10^{-6}$ | [0.001287,0.002104] |
| | 0.01 | 0.01681 | 0.015008 | 0.001797 | 6.8 | $3.96 \times 10^{-5}$ | [0.001244,0.002350] |
| | 0.05 | 0.021545 | 0.018772 | 0.002772 | 8.0 | $1.13 \times 10^{-5}$ | [0.002045,0.003499] |
| WQW | 0.005 | 0.016598 | 0.015962 | 0.000637 | 9.4 | $2.96 \times 10^{-6}$ | [0.000495,0.000778] |
| | 0.01 | 0.016940 | 0.016217 | 0.000723 | 9.9 | $7.31 \times 10^{-5}$ | [0.000570,0.000876] |
| | 0.05 | 0.017908 | 0.016955 | 0.000953 | 14.4 | $7.90 \times 10^{-8}$ | [0.000815,0.001092] |

Paired t-test between our proposed method and SVR

| Dataset | Noise level | AVG SVR MSE | AVG MCP MSE | AVG improvement | t-value | p-value | Confidence interval of AVG improvement |
|---|---|---|---|---|---|---|---|
| ASN | 0.005 | 0.015074 | 0.013495 | 0.001579 | 26.3 | $3.95 \times 10^{-10}$ | [0.001453,0.001704] |
| | 0.01 | 0.016072 | 0.014571 | 0.001501 | 16.2 | $2.87 \times 10^{-8}$ | [0.001307,0.001695] |
| | 0.05 | 0.019846 | 0.018631 | 0.001215 | 7.3 | $2.18 \times 10^{-5}$ | [0.000869,0.001562] |
| HOUS -ING | 0.005 | 0.015408 | 0.013825 | 0.001583 | 5.6 | $1.64 \times 10^{-4}$ | [0.000993,0.002172] |
| | 0.01 | 0.017457 | 0.015008 | 0.002449 | 7.0 | $3.23 \times 10^{-5}$ | [0.001714,0.003183] |
| | 0.05 | 0.022312 | 0.018772 | 0.003540 | 12.5 | $2.66 \times 10^{-7}$ | [0.002949,0.004131] |
| WQW | 0.005 | 0.016658 | 0.015962 | 0.000697 | 13.9 | $1.06 \times 10^{-7}$ | [0.000592,0.000801] |
| | 0.01 | 0.016952 | 0.016217 | 0.000734 | 12.0 | $3.89 \times 10^{-7}$ | [0.000606,0.000863] |
| | 0.05 | 0.017609 | 0.016955 | 0.000654 | 10.5 | $1.15 \times 10^{-6}$ | [0.000524,0.000784] |

0.05. Also, all the computed $t$-values are greater than the critical $t$-value. That means, the improvement of using the ADMM-MCP is statistically significant.

## 5   Conclusion

This paper devises a noise resistant training algorithm for the RBF model. The proposed algorithm can train RBF weights and select the RBF centers at the same time. There are three terms in the training objective of the proposed algorithm. The first term is the MSE value of the training set under the noise-free situation. The second term is the increase of MSE due to the presence of weight/node noise. The last term is a MCP term that can delete some unimportant node during training. Afterwards, we derived an ADMM algorithm to optimize the RBF weights to delete some unnecessary weights. We also presented the convergent properties of the proposed ADMM-MCP algorithm. Simulation showed that the proposed algorithm is superior to other comparison methods, including the fault tolerant OLS approach [7], the SVR algorithm [15], and the Homotopy method (HOM) [16].

**Acknowledgments.** The work was supported by a research grant from City University of Hong Kong (7004842).

# References

1. Poggio, T., Girosi, T.: Networks for approximation and learning. Proc. IEEE **78**(9), 1481–1497 (1990)
2. Haykin, S.: Neural Networks: A Comprehensive Foundation. Prentice Hall, Upper Saddle River (1998)
3. Gomm, J., Yu, D.: Selecting radial basis function network centers with recursive orthogonal least squares training. IEEE Trans. Neural Netw. **11**(2), 306–314 (2000)
4. Burr, J.B.: Digital neural network implementations. In: Neural Networks, Concepts, Applications, and Implementations, vol. 3, pp. 237–285. Prentice Hall (1995)
5. Han, Z., Feng, R., Wan, W.Y., Leung, C.S.: Online training and its convergence for faulty networks with multiplicative weight noise. Neurocomputing **155**, 53–61 (2015)
6. Bernier, J.L., Ortega, J., Ros, E., Rojas, I., Prieto, A.: A quantitative study of fault tolerance, noise immunity, and generalization ability of MLPs. Neural Comput. **12**(12), 2941–2964 (2000)
7. Leung, C.S., Wan, W.Y., Feng, R.: A regularizer approach for RBF networks under the concurrent weight failure situation. IEEE Trans. Neural Netw. Learn. Syst. **28**(6), 1360–1372 (2017)
8. Zhang, C.H.: Nearly unbiased variable selection under minimax concave penalty. Ann. Stat. **38**(2), 894–942 (2010)
9. Boyd, S., Parikh, N., Chu, E., Peleato, B., Eckstein, J.: Distributed optimization and statistical learning via the alternating direction method of multipliers. Found. Trends Mach. Learn. **3**(1), 1–122 (2011)
10. Breheny, P., Huang, J.: Coordinate descent algorithms for nonconvex penalized regression, with applications to biological feature selection. Ann. Appl. Stat. **5**(1), 232–253 (2011)
11. Donoho, D.L., Johnstone, I.M.: Ideal spatial adaptation by wavelet shrinkage. Biometrika **81**(3), 425–455 (1994)
12. Attouch, H., Bolte, J., Svaiter, B.F.: Convergence of descent methods for semi-algebraic and tame problems: proximal algorithms, forward-backward splitting, and regularized gauss-seidel methods. Math. Program. **137**(1–2), 91–129 (2013)
13. Wang, Y., Yin, W., Zeng, J.: Global convergence of ADMM in nonconvex nonsmooth optimization. J. Sci. Comput. (2015, accepted)

14. Lichman, M.: UCI machine learning repository (2013)
15. Zhang, Q., Hu, X., Zhang, B.: Comparison of $l_1$-norm SVR and sparse coding algorithms for linear regression. IEEE Trans. Neural Netw. Learn. Syst. **26**(8), 1828–1833 (2015)
16. Malioutov, D.M., Cetin, M., Willsky, A.S.: Homotopy continuation for sparse signal representation. In: Proceedings of the IEEE CASSP 2005, vol. 5, pp. 733–736. IEEE Press, New York (2005)

# Fault-Resistant Algorithms for Single Layer Neural Networks

Muideen Adegoke[1], Andrew Chi Sing Leung[1(✉)], and John Sum[2]

[1] Department of Electronic Engineering, City University of Hong Kong,
Kowloon Tong, Hong Kong
maadegoke2-c@my.cityu.edu.hk, eeleungc@cityu.edu.hk
[2] Institute of Technology Management, National Chung Hsing University,
Taichung, Taiwan
pfsum@dragon.nchu.edu.tw

**Abstract.** Incremental extreme learning machine (IELM), convex incremental extreme learning machine (C-IELM) and other variants of extreme learning machine (ELM) algorithms provide low computational complexity techniques for training single layer feed-forward networks (SLFNs). However, the original IELM and C-IELM consider faultless network situations only. This paper investigates the performance of IELM and C-IELM under the multiplicative weight noise situation, where the input weights and the output weights are contaminated by noise. In addition, we propose two incremental fault tolerant algorithms, namely weight deviation tolerant-IELM (WDT-IELM) and weight deviation tolerant convex-IELM (WDTC-IELM). The performance of the two proposed algorithms is better than that of the two original ELM algorithms. Moreover, the convergence properties of the proposed algorithms are presented.

**Keywords:** Weight noise · Extreme learning machine · Fault tolerance

## 1 Introduction

It is well known that single layer feed-forward neural networks (SLFNs) have the ability to act as universal approximators [1,2]. In the conventional gradient-based approach, such as the back-propagation concept, all the weights, including the weights from the input layer to the hidden nodes and the output weights, need to be adjusted. The drawbacks of this approach are local minimum and long training time. Huang et al. [3–5] showed that the universal approximation property of neural networks can still be retained even though the input weights and biases of the hidden nodes are generated randomly. Based on this discovery and the pseudo-inverse concept, they proposed extreme learning machine (ELM) [6]. In addition, two incremental learning techniques, namely incremental extreme learning machine (IELM) and convex incremental extreme learning machine (C-IELM), were proposed [3,4]. However, these two ELM algorithms were designed

© Springer Nature Switzerland AG 2018
L. Cheng et al. (Eds.): ICONIP 2018, LNCS 11302, pp. 680–689, 2018.
https://doi.org/10.1007/978-3-030-04179-3_60

for the faultless situation, where weights or nodes are implemented in a perfect way.

Without doubts, in the neural network's implementation, network fault occurs inevitably [7,8]. As an instance, when a finite precision technology is utilized to implement a trained network, additive or multiplicative weight noise in the input weights and output weights would be introduced. Many studies have shown that without proper fault or noise resistant procedure, the performance of the trained network is compromised. Many batch mode fault tolerant techniques for traditional neural network models, such as radial basis function networks were presented [9–11]. However, few results on weight deviation tolerant or noise resistant network on the incremental mode of ELM were reported.

This paper considers that a trained SLFN is affected by weight noise in the input layer (weights between the input and hidden nodes) and the output layer (weights between the hidden nodes and the output layer). We call this situation "weight deviation SLFNs" (WDSLFNs). We first present the training set error of WDSLFNs. Furthermore, we develop an algorithm, namely weight deviation tolerant IELM (WDT-IELM) to tolerate the fault in the network. Besides, to further enhance the performance of the network under this fault situation, we propose another algorithm, namely weight deviation tolerant convex IELM (WDTC-IELM). For the two proposed algorithms, we prove that in term of training objective, the proposed algorithms converge. Also, the superiority of the approaches is shown by comparing them to the IELM and C-IELM [4].

The rest of the paper is organized as follows. The background of ELM and the concept of WDSLFNs are presented in Sect. 2. In Sect. 3, the two proposed weight deviation ELM algorithms are developed. Section 4 presents the simulation results. This paper is then concluded in Sect. 5.

## 2    ELM and Weight Noise Model

In this paper, the nonlinear regression problem is considered. The training set is denoted as $\mathbb{D}_{train} = \{(\boldsymbol{x}_i, t_i) : \boldsymbol{x}_i \in \mathbb{R}^K, t_i \in R, \quad i = 1, \ldots, N\}$, where $K$ is the number of input features, $N$ is the number of training samples, and $\boldsymbol{x}_i$ and $t_i$ are the inputs and target outputs of the $i$th sample, respectively. Similarly, the test set is denoted as $\mathbb{D}_{test} = \{(\boldsymbol{x}'_{i'}, t'_{i'}) : \boldsymbol{x}'_{i'} \in \mathbb{R}^K, t'_{i'} \in R, \quad i' = 1, \ldots, N'\}$, where $N'$ is the number of samples in the test set. We utilize $m$ hidden nodes to model the input-output relationship. The output of a SLFN is given by

$$y_m(\boldsymbol{x}) = \sum_{j=1}^{m} \beta_j g_j(\boldsymbol{x}), \quad \text{where } g_j(\boldsymbol{x}) = \frac{1}{1 + \exp^{-(\boldsymbol{a}_j^T \boldsymbol{x} + b_j)}}, \tag{1}$$

where $\beta_j$ denotes the output weight of the $j$th hidden nodes, $g_j(\cdot)$ is the output of the $j$th hidden node, and $\boldsymbol{a}_j$ and $b_j$ are the input weight vector and input bias of the $j$th hidden node, respectively. By grouping the input weight vector and input bias together, the hidden node output is rewritten as

$$g_j(\boldsymbol{x}) = \frac{1}{1 + \exp^{-(\boldsymbol{a}_j^T \boldsymbol{o})}}, \tag{2}$$

where $\boldsymbol{w}_j = [\boldsymbol{a}_j^{\mathrm{T}}, b_j]^{\mathrm{T}} = [w_{j1}, \ldots, w_{j(K+1)}]^{\mathrm{T}} \in \mathbb{R}^{K+1}$ and $\boldsymbol{o} = [\boldsymbol{x}^{\mathrm{T}}, 1]^{\mathrm{T}}$. The training set error is given by

$$\xi = \sum_{i=1}^{N}(t_i - y_m(\boldsymbol{x}_i))^2 = \left\| \boldsymbol{t} - \sum_{j=1}^{m} \beta_j \boldsymbol{g}_j \right\|_2^2, \tag{3}$$

where $\boldsymbol{t} = [t_1, \ldots, t_N]^{\mathrm{T}}$, and $\boldsymbol{g}_j = [g_j(\boldsymbol{x}_i), \ldots, g_j(\boldsymbol{x}_N)]^{\mathrm{T}}$. In the traditional IELM [3] and C-IELM [4], the bias and the input weights are randomly generated, therefore only the output weights $\beta_j$'s need to be estimated during the network training.

When we implement a trained network, weight deviation or weight noise are inevitable. In the digital implementation, finite precision can be modelled as multiplicative weight noise [7]. Given the $i$th input vector, when noise occur in the input weights and input biases, the node output can be modelled as

$$\tilde{g}_j(\boldsymbol{o}_i) = \frac{1}{1 + \exp^{-(\tilde{\boldsymbol{w}}_j^{\mathrm{T}} \boldsymbol{o}_i)}} \quad \forall i = 1, \ldots, N \quad \text{and} \quad \forall j = 1, \ldots, m. \tag{4}$$

where $\tilde{\boldsymbol{w}}_j = [\tilde{\boldsymbol{a}}_j^{\mathrm{T}}, \tilde{b}_j]^{\mathrm{T}} = [\tilde{w}_{j1}, \ldots, \tilde{w}_{j(K+1)}]^{\mathrm{T}}$, $\tilde{\boldsymbol{a}}_j = \boldsymbol{a}_j + \Delta \boldsymbol{a}_j$, $\tilde{b}_j = b_j + \Delta b_j$, $\Delta \boldsymbol{a}_j$ and $\Delta b_j$ are small changes in the input weights and biases, respectively. From the notation of $\tilde{\boldsymbol{w}}_j$'s and $\tilde{\boldsymbol{a}}_j$'s, the implemented weights can be expressed as $\tilde{\boldsymbol{w}}_j = \boldsymbol{w}_j + \Delta \boldsymbol{w}_j$, where $\Delta \boldsymbol{w}_j = [\Delta \boldsymbol{a}_j^{\mathrm{T}}, \Delta b_j]^{\mathrm{T}} = [\Delta w_{j1}, \ldots, \Delta w_{j(K+1)}]^{\mathrm{T}}$. Here, we assume that the input weights and input biases are contaminated by multiplicative noise, given by

$$\Delta w_{jk} = \delta_{jk} w_{jk}, \tag{5}$$

where $\delta_{jk}$'s are independent and identically distributed (i.i.d.) random variables with zero mean and variance equal to $\sigma_w^2$. Furthermore, we consider that the output weights $\beta_j$'s are contaminated by multiplicative noise too. Hence, the implemented output weights $\beta_j$'s are given by

$$\tilde{\beta}_j = (1 + \gamma_j)\beta_j \quad \forall j = 1, \ldots, m. \tag{6}$$

where $\gamma_j$'s are i.i.d. zero-mean random variable with variance $\sigma_\beta^2$. Combining (4) and (6), we obtain the weighted output of a hidden node, given by

$$\tilde{\beta}_j \tilde{g}_j(\boldsymbol{o}_i) = (1 + \gamma_j)\beta_j \tilde{g}_j(\boldsymbol{o}_i), \quad \forall i = 1, \ldots, N \quad \text{and} \quad \forall j = 1, \ldots, m. \tag{7}$$

In order to expand (7) further, we use the Taylor series concept to linearize $\tilde{g}_j(o_i)$ around the $\mathbf{w}_j$, hence

$$\tilde{g}_j(\boldsymbol{o}_i) \simeq g_j(\boldsymbol{o}_i) + \sum_{k=1}^{K+1} \Delta w_{jk} \frac{\partial \tilde{g}_j(\boldsymbol{o}_i)}{\partial \tilde{w}_{jk}}\bigg|_{\tilde{w}_{jk} = w_{jk}} = g_j(\boldsymbol{o}_i) + \sum_{k=1}^{K+1} \delta_{jk} w_{jk} \frac{\partial \tilde{g}_j(\boldsymbol{o}_i)}{\partial \tilde{w}_{jk}}\bigg|_{\tilde{w}_{jk} = w_{jk}}. \tag{8}$$

From (7) and (8), the weighted output of a hidden node is given by

$$\tilde{\beta}_j \tilde{g}_j(\boldsymbol{o}_i) = (1 + \gamma_j)\beta_j \Delta G \quad \forall\, i = 1, \ldots, N \quad \text{and} \quad \forall\, j = 1, \ldots, m. \tag{9}$$

where $\Delta G_{ijk} = g_j(\boldsymbol{o}_i) + \sum_{k=1}^{K+1} \delta_{jk} w_{jk} \frac{\partial \tilde{g}_j(\boldsymbol{o}_i)}{\partial \tilde{w}_{jk}}\Big|_{\tilde{w}_{jk}=w_{jk}}$ . Using the statistics of $\delta_{wj}$'s and $\gamma_j$'s, we obtain the expectations of weighted outputs of the faulty network as

$$\left\langle \tilde{\beta}_j \tilde{g}_j(\boldsymbol{o}_i) \right\rangle = \beta_j g_j(\boldsymbol{o}_i), \tag{10}$$

$$\left\langle \tilde{\beta}_j \tilde{g}_j(\boldsymbol{o}_i)\tilde{\beta}_{j'}\tilde{g}_{j'}(\boldsymbol{o}_i) \right\rangle = \beta_j \beta_{j'} g_j(x_i) g_{j'}(x_i), \quad \forall\, j \neq j', \tag{11}$$

$$\left\langle \tilde{\beta}_j^2 \tilde{g}_j^2(\boldsymbol{o}_i) \right\rangle = (1 + \sigma_\beta^2)\left( \beta_j^2 g_j^2(\boldsymbol{o}_i) + \beta_j^2 \sum_{k=1}^{K+1} \sigma_w^2 w_{jk}^2 \Delta H_{ijk}^2 \right). \tag{12}$$

where $\Delta H_{ijk} = \frac{\partial \tilde{g}_j(\boldsymbol{o}_i)}{\partial \tilde{w}_{jk}}\Big|_{\tilde{w}_{jk}=w_{jk}}$ and $\langle \cdot \rangle$ denotes the expectation operator.

## 3    Weight Deviation Tolerant Learning

### 3.1    WDT-IELM Algorithm

In this subsection, we derive the WDT-IELM algorithm which inserts one new hidden node into network at each iteration. It does not modify the previous learned output weights. Taking the expectation over all possible noise patterns, we obtain

$$\left\langle \tilde{\xi} \right\rangle = \left\langle \sum_{i=1}^{N} \left( t_i - \sum_{j=1}^{m} \tilde{\beta}_j \tilde{g}_j(\boldsymbol{o}_i) \right)^2 \right\rangle. \tag{13}$$

Using (11)–(12), (13) becomes

$$\left\langle \tilde{\xi} \right\rangle = \left\| \boldsymbol{t} - \sum_{j=1}^{m} \beta_j \boldsymbol{g}_j \right\|_2^2 + \sigma_\beta^2 \sum_{j=1}^{m} \beta_j^2 \|\boldsymbol{g}_j\|_2^2$$

$$+ (1 + \sigma_\beta^2) \left( \sum_{j=1}^{m} \beta_j^2 \left( \sum_{k=1}^{K+1} \sigma_w^2 w_{jk}^2 \|\boldsymbol{\Delta H}_{jk}\|_2^2 \right) \right), \tag{14}$$

where $\boldsymbol{\Delta H}_{jk} = [\Delta H_{1jk}, \ldots, \Delta H_{Njk}]^{\mathrm{T}}$. In the incremental learning concept, we add hidden nodes one-by-one into the network. At the $m$th iteration, the training objective is given by

$$J_m = \left\| \boldsymbol{t} - \sum_{j=1}^{m} \beta_j \boldsymbol{g}_j \right\|_2^2 + \sigma_\beta^2 \sum_{j=1}^{m} \beta_j^2 \|\boldsymbol{g}_j\|_2^2$$

$$+ (1 + \sigma_\beta^2)\sigma_w^2 \left( \sum_{j=1}^{m} \beta_j^2 \left( \sum_{k=1}^{K+1} w_{jk}^2 \|\boldsymbol{\Delta H}_{jk}\|_2^2 \right) \right). \tag{15}$$

In order to define the incremental learning algorithm, we define the following equations:

$$F_m = \sum_{j=1}^{m} \beta_j g_j, \quad H_m = \sum_{j=1}^{m} \beta_j^2 \left( \sum_{k=1}^{K+1} w_{jk}^2 \|\Delta H_{jk}\|_2^2 \right) \tag{16}$$

$$L_m = \sum_{j=1}^{m} \beta_j^2 \|g_j\|_2^2, \quad e_m = t - \sum_{j=1}^{m} \beta_j g_j. \tag{17}$$

From (16) and (17), (15) becomes

$$J_m = \|e_m\|_2^2 + \sigma_\beta^2 L_m + (1 + \sigma_\beta^2)\sigma_w^2 H_m. \tag{18}$$

In addition, the reduction in the training objective, $D_m = J_m - J_{m-1}$, is given by

$$D_m = -2\beta_m e_{m-1}^T g_m + (1 + \sigma_\beta^2)\|\beta_m^2 g_m\|_2^2 + (1 + \sigma_\beta^2)\sigma_w^2 \beta_m^2 \sum_{k=1}^{K+1} w_{mk}^2 \|\Delta H_{mk}\|_2^2. \tag{19}$$

From (19), it can be seen that $D_m$ is a quadratic function of $\beta_m$ which has its minimum value equal to a negative value. Hence the optimal value $\beta_m^*$ of $\beta_m$ to maximize the reduction is given by

$$\beta_m^* = \frac{e_{m-1}^T g_m}{(1 + \sigma_\beta^2)\|g_m\|_2^2 + (1 + \sigma_\beta^2)\sigma_w^2 \sum_{k=1}^{K+1} w_{mk}^2 \|\Delta H_{mk}\|_2^2}. \tag{20}$$

With the optimal value $\beta_m^*$, the change in the objective value is given by

$$D_m = -\frac{\left(e_{m-1}^T g_m\right)^2}{\left(1 + \sigma_\beta^2\right)\left(\|g_m\|_2^2 + \sigma_w^2 \sum_{k=1}^{K+1} w_{mk}^2 \|\Delta H_{mk}\|_2^2\right)}. \tag{21}$$

From (21), after adding a new hidden node into the network, the training objective decreases. This indicates that the proposed WDT-IELM algorithm converges. The proposed WDT-IELM algorithm is shown in Algorithm 1. At the $m$th iteration, the computational complexity of the proposed WDT-IELM algorithm is $O(N) + O(NK)$ and it is the same as that of the original IELM algorithm.

## 3.2    WDTC-IELM Algorithm

It was shown in [4], that in term of training set error, the C-IELM algorithm [4] outperforms the IELM [3] algorithm under the noiseless situation. However, the original C-IELM algorithm has a very poor noise resistant ability under the weight deviation cases (shown in Sect. 4 of this paper). Hence it is fascinating to develop a weight deviation tolerant version of the C-IELM algorithm, namely

---

**Algorithm 1.  WDT-IELM**

---

**Require:** Set $m$ equal to zero ($m = 0$), $e_0 = t$ and $F_0 = 0$
    **while** $m \leq m_{max}$ **do** $m = m + 1$
    Insert a new hidden node into the network.
    Compute the hidden node output vector $g_m$ (over all training input vectors) of the
    newly inserted hidden node.
    Compute the output weight $\beta_m$ of the newly inserted node:
$$\beta_m = \frac{e_{m-1}^{\mathrm{T}} g_m}{(1+\sigma_\beta^2)\|g_m\|_2^2 + (1+\sigma_\beta^2)\sigma_w^2 \sum_{k=1}^{K+1} w_{mk}^2 \|\Delta H_{mk}\|_2^2} .$$
    Update $F_m = F_{m-1} + \beta_m g_m$ and $e_m = t - F_m$
    **end while**

---

WDTC-IELM. In the development of WDTC-IELM, after estimating the output weight $\beta_m$ at the $m$th iteration, all the previous trained weights are updated by

$$\beta_j = (1 - \beta_m)\beta_j \forall j = 1, \cdots, m - 1. \tag{22}$$

Thus, the recursive definitions for $F_m$, $e_m$, $L_m$ and $H_m$ are obtained as:

$$F_m = \beta_m g_m + (1 - \beta_m)F_{m-1}, \quad L_m = \beta_m^2 \|g_m\|_2^2 + (1 - \beta_m)^2 L_{m-1}$$

$$H_m = (1 - \beta_m)^2 H_{m-1} + \beta_m^2 \left( \sum_{k=1}^{K+1} w_{mk}^2 \|\Delta H_{mk}\|_2^2 \right), \quad e_m = t - F_m.$$

where $F_0 = 0$, $e_0 = t$, $L_0 = 0$ and $H_0 = 0$. Using this new updating scheme (22), from (15), the change in the objective value, $D_m = J_m - J_{m-1}$, is given by

$$D_m = -2\beta_m \left( e_{m-1}^{\mathrm{T}} q_m + \sigma_\beta^2 L_{m-1} + (1 + \sigma_\beta^2)\sigma_w^2 H_{m-1} \right) + \beta_m^2 \Lambda, \tag{23}$$

where

$$q_m = g_m - F_{m-1}, \tag{24}$$

$$\Lambda = \|q_m\|_2^2 + \sigma_\beta^2 \left( L_{m-1} + \|g_m\|_2^2 \right)$$

$$+ (1 + \sigma_\beta^2)\sigma_w^2 \left( H_{m-1} + \left( \sum_{k=1}^{K+1} w_{mk}^2 \|\Delta H_{mk}\|_2^2 \right) \right). \tag{25}$$

In order to maximize the decrease in the training objective values, $\beta_m$ should be given by

$$\beta_m = \frac{\left( e_{m-1}^{\mathrm{T}} q_m + \sigma_\beta^2 L_{m-1} + (1 + \sigma_\beta^2)\sigma_w^2 H_{m-1} \right)}{\Lambda}. \tag{26}$$

With (26), the change in the training objective value is given by

$$D_m = -\frac{\left( e_{m-1}^{\mathrm{T}} q_m + \sigma_\beta^2 L_{m-1} + (1 + \sigma_\beta^2)\sigma_w^2 H_{m-1} \right)^2}{\Lambda}. \tag{27}$$

From (27), it can be seen that when a new hidden node is inserted, the objective value decreases. Hence in terms of the objective value, the WDTC-IELM algorithm converges as well. The proposed WDTC-IELM algorithm is shown in Algorithm 2. From the algorithm, at the $m$th iteration, the computational complexity of the proposed WDTC-IELM algorithm is $O(N) + O(NK) + O(m)$ and it is the same as that of the original C-IELM algorithm.

---

**Algorithm 2.   WDTC-IELM**

---

**Require:** Set $m$ equal to zero $(m = 0)$, $e_0 = t$, $F_0 = 0$, $L_0 = 0$ , and $H_0 = 0$.
   while $m \leq m_{max}$ do $m = m + 1$
Insert a new hidden node into the network.
Compute the output vector $g_m$ (for all training samples) of this new hidden node.
Compute $q_m = g_m - F_{m-1}$ and $\Lambda$.
Compute the new output weight $\beta_m$ of the newly added node:
$\beta_m = \frac{(e_{m-1}^T q_m + \sigma_\beta^2 L_{m-1} + (1+\sigma_\beta^2)\sigma_w^2 H_{m-1})}{\Lambda}$.
$F_m = \beta_m g_m + (1 - \beta_m) F_{m-1}$.
$e_m = t - F_m$
$L_m = \beta_m^2 \|g_m\|_2^2 + (1 - \beta_m)^2 L_{m-1}$.
$H_m = (1 - \beta_m)^2 H_{m-1} + \beta_m^2 \left( \sum_{k=1}^{K+1} w_{mk}^2 \|\Delta H_{mk}\|_2^2 \right)$.
$\beta_j = (1 - \beta_m)\beta_j, \quad \forall j = 1, \dots, m - 1$.
**end while**

---

**Table 1.** Details of the data-sets

| Data-set | Training set size | Test set size | Number of features |
|---|---|---|---|
| Concrete | 500 | 530 | 9 |
| Abalone | 2000 | 2177 | 9 |
| Housing price | 300 | 206 | 14 |

# 4   Simulation

## 4.1   Setting

Three real-life data sets are obtained from the University of California Irvine (UCI) regression data repository [12]. The datasets includes Housing price, Concrete Compressive Strength, and Abalone. The details of these three data sets are shown in Table 1. The data are first pre-processed. The input features of the datasets are normalized to the range of $[-1, 1]$ while the target outputs of the data-sets are normalized to the range of $[0, 1]$. Furthermore, the input weights and the biases of the hidden nodes of the network are generated randomly which are set between the range $[-1, 1]$.

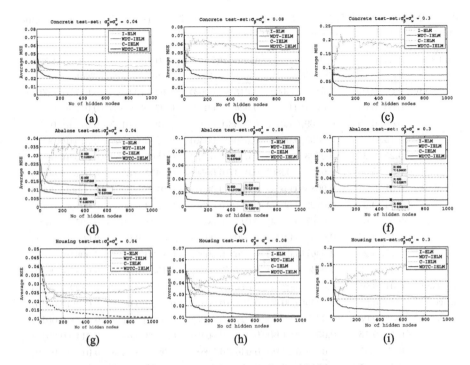

**Fig. 1.** Test set error versus number of hidden nodes.

## 4.2   Test Set Error Versus the Number of Hidden Nodes

We compare four incremental algorithms: IELM, C-IELM, the proposed WDT-IELM and the proposed WDTC-IELM. Three noise levels are considered: $\{\sigma_\beta^2 = \sigma_w^2 = 0.04, \sigma_\beta^2 = \sigma_w^2 = 0.08, \sigma_\beta^2 = \sigma_w^2 = 0.3.$ Figure 1 shows the test set mean squared error (MSE) of faulty networks versus the number of hidden nodes. In order to obtain the test set MSE of faulty networks, we generate 100 faulty patterns for each trained networks.

From the figure, it can be seen that the two proposed algorithms outperform the IELM and C-IELM. Especially, when the noise level is large, the improvement is significant. Also, around 200 to 500 hidden nodes are enough for the three datasets. Further increasing the number of hidden nodes does not significantly reduce the MSE values. The performance of the proposed WDT-IELM is better than that of IELM. Moreover, the original C-IELM algorithm performs poorly under the imperfect weight cases considered. The performance of the proposed WDTC-IELM is the best. For instance, for the Abalone dataset shown in Figs. 1(d)–(f), when the noise level is 0.04, the MSE values of IELM and C-IELM are 0.01246 and 0.03314, respectively. Clearly, the performance of C-IELM is very poor. With WDT-IELM, the MSE value reduces to 0.01209. With WDTC-IELM, the MSE value further reduces to 0.0070705. For high noise level, the improvements of using WDT-IELM and WDTC-IELM are more significant.

**Table 2.** Average MSE for Test data-sets of the faulty Network. The average values are taken over 100 trials. There are 500 hidden nodes.

| Data set | Noise level $\sigma_\beta^2, \sigma_w^2$ | IELM mean(std) | WDT-IELM mean(std) | C-IELM mean(std) | WDTC-IELM mean(std) |
|---|---|---|---|---|---|
| Concrete | 0.01, 0.01 | 0.0187(0.0003) | 0.0186(0.0003) | 0.0295(0.0024) | 0.0171(0.0002) |
| | 0.04, 0.04 | 0.0251(0.0011) | 0.0240(0.0013) | 0.0730(0.0070) | 0.0179(0.0001) |
| | 0.08, 0.08 | 0.0339(0.0024) | 0.0308(0.0015) | 0.1276(0.0165) | 0.0184(0.0002) |
| | 0.15, 0.15 | 0.0498(0.0048) | 0.0408(0.0030) | 0.2332(0.0267) | 0.0194(0.0003) |
| | 0.25, 0.25 | 0.0703(0.0067) | 0.0505(0.0047) | 0.3868(0.0462) | 0.0207(0.0002) |
| | 0.30, 0.30 | 0.0824(0.0086) | 0.0537(0.0050) | 0.4451(0.0542) | 0.0213(0.0003) |
| Abalone | 0.01, 0.01 | 0.0088(0.0002) | 0.0087(0.0002) | 0.0105(0.0006) | 0.0066(0.0001) |
| | 0.04, 0.04 | 0.0122(0.0005) | 0.0119(0.0006) | 0.0243(0.0028) | 0.0070(0.0001) |
| | 0.08, 0.08 | 0.0167(0.0014) | 0.0154(0.0012) | 0.0431(0.0040) | 0.0073(0.0001) |
| | 0.15, 0.15 | 0.0249(0.0023) | 0.0202(0.0017) | 0.0762(0.0075) | 0.0078(0.0001) |
| | 0.25, 0.25 | 0.0367(0.0051) | 0.0263(0.0026) | 0.1320(0.0140) | 0.0084(0.0001) |
| | 0.30, 0.30 | 0.0444(0.0053) | 0.0278(0.0026) | 0.1534(0.0196) | 0.0087(0.0002) |
| Housing price | 0.01, 0.01 | 0.0151(0.0022) | 0.0151(0.0024) | 0.0158(0.0029) | 0.0122(0.0018) |
| | 0.04, 0.04 | 0.0225(0.0048) | 0.0214(0.0036) | 0.0259(0.0067) | 0.0130(0.0019) |
| | 0.08, 0.08 | 0.0325(0.0067) | 0.0275(0.0038) | 0.0397(0.0111) | 0.0138(0.0017) |
| | 0.15, 0.15 | 0.0507(0.0149) | 0.0372(0.0050) | 0.0603(0.0166) | 0.0146(0.0018) |
| | 0.25, 0.25 | 0.0803(0.0338) | 0.0464(0.0068) | 0.0983(0.0208) | 0.0159(0.0019) |
| | 0.30, 0.30 | 0.0946(0.0620) | 0.0501(0.0067) | 0.1277(0.0485) | 0.0164(0.0021) |

When the noise level is 0.3, the MSE value of IELM is 0.04431 and the MSE level is extreme large. With WDT-IELM, the MSE value reduces to 0.02671. With WDTC-IELM, the MSE value further reduces to 0.008138.

### 4.3  Comparison

To further test the performance of the proposed algorithms. we repeat the experiment 100 times with different sets of random nodes. Since using 500 hidden nodes (see Fig. 1) is enough for these three datasets, we fix the number of hidden nodes to 500. The results are summarized in Table 2.

From the table, it can be seen that the performance of C-IELM is very poor, and that the performance of the proposed WDT-IELM algorithm is better than that of IELM. Besides, the WDTC-IELM algorithm is the best. Furthermore, as the noise level increases, the proposed WDTC-IELM algorithm greatly resists weight noise effect. For instance, from Table 2, for the Abalone data-set, when the noise level is $\sigma_\beta^2 = \sigma_w^2 = 0.01$, the test set error of the WDTC-IELM algorithm is equal to 0.0066. When the noise level increases to 0.3, the test set error slightly increases to 0.0087 only. This shows that the noise resistance ability of WDTC-IELM algorithm is much better than that of the other three algorithms. Hence, a fascinating characteristic of WDTC-IELM is that its test set error is insensitive to the weight noise level.

# 5   Conclusion

In this work, two incremental learning algorithms are presented for ELM under the input and out weight noise situations. The proposed algorithms, namely WDT-IELM and WDTC-IELM, has ability to resist the effect of weight noise. The algorithms work by inserting the randomly generated hidden nodes, one after the other into the network. Moreover, we presented the convergence properties of the two proposed algorithms.

Besides, the WDT-IELM algorithm only changes the output weight of the newly inserted hidden node. Its performance under noisy network is better than that of the original IELM algorithm and the original C-IELM.

Furthermore, we proposed the WDTC-IELM algorithm. It adjusts the output weight of the newly added node and modifies all the previous weight based on a single technique. Its performance under noisy network is better than that of IELM, C-IELM, and WDT-IELM.

**Acknowledgment.** The work was supported by a research grant from Government of the Hong Kong Special Administrative Region of the People's Republic of China (CityU 11259516).

# References

1. Hornik, K.: Approximation capabilities of multilayer feedforward networks. Neural Netw. **4**(2), 251–257 (1991)
2. Witten, I.H., Frank, E., Hall, M.A., Pal, C.J.: Data Mining: Practical machine learning tools and techniques. Morgan Kaufmann (2016)
3. Huang, G.B., Chen, L., Siew, C.K.: Universal approximation using incremental constructive feedforward networks with random hidden nodes. IEEE Trans. Neural Netw. **17**(4), 879–892 (2006)
4. Huang, G.B., Chen, L.: Convex incremental extreme learning machine. Neurocomputing **70**(16–18), 3056–3062 (2007)
5. Huang, G.B., Zhu, Q.Y., Siew, C.K.: Extreme learning machine: theory and applications. Neurocomputing **70**(1), 489–501 (2006)
6. Huang, G.B., Zhu, Q.Y., Siew, C.K.: Extreme learning machine: a new learning scheme of feedforward neural networks. In: 2004 IEEE International Joint Conference on Neural Networks, vol. 2, pp. 985–990 (2004)
7. Burr, J: Digital neural network implementations. In: Neural Networks, Concepts, Applications, and Implementations, pp. 237–285. Prentice Hall (1995)
8. Zhang, D.D., Zukowski, J.: Parallel VLSI Neural System Design. Springer-Verlag New York, Inc, New York (1997)
9. Leung, C.S., Sum, J.P.F.: A fault-tolerant regularizer for RBF networks. IEEE Trans. Neural Netw. **19**(3), 493–507 (2008)
10. Leung, C.S., Wan, W.Y., Feng, R.: A regularizer approach for RBF networks under the concurrent weight failure situation. IEEE Trans. Neural Netw. Learn. Syst. **28**(6), 1360–1372 (2017)
11. Mahdiani, H.R., Fakhraie, S.M., Lucas, C.: Relaxed fault-tolerant hardware implementation of neural networks in the presence of multiple transient errors. IEEE Trans. Neural Netw. Learn. Syst. **23**(8), 1215–1228 (2012)
12. Lichman, M.: UCI machine learning repository (2013). http://archive.ics.uci.edu/ml

# Adversarial Minimax Training for Robustness Against Adversarial Examples

Ryota Komiyama[✉] and Motonobu Hattori

Interdisciplinary Graduate School of Medicine, Engineering and Agriculture,
University of Yamanashi, Kofu, Yamanashi, Japan
{g18tk006,m-hattori}@yamanashi.ac.jp

**Abstract.** In this paper, we propose a novel method to improve robustness against adversarial examples. In conventional methods, in order to take measures against adversarial examples, a classifier is learned with adversarial examples generated in a specific way. However, this method can defend against only limited types of adversarial examples. In the proposed method, in order to deal with a wide range of adversarial examples, two networks are used: a generator network and a classifier network. The generator network generates noise to make an adversarial example and the classifier network acquires robustness by learning the adversarial example. Computer simulation results show that the proposed method is more robust against adversarial examples generated by some different methods in black box attacks than the conventional adversarial training methods.

**Keywords:** Adversarial examples · Adversarial training
Black box attack

## 1 Introduction

Recent research has revealed that machine learning methods including deep learning has a vulnerability called adversarial examples. An adversarial example is an image generated by adding malicious noise to the original image which is too small to be seen by human eyes [1]. Since there is little difference between the original image and adversarial examples, human can recognize the image correctly, whereas machine learning methods can not. This phenomenon is contrary to intuition and it becomes a big problem in environments where security is required, such as face recognition systems, road sign recognition systems in automatic driving, and so on.

Numerous studies have been done to obtain robustness against adversarial examples. Each method can be categorized into four groups. (1) adversarial training that trains not only original images but also adversarial examples at the same time [2], (2) using a network to determine whether the input image is an original image or an adversarial example [3,4], (3) a method of removing noise contained in the input image so that the neural network can correctly

© Springer Nature Switzerland AG 2018
L. Cheng et al. (Eds.): ICONIP 2018, LNCS 11302, pp. 690–699, 2018.
https://doi.org/10.1007/978-3-030-04179-3_61

recognize it [4], (4) a method of distilling a network and acquiring robustness against adversarial examples [5]. Although many methods of defense have been proposed, dealing with adversarial examples is still a big problem.

In this paper, we focused on adversarial training where numerous studies exist. In the conventional methods, in order to acquire robustness against adversarial examples, a network is trained by using not only original images but also adversarial examples with the same teacher label. However, since this method uses adversarial examples generated by a specific method, only limited robustness can be obtained. Therefore, we need to consider an adversarial training method with which the classifier can learn a wide range of adversarial examples.

In this research, we propose an adversarial training method which does not fix the generation method of adversarial examples. In order to generate a wide range of adversarial examples, we use a generator network that generates adversarial noise. This network is trained to generate noise for an adversarial example so that the other classifier network misrecognizes it. On the other hand, the classifier network is trained to recognize it correctly. By repeating this, the classifier network can acquire robustness to a wide range of adversarial examples.

The rest of this paper is organized as follows. In Sect. 2, we explain the conventional generation methods of adversarial examples and adversarial attack scenarios. We explain the proposed method in Sect. 3 and evaluate the performance of the proposed method in Sect. 4. Finally we conclude in Sect. 5.

## 2    Adversarial Attack

### 2.1    Generation Algorithms of Adversarial Examples

A number of algorithms to generate adversarial examples have been proposed so far. Here, we introduce some representative methods.

**Fast Gradient Sign Method (FGSM).** FGSM [2] is a generation method for adding noise to an image so that the network error increases. In this method, the distance between each element of the original image and that of the adversarial example is equal, and it is optimized for $L_\infty$ [2]. Specifically, noise is added as follows:

$$x_{adv} = x + \epsilon \cdot \text{sign}(\nabla_x J(x)) \tag{1}$$

where $x$ is the input image, $\epsilon$ is the noise size added to each feature, $\text{sign}(\cdot)$ is the sign function, $J(\cdot)$ is the error function of the target network. This algorithm generates an adversarial example to be misrecognized in a class different from the teacher label, and we can not specify the class misrecognized. So, this is called a non-targeted attack. Although this method operates fast because it updates the noise only once, it does not always produce optimum noise.

**Jacobian Based Saliency Map Approach (JSMA).** JSMA [6] is a method to minimize $L_0$ (a measure representing the number of changed features). Briefly, this is a method of selecting input feature(s) that gives the greatest influence on the output result and adding relatively large noise to the input feature repeatedly. In JSMA, assuming that positive noise will be added, the feature $ind$, which has the greatest influence on the output result, is selected as follows:

$$a_i = \frac{\partial F_{tr}(\boldsymbol{x_{adv}})}{\partial x_i} \tag{2}$$

$$b_i = \sum_{j \neq tr} \frac{\partial F_j(\boldsymbol{x_{adv}})}{\partial x_i} \tag{3}$$

$$c_i = \begin{cases} 0 & a_i < 0 \text{ or } b_i > 0 \\ -a_i \times b_i & \text{otherwise} \end{cases} \tag{4}$$

$$ind = \arg\max_i(c_i) \qquad \text{for } i = 1, \cdots, n$$

where $F_j$ is the $j$th output in the output layer of the target network, $tr$ is the target class to be misrecognized, and $n$ is the number of input features. $a_i$ represents the amount of change in the target class when the $i$th input feature is changed, and $b_i$ represents the total amount of change in the classes excluding the target one. Therefore, it means that the adverse influence increases as $c_i$ increases. By using $\boldsymbol{i} = (i_1, i_2, \cdots)$ instead of $i$, we can calculate the effect of changes in multiple input features at the same time.

JSMA is a method to select input features that affect the output until the generated image is misrecognized. Therefore, when the output of the target class $tr$ becomes dominant, the noise adding process ends. In addition to the original JSMA, we also used JSMA considering the classification probability in our simulation. JSMA considering the classification probability is a method to select features until the classification probability exceeds the threshold $p$. So, it is possible to generate more powerful adversarial examples compared with the original JSMA.

**L2 Optimization Attack of Carlini and Wagner Attack (L2CW).** L2CW [7] is a method to minimize a specific metric by using a logit (the input of an output neuron). Since logits give more information about a neural network compared to the outputs through the activation function such as the softmax function, it is possible to add more powerful and unrecognizable noise to the image. We generate adversarial examples by using $L_2$ as a metric and minimize the following objective function:

$$\min_{\boldsymbol{\delta}}(||\boldsymbol{\delta}||_2 + c \cdot f(\boldsymbol{x} + \boldsymbol{\delta})) \tag{5}$$

$$\text{such that } \boldsymbol{x} + \boldsymbol{\delta} \in [0, 1]^n$$

where $\boldsymbol{\delta}$ denotes noise to be added, $f(\boldsymbol{u})$ is a function that takes a non-positive value if and only if $\boldsymbol{u}$ is misrecognized, $c$ is a constant indicating the importance between the noise size and the degree of misrecognition. Although there are many choices for $f(\cdot)$, we use $f(x) = \max(0, Z_{tch}(x) - \max_{i \neq tch}(Z_i(x)) + k)$ because we can adjust the classification probability, where $Z_i(\cdot)$ is the $i$th logit, $k$ is a parameter to adjust the size of the logit, and $tch$ is the teacher label. The higher the value of $k$, the higher the classification probability of the misrecognized class in adversarial examples generated by this method. There is a way to adjust the hyper-parameter, $c$ efficiently. Since the ideal value of $c$ is the smallest value in which the target network is misrecognized, it is possible to search it by using the binary search. In our simulation, the binary search was performed 50 times, and minimization of the function (5) in each $c$ was carried out 100 times. Although this method is available for both targeted attacks and non-targeted attacks, we used non-targeted attacks in our simulation.

## 2.2   Adversarial Attack Scenarios

There are two scenarios for adversarial attacks: a white box attack and a black box attack. A white box attack is a situation in which all the information necessary for generating adversarial examples, such as the network structure, weights and dataset for training, is available. However, this is not a realistic scenario, because it is extremely difficult to obtain weights. On the other hand, a black box attack is a scenario where attackers generate adversarial examples in the situation where limited information is available. There are two kinds of situations for black box attacks. The first situation is that the target network recognizes an image and attackers can use the result (black box attack with query). In this situation, the attackers can train the replica of the target network by using results obtained from the target, and use the replica to generate adversarial examples. Since adversarial examples have high transferability, this attack succeeds in many cases. The second situation is that attackers can not use any information including recognition results (black box attack without query) [8]. In order to raise the success rate of the attack under this situation, attackers need to use multiple networks with different structures to generate adversarial examples.

# 3   Adversarial Minimax Training

In the conventional methods of adversarial training, only limited types of adversarial examples can be learned even if multiple methods to generate adversarial examples are used during training. Moreover, it may occur that a new method to generate adversarial examples is proposed, then its countermeasure is taken, after that another new generation method is developed, .... It seems like a see-saw game and there is no end. Therefore, we need a method to fix weaknesses of the network by itself without using specific generation methods during training. Although there are several studies to generate adversarial examples using neural networks [9,10], there are no studies used for training. Inspired by these studies, we propose Adversarial Minimax Training.

In the proposed method, we use two types of networks, a classifier and a generator. The generator network is trained to generate noise for adversarial examples and to make the classifier network misrecognized by the adversarial examples. The classifier network is trained to be able to recognize the adversarial examples correctly. Specifically, the classifier minimizes the objective function (6) and the generator maximizes the objective function (7).

$$\min_J(J(\boldsymbol{x}) + J(\boldsymbol{x} + G(\boldsymbol{x}))) \tag{6}$$

$$\max_G(J(\boldsymbol{x} + G(\boldsymbol{x}))) \tag{7}$$

$$\text{such that } G(\boldsymbol{x}) \in [-ns, ns]^n$$

where $J(\cdot)$ is the error function of the classifier such as the cross entropy, $G(\cdot)$ is the output of the generator (i.e. adversarial noise), and $ns$ is the upper bound of noise added to each pixel. Therefore, $L_\infty$ in the generated noise is less than or equal to $ns$. For a wide range of training, we change $ns$ according to the progress of training. The flow of the whole learning is based on GAN (Generative Adversarial Network) [11], and details are as follows:

(1) $iter \leftarrow 0$, $ns \leftarrow 0$, determine the upper limit of training iterations, $iter_{max}$.
(2) Repeat the following operations while $iter \leq iter_{max}$.
  (a) Maximize the function (7) using the hill climbing method.
  (b) Minimize the function (6) using the gradient descent method.
  (c) $ns \leftarrow s(ns, iter)$
  (d) $iter \leftarrow iter + 1$

where $s(ns, iter)$ is a function to determine the next $ns$ from the current $ns$ and $iter$. The above procedure shows just a framework, and there may be some ingenious ways to improve the performance. In this paper, we try to improve the performance by combining the proposed method with adversarial training, and also investigate whether the proposed method can coexist with other methods.

## 4     Computer Simulation

### 4.1     Setup

In order to investigate robustness of the proposed method against adversarial examples, we generated multiple adversarial examples using the CleverHans [12] library and attacked the proposed method (more precisely, the classifier network trained by the proposed method). The comparison methods include a network without measures against adversarial examples (Normal), a network where training data was extended by adding random noise (Random), a network trained by adversarial training using FGSM (AT), a network trained by FGSM while changing noise ($\epsilon$) (ATd). The proposed method includes a network trained by Adversarial Minimax Training (AMT), and a network trained by Adversarial

Minimax Training and adversarial training using FGSM (AMT+AT). We used three methods to generate adversarial examples, FGSM, JSMA, and L2CW. We varied the magnitude of noise in each method. Table 1 shows parameters to adjust the noise size and the classification probability used in the simulation. We used large and small noise denoted as "L" and "s", respectively. The datasets used in the simulation were MNIST and CIFAR10, and we adopted white box attacks and black box attacks with query as attack scenarios. In the black box attacks, it was assumed that each target network was attacked under the worst case scenario, that is, all the information except weights was regarded as accessible. So, the structure of the replica was set to the same one as the target network, and we used all the same for the optimization method, training dataset, and hyper-parameters in the replica as those used in the target network. Moreover, we used the actual output of the target network corresponding to each input image for the learning of the replica.

**Table 1.** Amount of attack noise and probability control parameters

| Dataset | Setting | FGSM ($\epsilon$) | | JSMA ($p$) | | L2CW ($k$) | | |
|---------|---------|---------|---------|--------|--------|--------|--------|--------|
| MNIST | White | 0.15(s) | 0.25(L) | 0(s) | — | 0 | — | — |
| | Black | 0.15(s) | 0.25(L) | 0(s) | 0.9(L) | — | 10(s) | 20(L) |
| CIFAR10 | White | 0.05(s) | 0.15(L) | 0(s) | — | 0 | — | — |
| | Black | 0.05(s) | 0.15(L) | 0(s) | 0.9(L) | — | 20(s) | 40(L) |

The network structure was changed according to the datasets. The structure of the classifier is shown in Table 2. We used the widely used structure in MNIST and we used the distilled model used in [5] in CIFAR10.

**Table 2.** Architecture (Classifier)

| Layer Name | MNIST | CIFAR10 |
|------------|-------|---------|
| Input | 28 × 28 pixels, 1 channel | 32 × 32 pixels, 3 channels |
| Convolution + ReLU | 5 × 5 filters, 20 channels | 3 × 3 filters, 64 channels |
| | | 3 × 3 filters, 64 channels |
| Max Pooling | 2 × 2 filters, 2 strides, no padding | |
| Convolution + ReLU | 5 × 5 filters, 50 channels | 3 × 3 filters, 128 channels |
| | | 3 × 3 filters, 128 channels |
| Max Pooling | 2 × 2 filters, 2 strides, no padding | |
| Full Connection + ReLU | 500 neurons | 256 neurons |
| | | 256 neurons |
| Softmax | 10 neurons | 10 neurons |

The structure of the generator is shown in Table 3. We referred to the generators used in GAN [11] for the structures. In order to limit the output range of the output layer, we used the hyperbolic tangent and converted the output range from $[-\infty, \infty]$ to $[-1, 1]$. The output was multiplied by $ns$ and used as noise to make an adversarial example.

**Table 3.** Architecture (Generator)

| Layer Name | MNIST | CIFAR10 |
|---|---|---|
| Input | 784 neurons | 3072 neurons |
| Full Connection + ReLU | 1200 neurons | 8000 neurons |
| | 1200 neurons | 8000 neurons |
| | | 8000 neurons |
| Full Connection + tanh | 784 neurons | 3072 neurons |

The parameters used in the simulation are shown in Table 4. The classifier and the generator were trained alternately with one iteration and the generator used a learning rate ten times larger than that of the classifier. Noise $(ns, \epsilon)$ used for learning was 0.2 for MNIST and 0.1 for CIFAR10. In ATd, AMT, and AMT of AMT+AT, the magnitude of noise was changed from 0.0 to 0.2 with equally spaced 100 gradations for MNIST. Similarly, it was changed from 0.0 to 0.1 for CIFAR10. In AT, the magnitude of noise was fixed at 0.2 for MNIST and 0.1 for CIFAR10.

**Table 4.** Parameters used in training and attack

| Parameter | MNIST | CIFAR10 |
|---|---|---|
| Learning rate of classifier | $0.01 \rightarrow 0.006$ | $0.01(\times 0.95$ per 4000 iter) |
| Learning rate of generator | (Learning rate of classifier) $\times$ 10 | |
| Momentum | 0.9 | $0.9(\times 0.5$ per 4000 iter) |
| Batch size | 128 per 1 iter | |
| Dropout rate (used in FC) | no dropout | 0.5 |
| No. of iterations | 10000 | 20000 |
| Max amounts $(L_\infty)$ of noise $(ns, \epsilon)$ | 0.2 | 0.1 |
| Added JSMA noise | 1.0 | 0.4 |

We evaluated the performance with the accuracy for the test data and the attack success rate. We generated 100 adversarial examples by each generation method, and gave them to each defense method. Then, we counted the number of adversarial examples which were not recognized as their original correct classes. Since this number shows the success of attacks, we used the ratio as the attack success rate.

## 4.2   Results

Table 5 shows the accuracy for the test data and the attack success rate of each generation algorithm under the white box attack scenario. Comparing the proposed AMT with AT and ATd, we can see that AMT has not acquired robustness for all three generation methods. However, comparing AMT with Random, AMT gained more robustness, and further robustness was obtained by combining AMT with adversarial training using FGSM (AMT+AT). These results show that the generator learns weak points of the classifier to some extent and the classifier overcomes them through training.

**Table 5.** Accuracy (%) and attack success rate (%) in white box setting

| Dataset | Method | Accuracy | FGSM(s) | FGSM(L) | JSMA(s) | L2CW(0) |
|---------|--------|----------|---------|---------|---------|---------|
| MNIST   | Normal | 99.14 | 42.3 | 74.1 | 96.7 | 100.0 |
|         | Random | 99.14 | 39.4 | 68.3 | 93.6 | 100.0 |
|         | AT | 99.15 | 8.2 | **1.2** | **61.0** | **99.1** |
|         | ATd | 99.26 | 2.6 | 13.9 | 88.5 | 100.0 |
|         | AMT | 99.11 | 30.7 | 62.5 | 90.6 | 100.0 |
|         | AMT+AT | 99.23 | **1.8** | 7.6 | 69.7 | 99.7 |
| CIFAR10 | Normal | 77.21 | 87.8 | 89.8 | 90.1 | 100.0 |
|         | Random | 76.56 | 90.3 | 90.3 | 90.8 | 100.0 |
|         | AT | 77.04 | 59.4 | 56.1 | **57.0** | 100.0 |
|         | ATd | 76.57 | **58.1** | 72.9 | 63.1 | 100.0 |
|         | AMT | 76.34 | 80.9 | 80.7 | 87.6 | 100.0 |
|         | AMT+AT | 76.76 | 68.5 | **42.0** | 76.6 | 100.0 |

Table 6 shows the attack success rate of each generation algorithm under the black box attack scenario. As for the proposed AMT, it is inferior in terms of robustness against FGSM attacks to the networks trained by FGSM (AT and ATd) in both datasets. However, in JSMA and L2CW attacks in CIFAR10 dataset, AMT shows better performance than AT and ATd. Moreover, like the results with the white box attacks, the combination of Adversarial Minimax Training and adversarial training (AMT+AT) has acquired the best robustness against adversarial examples in total. The results show that the proposed AMT+AT can learn a wide range of adversarial examples. Even if a specific weakness exists in the network like the FGSM of this simulation, the classifier can learn its weak point by adversarial training in combination with Adversarial Minimax Training, and it is possible to expect overall robustness acquisition.

**Table 6.** Attack success rate (%) in black box setting

| Dataset | Method | FGSM(s) | FGSM(L) | JSMA(s) | JSMA(L) | L2CW(s) | L2CW(L) |
|---------|--------|---------|---------|---------|---------|---------|---------|
| MNIST | Normal | 26.5 | 69.3 | 18.1 | 43.7 | 97.8 | 99.1 |
| | Random | 29.0 | 66.7 | 15.7 | 34.8 | 94.9 | 99.6 |
| | AT | **0.1** | 7.8 | 16.2 | 28.7 | 42.3 | 94.1 |
| | ATd | 1.6 | 10.4 | **12.2** | 25.7 | 53.0 | 95.8 |
| | AMT | 20.8 | 60.6 | 18.5 | 37.3 | 88.1 | 99.4 |
| | AMT+AT | 0.4 | **6.3** | 12.9 | **25.0** | **28.6** | **91.7** |
| CIFAR10 | Normal | 76.1 | 86.4 | 4.2 | 12.0 | 83.8 | 98.6 |
| | Random | 70.4 | 90.0 | 2.9 | 7.9 | 66.0 | 92.8 |
| | AT | 42.6 | 52.0 | 1.6 | 6.8 | 81.2 | 86.4 |
| | ATd | **32.8** | 69.1 | 2.2 | 8.0 | 82.7 | 80.9 |
| | AMT | 59.0 | 86.5 | 1.2 | 5.7 | **41.3** | 78.8 |
| | AMT+AT | 44.8 | **47.9** | **1.1** | **3.9** | 48.5 | **76.1** |

## 5 Conclusions

In this paper, we have proposed Adversarial Minimax Training as a method to improve robustness against adversarial examples. Like GAN, the proposed method uses two networks: a generator network and a classifier network. The generator network generates noise to make adversarial examples, and the classifier network learns the adversarial examples with original images. Computer simulation results show that (1) the generator learned weak points of the classifier, (2) the classifier obtained robustness against adversarial examples by training the weak points, and (3) the classifier was able to acquire further robustness by combining Adversarial Minimax Training with adversarial training.

For the future work, we would like to make it possible for classifiers to acquire further robustness using only Adversarial Minimax Training.

**Acknowledgments.** This work was supported in part by JSPS KAKENHI (Grant no. 16K00329).

## References

1. Szegedy, C., et al.: Intriguing properties of neural networks. arXiv:1312.6199 (2014)
2. Goodfellow, I., Shlens, J., Szegedy, C.: Explaining and harnessing adversarial examples. arXiv:1412.6572 (2015)
3. Carrara, F., Falchi, F., Caldelli, R., Amato, G., Fumarola, R., Becarelli, R.: Detecting adversarial example attacks to deep neural networks. In: 15th International Workshop on Content-Based Multimedia Indexing, no. 38, pp. 1–7. ACM (2017)
4. Meng, D., Chen, H.: MagNet: a two-pronged defense against adversarial examples. In: 2017 ACM SIGSAC Conference on Computer and Communications Security, pp. 135–147. ACM (2017)

5. Papernot, N., McDaniel, P., Wu, X., Jha, S., Swami, A.: Distillation as a defense to adversarial perturbations against deep neural networks. In: 37th IEEE Symposium on Security and Privacy, pp. 582–597. IEEE Press (2016)
6. Papernot, N., McDaniel, P., Jha, S., Fredrikson, M., Celik, B.Z., Swami, A.: The limitations of deep learning in adversarial settings. In: 1st IEEE European Symposium on Security and Privacy, pp. 372–387. IEEE Press (2016)
7. Carlini, N., Wagner, D.: Towards evaluating the robustness of neural networks. In: IEEE Symposium on Security and Privacy, pp. 39–57. IEEE Press (2017)
8. Kurakin, A., et al.: Adversarial Attacks and Defences Competition. arXiv:1804.00097 (2018)
9. Xiao, C., Li, B., Zhu, J., He, W., Liu, M., Song, D.: Generating Adversarial Examples with Adversarial Networks. arXiv:1801.02610 (2018)
10. Baluja, S., Fischer, I.: Adversarial Transformation Networks: Learning to Generate Adversarial Examples. arXiv:1703.09387 (2017)
11. Goodfellow, I., et al.: Generative adversarial nets. In: Advances in Neural Information Processing Systems 27, pp. 2672–2680. NIPS (2014)
12. Papernot, N., et al.: cleverhans v2.0.0: an adversarial machine learning library. arXiv:1610.00768 (2017)

# SATB-Nets: Training Deep Neural Networks with Segmented Asymmetric Ternary and Binary Weights

Shuai Gao[1]([✉]), JunMin Wu[2]([✉]), Da Chen[1], and Jie Ding[2]

[1] School of Software Engineering, University of Science and Technology of China,
SuZhou, China
{gshuai16,soloda}@mail.ustc.edu.cn
[2] Department of Computer Science and Technology, University of Science
and Technology of China, HeFei, China
jmwu@ustc.edu.cn, dj1993@mail.ustc.edu.cn

**Abstract.** Deep convolutional neural networks (CNNs) are both computationally intensive and memory intensive, making them difficult to deploy on embedded systems with limited hardware resources efficiently. To address this limitation, we introduce SATB-Nets, a method which trains CNNs with segmented asymmetric ternary weights for convolutional layers and binary weights for the fully-connected layers. We compare SATB-Nets with previous proposed ternary weight networks (TWNs), binary weight networks (BWNs) and full precision networks (FPWNs) on CIFAR-10 and ImageNet datasets. The result shows that our SATB-Nets model outperforms full precision model VGG16 by 0.65% on CIFAR-10 and achieves up to 29× model compression rate. On ImageNet, there is 31× model compression rate and only 0.15% accuracy degradation over the full-precision AlexNet model of Top-1 accuracy.

**Keywords:** Deep convolutional neural networks
Segmented asymmetric ternary and binary weights
Model compression · Embedded efficient neural networks

## 1 Introduction

Deep Neural Networks (DNNs) have inexorably pushed the amazing performances in lots of application domains including but not limited to the speech recognition [1,2] and computer vision, mainly including object recognition [3,4,6,23] and object detection [7,8,10]. A particular type of networks, named Convolution Neural Networks (CNNs), are being deployed to real world applications on smart phones and other embedded devices. However, it is difficult to deploy these computationally intensive and memory-intensive CNNs on embedded devices which are both computational resources limited and storage limited.

© Springer Nature Switzerland AG 2018
L. Cheng et al. (Eds.): ICONIP 2018, LNCS 11302, pp. 700–710, 2018.
https://doi.org/10.1007/978-3-030-04179-3_62

## 1.1   Binary Weight Networks and Model Compression

To address the storage and computational issues [5,21], methods that seek to binarize weights or activations in DNNs models have been proposed. BinaryConnect [11] binarizes the weights to $\{+1, -1\}$ with a single sign function. Binary Weight Networks [12] improve the models' capacity by adding an extra scaling factor on the basis of the previous method. BinaryNet [12] and XNOR-Net [13] binarize not only weights but also activations as extensions of the previous methods. These models eliminate most of the multiplication operations in the forward and backward propagations [16] and model compression rate achieves up to 32×, but there are also considerable accuracy loss.

## 1.2   Ternary Weight Networks and Model Compression

Nowadays, more and more researchers are engaged in the quantization of 2-bit neural networks especially the ternary weights quantization. Ternary weights networks (TWNs) [14] were introduced with the weights constrained to $\{-1, 0, +1\}$ to maximize scale model compression and minimize the precision loss of the model as far as possible. Compared with the previous binary quantization network, the accuracy loss has been reduced obviously because of the increased weights precision. However, there are also some tricks to improve the capacity of ternary weights networks with the different scaling factors for positive and negative weights.

We optimize the previous methods [14,20] by proposing Segmented Asymmetric Ternary and Binary Weights Networks (SATB-Nets) to explore higher model capacity and model compression rate. For each layer, we segment the weights vector space into many disjoint subspaces. In each subspace, we confine weights to three values $\{+W_{ls}^{pt}, 0, -W_{ls}^{nt}\}$ for convolutional (CONV) layers and two values $\{+W_{ls}^{pb}, -W_{ls}^{nb}\}$ for fully-connected (FC) layers, which can also be encoded with two bits and a single bit. Compared with TWNs [14] and BWNs [11] quantization method, our SATB-Nets are able to explore the local redundancy structure better and gain more stronger expressive abilities leading to better performance. In addition, the fixed scaling factors $\{+W_{ls}^{p*}, 0, -W_{ls}^{n*}\}$ provide more possibilities for computing acceleration.

## 2   Segmented Asymmetric Ternary and Binary Weights Networks

We will detailedly introduce how to obtain Segmented Asymmetric Ternary and Binary Weights Networks (SATB-Nets) and train them efficiently in this section.

### 2.1   Segmentation

Product quantization (PQ) [18] partitions the vector space into many disjoint subspaces to explore the redundancy of structures in vector space. Authors of

[9] proposes the segmentation of the weight matrix and then the performance of quantization in each subspace. Similarly, we partition weight matrix into several submatrices to improve the expression ability of the quantized networks:

$$W = [W^1, W^2, ..., W^k], \tag{1}$$

Where $W \in R^{m*n}$ and $W^i \in R^{m*(n/k)}$ assuming $n$ is divisible by $k$. We can quantify each submatrix $W^i$ with ternary and binary value. More segments lead to higher model capacity but will aggressively increase the codebook size. So, by using the same trick as described in [9], we fixed the number of segments $k$ to 8 to keep a satisfying balance between compression rate and output precision loss of the networks.

## 2.2   Asymmetric Binary Weights for Fully-Connected(FC) Layers

We constrain the full precision weights $W_{lsi}$ (lth layers, sth segments and ith parameters) to binary weights with values belong to $\{+W_{ls}^{pb}, -W_{ls}^{nb}\}$. The quantization function is shown in (2).

$$w_{lsi}^b = f_b(w_{lsi}^b) = \begin{cases} +W_{ls}^{pb} & w_{lsi} \geq 0 \\ -W_{ls}^{nb} & w_{lsi} < 0 \end{cases} \tag{2}$$

Here 0 is threshold and $\{W_{ls}^{pb}, W_{ls}^{nb}\}$ are the scaling factors. In order to get as well performance as possible, the minimization of Euclidian distance between the floating-point weights $W_{ls}$ and binary weights $W_{ls}^b$ is adopted and the optimization problem is transformed to (3):

$$\begin{cases} W_{ls}^* = arg\min J(W_{ls}^*) = arg\min \left\|W_{ls} - W_{ls}^b\right\|_2^2 \\ s.t. W_{ls}^* > 0; w_{lsi}^b \in \{+W_{ls}^{pb}, 0, -W_{ls}^{nb}\}; i = 1, 2, ..., n; s = 1, 2, ..., k \end{cases} \tag{3}$$

Substitute the binary function (2) into the formula (3), we can get the expression as (4):

$$J(W_{ls}^*) = \left\|W_{ls} - W_{ls}^b\right\|_2^2 = \sum_{i \in I_*}^{n_s} ||w_{lsi}| - W_{ls}^*|^2 \tag{4}$$

where $I_* = \{I_p, I_n\}$, $I_p = \{i | w_{lsi} \geq 0\}$, $I_n = \{i | w_{lsi} < 0\}$. According to (4), It is not complicated to obtain binary weights from the floating-point weights as (5):

$$W_{ls}^* = \frac{1}{|I_*|} \sum_{i \in I_*}^{n_s} |w_{lsi}| \tag{5}$$

where $|I_*|$ denotes the number of elements in $I_*$ in each segment.

## 2.3 Asymmetric Ternary Weights for Convolutional (CONV) Layers

Similarly, we also constrain the floating-point weights $W_{lsi}$ ($l$th layers, $s$th segments and $i$th parameters) to ternary weights with values belong to $\{+W_{ls}^{pt}, 0, -W_{ls}^{nt}\}$. The quantization function is shown in (6).

$$w_{lsi}^t = f_t(w_{lsi}^t | \Delta_{ls}^p, \Delta_{ls}^n) = \begin{cases} +W_{ls}^{pt} & w_{lsi} > \Delta_{ls}^p \\ 0 & -\Delta_{ls}^n \leq w_{lsi} \leq \Delta_{ls}^p \\ -W_{ls}^{nt} & w_{lsi} < -\Delta_{ls}^n \end{cases} \quad (6)$$

Here $\{\Delta_{ls}^p, \Delta_{ls}^n\}$ are the threshold and $\{W_{ls}^{pt}, W_{ls}^{nt}\}$ are the scaling factors. The optimization problem is formulated as (7):

$$\begin{cases} W_{ls}^* = \arg\min J(W_{ls}^*) = \arg\min \|W_{ls} - W_{ls}^t\|_2^2 \\ s.t. W_{ls}^* > 0; W_{lsi}^t \in \{+W_{ls}^{pt}, 0, -W_{ls}^{nt}\}; i = 1, 2, ..., n; s = 1, 2, ..., k \end{cases} \quad (7)$$

Substitute the ternary function (6) into the formula (7), we can get the expression as (8):

$$J(W_{ls}^*) = \|W_{ls} - W_{ls}^t\|_2^2 = |I_{\Delta_{ls}^*}| * (W_{ls}^*)^2 - 2 * (\sum_{i|i \in I_{\Delta_{ls}^*}}^{n_s} |w_{lsi}|) * (W_{ls}^*) + C \quad (8)$$

Where $I_{\Delta_{ls}^*} = \{I_{\Delta_{ls}^p}, I_{\Delta_{ls}^n}\}$, $I_{\Delta_{ls}^p} = \{i | w_{lsi} > \Delta_{ls}^p\}$, $I_{\Delta_{ls}^n} = \{i | w_{lsi} < -\Delta_{ls}^n\}$ and $|I_{\Delta_{ls}^*}|$ denotes the number of elements in $I_{\Delta_{ls}^*}$ in each segment. $\Delta_{ls}^p$ and $\Delta_{ls}^n$ are independent together. $C = \sum_i^{n_s} |w_{lsi}|^2$ is a $\{W_{ls}^{pt}, W_{ls}^{nt}\}$-independent constant. Therefore, our scaling factors $\{W_{ls}^{pt}, W_{ls}^{nt}\}$ can be simplified to:

$$W_{ls}^* = \arg\min J(W_{ls}^*) = \arg\min(|I_{\Delta_{ls}^*}| * (W_{ls}^*)^2 - 2 * (\sum_{i|i \in I_{\Delta_{ls}^*}}^{n_s} |w_{lsi}|) * (W_{ls}^*)) \quad (9)$$

According to (9), It is not complicated to obtain tenary weights from the floating-point weights as (10):

$$W_{ls}^* = \frac{1}{|I_{\Delta_{ls}^*}|} \sum_{i|i \in I_{\Delta_{ls}^*}}^{n_s} |w_{lsi}| \quad (10)$$

Here $\{\Delta_{ls}^p, \Delta_{ls}^n\}$ are both positive values. There is no straightforward solutions to figure out $\Delta_{ls}^p$ and $\Delta_{ls}^n$ as [17]. But values are generated from uniform or normal distribution empirically, adopting the method mentioned in [14], the thresholds are as following:

$$\Delta_{ls}^* \approx 0.7 * \frac{1}{|I^*|} \sum_{i|i \in I^*}^{n_s} |w_{lsi}| \quad (11)$$

Where $I^* = \{I^p, I^n\}$, $I^p = \{i|w_{lsi} \geq 0|i = 1, 2, ..., n_s\}$, $I^n = \{i|w_{lsi} < 0|i = 1, 2, ..., n_s\}$. Finally, by substituting (10) and (11) to (6), Ternary weights can be easily obtained from the floating-point weights.

### 2.4    Heterogeneous Quantized Weights Structure

In order to achieve a good balance between compression rate and accuracy, we train CNNs with ternary weights for convolutional layers and binary weights for the fully-connected layers. On the one hand, these densely and highly redundancy fully-connected layers take up most of the parameters, binarization is more helpful in removing redundancy and higher proportion of compression. On the other hand, [21] shows that convolutional layers require more bits of precision than fully-connected layers, so ternary weights for convolutional layers improve the expression capacity. In addition, the quantization values of zero for convolutional layers reduce the calculation of the multiplication to accelerate the networks.

### 2.5    Train the SATB-Nets with Stochastic Gradient Descent (SGD) Method

Stochastic Gradient Descent (SGD) algorithm is used as the training algorithm for SATB-Nets, about which more detail is shown in Algorithm 1.

The whole training process is almost the same as normal training method, except that segmented asymmetric ternary weights for convolutional (CONV) layers and binary weights for the fully-connected (FC) layers are used in forward propagation (**step 1**) and backward derivation (**step 2**), which is similar to training method as BinaryConnect [11]. In order to overcome the difficulty of convergence of models using quantized weights, we reserved the full precision floating-point weights to update weights to obtain the tiny changes in each iteration (**step 3**).

In addition, Batch Normalization (BN) [24] and learning rate scaling, as two useful tricks, are adopted. We also use momentum for acceleration.

## 3    Experiments

In this section, we benchmark SATB-Nets with full precision weights networks (FPWNs), binary weights networks (BWNs) and Ternary Weights Networks (TWNs) on the small scale datasets (CIFAR-10) and the large scale dataset (ImageNet datasets). We adopt the VGG [6] networks on Cifar-10 and the AlexNet [3] on ImageNet. To be fair, the following terms are identical: network architecture, learning rate scaling procedure (multi-step), optimization method (SGD with momentum) and regularization method (L2 weight decay). We conjecture that SATB-Nets have sufficient expressiveness in the depth networks and adopt the data augmentation and the sparse weights like dropout [15] to prevent overfitting. In addition, all the neural networks are deployed on framework of Caffe [25]. For more detailed configurations, we can see Table 1.

**Algorithm 1.** Training a DNNs with SATB-Nets. **L** is the number of layers and **K** is the number of segments. Full precision weights for layer $l$ and segment $s$ are $W_{ls}$ and $b_{ls}$ and the output is $a_{ls}$. **J** is the cost function of networks and **Binary**$(W_{ls}^b)$ and **Ternary**$(W_{ls}^t)$ mean to quantize weights to binary in fully connected layers (FC) and ternary values in convolution layers (CONV).

---

**Begin**

  **1. Forward propagation:**

  **for** $l \leftarrow 1$ to $L-1$ **do**

    **for** $s \leftarrow 1$ to $K-1$ **do**

      **if** $CONV$ **then**

        $W_{ls}^t \leftarrow$ **Ternary**$(W_{ls})$

        $a_{l+1s} \leftarrow f\left(W_{ls}^t * a_{ls} + b_{ls}\right)$

      **end if**

      **if** $FC$ **then**

        $W_{ls}^b \leftarrow$ **Binary**$(W_{ls})$

        $a_{l+1s} \leftarrow f\left(W_{ls}^b * a_{ls} + b_{ls}\right)$

      **end if**

    **end for**

  **end for**

  **2. Backward derivation:**

  **for** $l \leftarrow 1$ to $L-1$ **do**

    **for** $s \leftarrow 1$ to $K-1$ **do**

      **if** $CONV$ **then**

        $\frac{\partial J}{\partial a_{ls}} \leftarrow \left(\left(W_{ls}^t\right)^T * \frac{\partial J}{\partial a_{l+1s}}\right) \circ f'$

        // $\circ$ means element-wise product

      **end if**

      **if** $FC$ **then**

        $\frac{\partial J}{\partial a_{ls}} \leftarrow \left(\left(W_{ls}^b\right)^T * \frac{\partial J}{\partial a_{l+1s}}\right) \circ f'$

      **end if**

      $\frac{\partial J}{\partial W_{ls}} \leftarrow \frac{\partial J}{\partial a_{l+1s}} * \left(a_{ls}\right)^T$

      $\frac{\partial J}{\partial b_{ls}} \leftarrow \frac{\partial J}{\partial a_{l+1s}}$

    **end for**

  **end for**

  **3. Weights update:**

  **for** $l \leftarrow 1$ to $L-1$ **do**

    $W_{ls} \leftarrow W_{ls} - \eta * \frac{\partial J}{\partial W_{ls}}$

    $b_{ls} \leftarrow b_{ls} - \eta * \frac{\partial J}{\partial b_{ls}}$

  **end for**

**End**

---

## 3.1   VGGNets on CIFAR-10

CIFAR-10 is a benchmark image classification dataset which consists of 60K $32 \times 32$ color images and Five sixths of them belong to a training set and the rest belong to the test set. To prevent over-fitting while training VGG [6] networks, data-augmentation is used following [4]. A random $32 \times 32$ crop is from the padded images on which 4 pixels are padded each side. The cropped images

**Table 1.** Networks and some hyper-parameters of them on datasets

|  | Cifar-10 | | | ImageNet |
|---|---|---|---|---|
| Networks | VGG10 | VGG13 | VGG16 | AlexNet |
| Weight decay | $5 \times 10^{-4}$ | $5 \times 10^{-4}$ | $5 \times 10^{-4}$ | $5 \times 10^{-4}$ |
| Momentum | 0.9 | 0.9 | 0.9 | 0.9 |
| Mini-batch size of BN | 100 | 100 | 100 | 256 |
| Learning rate | 0.01 | 0.01 | 0.01 | 0.0001 |
| Learning rate decay epochs (0.1) | 50, 100, 150, 200 | | | 50, 60 |

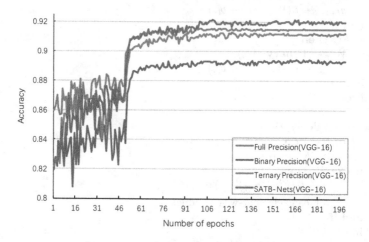

**Fig. 1.** Validation accuracy curves of VGG16 on Cifar-10

are used for training while original images are for testing. We adopt VGG16 [6] architecture for the experiment firstly. Beside, in order to solve the difficulty of training so deep neural network, we initialize these networks with full-trained full precision model.

We compare SATB-Nets with the FPWNs, BWNs and TWNs. The result (Fig. 1 and Table 2) shows that SATB-Nets from VGG16 outperforms BWNs, TWNs and FPWNs by 2.52%, 1.09% and 0.65% respectively. In the meanwhile, SATB-Nets from VGG10, VGG13 and VGG16 are always outperforming BWNs and TWNs.

To our surprise, the SATB-Nets constrained from VGG13 and VGG16 outperform the full precision weights networks. According to our analysis, we conjecture that our SATB-Nets have adequate capacity for expression and the sparse weights networks prevent over-fitting like dropout [15].

For the more sufficient experimental verification, we expand the experiment to VGG13 removing last 3 convolutional layers of VGG16 [6] and VGG10

**Table 2.** Validation accuracy of VGGNet on CIFAR-10 (%)

| Model | FPWNs | BinaryNet | TWNs | SATB-Nets | Improvements(%) |
|-------|-------|-----------|------|-----------|-----------------|
| VGG-10 | **90.45** | 88.76 | 89.50 | **90.40** | $-0.05/2.64/0.90$ |
| VGG-13 | **91.25** | 89.09 | 90.60 | **91.96** | $+0.71/2.87/1.36$ |
| VGG-16 | **91.50** | 89.63 | 91.06 | **92.15** | $+0.65/2.52/1.09$ |

**Table 3.** Compression ratio for VGG-16 (Byte)

| Layer | Full-weights | BinaryNet | TWNs | SATB-Nets |
|-------|--------------|-----------|------|-----------|
| CONV | 58.84 M | 1.84 M | 3.68 M | 3.68 M |
| FC | 494.52 M | 15.44 M | 30.92 M | 15.44 M |
| Total | 553.36 M | 17.28 M ($32\times$) | 34.24 M ($16\times$) | 19.12 M ($\mathbf{29\times}$) |

removing last 6 convolutional layers. The results met our expectations which are listed in Table 2. In the meanwhile, Table 3 shows the compression ratio of VGG-16.

## 3.2 AlexNet on ImageNet

We further examine the performance of SATB-Nets on the ImageNet ILSVRC-2012 dataset, which has over $1.2M$ training examples and $50K$ validation examples. We use the AlexNet Caffe model [26] as the reference model. Beside, in order to solve the difficulty of training so deep neural network, we initialize these networks with full-trained full precision model.

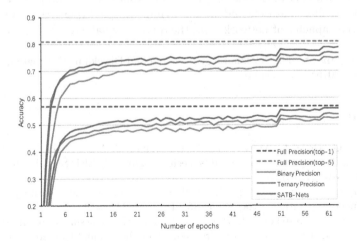

**Fig. 2.** Validation accuracy curves of AlexNet on ImageNet

**Table 4.** Validation accuracy of AlexNet on ImageNet(%)

| Accuracy | FPWNs | BWNs | TWNs | SATB-Nets |
|----------|-------|-------|-------|-----------|
| Top-1 | **56.72** | 52.65 | 54.01 | **56.57** |
| Top-5 | **80.17** | 74.89 | 76.43 | **78.76** |

**Table 5.** Compression statistics for AlexNet (Byte)

| Layer | Full-weights | BinaryNet | TWNs | SATB-Nets |
|-------|--------------|-----------|------|-----------|
| CONV | 9.32 M | 0.28 M | 0.60 M | 0.60 M |
| FC | 234.52 M | 7.32 M | 14.64 M | 7.32 M |
| Total | 243.84 M | 7.32 M (32×) | 15.24 M (16×) | 7.92 M (**31×**) |

Our training curves are shown in Fig. 2, the complete result (Fig. 2 and Table 4) shows that SATB-Nets reaches the top-1 validation accuracy of 56.57% which has only 0.15% accuracy degradation over full precision counterpart.

Tables 3 and 5 show the compression ratio of VGG-16 and AlexNet. SATB-Nets achieve up to 29× and 31× model compression rate respectively which are closed to the binary weights compression without impacting accuracy.

## 4    Conclusion

In this paper, we propose ternary and binary weights networks optimization problems. Next, We propose SATB-Nets which nearly achieve up to binary compression ratio. Meanwhile, experiments show that benchmarks demonstrate the superior performance of the method which we proposed. Next step, we will apply the method to more datasets and models to more deeply explore the relationships between the capacity of networks and the quantized values.

**Acknowledgment.** The National Key Research and Development Program of China (Grants No. 2016YFB1000403).

## References

1. Hinton, G., Deng, L., Yu, D., Dahl, G.E., Mohamed, A.R., Jaitly, N.: Deep neural networks for acoustic modeling in speech recognition. IEEE Signal Process. Mag. **29**(6), 82–97 (2012)
2. Lecun, Y., Bottou, L., Bengio, Y., Haffner, P.: Gradient-based learning applied to document recognition. Proc. IEEE **86**(11), 2278–2324 (1998)
3. Krizhevsky, A., Sutskever, I., Hinton, G.E.: ImageNet classification with deep convolutional neural networks. In: Proceedings of the Annual Conference on Neural Information Processing Systems, pp. 1097–1105 (2012)

4. He, K., Zhang, X., Ren, S., Sun, J.: Deep residual learning for image recognition. In: IEEE Conference on Computer Vision and Pattern Recognition, pp. 770–778. IEEE Computer Society (2016)

5. Esser, S.K., Merolla, P.A., Arthur, J.V., Cassidy, A.S., Appuswamy, R., Andreopoulos, A.: Convolutional networks for fast, energy-efficient neuromorphic computing. In: Proceedings of the National Academy of Sciences of the USA, pp. 1441–1446 (2016)

6. Simonyan, K., Zisserman, A.: Very deep convolutional networks for large-scale image recognition (2014). arXiv preprint arXiv:1409.1556

7. Girshick, R., Donahue, J., Darrell, T., et al.: Rich feature hierarchies for accurate object detection and semantic segmentation. In: Proceedings of the IEEE Conference on Computer Vision and Pattern Recognition, pp. 580–587 (2014)

8. Ren, S., He, K., Girshick, R., Sun, J.: Faster R-CNN: towards real-time object detection with region proposal networks. In: International Conference on Neural Information Processing Systems. vol. 39, pp. 91–99. MIT Press, Cambridge (2015)

9. Gong, Y., Liu, L., Yang, M., et al.: Compressing deep convolutional networks using vector quantization (2014). arXiv preprint arXiv:1412.6115

10. Girshick, R.: Fast r-cnn. In: Proceedings of the IEEE Conference on Computer Vision and Pattern Recognition, pp. 1440–1448 (2015)

11. Courbariaux, M., Bengio, Y., David, J.P.: Binaryconnect: Training deep neural networks with binary weights during propagations. In: Proceedings of the Annual Conference on Neural Information Processing Systems, pp. 3123–3131 (2015)

12. Hubara, I., Courbariaux, M., Soudry, D., et al.: Binarized neural networks. In: Proceedings of the Annual Conference on Neural Information Processing Systems, pp. 4107–4115 (2016)

13. Rastegari, M., Ordonez, V., Farhadi, A., et al.: XNOR-Net: Imagenet classification using binary convolutional neural networks. In: Proceedings of the European Conference on Computer Vision, pp. 525–542 (2016)

14. Li, F., Zhang, B., Liu, B.: Ternary weight networks (2016). arXiv preprint arXiv:1605.04711

15. Srivastava, N., Hinton, G., Krizhevsky, A., Sutskever, I., Salakhutdinov, R.: Dropout: a simple way to prevent neural networks from overfitting. J. Mach. Learn. Res. **15**(1), 1929–1958 (2014)

16. Lin, Z., Courbariaux, M., Memisevic, R., Bengio, Y.: Neural networks with few multiplications (2015). arXiv preprint arXiv:1510.03009

17. Hwang, K., Sung, W.: Fixed-point feedforward deep neural network design using weights $+1$, 0, and $-1$ (2014). arXiv preprint arXiv:1405.3866

18. Jgou, H., Douze, M., Schmid, C.: Product quantization for nearest neighbor search. IEEE Trans. Pattern Anal. Mach. Intell. **33**(1), 117–128 (2010)

19. Lee, C.Y., Xie, S., Gallagher, P., Zhang, Z., Tu, Z.: Deeply-supervised nets. In: In Proceedings of the Eighteenth International Conference on Artificial Intelligence and Statistics, pp. 562–570 (2015)

20. Ding, J., Wu, J.M., Wu, H.: Asymmetric ternary networks. In: International Conference on TOOLS with Artificial Intelligence IEEE Computer Society, pp. 61–65 (2017)

21. Han, S., Mao, H., Dally, W.J.: Deep compression: Compressing deep neural networks with pruning, trained quantization and huffman coding (2015). arXiv preprint arXiv:1510.00149

22. Deng, J., Dong, W., Socher, R., et al.: Imagenet: a large-scale hierarchical image database. In: Proceedings of the IEEE Conference on Computer Vision and Pattern Recognition, pp. 248–255 (2009)

23. Szegedy, C., Liu, W., Jia, Y., et al.: Going deeper with convolutions. In: Proceedings of the IEEE Conference on Computer Vision and Pattern Recognition, pp. 1–9 (2015)
24. Ioffe, S., Szegedy, C.: Batch normalization: accelerating deep network training by reducing internal covariate shift. In: Proceedings of the International Conference on Machine Learning, pp. 448–456 (2015)
25. Jia, Y., Shelhamer, E., Donahue, J., et al.: Caffe: convolutional architecture for fast feature embedding. In: Proceedings of the International Conference on Multimedia Retrieval, pp. 675–678 (2014)
26. BVLC: Caffe model zoo. http://caffe.berkeleyvision.org/model_zoo

# Asynchronous Methods for Multi-agent Deep Deterministic Policy Gradient

Xuesong Jiang$^{(\boxtimes)}$, Zhipeng Li, and Xiumei Wei

College of Information, Qilu University of Technology
(Shandong Academy of Sciences), Jinan 250353, China
{jxs,wxm}@qlu.edu.cn, lizhipengqilu@gmail.com

**Abstract.** We propose a variant framework for optimizing the deep neural network controller using asynchronous gradient descent method for the Multi-Agent Deep Deterministic Policy Gradient (MADDPG) algorithm. Using CPU's multicore to create multiple parallel environments, each thread interacts with its own environment replica. Each copy uses prioritized batch data. The evaluation method of Critic was adjusted, and advantage was used as the evaluation of action. The batch data processed by multiple copies is collected and the loss values of each copy are calculated. Using batch data with maximum loss as sampling for global network. In addition, we show the successful application of multi-agent collaboration based on asynchronous methods. The results show that the mean episode reward is higher than the reward obtained by previous algorithm.

**Keywords:** MADDPG · Asynchronous methods · Multithreading
Prioritized batch data

## 1 Introduction

In recent years, with the use of deep neural networks in reinforcement learning, more and more reinforcement learning algorithms use Deep Neural Networks (DNNs) as a function approximator to solve problems in the real world. Q-learning is the most classical algorithm for reinforcement learning. The Q-learning algorithm has been applied in some fields. In Conventional Q-learning algorithm, the state and the action space are discrete and the dimension is not high. When you use the Q table, you can store the Q value of each state action pair. When the state and the action space are high-dimensional continuous, using a Q-Table is not practical. In February 2015, Google published a paper Human in Human-level control through deep reinforcement learning, which describes how to let the computer learn to play Atari 2600 electronic games. The input of the Google algorithm is only the image of the game screen and the score of the game. Without human intervention, the computer learned the game play itself and broke the record of the human player in 29 games. DQN (Deep Q-Learning) [1] is a pioneering work of Deep Reinforcement Learning (DRL). It is a combination of deep learning and reinforcement learning to achieve end-to-end learning from Perception to action. New algorithms to promote the vigorous development of the field of learning.

© Springer Nature Switzerland AG 2018
L. Cheng et al. (Eds.): ICONIP 2018, LNCS 11302, pp. 711–721, 2018.
https://doi.org/10.1007/978-3-030-04179-3_63

When we build the DQN agent, we use the neural networks to estimate the value of the Q(s, a) function. In another paper, the author uses different methods to do. Since policy is a function of state, then we can directly estimate the choice of policy based on the input of the state. Therefore, the policy gradient [2] is proposed, and a probability distribution is output according to the state of the input. In reality. Due to lack of sampling, large variance, etc., the policy gradient can easily collapse to a local solution. The actor-critic algorithm combines value-based and policy-based methods. The policy network is an actor that outputs actions. The value network is a critic to evaluate the quality of the selected action of the actor network and generate a td_error signal to guide the update of the critic network and the actor network. The input of the actor network is state, output action, function fitting with DNNs, tanh or sigmod for the continuous action neural networks output layer, and the probability output for the discrete action with softmax as the output layer. Critic network input is state and action, and output is Q value. The input to the actor network is state and the output is action. For continuous actions, the output layer of the neural network can use tanh or sigmod, while the discrete action uses softmax as the output layer to achieve the effect of the probability output. The critic network's input is state and action, and the output is Q value.

DQN is a discrete-oriented control algorithm, that is, the output action is discrete. Corresponding to the Atari game, only a few discrete keyboard or joystick buttons are needed for control. However, the control problem is continuous in practice, so the high-dimensionality cannot be solved by the traditional DQN method. Deep Deterministic Policy Gradient (DDPG) [3] absorbs the feature of Actor-critic that allows policy gradients to be updated step by step, and also absorbs the essence of DQN (using a experience replay [4] and two sets of neural networks with the same structure but different update frequency of parameters). Last July, researchers from OpenAI company invented a new algorithm, Multi-Agent Deep Deterministic Policy Gradient (MADDPG) [3], which is suitable for centralized learning and decentralized execution in multi-agent environments, and allows agents to learn to cooperate and compete.

In order to alleviate the instability of the traditional policy gradient method combined with the neural network, all kinds of depth policy gradient methods (such as DDPG, SVG, etc.) have adopted an experience replay mechanism to eliminate the correlation between training data. However, there are two problems with the experience replay mechanism. Each real-time interaction between the agent and the environment requires a lot of memory and computing power. The experience playback mechanism requires the agent to use the off-policy method to learn, and the off-policy method can only be updated based on the data generated by the old policy. To solve these problems, Mnih et al. proposed a lightweight DRL framework based on the idea of Asynchronous Reinforcement Learning (ARL) [5], which use the idea of multi-threading, Performing multiple agents asynchronously, and removing the correlation between state transition samples generated during training through the different states experienced by parallel agents. Only a standard multi-core CPU is required to implement the algorithm, in terms of effect. Both time and resource consumption are superior to traditional methods. At present, another algorithm is reinforcement learning with unsupervised auxiliary tasks [6]. The algorithm is based on Asynchronous Advantage Actor-Critic (A3C) and adds many auxiliary tasks. The A3C algorithm is on policy, while the auxiliary task that uses the experience replay in A3C is off policy. The efficiency of this algorithm has been greatly improved.

Combining with previous research, this paper combines asynchronous thought and multi-agent cooperation algorithm Multi-Agent Deep Deterministic Policy Gradient (MADDPG) [7], then proposes a new algorithm structure, Asynchronous Advantage Multi-Agent Deep Deterministic Policy Gradient (AMADDPG). Different from the algorithm structure in A3C, this algorithm uses an experience replay [8]. The function of the experience replay is mainly to solve the problems of correlation and non-static distribution. The specific approach is to store the transfer samples $(x^j, a^j, r^j, x^{\prime j})$ obtained by the interaction between each time step agent and the environment into the experience replay unit, and then randomly take the training samples. Taking some minibatch training. It has a central brain and multiple workers. Multiple workers are copies of the central brain. The central brain and workers have multiple centralized Critic networks. The data in the experience replay is divided into N size at first, and then the worker samples from the each size to perform operations on a single machine through multiple threads of a CPU to achieve multiple workers running simultaneously. Advantage is used as the evaluation method of action to calculate the maximum loss in each worker network. The batch data of the worker network with the largest loss is selected as the sampling data of the central brain batch, and the Q value of critic is updated by one-step. By calculating the mean episode reward through experiments, we find that our algorithm structure is higher than the mean episode reward obtained by the previous algorithm structure, and the processing time is faster than before.

## 2  Related Work

The latest advances in deep reinforcement learning stem from the constant optimization of new algorithms and related systems. The study of algorithmic space seems to be the most common method. Traditional non-deep learning function fitting is more of an artificial feature and a linear model fitting. These years have been accompanied by profound success in supervised learning in recent years. Deep neural networks are used to fit Q-values end-to-end, that is DQN. DQN uses experience replay to remove correlation by random sampling, but at the same time brings with it computational cost and memory problems. The idea of Double Q-learning is to separate the networks that select and evaluate the action, which can reduce the overestimation of DQN. Comparing with the previous method, Dueling DQN's [9] innovation lies in changing the network structure into two streams, which is the state value function and the advantage function. Dueling net combines DDQN [10] and Prioritized Experience Replay training methods. DDQN can estimate more accurate Q values and achieve more stable and effective strategies in some Atari2600 games.

The Asynchronous Advantage Actor-Critic (A3C) [5] algorithm is a better and more general depth-enhanced learning algorithm proposed by DeepMind in 2015 than DQN. The A3C algorithm completely uses the Actor-Critic framework, and introduces the idea of asynchronous training, which greatly speeds up the training while increasing performance. The basic idea of the A3C algorithm, is to evaluate the outcome of the output. If the action is considered as advantage, then the Actor Network is adjusted to increase the likelihood of the action. On the other hand, if the action is considered as disadvantage, the possibility of the action appearing is reduced. Through repeated training, constantly adjust the mobile network to find the best action. AlphaGo's [11] self-learning is also based on this idea. The A3C algorithm uses the idea of asynchronous training to increase the training speed. That is, multiple training environments are started at the same time, sampling is performed at the same time, and the collected samples are directly used for training. Compared to the DQN algorithm, the A3C algorithm does not need to use an experience replay to store historical samples, which save storage space, and using asynchronous training, which greatly accelerate the data sampling rate. So those methods improve the training speed. At the same time, A3C algorithm use multiple different training environments to collect samples. The distribution of the samples is more uniform, which is more conducive to the training of neural networks. The A3C algorithm has made improvements in the above aspects, making its average score on the Atari game four times that of the DQN algorithm. It has achieved a tremendous improvement, and the training speed has also multiplied. Therefore, the A3C algorithm replaces DQN to become a better depth enhancement learning algorithm.

The Unsupervised Reinforcement and Auxiliary Learning (UNREAL) [12] algorithm is the latest deep learning algorithm proposed by DeepMind in November 2016. The UNREAL algorithm is based on the A3C algorithm. It uses another method to improve the algorithm by training multiple auxiliary tasks while training A3C. The UNREAL algorithm further improves performance and speed based on the A3C algorithm, achieving 8.8 times human performance on the Atari game, and reaching a human level of 87% in the 3D maze Labyrinth of the first perspective. Become the best depth deep reinforcement learning algorithm.

Babaeizadeh M et al. proposed a hybrid CPU/GPU version of the A3C,which is Reinforcement Learning through Asynchronous Advantage Actor-Critic on a GPU (GA3C) [13]. GA3C is similar to A3C in that it collects samples but does not need to copy a model. It only needs to add the current State as a request to the Prediction Queue before each action is selected. After performing n steps, it will backtrack and calculate the total return for each step, and add the n states to the training queue. Predictor populates the request sample in the prediction queue as a minibatch into the GPU's network model, and returns the model's predicted action to the respective Agent. To reduce latency, you can use multiple threads to parallel multiple predictors. Compared with the A3C algorithm, the GA3C algorithm adds GPUs to operations, which makes the training speed of the GA3C significantly faster and reduces memory consumption.

## 2.1    Reinforcement Learning

Reinforcement learning is a branch of machine learning, emphasizing how to take action based on the environment in order to maximize the expected benefits. Its essence is to solve a series of decision problems. The source of reinforcement learning is the Markov decision process, which is a discrete-time random process described by six-element tuple {S, A, D, P, r, J} [14]. In this six tuple, S is the state space of the limited environment; A is the limited system action space; D is the probability distribution of the initial state. $P(s, a, s) \in [0, 1]$ is to select a in the state s. The probability that the state changes to $s_{t+1}.r(s, a, s_{t+1})$ is an immediate reward for the learning system to perform action a from state s to state $s_{t+1}$ [15, 16].

The core issue of Markov Decision Process (MDP) is to find a policy for decision makers. Once the Markov decision process is combined with the policy in this way, the behavior of each state can be solved, and the resulting combined behavior behaves like a Markov chain. The MDP goal is to choose a policy that will maximize some cumulative function of random rewards, usually the sum of expected discounts over a potentially infinite time frame. After defining the MDP policy, the corresponding value function of the MDP can be divided into a state value function and an action value function. The state-valued function $v_\pi(s)$ starts from the state s and is expected to behave according to the policy $\pi$. The state-valued function can be used to evaluate the state.

The Q-Learning algorithm is a classical algorithm in the field of reinforcement learning. Its goal is to achieve the target state and obtain the highest reward. Once it reaches the target state, the final reward remains unchanged. Therefore, the target state is also called the absorption state. The agent under the Q-Learning algorithm does not know the overall environment but knows which actions can be selected in the current state. In general, we need to construct an instant reward matrix R that represents the action reward value from the state s to the next state $s_{t+1}$. The Q matrix that guides the agent's actions is calculated from the instant reward matrix R. The Q matrix is the brain of the agent. Initially, Q matrix elements are all initialized to 0, indicating that the current agent brain is blank and does not know anything. Q-learning is a very powerful algorithm, but its main drawback is its lack of versatility. If you understand Q-learning as updating a number in a two-dimensional array (action space, state space), it is actually similar to dynamic programming. This shows that the Q-learning agent does not know what action to take on the unseen state. In other words, the Q-learning agent has no ability to value the unseen state. To solve this problem, DQN introduces neural networks to get rid of two-dimensional arrays. DQN has made some modifications to Q-learning. DQN uses a deep convolutional neural network to approximate the value function. DQN uses an empirical playback training to enhance the learning process. DQN sets the target network independently to handle the TD bias in the time difference algorithm. Actor- Critic incorporates two types of reinforcement learning algorithms based on values (such as Q learning) and based on the probability of actions (such as Policy Gradients). Actor's predecessor was Policy Gradients. This method can select appropriate actions in continuous action. Critic's predecessor is Q-learning or other value-based learning method. It can perform single-step update, but it cannot handle continuous action, while traditional Policy Gradients are round updates. In order to handle both continuous and single-step updates, DDPG appeared. Actor network's input is state, and output is action. Critic network's input is state and action, and output is Q value. Therefore, Actor Critic is formed. However, the traditional Actor-Critic has high correlation and learning, so people have learned the methods in DQN. They use a experience replay and two neural networks with the same structure but different update frequency of parameters to train. Deepmind company proposed Deep Deterministic Policy Gradient (DDPG) in 2016, which is a policy learning method that integrates deep learning neural network into Deterministic Policy Gradient (DPG) [17]. The core improvement relative to DPG is the use of convolutional neural networks as a simulation of the strategy function μ and Q functions. i.e., the policy network and the Q network, then the deep learning method is used to train the above neural networks. The algorithm makes use of the behavior evaluation method in DQN to establish a neural network with a behavioral function network and an evaluation function network. The emergence of DDPG algorithm solves the problem of continuous decision well.

## 2.2    Multi-agent Deep Deterministic Policy Gradient

It has been an important research topic in the field of artificial intelligence for the agent society to cooperate and compete. It is also a necessary condition for the realization of universal artificial intelligence. Lowe et al. proposed a multi-agent deep deterministic policy gradient (MADDPG) [7] for cooperative and competitive hybrid environments. MADDPG extends the DDPG reinforcement learning algorithm to achieve centralized learning and distributed execution of multi-agents, allowing agents to learn and cooperate with each other. MADDPG outperforms DDPG in multiple testing tasks. It is a simple extension of the actor-critic policy gradient method. Critic adds additional information on other agent's policy.

Critic's network structure is inputting observations and actions, and outputting the Q value of each agent. Each actor has a crisp that collects the observations and actions of all agents and outputs the Q of a single actor to update the actor. For each agent, it samples a random mini batch of S samples $(x^j, a^j, r^j, x'^j)$ from the reply buffer D to the Critic network. The Q value of the output is fed back to the Actor, whose objective function is:

$$y^j = r_i^j \gamma Q_i^{\mu'}\left(x'^j, a_1', \ldots, a_N'\right)\big|_{a_k' = \mu_k'(o_k^j)} \tag{1}$$

Q in the formula is output by the above mentioned Critic network. r is a single agent reward and is used to update the loss of the Critic network. The loss formula is

$$\mathcal{L}\left(\emptyset_i^j\right) = -E_{o_j, a_j}\left[\log \hat{\mu}_i^j(a_j|o_j) + \lambda H\left(\hat{\mu}_i^j\right)\right] \tag{2}$$

Where H is the entropy of the policy distribution. Each agent can additionally maintain an approximation $\hat{\mu}_{\phi_i^j}$ relating to the real policy of the agent j, $\mu_j$ (where φ are approximate parameter). After the Actor receives the Q value, it uses its own network to calculate the strategy gradient for optimizing the parameters. The optimization formula is

$$\nabla_{\theta_i} J \approx \frac{1}{S}\sum_j \nabla_{\theta_i \mu_i}\left(o_i^j\right)\nabla_{a_i} Q_i^{\mu}\left(x^j, a_1^j, \ldots, a_N^j\right)\big|_{a_i = \mu_i(o_i^j)} \tag{3}$$

For more details on MADDDPG algorithm, see [7].

## 3  Asynchronous Advantage Multi-agent Deep Deterministic Policy Gradient

The A3C algorithm uses the idea of asynchronous training to increase the training speed. That is, multiple training environments are started at the same time, sampling is performed at the same time, and the collected samples are directly used for training. Compared to the DQN algorithm, the A3C algorithm does not need to use an experience repay to store historical samples, therefore it can save storage space. Otherwise, it uses asynchronous training, which greatly doubles the data sampling rate, and thus improves the training speed. At the same time, using multiple different training environments to collect samples, the distribution of the samples is more uniform, which is more conducive to the training of neural networks. The A3C algorithm has made improvements in the above aspects, making its average score on the Atari game four times bigger than that of the DQN algorithm, and has achieved a tremendous improvement, and the training speed has also multiplied.

Based on the basic idea of Actor-Critic, the value network of Critic's judging module (Value Network) can be updated using the DQN method. How to construct the loss function of the mobile network and realize the training of the network is the key to the algorithm. The general action network has two output modes: one is a probability mode, that is, the probability of outputting one action; the other is a deterministic mode, that is, outputting a specific action. AMADDPG uses a probabilistic output method. Therefore, we get a good or bad evaluation of the action from the Critic evaluation module, i.e. the value network. In order to achieve better results, how to accurately evaluate the performance of the algorithm is also the key to the algorithm. The AMADDPG algorithm uses the advantage A as the evaluation of movement on the basis of action value Q. Advantage A refers to the advantage of action a relative to other actions in state x. Using output action Log Likelihood multiplied by action evaluation as the loss function of the action network. The goal of the action network is to maximize this loss function. If the action is evaluated positive, it increases its probability and vice versa. This is consistent with the basic idea of Actor-Critic. With the loss function of the mobile network, it is also possible to update the parameters by means of a random gradient descent. The updating method of the AMADDPG algorithm can be considered as

$$\nabla_{\theta_i} J = \frac{1}{S} \sum_j \nabla_{\theta_i \mu_i} \left( o_i^j \right) \nabla_{\theta_i'} \log \pi \left( a_1^j, \ldots, a_i, \ldots, a_N^j | x^j; \theta_i' \right) A \left( x_i, a_i; \theta_i, \theta_{i,v} \right) \quad (4)$$

Where

$$A \left( x_i, a_i; \theta_i, \theta_{i,v} \right) = r_i^j + \gamma V \left( x^j; \theta_{i,v} \right) - V \left( x^j; \theta_{i,v} \right) \quad (5)$$

$$\theta_{i,v} = a_1^j, \ldots, a_i^j, \ldots, a_N^j \quad (6)$$

The policy parameters $\theta_i$ and value parameters $\theta_{i,v}$ share some parameters in practice.

In order to improve the training speed, the AMADDPG algorithm also adopts the idea of asynchronous training. Starting multiple training environments simultaneously. We adopt this global network and multi-worker architecture in our work. Unlike the asynchronous methods in A3C, we use the experience replay. The worker randomly samples M size of batch data from replay memory, and we divide the batch data into K parts. Then each worker will get M/K size of batch data and prioritize these batches. At the same time the master (which is the global network in Fig. 1) using one-step update of the batch data with the maximization loss (Eq. 4).

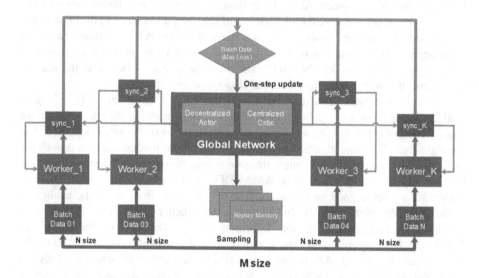

**Fig. 1.** The structure of AMADDPG

The global network and worker consists of a centralized actor network and a decentralized reviewer network. When the worker requests resources from the master, the master and the worker maintain the thread security through synchronization. After the master allocates resources to each worker, multiple agents are executed asynchronously. Through the different states experienced by parallel agents, the correlation between state transition samples generated during training is removed. The detailed algorithm structure is shown in Fig. 1.

## 4 Experiments

In order to implement our experiments, we used the grounded communication environment proposed in [7]. It consists of N agents and L landmarks. It exists in a two-dimensional world with continuous space and discrete time. In order to compare the performance of the algorithm, we took the pursuit of escape as an example, selected three predator agents, and completed the encircling of prey by cooperation (see Fig. 2).

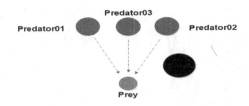

**Fig. 2.** Multi-agent cooperation based on AMADDPG algorithm (In the figure, red represents the predator and blue represents the prey.) (Color figure online)

In the experiment, the Critic network has 15 dimensional observations. The observations pass through neural network which contains two hidden layers. They have 500, 200 neurons respectively. The hidden layer activation function is relu. Then the output layer activation function is sigmod, and the final output has 5 dimensional actions. The critic network's input has 15 dimension action values and 15 dimensional observations. 15 dimensional action values pass through two hidden layers, which have 500 and 200 neurons respectively. Two hidden layer activation function is Relu. The action passes the output of the hidden layer with 300 neurons and the observations pass through both hidden layer outputs as input. After going through 300 neurons, the advantage A whose output is to guide the action of the actor network. In our experiment, the learning rate of our actor and Critic network is 0.1. We set the prioritized batch data of the experience replay to 255. We use multithreading to start three workers at the same time. Other parameters are default. At the beginning, we find that the agents are disorderly. Through training, the three predators gradually learn policy and eventually learn to collaborate to circumvent prey.

In the experiment, We compare the AMADDPG with the MADDPG algorithm. Through training, we plotted the mean episode reward and episode curves (see Fig. 3). As can be seen from the figure, It was not stable at first. With the increase in the number of training episodes, the mean episode reward obtained by the AMADDPG algorithm framework are higher than the original algorithm. Therefore, the experiment proves the effectiveness of the algorithm framework proposed in this paper.

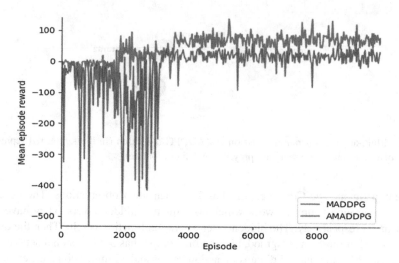

**Fig. 3.** The mean episode reward of two algorithm

## 5  Conclusions and Future Work

We have proposed an Asynchronous Advantage Multi-Agent Deep Deterministic Policy Gradient (AMADDPG) algorithm. We used asynchronous methods to create multiple copies using multithreading. Using advantage as an evaluation of actor in actor-critic. Speeding up training speed with priority batch processing. Experimentally, our improved AMADDPG algorithm outperforms other traditional RL algorithms. However, there are also some defects. For instance, we only use the predator-prey environment, increasing the number of predators and prey is not very effective. We will research the cooperation between agents and increase the number of agents in our further study.

**Acknowledgments.** This work was supported by Key Research and Development Plan Project of Shandong Province, China (No. 2017CXGC0614).

## References

1. Mnih, V., et al.: Human-level control through deep reinforcement learning. J. Nature. **518**, 529–533 (2015)
2. Peters, J.: Policy gradient methods. J. Scholarpedia **5**(11), 3698 (2010)
3. Lillicrap, T.P., et al.: Continuous control with deep reinforcement learning. In: International Conference on Learning Representations 2016, San Juan (2016)
4. Schaul, T., Quan, J., Antonoglou, I., Silver, D.: Prioritized experience replay. In: International Conference on Learning Representations 2016, San Juan (2016)
5. Mnih, V., et al.: Asynchronous methods for deep reinforcement learning. In: International Conference on Learning Representations 2016, San Juan (2016)

6. Jaderberg, M., et al.: Reinforcement learning with unsupervised auxiliary tasks. In: 5th International Conference on Learning Representations, Toulon (2016)
7. Lowe, R., Wu, Y., Tamar, A., Harb, J., Abbeel, P., Mordatch, I.: Multi-agent actor-critic for mixed cooperative-competitive environments. In: 31st Conference on Neural Information Processing Systems (NIPS 2017), Long Beach (2017)
8. Riedmiller, M.: Neural fitted Q iteration – First experiences with a data efficient neural reinforcement learning method. In: Gama, J., Camacho, R., Brazdil, Pavel B., Jorge, A.M., Torgo, L. (eds.) ECML 2005. LNCS (LNAI), vol. 3720, pp. 317–328. Springer, Heidelberg (2005). https://doi.org/10.1007/11564096_32
9. Wang, Z., et al.: Dueling network architectures for deep reinforcement learning. In: ICML 2016 Proceedings of the 33rd International Conference on International Conference on Machine Learning, pp. 1993–2005. JMLR Press. New York (2016)
10. Van Hasselt, H., Guez, A., Silver, D.: Deep reinforcement learning with double Q-learning. J. Comput. Sci. (2015)
11. Silver, D., et al.: Mastering the game of Go without human knowledge. J. Nature **550**, 354–359 (2017)
12. Coppens, Y., Shirahata, K., Fukagai, T., Tomita, Y., Ike, A.: GUNREAL: GPU-accelerated UNsupervised REinforcement and Auxiliary Learning. In: Fifth International Symposium on Computing and Networking, pp. 330–336, Shanghai (2017)
13. Babaeizadeh, M., Frosio, I., Tyree, S., Clemons, J., Kautz, J.: Reinforcement learning through asynchronous advantage actor-critic on a GPU. In: Fifth International Conference on Learning Representations, San Juan (2016)
14. Sutton, R.S., Barto, A.G.: Reinforcement learning: an introduction. MIT Press, Cambridge (2015)
15. Wang, X., Zhu, M., Cheng, Y.: The Principle of Reinforcement Learning and Its Application. Science Press, Beijing (2014)
16. Schwartz, H.M.: Multi-agent Machine Learning: A Reinforcement Approach, 458 p. Wiley, US (2014)
17. Silver, D., Lever, G., Heess, N., Degris, T., Wierstra, D., Riedmiller, M.: Deterministic policy gradient algorithms. In: The 31st International Conference on Machine Learning (ICML 2014), pp. 387–395, Beijing (2014)

# A Revisit of Reducing Hidden Nodes in a Radial Basis Function Neural Network with Histogram

Pey Yun Goh[✉], Shing Chiang Tan, and Wooi Ping Cheah

Multimedia University, Jln. Ayer Keroh Lama, 75450 Melaka, Malaysia
{pygoh,sctan,wpcheah}@mmu.edu.my

**Abstract.** In previous work [1], an incremental radial basis function network trained by a dynamic decay adjustment algorithm (RBFNDDA) was integrated with histogram to reduce redundant hidden neurons (or simply neurons). In order to remove unnecessary neurons, a weight-based indicator was utilized [1]. This hybrid model (RBFNDDA-HIST1) can reduce unnecessary neurons and maintain classification accuracy satisfactorily. However, another aspect of noises, i.e., overlapping among neurons of different classes in RBFNDDA-HIST1 and RBFNDDA, is not tackled fully for solutions. To close this research gap, another version of RBFNDDA-HIST (i.e., RBFNDDA-HISTR) is developed whereby the radius of a neuron (that overlaps with neurons of other classes) is checked before removing it from the network. Several public data sets that have a high level of overlapping records according to an overlapping indicator are used to evaluate the performance of RBFNDDA-HISTR in terms of number of neurons and classification accuracy. A performance comparison among RBFNDDA, RBFNDDA-HISTR and RBFNDDA-HIST1 are made. The results show that the proposed RBFNDDA-HISTR can reduce the number of neurons from RBFNDDA-HIST1 without deteriorating classification accuracy.

**Keywords:** RBFNDDA · Histogram · Weight · Radius · Pruning

## 1 Introduction

Similar to a typical radial basis function network (RBFN), the radial basis function network with dynamic decay adjustment (RBFNDDA) has three layers, i.e. the input, the hidden and the output layers. The learning behavior of RBFNDDA is incremental, dynamic and flexible where hidden neurons are created and adapted automatically on provision of data. RBFNDDA requires shorter execution time, performs with better classification accuracy than typical RBFN, and it is steady and well tested in implementation [2, 3]. It has been successfully applied in medical [4], software defects prediction [5] and recently, as a data generator [3]. However, the incremental behavior of RBFNDDA could over fit the data on hand as a result of absorbing information from the data using many hidden neurons. As such, RBFNDDA will have a huge and complicated network structure [6, 7].

Finding a suitable size of a neural network is always a challenge [8]. One of the famous and effective techniques to produce a suitable network structure is pruning.

© Springer Nature Switzerland AG 2018
L. Cheng et al. (Eds.): ICONIP 2018, LNCS 11302, pp. 722–731, 2018.
https://doi.org/10.1007/978-3-030-04179-3_64

Various pruning techniques are available in the literature, such as sensitivity approach [9, 10], magnitude approach [11, 12], clustering approach [13, 14], mutual information approach [15, 16] and statistical approach [17, 18]. In this study, a simple and cost effective method from a statistical approach is applied to reduce the network size of RBFNDDA. Specifically, RBFNDDA is integrated with histogram to remove overlapping neurons in its hidden layer. Histogram has a long history with these advantages: simple, does not incur heavy computational cost, approximate distribution of data [19] and does not require data to fit in a probability distribution. Histogram has been applied as a pruning technique in some domains such as word recognition [20, 21] and signal processing [22], but its application for pruning information learned from neural networks is relatively new.

In the works of Mart'in-Albo et al. [20] and Steinbiss et al. [21], histogram is used to limit the searching space in word recognition. Histogram is generated based on the frequency of active hypotheses within a time frame. Searching is time consuming especially when the number of active hypotheses in the search space is huge. Thus, pruning is applied when the number of active hypotheses exceeds a threshold. For signal processing, histogram pruning is applied to enhance the similarity search between query and store signals in audio and video. Histogram is generated based on the frequency of attribute vector occurrence within a signal time duration [22, 23]. The pruning indicator is the similarity degree. When the degree of query signal is less than a threshold, then pruning is applied.

Based on the related works [20–23] in word recognition and signal processing, hypothesis score or similarity degree is used as the pruning indicator. A question is raised before using histogram to prune hidden neurons from a neural network: what criterion should be considered as the pruning indicator? The histogram pruning with weight as the pruning indicator has been proposed in [1], i.e. RBFNDDA-HIST1. In RBFNDDA-HIST1, frequency distribution of weight values from RBFNDDA is computed to determine which bin should be removed and hence prunes the unnecessary neurons. However, it does not consider the overlapping among neurons of different classes. Overlapping reflects the degree of similarity between the neurons of two classes. Overlapping region between the neurons of different classes could lead to a prediction error [24]. It is claimed that when a data set consists of overlapping characteristic, it will cause RBFNDDA to create overlap neurons. Even though RBFNDDA can shrink the radius of neurons when overlapping between different classes occurs, the neurons may still suffer from the overlapping problem. The reason is, the original design of RBFNDDA is not equipped a pruning mechanism to remove these overlapping neurons. As such, in this paper, histogram is used to provide the distribution of all neurons learned by RBFNDDA and a pruning indicator based on the radii magnitude of a neuron is proposed.

The organization of this paper is divided into four sections, as follows: after introduction in Sect. 1, the building blocks for constructing RBFNDDA-HISTR are described in Sect. 2; experimental study is explained, results are compared and analyzed in Sect. 3; and, a summary and suggestions for future work are pointed out in Sect. 4.

## 2 The Methods

### 2.1 The RBFNDDA

Each hidden neuron of RBFNDDA is defined by a radial basis function, *Rad*. Two user-defined thresholds are applied in RBFNDDA to govern the dynamics of RBFNDDA, they are the positive $th^+$ and the negative threshold $th^-$. The lowest correct classification probability of the correct class is determined by $th^+$ whereas the highest misclassification probability is determined by $th^-$. These two parameters are not critical [25] and the default values are set as $th^+ = 0.4$ and $th^- = 0.2$ [6]. The training procedure of RBFNDDA in a single epoch is described as below:

1. Consider all hidden neurons, $Neu_i^c$ where $i = 1, 2, \ldots I_c, c = 1, 2, \ldots C$, the weight of each hidden neuron is set to zero, i.e. $w_i^c = 0$.
2. Assume that a training sample $\mathbf{x}$ of a class $c$ and a number of hidden neurons has been inserted into the network.
   (a) If activation through radial basis functions, $Rad^c$ of some hidden neurons are greater than $th^+$, increase the weight of the $i$-th hidden neuron, $Neu_i^c$ that has the highest $Rad_i^c$ by one: $w_i^c = w_i^c + 1$.
   (b) Otherwise, if none of the existing hidden neurons can classify $\mathbf{x}$ correctly, then perform the following procedure:

   {

   Increase the total number of hidden neuron by one: $I_c = I_c + 1$
   Add a new neuron, $Neu_{I_c}^c$ by setting the following:
   Set $w_{I_c}^c = 1$; Set the center of $Neu_{I_c}^c$, $\mathbf{z}_{I_c}^c = \mathbf{x}$;
   Set the radius $\sigma_{I_c}^c = \min\limits_{\substack{j \neq c \\ 1 \leq a \leq I_j}} \left\{ \sqrt{-\frac{\|\mathbf{z}_a^j - \mathbf{z}_{I_c}^c\|^2}{\ln(th^-)}} \right\}$

   }
   (c) Shrink the radius of neurons of conflicting classes where $j \neq c, 1 \leq a \leq I_c$,
   $$\sigma_a^j = \min\left\{ \sigma_a^j, \sqrt{-\frac{\|\mathbf{x} - \mathbf{z}_a^j\|^2}{\ln(\theta^-)}} \right\}$$
   (d) RBFNDDA is trained with another input sample $\mathbf{x}$ by repeating step 2. A training epoch is completed when all input samples are sent to RBFNDDA.

### 2.2 The Histogram

In this study, the proposed histogram (HIST) by Shimazaki and Shinomoto [26] is used. There is no systematic way in determining a suitable bin size in the typical histogram. Thus, Shimazaki and Shinomoto [26] has proposed HIST which has the strength of optimizing the number of bins and bin width to the data on hand. This eases the problem of having too many bins or too little bins which refrains the histogram from representing the distribution of data correctly. This HIST was proposed with an intention to estimate the spike rate (the rate of potential instantaneous activity of an individual neuron). This algorithm obtains the optimal bin size, $N$ through minimizing the cost function, $Cost_s(\Delta)$. The procedure is described below:

1. Divide the observation, *Obs* of period $T$ into $N$ bins with width $\Delta$. The setting of $N$ is a discrete value between 2 and 50.

$$\Delta = \frac{Obs_{max} - Obs_{min}}{N} \tag{1}$$

2. For each $s$ sequence (where $s$ is the number of sequence to obtain the spike rate), compute the number of spikes, $k_i$ that enter the $i$-th bin.

   a. Obtain the mean, $\bar{k}$ and variance, *var* of number of spikes by using the formulas as below:

$$\bar{k} = \frac{1}{N} \sum_{i=1}^{N} k_i \tag{2}$$

$$var = \frac{1}{N} \sum_{i=1}^{N} (k_i - \bar{k})^2 \tag{3}$$

   b. Compute the cost function, $Cost_s$:

$$Cost_s(\Delta) = \frac{2\bar{k} - var}{(s\Delta)^2} \tag{4}$$

3. Repeat step 1 till 2 by changing the value of $N$ which then change the bin width, $\Delta$ to search for optimal number of bin and bin width that minimize $Cost_s$.

### 2.3 The Hybrid of RBFNDDA and HIST with Radius as the Pruning Indicator

In RBFNDDA, radius of neurons is shrunk to reduce the overlapping among neurons of different classes. In order to remain high weight neurons, neurons in each bin are separated according to weight value into 2 groups: (1) neurons with weight $w > \beta$ where $\beta$ is a threshold and (2) $w \leq \beta$. Then, smallest radius, $\sigma_{min}$ among all neurons from group 1 is selected as a pruning threshold. Neurons from group 2 with radius $\sigma \leq \sigma_{min}$ are pruned. This allows each bin has a different threshold setting of $\sigma_{min}$ which is based on the distribution of neurons. In RBFNDDA-HISTR, neurons of all classes are organized together by using one histogram. Pruning is done after each learning epoch of RBFNDDA. The procedure of RBFNDDA-HISTR of one epoch pruning is as below:

1. After the learning process of RBFNDDA (refer Sect. 2.1) has been completed, the hidden layer has a group of neurons, where each $i$-th neuron is defined as:

$$Neu_i = \left\langle \mu_{i_z}, w_{i_z}, \sigma_{i_z} \right\rangle \tag{5}$$

where $\mu_i$ is the center, $w_i$ is the weight and $\sigma_i$ is the radius.

2. Center $\mu$ of neurons consists of multi-dimensions but HISTR accepts input with one dimension. Aggregate sum is used to transform $\mu$ to one single representative, i.e. $Obs$ (Eq. 6):

$$Obs_i = \sum_{m=1}^{M} \mu_{m_i} \tag{6}$$

where $m$ represents $m$-th attribute and $M$ is the total number of attributes.

3. Then, HIST procedure (refer Sect. 2.2) is executed to obtain optimum bin size.

4. For each bin $z$ with neurons, $Neu = \left\langle \mu_{1,\cdots,I_z}, w_{1,\cdots,I_z}, \sigma_{1,\cdots,I_z} \right\rangle$ where $I_z$ = the total number of neurons in the $z$-th bin:
   a. Separate neurons according to group 1: weight $w > \beta$ and group 2: $w < = \beta$.
   b. Select the smallest radius, $\sigma_{min}$ among the neurons from group 1.
   c. Neurons from group 2 with $\sigma < \sigma_{min}$ are pruned.

# 3  Experiment

## 3.1  Overlapping Computation

If the data samples of different classes in a data set are overlapped, such intrinsic association among data samples could be carried by hidden neurons during the incremental learning process of RBFNDDA. At the data level, to measure the degree of overlapping among the data samples of two classes, a Simpson coefficient [27] is applied:

$$\text{Overlap}(X, Y) = \frac{|X \cap Y|}{\min(|X|, |Y|)} \tag{7}$$

where $|X \cap Y|$ indicates the number of similar elements between class $X$ and class $Y$; $|X|$ and $|Y|$ indicates the number of elements of class $X$ and $Y$ respectively; min is a minimum operator. Equation 7 is applicable for measuring the degree of overlapping of the records from two classes at one input attribute. If a binary data set has more than one input attributes, Eq. 7 is extended to Eq. 8 where the degree of overlapping of data samples from two classes of all input attributes are computed, as follows:

$$\text{Avg}(OVL) = \frac{\sum_{i=1}^{M} \text{Overlap}_i(X, Y)}{M} \tag{8}$$

where $i = 1, 2, 3....M$, and $M$ is the total number of attributes. For a data set having more than two classes, a *one-versus-rest* classification is applied to re-label the data samples of one class as positive whereas the data samples of all other classes are labelled as negative. Equation 8 is extended to

$$\phi = \frac{\sum_{k=1}^{B} \text{Avg}_k(OVL)}{C} \tag{9}$$

where $\phi$ is the final overlapping coefficient, $k$ represents the $k$-th re-label class and $C$ is the total number of classes.

## 3.2    Benchmarked Data Sets and Experimental Setup

The result of RBFNDDA-HISTR is compared with RBFNDDA-HIST1 [1] in terms of number of neurons and number of classification accuracy. A group of data sets from UCI repository [28], i.e. blood transfuse (BldT), Indian liver (InL), diabetes (Dia), Haberman (Hab) and car evaluation (CarE) are used to evaluate the performance of RBFNDDA-HISTR. These data sets are measured with the overlapping coefficient, $\phi$. The detail characteristics of all data sets in terms of overlapping coefficient, number of samples, input attributes and classes are shown in Table 1. Notably, the overlapping coefficient of these five data sets are greater than 0.80. Table 1 also lists the size of training and test sets used in the experiment.

**Table 1.** Characteristics of data sets

| Data sets | No. of class | Sample size | No. of attributes | No. training | No. testing | $\phi$ |
|-----------|--------------|-------------|-------------------|--------------|-------------|--------|
| BldT | 2 | 748 | 5 | 374 | 374 | 0.83 |
| InL | 2 | 583 | 8 | 292 | 291 | 0.86 |
| Dia | 2 | 768 | 8 | 384 | 384 | 1.00 |
| Hab | 2 | 306 | 3 | 153 | 153 | 0.90 |
| CarE | 4 | 1728 | 6 | 864 | 864 | 1.00 |

The hardware specifications in running the experiment are: Windows 7, 4.0 GB RAM, Intel Core (TM)'s processor, i.e. i5-2410 M. In terms of parameter setting, default parameters are used for RBFNDDA whereas for RBFNDDA-HISTR, besides the initial parameters from RBFNDDA, the remaining parameters are based on the best setting (see Table 2). The experiment is repeated for 30 runs. The results are averaged.

## 3.3    Results and Discussions

The number of neurons of RBFNDDA, RBFNDDA-HIST1 and RBFNDDA-HISTR is shown in Table 3. The reduction ratio in Table 4 is computed with the following formula:

**Table 2.** Parameters setting

| Models | Parameters setting | |
|---|---|---|
| RBFNDDA | $th^+$ | 0.4 |
| | $th^-$ | 0.2 |
| | epoch | 6 |
| RBFNDDA-HIST1 | $th^+$ | 0.4 |
| | $th^-$ | 0.2 |
| | epoch | 6 |
| | Number of bins, $BIN$ | between 3 and 50 |
| | Sequence, $x$ | 1 |
| | Pruning threshold, $\eta$ | 0.25 |
| RBFNDDA-HISTR | $th^+$ | 0.4 |
| | $th^-$ | 0.2 |
| | epoch | 6 |
| | Number of bins, $BIN$ | between 3 and 50 |
| | Sequence, $x$ | 1 |
| | Threshold of weight, $\beta$ | 1 |

**Table 3.** Number of neurons

| Data set | Model | | | | | |
|---|---|---|---|---|---|---|
| | RBFNDDA | | RBFNDDA-HIST1 | | RBFNDDA-HISTR | |
| | # Neu. | Std. Dev. | # Neu. | Std. Dev. | # Neu. | Std. Dev. |
| BldT | 654.00 | 55.00 | 329.00 | 37.00 | 183.65 | 20.85 |
| InL | 262.62 | 5.42 | 213.36 | 7.41 | 179.63 | 22.99 |
| Dia | 288.50 | 6.10 | 202.90 | 26.20 | 154.33 | 22.62 |
| Hab | 119.37 | 11.96 | 79.41 | 26.35 | 76.73 | 12.82 |
| CarE | 471.06 | 10.95 | 449.63 | 28.04 | 258.11 | 28.10 |

#Neu = number of neurons; Std. Dev. = standard deviation

**Table 4.** Reduction ratio

| Data set | Model | |
|---|---|---|
| | RBFNDDA-HIST1 | RBFNDDA-HISTR |
| BldT | 0.50 | 0.72 |
| InL | 0.19 | 0.32 |
| Dia | 0.30 | 0.47 |
| Hab | 0.33 | 0.36 |
| CarE | 0.05 | 0.45 |

$$\frac{\#\text{Neu in RBFNDDA} - \#\text{Neu in Extended Model}}{\#\text{Neu in RBFNDDA}} \qquad (10)$$

where #Neu = number of neurons; and extended model is referred to either RBFNDDA-HIST1 or RBFNDDA-HISTR. The higher the ratio, the more neurons the model can reduce. From the results in Tables 3 and 4, both RBFNDDA-HIST1 and RBFNDDA-HISTR can reduce the number of neurons from RBFNDDA. RBFNDDA-HISTR shows better reduction than RBFNDDA-HIST1 in all the data sets where the lowest ratio is 0.32 and the highest ratio is 0.72. These results show that RBFNDDA-HISTR is good in dealing with data samples that are overlapped between classes.

**Table 5.** Classification accuracy

| Data set | Model | | | | | |
|---|---|---|---|---|---|---|
| | RBFNDDA | | RBFNDDA-HIST1 | | RBFNDDA-HISTR | |
| | Accuracy | Std. Dev. | Accuracy | Std. Dev. | Accuracy | Std. Dev. |
| BldT | 61.56 | 2.68 | 62.36 | 3.06 | 65.70 | 3.60 |
| InL | 74.19 | 2.25 | 74.39 | 2.33 | 74.13 | 2.21 |
| Dia | 74.35 | 0.97 | 72.85 | 1.24 | 76.07 | 0.88 |
| Hab | 70.08 | 3.04 | 68.09 | 10.75 | 70.76 | 2.75 |
| CarE | 73.72 | 1.72 | 73.85 | 1.80 | 75.47 | 1.89 |

Std. Dev. = standard deviation

In terms of classification accuracy (see Table 5), RBFNDDA-HISTR outperforms RBFNDDA and RBFNDDA-HIST1 in BldT, Dia and CarE and it maintains the accuracy in InL and Hab.

In general, both RBFNDDA-HIST1 and RBFNDDA-HISTR can reduce the number of neurons in RBFNDDA. However, when a data set is highly overlapped, RBFNDDA-HIST1 could face information loss. This could be observed from the result in Dia and Hab. RBFNDDA-HISTR can classify the data with better or at least maintain the classification accuracy and better reduce the number of neurons than RBFNDDA-HIST1 when overlapping is present in the data set.

# 4 Conclusion

In this study, a neuron reduction method through the hybrid of RBFNDDA and histogram, i.e. RBFNDDA-HISTR is proposed. Experiment shows that this method can reduce neurons well and has a better prediction capability when the on hand problems consist of overlapping information. In future, it is interesting to extend the work where both weight and radius pruning criteria are integrated together into one algorithm. With such a combination, it is believed that better reduction power and good classification accuracy could be achieved.

# References

1. Goh, P.Y., Tan, S.C., Cheah, W.P., Lim, C.P.: Reducing the complexity of an adaptive radial basis function network with a histogram algorithm. Neural Comput. Appl. **28**(1), 365–378 (2017)
2. Du, K.-L., Swamy, M.N.S.: Neural Network and Statistical Learning. Springer-Verlag, London (2014). https://doi.org/10.1007/978-1-4471-5571-3
3. Robnik-Šikonja, M.: Data generators for learning systems based on RBF networks. IEEE Trans. Neural Networks Learn. Syst. **27**(5), 926–938 (2016)
4. Kurjakovic, S., Svennberg, A.: Implementing an RBF in VHDL, Kungl Tekniska Hogskolan. Royal Institute of Technology), Swedish (2002)
5. Bezerra, M.E.R., Oliveiray, A.L.I., Adeodatoz, P.J.L.: Predicting software defects: a cost-sensitive approach. In: 2011 IEEE International Conference on Systems, Man, and Cybernetics (SMC), pp. 2015–2522. IEEE, Anchorage (2011)
6. Paetz, J.: Reducing the number of neurons in radial basis function networks with dynamic decay adjustment. Neurocomputing **62**, 79–91 (2004)
7. Reed, R.: Pruning algorithms-a survey. IEEE Trans. Neural Networks **4**(5), 740–747 (1993)
8. Wang, J., Xu, C., Yang, X., Zurada, J.M.: A novel pruning algorithm for smoothing feed forward neural networks based on group lasso method. IEEE Trans. Neural Networks Learn. Syst. **29**(5), 2012–2024 (2018)
9. Zhai, J., Shao, Q., Wang, X.: Architecture selection of ELM networks based on sensitivity of hidden nodes. Neural Process. Lett. **44**(2), 471–489 (2016)
10. Wang, H., Yan, X.: Improved simple deterministically constructed cycle reservoir network with sensitive iterative pruning algorithm. Neurocomputing **145**, 353–362 (2014)
11. Medeiros, C.S., Barreto, G.: A novel weight pruning method for MLP classifiers based on the MAXCORE principle. Neural Comput. Appl. **22**(1), 71–84 (2013)
12. Leung, C., Tsoi, A.: Combined learning and pruning for recurrent radial basis function networks based on recursive least square algorithms. Neural Comput. Appl. **15**(1), 62–78 (2005)
13. Kusy, M., Kluska, J.: Probabilistic neural network structure reduction for medical data classification. In: Rutkowski, L., Korytkowski, M., Scherer, R., Tadeusiewicz, R., Zadeh, Lotfi A., Zurada, Jacek M. (eds.) ICAISC 2013. LNCS (LNAI), vol. 7894, pp. 118–129. Springer, Heidelberg (2013). https://doi.org/10.1007/978-3-642-38658-9_11
14. Xie, Y., Fan, X., Chen, J.: Affinity propagation-based probability neural network structure optimization. In: 2014 Tenth International Conference on Computational Intelligence and Security, pp. 85–89. IEEE, Kunming (2014)
15. Xing, H.J., Hu, B.G.: Two-phase construction of multilayer perceptrons using information theory. IEEE Trans. Neural Netw. **20**(4), 715–721 (2009)
16. Deco, G., Finnoff, W., Zimmermann, H.G.: Elimination of overtraining by a mutual information network. In: Gielen, S., Kappen, B. (eds.) ICANN '93, pp. 744–749. Springer, London (1993). https://doi.org/10.1007/978-1-4471-2063-6_208
17. Thomas, P., Suhner, M.-C.: A new multilayer perceptron pruning algorithm for classification and regression applications. Neural Process. Lett. **42**(2), 437–458 (2015)
18. Lo, J.T.: Pruning method of pruning neural networks. In: International Joint Conference on Neural Networks, IJCNN 1999, Washington, DC, vol. 3, pp. 1678–1680 (1999)
19. Ioannidis, Y.: The history of histogram. In: Freytag, J-C., Lockemann, P., Abiteboul, S., Carey, M., Selinger, P., Heuer, A. (eds.) Proceedings of the 29th International Conference on Very Large Data Bases, Berlin, Germany, vol. 29, pp. 19–30 (2003)

20. Martín-Albo, D., Romero, V., Vidal, E.: An experimental study of pruning techniques in handwritten text recognition systems. In: Sanches, João M., Micó, L., Cardoso, Jaime S. (eds.) IbPRIA 2013. LNCS, vol. 7887, pp. 559–566. Springer, Heidelberg (2013). https://doi.org/10.1007/978-3-642-38628-2_66

21. Steinbiss, V., Tran, B.-H., Ney, H.: Improvement in beam search. In: Third International Conference on Spoken Language Processing, Yokohama, Japan, pp. 2143–2146 (1994)

22. Zhang, W.-Q., Liu, J.: Two-stage method for specific audio retrieval. In: IEEE International Conference on Acoustics, Speech and Signal Processing, Honolulu, HI, vol. 4, pp. IV–85–IV–88 (2007)

23. Kashino, K., Kurozumi, T., Murase, H.: A quick search method for audio and video signals based on histogram pruning. IEEE Trans. Multimed. 5(3), 348–357 (2003)

24. Tang, W., Mao, K.Z., Mak, L.O., Ng, G.W.: Classification for overlapping classes using optimized overlapping region detection and soft decision. In: 13th International Conference on Information Fusion, pp. 1–8. IEEE, Edinburgh (2010)

25. Berthold, M.R., Diamond, J.: Boosting the performance of RBF networks with dynamic decay adjustment. In: NIPS 1994 Proceedings of the 7th International Conference on Neural Information Processing Systems, pp. 521–528. MIT Press, Cambridge (1995)

26. Shimazaki, H., Shinomoto, S.: A method for selecting the bin size of a time histogram. J. Neural Comput. 19(6), 1503–1527 (2007)

27. Hayashi, Y., Tsuji, H., Saga, R.: Visualizing method based on item sales records and its experimentation. In: 2009 IEEE International Conference on Systems, Man and Cybernetics, pp. 524–528. IEEE, San Antonio (2009)

28. Bache, K., Lichman, M.: UCI Machine Learning Repository, University of California, School of Information and Computer Science (2013). http://archive.ics.uci.edu/ml

# Author Index

Printed in the United States
By Bookmasters